ELASTIC-PLASTIC FRACTURE: SECOND SYMPOSIUM, VOLUME I— INELASTIC CRACK ANALYSIS

A symposium
sponsored by ASTM
Committee E-24 on
Fracture Testing
Philadelphia, Pa., 6–9 Oct. 1981

ASTM SPECIAL TECHNICAL PUBLICATION 803

C. F. Shih, Brown University, and
J. P. Gudas, David Taylor Naval Ship
R&D Center, editors

ASTM Publication Code Number (PCN)
04-803001-30

ASTM 1916 Race Street, Philadelphia, Pa. 19103

Library of Congress Catalog Card Number: 82-83520

NOTE

The Society is not responsible, as a body, for the statements and opinions advanced in this publication.

Printed in Baltimore, Md. (b)
November 1983

Foreword

The Second International Symposium on Elastic-Plastic Fracture Mechanics was held in Philadelphia, Pennsylvania, 6–9 Oct. 1981. This symposium was sponsored by ASTM Committee E-24 on Fracture Testing. C. F. Shih, Brown University, and J. P. Gudas, David Taylor Naval Ship Research and Development Center, presided as symposium chairmen. They are also editors of this publication.

Related ASTM Publications

Fracture Mechanics (13th Conference), STP 743 (1981), 04-743000-30

Fractography and Materials Science, STP 733 (1981), 04-733000-30

Crack Arrest Methodology and Applications, STP 711 (1980), 04-711000-30

Fracture Mechanics (12th Conference), STP 700 (1980), 04-700000-30

Elastic-Plastic Fracture, STP 688 (1979), 04-688000-30

A Note of Appreciation to Reviewers

The quality of the papers that appear in this publication reflects not only the obvious efforts of the authors but also the unheralded, though essential, work of the reviewers. On behalf of ASTM we acknowledge with appreciation their dedication to high professional standards and their sacrifice of time and effort.

ASTM Committee on Publications

ASTM Editorial Staff

Acknowledgments

The editors would like to acknowledge the assistance of Professor G. R. Irwin, Dr. J. D. Landes, Professor P. C. Paris, and Mr. E. T. Wessel in planning and organizing the symposium. We are grateful for the support provided by the ASTM staff, particularly Ms. Kathy Greene and Ms. Helen M. Hoersch. The timely submission of papers by the authors is greatly appreciated. Finally, this publication would not have been possible without the tremendous effort and dedication that was put forth by the many reviewers. Their high degree of professionalism ensured the quality of this publication. The editors also wish to acknowledge the diligent assistance of Ms. Susan Beigquist, Ms. Ann Degnan, Mr. Steven Kopf, and Mr. Mark Kirk in preparing the index.

J. P. Gudas
C. F. Shih

Contents

Viscoplastic Analysis and Correlation

Introduction

In October 1981, ASTM Committee E24 sponsored the Second International Symposium on Elastic-Plastic Fracture Mechanics which was held in Philadelphia, Pennsylvania. The objective of this meeting was to provide a forum for review of recent progress and introduction of new concepts in this field. The impetus for this symposium was the historical development of elastic-plastic fracture technology. Concepts such as the J-Integral, COD, and HRR field generated tremendous interest which led to the First International Symposium held in 1977. The presentation and publication of works on such topics as J-controlled crack growth, tearing instability, and numerical description of crack tip fields, among others, led to major, broad-based research activities and application-oriented developments. This sustained growth provided the motivation for another meeting devoted solely to elastic-plastic fracture.

The call for papers for the Second International Symposium generated an overwhelming response. This was reflected by the number of papers presented, and the attendance which exceeded 300 participants. The papers submitted to this symposium underwent rigorous review. The works contained in these two volumes reflect the high degree of interest in this subject and the quality of the efforts of the individual authors.

In the first ASTM publication devoted to elastic-plastic fracture (ASTM STP 668), there were three major groupings including elastic-plastic fracture criterion and analysis, experimental test techniques and fracture toughness data, and applications of elastic-plastic methodology. The present collection of papers shows substantial growth in theoretical and analytical areas which now include topics ranging from fundamental analysis of crack growth under static and dynamic conditions, finite strain effects at the crack tip, elevated temperature effects, visco-plastic crack analysis, and tractable treatments of fully plastic crack problems and surface flaws. These theoretical and analytical developments, combined with progress in test method development and ductile fracture toughness characterization led to substantial growth in engineering application of elastic-plastic fracture methodologies as evidenced by the large selection of papers on this topic.

The papers in these two volumes have been grouped into six topic areas including elastic-plastic crack analysis, fully plastic crack and surface flaw analysis, visco-plastic crack analysis and correlation, engineering applications, test methods and geometry effects, and cyclic plasticity effects and material characterization. The first grouping contains papers on crack propagation under static and dynamic conditions, crack growth and fracture criteria, finite strain effects on crack tip fields, and plasticity solutions for important crack geometries and structural configurations. Fully plastic crack solutions, elastic-plastic line-spring models, approximate treatment of surface flaws, and surface flaw crack growth correlations are the focus of the second grouping of papers. An area not addressed in the first symposium and which has since attracted significant attention is crack growth at elevated temperature, and time dependent effects. Papers on this topic address theoretical aspects of creeping cracks, computational procedures, microstructural modelling, creep crack growth correlations, and materials characterization.

The second volume of this publication begins with the major section on engineering applications. This includes several papers on tearing instability, J-based design curves, evaluations of several fracture criteria, further developments of fracture analysis diagrams, and flawed pipe analyses. This is followed by a series of papers on test methods and geometry relationships. A majority of the papers focus on J_{Ic} and J_I-R-curve test procedures and computations, and several papers address the test specimen geometry dependence of these parameters. In the last section several papers are devoted to prior load history effects, crack growth in the elastic-plastic regime under cyclic loading, and micromechanism studies of the fracture process.

The collections of papers from this and the previous symposium contain many of the major works in the rapidly evolving subject of elastic-plastic fracture. It is hoped that these two volumes will serve to stimulate further progress in this field.

J. P. Gudas

David Taylor Naval Ship Research and Development Center, Annapolis, Md. 21401; symposium chairman and editor.

C. F. Shih

Division of Engineering, Brown University, Providence, R.I. 02912; symposium chairman and editor.

Elastic-Plastic Crack Analysis

L. B. Freund[1] *and A. S. Douglas*[2]

Dynamic Growth of an Antiplane Shear Crack in a Rate-Sensitive Elastic-Plastic Material

REFERENCE: Freund, L. B. and Douglas, A. S., **"Dynamic Growth of an Antiplane Shear Crack in a Rate-Sensitive Elastic-Plastic Material,"** *Elastic-Plastic Fracture: Second Symposium, Volume I—Inelastic Crack Analysis, ASTM STP 803*, C. F. Shih and J. P. Gudas, Eds., American Society for Testing and Materials, 1983, pp. I-5–I-20.

ABSTRACT: A numerical study of the dynamic steady-state antiplane shear (Mode III) crack growth process is described. The study is based on continuum mechanics, and the material is modeled as being elastic-viscoplastic. Using the small-scale yielding concept of fracture mechanics and allowing only for steady-state crack growth, but including the effects of material inertia explicitly, a full-field solution for the deformations is obtained by means of an iteration procedure involving the finite-element method. Numerical results for the strain distribution in the active plastic zone are presented. The fracture criterion for materials which fail in a locally ductile manner proposed by McClintock and Irwin [*1*],[3] which stipulates that crack growth will proceed such that a critical strain level is maintained at a characteristic distance ahead of the crack tip, is adopted. This criterion is coupled with the results of numerical calculations to develop theoretical dynamic fracture toughness versus crack tip speed relationships for two levels of critical strain. For rate-dependent materials a range of toughness-versus-speed relations is manifested, ranging from that similar to rate-independent materials for low rate sensitivity to a relationship which rises dramatically for even a small increase in crack speed at low crack speeds, but which levels off for higher speeds for extremely rate-sensitive materials.

KEY WORDS: fracture (materials), elastic-plastic crack propagation, dynamic crack propagation, crack propagation, small scale yielding, finite element method, elastic-plastic fracture

The in-plane modes of deformation (Modes I and II) are of main interest in fracture mechanics, but mathematical difficulties encountered in their analysis have prevented complete descriptions of their features, especially for dy-

[1]Professor, Division of Engineering, Brown University, Providence, R.I. 02912.

[2]Assistant professor, Department of Mechanical Engineering, Johns Hopkins University, Baltimore, Md.; formerly, research assistant, Brown University, Providence, R.I. 02912.

[3]The italic numbers in brackets refer to the list of references appended to this paper.

namic or elastic-plastic systems. The difficulties for the antiplane mode of deformation (Mode III) are not as severe, and several features of in-plane crack deformation have been anticipated from Mode III solutions. Here, dynamic steady-state crack growth in the antiplane shear mode through a strain-rate-sensitive elastic-plastic material is studied. The inertial resistance of the material to motion is included explicitly but for this steady-state situation (in which the crack-tip speed is constant, and has been for all time), the deformation is time independent as viewed by an observer fixed at the moving edge of the crack. The full deformation field is determined by means of an iteration procedure involving the numerical finite-element method, under conditions of the small-scale yielding hypothesis of fracture mechanics [*2*]. According to this hypothesis, the elastic-plastic field in the crack-tip region is controlled by the surrounding elastic field. A useful measure of this surrounding elastic field, for any crack tip speed v, is the linear elastic stress-intensity factor k, which is assumed to be known from the solution of the relevant elastic crack problem.

The same problem as that being considered here, but for rate-independent material response and with inertial effects neglected, was studied in some detail by Rice [*3*] and Chitaley and McClintock [*4*]. Each employed incremental plastic stress-strain relations based on the associated flow law and a yield condition suitable for antiplane shear in a nonhardening material. They showed that the principal shear lines in the active plastic zone are straight, and Rice [*3*] determined the distribution of plastic strain on the line directly ahead of the crack tip in terms of the unknown distance between the tip and the elastic-plastic boundary. Chitaley and McClintock [*4*] went on to integrate numerically the field equations. In a more recent study of the same problem, Dean and Hutchinson [*5*] noted some discrepancies between their finite-element calculations and the numerical results of Chitaley and McClintock. Whereas the latter authors assumed that the principal shear lines in the active plastic zone were all members of a centered fan, the results of Dean and Hutchinson suggest that, in the portion of the active plastic zone adjacent to the unloading boundary, the shear lines do not form a centered fan.

Progress on the corresponding steady-state dynamic problem, with material inertia taken into account but for rate-independent material response, has been reported by Douglas, Freund, and Parks [*6*] and by Freund and Douglas [*7*]. In Ref *6*, a numerical procedure similar to that developed by Parks, Lam, and McMeeking [*8*] and by Dean and Hutchinson [*5*] was used to find the complete deformation field for several values of crack propagation speed. It was found that the amount of plastic strain at a fixed distance ahead of the moving crack tip decreased with increasing crack-tip speed, for a fixed level of remote stress-intensity factor. An analysis leading to an exact expression for the strain distribution on the crack line within the active plastic zone for any crack propagation speed was presented in Ref *7*, and the strain distributions for selected speeds obtained previously by means of the numerical method compared very well with the exact results. Furthermore, the analytical and numerical results were combined with the "critical plastic strain at a characteristic distance" crack growth criterion in Ref *7* to generate theoretical fracture toughness versus crack speed relationships. The results showed a strong dependence of fracture toughness on crack speed for even moderate crack growth rates. Because the material response was independent of strain rate in the material model used, the results suggested that the influence of inertia on fracture resistance is quite significant.

During rapid crack propagation, material particles very near to the crack tip experience very high strain rates. Thus, for materials for which the flow stress has a significant dependence on strain rate, it is expected that strain-rate effects will also influence the fracture toughness versus crack speed relationship. To quantify the level of this influence, the numerical method which has been used in our earlier work on dynamic elastic-plastic crack growth has been modified to accommodate a particular model of elastic-plastic material response in which the flow stress depends on the instantaneous strain rate. The features of the material model and the results of some calculations are described in the following sections.

Governing Equations

All variables are referred to a translating set of Cartesian axes labelled (x_1, x_2, x_3) in the body. The x_3-axis coincides with the crack edge, and this edge moves in the x_1-direction with speed v. The displacement vector is in the x_3-direction and its magnitude u_3 depends only on position (x_1, x_2). The equation of motion is

$$\frac{\partial \sigma_{13}}{\partial x_1} + \frac{\partial \sigma_{23}}{\partial x_2} = \rho \ddot{u}_3 \tag{1}$$

where σ_{13} and σ_{23} are the nonzero stress components and ρ is the mass density of the material. The dot denotes the material time derivative. Because only steady-state solutions are sought, the material time derivative may be replaced by a spatial gradient as follows

$$\dot{u}_3 = -v \frac{\partial u_3}{\partial x_1} \tag{2}$$

The shear strain components corresponding to the stress components σ_{13} and σ_{23} are ϵ_{13} and ϵ_{23}, respectively. It is convenient to introduce the following nondimensional quantities:

$$x = \frac{x_1 \tau_0^2}{k^2} \qquad y = \frac{\alpha x_2 \tau_0^2}{k^2} \qquad w = \frac{\mu u_3 \tau_0}{k^2}$$

$$\gamma_x = \frac{2\mu\epsilon_{13}}{\tau_0} \qquad \gamma_y = \frac{2\mu\epsilon_{23}}{\tau_0}$$
$$\tau_x = \frac{\sigma_{13}}{\tau_0} \qquad \tau_y = \frac{\sigma_{23}}{\tau_0} \qquad m = \frac{v}{c} \tag{3}$$

where

τ_0 = flow stress of the material for very slow loading in shear,
μ = elastic shear modulus,
$c = \sqrt{\mu/\rho}$ = elastic wave speed, and
$\alpha = \sqrt{1 - v^2/c^2}$.

The strain-displacement relations in nondimensional form are

$$\gamma_x = \frac{\partial w}{\partial x} \qquad \gamma_y = \alpha \frac{\partial w}{\partial y} \tag{4}$$

The relationship between the stress and elastic strain, and the additivity of elastic and plastic strain components, may be expressed as

$$\tau_\beta = \gamma_\beta^{\,e} = \gamma_\beta - \gamma_\beta^{\,p}, \qquad \beta = x \quad \text{or} \quad y \tag{5}$$

Using Eqs 3 and 4, the equation of motion may be written in the form

$$\frac{\partial^2 w}{\partial x^2} + \frac{\partial^2 w}{\partial y^2} = \frac{1}{\alpha^2}\frac{\partial \gamma_x^{\,p}}{\partial x} + \frac{1}{\alpha}\frac{\partial \gamma_y^{\,p}}{\partial y} \tag{6}$$

The asymptotic behavior of the stress components as $r = \sqrt{x^2 + y^2} \rightarrow \infty$ is chosen to be consistent with the small-scale yielding hypothesis. Thus, the stress components are required to approach the elastic near-field limits, that is

$$\tau_x \rightarrow \frac{\sin(\theta/2)}{\alpha\sqrt{2\pi r}} \qquad \tau_y \rightarrow \frac{\cos(\theta/2)}{\sqrt{2\pi r}} \tag{7}$$

where $\theta = \arctan(y/x)$ as $r \rightarrow \infty$. The crack faces are traction-free.

The constitutive model used is of the overstress type which was introduced by Malvern [*9*] in his study of rate effects in uniaxial stress wave propagation. Malvern's stress-strain rate relation was subsequently generalized to general stress states by Perzyna [*10*], and these results were embedded in a general framework for time-dependent plastic response by Rice [*11*]. It is assumed here that the material responds isotropically, that the strain rate vector is co-linear with the stress vector, and that the magnitude of the strain rate is proportional to the difference between the current stress magnitude and the initial yield stress magnitude raised to some power. In terms of nondimensional variables, the plastic strain rate is given by

$$-\frac{\partial \gamma_\beta{}^p}{\partial x} = H(\tau - 1)^n \frac{\tau_\beta}{\tau}, \qquad \tau^2 = \tau_x{}^2 + \tau_y{}^2 \tag{8}$$

where H and n are material parameters. In terms of previously defined quantities, H is given by

$$H = \frac{2\mu h k^2}{v \tau_0{}^3} \tag{9}$$

where h is the actual strain rate sensitivity parameter. The latter is defined as that strain rate at which the instantaneous flow stress magnitude is twice the slow-loading flow stress τ_0. The response is governed by Eq 8 as long as the strain rate is high enough to result in a positive stress rate $\dot{\tau}$. If the strain rate magnitude falls below the level required to meet this criterion, but the strain magnitude continues to increase, then the deformation proceeds with $\dot{\tau} = 0$. If the strain magnitude begins to decrease, then the material responds elastically. This description of behavior seems to be consistent with the limited experimental data obtained for changes in strain rate at the very high strain rates which are anticipated in the crack-tip region during rapid crack propagation (see Lipkin, Campbell, and Swearengen [*12*] and Chiem and Duffy [*13*]). The essential features of the material model are shown for uniaxial response in Fig. 1.

Numerical Procedure

In order to develop the numerical procedure used, the governing field Eq 6 is recast into variational form. Both sides of Eq 6 are multiplied by an arbitrary function, say w^*, which has the same smoothness properties and satisfies the same kinematic constraints as w, and are then integrated over any portion A of the x,y-plant not containing the crack itself. The result of applying the divergence theorem to the equation obtained is

$$\int_A \nabla w^* \cdot \nabla w \, dA = -\int_A w^* \left[\frac{1}{\alpha^2} \frac{\partial \gamma_x{}^p}{\partial x} + \frac{1}{\alpha} \frac{\partial \gamma_y{}^p}{\partial y} \right] dA + \int_{\partial A} w^* (\mathbf{n} \cdot \nabla w) \, dS \tag{10}$$

where

∂A = closed boundary of A,
∇ = two-dimensional gradient operator, and
$\mathbf{n}$ = unit normal to ∂A pointing out of the region A.

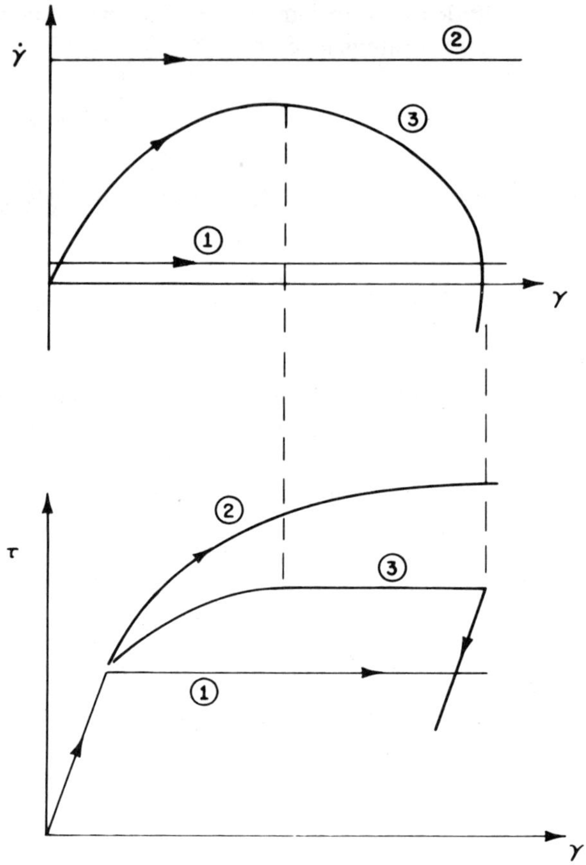

FIG. 1—*Schematic diagram of uniaxial material response for representative strain rate histories:* (1) *uniform "low" strain rate,* (2) *uniform "high" strain rate, and* (3) *nonuniform strain rate.*

The rectangular region $-X < x < X$, $0 < y < Y$ (see Fig. 2) is then divided into finite elements, and the displacement field is approximated by discrete interpolation functions ϕ_i, that is,

$$w = \phi_i \cdot \delta_i \tag{11}$$

where summation is implied and δ_i are the values of the displacement at the N discrete node points. Likewise,

$$w^* = \phi_i \cdot \delta_i^* \tag{12}$$

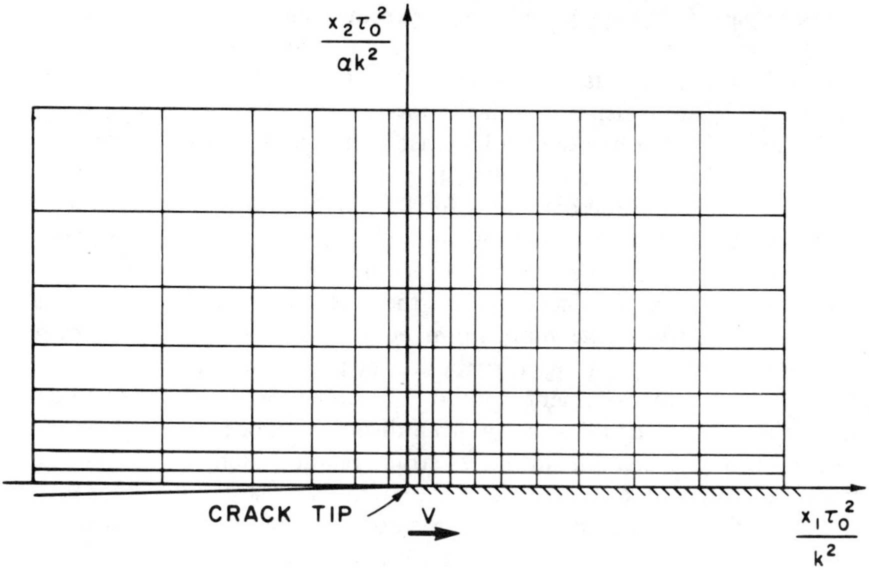

FIG. 2—*Plane of deformation, showing a coarse representation of the finite-element mesh.*

With the finite-element formulation and with δ_i^* arbitrary, the variational statement of the governing Eq 10 may be reduced to the system of linear equations

$$K_{ij}\delta_j = R_i(\gamma_x{}^p, \gamma_y{}^p) \tag{13}$$

The N by N stiffness matrix K_{ij} in Eq 13 for any region A is then

$$K_{ij} = \int_A \nabla\phi_i \cdot \nabla\phi_j dA \tag{14}$$

The right-hand side of Eq 13 is

$$R_i = \int_A \left[\frac{\gamma_x{}^p}{\alpha^2}\frac{\partial\phi_i}{\partial x} + \frac{\gamma_y{}^p}{\alpha}\frac{\partial\phi_i}{\partial y}\right] dA + \int_{L_p} \phi_i \left[\frac{m^2}{\alpha^2}\gamma_x{}^p - \tau_x\right] dy + \int_{L_e} \frac{\phi_i}{\alpha\sqrt{2\pi r}}[n_y \cos(\theta/2) - n_x \sin(\theta/2)]dL \tag{15}$$

where the path L_e is the perimeter of the rectangular region shown in Fig. 2 except for the portion of the left-hand boundary which is in the wake region and which is identified as L. As is evident in Eq 15, the boundary tractions imposed on L_e are consistent with the far-field small-scale yielding stresses (Eq 7). For this identification to be reasonable, points on L_e should be far from the crack tip when compared with plastic zone size.

In order to solve these equations for the unknown displacements δ_i, the plastic strains must be known. These are, however, unknown *a priori*, and an iterative procedure must be adopted. There are two main steps to the iteration procedure used here. One step involves the determination of the total strain distribution satisfying the momentum equation, subject to the specified boundary conditions, for *any* distribution of plastic strain. This is accomplished with the finite-element method. The other step involves the calculation of the stress distribution for a given distribution of total strain through integration of the incremental stress-strain relation along a y = constant "pathline" from $x = X$ to $x = -X$. If the strain rate is large enough for Eq 8 to hold, then the stress at an integration station is determined by straightforward application of Newton's rule to Eq 8 at the previous integration station on the same pathline. If the strain rate falls below the level required for Eq 8 to hold, but plastic flow continues, then the stress is determined by sequential application of the method of Rice and Tracey [*14*] at integration stations along a pathline. The details of this numerical integration procedure are described in Ref *15*.

The stress and deformation fields for a range of values of the parameter H are obtained by starting with the value $H = 0$, which reproduces the well-known elastic results. This parameter is then increased a small amount and the iteration procedure, starting from the final results for the previous value of H as an initial guess, is followed until convergence is achieved. The steps in the numerical procedure may be summarized as follows:

1. Calculate the elastic displacements (that is, those for $H = 0$) at the nodes. Use these as the first estimate, say $\delta_j(1)$.
2. Increase the value of H a small amount.
3. Calculate the total strains $\gamma_\beta(i)$ from the ith estimate of displacement $\gamma_j(i)$ by means of the strain-displacement relations.
4. Determine whether or not the stress response at each integration station is rate-sensitive or slow-loading; if rate-sensitive, go to Step 5; if slow-loading, go to Step 6.
5. Integrate Eq 8 to calculate the stresses. Go to Step 7.
6. Use the method of Rice and Tracey [*14*] to determine the stresses. Go to Step 7.
7. Determine the plastic strains $\gamma_\beta{}^P(i)$ according to Eq 5.
8. Calculate the $(i + 1)$th estimate of displacement $\delta_j(i + 1)$ from Eq 13.
9. Compare $\delta_j(i + 1)$ with $\delta_j(i)$. If the difference is sufficiently small, go to Step 8; otherwise, increment the index i by 1 and return to Step 3.

10. If H is large enough, the process is complete; otherwise, reset $i = 1$ and return to Step 2.

The finite-element calculations were based on an array of 2800 rectangular elements, each comprising four constant strain triangles formed by the diagonals. Nodal points were concentrated near the crack tip and element size increased with distance from the crack tip. The size of the smallest element was approximately 0.2 percent of the maximum extent of the plastic zone. The distance to the remotely applied elastic stress field, that is, the value of X and Y, was chosen to be at least ten times the maximum plastic zone size. A coarse representation of a typical mesh is shown in Fig. 2. This choice of element array led to a few elements with very small aspect ratios but they gave rise to no numerical difficulties, perhaps because they occurred only in regions where all fields had a high degree of smoothness.

Results

To illustrate the accuracy of the numerical method, the distribution of strain on the crack line within the active plastic zone for a rate-independent material has been computed for several values of crack-tip speed and the results are compared with the exact analytical results in Fig. 3. These data are reproduced from Ref 7. The general features of the computed strain distributions, including the very large strains at the crack tip, are consistent with the exact distributions and the computed values are approximately correct. The strain distributions shown in Fig. 3 are also obtained from the calculation for a rate-dependent material if a very large value of H is assumed, which would correspond to a relatively rate-insensitive material ($h \to \infty$) or to a very slow crack speed ($v \to 0$).

The computed strain distributions on the crack line within the active plastic zone for a rate-dependent material are shown in Figs. 4 and 5 for the nondimensional crack speeds $m = 0.3$ and $m = 0.5$, respectively. The calculations were done with the value of the exponent $n = 5$ and with the various values of the rate-sensitivity parameter H shown in the graphs. For purposes of comparison, the corresponding strain distributions for a rate-independent material are also shown in the same figures. No analytical results are available with which these computed results may be compared for rate-dependent material response. From these figures, it appears that the strain distributions for the rate-dependent materials are more strongly singular at the crack tip, and that the size of the region over which the stronger singularity applies increases with increasing speed. The strength of the singularity may be estimated from an examination of the equation of motion (Eq 6) and the stress-strain relation as represented by Eqs 5 and 8. Suppose that the strain varies with distance ahead of the crack tip x as an inverse power, say x^{-s}, and consider the possibility that the elastic strain dominates the plastic strain. Then the governing equations are essentially the elastic equations and s must have the value ½.

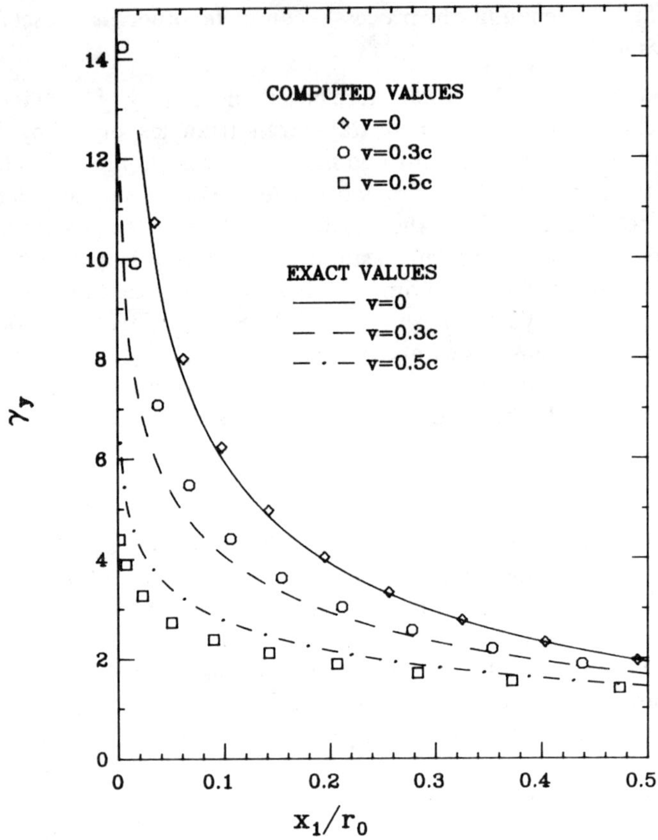

FIG. 3—*Normalized shear strain versus distance* x_1 *along the crack line in the active plastic zone for a rate-independent elastic-ideally plastic material;* r_0 *is the distance along the crack line from the crack tip to the elastic-plastic boundary.*

But if $s = ½$ then the elastic strain can dominate the plastic strain at the crack tip only if $n < 3$. Next, consider the possibility that the plastic strain dominates the elastic strain at the crack tip. The equation of motion suggests that the strength of the singularities in stress and total strain are the same. Then, in the stress-strain relation, there is no singularity stronger than the elastic singularity to balance the plastic strain singularity and, therefore, the magnitude of the plastic strain singularity is zero. For $n > 3$, this leaves as the only possibility that the strengths of the elastic and plastic strain singularities are the same, which implies that $s = 1/(n - 1)$. If this reasoning is correct, then the strain distributions in Figs. 4 and 5 are both singular at the crack tip as $x^{-1/4}$ for the case $n = 5$. It is interesting to note that the strain singularity anticipated here for $n > 3$ is the same as that predicted by Hui and Riedel [*16*] in their study of the asymptotic strain field near the tip of a growing crack

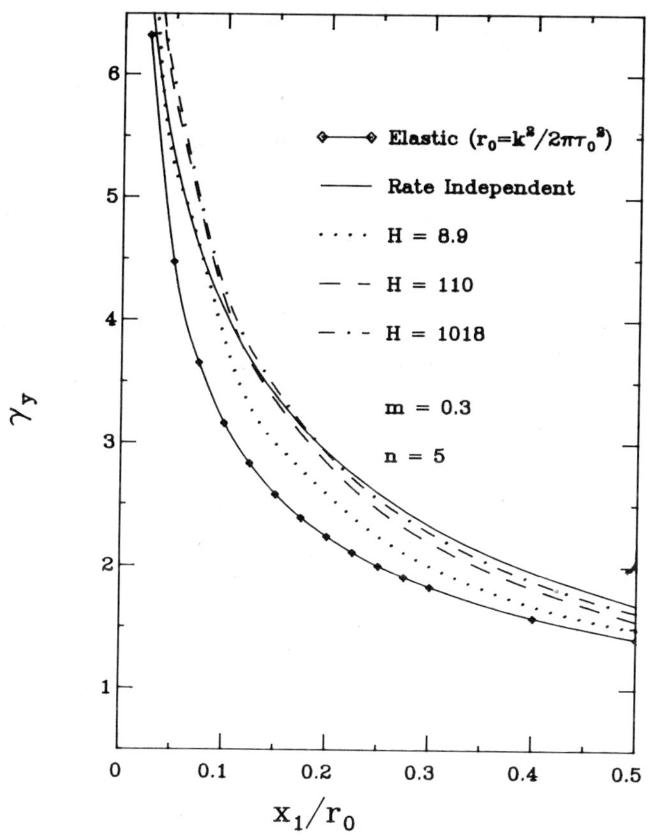

FIG. 4—*Normalized shear strain versus distance* x_1 *along the crack line in the active plastic zone for rate-sensitive material;* r_0 *is the distance along the crack line from the crack tip to the elastic-plastic boundary.* H *and* n *are defined through Eq 8.*

under creep conditions. However, because of the presence of the inertia term in the governing equations, it appears that the arguments concerning the asymptotic field cannot be made in the same concise way in which they are presented in Ref *16*. The range over which such an asymptotic field might be valid cannot be established without a more complete solution of the problem. Such a complete solution was presented for the rate-independent material in Ref *7*, where it was shown that the range of validity of the asymptotic solution did indeed increase with increasing crack tip speed.

To determine the level of the remotely applied stress-intensity factor k required to sustain crack growth at the predetermined speed v, a ductile fracture criterion must be introduced. The fracture criterion adopted is that proposed in Ref *1*, namely, that a crack will grow if a critical level of plastic strain occurs at a point on the line directly ahead of the crack at a characteristic

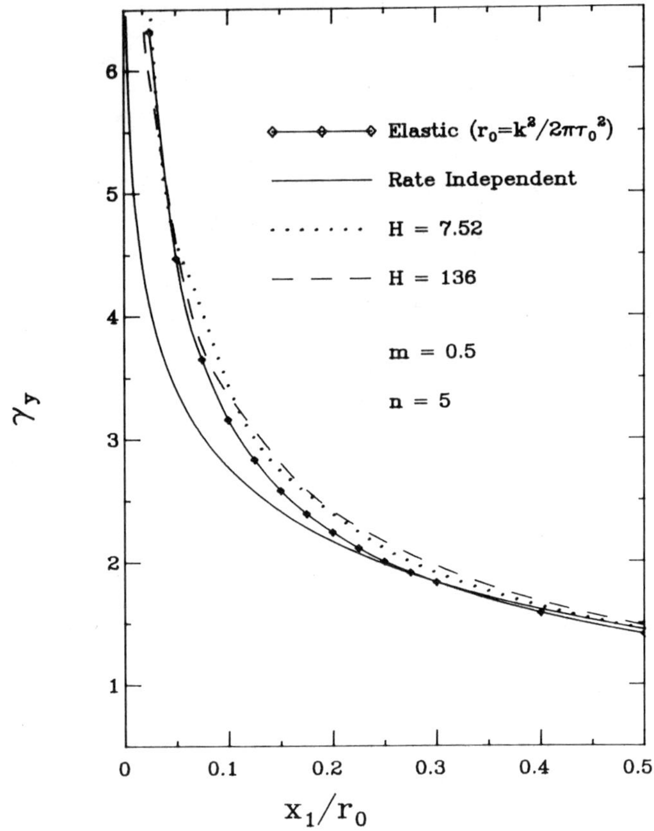

FIG. 5—*See Fig. 4.*

distance from the tip. If γ_f and d_f represent the critical (nondimensional) plastic strain and the characteristic length, respectively, then the crack will grow with $\gamma_y{}^p = \gamma_f$ at $x_1 = d_f$ on $x_2 = 0$. For levels of plastic strain below γ_f at $x_1 = d_f$ the crack cannot grow, and strains above γ_f at $x_1 = d_f$ are inaccessible. For present purposes, it is useful to eliminate d_f in favor of the critical stress-intensity factor k_c (assuming small-scale yielding). The value of k_c is defined as that stress-intensity factor required to satisfy the fracture criterion for a *stationary* crack in an elastic-ideally plastic material under slow-loading conditions. Thus, according to Rice [2], these material parameters are related by

$$d_f(\gamma_f + 1) = (k_c/\tau_0)^2 \tag{16}$$

which, on using Eqs 3, gives the nondimensional characteristic distance in terms of the failure strain and the ratio of the stress-intensity factor k needed to sustain crack advance at speed v to the static stress-intensity factor k_c

$$x_f = \frac{d_f \tau_0^2}{k^2} = \left(\frac{k_c}{k}\right)^2 \frac{1}{\pi(\gamma_f + 1)} \tag{17}$$

For a given critical strain γ_f, the stress-intensity factor (for the speed v) can be found from the strain versus nondimensional distance ahead of the crack tip x_f through Eq 17. This allows construction of a fracture toughness versus crack speed relationship for a material which fails in a locally ductile manner.

The next objective is to generate theoretical fracture toughness versus crack speed relationships based on the critical plastic strain criterion. Whereas such a relationship should be obtained for a fixed set of material parameters, the calculated strain distributions were obtained for fixed values of H. It is clear from its definition (Eq 9) that H depends on k and v as well as on material parameters. However, in view of Eqs 16 and 17, H may alternatively be expressed by

$$H = \frac{2\mu h d_f}{c\tau_0} \frac{1}{m x_f} = \frac{h^*}{m x_f} \tag{18}$$

where h^* depends only on the material parameters, including the rate sensitivity parameter h. In order to obtain toughness values for a range of values of crack speed m, but for *fixed* values of h^*, γ_f, and d_f, it is necessary to guess a value for H for each particular crack speed ratio m and critical strain γ_f. The value for h^* is then given by Eq 18, where x_f is obtained from the strain versus nondimensional distance ahead of the crack x for the particular critical strain γ_f. In order to obtain results for values of h^* within 1 percent of a desired value, typically three runs for each crack velocity and critical strain were needed. However, with information gained from previous calculations, good estimates of the H required to give a particular h^* can be made.

Estimates of the driving force ratio k/k_c required to sustain crack growth were obtained for $n = 5$ and two values of the nondimensional critical plastic strain, $\gamma_f = 2$ and 5. Calculations were performed for crack velocities of 1, 5, 10, 20, 30, 40, and 50 percent of the elastic shear wave speed c. Values of H were selected to result in graphs of k/k_c versus m for $h^* = 0.02$, 0.1, 1.0, and 10.0. The results are shown in Figs. 6 and 7. The results suggest that the influence of strain rate sensitivity is greatest at the lower crack speeds, where the influence of inertia is least. Although the calculated toughness continues to increase with increasing speed for all levels of rate sensitivity, the rate of increase is lower than for the corresponding rate-independent material for the higher values of speed. In fact, for the case of $\gamma_f = 5$, the curves seem to have a common crossover point at about $m = 0.45$. For speeds beyond the crossover point, the results shown an inverse dependence of calculated toughness on crack propagation speed. This feature may be connected with the increasing domain of validity of the strong algebraic singularity in strain at the crack tip with increasing speed.

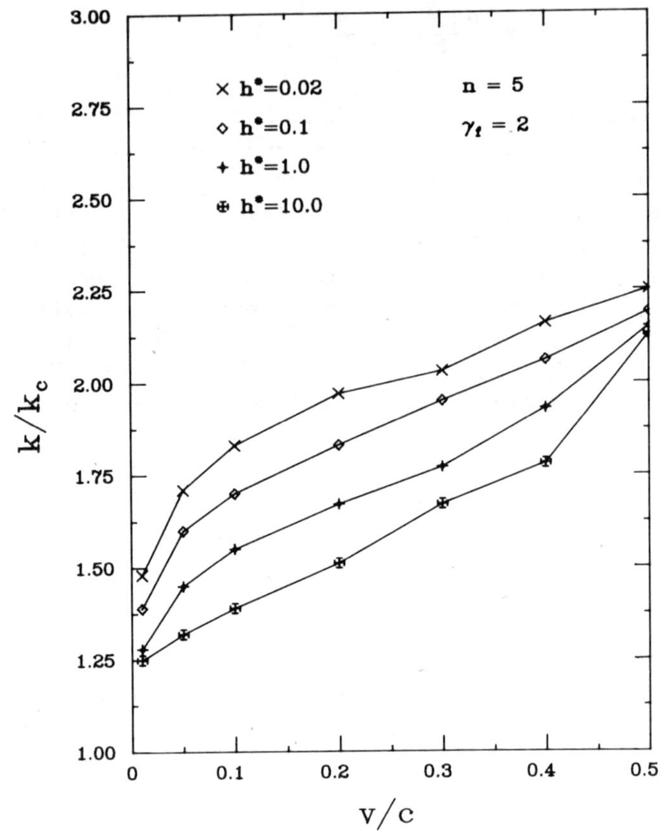

FIG. 6—*Remote stress-intensity factor required to sustain steady crack growth according to critical plastic strain criterion versus crack-tip speed. The material parameter* h* *is defined in Eq 18.*

The calculations have been performed with a critical plastic strain γ_f which is independent of crack tip speed or strain rate. If it were assumed that the value of γ_f increased with crack-tip speed, then the toughness would increase more rapidly with speed and the crossover point just mentioned would be moved to the right or even eliminated. On the other hand, if it were assumed that the value of γ_f decreased with crack-tip speed, then the toughness would increase less rapidly with speed. In this situation, it would be possible to generate toughness versus speed relationships which have local maxima and minima. If the value of k at a local minimum should fall near or below the value of k_c, it would have important implications in considering the fast fracture arrest capabilities of the material. A thorough discussion of this issue has been presented recently by Dantam and Hahn [*17*].

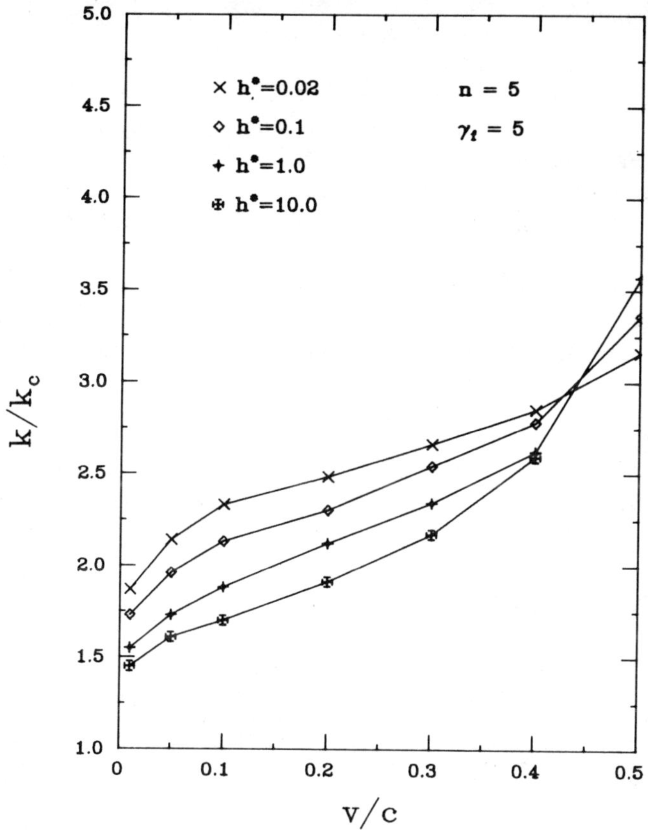

FIG. 7—*See Fig. 6.*

For the sake of brevity, attention was restricted here to the case of $n = 5$. Corresponding results for the case of $n = 1$ have also been obtained and they are reported in Ref *15*.

Finally, it is noted that a large amount of experimental data on fracture toughness versus crack speed for metals which fracture in a locally ductile manner are presented in Ref *17*. In all cases for which data are available, the toughness is a monotonically increasing function of crack speed. Furthermore, materials with significant strain rate sensitivity show the reversed curvature behavior exhibited in Figs. 6 and 7. The range of values of h^* for which calculations were done seems to include realistic values. For example, if the rate sensitivity parameter is in the range $10^3/\text{s} < h < 10^5/\text{s}$, the yield strain of the material is $\tau_0/\mu = 0.002$, and the critical distance $d_f = 10^{-4}$ m, then the value of h^* is in the range $0.02 < h^* < 2.2$.

Acknowledgments

The project described here is supported by the National Science Foundation, Solid Mechanics Program, Grant No. CME 77-15564. All computations were done in the Brown University, Division of Engineering VAX-11/780 Computer Facility. The acquisition of this computer was made possible by grants from the National Science Foundation (Grant No. ENG78-19378), the General Electric Foundation, and the Digital Equipment Corp.

Helpful discussions with Professor D. M. Parks of Massachusetts Institute of Technology concerning the numerical finite-element method and with Dr. K. K. Lo of Chevron Research Laboratories concerning the crack-tip strain singularity for rate-sensitive materials are gratefully acknowledged.

References

[*1*] McClintock, F. A. and Irwin, G. R. in *Fracture Toughness Testing and Its Applications, ASTM STP 381*, American Society for Testing and Materials, 1965, pp. 84–113.
[*2*] Rice, J. R. in *Fatigue Crack Propagation, ASTM STP 415*, American Society for Testing and Materials, 1967, pp. 247–311.
[*3*] Rice, J. R. in *Fracture: An Advanced Treatise*, H. Liebowitz, Ed., Academic Press, New York, 1968, pp. 191–311.
[*4*] Chitaley, A. D. and McClintock, F. A., *Journal of the Mechanics and Physics of Solids*, Vol. 19, 1971, pp. 147–163.
[*5*] Dean, R. H. and Hutchinson, J. W. in *Fracture Mechanics: Twelfth Conference, ASTM STP 700*, American Society for Testing and Materials, 1980, pp. 383–405.
[*6*] Douglas, A. S., Freund, L. B., and Parks, D. M. in *Proceedings*, Fifth International Conference on Fracture, Cannes, France, D. Francois, Ed., Pergamon Press, New York, 1981, pp. 2233–2240.
[*7*] Freund, L. B. and Douglas, A. S., *Journal of the Mechanics and Physics of Solids*, Vol. 30, 1982, pp. 59–74.
[*8*] Parks, D. M., Lam, P. S., and McMeeking, R. M. in *Proceedings*, Fifth International Conference on Fracture, Cannes, France, D. Francois, Ed., Pergamon Press, New York, 1981, pp. 2607–2614.
[*9*] Malvern, L. E., *Journal of Applied Mechanics*, Vol. 18, 1951, pp. 203–208.
[*10*] Perzyna, P., *Quarterly of Applied Mathematics*, Vol. 20, 1963, pp. 321–331.
[*11*] Rice, J. R., *Journal of Applied Mechanics*, Vol. 37, 1970, pp. 728–737.
[*12*] Lipkin, J., Campbell, J. D., and Swearengen, J. C., *Journal of the Mechanics and Physics of Solids*, Vol. 26, 1978, pp. 251–268.
[*13*] Chiem, C. Y. and Duffy, J., "Strain Rate History Effects in Aluminum Single Crystals During Dynamic Loading in Shear," Division of Engineering Technical Report, Brown University, Providence, R.I., 1981.
[*14*] Rice, J. R. and Tracey, D. M. in *Numerical and Computer Methods in Structural Mechanics*, S. J. Fenves, Ed., Academic Press, New York, 1973, pp. 585–623.
[*15*] Douglas, A. S., "Dynamic Fracture Toughness of Ductile Materials in Antiplane Shear," Ph.D. Thesis, Division of Engineering, Brown University, Providence, R.I., Aug. 1981.
[*16*] Hui, C. Y. and Riedel, H., "The Asymptotic Stress and Strain Field Near the Tip of a Growing Crack Under Creep Conditions," Division of Engineering Technical Report, Brown University, Providence, R.I., 1979.
[*17*] Dantam, V. and Hahn, G., "Evaluation of the Crack Arrest Capabilities of Ductile Steels," presented at the American Society of Mechanical Engineers Pressure Vessel and Piping Conference, Denver, Colo., June 1981.

Albert S. Kobayashi[1] *and Ouk S. Lee*[1]

Elastic Field Surrounding a Rapidly Tearing Crack

REFERENCE: Kobayashi, A. S. and Lee, O. S., **"Elastic Field Surrounding a Rapidly Tearing Crack,"** *Elastic-Plastic Fracture: Second Symposium, Volume I—Inelastic Crack Analysis, ASTM STP 803*, C. F. Shih and J. P. Gudas, Eds., American Society for Testing and Materials, 1983, pp. I-21–I-38.

ABSTRACT: The transient elastic stress field surrounding a rapidly tearing crack in a single-edged notch (SEN) specimen is analyzed by dynamic photoelasticity using a 16-spark gap Cranz-Schardin camera system. The 1.6-mm-thick annealed polycarbonate SEN specimens produced rapidly tearing cracks with approximately 11-mm-long necked regions which were modeled as Dugdale strip yield zones. In addition, the wake of the residual compressive stress along the rapidly tearing crack was modeled statically by a single force parallel to the crack and acting close to the physical crack tip. Theoretical isochromatics generated by this residual-stress-corrected, propagating Dugdale model of a rapidly tearing crack agreed reasonably well with those obtained experimentally by dynamic photoelasticity.

The experimental analysis showed that the length of Dugdale strip yield zone remained essentially constant through much of the rapid tearing. The crack-tip opening angle also remained constant, while the crack-mouth opening displacement increased during the terminal stage of crack propagation.

KEY WORDS: dynamic fracture, rapid tearing, ductile fracture, Dugdale strip yield zone, dynamic photoelasticity, elastic-plastic fracture

Unlike the efforts in elastodynamic fracture, little information is available on the transient state involved in dynamic ductile fracture, particularly in the presence of large-scale yielding. Earlier investigations on dynamic ductile fracture, including those of Kanninen et al [*1*],[2] Hahn et al [*2*], and Ogasawara et al [*3*] as well as by Koshiga et al [*4*] and Nora et al [*5*] involved phenomenological modelings of specific experimental observations. While these results are useful in predicting dynamic ductile fracture responses in specific

[1] Professor and graduate student, respectively, Department of Mechanical Engineering, University of Washington, Seattle, Wash. 98195.

[2] The italic numbers in brackets refer to the list of references appended to this paper.

structures, the lack of data on the near-field states of stress and strain makes it difficult to extract a dynamic ductile fracture criterion from these studies.

A major obstacle in experimental research in dynamic ductile fracture studies is the lack of a suitable experimental technique with which the dynamic states of stress, strain, plastic yield zone, etc., can be measured directly for detailed scrutiny. In contrast, dynamic photoelasticity and dynamic caustics provide an optical image of the elastodynamic field surrounding the running crack from which the dynamic stress-intensity factor, K_I^{dyn}, among others, can be extracted for elastodynamic fracture analysis. The plasticity counterpart of these optical techniques, that is, dynamic photoplasticity and dynamic caustics with plasticity, cannot be used because the theoretical dynamic near-field states of stress, strain, and displacement, which are needed for data interpretation, are not explicitly known at this time [*6, 7*]. Compounding this difficulty are the uncertainties in the theory of static photoplasticity [*8–10*] and hence in the theory of dynamic photoplasticity. The science of caustics in the presence of plastic yielding is in a similar state [*11*] although efforts have been made to model the plasticity effect by a propagating Dugdale-type strip yield zone [*12*].

Despite the aforementioned uncertainties, dynamic photoelasticity does provide information on the surrounding elastic field from which an estimate of the influence of a propagating crack-tip yield zone can be made. For example, a dynamic extension of Irwin's plasticity correction factor [*13*] can be extracted from this surrounding elastic field and used to correct the otherwise elastic dynamic stress-intensity factor, K_I^{dyn}. The experimentally determined elastodynamic field can also be used as a prescribed boundary condition for an elastoplastic dynamic finite-element analysis of the interior plastic field surrounding the propagating crack tip. Alternatively, the surrounding elastic field can be used to develop a mathematical model of the propagating plastic yield zone for the purpose of extracting fracture parameters which could relate to the elastoplastic dynamic fracture process.

The purpose of this paper is to develop such an elastic-plastic model using the dynamic isochromatics associated with a rapidly tearing crack. The Dugdale strip yield zone is used in this paper for its mathematical simplicity as well as for its apparent simulation of the observed strip yield zone ahead of the rapidly tearing crack in polycarbonate fracture specimens. The polycarbonate sheet, 1.6 mm thick, used in this investigation exhibited negligible strain-hardening prior to necking and thus the strain-hardening effect was not incorporated in the Dugdale strip yield zone.

Theoretical Background

Moving Dugdale Strip Yield Zone

The static and dynamic near-field states of stress in the vicinity of a Dugdale strip yield zone shown in Fig. 1 were derived in Ref *14* and Refs *12*

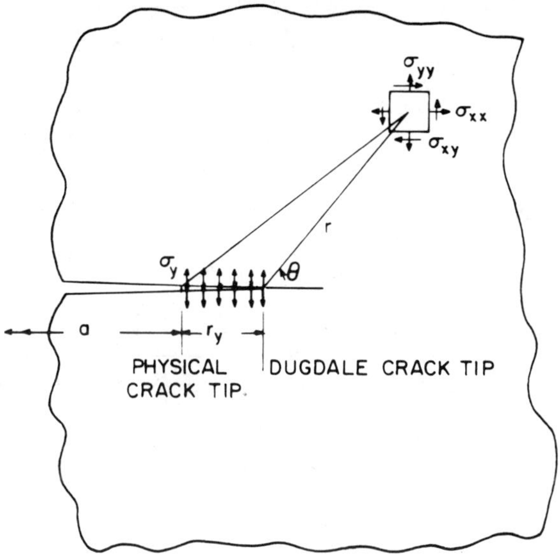

FIG. 1—*Dugdale strip yield zone at the crack tip of a semi-infinite crack.*

and *15*, respectively. In terms of the local polar coordinate system shown in Fig. 1, the dynamic maximum shear stress, τ_m, can be represented in the three stress components σ_{xx}, σ_{yy}, and σ_{xy} which were derived from Ref *16* as

$$\tau_m = \left[\left(\frac{\sigma_{yy} - \sigma_{xx}}{2}\right)^2 + \sigma_{xy}{}^2\right]^{1/2} \tag{1a}$$

where

$$\frac{\sigma_{yy} - \sigma_{xx}}{2} = -\frac{\sigma_{0x}}{2} - \sigma_y \left[B_1(1 + \beta_1{}^2) \tan^{-1} \frac{2\sqrt{\frac{r_y}{r_1}} \cos \frac{\theta_1}{2}}{1 - \frac{r_y}{r_1}} + B_2(1 + \beta_2{}^2) \tan^{-1} \frac{2\sqrt{\frac{r_y}{r_2}} \cos \frac{\theta_2}{2}}{1 - \frac{r_y}{r_2}} \right] \tag{1b}$$

$$\sigma_{xy} = \sigma_y \left[\beta_1 B_1 \log \frac{1 + \frac{r_y}{r_1} + 2\sqrt{\frac{r_y}{r_1}} \sin \frac{\theta_1}{2}}{1 + \frac{r_y}{r_1} - 2\sqrt{\frac{r_y}{r_1}} \sin \frac{\theta_1}{2}} \right.$$

$$+ \frac{(1+\beta_2{}^2)^2}{4\beta_2} B_2 \log \frac{1 + \dfrac{r_y}{r_2} + 2\sqrt{\dfrac{r_y}{r_2}} \sin \dfrac{\theta_2}{2}}{1 + \dfrac{r_y}{r_2} - 2\sqrt{\dfrac{r_y}{r_2}} \sin \dfrac{\theta_2}{2}} \Bigg] \tag{1c}$$

$$r_1 = r(\cos^2\theta + \beta_1 \sin^2\theta)^{1/2} \qquad r_2 = r(\cos^2\theta + \beta_2 \sin^2\theta)^{1/2} \tag{1d}$$

$$\tan\theta_1 = \beta_1 \tan\theta \qquad \tan\theta_2 = \beta_2 \tan\theta \tag{1e}$$

$$B_1 = \frac{1+\beta_2{}^2}{\pi D} \qquad B_2 = -\frac{4\beta_1\beta_2}{\pi D(1+\beta_2{}^2)} \tag{1f}$$

$$D = 4\beta_1\beta_2 - (1+\beta_2{}^2)^2 \tag{1g}$$

$$\beta_1{}^2 = 1 - \left(\frac{C}{C_1}\right)^2 \qquad \beta_2{}^2 = 1 - \left(\frac{C}{C_2}\right)^2 \tag{1h}$$

and

σ_{ox} = remote stress component,
σ_y = yield stress in uniaxial tension,
r_y = length of the Dugdale strip yield zone, and
C, C_1, C_2 = crack, dilitational, and distortional wave velocities, respectively.

Residual Stress Field in the Wake of a Propagating Crack

The effect of residual stresses in the wake of the propagating ductile crack cannot be ignored when large-scale yielding is involved. The residual plane-stress deformation field in the wake of a moving Dugdale strip yield zone has been analyzed by Budiansky and Hutchinson [*17*]. The σ_{xx} component in the wake of this Dugdale model vanishes while experimental results, as shown later, indicated the existence of a substantial residual σ_{xx} component in the unload region immediately behind the moving crack tip.

As a simplified modeling of the residual stress effect in the wake, the state of stress can be approximated as a varying uniaxial stress along the crack surface. An estimate of this normal stress distribution can be made by observing the termini of the isochromatics at the wake of the propagating crack in photoelastic specimens. Further simplification can be made by replacing the residual normal stress distribution by the average stress in the wake of the crack and subsequently with a concentrated force Q, acting close to the physical crack tip, that is, at a distance r_y from the Dugdale crack tip, as shown in Fig. 2. The *static* near-field state of stresses at the Dugdale crack tip due to this concentrated force can then be derived from a solution in Ref *18*. The

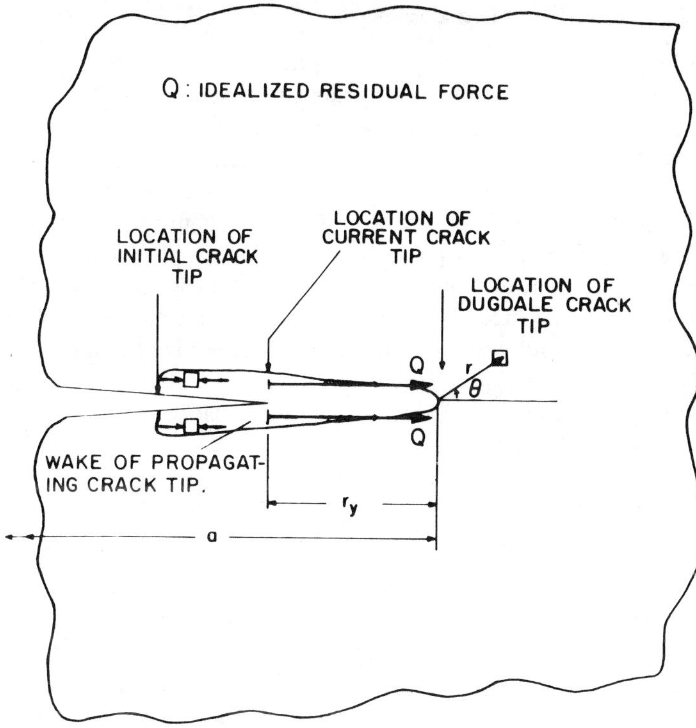

FIG. 2—*Residual stress field in wake of crack propagation.*

sum of the two normal stresses for this static boundary-value problem for $r_y \ll a$ becomes

$$\sigma'_{xx} = -\frac{2Q(r_y + r\cos\theta)^3}{\pi(r_y{}^2 + 2r_y r\cos\theta + r^2)^2} \tag{2a}$$

$$\sigma'_{yy} = -\frac{2Qr^2\sin^2\theta\,(r_y + r\cos\theta)}{\pi(r_y{}^2 + 2r_y r\cos\theta + r^2)^2} \tag{2b}$$

$$\sigma'_{xy} = -\frac{Q\sqrt{r_y r}\sin\theta\left(r_y{}^2\sin\frac{\theta}{2} + 2r_y r\sin\frac{3\theta}{2} + r^2\sin\frac{5\theta}{2}\right)}{\pi(r_y{}^2 + 2r_y r\cos\theta + r^2)^2}$$

$$-\frac{Q\sqrt{r_y r}\left(\frac{\sin\theta}{2}\cdot\sin\frac{5\theta}{2} - \cos\frac{3\theta}{2}\right)}{\pi(r_y{}^2 + 2r_y r\cos\theta + r^2)} \tag{2c}$$

where a is the half crack length of an embedded crack.

Moving Dugdale Strip Yield Zone with a Trailing Wake

If the Dugdale strip yield zone is propagating slowly, that is, at a crack velocity of $C/C_1 < 0.1$, then for all practical purposes the stress wave effect can be ignored and the static and dynamic crack-tip stress fields coincide. Equations 1*b*, 1*c*, and 2 can then be superimposed and, with Eqs 1*d* to 1*h*, be used through Eq 1*a* to construct near-field isochromatics at the tip of a propagating Dugdale strip yield zone. The validity of this model is then verified by comparing the experimentally determined dynamic photoelastic fringe patterns with those generated by Eqs 1 and 2 using experimentally determined parameters r_y and Q. Once the moving Dugdale strip yield zone is verified as a viable model of a rapidly tearing crack, then the associated parameters such as J and crack-opening displacement δ_t [*19*] can be studied in detail as plausible fracture parameters associated with rapid tearing.

Experimental Analysis

Polycarbonate SEN Specimens

The experimental program consisted of dynamic photoelasticity using a thin, that is, 1.6 mm thick, polycarbonate single-edged notch (SEN) specimen which is shown in Fig. 3. The annealed polycarbonate, from which the specimens were machined, has a well-defined static yield point of 44 MPa, flows under high stresses, and exhibits tensile instability with accompanying Lueder's bands. Under high strain rate loading, that is, about 10^{-1} 1/s, such as in the vicinity of a propagating crack tip, the strain-rate-sensitive polycarbonate yields at about 60 ~ 65 MPa [*19*], and fractures in a ductile mode with a 100 percent shear lip together with an apparent strip yield zone preceding the propagating crack tip. Details on this ductile brittle transition in dynamic loading of polycarbonate is discussed in Ref *20*.

In the as-received condition, the polycarbonate sheet exhibits considerable residual stress distribution and thus the sheets were annealed after rough cutting. This annealing, which caused some distortion and shrinkage, consisted of overnight heating at 160°C, followed by gradual cooling at the rate of 5°C per hour. The starter crack consisted of a 0.4-mm-wide edge crack approximately 25 mm in length which was machined and then chiseled to simulate a somewhat blunt crack tip. Dynamic stress-fringe constant, dynamic elastic modulus, dynamic yield stress, and Poisson's ratio at various strain rates were determined by a standard split Hopkinson bar system described in previous publications. A static modulus of elasticity of $E = 2.00$ GPa and yield strength of 62 MPa at a strain rate of 10^{-1} were determined separately for this annealed polycarbonate sheet.

The polycarbonate SEN specimens with blunt starter cracks of about 25 mm in length were loaded to failure and the dynamic photoelastic patterns

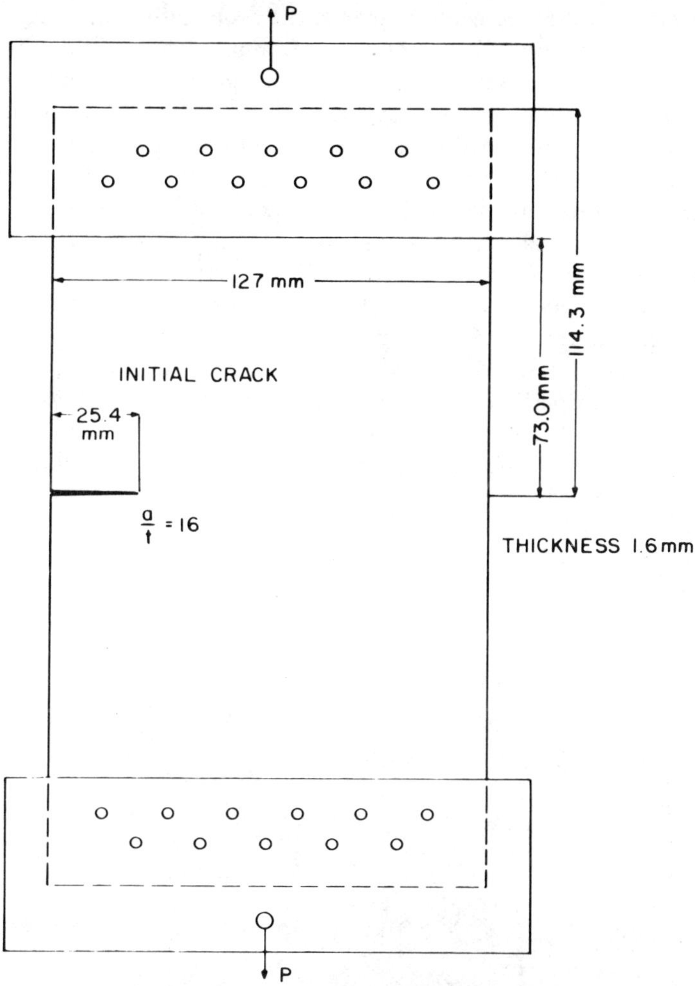

FIG. 3—*Polycarbonate single-edged notch specimen.*

surrounding the propagating crack tips were recorded. The 16-spark gap Cranz-Schardin camera and the associated dynamic photoelastic systems used to record 16 discrete dynamic photoelastic patterns has been described in many previous papers and thus will not be repeated here.

Data Reduction Procedure

Previous experimental investigations [*14,15*] showed that annealed thin polycarbonate fracture specimens can exhibit either cleavage or ductile fracture with shear lips depending on the loading conditions. For dynamic ductile fracture with about 50 percent shear lips, the propagating crack tip was preceded by a small strip yield zone of $r_y = 1.3$ mm in length. In these previous experimental investigations, a propagating Dugdale model was postulated and r_y was computed from the experimentally determined isochromatics. These computed r_y coincided with the lengths of the darkened apparent strip yield zones preceding the propagating crack. Thus, the length of this apparent strip yield zone represented by the darkened strip zone in the dynamic photoelastic record was used as the measured r_y in this investigation.

As mentioned previously, the elastic region surrounding the moving yield zone provides the elastodynamic state which is used to extract the apparent elastic fracture parameter for characterizing dynamic ductile fracture. In the static [*14*] and dynamic [*15*] procedures used in previous investigations, theoretical near-field isochromatics with disposal parameters, such as r_y and σ_{ox}, were least-squares fitted to the experimentally determined isochromatics and numerical values for these disposal parameters were extracted. For this inves-

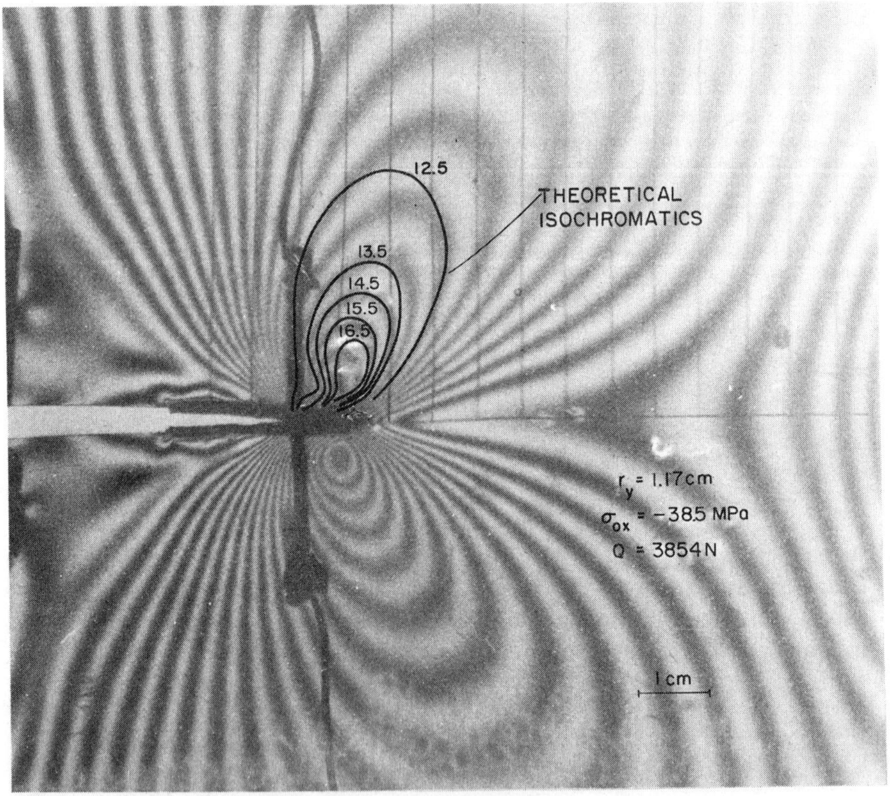

FIG. 4*a*—*Dynamic isochromatics of rapidly tearing polycarbonate SEN specimen (Specimen No. 70981P-OSL5, Frame No. 1).*

tigation, only σ_{ox} remained unknown since r_y was determined directly from experimental results.

The second disposal parameter in this investigation was the force Q which models the effect of the compressive residual stress in the trailing wake of the propagating crack shown in Fig. 2. Figures 4 and 5 show typical dynamic photoelastic patterns recorded in two rapidly tearing polycarbonate SEN specimens. The isochromatic loops which terminate at the trailing wake of the crack are approximately one half of σ_{xx} at this terminus since $\sigma_{yy} = 0$ on the crack surface. From these terminating isochromatics, an estimate of the residual stress distribution in the trailing wake can be obtained. Q can then be approximated by the average of the total compressive σ_{xx} distribution in the trailing wake and in part of the strip yield zone. Q is also assumed to act in a region adjacent to the physical crack tip for the reason described later in the Results section. In the actual data reduction procedure, Q was set as a disposal parameter which was determined together with r_y from the recorded isochro-

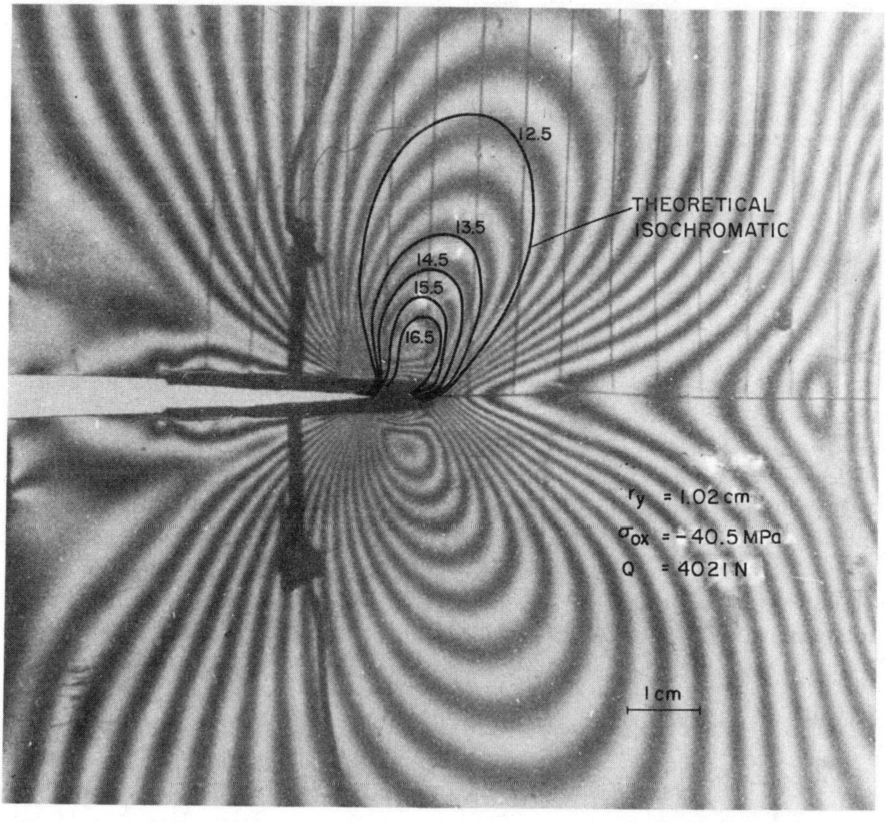

FIG. 4*b*—*Dynamic isochromatics of rapidly tearing polycarbonate SEN specimen (Specimen No. 70981P-0SL5, Frame No. 6).*

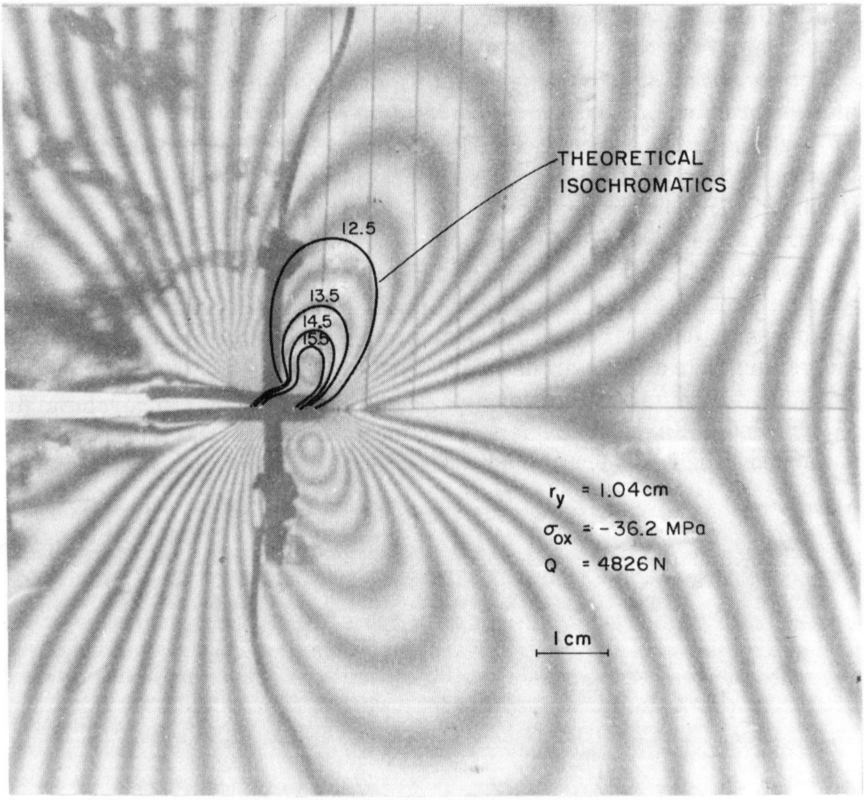

FIG. 5a—*Dynamic isochromatics of rapidly tearing polycarbonate SEN specimen (Specimen No. 70981P-0SL4, Frame No. 1).*

matic loops. The computed Q was then checked against Q estimated from the experimental results following the procedure described in the foregoing.

Having determined r_y directly from the dynamic photoelastic results, the overdeterministic least-squares method [*21*] was used to determine two unknowns, σ_{ox} and Q, in Eqs 1 and 4. This data reduction procedure for static and dynamic photoelasticity has been described by the authors and others in many previous papers and thus will not be elaborated on here.

Results

Crack Velocities

A total of four polycarbonate SEN specimens was tested and the crack velocities were determined by measuring the recorded instantaneous crack lengths at discrete time intervals. Measured crack velocities as shown in Fig. 6

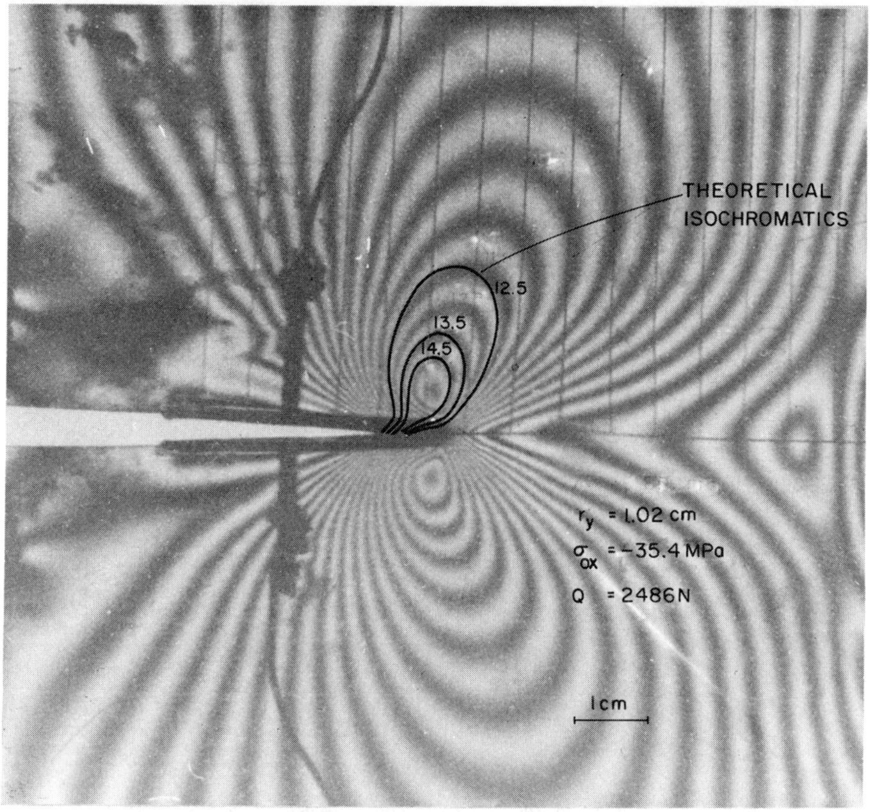

FIG. 5b—*Dynamic isochromatics of rapidly tearing polycarbonate SEN specimen (Specimen No. 70981P-0SL4, Frame No. 5).*

were at the most 0.05 C_1 and thus the computed dynamic effect in Eq 1 was negligible [*15*]. These results justify the superposition of the static results of Eq 4 onto the dynamic equations of Eq 1. In retrospect, a static near-field solution involving higher-order terms [*22*] could have been used together with Eq 4 for fitting larger isochromatic lobes.

Dugdale Strip Yield Zone Length, r_y

Figure 7 shows the variations in Dugdale strip yield zone lengths, r_y, with crack propagation. As mentioned previously, r_y was measured directly by the darkened yield zones ahead of the propagating crack tip in the photoelastic records. Figure 7 shows that r_y remained relatively constant throughout the entire crack propagation history. The measured r_y were then used as one of the input data in Eq 1.

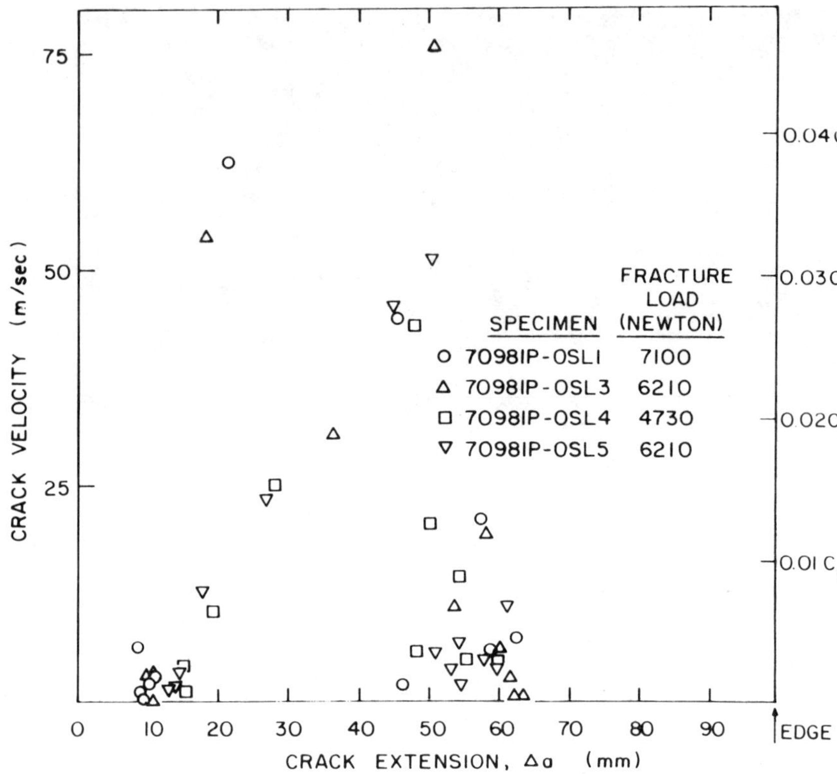

FIG. 6—*Crack velocity in four rapidly tearing polycarbonate SEN specimens.*

Isochromatics Surrounding the Crack Tip

Figures 4 and 5 show two typical dynamic photoelastic patterns in two tests. Isochromatic loops with a fringe order of $N = 13.5$ were used in the least-squares method of data reduction to compute σ_{ox} and Q. This fringe order was selected as the nearest isochromatic loop to the elastic-plastic boundary which corresponds to a fringe order of $N = 13.5$ based on a maximum in-plane shear stress of 31 MPa for yielding. This in-plane shear stress happens to be the maximum shear stress as will be shown in the Discussion section. The σ_{ox} and Q, obtained from the isochromatics, together with the measured r_y were then used to generate theoretical isochromatics. The measured r_y as well as the computed σ_{ox} and Q for each of the four dynamic photoelastic records in Figs. 4 and 5 are shown in the legend. The percent difference between these computed Q's and those measured directly from the isochromatics ranged from a low of 16 percent for Fig. 4*a* to a high of 28 percent for Fig. 5*a*. Nonlinearity in the stress optical coefficient for $N > 13.5$ was accounted for in regenerating the isochromatic loops using Eq 1. These computed iso-

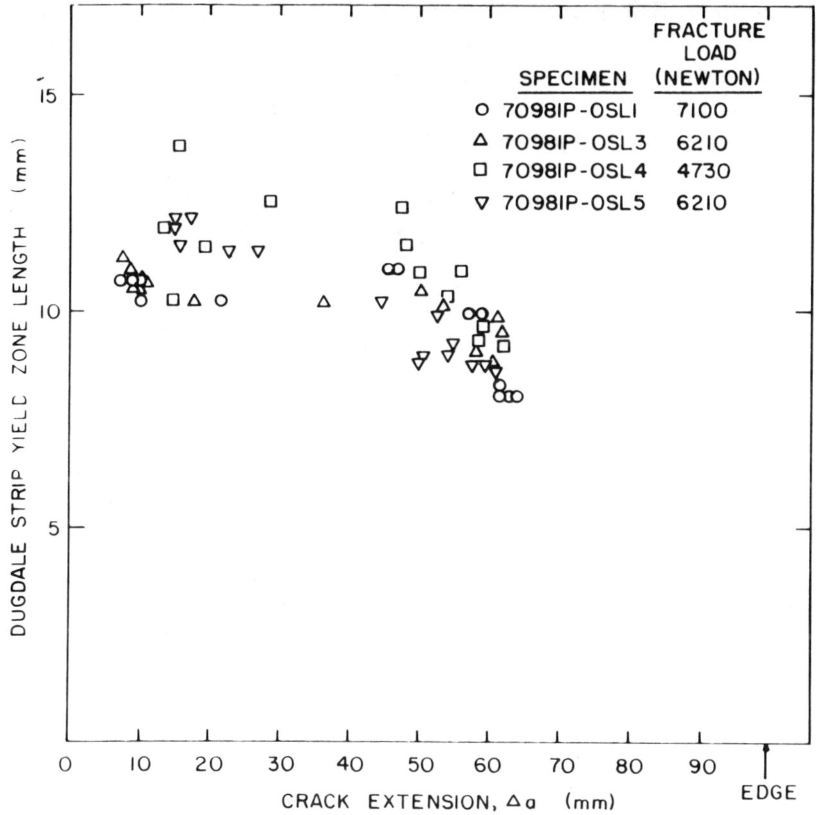

FIG. 7—*Dugdale strip yield zone length,* r_y, *in rapidly tearing polycarbonate SEN specimens.*

chromatics were then superimposed onto the recorded isochromatics as shown in black-inked curves in Figs. 4 and 5. Other than the obvious agreement between the fitted and experimental isochromatics of fringe order $N =$ 13.5, the closeness of the experimental and analytical isochromatics for $N -$ 14.5 and 15.5 is noted. The noted differences between theoretical and experimental isochromatics for $N = 12.5$ could be due to the effect of the finite boundary where the two-term representation for the near-field state of stress, that is, r_y and σ_{ox}, in Eq 1, is no longer sufficient to characterize the stress field which is obviously affected by the boundary. A higher-order static stress field [22] would have been a better choice for data reduction for these larger isochromatics.

The foregoing agreement between the theoretical and experimental isochromatics, that is, of fringe orders of $N = 13.5$ and 14.5, indicates that the propagating Dugdale strip yield zone with a trailing wake of residual stress field is a reasonable model of the rapidly tearing crack in the polycarbonate

SEN specimen. This model was then used to compute the three stresses, that is, σ_{xx}, σ_{yy}, and σ_{xy}, in the crack-tip region and a search was made for the yield zone, which is based on Treaca's yield criterion, surrounding the propagating Dugdale strip yield zone. For a dynamic yield stress of $\sigma_y = 62$ MPa, the yield zone could be approximated by the isochromatic of a fringe order of $N = 13.5$. Since $\sigma_{yy} = 62$ MPa along the Dugdale strip yield zone, the isochromatics of $N = 13.5$ must merge with the observed strip yield zone. Fig-

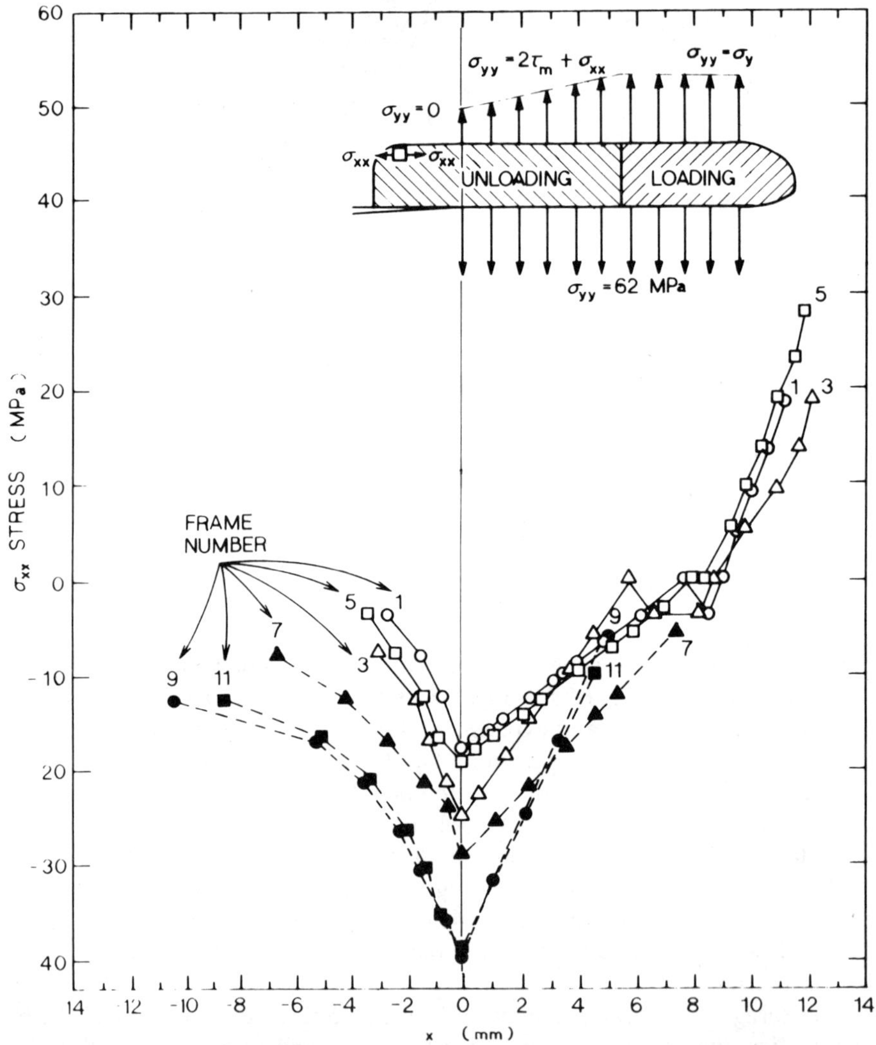

FIG. 8—σ_{xx} *distribution along the Dugdale strip yield zone and its wake of rapidly tearing crack (Specimen No. 70981P-0SL4).*

ures 4 and 5 show that the frontal portion of the $N = 13.5$ isochromatic does run into the frontal upper and lower boundaries of the strip yield zone, while the trailing portion of the $N = 13.5$ intersects with the strip yield zone ahead of the crack tip. The latter point of intersection is presumed to be the boundary between the loading and unloading regions [6, 7] in the strip yield zone as illustrated schematically in the legend of Fig. 8.

The higher fringe order of $N = 14.5$ and 15.5 is contained within the elastic-plastic boundary of $N = 13.5$. The prominent rotation of these higher-order isochromatics near the strip yield zone is due to the residual stresses in the trailing wake which are modeled reasonably well by Eq 4 as evidenced by the quantitative agreements between the theoretical and analytical isochromatics in Figs. 4 and 5.

Figure 8 shows the distribution of σ_{xx} along the wake and the strip yield zone. σ_{xx} along the wake was estimated by assuming $\sigma_{yy} = 0$ throughout the wake along the traction-free crack surface. Continuity of σ_{xx} behind and ahead of the propagating crack tip together with the isochromatics of fringe order less than $N = 13.5$ required that σ_{yy} be less than the yield stress of $\sigma_y =$ 62 MPa and thus a linearly increasing σ_{yy} was assumed in this unloading re-

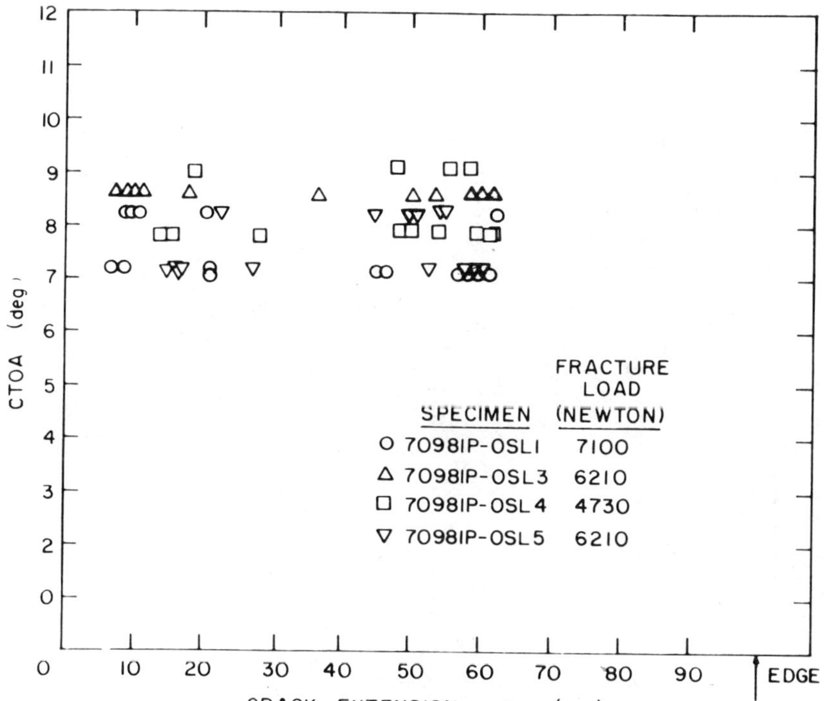

FIG. 9—*CTOA in four rapidly tearing polycarbonate SEN specimens.*

gion. The remaining σ_{xx} distribution in the strip yield zone was then constructed from the isochromatic distribution based on this model. The sharp change in σ_{xx} at the physical crack tip justifies the application of force Q at this point to simulate the effect of the trailing wake of the propagating strip yield zone.

CTOA and CMOD

The crack-tip opening angle (CTOA) and the crack-mouth opening displacement (CMOD) can be measured directly from the photoelastic record. CTOA is defined as the crack-tip angle subtending the crack opening displacement at a distance r_m from the crack tip. This small distance, r_m, which should be comparable to the initial fracture process zone [*23*], was equated to the initial crack-tip blunting. CMOD is defined as the crack-tip angle subtending the crack opening displacement at the initial crack tip. Figures 9 and 10 show the variations in measured CTOA and CMOD, respectively. The CTOA remained relatively constant through the crack propagation. Sharp increases in CMOD as the crack approaches the edge of the specimen indicate that CMOD will not qualify as a possible fracture parameter in rapid tearing.

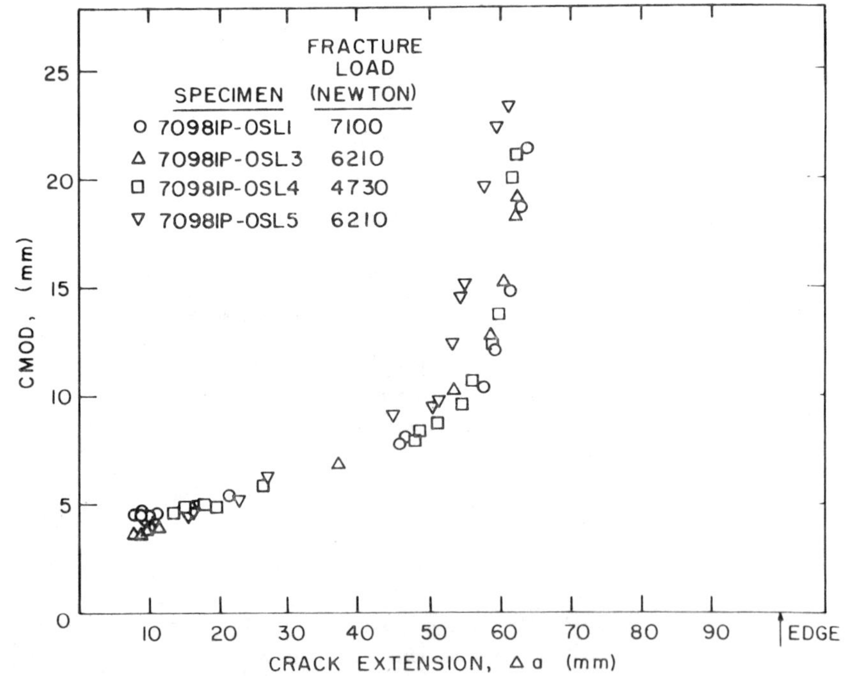

FIG. 10—*CMOD in four rapidly tearing polycarbonate SEN specimens.*

Conclusions

1. A quasi-static analysis of rapidly tearing thin polycarbonate SEN specimens indicates that the moving Dugdale strip yield zone with a trailing wake of residual stress field models reasonably well the stress field surrounding the propagating crack tip.

2. The length of the Dugdale strip yield zone, r_y, and the crack-tip opening angle, CTOA, remained relatively constant through the rapid tearing process.

3. CMOD increased sharply as the crack approached the specimen edge and thus may not be a suitable candidate as a fracture parameter.

Acknowledgments

The results of this investigation were obtained through a research grant funded by the National Science Foundation (NSF) Grant No. CME-792507. The authors wish to acknowledge the support and encouragement of Dr. C. J. Astill of NSF during the course of this investigation.

References

[*1*] Kanninen, M. F., Mukherjee, A. K., Rosenfield, A. R., and Hahn, G. T. in *Mechanical Behavior of Materials Under Dynamic Loads*, U. S. Lindholm, Ed., Springer-Verlag, Berlin, 1968, p. 96.

[*2*] Hahn, G. T., Sarate, M., Kanninen, M. F., and Rosenfield, A. R., *International Journal of Fracture*, Vol. 9, June 1973, pp. 209–222.

[*3*] Ogasawara, M., Mimura, M., and Ishiguro, K. in *Proceedings*, International Symposium on Crack Propagation in Pipelines, Newcastle-Upon-Tyne, U.K., March 1974.

[*4*] Koshiga, F., Kurita, Y., Akiyama, T., and Kawahara, M., *Transactions*, American Society of Mechanical Engineers, *Journal of Pressure Vessel Technology*, Vol. 100, Feb. 1980, pp. 18–23.

[*5*] Nora, Y., Fukada, M., Nogaki, N., Koga, T., and Takeuchi, I., "Study on Resisting of Various Types of Steels Against Propagating Shear Fracture by Modified West Jefferson Type Burst Test," ASME Preprint 78-PVP-71, American Society of Mechanical Engineers, 1978.

[*6*] Achenbach, J., Kanninen, M. F., and Popelar, C. H., *Journal of the Mechanics and Physics of Solids*, Vol. 29, No. 3, 1981, pp. 211–225.

[*7*] Achenbach, J. and Dunayevsky, V., *Journal of the Mechanics and Physics of Solids*, Vol. 29, No. 4, 1981, pp. 283–303.

[*8*] Zachary, L. W. and Riley, W. F., *Experimental Mechanics*, Vol. 17, No. 9, Sept. 1977, pp. 321–326.

[*9*] Doyle, J. F., *Experimental Mechanics*, Vol. 20, No. 2, Feb. 1980, pp. 65–67.

[*10*] Freire, J. L. F. and Riley, W. F., *Experimental Mechanics*, Vol. 20, No. 4, April 1980, pp. 118–125.

[*11*] Theocaris, P. S., *Transactions*, American Society of Mechanical Engineers, *Journal of Applied Mechanics*, Vol. 37, Series E, No. 2, June 1970, pp. 409–415.

[*12*] Rosakis, A. J. and Freund, L. B., *Transactions*, American Society of Mechanical Engineers, *Journal of Applied Mechanics*, Vol. 48, No. 2, June 1981, pp. 302–308.

[*13*] Irwin, G. R. in *Proceedings*, Sagamore Ordinance Materials Research Conference, Report ME TE 661-611/F, Syracuse Research Institute, Syracuse, N.Y., 1960, pp. 63–78.

[*14*] Kobayashi, A. S., Ramulu, M., and Kang, B. S.-J. in *Proceedings*, 2nd International Conference on Numerical Methods in Fracture, D. R. J. Owens and A. R. Luxmore, Eds., University College of Swansea, Swansea, U.K., 7–11 July 1980, pp. 535–538.

[15] Betser, A. A., Kobayashi, A. S., Lee, O. S., and Kang, B. S.-J., "Crack Tip Dynamic Isochromatics in the Presence of Small Scale Yielding," *Experimental Mechanics*, Vol. 22, April 1982, pp. 132–138.
[16] Kanninen, M. F., *Journal of the Mechanics and Physics of Solids*, Vol. 16, 1968, pp. 215–225.
[17] Budiansky, B. and Hutchinson, J. W., *Transactions*, American Society of Mechanical Engineers, *Journal of Applied Mechanics*, Vol. 45, No. 2, June 1978, pp. 267–276.
[18] Erdogan, F. in *Proceedings*, 4th U.S. National Congress on Applied Mechanics, 1962, pp. 547–553.
[19] Rice, R. C., *Transactions*, American Society of Mechanical Engineers, *Journal of Applied Mechanics*, Vol. 35, June 1968, pp. 379–386.
[20] Yee, A. F., "The Yield and Deformation Behavior of Some Polycarbonate Blend," General Electric Co. Technical Information Series, 76 CRD 130, Schenectady, N.Y., Aug. 1976.
[21] Kobayashi, A. S. and Ramulu, M., *Experimental Mechanics*, Vol. 21, Jan. 1981, pp. 41–48.
[22] Bradley, W. B. and Kobayashi, A. S., *Engineering Fracture Mechanics*, Vol. 3, 1971, pp. 317–332.
[23] Rice, J. R., Drugan, W. J., and Sham, T-L. in *Fracture Mechanics, ASTM STP 700*, American Society for Testing and Materials, July 1980, pp. 189–221.

Rick H. Dean[1]

Elastic-Plastic Steady Crack Growth in Plane Stress

REFERENCE: Dean, R. H., **"Elastic-Plastic Steady Crack Growth in Plane Stress,"** *Elastic-Plastic Fracture: Second Symposium, Volume I—Inelastic Crack Analysis, ASTM STP 803,* C. F. Shih and J. P. Gudas, Eds., American Society for Testing and Materials, 1983, pp. I-39–I-51.

ABSTRACT: Results are presented for a crack which is growing quasi-statically under plane-stress conditions. Crack-tip stress and strain fields have been computed by the finite-element method for a crack which is propagating steadily with symmetric small-scale yielding boundary loads (Mode I). The analysis assumes that the material is a standard J_2-flow theory elastic-plastic solid exhibiting isotropic-hardening behavior. Results are discussed for various levels of material hardening.

KEY WORDS: small-scale yielding, quasi-static, elastic-plastic fracture, plane stress, Mode I, steady state, crack propagation

Substantial amounts of stable crack growth have been observed in plates possessing various crack geometries [*1–3*].[2] Since stable crack growth occurs frequently, and is often extensive, it is important that the entire crack growth process be thoroughly understood. This paper is concerned with the steady-state stress and strain fields which develop at the tip of a crack which is growing quasi-statically in an elastic-plastic solid.

There have been a number of recent publications in the area of quasi-static steady crack growth in elastic-plastic solids. Amazigo and Hutchinson [*4*] calculated the asymptotic steady-state solutions for Mode I loading conditions in plane stress and plane strain and Mode III loading conditions in antiplane shear for a bilinear hardening solid (a material whose uniaxial stress-strain curve is composed of two straight-line segments). Yu-chen and Keh-chih [*5*] performed an analysis similar to that of [*4*] for a power-hardening material. Numerical studies of quasi-static steady crack growth include the works of

[1]Senior reservoir engineer, Arco Oil and Gas Co., Dallas, Tex. 75075.

[2]The italic numbers in brackets refer to the list of references appended to this paper.

Chitaley and McClintock [6] for the Mode III antiplane shear problem, Andersson [7] on the Mode I plane stress and Mode III antiplane shear problems, Dean and Hutchinson [8] on Mode I plane strain and Mode III antiplane shear, and Parks, Lam, and McMeeking [9] on Mode I in plane strain.

This paper presents steady-state results for a bilinear hardening material in plane stress. The finite-element method has been used to compute the steady-state solutions for a growing crack under plane-stress Mode I conditions where a diffuse plastic zone develops at the crack tip. Small scale yielding assumptions (SSY) are made throughout and all strain-rate sensitivity and inertial effects are neglected. Sources of numerical error are carefully examined and comparisons are made between the present results and those available in the current literature.

Solution Procedure

In SSY, the crack is taken to be semi-infinite and the stresses are given by

$$\sigma_{ij} \rightarrow \frac{K}{\sqrt{2\pi r}} f_{ij}(\theta) \tag{1}$$

as $r \rightarrow \infty$, where K is the stress-intensity factor and f_{ij} a function of the angle θ determined by the Mode I plane-stress loading conditions. If the load is slowly increased on a stationary crack, the crack will begin to grow at some value of K called K_c as shown in Fig. 1. An increasing K-value is then required

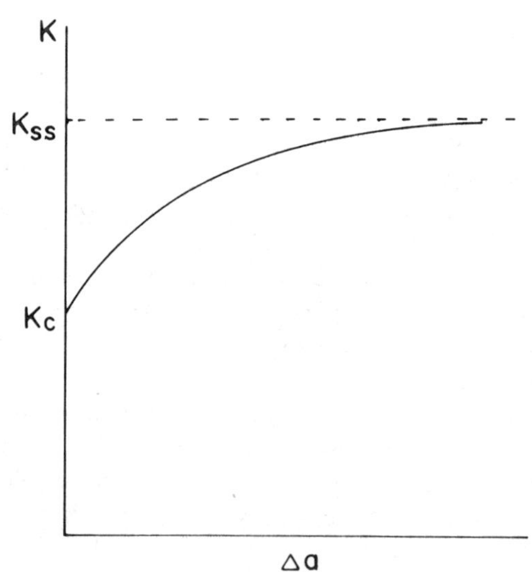

FIG. 1—*Crack growth resistance curve.*

to further propagate the crack. The K-value asymptotes to a steady-state value (K_{ss}) after a sufficient amount of crack growth. A typical steady-state field which might develop is shown in Fig. 2 where there is an active plastic zone ahead of the crack tip with a trailing wake of residual plastic strains and a possible reloading zone along the crack face.

The elastic-plastic solid is modeled by a bilinear hardening flow rule from a simple tension test for which the constitutive equation is given by

$$\dot{\sigma}_{ij} = \frac{E}{1+\nu}\left\{\dot{\epsilon}_{ij} + \frac{\nu}{1-2\nu}\,\dot{\epsilon}_{KK}\delta_{ij} - \beta\,\frac{(1-\alpha)}{1+\alpha(2\nu-1)/3}\,\mu_{ij}\mu_{kl}\dot{\epsilon}_{kl}\right\} \quad (2)$$

where

β = 0 (elastic) or 1 (plastic),
$\alpha = E_t/E$ (nondimensional tangent modulus),
$\mu_{ij} = S_{ij}/\sqrt{S_{kl}S_{kl}}$ (unit normal to the yield surface), and
$S_{ij} = \sigma_{ij} - 1/3\sigma_{KK}\delta_{ij}$ (stress deviator).

The material is assumed to be initially isotropic with a classical Huber-von Mises initial yield surface given by $J_2 = 1/2S_{ij}S_{ij} = 1/3\sigma_0^2$ where σ_0 is the yield stress in simple tension. The yield surface is made to expand uniformly (isotropic hardening) throughout the loading history.

The steady-state solution propagates with the crack as shown in Fig. 2. Thereby, derivatives of stress and strain in Eq 2 can be replaced by derivatives taken in the direction of the propagation of the crack (x-direction). The loading history of a particular material point is then determined entirely by the stress history parallel to the x-axis.

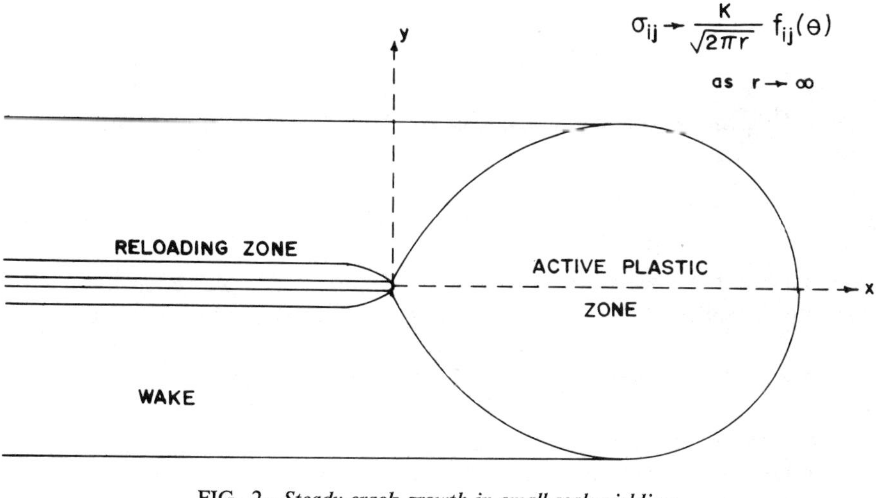

FIG. 2—*Steady crack growth in small-scale yielding.*

The field equations are nondimensionalized in order to reduce the number of material parameters. The original displacements, positions, stresses, and strains are replaced by

$$\begin{Bmatrix} \bar{U}_x \\ \bar{U}_y \end{Bmatrix} = \frac{E\sigma_0}{K^2} \begin{Bmatrix} U_x \\ U_y \end{Bmatrix} \qquad \bar{r} = \left(\frac{\sigma_0}{K}\right)^2 r \tag{3}$$

$$\bar{\sigma}_{ij} = \sigma_{ij}/\sigma_0 \quad \text{and} \quad \bar{\epsilon}_{ij} = E\epsilon_{ij}/\sigma_0$$

where σ_0 is the yield stress in simple tension and all barred variables are now nondimensional. The form of Eq 1 remains unchanged in the nondimensional system except the K is no longer present. The stress-intensity factor does not enter into any of the finite-element calculations. The steady-state value of K is determined by the ratio of K_{ss} to K_c when a fracture criterion is finally used in comparing solutions for the stationary and steadily growing crack problems. The only material parameters present in the calculations are the nondimensional tangent modulus E_t/E and Poisson's ratio ν (which is taken to be 0.3).

An iterative finite-element technique used in Refs *8, 9,* and *10* was implemented to compute the steady-state solutions. The principal of virtual work (PVW) may be written in vector form as

$$\int_A \vec{\sigma}^T \, \delta \, \vec{\epsilon} dA = \int_S \vec{F}^T \, \delta \, \vec{u} ds \tag{4}$$

where dA is an area element and the traction vector $\vec{F}$ is applied along a boundary S far removed from the crack tip. The vector $\vec{F}$ is determined from the elastic asymptotic stresses given in Eq 1. The strains and displacements within an element are given by

$$\vec{\epsilon} = \mathbf{B}\vec{U} \quad \text{and} \quad \vec{u} = \mathbf{R}\vec{U} \tag{5}$$

where $\vec{U}$ is the vector of nodal displacements. The total stresses and strains are connected by

$$\vec{\sigma} = \mathbf{D}^e(\vec{\epsilon} - \vec{\epsilon}^p) \tag{6}$$

where $\mathbf{D}^e$ is the elastic constitutive matrix and $\vec{\epsilon}^p$ the plastic strain vector. Writing the elastic stiffness matrix as $\mathbf{K}^e = \int_A \mathbf{B}^T\mathbf{D}^e\mathbf{B}dA$ the PVW becomes

$$\mathbf{K}^e\vec{U}_{i+1} = \int_S \mathbf{R}^T\vec{F}ds + \int_A \mathbf{B}^T\mathbf{D}^e\vec{\epsilon}_i^p dA \tag{7}$$

where the i denotes the ith iteration.

The iterative procedure begins with the initial plastic strain vector set equal to zero ($\vec{\epsilon}_0^p = 0$). Equation 7 is solved for $\vec{U}_1$ and then $\vec{\epsilon}_1$ is calculated for any element of interest using Eq 5. After the strains are found everywhere, the stresses are then determined by integrating Eq 2 along horizontal lines in the negative x-direction where the elastic-plastic constitutive matrix in Eq 2 is calculated in the form described by Rice and Tracey [*11*]. The plastic strains are then calculated using Eq 6 and then plugged into Eq 7 to calculate $\vec{U}_2$. The iterations proceed in the manner just described until the relative difference between iterations is less than 0.01 percent. Each steady-state solution required 10 to 90 iterations, depending on the degree of material hardening, where each iteration consumed two seconds of central processing unit time on an Amdahl 470 V/6-II for a system with 3218 degrees of freedom.

Error Analysis

This section discusses the sources of numerical errors which are present in a finite-element approximation to a continuous solid. No attention will be given to the errors inherent in representing a real material with a bilinear hardening constitutive equation. There are several forms of numerical error which arise during the solution procedures. Sources of error originate because of the discretization process, computation round-off, the representation of an infinite SSY domain by a finite region, and due to a finite number of iterative steps for a nonlinear problem.

The computational round-off errors were assessed by first solving the elastic plane-stress SSY problem in single and double precision with corresponding real number lengths of 7.2 and 16.8 decimal digits, respectively. The single-precision stresses, strains, and displacements were slightly larger in magnitude than the corresponding analytical results over most of the region outside the crack tip. The double-precision results were in better agreement with the analytical solution, and the corresponding hypothetical plastic boundaries ($J_2 = 1/2S_{ij}S_{ij} = 1/3\sigma_0{}^2$) are shown in Fig. 3. The elastic-perfectly plastic solution was also computed in single- and double-precision arithmetic. However, the two solutions for the elastic-plastic case were in better agreement than the two solutions for the purely elastic problem. The decomposition of the stiffness matrix in Eq 7 is a major source of round-off error. In addition, the elastic stiffness matrix has a stronger influence on the purely elastic problem than on the elastic-plastic problem and this may explain the reason we see a larger round-off error for the purely elastic solution. All results reported in this paper were computed using double-precision arithmetic.

A second source of error arises when an infinite domain is approximated by a finite region. The asymptotic loads determined by Eq 1 were applied at a distance R_b from the crack tip. The elastic-perfectly plastic problem was then solved for various values of the nondimensional distance $R_b/(K/\sigma_0)^2$ where each solution predicted a slightly different plastic zone. Let R_p be the length of the plastic zone ahead of the crack tip. Figure 4 shows the length R_p as a function of R_b where R_p appears to approach an approximate value of 0.305

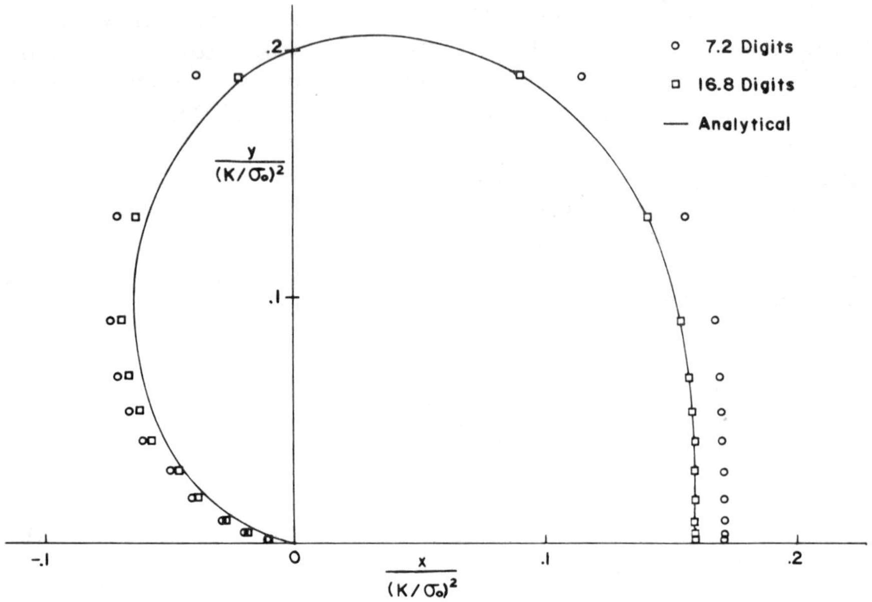

FIG. 3—*Single- and double-precision elastic-plastic boundaries for the purely elastic computations.*

$(K/\sigma_0)^2$ as $R_b \rightarrow \infty$. All results reported in this paper were computed for an R_b to R_p ratio of approximately 30 to 1.

Another source of numerical error is associated with the discretization process. A continuous domain is represented by a discrete number of constant-strain triangles and condensed quadrilaterals (each quadrilateral is composed of four constant-strain triangles with the center node condensed out before forming the global stiffness matrix). A coarse representation of the finite-element mesh is shown in Fig. 5. The rectangular region was required along the crack plane to be able to integrate the loading histories as discussed earlier. The mesh was adjusted so that an adequate solution was obtained for the stress and strain singularities in the purely elastic case. The same mesh was then used for the elastic-plastic solutions. The ratio of plastic zone size (R_p) to smallest quadrilateral was approximately 34 to 1. The mesh consisted of 3218 degrees of freedom (DOF) before nodal condensation and 1647 DOF after condensation.

A final source of numerical error results from the iterative approach to a nonlinear problem. The iterations were terminated when the relative difference between iterates was less than 0.01 percent. Given the other possible forms of numerical error, it was felt that this small difference between iterations was sufficient for the accuracy desired.

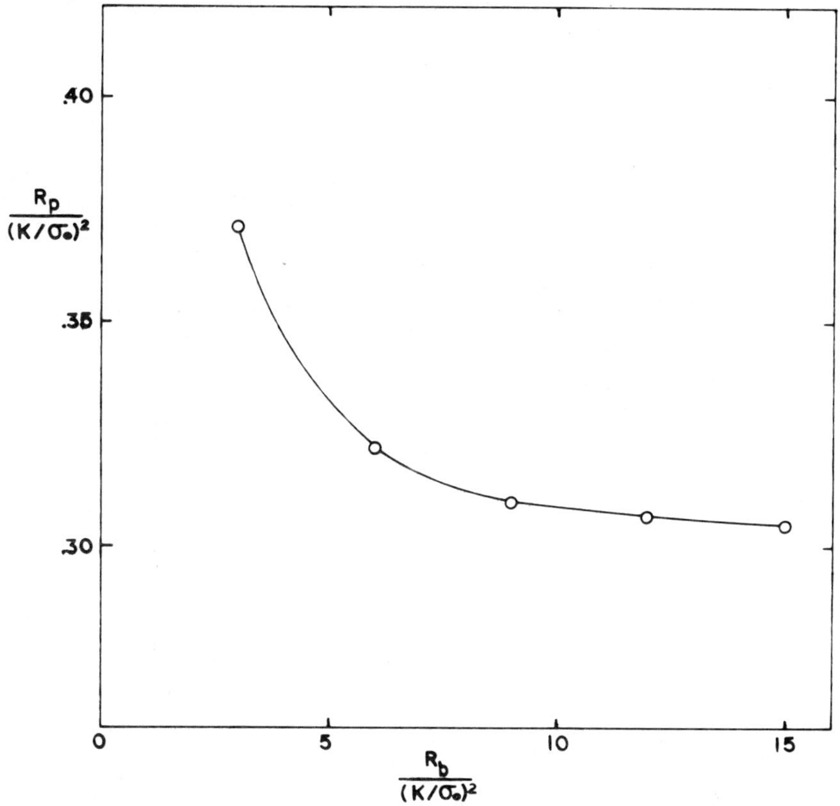

FIG. 4—*Boundary load influence on elastic-perfectly plastic plastic zone.*

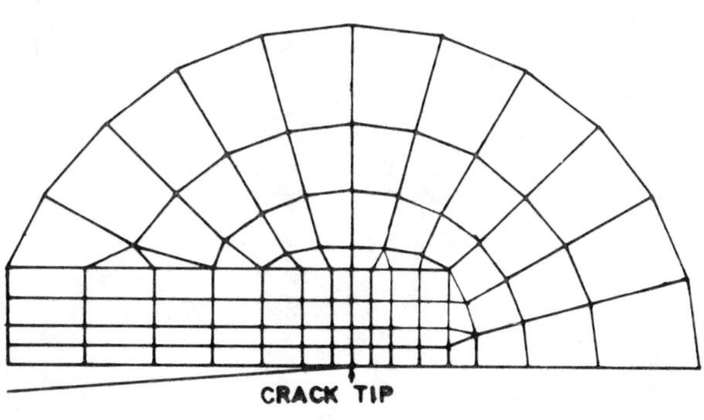

FIG. 5—*Coarse representation of finite-element mesh.*

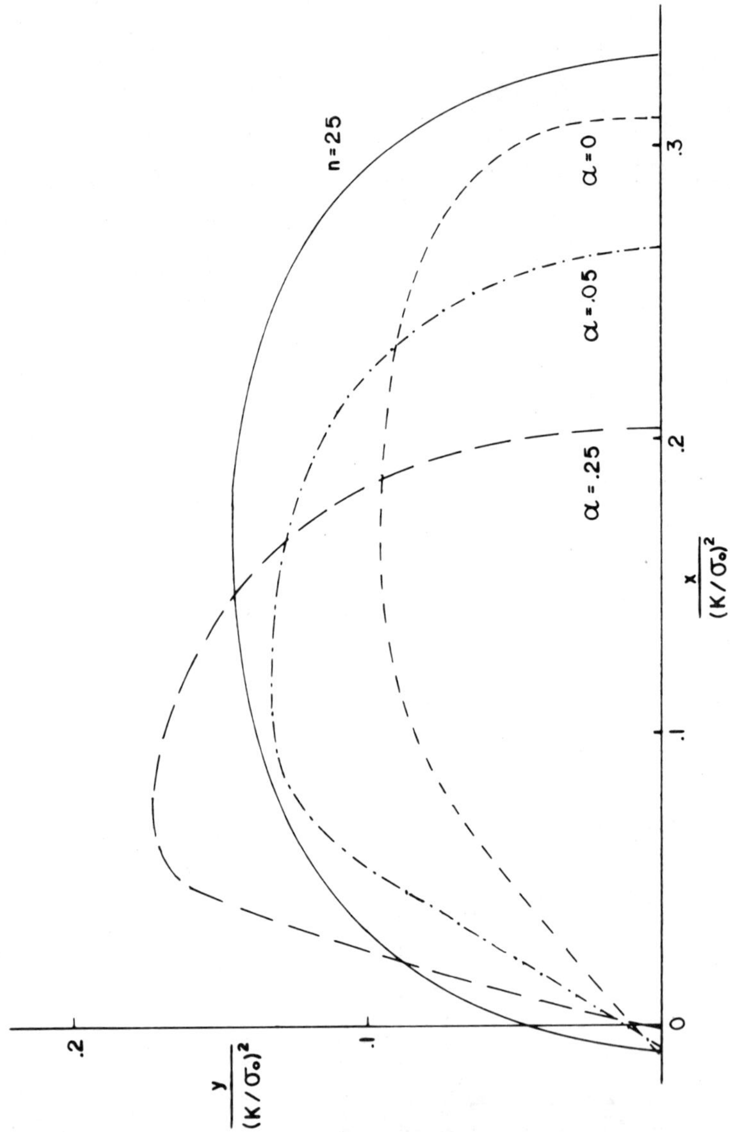

FIG. 6—*Stationary and steady-state plastic zones.*

Results

Steady-state solutions were computed for values of the nondimensional tangent modulus, α in Eq 2, equal to 1.0, 0.25, 0.05, and 0 (elastic-perfectly plastic). Plastic zones are shown in Fig. 6 for the steadily growing crack along with the plastic zone for a stationary crack in a power-hardening material. The stationary crack was analyzed by Hutchinson and Shih [*12*] for a power-hardening solid with a flow rule of the form $\epsilon^p \sim \alpha(\sigma/\sigma_0)^{n-1}\,\sigma/E$ where the stationary result in Fig. 6 is for an exponent of $n = 25$. No reloading zone was seen along the crack face for the steadily growing crack for any level of hardening α.

The stress distribution surrounding the crack tip is shown in Fig. 7 for a crack which is growing steadily in an elastic-perfectly plastic solid. Hutchinson [*13*] derived the asymptotic stress field for a stationary crack in an elastic-perfectly plastic material. The present results are fairly similar to those presented in Ref *13* except for the σ_{rr} term for the angle $\theta > 150$ deg. The present σ_{rr} results go to zero as θ approaches 180 deg while σ_{rr} in Ref *13* goes to -1.

The steady-state crack-opening profiles are shown in Fig. 8 for selected values of the hardening parameters α. The crack-opening profiles vary considerably with the hardening level, which also occurred in the Mode III anti-

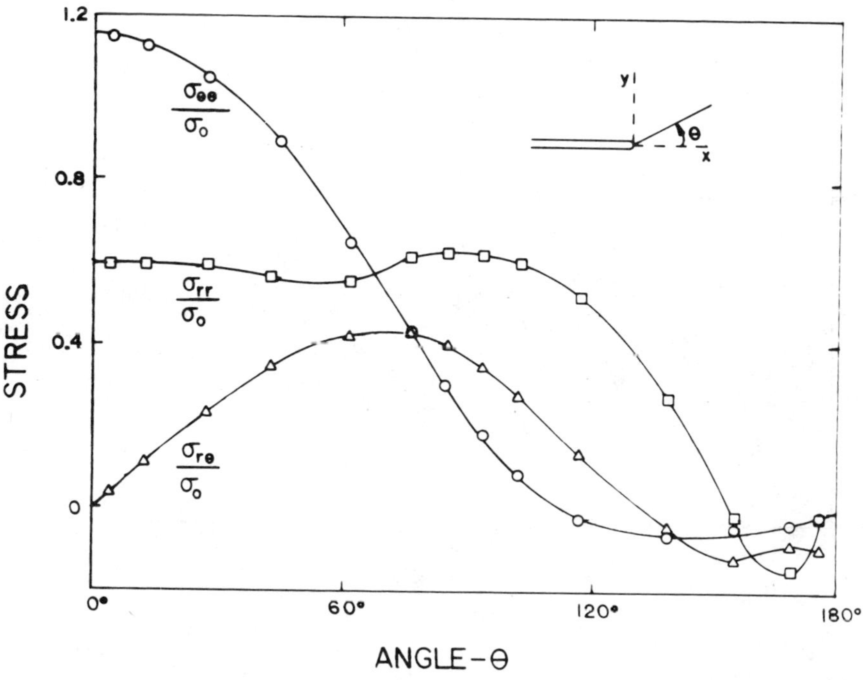

FIG. 7—*Stress distribution at the tip of a moving crack in an elastic-perfectly plastic solid.*

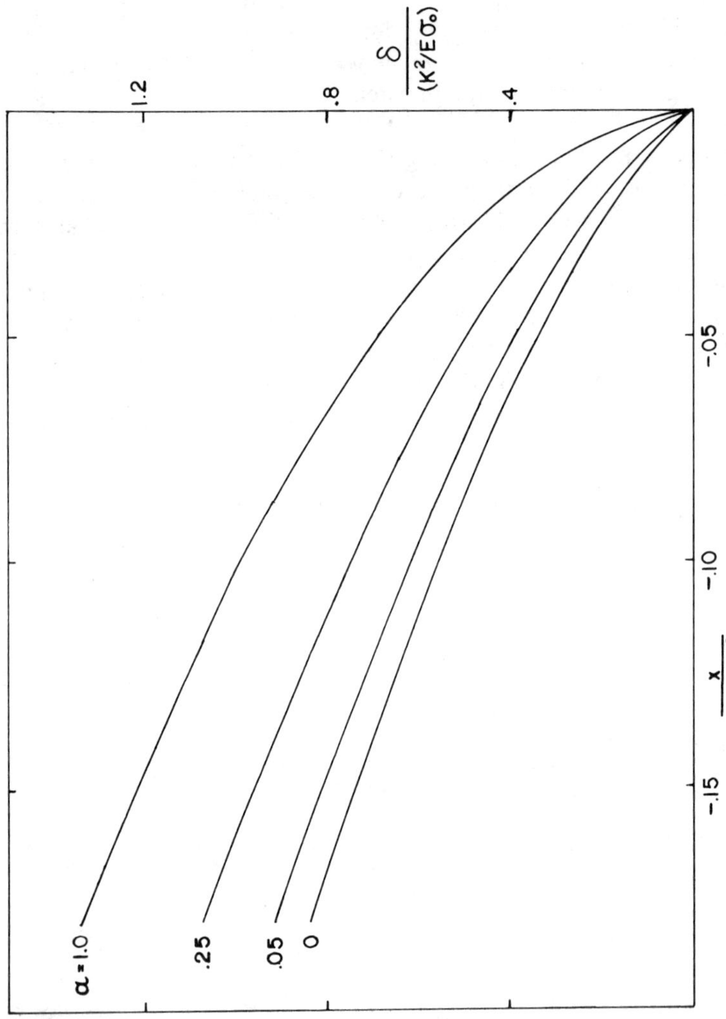

FIG. 8—*Crack-opening profiles for a bilinear hardening material.*

plane shear case in Ref *8*. For comparison, the crack-opening displacement δ for a stationary crack in an elastic-perfectly plastic solid is equal to $K^2/E\sigma_0$.

Amazigo and Hutchinson [*4*] calculated the asymptotic solutions for a crack which was growing steadily in a bilinear hardening solid. They found that the crack-opening profiles could be written asymptotically as $\delta = Ar^{\beta}$. Using a linear least-squares fit for nodes near the crack tip, the present finite-element results predict values of the exponent β equal to 0.52, 0.62, and 0.69 when the hardening level α is 1.0, 0.25, and 0.05, respectively. Values of β in Ref *4* are 0.5, 0.67, and 0.82. The finite-element results may give only a rough estimate of the exponent β because the region over which the most singular term dominates shrinks as α approaches zero.

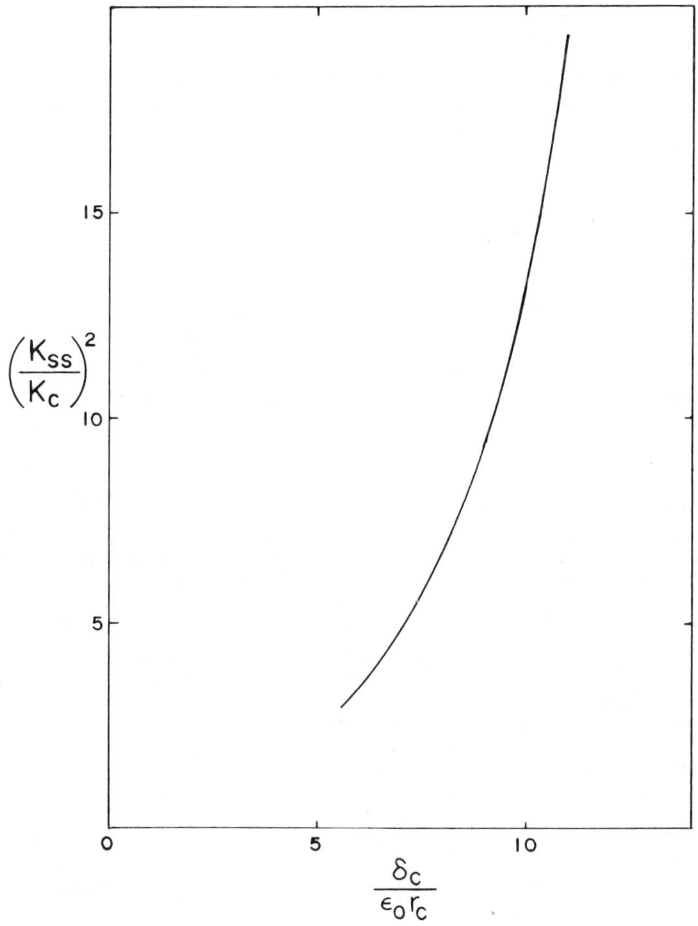

FIG. 9—*Elastic-perfectly plastic* K_{ss}/K_c *using a critical crack-opening angle criterion.*

The results for the stationary crack provided by Shih [*14*] can be combined with the present steady-state results, as was done in Ref *8*, to calculate the ratio of K_{ss} to K_c where K_{ss} and K_c are the stress-intensity factors for steady growth and crack initiation, respectively. Here a crack-opening displacement criterion will be used to derive K_{ss}/K_c as a function of ductility. Assume that the crack grows when a critical displacement δ_c is maintained at some given distance r_c behind the crack tip. This is essentially a constant crack-opening angle (COA) criterion. In actual specimens the COA has been found to vary during the initial stages of crack growth but tends to level off to a constant value fairly rapidly [*15–17*]. Applying the COA criterion to the elastic-perfectly plastic solid, the curve in Fig. 9 can be derived in the manner described in Ref *8*. Figure 9 predicts that substantial amounts of stable crack will occur in plane stress for modest values of $\delta_c/\epsilon_0 r_c$ where $\epsilon_0 = \sigma_0/E$. In Fig. 9, $(K_{ss}/K_c)^2$ is approximately equal to 13 when $\delta_c/\epsilon_0 r_c$ is equal to 10 for the plane-stress problem while the corresponding value of $(K_{ss}/K_c)^2$ was approximately 1 in plane strain [*8*]. This large difference between the plane-stress and plane-strain predictions should not be surprising since cracks have been found to grow more stably in plane stress.

Acknowledgments

The author would like to thank C. F. Shih for providing his data on the stationary elastic-plastic problem. This work was completed while the author was an assistant professor at The Ohio State University. This work was supported in part by The Ohio State University Small Research Grants Program and all computational funds were supplied by the Instruction and Research Computer Center at The Ohio State University.

References

[*1*] Broek, D., *International Journal of Fracture Mechanics*, Vol. 4, 1968, pp. 19–29.

[*2*] Green, G. and Knott, J. F., *Journal of the Mechanics and Physics of Solids*, Vol. 23, 1975, pp. 167–183.

[*3*] Yoder, G. R. and Griffis, C. A. in *Mechanics of Crack Growth, ASTM STP 590*, American Society for Testing and Materials, 1976, pp. 61–81.

[*4*] Amazigo, J. C. and Hutchinson, J. Q., *Journal of the Mechanics and Physics of Solids*, Vol. 25, 1977, pp. 81–97.

[*5*] Yu-chen, G. and Keh-chich, H. in *Proceedings*, Fifth International Conference on Fracture, Cannes, France, 1981, pp. 669–682.

[*6*] Chitaley, A. D. and McClintock, F. A., *Journal of the Mechanics and Physics of Solids*, Vol. 19, 1971, pp. 147–163.

[*7*] Andersson, H., *Journal of the Mechanics and Physics of Solids*, Vol. 22, 1974, pp. 285–308.

[*8*] Dean, R. H. and Hutchinson, J. W. in *Fracture Mechanics, Twelfth Conference, ASTM STP 700*, American Society for Testing and Materials, 1980, pp. 383–405.

[*9*] Parks, D. M., Lam, P. S., and McMeeking, R. M. in *Proceedings*, Fifth International Conference on Fracture, Cannes, France, 1981.

[*10*] Douglas, A. S., Freund, L. B., and Parks, D. M. in *Proceedings*, Fifth International Conference on Fracture, Cannes, France, 1981.

[11] Rice, J. R. and Tracey, D. M. in *Numerical and Computer Methods in Structural Mechanics*, Academic Press, New York, 1973, pp. 585–623.
[12] Hutchinson, J. W. and Shih, C. F., *Recent Advances in Engineering Science*, Vol. 6, 1975, pp. 245–253.
[13] Hutchinson, J. W., *Journal of the Mechanics and Physics of Solids*, Vol. 16, 1968, pp. 337–347.
[14] Shih, C. F., "Elastic-Plastic Analysis of Combined Mode Crack Problems," Ph.D. Thesis, Harvard University, Cambridge, Mass., 1973.
[15] deKoning, A. U. in *Proceedings*, Fracture 1977, Waterloo, Ont., Canada, Vol. 3, 1977, pp. 25–31.
[16] Shih, C. F., deLorenzi, H. G., and Andrews, W. R. in *Elastic-Plastic Fracture, ASTM STP 668*, J. D. Landes, J. A. Begley, and G. A. Clarke, Eds., American Society for Testing and Materials, 1979, pp. 65–120.
[17] Kanninen, M. F. et al, "The Development of a Plastic Fracture Methodology," Battelle Columbus Laboratories Final Report to EPRI on RP601-1, Columbus, Ohio, 1980.

Ting-Leung Sham[1]

A Finite-Element Study of the Asymptotic Near-Tip Fields for Mode I Plane-Strain Cracks Growing Stably in Elastic-Ideally Plastic Solids

REFERENCE: Sham, T-L., **"A Finite-Element Study of the Asymptotic Near-Tip Fields for Mode I Plane-Strain Cracks Growing Stably in Elastic-Ideally Plastic Solids,"** *Elastic Plastic-Fracture: Second Symposium, Volume I—Inelastic Crack Analysis, ASTM STP 803,* C. F. Shih and J. P. Gudas, Eds., American Society for Testing and Materials, 1983, pp. I-52–I-79.

ABSTRACT: The asymptotic near-tip stress and deformation fields for a Mode I plane-strain crack growing quasi-statically in an elastic-ideally plastic solid under small-scale yielding conditions are analyzed numerically by the finite-element method. The asymptotic analyses of Rice and Sorensen; Rice, Drugan, and Sham; and Drugan, Rice, and Sham predict that the crack opening rate $\dot{\delta}$ at a small distance r from the growing crack tip under contained yielding conditions is given by

$$\dot{\delta} = \alpha \frac{\dot{J}}{\sigma_0} + \beta \frac{\sigma_0}{E} \dot{a} \ln \frac{R}{r}, \qquad \text{as } r \to 0$$

where a is the crack length, σ_0 the tensile yield strength, E Young's modulus, J the far-field value of the J-integral evaluated on contours in the elastic region, $\beta = 5.462$ (Drugan et al) for Poisson ratio $\nu = 0.3$, and the parameters α and R are undetermined by the asymptotic analysis. The finite-element solutions generated are very detailed, with the maximum extent of the plastic zone being about 100 times the smallest element size. This high resolution in the finite-element solution enables correlation of the finite-element results with the asymptotic crack opening rate to be made and the expressions for the parameters α and R to be determined. Various crack growth histories are simulated by relaxing the nodal force at the crack tip and simultaneously increasing the external applied load in order to investigate the possible dependence of α and R on the crack growth history. The maximum amount of crack growth simulated is about 20 percent of the maximum extent of the plastic zone size. The numerical results reveal that a well-defined elastic unloading sector develops and moves with the advancing crack tip and its location coincides well with that predicted by the asymptotic analysis. Estimation from the nu-

[1]Assistant professor, Department of Mechanical Engineering, Aeronautical Engineering and Mechanics, Rensselaer Polytechnic Institute, Troy, N.Y. 12181; formerly, research assistant, Division of Engineering, Brown University, Providence, R.I. 02912.

merical results gives $\beta = 5.46$, which is in very good agreement with the asymptotic result. The parameter α is found to be independent of the crack growth histories examined and has a value of 0.58. An attempt to relate R to the quantity EJ/σ_0^2 is made. However, it is found that the parameter s, defined by $s \equiv R/(EJ/\sigma_0^2)$, varies in a range 0.113 to 0.133 for the crack growth histories simulated. Thus, the precise definition for R remains unclear.

KEY WORDS: elastic-plastic crack growth, ductile tearing fracture, asymptotic near-tip fields, embedded finite elements, plane strain, small-scale yielding, fractures (materials), quasi-static crack propagation, elastic-plastic fracture

Due to the earlier investigations of plane-strain stable crack growth, the nature of the elastic-plastic strain singularity in the centered fan sector moving with a growing crack has been quite well established. In an elastic-ideally plastic solid the near-tip strain field changes from a strong $(1/r)$ singularity for a stationary crack, where r is the distance measured from the crack tip, to a weaker $\ell n\ (1/r)$ singularity for a growing crack. These results have been discussed by Rice [*1,2*],[2] Cherepanov [*3*], and Rice and Sorensen [*4*] and slip-line methods were used in these analyses. The asymptotic near-tip fields for growing cracks in an elastic-ideally plastic solid under plane-strain conditions were first obtained by Slepyan [*5*] for a Tresca material, and later, independently, by Gao [*6*] and by Rice et al [*7*] for a Huber-Mises material. In these analyses, the near-tip Prandtl stress field configuration for a stationary crack was shown to be inadmissible for the growing crack because of negative plastic work dissipation along a ray emanating from the moving crack tip at 135 deg from the crack line ahead of the crack. For an incompressible solid (Poisson ratio $\nu = 0.5$), this necessitates the addition of an elastic unloading sector to the basic Prandtl configuration.

Subsequently, Rice [*8*] has analyzed the whole class of asymptotic structures of the near-tip stress and deformation fields for cracks growing in elastic-ideally plastic solids under any of the three modes of fracture and has given a general formulation for materials of arbitrary yield condition and associated flow rule, including anisotropic response. Rice [*8*] has also made some corrections to the work of Rice et al [*7*] concerning a limit procedure used in their analyses. Specializing Rice's [*8*] general formulation, Drugan et al [*9*] have constructed an exact solution for the plane-strain near-tip stress field which corrects and extends the results in the analysis of Rice et al [*7*] when compressibility is admitted ($\nu \neq 0.5$).

In the analysis of Drugan et al [*9*] the full three-dimensional Mises yield condition is employed for plastic sectors which do not produce singular strain and the resulting configuration of the near-tip stress field is shown in Fig. 1 for $\nu = 0.3$. In Fig. 1, A and B are, respectively, plastic sectors of the constant stress and centered fan type in which the out-of-plane deviatoric stress,

[2]The italic numbers in brackets refer to the list of references appended to this paper.

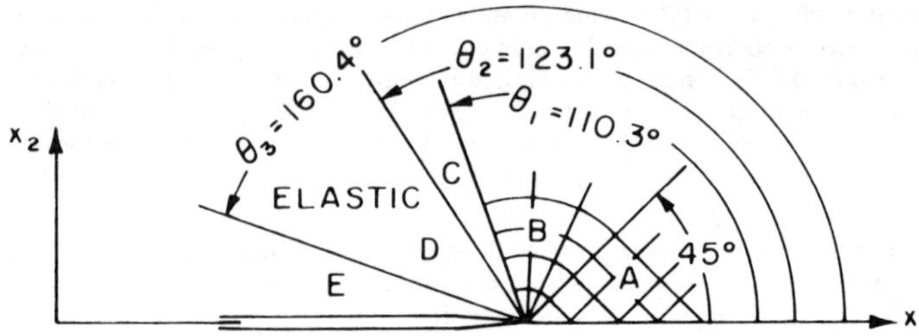

FIG. 1—*Configuration of the near-tip stress field for a growing crack with Poisson ratio* ν = *0.3 (Drugan et al [9]). Sector A is an* $s_{33} = 0$ *plastic sector of the constant stress type and Sector B is an* $s_{33} = 0$ *plastic sector of the centered fan type. Sectors C and E are* $s_{33} \neq 0$ *plastic sectors in which the plastic strains are nonsingular at the crack tip, and Sector D is an elastic unloading sector.*

s_{33}, is zero; C and E are $s_{33} \neq 0$ plastic sectors in which the plastic strains are bounded at the crack tip; D is an elastic unloading sector.

The result in Ref *9* shows that the plastic strains in the centered fan sector have the singular form

$$\epsilon_{ij}{}^{P} = \frac{5-4\nu}{2\sqrt{6}}\frac{\sigma_0}{E} G_{ij}(\theta)\, \ell\text{n}\frac{\overline{R}}{r} + H_{ij}(\theta), \qquad \text{as } r \to 0 \tag{1}$$

Here, the origin of the polar coordinate system (r, θ) is located at the moving crack tip, σ_0 is the tensile yield strength, E Young's modulus, $G_{ij}(\theta)$ are known functions of angle θ, and $\overline{R}$ and $H_{ij}(\theta)$ are undetermined by the asymptotic analysis. However, the plastic strain accumulated by a material point passing through the constant-stress sector ahead of the growing crack tip is at most $\ell n(\ell n(r))$ singular as $r \to 0$. Gao [*10*] has also constructed a near-tip stress field which includes compressibility and it has the same configuration as that shown in Fig. 1 but with different boundary locations for the sectors. Nevertheless, Gao [*10*] has concluded erroneously (see discussion by Drugan et al [*9*]) that the plastic strain in the constant-stress sector ahead of a growing crack is $\ell n(r)$ singular as $r \to 0$.

Drugan et al [*9*] have also given an expression for the rate of crack opening, $\dot{\delta}$, as

$$\dot{\delta} = \alpha\frac{\dot{J}}{\sigma_0} + \beta\frac{\sigma_0}{E}\dot{a}\, \ell\text{n}\frac{R}{r}, \qquad \text{as } r \to 0 \tag{2}$$

where

J = far-field value of the J-integral evaluated on contours in the elastic region,
$\dot{a}$ = crack growth rate, and
β = 5.462 for ν = 0.3.

Again, α and R are undetermined by the asymptotic analysis. This expression for the crack opening rate is of the same form as that originally suggested by Rice and Sorensen [4] and Rice et al [7] but with a revised value for β.

It is seen that the asymptotic analysis gives complete information about the near-tip stress state and also the singular form of the plastic strains. However, in an elastic-ideally plastic solid, the asymptotic analyses show that the stress state directly ahead of a stationary crack and a stably growing crack are approximately the same and that the deformation field alone characterizes the state at the crack tip. Thus a reasonable crack growth criterion would be based on the deformation field. One form of such a criterion is that discussed by Rice and Sorensen [4] which requires that a critical opening be maintained at a small distance behind the crack tip once the crack has grown a distance comparable to the fracture process zone. This growth criterion has been rephrased by Rice et al [7] in a form that does not involve microscale parameters. They show that the near-tip crack opening can be written as

$$\delta = \beta \frac{\sigma_0}{E} r \ln \frac{\rho}{r}$$

where

$$\rho \equiv R \exp(1 + \alpha T/\beta)$$

and $T(\equiv (E/\sigma_0{}^2)(dJ/da))$ is the Paris tearing modulus. Hence, a crack growth criterion which requires a geometrically similar profile near the crack tip can be stated as the requirement that ρ be constant during growth. It is emphasized that these equivalent forms of the crack growth criterion are theorized to hold for arbitrary amounts of stable crack growth once the crack has grown past a distance of the order of the fracture process zone.

The finite-element results of Sham [11] cited in Ref 7 constituted the first numerical solution that showed elastic unloading of finite elements behind a growing crack. Although in the Sham [11] solution the position and size of the elastic unloading sector is far from exact compared with the asymptotic result, the presence of this sector is clearly evident. Sham [11] has also made estimates for the parameters α and R. Although these estimates are not definitive, the numerical solution together with the asymptotic result provides a fairly complete description of the near-tip deformation state as discussed by Rice et al [7]. In view of the lack of precision of the α and R estimates in the Sham [11] solution, the present investigation was carried out in an effort to determine more accurately the expressions for α and R and their possible dependence on crack growth history.

Details of Finite-Element Computation

A boundary-layer formulation, as originally introduced by Rice [1,12], is followed to study the problem of plane-strain stable crack growth unuer small-scale yielding conditions. This formulation is a very useful tool of analysis in fracture mechanics. It stipulates that the small-scale yielding solution for any flawed structure can be obtained by considering an auxiliary problem of a semi-infinite crack with the asymptotic boundary conditions

$$\sigma_{ij} \rightarrow \frac{K}{\sqrt{2\pi r}} g_{ij}(\theta), \qquad \text{as } r \rightarrow \infty \tag{3}$$

where

K = stress-intensity factor which incorporates the effects of external loading and geometry,
$g_{ij}(\theta)$ = set of universal functions of angle θ, and
r = distance measured from the crack tip.

The techniques for applying this formulation to finite-element analyses are well established. As the boundary conditions (Eq 3) are asymptotic in nature, Rice and Tracey [*13*] implemented the boundary-layer formulation to a large but finite region and restricted the size of the plastic zone to be about one tenth of the maximum radius in their finite-element mesh in order not to violate the small-scale yielding conditions. Hilton and Hutchinson [*14*] had worked with an infinite region and solved the problem by using a modified variational principle in their finite-element implementation. This approach (see Appendix) is adopted in the present investigation, because the plastic zone can then be allowed to spread very close to the outer boundary of the finite-element mesh without violating the small-scale yielding conditions. Thus all the available finite-element degrees of freedom, which are limited by the computer capability, can be concentrated inside and near the plastic zone.

The finite-element formulation used in this work models the material, within the framework of small-strain theory, as isotropic, elastic-ideally plastic, satisfying the Prandtl-Reuss flow rule and Mises yield condition. No account is taken of the possible development of a vertex on the yield surface due to the strongly nonproportional stressing paths experienced near a growing crack tip. Dean and Hutchinson [*15*] and Parks et al [*16*] have discussed the effects of this on the behavior of the near-tip fields in Modes III and I, respectively. For the element integration station at which the stress state is currently at yield, the partial stiffness approach of Marcal and King [*17*] as modified by Rice and Tracey [*13*] is used to determine the best representative constitutive relation within a load increment. Four-noded constant-dilatation elements, as suggested by Nagtegaal et al [*18*], were used in the calculations to free the artificial mesh-locking effect under nearly incompressible conditions.

Crack growth under continuously increasing external loading was simulated by relaxing the nodal reaction force at the current crack-tip node incrementally. In the scheme of Rice and Tracey [*13*], a chord approximation for the yield surface is made within each load increment, and thus it is desirable to limit the distance that a material point can traverse on the yield surface in an increment to as small an amount as possible. Twenty steps were chosen to relax the nodal reaction force of one finite element. This corresponds to limiting the maximum angle of rotation undergone by a material point on the yield surface to about 6.5 deg in each step. The external load was increased simultaneously during the nodal release procedure in order to model the crack growth process in a more realistic manner.

Under small-scale yielding conditions, the appropriate loading parameter is the stress-intensity factor, K. However, the far-field value of the J-integral evaluated on contours in the elastic region has a unique relation to K, namely

$$J = (1 - \nu^2)K^2/E$$

Thus, some of the results will be reported in terms of the far-field value of the J-integral. The load versus crack length histories are depicted in Fig. 2. A load, J_i, required to cause incipient yielding at one finite-element integration station was applied and the yielding was allowed to spread gradually to other elements at fixed crack length by increasing the load in equal increments of $0.35\, J_i$ to J_0 ($=176\, J_i$). At this stage, the maximum extent of the plastic zone is about 100 times the smallest element size. Then two growth histories, each consisting of 20 one-element crack growth steps, were simulated corresponding to crack growth at two constant values of the Paris tearing modulus, T, namely

$$T = 5 \quad \text{and} \quad T = 15$$

Figure 3 shows the finite-element mesh used in the calculation. Since there is no characteristic length dimension in the configuration of a semi-infinite crack, all coordinates of the nodes in the finite-element mesh are expressed in terms of an arbitrary length measure. There is a total of 1725 four-noded constant-dilatation elements and 1796 nodes in the mesh. Material properties used were $\nu = 0.3$ and $\sigma_0/E = 0.005$.

Finite-Element Results

In the present finite-element calculations, no singularity elements are used to model properly either the crack-tip singular behavior of a stationary crack or that of a stably growing crack during the entire loading histories. Instead, a refined finite-element mesh with regular finite elements at the crack-tip region is employed, so that the crack-tip singularities can be *approximated* as closely as possible. Thus, useful results can be obtained in the neighborhood of the crack tip. Data can then be extrapolated back to the crack tip to reveal the singular nature, or at least the structure, of the near-tip field of a stably growing crack. Indeed, such a procedure has been employed successfully by

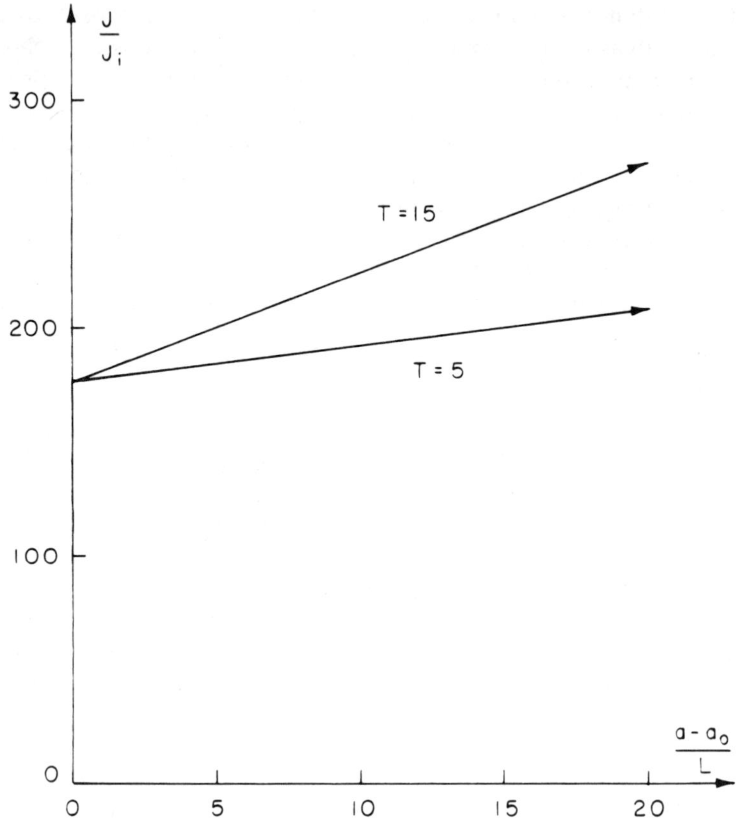

FIG. 2—*Load versus crack growth histories simulated for crack growth under constant Paris tearing modulus* T = *5, 15. Here* J *is the value of the J-integral evaluated on contours in the elastic region and* J_i *is the value of* J *at incipient yielding;* $a - a_0$ *is the amount of crack growth and* L *is the size of the smallest finite element near the crack tip.*

Chan et al [*19*] to determine the stress-intensity factors of various test specimens in linear elastic fracture mechanics (LEFM). In order to establish better confidence in the present results, the numerical results for the stationary crack are investigated in some detail before those for the growing crack are presented since both theoretical and numerical solutions for a stationary crack in elastic-ideally plastic materials are well known [*1,12,13,20,21*].

Crack-Tip Opening Displacement for a Stationary Crack

The crack-tip opening for a stationary crack under monotonic loading can be obtained by setting $\dot{a} = 0$ in Eq 2 to give

$$(\delta)_{r=0} = \alpha \frac{J}{\sigma_0} \tag{4}$$

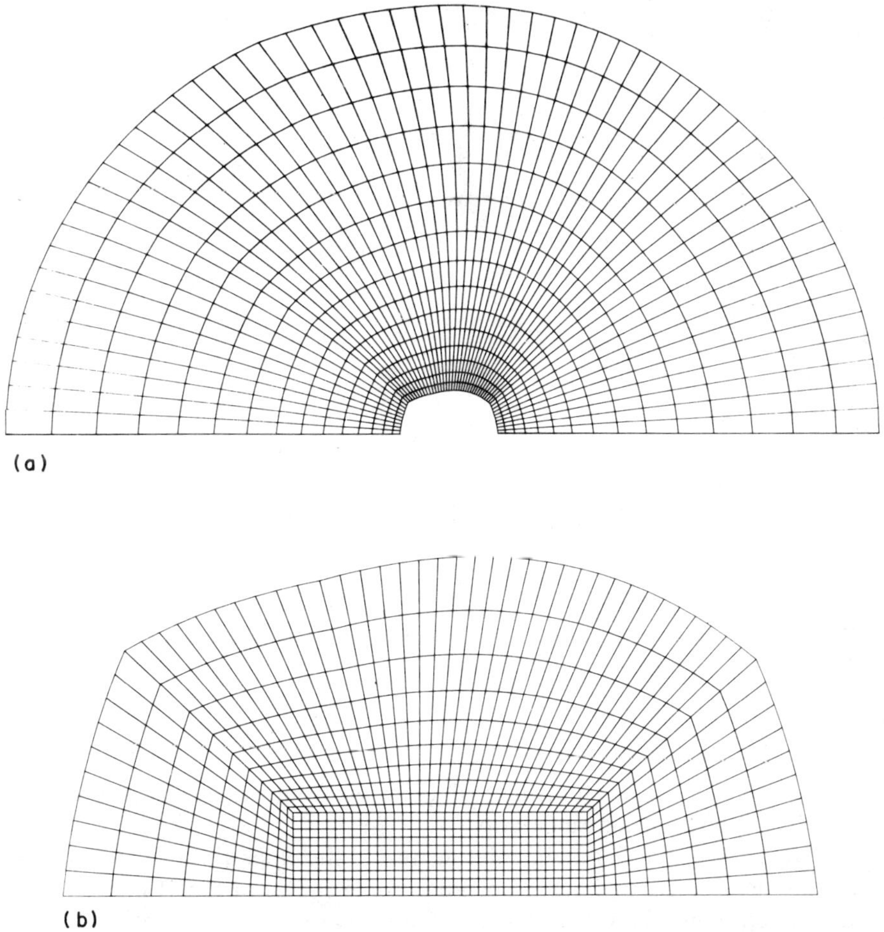

FIG. 3—*Finite-element discretization of Region B (see Appendix). The outer mesh is shown in* (a) *and the details of the mesh near the crack tip are shown in* (b). *There are 1725 elements and 1796 nodes. The finite-element mesh is embedded in the analytically simulated continuum.*

where α, by dimensional considerations, must be a constant for small-scale yielding.

A plot of the nondimensional crack opening profile, $\delta/(J/\sigma_0)$, obtained from the finite-element results is shown in Fig. 4. The strong $(1/r)$ plastic strain singularity associated with a stationary crack at small-scale yielding is reflected in the immense amount of shearing found in the finite elements next to the crack tip. Since no singularity elements were used in the computation, the interpolation functions employed in the finite-element formulation could not possibly reproduce the type of deformation demanded by the true solution. Thus, in order to accommodate the large shearing present, the kinematics of the finite-element mesh produces a "kink" on the near-tip crack

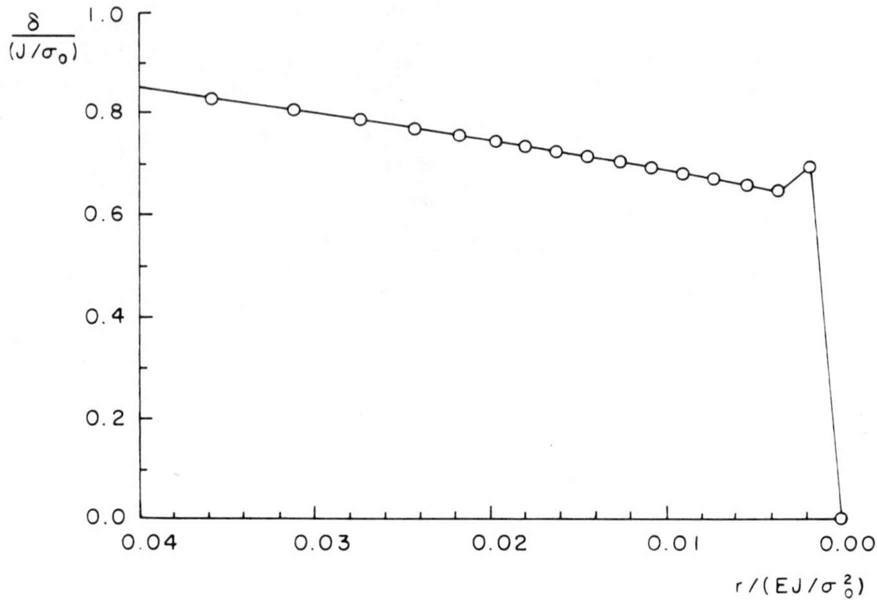

FIG. 4—*Normalized crack opening for a stationary crack (*r *is the distance measured from the crack tip).*

opening profile. This is an artifact which arises from the incorrect modeling of the $(1/r)$ plastic strain singularity.

So, by ignoring the first data point nearest to the crack tip, one can extrapolate the nondimensional crack opening profile shown in Fig. 4 to the tip in order to determine the value of α. It is found that α has a value of 0.62. This falls in the range of acceptable values obtained by various workers with different finite-element formulations and different material parameters, σ_0/E. A summary of these results is given in the works of Rice and Sorensen [*4*] and Shih [*22*].

Stress Distributions Ahead of a Stationary Crack

From an asymptotic analysis of a mathematically sharp crack in elastic-ideally plastic solids under plane-strain conditions, Rice [*1,12*] and Rice and Tracey [*13*] have shown, within the small-strain theory, that the limiting near-tip stress state corresponds to that of a Prandtl field with

$$\sigma_{11}/\sigma_0 = \pi/\sqrt{3} = 1.814, \qquad \sigma_{22}/\sigma_0 = (2+\pi)/\sqrt{3} = 2.968,$$

$$\sigma_{12}/\sigma_0 = 0.00, \qquad \sigma_{33}/\sigma_0 = (1+\pi)/\sqrt{3} = 2.391$$

ahead of the crack tip. This result is obtained by assuming that yielding completely surrounds the crack tip, together with the conditions that the out-of-plane deviatoric stress vanishes at all values of angle θ and all stress components are continuous. The near-tip stresses, σ_{11}, σ_{12}, σ_{22}, and σ_{33}, normalized by the yield stress, have been plotted in Fig. 5 for material points lying directly ahead of the stationary crack tip. The position of the material point measured from the crack tip is normalized by the similarity quantity $(K/\sigma_0)^2$. The data points shown in Fig. 5 are the values of the stress components averaged over the integration stations of an element. The stress state shown corresponds to the last load step of the stationary crack. One can see from Fig. 5 that the same type of disturbance arises near the crack tip due to the incorrect modeling of the $(1/r)$ plastic strain singularity. However, if the first four data points nearest to the crack tip are ignored, extrapolation of the data back to the crack tip would lead to a stress state given by

$$\sigma_{11}/\sigma_0 = 1.77 \qquad \sigma_{22}/\sigma_0 = 2.96$$

$$\sigma_{12}/\sigma_0 = 0.00 \qquad \sigma_{33}/\sigma_0 = 2.13$$

at the crack tip.

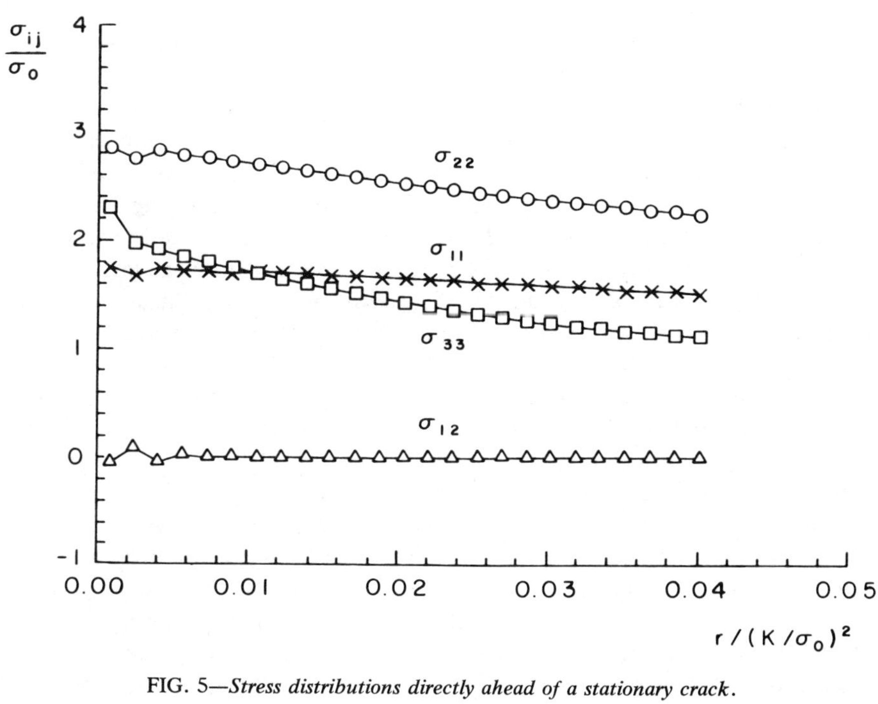

FIG. 5—*Stress distributions directly ahead of a stationary crack.*

The Crack Size Plasticity Adjustment of a Stationary Crack

The notion of a plastic zone correction, as suggested by Irwin [*23*] and discussed in detail by Irwin and Paris [*24*], provides a very useful concept in extending LEFM to the analysis of crack problems under small-scale yielding conditions. Irwin argued that the presence of plasticity at the near-tip region of a crack of length a would cause the material far from the plastic zone to behave as if it were responding to a sharp crack of length $a + c$ in a linear elastic solid and c was estimated to be 0.053 $(K/\sigma_0)^2$ under plane-strain conditions. In the present computation, the stress and deformation states in the elastic region far away from the crack tip are represented by convergent series in terms of unknown coefficients which were determined at each load step. The plastic zone correction, c, can be correlated to these coefficients (see Appendix) and it is found that

$$c = 0.025\,(K/\sigma_o)^2$$

Stress Distributions for a Growing Crack

The stress distributions ahead of a growing crack tip for tearing modulus $T = 5$ and $T = 15$ have been plotted in Figs. 6*a* and 6*b*, respectively, in terms of the same normalizing quantities as were used in the plotting of the stress distributions ahead of the stationary crack tip. These stress distributions are taken from the end of the load versus crack length histories which correspond to 20 one-element crack growth steps for the two values of T. Oscillations in the stress distributions are again noticed within a region of three to four finite elements from the crack tip. It is observed that the curves traced out by the points in Figs. 6*a* and 6*b* superpose on one another very well. This suggests that the stress distributions directly ahead of a growing crack are self-similar for different values of T when distances are measured in terms of $(K/\sigma_0)^2$, or equivalently in terms of EJ/σ_0^2. This result is not a surprising one since, from dimensional considerations, $(K/\sigma_0)^2$ or EJ/σ_0^2 is an appropriate length scale for both stationary and growing cracks under small-scale yielding conditions.

With the type of mesh refinement at the near-tip region used in the present computations, the angular stress distributions around a growing crack tip can be investigated in some detail. As the crack advances, one can consider a contour which is fixed relative to the crack tip and moves with the tip through the material. The material points on this contour are, of course, ever-changing. An observer is imagined to be moving with the advancing crack tip and observes the state of stress of those material points which coincide, instantaneously, with the time-dependent contour. Thus, the angular variation of the stresses around a moving crack tip can be obtained by plotting stresses along this contour. This is done in Figs. 7*a* and 7*b* for the two values of T consid-

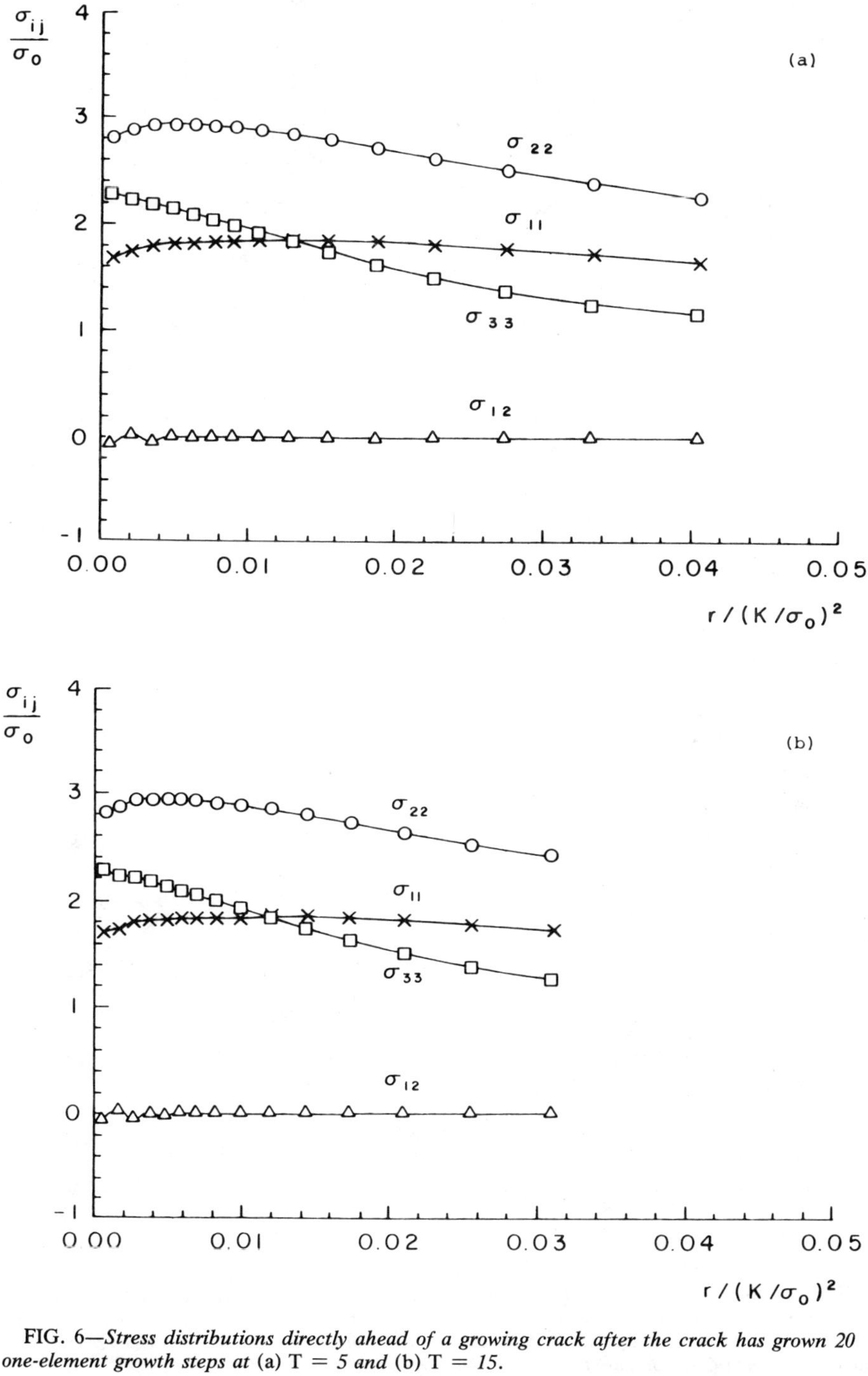

FIG. 6—*Stress distributions directly ahead of a growing crack after the crack has grown 20 one-element growth steps at* (a) T = *5 and* (b) T = *15.*

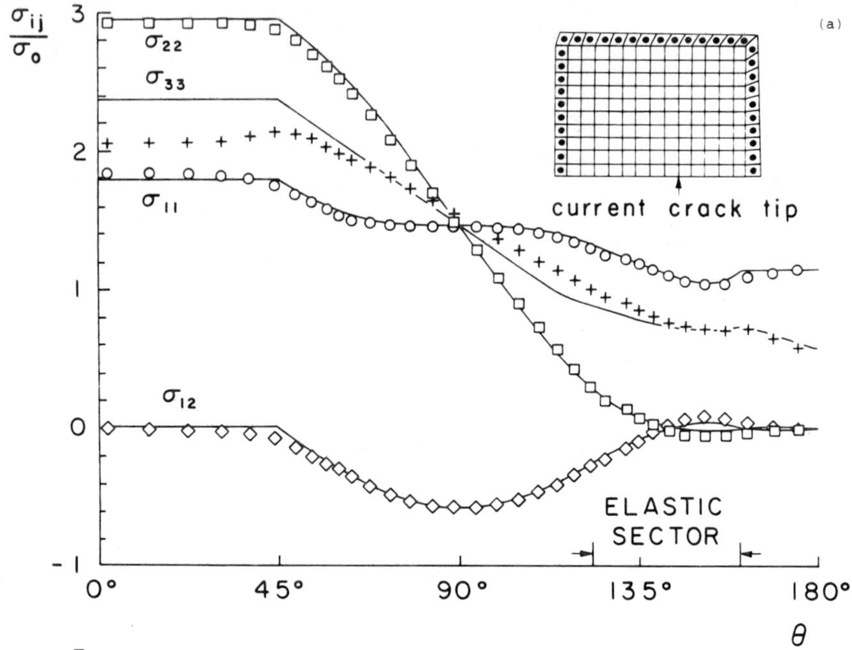

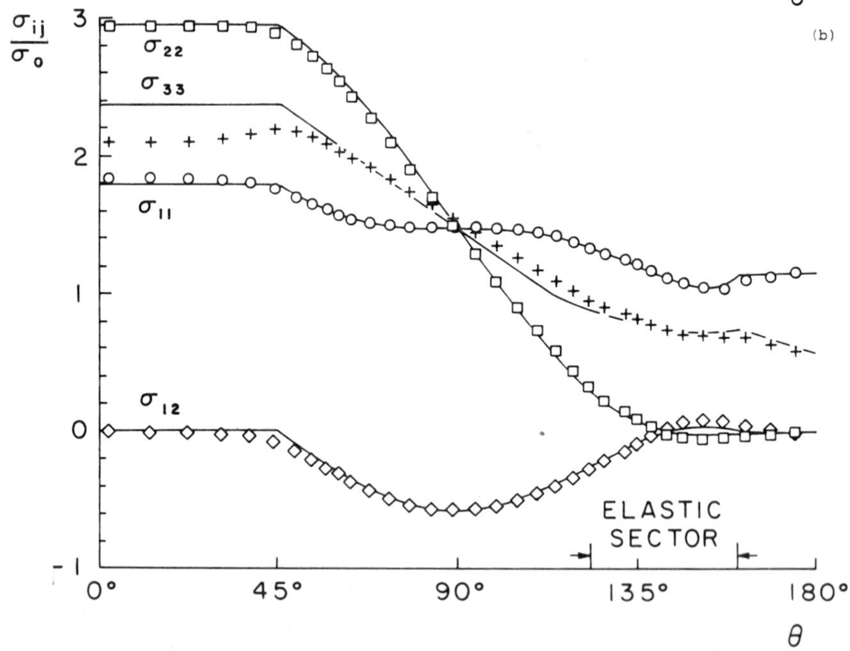

FIG. 7—*Angular stress distributions around the crack tip for a growing crack after the crack has grown 20 one-element growth steps at* (a) T = *5 and* (b) T = *15. The data are taken on the path shown in the inset in Fig. 7*a. *The solid lines are the exact asymptotic result of Drugan et al [9] for* v = *0.3 and the symbols are the finite-element results.*

ered and the two stress states are also taken from the end of the two growth histories simulated. In Figs. 7*a* and 7*b*, the stresses, normalized by the yield stress, are plotted along a contour as shown in the inset diagram in Fig. 7*a*. The solid lines are the exact asymptotic result of Drugan et al [9] for $\nu = 0.3$, and the angular position of the unloading sector predicted by the asymptotic analysis is also indicated in these two figures. It is seen that the numerically computed values of σ_{11}, σ_{12}, and σ_{22} agree well with the asymptotic results. Some drifting of the data points from the asymptotic results is noticed in the angular range 90 to 140 deg for the σ_{33} component and, in addition, its values are quite low compared with the asymptotic results in the constant-stress sector (0 deg $\leq \theta \leq$ 45 deg). However, the σ_{33} distributions directly ahead of the crack tip shown in Figs. 6*a* and 6*b* suggest that closer agreement with the result of Drugan et al [9] would be expected as the data points are taken nearer to the crack tip.

Plastic Zone Shape for Stationary Crack

The plastic zone obtained at the last load step of the monotonic loading history of the stationary crack is shown in Fig. 8. Each point shown in this figure corresponds to the position of a finite-element integration station at which the stress state is at yield. All coordinates are normalized by the similarity quantity $(K/\sigma_0)^2$. These points trace out the familiar butterfly-shaped plastic zone for a Mode I stationary crack under small-scale yielding conditions. The maximum extent of the plastic zone, which occurs at about 70 deg from the crack line ahead of the tip, is 0.163 $(K/\sigma_0)^2$, and the extent of the plastic zone ahead of the tip on the crack line is 0.036 $(K/\sigma_0)^2$.

Active Plastic Zone Shapes for Growing Cracks

The near-tip active plastic zone of a growing crack has a character different from that of a stationary crack. The asymptotic analysis of Drugan et al [9] for a growing crack predicts that there is an elastic unloading sector, which lies between 123.1 and 160.4 deg from the crack line ahead of the growing crack for $\nu = 0.3$, emanating from the crack tip. The finite-element mesh at the near-tip region is shown in Figs. 9*a* and 9*b* where each point marked on the mesh is the position of an integration station at which the stress state is currently at yield. These data are also taken from the end of the crack growth histories simulated at $T = 5$ and $T = 15$. It is seen in these two figures that the elastic unloading sectors near the crack tip are very prominent. In fact, the elastic unloading sector develops this well-defined shape once the crack has grown for a distance of four finite elements. However, the elastic unloading sectors do not focus at the crack tip but, rather, cut onto the crack surfaces at two finite elements behind the crack tip. The reason for this feature is thought to be twofold. First, the kinematics of the mesh pro-

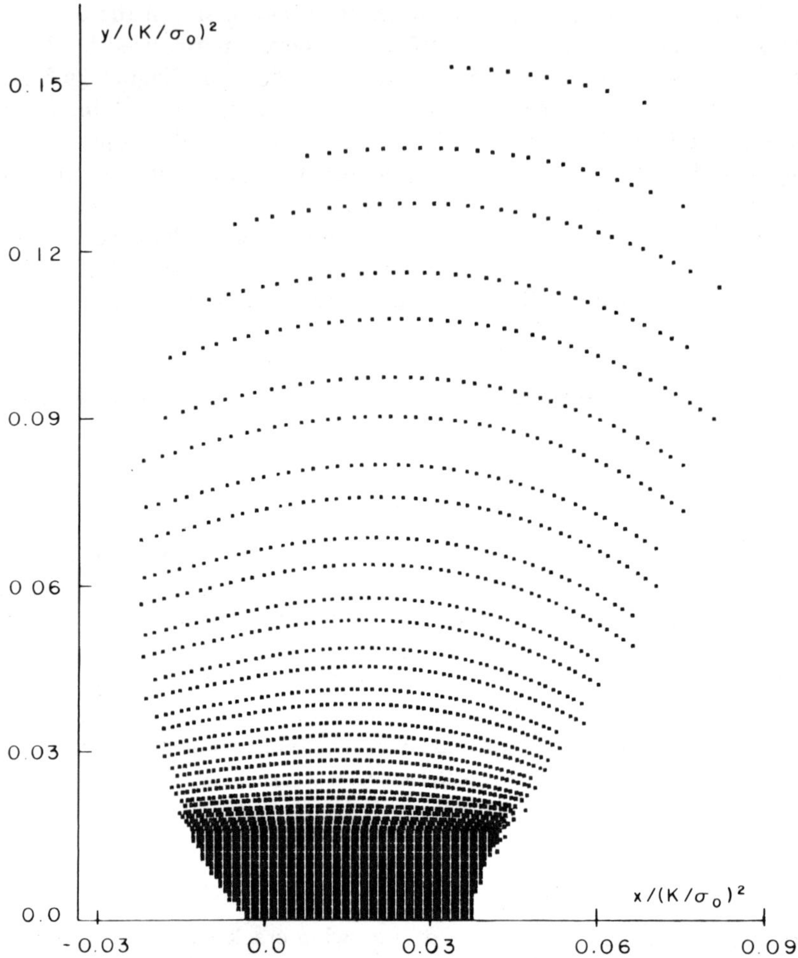

FIG. 8—*Shape of the plastic zone for a stationary crack. Each point shown corresponds to the position of a finite-element integration station at which the stress state is at yield and the position of these points are normalized by the similarity quantity* $(K/\sigma_0)^2$. *The crack tip is at the origin of the axes.*

duces an immense amount of shearing in the finite elements surrounding the moving crack tip in order to accommodate the milder $\ell n(1/r)$ plastic strain singularity in the centered fan sector of a growing crack. Secondly, due to the finite-element discretization procedure and the particular nodal release technique employed to simulate crack growth, there is an ambiguity, to within one to two finite elements, in the exact position of the crack tip as sensed by the numerical solution.

The size of the unloading sector obtained from the numerical solution is

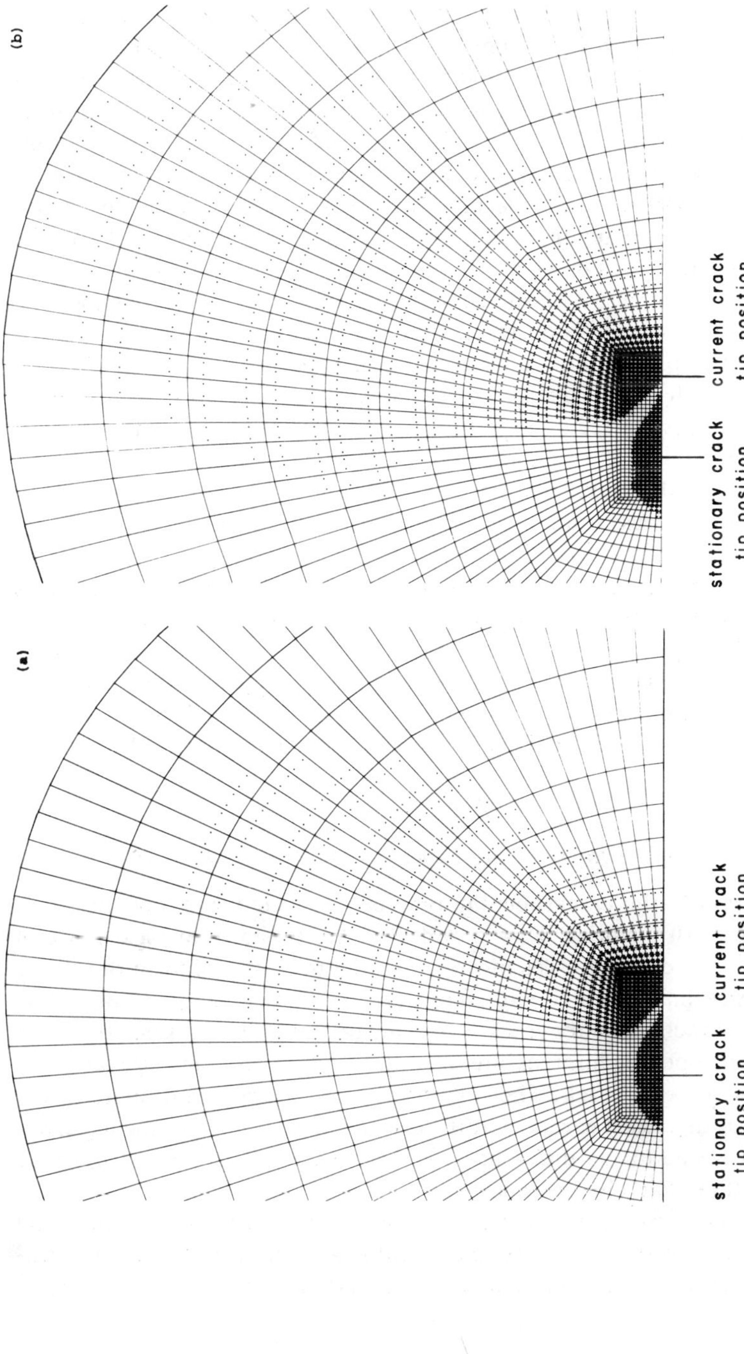

FIG. 9—*Finite-element mesh at the near-tip region of a growing crack. Each point shown corresponds to the position of a finite-element integration station at which the stress state is currently at yield. These points trace out the shape of the active plastic zone after the crack has grown 20 one-element growth steps at* (a) T = 5 *and* (b) T = 15.

small compared with the result of Drugan et al [*9*] even if the crack-tip position sensed by the numerical solution is taken as being two finite elements behind the current crack tip shown in Figs. 9*a* and 9*b*. However, examination of the near-tip stress field of Drugan et al [*9*] indicates that the stress state is less than 2 percent below yield from about 123.1 to 133.3 deg. This is the angular range where the numerical solution has taken the stress state to be at yield. In light of this observation, it is not surprising that the present finite-element solution did not show elastic unloading of finite elements in this angular range. The result of Drugan et al [*9*] also shows that the stress state is below yield at a more noticeable level, ranging from 2 to 10 percent, in the remaining part of the elastic unloading sector. This is the part of the elastic unloading sector which was detected by the finite-element solution.

The shapes of the active plastic zone have been monitored throughout the crack growth histories. It is observed that the size and shape of the leading part of the active plastic zones measured with respect to the crack-tip moving coordinates, which are normalized by $(K/\sigma_0)^2$, do not change significantly as the crack advances. In fact, the butterfly-shaped plastic zone of the stationary crack coincides well with these active plastic zones in the angular range 0 to 70 deg from the crack line ahead of the tip. This similarity occurs because the near-tip stress state predicted by the asymptotic analysis for a growing crack is very close to the Prandtl stress field of a stationary crack in this angular range. In contrast, the trailing part of the active plastic zone exhibits a much different shape from that of a stationary crack. Apart from the presence of the elastic unloading sector, an active plastic wake is found trailing behind the growing crack. It extends beyond the original stationary crack tip position and encompasses material points which have been processed neither by the centered fan sector of the stationary crack nor by that of the growing crack.

Crack Opening Profiles for Growing Cracks

The evolution of the crack opening profiles at selected stages of the simulated crack growth histories, $T = 5$ and $T = 15$, is shown in Figs. 10*a* and 10*b*. The growing crack profiles in these two figures correspond to the stages where the crack has grown 10 and 20 one-element growth steps, respectively. The profile for the stationary crack is also included for comparison. It is seen that the crack opening profiles are very much different for the stationary crack and the growing crack. The stationary crack possesses a discrete crack-tip opening, as discussed earlier, but the opening at the tip of a growing crack is zero. This is due to the different deformation states experienced near the crack tip in these two circumstances.

With the crack opening of a growing crack obtained from the numerical solution, estimation of the parameters α and R appearing in the crack opening rate expression (Eq 2) can be made. This is done in the following section.

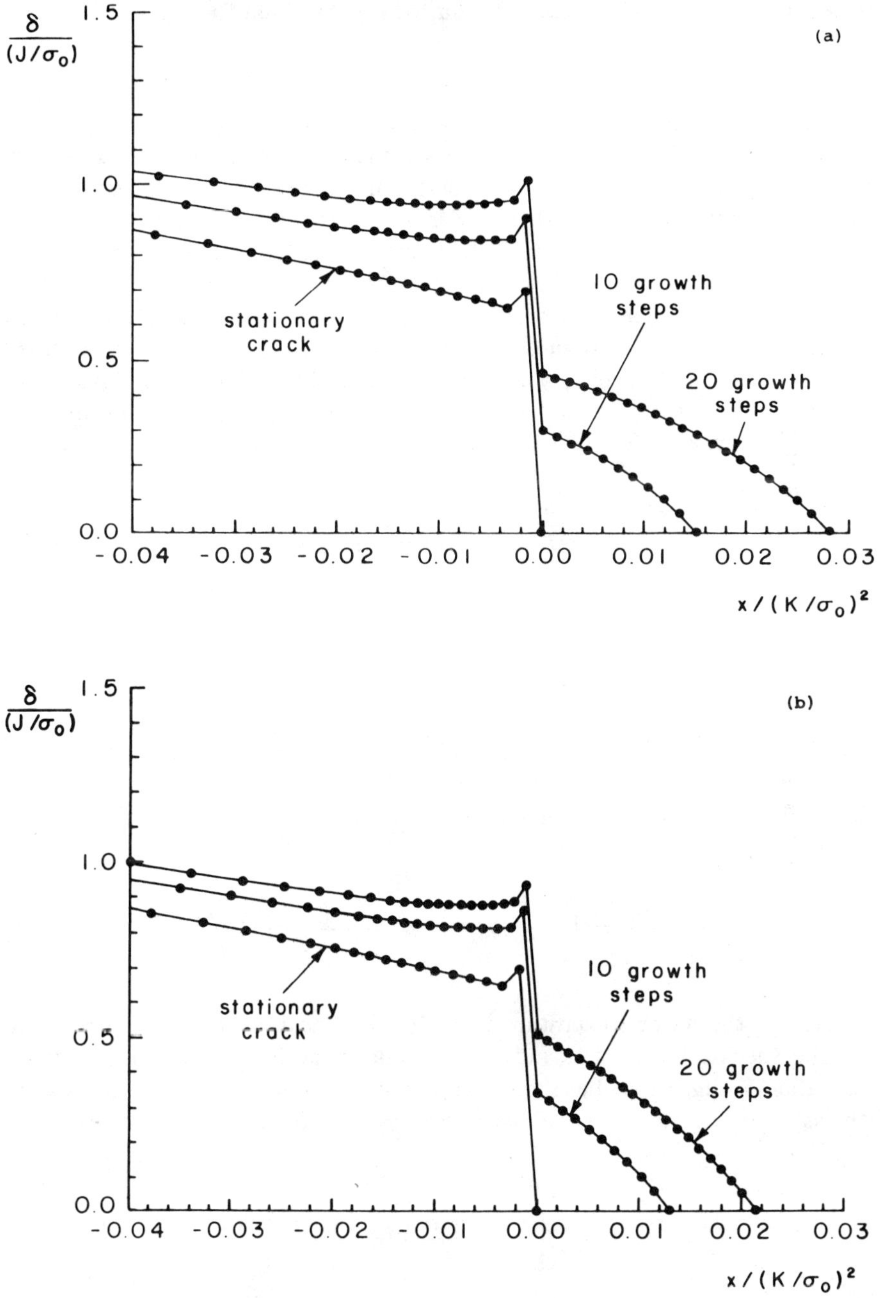

FIG. 10—*Evolution of the crack opening profiles—the stationary crack, the growing crack after the crack has grown 10 one-element growth steps and 20 one-element growth steps at* (a) T = *5 and* (b) T = *15.*

Correlation of Finite-Element Solution with Asymptotic Result

Estimation of β

Based on dimensional considerations, Rice and Sorensen [*4*] and Sham [*11*] have taken R to be in scale with the plastic zone size in estimating the parameters α and R. They have assessed the accuracy of the α and R estimates by determining the value of β in Eq 2 from their numerical solutions and comparing it with the theoretical result. The same comparison for the value of β is done here. However, in contrast to the previous attempts [*4,11*], the value of β is determined from the present numerical solution without making any *a priori* assumption about the functional forms of these parameters. To this end, an arbitrary constant of length dimension is first introduced and this constant is taken to be the smallest element size, L, for convenience; then the expression for the crack opening rate (Eq 2) can be rewritten as

$$\frac{E}{\sigma_0}\dot{\delta} = \dot{a}\left[f\left(\frac{a}{L}\right) + \beta\,\ell\mathrm{n}\left(\frac{L}{r}\right)\right], \qquad \text{as } r \to 0 \tag{5}$$

where

$$f\left(\frac{a}{L}\right) \equiv \alpha T + \beta\,\ell\mathrm{n}\left(\frac{R}{L}\right) \tag{6}$$

Here, α, T, and R are taken to be unknown functions of the crack length a. Taking Eq 5 to relate increments of crack opening to increments of crack length near the crack tip, one can integrate Eq 5 for a fixed material point x to obtain

$$\frac{E}{\sigma_0}\frac{\delta\,(x;a)}{L} = \int_{x/L}^{a/L} f(\zeta)d\zeta + \beta\left(\frac{a-x}{L}\right)\ell\mathrm{n}\left(\frac{eL}{a-x}\right) \tag{7}$$

where e is the natural logarithm base. It is emphasized that no assumptions on the specific forms of α and R are made in performing this integration. Equation 7 suggests a basis for correlating the finite-element solution with the asymptotic result. For two successive crack lengths a_1 and a_2, with $a_2 > a_1$, Eq 7 gives

$$\frac{E}{\sigma_0}\frac{\Delta\delta}{L} = \Delta F + \beta\left[\left(\frac{a_2-x}{L}\right)\ell\mathrm{n}\left(\frac{eL}{a_2-x}\right) - \left(\frac{a_1-x}{L}\right)\ell\mathrm{n}\left(\frac{eL}{a_1-x}\right)\right] \tag{8}$$

where

$$\Delta\delta \equiv \delta(x; a_2) - \delta(x; a_1)$$

$$\Delta F \equiv \int_{a_1/L}^{a_2/L} f(\zeta) d\zeta$$

For given a_1 and a_2, ΔF is a number; thus one can plot

$$\frac{E}{\sigma_0}\frac{\Delta\delta}{L} \quad \text{versus} \quad \left(\frac{a_2 - x}{L}\right)\ell\text{n}\left(\frac{eL}{a_2 - x}\right) - \left(\frac{a_1 - x}{L}\right)\ell\text{n}\left(\frac{eL}{a_1 - x}\right)$$

and determine β and ΔF from the slope and axis-intercept of the graph, respectively. This procedure has been carried out for the successive one-element growth steps 10 to 20 for both $T = 5$ and $T = 15$. The best-fit values of β have an average value of 5.46 and they vary within a ± 0.3 percent range. From the result of Drugan et al [9] for $\nu = 0.3$, β has a theoretical value of 5.462 and the average value for the numerically estimated β is in very good agreement with this theoretical value. In light of this close agreement, the theoretical slope $\beta = 5.462$ was used to fit the numerical data and to obtain the values of ΔF. The resulting values of ΔF can be interpreted as the value of $f(a/L)$ at $a/L = (a_1 + a_2)/(2L)$, multiplied by $(a_2 - a_1)/L$.

Estimation of α *and* R

From the asymptotic analyses of Rice et al [7] and Drugan et al [9], the crack opening rate $\dot{\delta}$ is homogeneous of degree one in $\dot{a}$ and $E\dot{J}/\sigma_0^2$. Thus $\dot{\delta}$ can be written as

$$\dot{\delta} = \dot{a}\, B(T) \tag{9}$$

where B is a dimensionless function of the Paris tearing modulus T. In order to examine the dependence of $\dot{\delta}$ on T, two one-element growth steps were simulated for each of $T = 0, 5, 10, 15$, after 15 one-element growth steps had been simulated at $T = 5$. This is shown schematically in Fig. 11. From dimensional considerations, the two appropriate length scales in the problem are the amount of growth $a - a_0$ and the quantity EJ/σ_0^2. Now, from this observation, R can be written as

$$R = s\frac{EJ}{\sigma_0^2} \tag{10}$$

Here s is a constant for a given amount of crack growth but it could vary with the amount of crack growth. With this definition for R, Eq 6 becomes

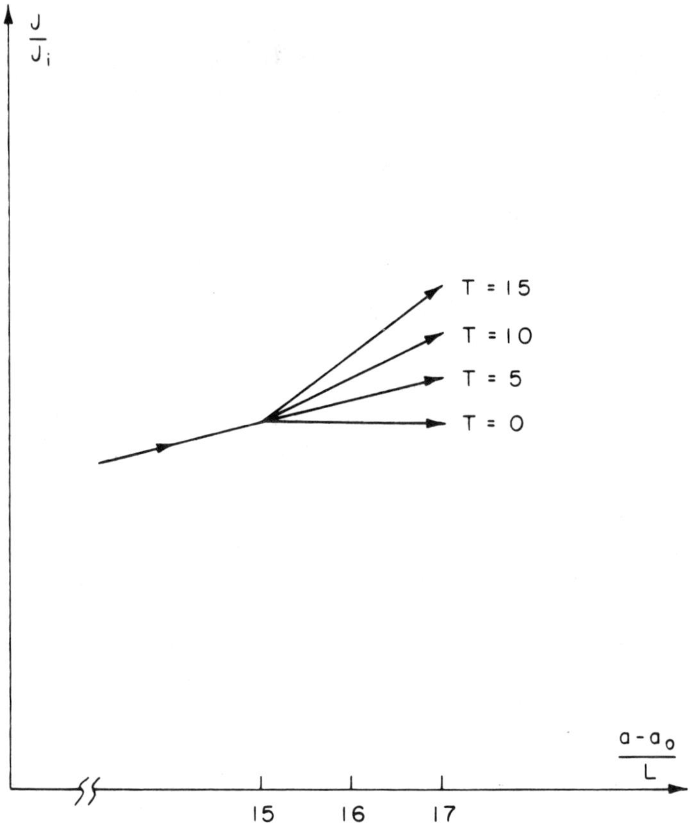

FIG. 11—*A schematic representation of the growth histories simulated to examine the dependence of $\dot{\delta}$ on* T. *The crack is taken to grow at* T = *5 for 15 one-element growth steps, followed by 2 one-element growth steps at each of* T = *0, 5, 10, and 15.*

$$f + \beta \,\ell\text{n}\left(\frac{L}{EJ/\sigma_0^2}\right) = \alpha T + \beta \,\ell\text{n}\,(s) \tag{11}$$

The procedure outlined in the foregoing to determine $f(a/L)$ by using the theoretical slope $\beta = 5.462$ to fit the numerical data was performed for each of these four load versus crack length paths. Using these values for f, the left-hand side of Eq 11 can be computed, and the quantity $\alpha T + \beta \,\ell\text{n}(s)$ is plotted versus the Paris tearing modulus T in Fig. 12 for the simulated growth steps depicted in Fig. 11. It is found that a straight line can be fitted through the data points, and the best-fit values for α and s are 0.584 and 0.125, respectively. This confirms that for a given amount of crack growth, α and s are constants and they have no dependence on the instantaneous value of T, at

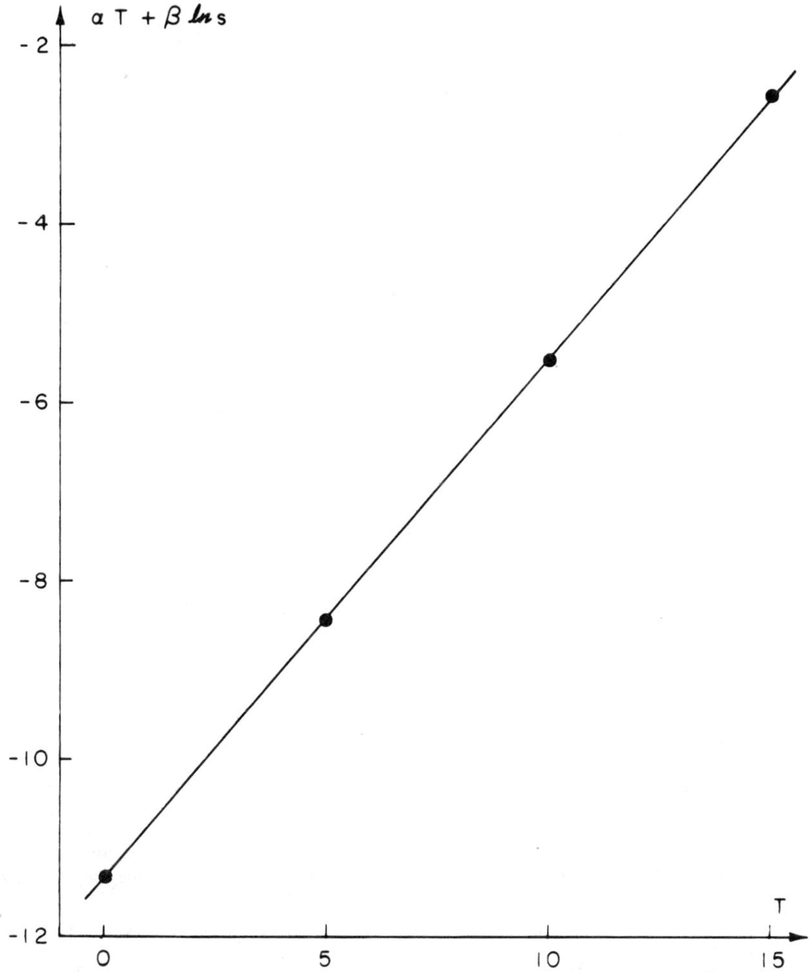

FIG. 12—*Plot of* αT + β ℓn(s) *versus* T *for the growth histories depicted in Fig. 11.*

least for the limited range of T examined. Thus the crack opening rate $\dot{\delta}$ is linear in the instantaneous T.

Once this linearity is established, the dependence of α and s on the amount of crack growth and their possible dependence on the prior values of T can be investigated. This is done by simulating various one-element growth steps that branch off the principal growth histories at different stages; these paths are shown schematically in Fig. 13. The methodology employed in the preceding to determine α and s is applied repeatedly to each set of these branching paths, and Fig. 14 shows graphs of $\alpha T + \beta \ell n(s)$ versus T which correspond to the branching sequences indicated in Fig. 13. The best-fit values for

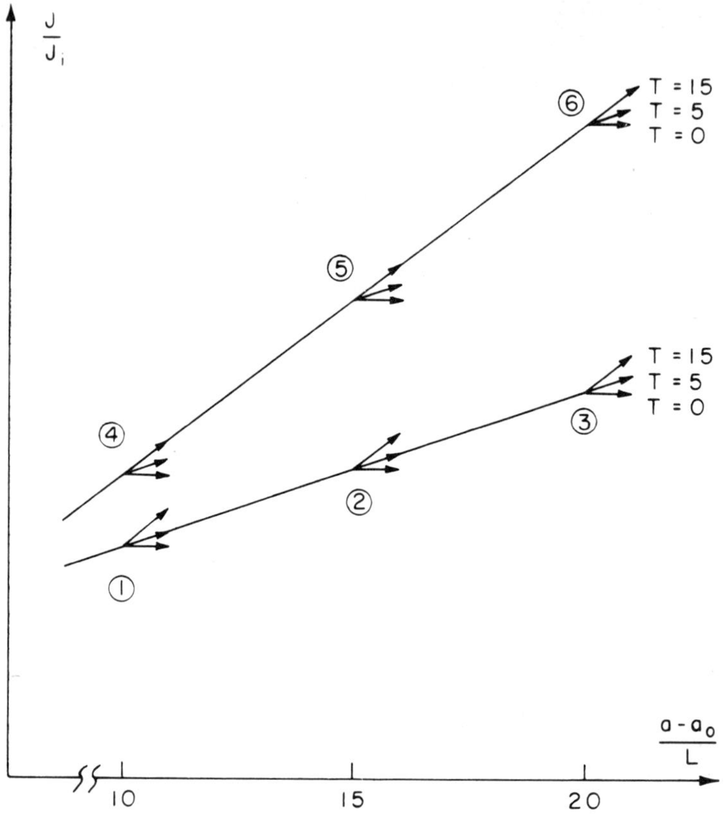

FIG. 13—*A schematic representation of the growth histories simulated to examine the dependence of* α *and* s *on the amount of crack growth and their possible dependence on prior history of* T.

α and s are shown in the inset in Fig. 14. The values for α appear to be fairly constant and they vary within a ± 2 percent range of the average α-value, which is taken to be 0.58. In the range of modest tearing moduli (0 to 15) and amounts of crack growth (20 percent of the maximum extent of the plastic zone) examined, α appears to be independent of crack growth history. This value of α for the growing crack is 6 percent less than that estimated for the stationary crack. On the other hand, the values for s show more variations which range from 0.113 to 0.133. With the limited data available, it is difficult to conjecture whether or not s would saturate to a constant value when the amount of crack growth is substantially more than that simulated here. Indeed, there is a wide scattering in the s estimates reported in the literature [*7,11,15,16*] and these are summarized in the work of Drugan et al [*9*]. Unfortunately, yet more work needs to be done to determine the precise meaning of R and its possible dependence on crack growth history.

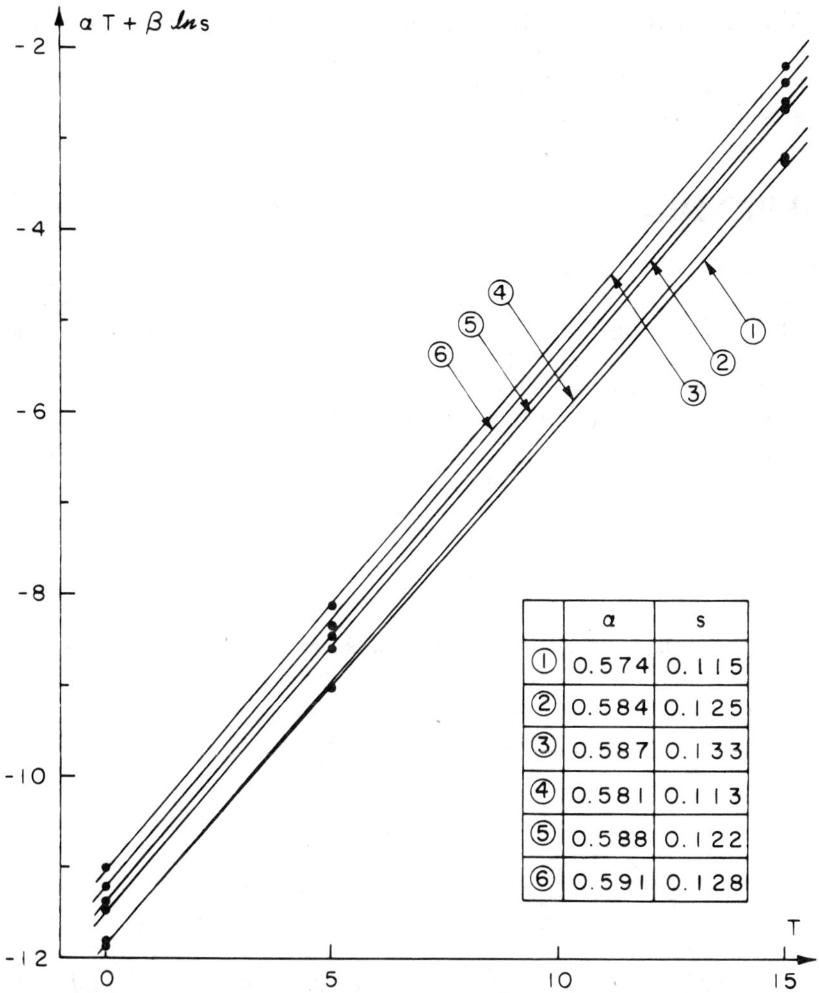

	α	s
1	0.574	0.115
2	0.584	0.125
3	0.587	0.133
4	0.581	0.113
5	0.588	0.122
6	0.591	0.128

FIG. 14—*Plot of* αT + β ℓn(s) *versus* T *for the growth histories depicted in Fig. 13. The keys to the graphs correspond to the branching sequences in Fig. 13. The inset shows the best fit of the* α *and* s*-values.*

Acknowledgments

The author gratefully acknowledges the support provided by the Division of Materials Science, Office of Basic Energy Sciences, U.S. Department of Energy under Contract AC02-80-ER10556 with Brown University. The computations were carried out on the Brown University, Division of Engineering, VAX-11/780 computer. The acquisition of this computer was made possible by grants from the U.S. National Science Foundation (Grant ENG78-19378), the General Electric Foundation, and the Digital Equipment Corp.

The author is also grateful to Professor J. R. Rice of Harvard University for his supervision of this work, and to Professors D. M. Parks of Massachusetts Institute of Technology and A. Needleman and C. F. Shih of Brown University for helpful discussions.

APPENDIX

Modified Variational Principle for Small-Scale Yielding

The configuration of a semi-infinite crack in the upper half space is shown in Fig. 15. Let Γ be a semi-circle with radius r_0 centered at the crack tip such that the plastic zone is always contained within it. The region outside and inside Γ is denoted as A and B, respectively. The unit normal **n** to Γ is defined in Fig. 15. The elastic singular stress field and its corresponding displacement field are

$$\sigma^{\infty} = \frac{K}{\sqrt{2\pi r}}\,\mathbf{g}(\theta)$$
$$\mathbf{u}^{\infty} = \frac{K}{\sqrt{2\pi}}\,\frac{\sqrt{r}}{2\mu}\,\mathbf{h}(\theta) \qquad (12)$$

where μ is the elastic shear modulus and **g** and **h** are known functions of the angle, θ.

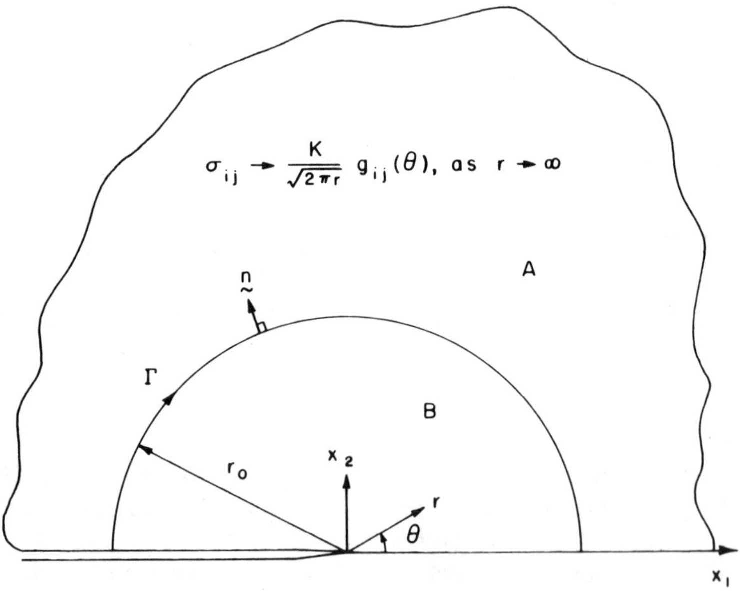

FIG. 15—*Division of the upper half space associated with the modified variational principle. In A, material response is always linear elastic and the stress and deformation fields are represented by convergent series. Region B is discretized into finite elements.*

The boundary conditions for this small-scale yielding problem are that

1. the traction rate at infinity is prescribed to be that corresponding to the elastic singular stress field,
2. the crack surface is traction free, and
3. the Mode I symmetry conditions are maintained on the crack line ahead of the crack tip.

The rate form of a modified potential energy functional, $\tilde{\pi}$, for small-scale yielding can be defined as (Hilton and Hutchinson [*14*])

$$\tilde{\pi}(\dot{\mathbf{u}}, \dot{\tilde{\mathbf{u}}}) = \int_A W_A(\dot{\tilde{\epsilon}})dS + \int_B W_B(\dot{\epsilon})dS - \int_\Gamma \mathbf{n}\cdot\dot{\sigma}^\infty\cdot\dot{\tilde{\mathbf{u}}}\, ds \tag{13}$$

with

$$\dot{\tilde{\mathbf{u}}} = 0\left(\frac{1}{\sqrt{r}}\right) \tag{14}$$

where

$W(\dot{\epsilon}) = \frac{1}{2}\, tr(\dot{\sigma}(\dot{\epsilon})\cdot\dot{\epsilon})$,
$\dot{\mathbf{u}}$ = velocity field,
$\dot{\sigma}$ = stress rate field, and
$\dot{\epsilon}$ = strain rate field.

The velocity $\dot{\tilde{\mathbf{u}}}$ is defined as

$$\dot{\tilde{\mathbf{u}}} = \dot{\mathbf{u}} - \dot{\mathbf{u}}^\infty$$

and $\dot{\tilde{\epsilon}}$ is the corresponding strain rate for $\dot{\tilde{\mathbf{u}}}$. It can be shown that the exact velocity fields $\dot{\mathbf{u}}$ in B and $\dot{\tilde{\mathbf{u}}}$ in A minimize the modified potential energy functional $\tilde{\pi}$. The implementation of this modified variational principle within the finite-element context is as follows.

First, it is noted that the stress field $\tilde{\sigma}$ and displacement field $\tilde{\mathbf{u}}$ in A can be expressed in terms of two analytic functions $\phi(z)$ and $\Omega(z)$ (for example Rice [*1*]) as

$$\begin{aligned} \tilde{\sigma}_{11} + \tilde{\sigma}_{22} &= 2[\phi'(z) + \overline{\phi'(z)}] \\ \tilde{\sigma}_{22} - \tilde{\sigma}_{11} + 2i\tilde{\sigma}_{12} &= 2[\Omega'(z) + (\bar{z} - z)\phi''(z) - \phi'(z)] \\ 2\tilde{\mu}(\tilde{u}_1 + i\tilde{u}_2) &= (3 - 4\nu)\phi(z) - \overline{\Omega(z)} + (\bar{z} - z)\overline{\phi'(z)} \end{aligned} \tag{15}$$

where $z = x_1 + i\,x_2$ with i being the imaginary unit. The analytic functions that satisfy the aforementioned boundary conditions and restriction (Eq 14) are given by

$$\phi'(z) = \Omega'(z) = \sum_n a_{n+1/2}\left(n + \frac{3}{2}\right)z^{n+1/2} + \sum_n a_n(n + 1)z^n \tag{16}$$

for $|z| \geq r_0$ with $n = -2, -3, -4, \ldots$ and $a_{n+1/2}$ and a_n being unknown real coefficients.

Thus, with $\tilde{\sigma}$ and $\tilde{\mathbf{u}}$ determined in A, the first and the third integral on the right-hand side of Eq 13 can be expressed in terms of a finite number of these unknown coefficients by truncating the series in Eq 16. In Region B, the usual procedure is employed to discretize it into finite elements. The nodal displacement increments along Γ are then taken to be those given by

$$\dot{\mathbf{u}}(\mathbf{x}) = \dot{\tilde{\mathbf{u}}}(\mathbf{x}) + \dot{\mathbf{u}}^{\infty}(\mathbf{x})$$

Hence, the modified potential energy functional, which is expressed in terms of the nodal displacement increments (excluding those along Γ) in B and the unknown coefficients, can be minimized with respect to these two sets of unknowns resulting in two sets of algebraic equations. Upon eliminating the set of unknown coefficients, one obtains a set of nonlinear finite-element equilibrium equations with the nodal displacement increments inside B as unknowns. The stiffness matrix obtained is symmetric and positive definite and the usual finite-element solution scheme can be used to determine the nodal displacement increments and hence the unknown coefficients.

Plastic Zone Correction

Here, a procedure for determining the plastic zone correction from the finite-element results is given. First, the problem of a semi-infinite crack in a linear elastic half space under Mode I loading and plane-strain conditions is considered. The crack tip is positioned at a distance c ahead of the origin of the Cartesian coordinates. The analytic functions $\phi'(z)$ and $\Omega'(z)$ for this problem are

$$\phi'(z) = \Omega'(z) = \frac{K}{2\sqrt{2\pi}} \frac{1}{\sqrt{z - c}} \tag{17}$$

These functions can be expanded for large z as

$$\phi'(z) = \Omega'(z) = \frac{K}{2\sqrt{2\pi z}} \left[1 + \frac{1}{2}\frac{c}{z} + \frac{3}{8}\left(\frac{c}{z}\right)^2 + \cdots\right] \tag{18}$$

Comparing the coefficients in Eq 16 with those in Eq 18 for the same order in $z^{-3/2}$ gives

$$\frac{c}{(K/\sigma_0)^2} = -2\sqrt{2\pi} \frac{\alpha_{-3/2}}{\sigma_0(K/\sigma_0)^3} \tag{19}$$

Thus, the plastic zone correction c is related to $a_{-3/2}$ which is computed at each stage of the load incrementation for the stationary crack. It is found that

$$c = 0.025\,(K/\sigma_0)^2$$

References

[1] Rice, J. R. in *Fracture: An Advanced Treatise*, Vol. 2, H. Liebowitz, Ed., Academic Press, New York, 1968, pp. 191-311.

[2] Rice, J. R. in "Mechanics and Mechanisms of Crack Growth" (*Proceedings*, Conference at Cambridge, England, April 1973), M. J. May, Ed., British Steel Corp. Physical Metallurgy Centre Publication, 1974, pp. 14-39.

[3] Cherepanov, G. P., *Mechanics of Brittle Fracture* (in Russian), Nauka, Moscow, 1974; English translation by A. L. Peabody; R. deWit and W. C. Cooley, Eds., McGraw-Hill, New York, 1979, pp. 284-289.

[4] Rice, J. R. and Sorensen, E. P., *Journal of the Mechanics and Physics of Solids*, Vol. 26, 1978, pp. 163–186.
[5] Slepyan, L. I., *Izvestiia Akademii Nauk S.S.S.R. Mekhanika Tverdogo Tela*, Vol. 9, 1974, pp. 57–67.
[6] Gao, Y-c., *Li-hsueh-hsueh-pao Acta Mechanica Sinica* (in Chinese), No. 1, 1980, pp. 48–56.
[7] Rice, J. R., Drugan, W. J., and Sham, T-L. in *Fracture Mechanics: Twelfth Conference, ASTM STP 700*, American Society for Testing and Materials, 1980, pp. 189–221.
[8] Rice, J. R. in *Mechanics of Solids (The Rodney Hill 60th Anniversary Volume)*, H. G. Hopkins and M. J. Sewell, Eds., Pergamon Press, Oxford, U.K., 1982, pp. 539–562.
[9] Drugan, W. J., Rice, J. R., and Sham, T-L., *Journal of the Mechanics and Physics of Solids*, Vol. 30, 1982, pp. 447–443, and erratum, Vol. 31, 1983, p. 191.
[10] Gao, Y-c., "The Influence of Compressibility on the Elastic-Plastic Field of a Growing Crack," presented at the American Society for Testing and Materials Second International Symposium on Elastic-Plastic Fracture Mechanics, Philadelphia, Pa., Oct. 1981.
[11] Sham, T-L., "A Finite Element Analysis of Quasi-Static Crack Growth in an Elastic-Perfectly Plastic Solid," Sc.M. Thesis, Brown University, Division of Engineering, Providence, R.I., March 1979.
[12] Rice, J. R., *Journal of Applied Mechanics*, Vol. 35, 1968, pp. 379–386.
[13] Rice, J. R. and Tracey, D. M. in *Numerical and Computer Methods in Structural Mechanics*, S. J. Fenves et al, Eds., Academic Press, New York, 1973, pp. 585–623.
[14] Hilton, P. D. and Hutchinson, J. W., *Engineering Fracture Mechanics*, Vol. 3, 1971, pp. 435–451.
[15] Dean, R. H. and Hutchinson, J. W. in *Fracture Mechanics: Twelfth Conference, ASTM STP 700*, American Society for Testing and Materials, 1980, pp. 383–405.
[16] Parks, D. M., Lam, P. S., and McMeeking, R. M. in "Advances in Fracture Research" (*Proceedings* of the Fifth International Conference on Fracture, Cannes, April 1981), D. Francis et al, Eds., Vol. 5, Pergamon Press, Oxford, U.K., 1981, pp. 2607–2614.
[17] Marcal, P. V. and King, I. P., *International Journal of Mechanical Sciences*, Vol. 9, 1967, pp. 143–155.
[18] Nagtegaal, J. C., Parks, D. M., and Rice, J. R., *Computer Methods in Applied Mechanics and Engineering*, Vol. 4, 1974, pp. 153–177.
[19] Chan, S. K., Tuba, I. S., and Wilson, W. K., *Engineering Fracture Mechanics*, Vol. 2, 1970, pp. 1–17.
[20] Parks, D. M., "Some Problems in Elastic-Plastic Finite Element Analysis of Cracks," Ph.D. Thesis, Brown University, Providence, R.I., 1975.
[21] McMeeking, R. M., "Large Plastic Deformation and Initiation of Fracture at the Tip of a Crack in Plane Strain," Ph.D. Thesis, Brown University, Providence, R.I., 1976.
[22] Shih, C. F., *Journal of the Mechanics and Physics of Solids*, Vol. 29, 1981, pp. 305–326.
[23] Irwin, G. R., *Applied Materials Research*, Vol. 3, 1964, pp. 65–81.
[24] Irwin, G. R. and Paris, P. C. in *Fracture: An Advanced Treatise*, Vol. 3, H. Liebowitz, Ed., Academic Press, New York, 1971, p. 1.

Alan Needleman[1] *and Viggo Tvergaard*[2]

Crack-Tip Stress and Deformation Fields in a Solid with a Vertex on Its Yield Surface

---

REFERENCE: Needleman, A. and Tvergaard, V., **"Crack-Tip Stress and Deformation Fields in a Solid with a Vertex on Its Yield Surface,"** *Elastic-Plastic Fracture: Second Symposium, Volume I—Inelastic Crack Analysis, ASTM STP 803*, C. F. Shih and J. P. Gudas, Eds., American Society for Testing and Materials, 1983, pp. I-80–I-115.

ABSTRACT: For deeply cracked specimens a finite strain numerical analysis is employed to investigate the development of the near-tip stress and strain fields during crack-tip blunting. The material is characterized by an elastic-plastic constitutive law with a vertex on subsequent yield surfaces at the point of loading, as suggested by physical models of polycrystalline aggregates. In contrast to classical smooth yield surface theory, this corner theory of plasticity permits shear band formation at achievable strain levels in work-hardening solids. Both for an edge-cracked bend specimen and for a center-cracked panel the corner theory description results in localization into shear bands in a small material region near the crack tip, so that very high strain peaks develop at an early stage of blunting. An edge-cracked bend specimen computation based on the classical Prandtl-Reuss plasticity theory, carried out for comparison purposes, shows a smooth variation of the strain field around the blunting crack tip. For the two edge-cracked bend specimens the stress and strain fields a few openings away from the tip are, however, in close agreement.

KEY WORDS: fracture, crack initiation, J-integral, plasticity, plastic properties, large-scale yielding, tip field dominance, elastic-plastic fracture

Analyses of local crack-tip stress and deformation fields, including the effects of the finite geometry changes which accompany crack-tip blunting, play a role in relating continuum stress analyses to micromechanical failure mechanisms in the prediction of criteria for crack extension. Such analyses also provide a framework for assessing the degree to which a single param-

[1]Professor of engineering, Division of Engineering, Brown University, Providence, R.I. 02912.
[2]Lecturer, Department of Solid Mechanics, Technical University of Denmark, 2800 Lyngby, Denmark.

eter characterization of the crack tip is appropriate, since the blunted crack-tip opening sets the scale over which large strains and high stress triaxiality develop.

When geometry changes are neglected, the stress and deformation states at a plane-strain tensile crack in a power-law hardening plastic solid are governed by the one-parameter singular field derived by Hutchinson [*1*][3] and Rice and Rosengren [*2*] (the HRR field). For lightly hardening solids, the HRR field gives rise to intense shearing above and below the crack tip and to a large hydrostatic stress elevation directly ahead of the tip.

Assessments of the influence of the finite geometry changes associated with crack-tip blunting on the near-tip stress and deformation states of plane-strain tensile cracks have been carried out by Rice and Johnson [*3*] and by McMeeking [*4*], within the context of small-scale yielding. Rice and Johnson [*3*] employed an approximate analysis for a smoothly blunting crack based on the local use of rigid-plastic slipline theory, with the remote boundary conditions taken to represent small-scale yielding. Rice and Johnson [*3*] found that consideration of finite geometry changes modifies the near tip fields in two significant respects. One is that blunting induces intense straining directly ahead of the crack tip and the other is that the stress-free boundary conditions on the deformed crack surface limit the hydrostatic stress elevation at the tip.

McMeeking [*4*] carried out finite-element analyses based on classical smooth yield surface plasticity theory and considered isotropically strain-hardening as well as ideally plastic solids. McMeeking [*4*] found that the initially smooth crack tip blunted into a smooth shape resembling the deformed shape obtained in Ref *3*. Also, the near-tip stress and deformation states obtained in the numerical solutions were in fairly good accord with those given by Rice and Johnson's approximate analysis [*3*]. Both the approximate analysis of Rice and Johnson [*3*] and the numerical results of McMeeking [*4*] indicate that the finite geometry changes associated with crack-tip blunting dominate over a range comparable in size to the crack-tip opening displacement (CTOD).

Subsequently, McMeeking and Parks [*5*] have conducted detailed plane-strain finite deformation analyses of deeply cracked center-cracked panels and single edge-cracked bend specimens using the finite-element method. Their calculations, also based on classical smooth yield surface plasticity theory with isotropic hardening, spanned the range from small-scale yielding to fully plastic behavior. The comparison between the center-cracked panel specimens and the edge-cracked bend specimens is of interest since McClintock [*6*] has observed the very different crack-tip stress and deformation states predicted for these configurations by ideally plastic slipline theory. The plastic limit state for the edge-cracked bend specimen gives rise to high triax-

[3]The italic numbers in brackets refer to the list of references appended to this paper.

ial tension ahead of the crack tip while that for the center-cracked panel does not. McMeeking and Parks [*5*] focused on the question of when the blunted crack-tip region is uniquely characterized by a single parameter such as Rice's J-integral [*7*], or, put another way, on the minimum specimen size requirements for a valid *J*-test. For the deeply cracked bend specimen good agreement of the near-tip fields with the corresponding small-scale yielding fields [*4*] was found even in the range when there was large-scale plasticity [*5*]. For the center-cracked panel specimens noticeable deviation from the small-scale yielding fields occurred much earlier [*5*].

The mode of blunting obtained in the numerical solutions of McMeeking and Parks [*5*] was the smooth crack-tip profile analyzed by Rice and Johnson [*3*] and McMeeking [*4*]. However, it is of interest to note that in the context of rigid perfectly plastic slipline theory, the blunted crack-tip shape is not unique. In addition to the smoothly blunting solution, solutions in which the crack blunts with two or more sharp vertices are also available [*6*].

In this paper we report on some initial studies of plane-strain crack-tip stress and deformation fields employing a plastic flow rule for a solid that develops a vertex on its yield surface. The occurrence of vertices on yield surfaces is predicted by models of the plastic flow properties of polycrystalline metals based on single crystal slip. The discreteness of slip systems in each grain leads to the prediction of a yield surface vertex, at least when the yield surface is defined for small offset plastic strains [*8,9*]. Although experimental evidence for yield surface vertices is ambiguous and conflicting, there is evidence in some cases for at least the development of a region of high curvature on the yield surface at the current stress point [*10,11*]. The question of whether a smooth yield surface plasticity theory or a cornered yield surface plasticity theory is an appropriate characterization of the plastic flow properties of structural metals has implications for the stress and deformation fields near a blunting crack tip. One immediate consequence of the presence of a vertex on the yield surface is that yield surface vertex effects permit shear band development at achievable strain levels in strain-hardening solids [*14,15*], whereas the classical smooth yield surface elastic-plastic solid is quite resistant to the localization of deformation into a shear band [*12,13*]. As our results will show, the highly localized deformation accompanying shear banding leads to a near-tip strain field very different from that predicted by the classical smooth yield surface model.

Another consequence of yield surface vertex effects arises from the fact that the notion of the J-integral as a path-independent characterizing parameter is based on a deformation theory of plasticity. Deformation plasticity theory is a small-strain nonlinear elastic (and, therefore, path independent) constitutive relation so that it is not an acceptable plasticity theory for nonproportional loading paths. However, the polycrystalline model based on the concept of single crystal slip considered by Hutchinson [*9*] does give closer agreement with deformation theory predictions than with smooth yield surface flow theory predictions for a nonproportional path consisting of small shear-stress increments superposed on monotonic tension. Indeed, such models, which have pointed yield surfaces, suggest that a deformation plasticity theory may be an appropriate model for the plastic flow properties of real materials for a limited range of loading histories that are nonproportional. Tests of the path independence of J have been based on smooth yield surface plasticity theory which coincides with deformation theory only for proportional loading paths. For a solid with a vertex on its yield surface the range of path independence of J is expected to be greater than indicated by analyses based on smooth yield surface plasticity theory.

We base our calculations on the phenomenological corner theory of plasticity, termed J_2 corner theory, proposed by Christoffersen and Hutchinson [*16*]. For nearly proportional loading, the instantaneous moduli are associated with those of a nonlinear elastic solid (a finite strain deformation plasticity theory). For increasing deviation from proportional loading the moduli stiffen smoothly until they coincide with the linear elastic moduli for stress increments directed along or within the corner of the yield surface.

Following McMeeking and Parks [*5*], we focus attention on deeply cracked center-cracked panel (CCP) specimens and edge-cracked bend (ECB) specimens. The specimen geometries are illustrated schematically in Fig. 1. We restrict attention to plane-strain deformations.

We carry out one J_2 corner theory computation for an ECB specimen and one for a CCP specimen employing identical material parameters in each case. For comparison purposes we also carry out a computation for the ECB specimen employing classical smooth yield surface plasticity theory, but with the specimen geometry and material stress-strain behavior otherwise identical to that employed in the corresponding corner-theory computation. The finite-element mesh and increment sizes employed in this smooth yield surface theory computation are also the same as those employed in the J_2 corner theory ECB calculation.

Computational Procedure

A Lagrangian convected coordinate formulation of the governing equations serves as the basis for the finite-element implementation employed in our analysis. This approach has been extensively used in previous finite deformation finite-element studies, for example, Needleman [*17*] and Tvergaard [*18*] for J_2 flow theory with isotropic hardening, and Tvergaard et al [*19*] and Triantafyllidis et al [*20*] for J_2 corner theory. This formulation will only be briefly outlined here.

Field Equations

The initial unstressed configuration is taken as reference and a material point is identified by its Cartesian coordinates X^i in the reference state. The

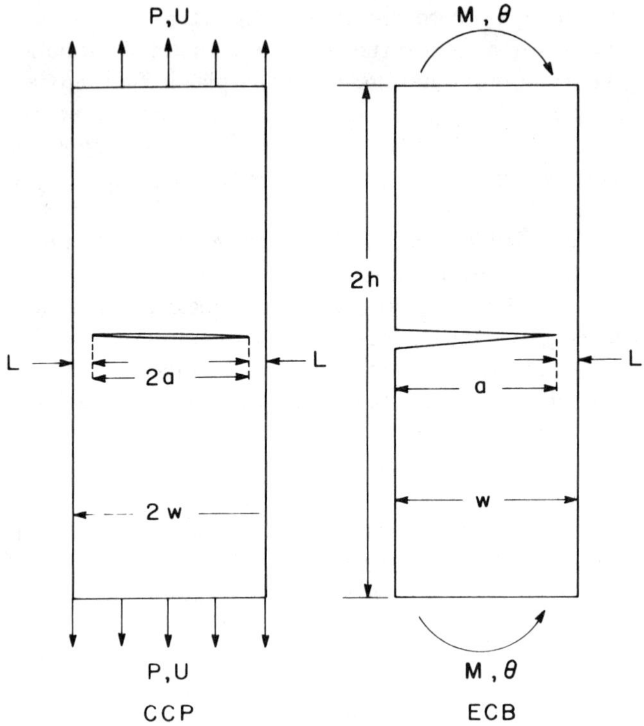

FIG. 1—*Schematic representation of specimen geometries and loadings, CCP (center-cracked panel in tension) and ECB (edge-cracked beam in bending). Attention is restricted to plane-strain deformations.*

reference state metric tensor is denoted by g_{ij}, with determinant g, while in the current configuration the metric tensor is G_{ij} and its determinant is G. The Lagrangian strain tensor is given by

$$\eta_{ij} = \frac{1}{2}(u_{i,j} + u_{j,i} + u_{,i}^{k} u_{k,j}) \tag{1}$$

where u^k are the displacement components on the reference base vectors and $(\ )_{,i}$ denotes covariant differentiation in the reference coordinate system.

Under plane-strain conditions, the incremental principle of virtual work on which the finite-element discretization is based is written as

$$\int_A [\dot{\tau}^{\alpha\beta}\delta\eta_{\alpha\beta} + \tau^{\alpha\beta}\dot{u}_{,\alpha}^{\gamma}\delta u_{\gamma,\beta}]dA$$

$$= \int_S \dot{T}^{\alpha}\alpha u_{\alpha}dS + \left[\int_S T^{\alpha}\delta u_{\alpha}dS - \int_A \tau^{\alpha\beta}\delta\eta_{\alpha\beta}dA\right] \tag{2}$$

Here, A and S denote, respectively, the cross-sectional area and perimeter of the plane of deformation in the reference configuration. The contravariant components of the Kirchhoff stress tensor on the embedded deformed coordinates, $\tau^{\alpha\beta}$, are related to the corresponding components of the Cauchy stress tensor $\sigma^{\alpha\beta}$ by

$$\tau^{\alpha\beta} = \sqrt{\frac{G}{g}}\,\sigma^{\alpha\beta} \tag{3}$$

and the nominal traction components, T^{α}, on a curve with normal n_{β} in the reference configuration are given by

$$T^{\alpha} = (\tau^{\alpha\beta} + \tau^{\gamma\beta} u_{,\gamma}^{\alpha}) n_{\beta} \tag{4}$$

In Eq 2, and subsequently, a superposed dot, $(\dot{\ })$, denotes an increment of a field quantity. The bracketed term on the right-hand side of Eq 2 is an equilibrium correction term included to inhibit drifting of the discretized solution from the true equilibrium path due to incrementation errors.

Constitutive Relations

The J_2 corner theory of Christoffersen and Hutchinson [*16*], was introduced in order to develop an analytically tractable phenomenological vertex theory of plasticity. The significance of vertex formation for bifurcation-related phenomena in the plastic range has been appreciated for some time in the context of plastic buckling applications [*21,22*] and more recently in the context of tensile bifurcation problems, in particular, in relation to the shear band bifurcations of concern here (for example, Rudnicki and Rice [*12*], Rice [*13*], Needleman and Rice [*14*], and Hutchinson and Tvergaard [*15*]).

In J_2 corner theory the instantaneous moduli for nearly proportional loading are chosen to be the moduli of J_2 deformation theory (the finite strain generalization of J_2 deformation theory employed here is a path-independent nonlinear elastic constitutive relation). With M_{ijkl}^{0} denoting the deformation theory compliances and $\mathfrak{M}_{ijkl}$ denoting the linear elastic compliances, the plastic compliances of deformation theory are

$$C_{ijkl} = M_{ijkl}^{0} - \mathfrak{M}_{ijkl} \tag{5}$$

so that

$$\dot{\eta}_{ij}^{p} = C_{ijkl}\,\overset{\nabla}{\tau}{}^{kl} \tag{6}$$

Here, $\dot{\eta}_{ij}^{p}$ is the plastic part of the strain rate and $\overset{\nabla}{\tau}{}^{kl}$ are the contravariant components of the Jaumann derivative of Kirchhoff stress.

The yield surface in the neighborhood of the current loading point is taken to be a cone in stress deviator space with the cone axis in the direction

$$\lambda^{ij} = s^{ij}(C_{mnpq}\, s^{mn}\, s^{pq})^{-1/2} \tag{7}$$

where

$$s^{ij} = \tau^{ij} - \frac{1}{3} G_{kl}\, \tau^{kl}\, G^{ij} \tag{8}$$

An angular measure of the stress rate direction relative to the cone axis is defined by

$$\cos\theta = C_{ijkl}\, \lambda^{ij}\, \overset{\nabla}{s}{}^{kl}(C_{mnpq}\, \overset{\nabla}{s}{}^{mn}\, \overset{\nabla}{s}{}^{pq})^{-1/2} \tag{9}$$

The stress rate potential at the vertex is taken as

$$W = \frac{1}{2}\, \mathfrak{M}_{ijkl}\, \overset{\nabla}{\tau}{}^{ij}\, \overset{\nabla}{\tau}{}^{kl} + \frac{1}{2} f(\theta)\, C_{ijkl}\, \overset{\nabla}{\tau}{}^{ij}\, \overset{\nabla}{\tau}{}^{kl} \tag{10}$$

The transition function $f(\theta)$ is unity throughout the total loading range, which is specified by $0 \leq \theta \leq \theta_0$ and is identically zero for $\theta_c < \theta \leq \pi$, where θ_c denotes the angle of the yield surface cone. In the transition regime, $\theta_0 \leq \theta \leq \theta_c$, $f(\theta)$ decreases monotonically and smoothly from unity to zero in a manner that ensures convexity of the incremental relation. The transition function employed here is the same as that employed in Refs *15*, *19*, and *20* and is specified by

$$f(\theta) = \frac{1}{g(\phi)\,[1 + l^2(\phi)]} \tag{11}$$

where

$$\theta(\phi) = \phi + \arctan[l(\phi)], \qquad l(\phi) = \frac{dg/d\phi}{2g(\phi)} \tag{12}$$

and

$$g(\phi) = \begin{cases} 1, & 0 \leq \phi \leq \theta_0 \\ \{1 - [(\phi - \theta_0)/(\theta_n - \theta_0)]^3\}^{-2}, & \theta_0 \leq \phi \leq \theta_n \end{cases} \tag{13}$$

with $\theta_n = \theta_c - \pi/2$. Of the transition functions considered by Christoffersen and Hutchinson [*16*], this transition function was found to most closely duplicate the moduli found by Hutchinson [*9*] using a self-consistent model of a polycrystalline aggregate.

From the potential function Eq 10, the strain rate is found to be

$$\dot{\eta}_{ij} = \frac{\partial^2 W}{\partial \overset{\triangledown}{\tau}^{ij} \partial \overset{\triangledown}{\tau}^{kl}} \overset{\triangledown}{\tau}^{kl} = M_{ijkl}(\theta)\, \overset{\triangledown}{\tau}^{kl} \tag{14}$$

Inverting Eq 14 gives the moduli $R^{ijkl}(\theta)$ relating the Jaumann rate of Kirchhoff stress and the strain rate, so that

$$\overset{\triangledown}{\tau}^{ij} = R^{ijkl}(\theta)\, \dot{\eta}_{kl} \tag{15}$$

The determination of the deformation theory moduli and compliances is outlined in Ref *19* and will not be repeated here.

With the transition function specified as in the preceding, the parameters characterizing the corner are θ_0, the angle limiting the total loading range, and θ_c, the angle of the yield surface cone. However, in J_2 corner theory a more convenient measure of the yield surface cone angle than θ_c is the angle β_c, where

$$\tan \beta_c = \frac{-\sigma_0}{(\sigma_e^2 - \sigma_0^2)^{1/2}} \tag{16}$$

Here, σ_e is the Mises effective stress, $\sigma_e^2 = 3/2\, s_{ij} s^{ij}$, and σ_0 is the initial tensile yield strength.

The angular measure θ_c and β_c are related by

$$\tan \theta_c = \sqrt{a} \tan \beta_c, \qquad a = \left| \frac{E}{E_s q} - 1 \right| \left| \frac{E}{E_t} - 1 \right|^{-1} \tag{17}$$

where

E = Young's modulus,
E_t = slope of the uniaxial Kirchhoff stress-logarithmic strain curve,
E_s = ratio of stress to strain on this stress strain curve, and

$$q = (\epsilon_1 - \epsilon_2) \coth (\epsilon_1 - \epsilon_2) \tag{18}$$

with ϵ_1 and ϵ_2 being the logarithms of the principal stretches.

The sharpness of the cone angle is limited by employing Eq 16 to give β_c for $\pi/2 \leq \beta_c \leq (\beta_c)_{max}$ and taking $\beta_c = (\beta_c)_{max}$ otherwise. In particular, in the computations carried out here we take $(\beta_c)_{max} = 3\pi/4$ (135 deg), and the angle limiting the total loading range, θ_0, is taken as $\theta_0 = \theta_c/2 - \pi/4$, with θ_c obtained from β_c via Eq 17. This vertex characterization has been employed in previous studies of shear band development [*15,19,20*].

In the incremental principle of virtual work, Eq 2, we need the moduli relating the contravariant components of the convected derivative of Kirchhoff stress, $\dot{\tau}^{\alpha\beta}$, to the strain rates, namely

$$\dot{\tau}^{\alpha\beta} = L^{\alpha\beta\gamma\delta} \dot{\eta}_{\gamma\delta} \tag{19}$$

The moduli **L** in Eq 19 are obtained from the moduli **R** in Eq 15 via the relation

$$\dot{\tau}^{\alpha\beta} = \overset{\nabla}{\tau}^{\alpha\beta} - G^{\alpha\gamma}\tau^{\beta\delta}\dot{\eta}_{\gamma\delta} - G^{\beta\gamma}\tau^{\alpha\delta}\dot{\eta}_{\gamma\delta} \tag{20}$$

For the classical smooth yield surface theory computation we employ the finite strain version of J_2 flow theory with isotropic hardening described by Hutchinson [*23*], based on a proposal of Budiansky (unpublished work). For this constitutive law the moduli **R** in Eq 15 are given by

$$R^{ijkl} = \frac{E}{1+\nu}\left[\frac{1}{2}(G^{ik}G^{jl} + G^{il}G^{jk}) + \frac{\nu}{1-2\nu}G^{ij}G^{kl} - \frac{3}{2}\alpha\frac{(E/E_t - 1)}{E/E_t - (1-2\nu)/3}\,\frac{s^{ij}s^{kl}}{\sigma_e^2}\right] \tag{21}$$

All quantities appearing in Eq 21 have been defined in the foregoing except for ν, which is Poisson's ratio, and α, which is unity for plastic loading and zero for elastic unloading.

The moduli, Eq 21, represent the same finite strain version of J_2 flow theory with isotropic hardening employed by McMeeking [*4*] and McMeeking and Parks [*5*].

The uniaxial Kirchhoff stress-logarithmic strain curve is represented by a piecewise power law of the form

$$\tau = \begin{cases} E\epsilon, & \text{for} \quad \tau < \sigma_0 \\ \sigma_0\left(\dfrac{\epsilon}{\epsilon_0}\right)^N, & \text{for} \quad \tau \geq \sigma_0 \end{cases} \tag{22}$$

and we take $N = 0.1$, $\epsilon_0 = \sigma_0/E = 0.002$, and $\nu = 0.3$ in the computations reported on here.

Finite-Element Implementation

All three computations were carried out using the finite-element mesh depicted in Fig. 2. For the CCP computation this mesh represents one quarter of the specimen and it represents one half of the specimen for the ECB computations. The dimensions of the specimens shown in Fig. 1 are taken as $a/w = 0.9$ (so that $L/w = 0.1$) and $h/w = 2.0$.

Figure 2*a* shows the outer mesh, while Fig. 2*b* depicts the mesh in the region left blank in Fig. 2*a*. The mesh very near the crack tip (represented by a semi circular notch) is shown in Fig. 2*c*. This mesh fits into the semicircular blank region in Fig. 2*b*. However, in Fig. 2*c* we have depicted the near-tip mesh on both sides of the line of the crack. Symmetry about the crack line is imposed so that only one-half the mesh shown in Fig. 2*c* is actually employed

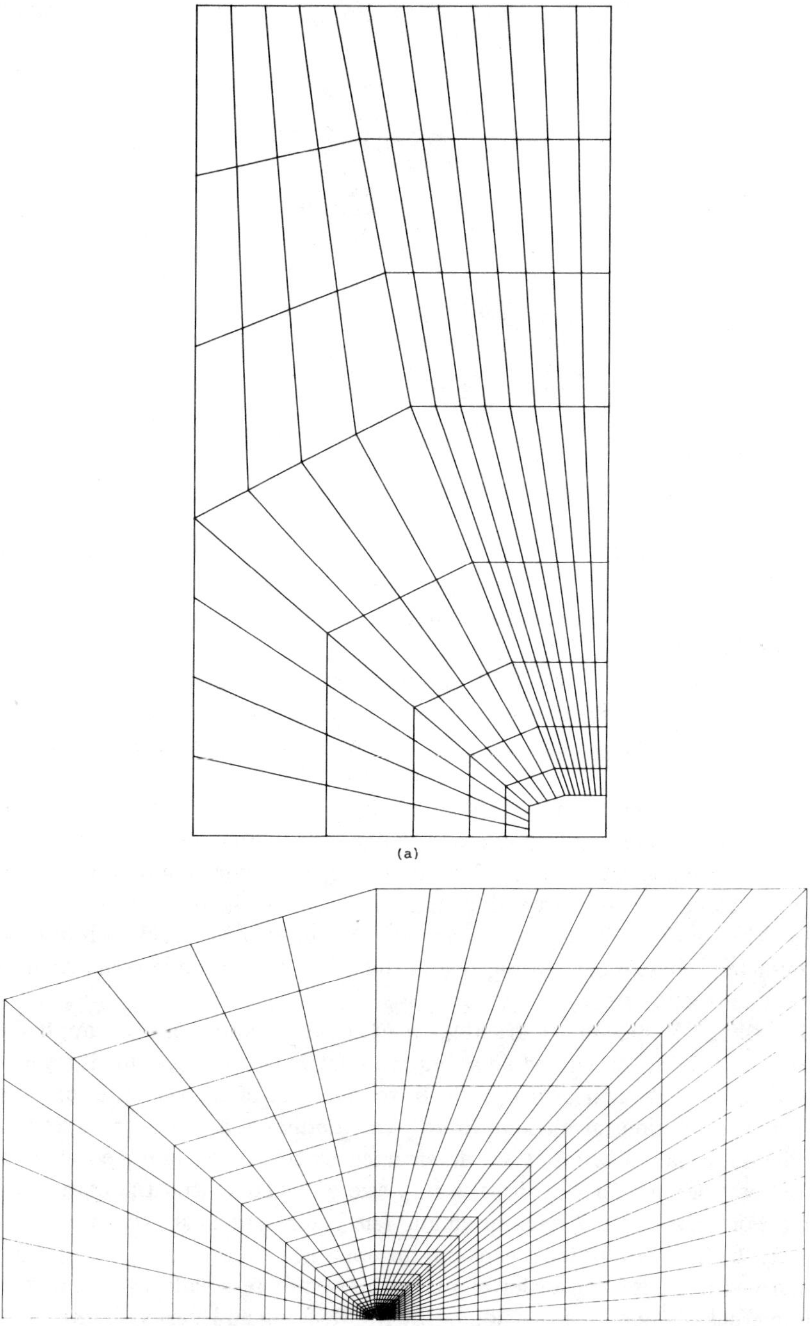

FIG. 2—*Finite-element mesh in initial configuration:* (a) *outer mesh,* (b) *mesh employed in region left blank in* (a), *and* (c) *near-tip mesh. In* (c) *Nodes* A *and* B *are subsequently employed to define current notch opening.*

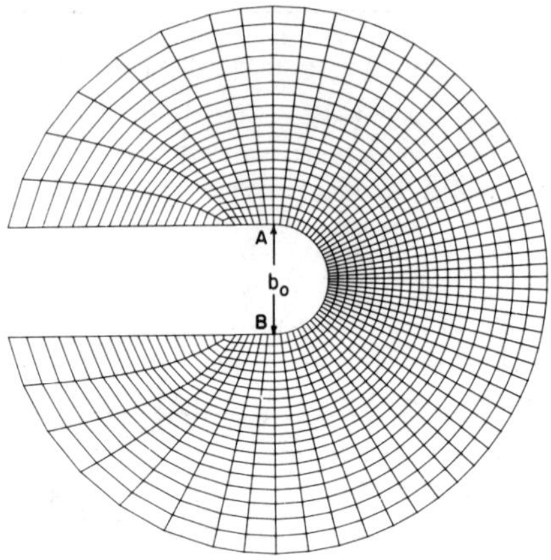

(c)

FIG. 2—*Continued.*

in the computations. The ratio of initial notch width, b_0, to uncracked ligament length, L, is 2×10^{-3}, with L measured from the center of the semicircular notch to the specimen boundary. This is the larger of the two b_0/L ratios considered by McMeeking and Parks [*5*].

The mesh is composed of 1403 quadrilateral elements. Each quadrilateral element consists of four crossed constant-strain triangles. This frees the mesh of any artificial constraint arising from nearly incompressible material behavior [*24*].

Traction-free boundary conditions are imposed on the notch surface and symmetry conditions are imposed on the line of the crack along the uncracked ligament. For the CCP calculation, symmetry boundary conditions are also imposed along the axis perpendicular to the crack at the midline of the specimen. Also, in the CCP computation a uniform tensile displacement is imposed as depicted schematically in Fig. 1 and the sides of the specimen are required to remain traction free.

For the ECB specimens, traction-free boundary conditions are applied at the sides of the specimen, while a rotation relative to the crack line is imposed (see Fig. 1), as described in Ref *20*. In the absence of a crack, the rotational boundary conditions give rise to finite pure bending of a plane-strain strip.

The deformation histories are determined by the linear incremental procedure described in Ref *19*, with the step size adjusted so that the equilibrium correction term in Eq 2 remains small. The deformations are continued until the crack tips are blunted to maximum openings about 2.5 times the initial notch width. This is an order of magnitude or more smaller than the maximum crack-tip openings considered by McMeeking and Parks [*5*]. As will be illustrated subsequently, for the corner-theory computations, shear band formation severely distorts the near-tip mesh and also leads to very high gradients where the mesh is most distorted. Hence, accuracy is lost in this region and therefore the computations are terminated. No such difficulty is encountered in the smooth yield surface computation for the range of crack-tip openings considered here, but since this computation is carried out for comparison purposes, it is arbitrarily terminated.

Results

Figure 3 displays the ratio of current notch opening, b, to initial notch opening, b_0, versus $J/\sigma_0 b_0$ for the three cases analyzed. The current notch opening, b, was identified with the current distance between Nodes A and B shown in Fig. 2c while J was evaluated using its finite deformation contour integral definition as given by Eshelby [25]

$$J = \int_{\Gamma} [W dX^2 - T^{\alpha} u_{\alpha,1}\, ds] \tag{23}$$

where

$$W = \int_0^{\eta_{\alpha\beta}} \tau^{\alpha\beta}\, d\eta_{\alpha\beta} \tag{24}$$

and T^{α} is given by Eq 4. In Eq 23, the crack line is assumed parallel to the X^1-axis in the initial state, Γ is some path, in the initial configuration, from the lower notch surface to the upper notch surface and ds is an element of arc length along the path Γ.

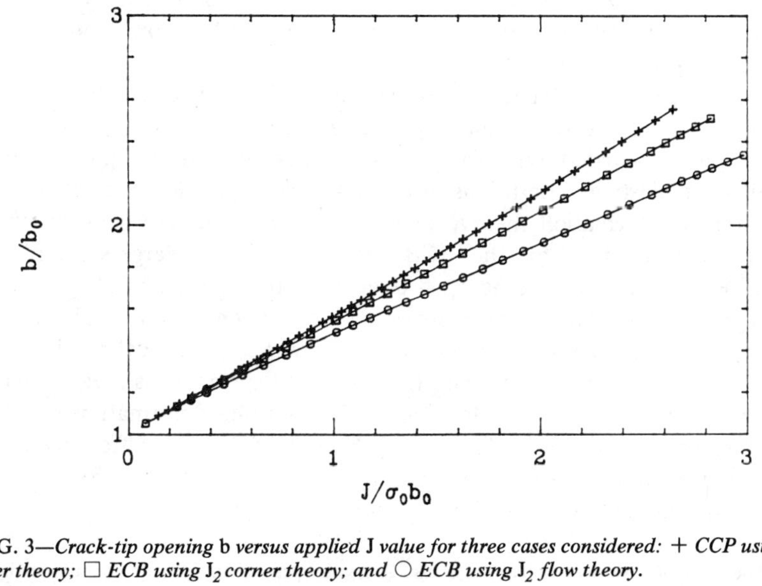

FIG. 3—*Crack-tip opening* b *versus applied* J *value for three cases considered:* + *CCP using* J_2 *corner theory;* □ *ECB using* J_2 *corner theory; and* ○ *ECB using* J_2 *flow theory.*

The path integral Eq 23 was evaluated along various contours, Γ, surrounding the crack tip. For contours sufficiently far from the tip, the value of J obtained numerically was virtually independent of the contour employed. Most values of J in Fig. 3 were obtained using a path through the ring of elements farthest from the notch tip in Fig. 2*b*.

As can be seen in Fig. 3, the curves of b/b_0 versus $J/\sigma_0 b_0$ are nearly linear for the two ECB specimens once $J/\sigma_0 b_0$ is greater than about unity. On the other hand, the curve for the CCP specimen is concave upwards. For the J_2 flow-theory ECB computation, the slope of the linear portion of this curve is 0.45 while for the J_2 corner-theory ECB computation $db/d(J/\sigma_0) = 0.53$. For the J_2 corner-theory CCP specimen, $db/d(J/\sigma_0)$ increases from 0.57 when $J/\sigma_0 b_0$ is around unity to 0.63 at the termination of the computation. This suggests, as will be illustrated in more detail subsequently, that an HRR field dominated region does not develop in the CCP computation.

The CTOD versus J curves obtained here can be compared with those obtained in previous investigations. Based on a small-strain analysis of a mathematically sharp crack, Shih [*26*] writes

$$\delta_t = d_n J/\sigma_0 \tag{25}$$

where δ_t is the tip opening displacement of the initially sharp crack, measured in Ref *26* at the intersection of 45 deg lines emanating from the deformed tip with the crack flanks. If we identify d_n in Eq 25 with the secant drawn from the leftmost point of a curve in Fig. 3 to the current point, then in the range $J/\sigma_0 b_0 > 1$, d_n decreases from 0.47 to 0.45 for the J_2 flow-theory ECB specimen, $d_n = 0.53$ in this range for the J_2 corner-theory ECB specimen, and d_n increases from 0.56 to 0.59 for the CCP computation which was based on J_2 corner theory.

Shih [*26*] reports that the HRR field gives $d_n = 0.50$ with $N = 0.1$ and $\sigma_0/E = 0.002$, the values employed here, while small-strain finite-element computations carried out using these values of material parameters for small-scale plasticity conditions give $d_n = 0.48$ for J_2 flow theory and $d_n = 0.52$ for J_2 deformation theory. These latter values agree very well with the values obtained here for the ECB specimens. Under large-scale plasticity conditions, Shih [*26*] obtains $d_n = 0.44$[4] for a fully plastic ECB specimen using J_2 flow theory which is also in very good agreement with the value obtained during the latter stages of loading in our J_2 flow-theory ECB calculation.

The value $d_n = 0.45$ obtained here for ECB specimen is greater than the value of 0.41 found by McMeeking [*4*] in his finite deformation small-scale yielding analysis with $N = 0.1$ and $\sigma_0/E = 0.00333$. The difference is probably somewhat larger than the apparent 10 percent since d_n is found by Shih [*26*] and McMeeking [*4*] to increase with σ_0/E.

[4]The value of 0.49 given in Table 1 of Ref *26* is a misprint (Shih, private communication).

We now focus attention, for each specimen analyzed, on four stages of the loading history. Figure 4 depicts the extent of the plastic zone at each of these four stages for the ECB specimens (the plastic zones are identical for the J_2 flow-theory and J_2 corner-theory computations), while Fig. 5 depicts the four plastic zone sizes for the J_2 corner-theory CCP computation. The portion of the mesh shown in Figs. 4 and 5 is the part displayed in Fig. 2*b*. The opening dis-

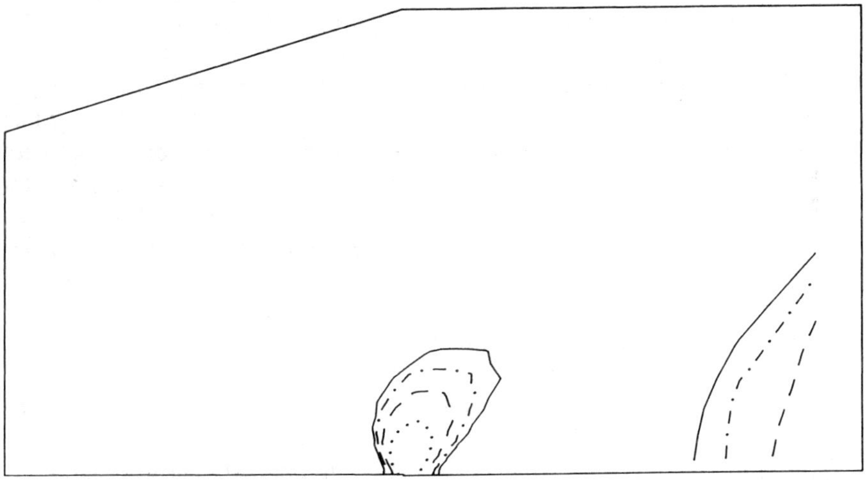

FIG. 4—*Plastic zone at four stages of deformation for the two ECB computations:* ··· $L\sigma_0/J =$ *495,* --- $L\sigma_0/J = $ *291,* -·- $L\sigma_0/J = $ *216, and* — $L\sigma_0/J = $ *177.*

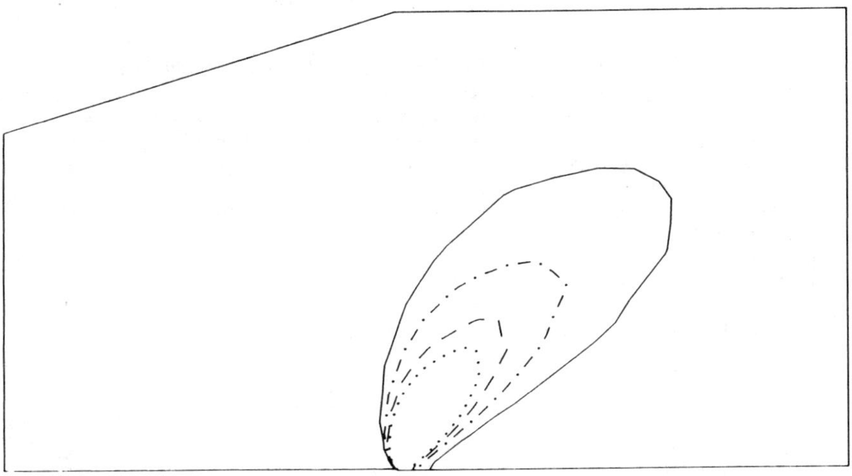

FIG. 5—*Plastic zone at four stages of deformation for CCP computation:* ··· $L\sigma_0/J = $ *482,* --- $L\sigma_0/J = $ *405,* -·- $L\sigma_0/J = $ *296, and* — $L\sigma_0/J = $ *195.*

placements and J-values at each of these loading stages are given in Table 1. Also given in this table is the plotting symbol employed in subsequent figures to denote field quantities (stresses, strains) at a given stage.

Crack-Tip Profiles

Figure 6 depicts the crack-tip profiles of the J_2 flow-theory ECB computation. This is the smoothly blunting profile analyzed by Rice and Johnson [*3*] and obtained from the finite-element analyses of McMeeking [*4*] and McMeeking and Parks [*5*].

The crack-tip profiles at the same $J/\sigma_0 b_0$-values for the J_2 corner-theory ECB computation are quite different. In Fig. 7*a*, the smooth notch tip has flattened out and in Fig. 7*b* the development of shear bands from the corners of this flattened region is quite clear. Subsequently, further opening occurs with increased shearing, Figs. 7*c* and 7*d*. Note the wedge-shaped region near the crack tip which undergoes relatively little deformation between Figs. 7*b* and 7*d*.

The blunted crack-tip profiles exhibited in Figs. 8*a* through 8*c* for the J_2 corner-theory CCP calculation show a similar development to that found in the J_2 corner-theory ECB case. However, in the latter stages of loading intense deformation occurs in the wedge-shaped region along the crack line at the crack tip, giving rise to the "V"-shaped profile in Fig. 8*d*.

The flattening out of a curved surface at the initiation of shear band development as exhibited in Figs. 7*a* and 8*a* has been encountered in previous calculations. This phenomenon was noted by Tvergaard et al [*19*] in their calculations of necking and shear band development in the plane-strain tensile test and was also observed by Larsson et al [*27*] in their analysis of flow

TABLE 1—*Values of crack opening displacement and* J *at four stages of deformation for each calculation (ECB = edge-cracked bend specimen; CCP = center-cracked panel; and* L *is defined in Fig. 1). Symbol is the plotting symbol used for a given stage in subsequent figures.*

Computation	Symbol	b/b_0	$J/\sigma_0 b_0$	$L\sigma_0/J$
J_2 flow-theory ECB	◇	1.49	1.01	495
J_2 flow-theory ECB	○	1.79	1.72	291
J_2 flow-theory ECB	□	2.05	2.32	216
J_2 flow-theory ECB	+	2.27	2.83	177
J_2 corner-theory ECB	◇	1.55	1.01	495
J_2 corner-theory ECB	○	1.92	1.72	291
J_2 corner-theory ECB	□	2.24	2.32	216
J_2 corner-theory ECB	+	2.51	2.83	177
J_2 corner-theory CCP	◇	1.59	1.04	482
J_2 corner-theory CCP	○	1.70	1.23	405
J_2 corner-theory CCP	□	1.97	1.69	296
J_2 corner-theory CCP	+	2.50	2.56	195

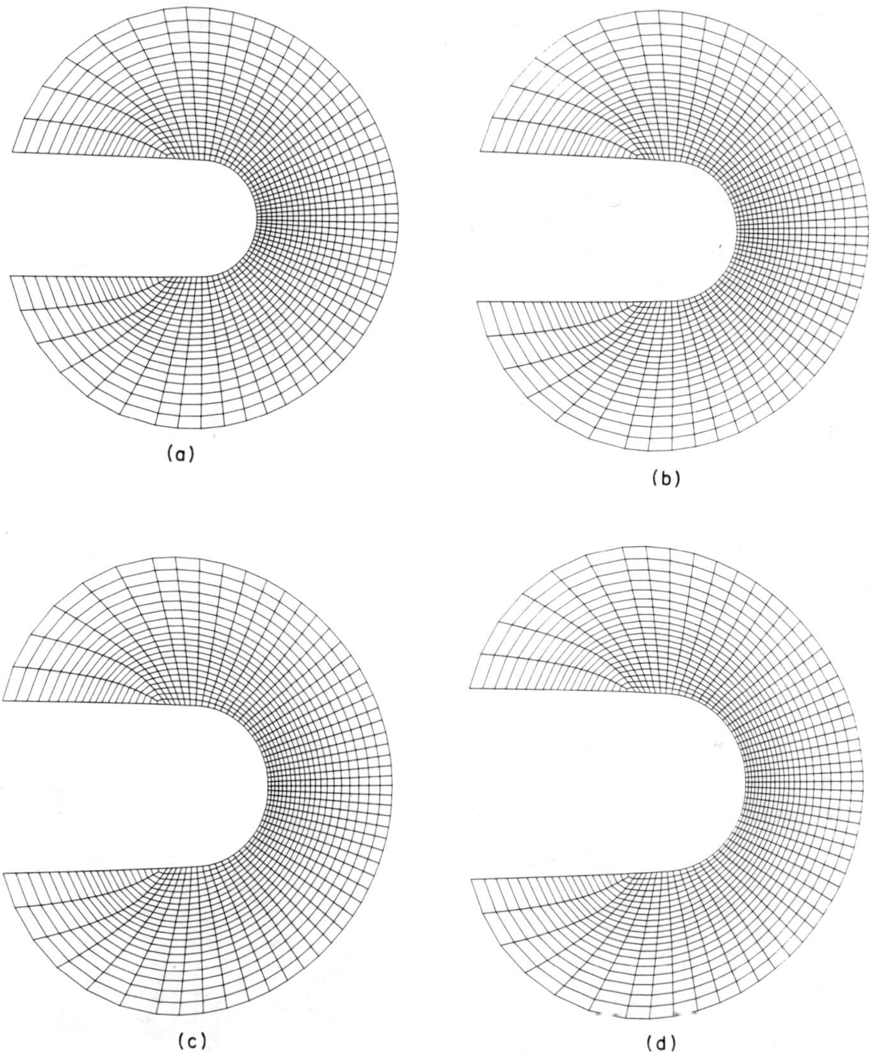

FIG. 6—*Deformed near-tip mesh at four stages of deformation for* J_2 *flow-theory ECB computation:* (a) $L\sigma_0/J$ = *495,* (b) $L\sigma_0 J$ = *291,* (c) $L\sigma_0 0/J$ = *216, and* (d) $L\sigma_0/J$ = *177.*

localization in internally pressurized cylinders. In particular, Tvergaard et al [*19*] in reference to their Figs. 5 and 6 note that this flattening is associated with the attainment of the critical strain for a short-wavelength surface instability at the free surface. For the material parameters employed here, the principal strain required for such a surface instability is 0.25 [*19*], which is reached quite early in the deformation history. At the deformation stages shown in Figs. 7*a* and 8*a*, the maximum principal strain is substantially

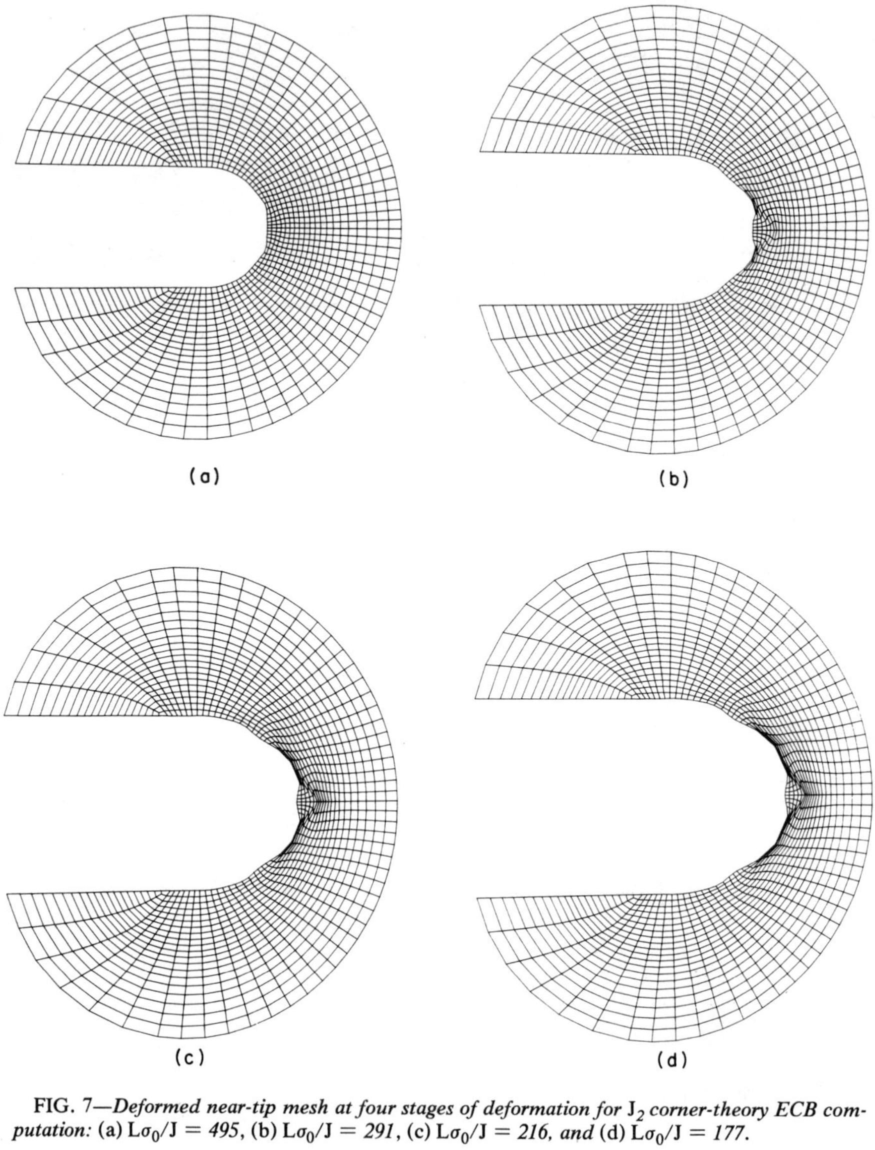

FIG. 7—*Deformed near-tip mesh at four stages of deformation for* J_2 *corner-theory ECB computation:* (a) $L\sigma_0/J = 495$, (b) $L\sigma_0/J = 291$, (c) $L\sigma_0/J = 216$, *and* (d) $L\sigma_0/J = 177$.

higher. In fact, the critical strain for shear band initiation, which in the present case is 0.32 [*19*], has already been exceeded locally in Figs. 7*a* and 8*a*.

Comparison with the HRR Field

In Figs. 9 through 11 the Cartesian component of stress normal to the line of the crack, σ_{22}, is plotted as a function of the distance R of a material point from

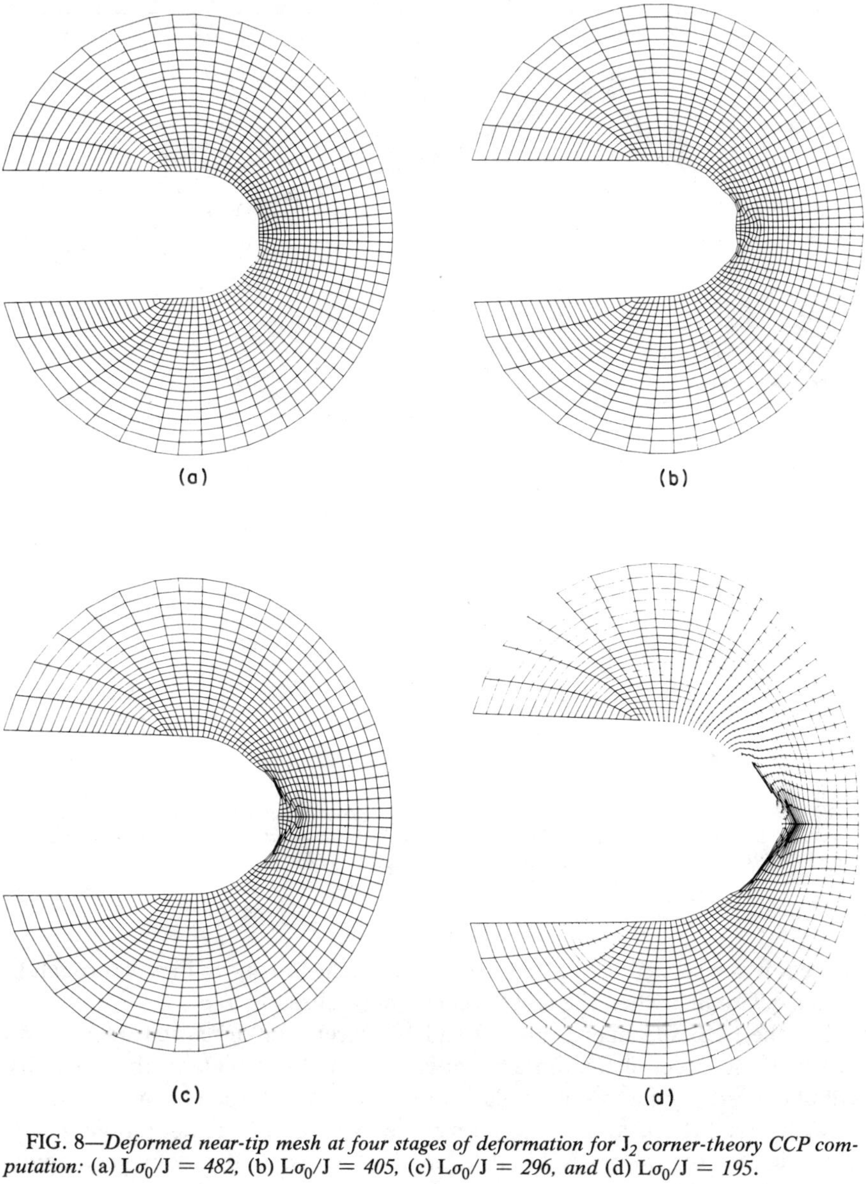

FIG. 8—*Deformed near-tip mesh at four stages of deformation for* J_2 *corner-theory CCP computation:* (a) $L\sigma_0/J = 482$, (b) $L\sigma_0/J = 405$, (c) $L\sigma_0/J = 296$, *and* (d) $L\sigma_0/J = 195$.

the tip in the initial configuration. Since the Kirchhoff stress differs from the Cauchy or true stress by a scale factor which differs from unity only due to elastic volume changes as shown in Eq 3, the distinction between Kirchhoff and Cauchy stress quantities will be neglected in the subsequent discussion. In these figures, the stress is normalized by the initial yield stress σ_0 and the dis-

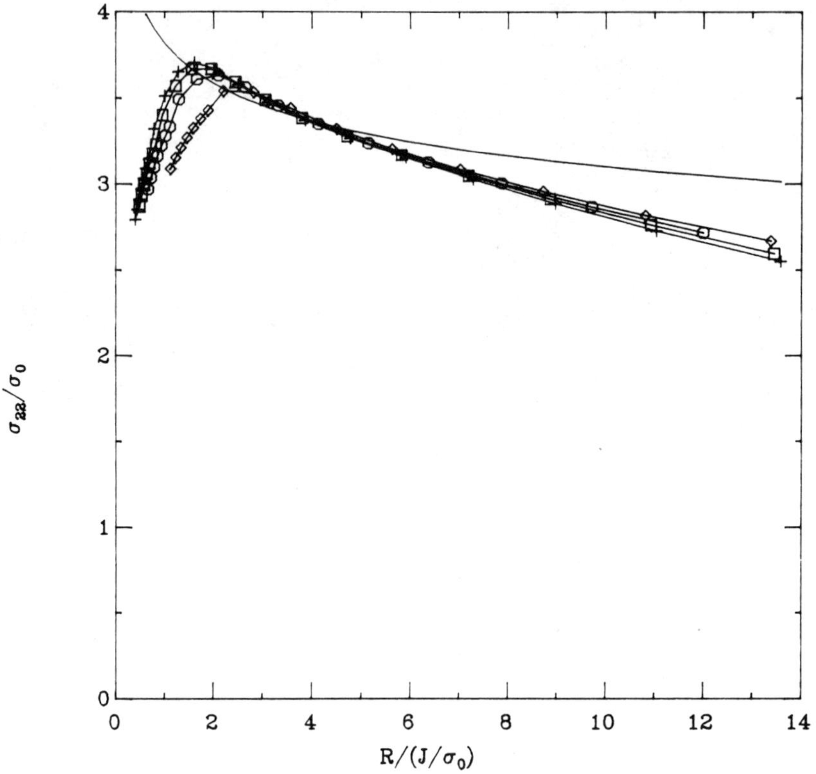

FIG. 9—*Normal stress on plane ahead of crack tip versus initial distance from tip for* J_2 *flow-theory ECB computation. The solid line is the HRR field value and* ◇ *is at* $L\sigma_0/J = 495$, ○ *is at* $L\sigma_0/J = 291$, □ *is at* $L\sigma_0/J = 216$, *and* + *is at* $L\sigma_0/J = 177$.

tance R is normalized by J/σ_0. The HRR field is the full line in Figs. 9 through 11 and is self-similar when the length scale is normalized by J/σ_0.

For the ECB specimens, Figs. 9 and 10, except for the initial stage shown in which the effect of the finite notch is still somewhat evident, the stress distributions are virtually self-similar. The stress distributions given by the J_2 flow-theory, Fig. 9, and the J_2 corner-theory, Fig. 10, computations nearly coincide over the range displayed, which is $R/(J/\sigma_0) > 0.4$. Very good agreement is found between the computed values of stress and the values given by the HRR field over the range $1.5 \leq R/(J/\sigma_0) \leq 5$. This upper limit to the size of the region over which the HRR stress distribution dominates in the ECB specimen is in good agreement with that observed by Shih and German [*28*] using a small-strain J_2 flow-theory formulation. The lower limit arises from the influence of the traction-free condition on the crack tip. We also note that McMeeking and Parks [*5*] have found that the self-similar charac-

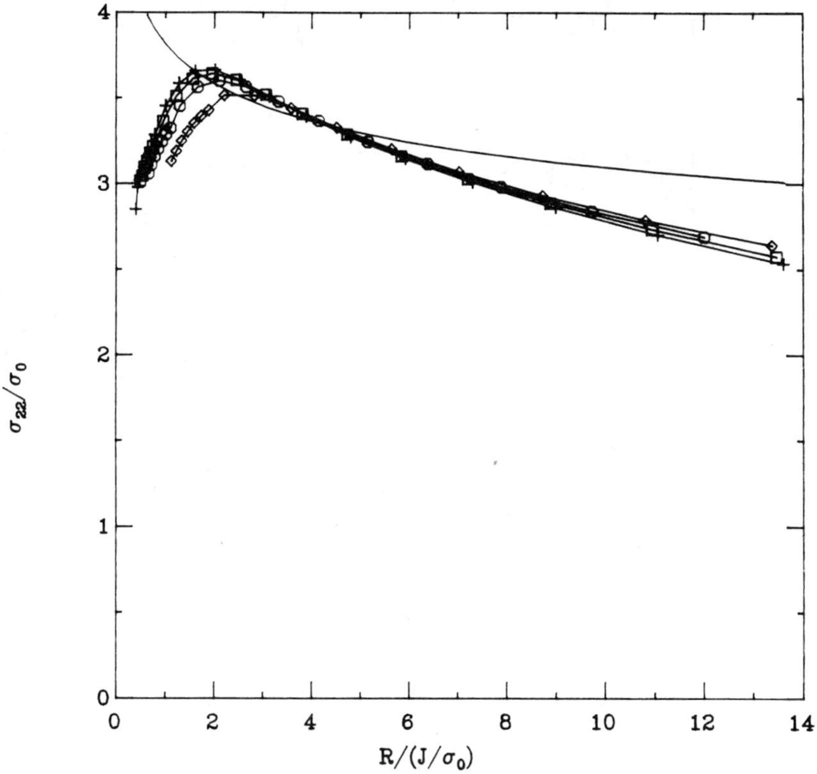

FIG. 10—*Normal stress on plane ahead of crack tip versus initial distance from tip for* J_2 *corner-theory ECB computation. The solid line is the HRR field value and* ◇ *is at* $L\sigma_0/J = 495$, ○ *is at* $L\sigma_0/J = 291$, □ *is at* $L\sigma_0/J = 216$, *and* + *is at* $L\sigma_0/J = 177$.

ter of the normal stress distribution ahead of the crack persists to much larger J-values than computed here.

The maximum normal stress ahead of the crack tip is approximately 3.7 σ_0 in both the J_2 flow-theory and J_2 corner-theory calculations. In the J_2 flow-theory case the stress maximum occurs at $R/(J/\sigma_0) = 1.6$ whereas in the J_2 corner-theory calculation it occurs at $R/(J/\sigma_0) = 2.0$. In each case, the location of the maximum stress is at about $R/b = 2$. Both the location of the maximum stress and the maximum stress value are in good agreement with the results obtained by McMeeking and Parks [5], in their J_2 flow-theory ECB computation employing $N = 0.1$, although they used $\sigma_0/E = 0.00333$ while we have employed $\sigma_0/E = 0.002$.

Figure 11 shows the normal stress acting on the plane ahead of the crack tip for the CCP specimen. The solid line is again the HRR field. The significant loss of triaxiality for this specimen, as also observed by McMeeking and

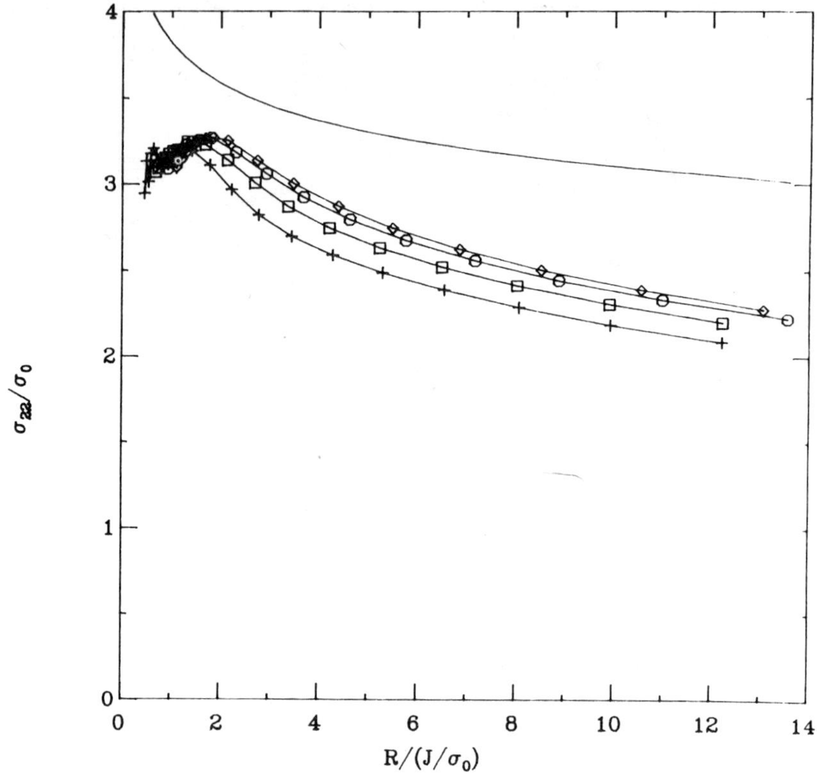

FIG. 11—*Normal stress on plane ahead of crack tip versus initial distance from tip for* J_2 *corner-theory CCP computation. The solid line is the HRR field value and* ◇ *is at* $L\sigma_0/J = 482$, ○ *is at* $L\sigma_0/J = 405$, □ *is at* $L\sigma_0/J = 296$, *and* + *is at* $L\sigma_0/J = 195$.

Parks [5], is evident. Note that the stress level here is dropping as deformation proceeds. The maximum normal stress along the line of the crack for the CCP specimen is approximately 3.3 σ_0 and is reached at about $R/(J/\sigma_0) =$ 1.8. Thus, as compared with the ECB specimens, the maximum of σ_{22}/σ_0 is smaller and occurs closer to the crack tip.

It can also be seen in Fig. 11 that, at the latest stage of deformation, a second local stress maximum has developed near the crack tip. In fact, in the intensely shearing region near the crack tip, the normal stress does not vary smoothly, but oscillates. This oscillation arises solely from the hydrostatic contribution to the normal stress (the effective stress varies smoothly) and is due to numerical difficulties associated with calculating the hydrostatic stress accurately near the shear band where strain gradients are very high. Such an oscillation in normal stress also occurs for the J_2 corner-theory bend specimen, but at values of $R/(J/\sigma_0)$ smaller than plotted in Fig. 10. However, the development of a local maximum in the normal stress at the shear band is to be expected since an increased hydrostatic tension develops during localization [15].

Figures 12 and 13 depict the total effective strain (the elastic part is not subtracted out) as a function of $R/(J/\sigma_0)$ at 90 deg to the crack. Here, the angle is measured from the center of the semicircle defining the notch tip and R is measured from the crack flank. Figure 12 displays values from both the J_2 flow-theory and J_2 corner-theory ECB computations, which nearly coincide for $R/(J/\sigma_0)$ greater than about one, and Fig. 13 shows the CCP effective strains. We have chosen to compare the effective strain distributions with the HRR field values at 90 deg, rather than 0 or 45 deg, since according to the HRR field, the effective strain reaches its maximum near 90 deg (at about 98 deg for $N = 0.1$).

The trend of the effective strain distribution for the ECB specimens is in excellent agreement with the HRR field distribution. In fact, except for a shift of approximately one J/σ_0, good quantitative agreement between the

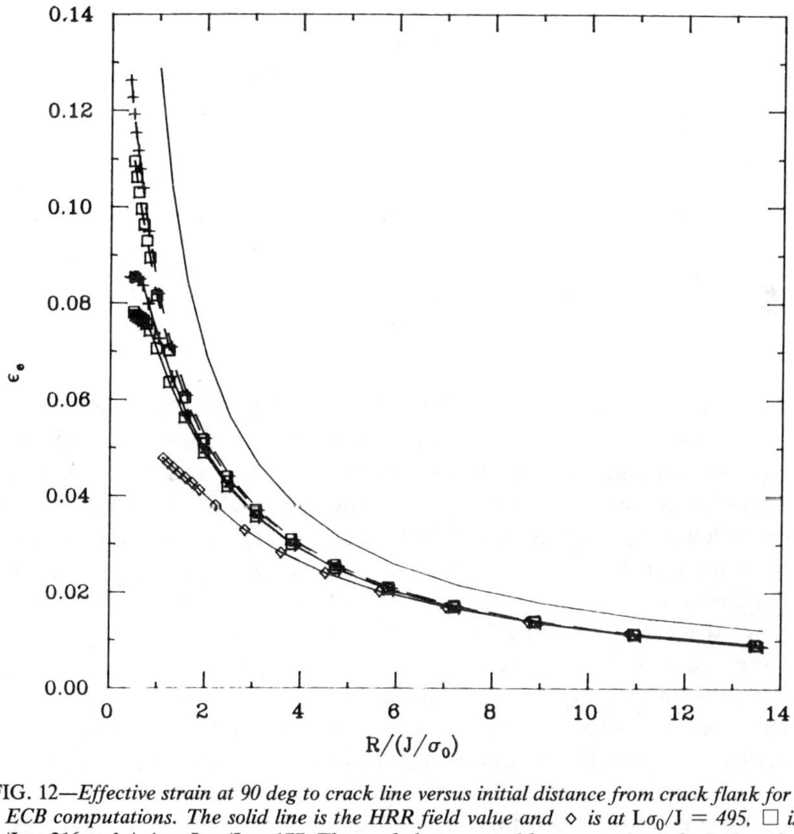

FIG. 12—*Effective strain at 90 deg to crack line versus initial distance from crack flank for the two ECB computations. The solid line is the HRR field value and* ◇ *is at* $L\sigma_0/J = 495$, □ *is at* $L\sigma_0/J = 216$ *and* + *is at* $L\sigma_0/J = 177$. *The symbols connected by --- are from the* J_2 *flow-theory computation and the symbols connected by —— are from the* J_2 *corner-theory computation.*

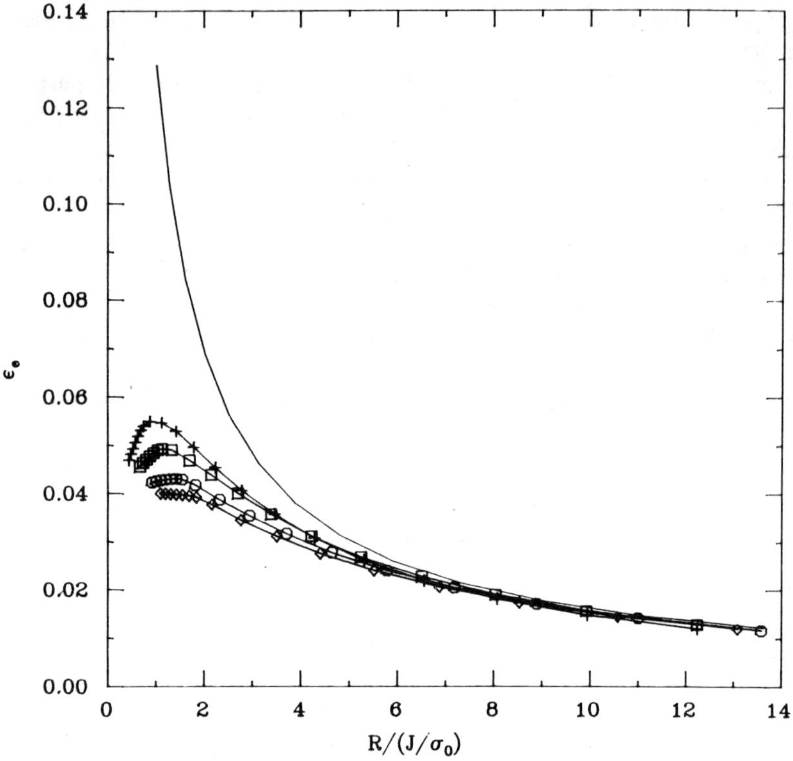

FIG. 13—*Effective strain at 90 deg to crack line versus initial distance from crack flank for the* J_2 *corner-theory CCP computation. The solid line is the HRR field value and* ◇ *is at* $L\sigma_0/J$ = *482,* ○ *is at* $L\sigma_0/J$ = *405,* □ *is at* $L\sigma_0/J$ = *296, and* + *is at* $L\sigma_0/J$ = *195.*

computed effective strains in Fig. 12 and the HRR field values holds in a similar range for which the computed and HRR field stresses agree in Figs. 9 and 10. We do note that the definition of R is somewhat arbitrary for the finite notch analyzed here; for the finite notch R could have been measured from the notch tip. However, this would not be consistent with the definition of angle used in these figures. Figure 13 displays, as does Fig. 11, the loss of dominance of the HRR field in the CCP specimen. For the CCP specimen the effective strain reaches a maximum at about $R/(J/\sigma_0) = 1$ and then falls off as the notch flank is approached.

The value of J as given by the contour integral Eq 23 was computed using various contours surrounding the notch tip. The contours were taken as straight-line segments connecting the centroids of various triangles within the quadrilaterals shown in Figs. 2*b* and 2*c*. Centroidal values of the integrand were employed in computing J. Figure 14 displays values of J so computed for every even-numbered ring of elements, except for rings very near

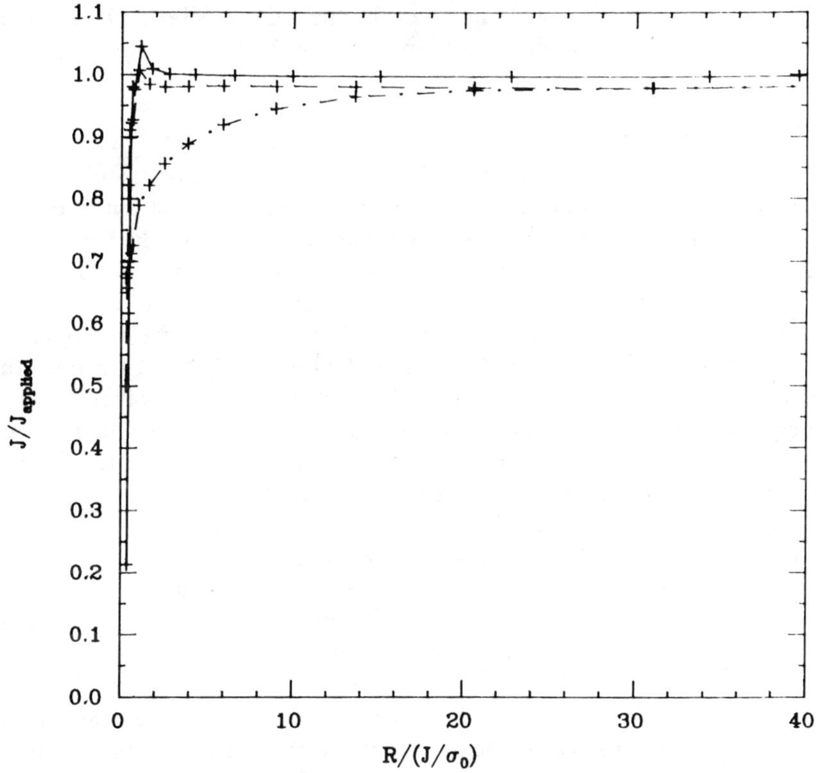

FIG. 14—*Value of* J, *computed as a contour integral, versus distance of contour from crack tip:* —— J_2 *corner-theory CCP calculation at* $L\sigma_0/J = 195$, --- J_2 *corner-theory ECB calculation at* $L\sigma_0/J = 177$, *and* -·· J_2 *flow-theory ECB calculation at* $L\sigma_0/J = 177$.

the crack tip where, in the J_2 corner-theory computations, intense shearing is taking place. The intense gradients in this region render the simple linear interpolation scheme for calculating J inaccurate. The distance R in Fig. 14 is the initial distance from the notch tip to the contour along the crack line and J is normalized by its remote value.

For the J_2 flow-theory ECB specimen the dependence of J on the distance R of the contour on which it is evaluated corresponds to that found by McMeeking [*4*]. For all contours with $R/(J/\sigma_0)$ greater than about 5 (corresponding to R/b greater than about 6), J is within 10 percent of its remote value. On the other hand, for both J_2 corner-theory specimens, J is within 2 percent of its remote value for $R/(J/\sigma_0) \geq 2$. Path independence of J, when the material is characterized by J_2 flow theory, only holds strictly when the loading is proportional. For J_2 corner theory, path independence strictly holds in the total loading regime and, as illustrated in Fig. 14, this gives rise to path independence much nearer the notch tip, even for the CCP specimen. In all three cases, however, J falls rapidly as the tip is approached.

Finite Strain Region

Figures 15 through 17 depict the extent of the finite strain region. In these figures contours of constant maximum principal logarithmic strain in the deformed configuration are shown for the three cases analyzed. In Fig. 15, the contours are plotted for the J_2 flow-theory ECB computation at $L\sigma_0/J = 177$. Figure 16 shows the corresponding contours for the J_2 corner-theory ECB computation at the same value of $L\sigma_0/J$, while in Fig. 17 the contours are plotted for the J_2 corner-theory CCP computation at $L\sigma_0/J = 195$. Since the material behavior is nearly incompressible, the value of the other in-plane principal strain is the negative of the one plotted and the contour lines are also contours of constant effective strain. The value of the effective strain is given by $2/\sqrt{3}$ times the value shown in the figures.

The 0.05 and 0.10 contours in Fig. 15 peak somewhat behind the semicircular notch, as would be anticipated from the HRR field, which gives the maximum effective (and, therefore, principal) strain at about 98 deg. For the J_2 corner-theory ECB, Fig. 16, these contour lines encompass a larger region and their peaks are shifted a bit toward the front of the crack. In Fig. 17, for the J_2 corner-theory CCP case, the peaks of the contour lines marked 0.05 and 0.10 extend farther into the body than for either ECB specimen and are ahead of the crack tip. The principal strain distribution in Fig. 17, which exhibits increased straining along a line at about 45 deg from the notch center,

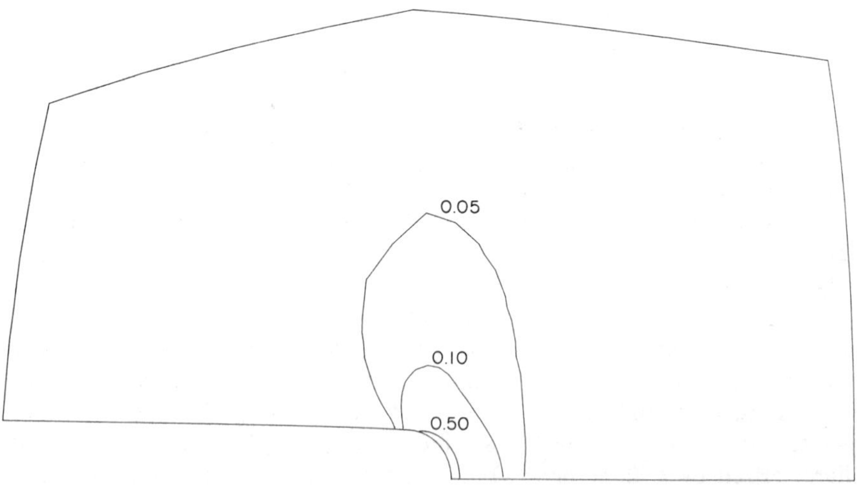

FIG. 15—*Contours of constant maximum principal logarithmic strain in the deformed configuration for the* J_2 *flow-theory ECB computation at* $L\sigma_0/J = 177$.

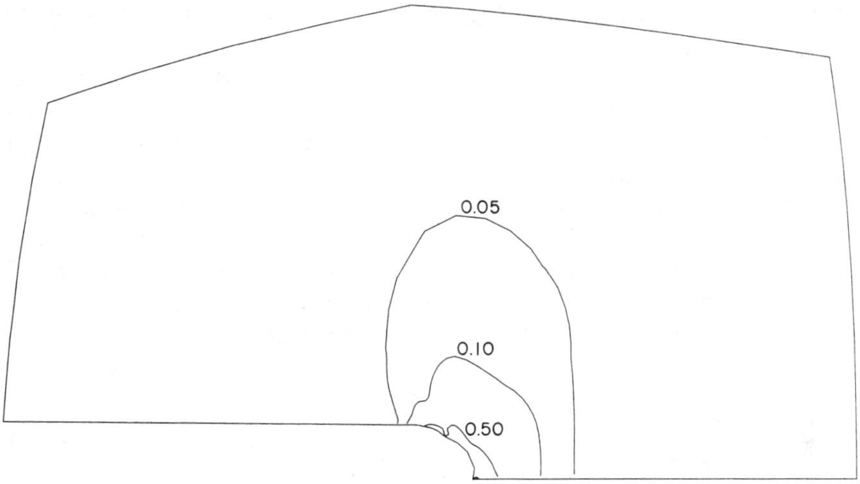

FIG. 16—*Contours of constant maximum principal logarithmic strain in the deformed configuration for the* J_2 *corner-theory ECB computation at* $L\sigma_0/J = 177$.

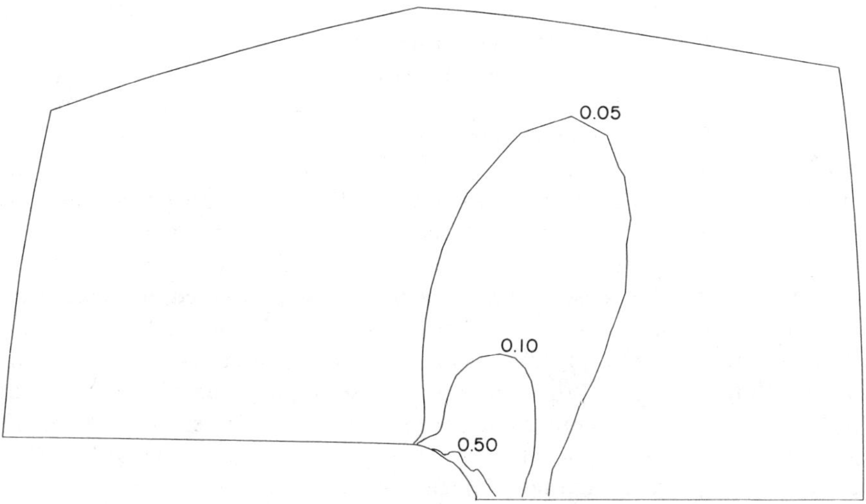

FIG. 17—*Contours of constant maximum principal logarithmic strain in the deformed configuration for the* J_2 *corner-theory CCP computation at* $L\sigma_0/J = 195$.

as indicated by the rigid-plastic slipline field, is another indication of the lack of dominance of the HRR field in the CCP computation.

If the extent of the 0.05 contour is taken arbitrarily to define the extent of the region in which finite deformation effects are significant, then, for the ECB specimens, this region extends about two opening displacements ($2b$)

into the body, while for the CCP specimen at the stage shown in Fig. 16, the extent is approximately $3b$ into the body. The maximum extent along the crack line is approximately b for the J_2 corner-theory ECB specimen and a little less than b in the other two cases. This extent along the crack line is slightly smaller than found by McMeeking and Parks [5], who, however, considered much larger openings than were obtained here.

In Fig. 18 contours of constant principal logarithmic strain are shown in the near-tip region for the J_2 corner-theory ECB computation at the four stages of deformation depicted in Fig. 7. The region depicted in this figure is the same as that depicted in Fig. 7, although here only the region above the line of the crack is shown. The greatest straining along the line of the crack occurs away from the crack tip. However, the maximum straining occurs along the notch surface off the crack line.

Figures 19 through 21 show the near-tip effective strain distributions along the line of the crack. Here, R, the distance of a material point from the tip in the initial configuration, is normalized by J/σ_0. In Fig. 19 the distribution obtained for the J_2 flow-theory ECB specimen is plotted. For the latter two stages of deformation the effective strain distribution is nearly self-similar right up to the tip of the crack. A little bit away from the crack tip, for $R/(J/\sigma_0)$-values greater than about 0.3, the effective strain distribution is nearly self-similar for all four stages of deformation shown.

Figure 20 shows the effective strain distribution ahead of the crack tip for the J_2 corner-theory ECB computation. Here, the self-similar character of the effective strain up to its maximum value on the crack line is evident. The strain then falls from this curve to the nearly constant value maintained right at the tip. Thus, although the near-tip effective strain distribution here is quite different from that given by small-strain J_2 deformation theory, the near-tip distribution is well characterized by the current value of J.

The reason for the stress oscillation near the crack tip noted previously can be seen in Fig. 20. As $R/(J/\sigma_0)$ decreases from 1.6, the normal stress monotonically decreases from its peak value. Then, at the intersection of the shear band with the crack line, the normal stress rises rapidly due to the enhanced hydrostatic tension accompanying localization [*15*]. For smaller values of $R/(J/\sigma_0)$ the normal stress will decrease as the crack tip is approached. The present mesh possesses too few elements in this region, where gradients are very high, to properly resolve this rapid rise and fall of the hydrostatic stress.

In Fig. 21, the effective strain distribution ahead of the crack is shown for the J_2 corner-theory CCP specimen. Also, for comparison purposes, the dashed line is the effective strain distribution for the J_2 corner-theory ECB specimen at $L\sigma_0/J = 177$, replotted from Fig. 20. The first three stages of deformation for the CCP calculation exhibit the same trend as for the ECB case. The self-similarity up to the maximum effective strain is not as well maintained as in the J_2 corner-theory ECB case, but these three curves and the ECB curve taken from Fig. 20 do fall within a well-defined band and do maintain nearly

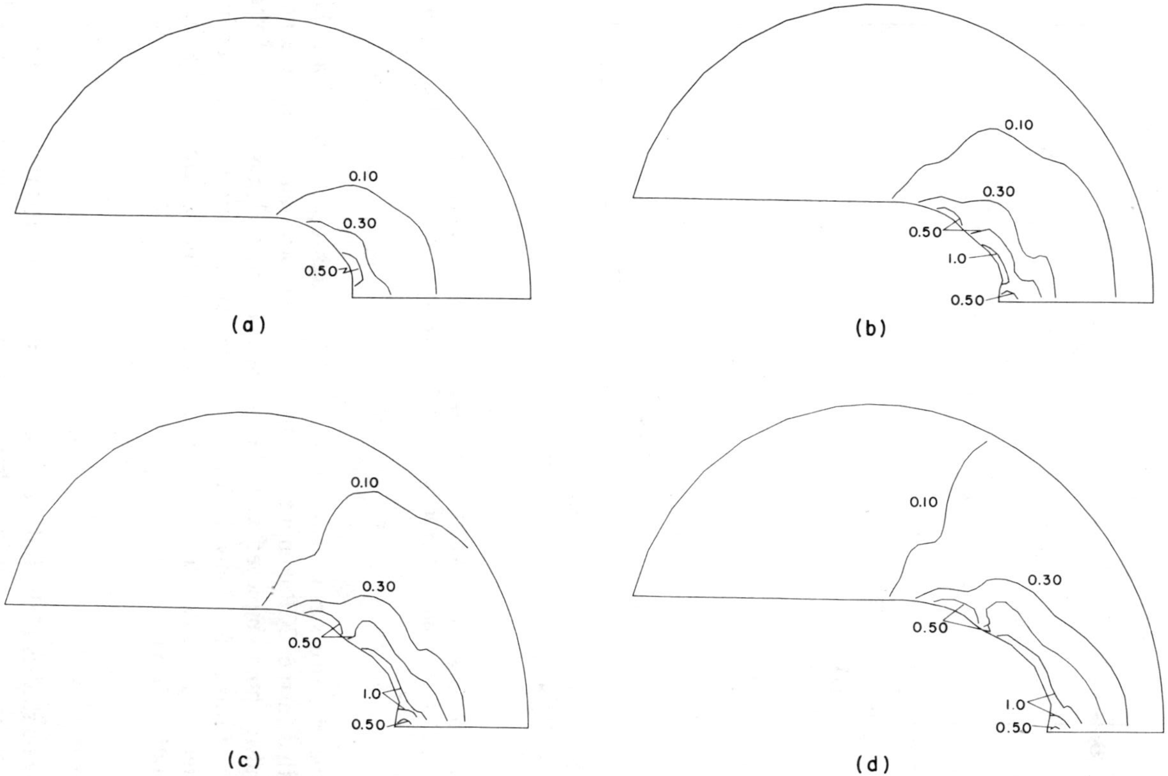

FIG. 18—*Contours of constant maximum principal logarithmic strain in the deformed configuration in the near-tip region for the* J_2 *corner-theory ECB computation:* (a) $L\sigma_0/J = 495$, (b) $L\sigma_0/J = 291$, (c) $L\sigma_0/J = 216$, *and* (d) $L\sigma_0/J = 177$.

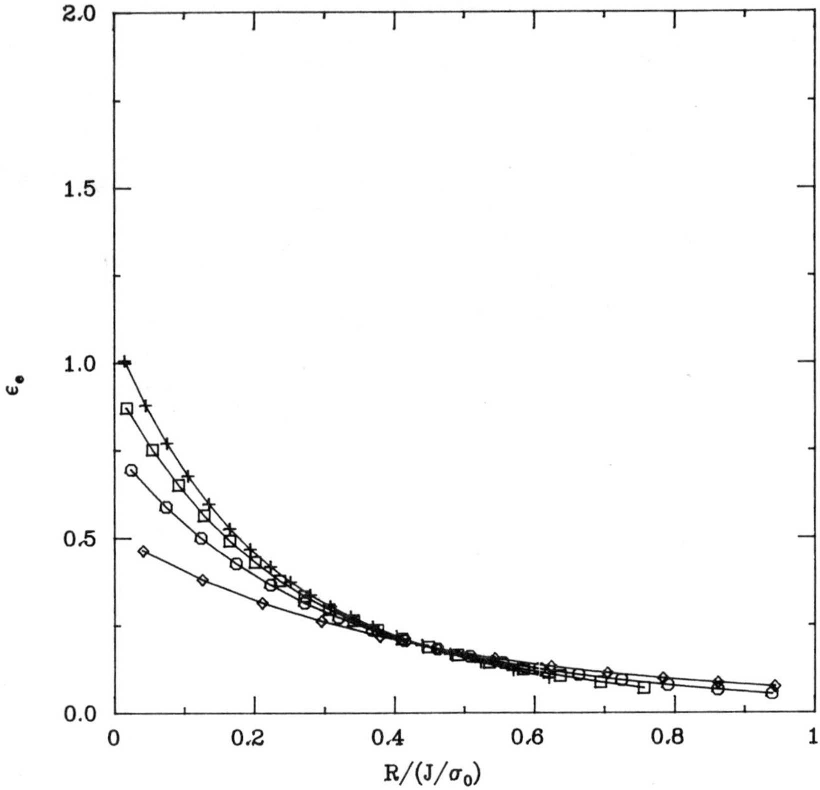

FIG. 19—*Effective strain on plane ahead of crack versus initial distance from tip for the* J_2 *flow-theory ECB computation:* ◇ *is at* $L\sigma_0/J = 495$, ○ *is at* $L\sigma_0/J = 291$, □ *is at* $L\sigma_0/J = 216$, *and* + *is at* $L\sigma_0/J = 177$.

the same effective strain value in front of the crack tip. It should also be noted that in the CCP computation the increment sizes employed were not identical to those employed in the ECB calculation, and how much of the breadth of this band, if any, is due to incrementation errors remains to be investigated. In any case, the loss of self-similarity for the last stage of the CCP computation shown is clear.

Discussion

Previous studies of the finite plastic deformations around the tip of a crack have been based on the classical Prandtl-Reuss plasticity theory with a smooth yield surface and isotropic hardening. Therefore, the main emphasis in the present investigation is on the differences induced by the presence of a vertex on subsequent yield surfaces. In particular, comparison with the crack-tip

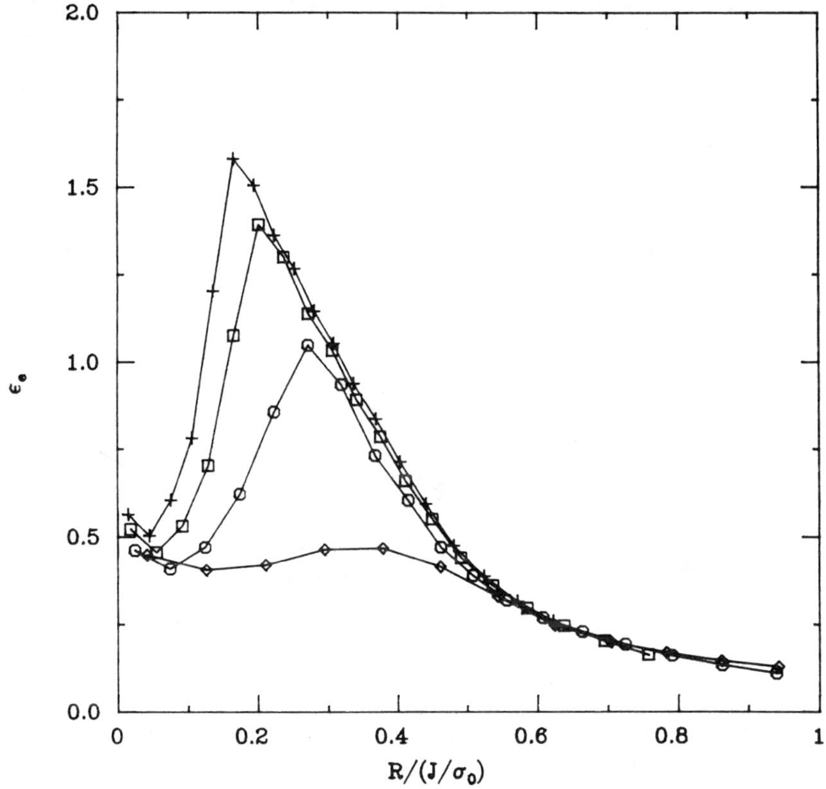

FIG. 20—*Effective strain on plane ahead of crack versus initial distance from the tip for the* J_2 *corner-theory ECB computation:* ◇ *is at* $L\sigma_0/J = 495$, ○ *is at* $L\sigma_0/J = 291$, □ *is at* $L\sigma_0/J = 216$, *and* + *is at* $L\sigma_0/J = 177$.

blunting results obtained by McMeeking and Parks [5] for ECB and CCP specimens is of interest.

The investigation of crack-tip blunting by McMeeking and Parks [5] mainly focussed on whether a single parameter, such as J or δ, characterizes the near-tip fields, not only as an amplitude of the asymptotic small-strain HRR field, but also in the fracture process zone, where large strains prevail. It was found for the ECB specimens that unique fields develop in the blunted crack-tip region, remaining virtually identical up to J-values as large as $\sigma_0 L/25$. For CCP specimens, on the other hand, the near-tip fields drift away from those of well-contained yielding at a smaller J-value, of the order of $\sigma_0 L/200$. Essentially, the corresponding results obtained in the present J_2 corner-theory computations are in good agreement with the smooth yield surface theory results of Ref 5. It should be emphasized, though, that the maximum crack openings considered in the present computations are $b/b_0 = 2.5$, whereas openings as

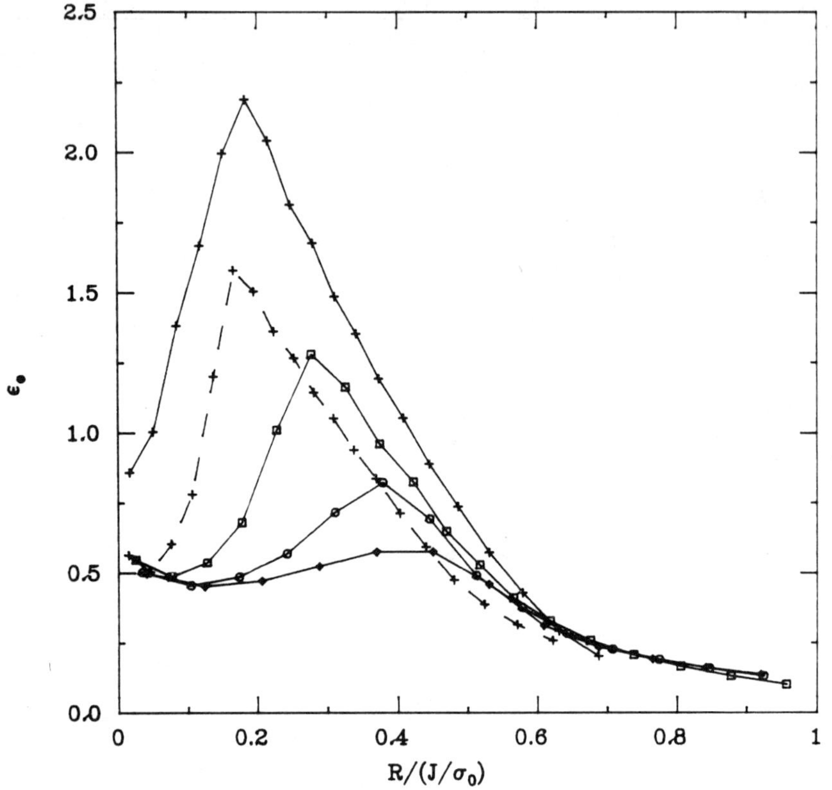

FIG. 21—*Effective strain on plane ahead of crack versus initial distance from tip for the* J_2 *corner-theory CCP computation:* ◇ *is at* $L\sigma_0/J = 482$, ○ *is at* $L\sigma_0/J = 405$, □ *is at* $L\sigma_0/J = 296$, *and* + *is at* $L\sigma_0/J = 195$. *The dashed line is the* J_2 *corner-theory ECB distribution at* $L\sigma_0/J = 177$ *from Fig. 20.*

large as $b/b_0 = 50$ were analyzed in Ref 5. Furthermore, the largest J-values considered here are $\sigma_0 L/177$.

For the ECB specimen the self-similar stress distributions given by J_2 flow theory and by J_2 corner theory are virtually identical (Figs. 9 and 10), and there is good agreement with the HRR field in a region of low straining outside the process zone. Also, the shapes of the strain contours (Figs. 15 and 16) are very similar, although the finite strain region is slightly larger for the corner-theory solution.

In the J_2 corner-theory CCP calculation the stresses remain well below the HRR field values (Fig. 11), even at the small J-values considered, and the stresses decay for increasing J. The decrease of triaxiality ahead of the crack and the amplification of plastic straining at 45 deg correspond to approaching the slipline theory solution for a nonhardening material. For strain-hardening solids, similar tendencies to those found here (Figs. 11 and 17) were also found in the context of classical plasticity theory [*5*].

The breakdown of J-dominance for the present corner-theory CCP calculation is also indicated by the slope db/dJ, which is initially 14 percent above the value $0.50/\sigma_0$ corresponding to the HRR field and which keeps increasing (Fig. 3). For the ECB calculations it is interesting to note that the slope obtained by corner theory is larger than that of J_2 flow theory, and that the corner-theory slope is very close to that obtained numerically by Shih [*26*] for small-strain J_2 deformation theory. This closer agreement with deformation theory rather than with flow theory is expected, since finite strain deformation theory moduli are used to characterize the J_2 corner-theory solid in the total loading range.

The most significant difference between the corner-theory results and those based on classical plasticity theory is the formation of localized shear bands at the crack tip, and the corresponding development of corners on the crack-tip profile. The occurrence of flow localization was expected, since large strains develop at a blunting crack tip, and since earlier corner-theory computations for nonhomogeneous fundamental strain fields [*19,20,27*] have resulted in shear bands as soon as the corresponding critical strain was exceeded locally. The location at which shear bands developed in the neck of a plane-strain tension specimen [*19*], or in the pure bending of a plate [*20*], varied considerably with small differences in the initial surface shapes. Likewise, for the blunting crack it is not *a priori* obvious that the shear bands should start at the surface points chosen by the solutions in Figs. 7, 8, and 18. In fact it is expected that an initial shape differing from the semicircular notch assumed here would result in a different shear band pattern.

The critical principal strain for shear band development under plane-strain conditions in the corner-theory material considered here is 0.32, and previous localization studies have shown that shear bands die out rapidly when reaching regions of lower straining. Thus, from the point of view of flow localization the large strain region is limited to $\epsilon_{max} > 0.3$, which is a rather small region of approximate thickness $b/2$ in front of the blunted crack tip, as can be seen in Figs. 15–18.

The main difference between the near-tip shear band patterns found for the ECB and CCP calculations, respectively (Figs. 7 and 8), is the intense straining of the wedge-shaped region at the root of the notch in the last stage of the CCP calculation. This could be connected with the changing field distributions during loss of J-dominance, but could also arise from the strong distortions of the mesh or from a saturation of the localized deformations. In fact the strain distributions for the ECB specimen (Fig. 19) show slowly increasing strains in this wedge-shaped region, suggesting that the same switch to intense straining may occur at a later stage of the deformation history.

The possibility that strain accumulation in a shear band saturates has been studied by Hutchinson and Tvergaard [*15*]. This effect is strongly dependent on the parameters characterizing the vertex, with the tendency to saturate increased by smaller values of $(\beta_c)_{max}$ or θ_0, corresponding to a more blunt vertex or a smaller total loading range, respectively. Thus, the localized deformations first observed at a blunting crack tip could saturate after some deformation, and subsequently formed bands of localized plastic flow could saturate also, etc., until localized deformations possibly continue growing in some set of bands.

In order to investigate this possibility, a more refined mesh than the one employed here would be required. In this regard it should be noted that in our calculations the finite-element discretization sets the minimum shear band width as well as the minimum separation between shear bands as discussed in Ref *19*. For real materials there are microstructural features which set these length scales, but such natural length scales are not included in the present formulation. Because of the severe distortion of the finite-element mesh in the near-tip region, another mesh effect is encountered here. Initially there is a tendency (Figs. 7*b* and 8*b*) for the bands meeting at the line of symmetry to cross one another and continue into the material, as occurred in a number of cases for shear bands in the neck of a tension specimen [*19*]. However, in the present problem the elements at the intersection point of the shear bands and the crack line have become stretched so that the mesh is no longer well suited for representing shear bands beyond the crossing.

Investigation of the issues discussed in the foregoing seems worthwhile since shear band formation in the fracture process zone can effect the rupture process in potentially significant ways. The actual ductile fracture mechanisms at a crack tip, which involve void nucleation, growth, and coalescence with the crack tip, have been discussed by Rice [*29*].

First of all, the picture of the fracture process mechanism that emerges from the smooth yield surface theory analyses [*3,4*] can be summarized as follows. The self-similar stress and strain fields that develop during crack-tip blunting sweep over the near-tip material so that a void-nucleating site at initial distance R from the tip is first reached by the stress peak when $R/(J/\sigma_0) \simeq 2$. Presumably, the high triaxial stress nucleates voids at weaker particles at this point, even though the plastic strains are small. Then, subsequently, the void is engulfed by the finite strain region and grows to coalescence with the crack tip. Rice and Johnson [*3*] presumed that voids on the crack line would first coalesce with the crack tip and, employing a criterion for coalescence between the void and the crack, computed the crack-tip opening displacement for fracture in terms of the ratio of void nucleating particle spacing to void nucleating particle size.

Green and Knott [*30*] studied several structural steels and found that the onset of crack growth occurred when the crack-tip opening displacement reached a magnitude of the order of the initial spacing between void nucleating particles. The trend of the data is captured by the Rice and Johnson [*3*] model, although, significantly, in one case in which the model overestimates the toughness, sectioning indicated that the failure process involved localized shearing [*30*].

McMeeking [*4*] noted that considering void growth at 45 deg to the crack tip and accounting for material strain hardening do not substantially alter the predictions of the Rice and Johnson [*3*] model. He compared data on high-strength steels and aluminum alloys with the model predictions and also noted that for materials prone to localization, the model overestimates the ductility. Indeed, tests on steels and aluminum alloys indicate that the interaction between the crack tip and an approaching void is, in some cases, terminated at an early stage by flow localization in a band from the tip to the void, particularly for low-hardening materials [*30–33*].

For a material prone to flow localization, the fracture process can be effected in several ways. Presuming that a relatively large weak particle is available to nucleate a void in the high-stress, small-plastic-strain region when $R/(J/\sigma_0) \simeq 2$, the additional strain concentration around such a void could effect the location of shear bands so that a shear band connecting the approaching void and the crack tip forms. This band could speed up coalescence considerably, either directly by slipping off, or by nucleating additional voids from populations of smaller particles, leading to rupture by a void sheet mechanism. Inside a shear band, conditions are favorable for the nucleation and growth of voids due to the intense straining and to the increased hydrostatic tension accompanying localization [*15*].

The additional strain concentration around a void is not necessary for shear band formation, which may be of relevance to relatively "clean" materials containing small second-phase particles. In such materials, the conditions several openings away from the crack tip (high hydrostatic tension, but small plastic strains) may not be sufficiently adverse to nucleate voids. However, in the shear band, the combination of large plastic strains and enhanced triaxial tension could promote void nucleation and rapid void growth.

In any case, the fracture process at a crack tip differs in significant respects from that in more homogeneously strained circumstances, such as in the neck of a tension specimen, as has been emphasized by Rice [*29*]. The stresses and strains near the crack tip, even at loads less than the fracture load, can be well above the levels at fracture under more homogeneous conditions. This has important implications for analyses such as the one carried out here, since initiation of shear band development occurs when the critical strain for a shear band bifurcation is achieved locally. In order to continue the calculation until conditions at the crack tip are representative of those at crack initiation, the numerical procedure must be able to accommodate much greater shearing (in one or more sets of shear bands) than is required to achieve conditions representative of fracture in more homogeneously strained specimens.

The analyses carried out in the present paper give an initial picture of shear band development at a blunting crack tip. More detailed studies of shear band initiation and development at a crack tip and the relation to microstructural mechanisms of fracture appear worthwhile. Actually accounting for the interaction of one or more voids with the crack-tip field to obtain quantitative predictions of the initiation of crack growth would be a useful further step. We conclude by noting that the localized shearing exhibited here arises from the presence of a vertex on the yield surface and thus is an inherent feature of the plastic flow process. In some circumstances, the weakening induced by void nucleation and growth may itself lead to localized deformation.

Acknowledgments

The work of the first author (A. N.) was supported by the Department of Energy through Contract No. DE-AC02-80ER 10556. We are pleased to acknowledge helpful discussions with Professor C. F. Shih of Brown University and with Professor D. M. Parks of the Massachusetts Institute of Technology. We are also indebted to Professor Shih for making his tabulated values of HRR field quantities available to us. The computations reported on here were carried out on the Brown University, Division of Engineering, VAX-11/780 computer. The acquisition of this computer was made possible by grants from the U.S. National Science Foundation (Grant ENG78-19378), the General Electric Foundation, and the Digital Equipment Corp.

References

[*1*] Hutchinson, J. W., *Journal of the Mechanics and Physics of Solids*, Vol. 16, 1968, pp. 13–31.
[*2*] Rice, J. R. and Rosengren, G. F., *Journal of the Mechanics and Physics of Solids*, Vol. 16, 1968, pp. 1–12.
[*3*] Rice, J. R. and Johnson, M. A. in *Inelastic Behavior of Solids*, M. F. Kanninen et al, Eds., McGraw-Hill, New York, 1970, pp. 641–671.
[*4*] McMeeking, R. M., *Journal of the Mechanics and Physics of Solids*, Vol. 25, 1977, pp. 357–381.
[*5*] McMeeking, R. M. and Parks, D. M. in *Elastic-Plastic Fracture, ASTM STP 668*, J. D. Landes, J. A. Begley, and G. A. Clarke, Eds., American Society for Testing and Materials, 1979, pp. 175–194.
[*6*] McClintock, F. A. in *Fracture: An Advanced Treatise*, H. Liebowitz, Ed., Vol. 3, Academic Press, New York, 1971, pp. 47–225.
[*7*] Rice, J. R., *Journal of Applied Mechanics*, Vol. 35, 1968, pp. 379–386.
[*8*] Hill, R., *Journal of the Mechanics and Physics of Solids*, Vol. 15, 1967, pp. 79–95.
[*9*] Hutchinson, J. W. in *Proceedings*, Royal Society of London, Vol. A318, 1970, pp. 247–272.
[*10*] Paul, B. in *Fracture: An Advanced Treatise*, H. Liebowitz, Ed., Vol. 2, Academic Press, New York, 1968, pp. 313–496.
[*11*] Hecker, S. S. in *Constitutive Equations in Viscoplasticity*, J. A. Stricklin and K. J. Saczalski, Eds., American Society of Mechanical Engineers, Applied Mechanics Division, Vol. 20, 1976, pp. 1–33.
[*12*] Rudnicki, J. W. and Rice, J. R., *Journal of the Mechanics and Physics of Solids*, Vol. 23, 1975, pp. 371–394.
[*13*] Rice, J. R. in *Proceedings*, 14th International Congress on Theoretical and Applied Mechanics, W. T. Koiter, Ed., Vol. 1, North-Holland, Amsterdam, 1976, pp. 207–220.

[*14*] Needleman, A. and Rice, J. R. in *Mechanics of Sheet Metal Forming*, D. P. Koistinen and N.-M. Wang, Eds., Plenum, 1978, pp. 237–265.
[*15*] Hutchinson, J. W. and Tvergaard, V., *International Journal of Solids and Structures*, Vol. 17, 1981, pp. 451–470.
[*16*] Christoffersen, J. and Hutchinson, J. W., *Journal of the Mechanics and Physics of Solids*, Vol. 27, 1979, pp. 465–487.
[*17*] Needleman, A., *Journal of the Mechanics and Physics of Solids*, Vol. 20, 1972, pp. 111–120.
[*18*] Tvergaard, V., *Journal of the Mechanics and Physics of Solids*, Vol. 24, 1976, pp. 291–304.
[*19*] Tvergaard, V., Needleman, A., and Lo, K. K., *Journal of the Mechanics and Physics of Solids*, Vol. 29, 1981, pp. 115–142.
[*20*] Triantafyllidis, N., Needleman, A., and Tvergaard, V., *International Journal of Solids and Structures*, Vol. 18, 1982, pp. 121–138.
[*21*] Batdorf, S. B., *Journal of the Aeronautical Sciences*, Vol. 16, 1949, pp. 405–408.
[*22*] Hutchinson, J. W., *Advances in Applied Mechanics,* Vol. 14, 1974, pp. 67–144.
[*23*] Hutchinson, J. W. in *Numerical Solution of Nonlinear Structural Problems*, R. F. Hartung, Ed., American Society of Mechanical Engineers, 1973, p. 17.
[*24*] Nagtegaal, J. C., Parks, D. M., and Rice, J. R., *Computer Methods in Applied Mechanics and Engineering*, Vol. 4, 1974, pp. 153–177.
[*25*] Eshelby, J. D. in *Inelastic Behavior of Solids*, M. F. Kanninen et al, Eds., McGraw-Hill, New York, 1970, pp. 77–115.
[*26*] Shih, C. F., *Journal of the Mechanics and Physics of Solids*, Vol. 29, 1981, pp. 305–326.
[*27*] Larsson, M., Needleman, A., Tvergaard, V., and Storåkers, B., *Journal of the Mechanics and Physics of Solids*, Vol. 30, 1982, pp. 121–154.
[*28*] Shih, C. F. and German, M. D., *International Journal of Fracture*, Vol. 17, 1981, pp. 27–43.
[*29*] Rice, J. R. in *The Mechanics of Fracture*, ASME Applied Mechanics Division, Vol. 19, American Society of Mechanical Engineers, 1976, pp. 23–53.
[*30*] Green, G. and Knott, J. F., *Transactions*, American Society of Mechanical Engineers, *Journal of Engineering Materials and Technology*, Vol. 98, 1976, pp. 37–46.
[*31*] Hahn, G. T., Kanninen, M. F., and Rosenfield, A. R., *Annual Review of Materials Science*, Vol. 2, 1972, pp. 381–404.
[*32*] Hahn, G. T. and Rosenfield, A. R., *Metallurgical Transactions*, Vol. 6A, 1975, pp. 653–668.
[*33*] Lautridou, J. C. and Pineau, A., *Engineering Fracture Mechanics*, Vol. 15, 1981, pp. 55–71.

Hiroshi Miyamoto,[1] *Kazuro Kageyama,*[2] *Masanori Kikuchi,*[1] *and Kenji Machida*[1]

The J_{ext}-Integral Based on the Concept of Effective Energy Release Rate

REFERENCE: Miyamoto, H., Kageyama, K., Kikuchi, M., and Machida, K., **"The J_{ext}-Integral Based on the Concept of Effective Energy Release Rate,"** *Elastic-Plastic Fracture: Second Symposium, Volume I—Inelastic Crack Analysis, ASTM STP 803,* C. F. Shih and J. P. Gudas, Eds., American Society for Testing and Materials, 1983, pp. I-116–I-129.

ABSTRACT: The J_{ext}-integral is introduced as the measure of the effective energy release rate in the process zone, and the expression of the two- and three-dimensional J_{ext}-integrals are obtained. The J_{ext}-integral is the measure of the inhomogeneities caused by the large deformation at the crack tip. The path-independent expression of the J_{ext}-integral is also introduced, which is useful when there are the body force and the eigen stress fields. For the two-dimensional case, some properties of the J_{ext}-integral are discussed and the relation to G^{Δ} is shown. It is pointed out that by considering a finite region like the process zone, J_{ext} can overcome the shortages of the energy concept as the fracture criterion. The J_{ext}-integral is evaluated for some basic problems.

KEY WORDS: J_{ext}-integral, effective energy release rate, process zone, fracture criterion, inhomogeneities, elastic-plastic fracture

Nomenclature

- F_i x_i-component of the body force
- J_{ext} extended J-integral
- J_{Ic} fracture toughness
- N strain-hardening exponent of Ramberg-Osgood-type materials

[1]Professor of Mechanical Engineering, Faculty of Science & Technology, and assistant professor and research assistant, Mechanical Engineering, respectively, Science University of Tokyo, Noda, Chiba, Japan.

[2]Researcher, Mechanical Engineering Laboratory, Agency of Industrial Science and Technology, MITI, Ibaraki, Japan.

r, θ	2 dimensional polar coordinate system
r_f	size parameter of the process zone
dS	surface element
ds	line element
T_i	x_i-component of the traction vector
δu_i^{el}	change of the x_i-component of the elastic displacement
δu_i^{pl}	change of the x_i-component of the plastic displacement
W^{el}	elastic strain energy density
x_i	rectangular coordinate system x_1, x_2, x_3
α_{ij}	dislocation density tensor
$\beta_{ji}^{el}, \beta_{ji}^{pl}, \beta_{ji}^*$	elastic, plastic, and eigen distortion tensor
$\epsilon_{ji}^{el}, \epsilon_{ji}^{pl}, \epsilon_{ji}^*$	elastic, plastic, and eigen strain
ϵ_{ikl}	permutation tensor
$\delta\Pi^{el}$	reversible energy change with crack growth
$\Omega_f, \lvert\Omega_f\rvert$	process zone and its boundary in three-dimensional cracked body
Γ_f	boundary of the process zone in two-dimensional cracked plate

Besides having the limitations that the constitutive equations are nonlinear elastic, the J-integral is inadequate as the fracture criterion because the energy release rate becomes zero for the elastic-plastic case as pointed out by Rice [*1*][3] and Yokobori [*2*]. The J_{ext}-integral proposed by Kageyama overcomes these shortcomings and can combine the energy concept with the submicroscopic mechanics of fracture. It is also shown that the J_{ext}-integral means the generalized force which is a measure of the amount of discontinuities in the process zone (the region in which the fracture process occurs), which is based on the concept of the continuously distributed dislocations by Kröner [*3*].

In this paper, the J_{ext}-integral is defined and its physical meaning is stated. Then it is treated as the fracture criterion and the relations to other criteria are discussed. Finally, the J_{ext}-integral is evaluated for some basic problems.

Concept of the Continuously Distributed Dislocations

In the present paper, the energy approach is applied to a small but finite fracture process zone near the crack tip. The fracture process zone is intensely deformed and has inhomogeneities such as voids, microcracks, and dislocations. From the viewpoint of the theory of the continuously distributed dislocations, these inhomogeneities are replaced by the incompatibility of the elastic strain in the homogeneous body, and the dislocation tensor is introduced.

[3]The italic numbers in brackets refer to the list of references appended to this paper.

By distortion we mean the sum of the strain and the corresponding rotation. Total distortion $\partial_j u_i$ is divided into elastic and the plastic parts

$$\partial_j u_i = \beta_{ji}^{\text{el}} + \beta_{ji}^{\text{pl}} \tag{1}$$

In the case of the incompatible distortion, the change of the elastic displacement is given as

$$\delta u_i^{\text{el}} = \beta_{ji}^{\text{el}} dx_j \tag{2}$$

and is not integrable. This means that a displacement field can be defined only in an infinitesimal neighborhood of a point. The incompatibility of the displacement field is represented by the dislocation tensor α_{ij}, and there is a simple relation between β_{ji} and α_{ij}

$$\alpha_{ij} = \epsilon_{ikl}\partial_k \beta_{lj}^{\text{el}} = -\epsilon_{ikl}\partial_k \beta_{lj}^{\text{pl}} \tag{3}$$

Total strain energy density W^{tot} is a sum of the elastic and the plastic ones

$$W^{\text{tot}} = W^{\text{el}} + W^{\text{pl}} \tag{4}$$

where

$$W^{\text{el}} = \int_0^{\epsilon_{ij}^{\text{el}}} \sigma_{ij} d\epsilon_{ij}^{\text{el}} = \int_0^{\beta_{ji}^{\text{el}}} \sigma_{ij} d\beta_{ji}^{\text{el}}$$
$$W^{\text{pl}} = \int_0^{\epsilon_{ij}^{\text{pl}}} \sigma_{ij} d\epsilon_{ij}^{\text{pl}} = \int_0^{\beta_{ji}^{\text{pl}}} \sigma_{ij} d\beta_{ji}^{\text{pl}} \tag{5}$$

W^{el} is reversible, and σ_{ij} is given as

$$\sigma_{ij} = \frac{\partial W^{\text{el}}}{\partial \epsilon_{ij}^{\text{el}}} = \frac{\partial W^{\text{el}}}{\partial \beta_{ji}^{\text{el}}} \tag{6}$$

Definition of J_{ext}-Integral Based on the Energy Concept

For the definition of the J_{ext}-integral, the following are assumed:

1. The fracture occurs in some finite region near the crack tip.
2. The energy released by the crack propagation is only the reversible one.
3. The fracture process is autonomous.

Based on these assumptions, the J_{ext}-integral is introduced from the macroscopic viewpoint and its expression is obtained in the form of the line integral.

Change of the Effective Energy

Referring to Fig. 1, consider the contour $|\Omega_f|$, the outer boundary of the process zone Ω_f, on which the traction force T_i acts. The coordinate axis at Point 0 is determined on the crack tip, as shown in Fig. 1.

When the crack propagates $d\xi_1$ along the x_1-axis, the change of the effective energy $d\Pi^{el}$, that is, the reversible energy which is neglected during a fracture process, as defined as

$$\delta\Pi^{el} = \int_{\Omega_f} \delta W^{el} dv - \int_{|\Omega_f|} T_i \delta u_i^{el} ds \tag{7}$$

where

W^{el} = elastic energy density,
T_i = traction force, and
δu_i^{el} = change of the elastic component of the displacement.

Assuming that only the elastic and reversible energy is available during the fracture process, we introduce a fracture parameter based on the effective energy change.

Definition of the J_{ext}-Integral

The J_{ext}-integral is defined as the effective energy release rate in Ω_f as

$$J_{ext} = -\left.\frac{\delta\Pi^{el}}{\delta\xi_1}\right|_{\Omega_f} \tag{8}$$

As the fracture process is assumed to be autonomous, the change of the elastic strain energy W^{el} in Ω_f with the crack growth $\delta\xi_1$ in the direction of

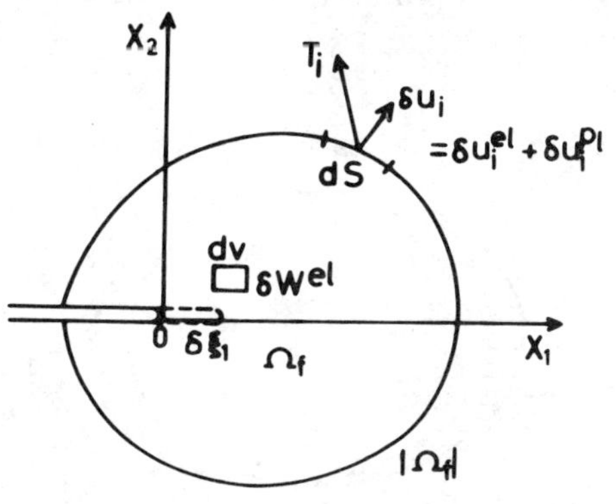

FIG. 1—*Finite fracture region Ω_f and evaluation of the effective energy.*

the positive x_1-axis can be obtained by replacing W^{el} (x_1, x_2, x_3) by W^{el} $(x_1 - \delta\xi_1, x_2, x_3)$. As shown in Fig. 2, the change of W^{el} can be measured on $|\Omega_f|$

$$\int_{\Omega_f} \delta W^{el} dv = -\delta\xi_1 \int_{|\Omega_f|} W^{el} n_1 dS \tag{9}$$

where n_i means the normal vector component on $|\Omega_f|$.

From the definition of the elastic distortion tensor $\beta_{ji}{}^{el}$, the change of the elastic displacement $\delta u_i{}^{el}$ is given as

$$\delta u_i{}^{el} = -\beta_{1i}{}^{el} \delta\xi_1 \tag{10}$$

By using Eqs 7–10, the effective energy release rate is written as

$$J_{ext} = \int_{|\Omega_f|} (W^{el} n_1 - T_i \beta_{1i}{}^{el}) dS \tag{11}$$

For the two-dimensional problem, Eq 11 is rewritten as

$$J_{ext} = \int_{\Gamma_f} (W^{el} dx_2 - T_i \beta_{1i}{}^{el} ds) \tag{12}$$

which coincides with the equation obtained by the authors in a previous paper [*4*].

The integrand in Eq 12 is identical in form to the path-dependent Q_1 integral proposed by Bilby [*5,6*]. Q_1 is the component on the axis x_1 of the resultant force on all dislocations in the region. On the other hand, J_{ext} is derived from the concept of the effective energy release rate, by considering the energy change in the special region Ω_f.

In the next section, J_{ext} is proved to be the generalized force acting on a

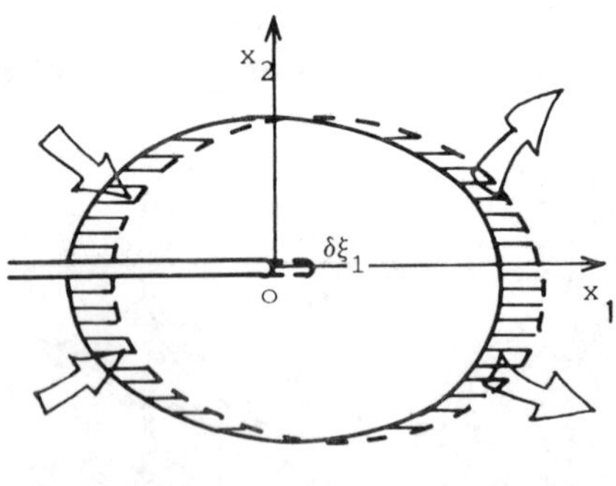

FIG. 2—*Change of the elastic energy in* Ω_f *during crack propagation.*

crack tip and all inhomogeneities in Ω_f, and a path-independent expression is obtained.

Interpretation of the J_{ext}-Integral from the Standpoint of Continuously Distributed Dislocations

As discussed by Mura [7] and others [8], the J_{ext}-integral can be interpreted as the parameter of micromechanics in the process zone on the basis of the continuously distributed dislocations.

The important point here is to see the applicability of Gauss's theorem. Discontinuities of integrands in Eq 12 cannot be inclined in the region. Therefore, Ω_f is interpreted here as the limit of the domain enclosed by $|\Omega_f|$, A^+, δA^+, A^-, and δA^-, as shown in Fig. 3, where δA is the crack surface increment due to the crack extension and δA^+ and δA^- are the upper and lower surfaces of δA, respectively. Then the next equation is obtained

$$\int_{|\Omega_f|+A^\pm+\delta A^\pm}(W^{el}n_1 - T_i\beta_{1i}{}^{el})dS = \int_{\Omega_f}\{\partial_1 W^{el} - \partial_j(\sigma_{ij}\beta_{1i}{}^{el})\}dv \quad (13)$$

On $A^\pm$, $T_i = n_1 = 0$, so J_{ext} is written as

$$J_{ext} = \int_{\delta A} T_i[\beta_{1i}{}^{el}]dS + \int_{\Omega_f}\{\partial_1 W^{el} - \partial_j(\sigma_{ij}\beta_{1i}{}^{el})\}dv \quad (14)$$

where [ ] represents multiple values on the crack surface and is defined as

$$[\beta_{ji}] = \beta_{ji}(\delta A^+) - \beta_{ji}(\delta A^-) \quad (15)$$

Using Eq 10,

$$[\beta_{1i}{}^{el}] = -\frac{1}{\delta\xi_1}[\delta u_i{}^{el}] \quad (16)$$

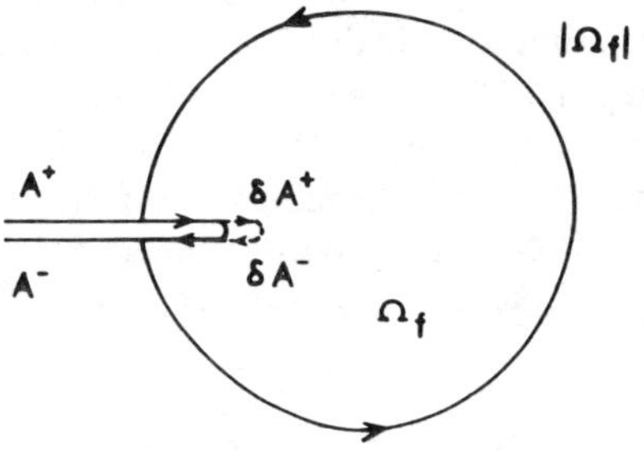

FIG. 3—*Application of Gauss's theorem.*

From Eq 6 and the equilibrium equation $\partial_j \sigma_{ij} = 0$, the integrand of the second term of Eq 14 is transformed

$$\begin{aligned}
&\int_{\Omega_f} \{\partial_1 W^{\text{el}} - \partial_j(\sigma_{ij}\beta_{1i}{}^{\text{el}})\} dv \\
&= \int_{\Omega_f} \left(\frac{\partial W^{\text{el}}}{\partial \beta_{ji}{}^{\text{el}}} \partial_1 \beta_{ji}{}^{\text{el}} - \partial_j \sigma_{ij} \beta_{1i}{}^{\text{el}} - \sigma_{ij} \partial_j \beta_{1i}{}^{\text{el}} \right) dv \\
&= \int_{\Omega_f} \sigma_{ij} (\partial_1 \beta_{ji}{}^{\text{el}} - \partial_j \beta_{1i}{}^{\text{el}}) dv \\
&= \int_{\Omega_f} \sigma_{ij} (\delta_{jl}\delta_{1k} - \delta_{jk}\delta_{1l}) \partial_k \beta_{li}{}^{\text{el}} dv \\
&= \int_{\Omega_f} \sigma_{ij} \epsilon_{1jh} \alpha_{hi} dv
\end{aligned} \tag{17}$$

where $\delta_{jl}\delta_{1k} - \delta_{jk}\delta_{1l} = \epsilon_{1jh}\epsilon_{hkl}$ and the dislocation density tensor $\alpha_{hi} = \epsilon_{hkl}\partial_k \beta_{li}{}^{\text{el}}$ are considered.

Due to the application of Gauss's theorem, the J_{ext}-integral is obtained as the sum of J_{ext} (crack) and J_{ext} (dis)

$$J_{\text{ext}} = J_{\text{ext}}\ (\text{crack}) + J_{\text{ext}}\ (\text{dis}) \tag{18}$$

$$J_{\text{ext}}\ (\text{crack}) = -\frac{1}{\delta\xi_1} \int_{\delta A} T_i [\delta u_i{}^{\text{el}}] dS \tag{19}$$

$$J_{\text{ext}}\ (\text{dis}) = \int_{\Omega_f} \epsilon_{ijh} \sigma_{ij} \alpha_{hj} dv \tag{20}$$

J_{ext} (crack) means the energy release rate by crack propagation alone, without moving the plastic strain. J_{ext} (dis) means the generalized force acting on dislocations in Ω_f, which is equivalent to the Peach-Koehler force acting on the continuously distributed dislocations. The inhomogeneities in the process zone—for example, dislocations, voids, and microcracks—can be transformed to the equivalent dislocation density, so the J_{ext}-integral is the measure of the magnitude of these inhomogeneities at the crack-tip region.

In the case of a blunted crack, Gauss's theorem is applied to Eq 21, by assuming that Ω_f is the domain enclosed by $|\Omega_f|$, A^+, $|\omega_f|$, and A^- as shown in Fig. 4*a*.

$$\int_{|\Omega_f| + A^{\pm} - |\omega_f|} (W^{\text{el}} n_1 - T_i \beta_{1i}{}^{\text{el}}) dS = \int_{\Omega_f} \{\partial_1 W^{\text{el}} - \partial_j(\sigma_{ij}\beta_{ji}{}^{\text{el}})\} dS \tag{21}$$

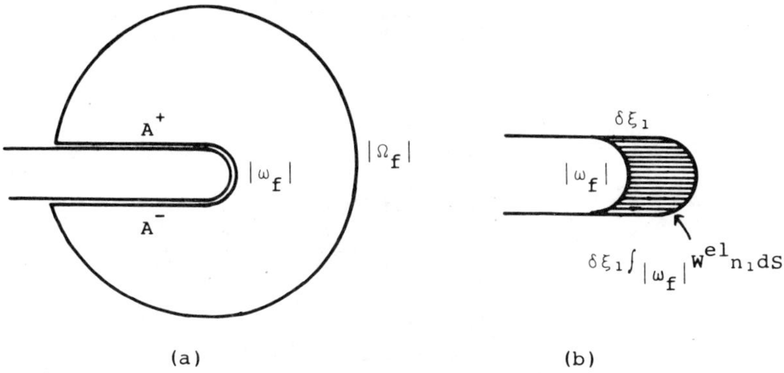

FIG. 4—J_{ext}-*integral applied to blunted crack.*

It has already been explained by Eq 17 that this integrand is transformed by using the dislocation density tensor. As $n_1 = 0$ and $T_i = 0$ on $A^{\pm}$ and $T_i = 0$ on $|\omega_f|$, then J_{ext} is given as

$$J_{ext} = \int_{\Omega_f} \sigma_{ij} \epsilon_{1jh} \alpha_{hi} dv + \int_{|\omega_f|} W^{el} n_1 dS \tag{22}$$

The first term of Eq 22 is identical to J_{ext} (dis) and means the Peach-Koehler force on the continuously distributed dislocations. The second term can be considered to be the released energy which is stored in the crack-tip region before crack propagation as shown in Fig. 4*b*.

As a result, the J_{ext}-integral can combine the macroscopic parameter with the submicroscopic parameter, that is, the energy release rate with the dislocation density in the process zone.

The Path-Independent Expression of the J_{ext}-Integral

As mentioned in the previous section, J_{ext} is defined on the special zone, the process zone, Ω_f. But this J_{ext} can be evaluated on an arbitrary zone Ω which is outside of the process zone (Fig. 5).

The region surrounded by $|\Omega|$, $|\Omega_f|$, and S in Fig. 5 is denoted by Ω_0. In order to obtain a path-independent expression of the J_{ext}-integral, consider the integral

$$I = \int_{|\Omega_0|} (W^{el} n_1 - T_i \beta_{1i}{}^{el}) ds \tag{23}$$

As $T_i = 0$ and $n_1 = 0$ on S, Eq 23 becomes

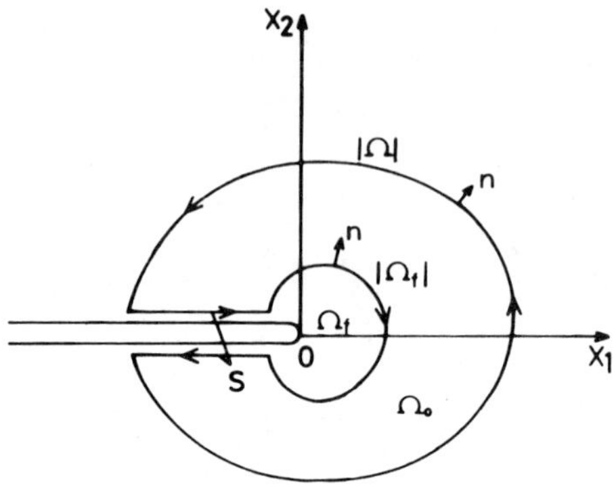

FIG. 5—*Path-independent expression of the J_{ext}-integral.*

$$I = \int_{|\Omega|} (W^{el} n_1 - T_i \beta_{1i}{}^{el}) ds - J_{ext} \tag{24}$$

By applying Gauss's theorem, Eq 23 is transformed into

$$I = \int_{\Omega_0} \{\partial_1 W^{el} - \partial_j (\sigma_{ij} \beta_{1i}{}^{el})\} dv \tag{25}$$

Due to Eqs 3–6 and the equilibrium condition

$$\partial_j \sigma_{ij} + F_i = 0 \tag{26}$$

where F_i is the x_i-component of the body force, we can obtain

$$I = \int_{\Omega_0} (\epsilon_{1jh} \sigma_{ij} \alpha_{hi} + F_i \beta_{1i}{}^{el}) dv \tag{27}$$

From Eqs 24 and 27 a path-independent expression of the J_{ext}-integral is obtained

$$J_{ext} = \int_{|\Omega|} (W^{el} n_1 - T_i \beta_{1i}{}^{el}) dS - \int_{\Omega_0} (\epsilon_{1jh} \sigma_{ij} \alpha_{hi} + F_i \beta_{1i}{}^{el}) dv \tag{28}$$

From Eq 28, we can calculate the J_{ext}-integral by using any contour far from the process zone. This is very useful for the numerical evaluation of the J_{ext}-value by finite-element analysis. The volume integral in Ω_0 is generally nonzero, for example, because of the body force F_i, the temperature distribution

θ, and the plastic distortion $\beta_{ji}{}^{pl}$. The distribution of the temperature θ can be replaced by the distribution of the eigen distortion tensor

$$\beta_{ji}{}^{*} = t_{ij}\theta \tag{29}$$

where t_{ij} is the coefficient of the thermal expansion. In Eq 26 α_{hi} is expressed as the sum of the plastic and the thermal deformation

$$\begin{aligned}\alpha_{hi} &= -\epsilon_{hkl}\partial_k(\beta_{li}{}^{pl} + \beta_{li}{}^{*}) \\ &= -\epsilon_{hkl}\partial_k\beta_{li}{}^{pl} - \epsilon_{hkl}t_{il}\partial_k\theta \end{aligned} \tag{30}$$

The J_{ext}-Integral as a Fracture Criterion

Hereinafter, only the two-dimensional problems are considered. It is assumed that fracture occurs when the following conditions are satisfied:

1. The strain ϵ_f in the process zone exceeds some critical value.
2. The effective energy release rate, J_{ext}, exceeds some critical value, γ_f, which provides the surface energy for the newly created surfaces.

In these statements, ϵ_f and γ_f are the material constants.

It is also assumed that the boundary of the process zone is given by (see Fig. 6)

$$r = \frac{r_f}{r_Y} R(\theta) \tag{31}$$

On Γ_f, the equivalent strain $\bar{\epsilon}$ is constant, $\bar{\epsilon} = \epsilon_f$, and the first condition of the fracture criterion is satisfied. The second condition is expressed by

$$J_{ext} = \gamma_f \tag{32}$$

For the plane-strain problem, by using Eq 41 (shown later), Eq 32 is rewritten as

$$J_{ext} = \gamma_f = J_{Ic} \bigg/ \left(\frac{\epsilon_f}{\epsilon_Y}\right)^{1-N} \tag{33}$$

where J_{Ic} is the fracture toughness, meaning the resistance to crack propagation, and N is the hardening exponent.

In Eq 33, ϵ_f/ϵ_Y is the measure of the ductility of the material. In general, the ductile material has large fracture toughness; therefore, for the material which has a large ϵ_f/ϵ_Y-value, J_{Ic} also becomes large. So these two factors cancel each other and the J_{ext}-integral is considered to be the more general fracture criterion.

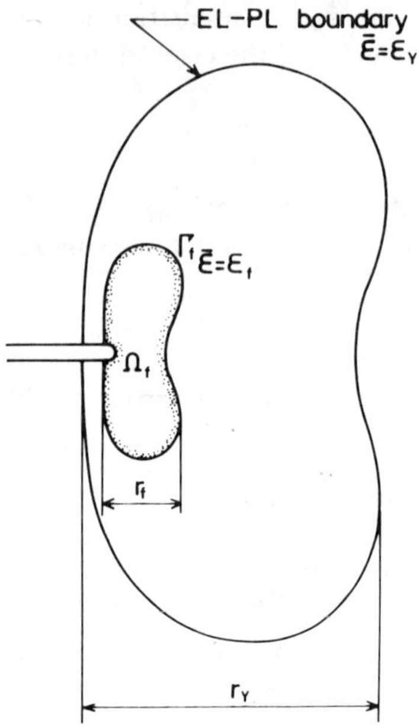

FIG. 6—*Application of* J_{ext} *fracture criterion.*

Evaluation of the J_{ext}-Integral

Bilby-Cottrell-Swinden (BCS) Model

Figure 7 shows the BCS model in Mode I. In the plastic region, $\sigma_{22} = \sigma_Y$ and $\alpha_{32} = \delta(x_2)\rho(x_1)$, where $\rho(x_1)$ is the dislocation density on the x_1-axis and σ_Y means the yield stress. In this problem, J_{ext} (crack) becomes zero because the stress singularity at the crack tip vanishes. The dimensions of Γ_f and the plastic region are denoted by r_f and r_Y, respectively. By using Eqs 18–20, the J_{ext}-integral is expressed by

$$J_{ext} = \sigma_Y\{\Phi(0) - \Phi(r_f)\} \tag{34}$$

$$\Phi(r) = \int_r^{r_Y} \rho(x_1)dx_1 \tag{35}$$

On the other hand, the J-integral is given as $J = \sigma_Y\Phi(0)$, which is equivalent to the case where r_f is equal to r_Y in Eq 34. For small-scale yielding, the dislocation density $\rho(x_1)$ is given as

$$\rho(x_1) = \frac{4\sigma_Y}{\pi E}\cosh^{-1}\left(\frac{2r_Y}{x_1} - 1\right) \tag{36}$$

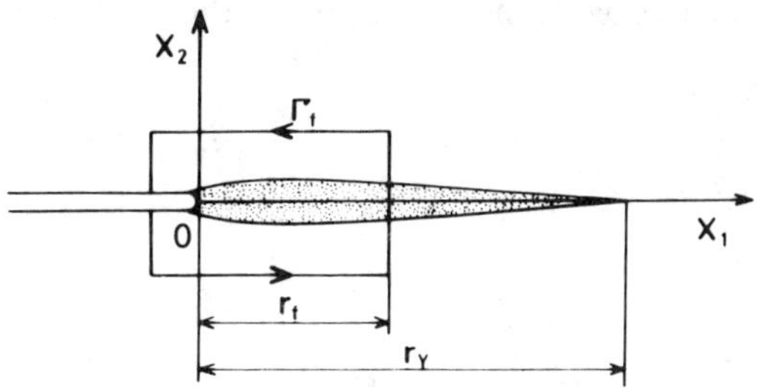

FIG. 7—*BCS model and a contour* Γ_f.

then putting $\xi = r_f/r_Y$

$$J_{\text{ext}} = J\left[\frac{\xi}{2}\left\{\cosh^{-1}\left(\frac{2}{\xi}-1\right) - \sqrt{\left(\frac{2}{\xi}-1\right)^2 - 1}\right\} + 1\right] \quad (37)$$

and for $\xi \ll 1$, Eq 37 becomes

$$J_{\text{ext}}/J = \frac{\xi}{2}\left\{\ell\text{n}\left(\frac{1}{\xi}\right) + \ell\text{n}\,4 + 1\right\} \quad (38)$$

Hutchinson-Rice-Rosengren (HRR) Singularity

Rice and Rosengren [*8*] and Hutchinson [*9*] showed the stress and strain fields at the crack tip for the Ramberg-Osgood-type materials. They are expressed as

$$\sigma_{ij} = K_\sigma(J)r^{-N/(1+N)}\Sigma_{ij}(\theta), \qquad \epsilon_{ij}{}^{\text{tot}} = K_\epsilon(J)r^{-1/(1+N)}E_{ij}(\theta) \quad (39)$$

K_σ and K_ϵ are the functions of J, and $\Sigma_{ij}(\theta)$ and $E_{ij}(\theta)$ give the shape of the stress and strain distributions. The equivalent stress and strain are given by

$$\bar{\sigma} = \sigma_Y\left\{\frac{R(\theta)}{r}\right\}^{N/(1+N)}, \qquad \bar{\epsilon} = \epsilon_Y\left\{\frac{R(\theta)}{r}\right\}^{1/(1+N)} \quad (40)$$

where $R(\theta)$ shows the elastic-plastic boundary approximately, which is obtained by Rice and Rosengren. Each term of the integrands of Eq 12 is expressed as

$$W^{\text{el}} = \frac{1}{2}\sigma_{ij}\epsilon_{ij}{}^{\text{el}} = \frac{K_\sigma{}^2}{E}r^{-2N/(1+N)}f(\theta), \qquad T_i\beta_{1i}{}^{\text{el}} = \frac{K_\epsilon{}^2}{E}r^{-2N/(1+N)}g(\theta) \quad (41)$$

The J_{ext}-value coincides with the J-value when the contour Γ_f coincides with $R(\theta)$, and we obtain

$$J_{ext} = J\left(\frac{r_f}{r_Y}\right)^{(1-N)/(1+N)} = J\Big/\left(\frac{\epsilon_f}{\epsilon_Y}\right)^{1-N} \tag{42}$$

Summary of the Theoretical Discussions

For the case of elastic-perfectly plastic material, $N = 0$. The relation of the J_{ext} to J depends on r_f/r_Y or ϵ_f/ϵ_Y. In the process zone, it is considered that the continuum theory is not applicable to describe the state of the material, but by setting $|\Omega_f|$ as the boundary of the process zone, the J_{ext} does not lose its meaning based on the continuum theory.

The discussion in the previous subsection is based on the deformation theory of plasticity, but for the incremental theory the J_{ext}-integral is also valid because the directions of the principal stresses do not change.

Relation Between J_{ext} and G^{Δ}

Kfouri and Rice [*10*] proposed a new fracture criterion, called G^{Δ}, which gives the energy release rate by some finite amount of crack propagation. G^{Δ} also aims to overcome the shortcoming mentioned earlier. They obtained the next equation by using the BCS model assuming that Δa is adequately small in comparison with the plastic region

$$G^{\Delta}/J = \frac{\xi}{4}\left\{\ell n\left(\frac{1}{\xi}\right) + \ell n\, 4 + \frac{3}{2}\right\} \tag{43}$$

For the same problem, J_{ext} is obtained by assuming $\epsilon = 2\xi = 2r_f/r_Y$ as

$$J_{ext}/J = \frac{\xi}{4}\left\{\ell n\left(\frac{1}{\xi}\right) + \ell n\, 8 + 1\right\} \tag{44}$$

By comparing Eq 44 with Eq 43, it is noticed that the first term of each equation is equal and the second term, the higher-order term, is similar because $\ell n 4 + 3/2 = 2.886$ and $\ell n\ 8 + 1 = 3.079$. The differences of both equations are very small; for example, for $\epsilon = 0.4$, the difference is under 5 percent. Though G^{Δ} and J_{ext} consider the different fracture processes, it is interesting that the two concepts result in nearly the same values as the fracture criterion.

Summary

The results are summarized as follows:

1. The J_{ext} concept is introduced as the measure of the effective energy release rate in the process zone, and expressions for the two- and three-dimensional J_{ext}-integrals are obtained.

2. The J_{ext}-integral is the measure of the inhomogeneities which occur by large deformation near the crack tip, and it is expressed in the form of the generalized force.

3. The path-independent expression of the J_{ext}-integral is also introduced, which is valid for the body force field and the eigen stress field.

4. For the two-dimensional case, some properties of the J_{ext}-integral are discussed and its relations to G^{Δ} are shown. By considering the finite region as the process zone, the J_{ext} concept overcomes the shortcomings of the energy concept as the fracture criterion.

References

[1] Rice, J. R. in *Proceedings*, 1st International Conference on Fracture, Sendai, Japan, Vol. 1, 1966, p. 269.

[2] Yokobori, T. and Ichikawa, M., *Reports of Research Institute for Strength and Fracture of Materials*, Tohoku University, Vol. 2, 1966, p. 21.

[3] Kröner, E., *Kontinuumstheorie der Versetzungen und Eigenspannungen*, Springer-Verlag, Berlin, 1958.

[4] Miyamoto, H. and Kageyama, K. in *Proceedings*, 1st International Conference on Numerical Methods in Fracture Mechanics, Swansea University, U.K., 1977, p. 479.

[5] Bilby, B. A. in *Proceedings*, 3rd International Conference on Fracture, PLI-111, Munich, 1973.

[6] Chell, G. G. in *Proceedings*, 4th International Conference on Pressure Vessel Technology, The Institution of Mechanical Engineers, London, Vol. 1, 1980, p. 117.

[7] Mura, T. in *Recent Research on Mechanical Behavior of Solids*, H. Miyamoto, Ed., University of Tokyo Press, 1979, p. 237.

[8] Rice, J. R. and Rosengren, G. F., *Journal of the Mechanics and Physics of Solids*, Vol. 16, 1968, p. 1.

[9] Hutchinson, J. W., *Journal of the Mechanics and Physics of Solids*, Vol. 16, 1968, p. 13.

[10] Kfouri, A. P. and Rice, J. R. in *Proceedings*, 4th International Conference on Fracture, Waterloo, Ont., Canada, Vol. 1, 1977, p. 43.

M. Saka,[1] *T. Shoji,*[2] *H. Takahashi,*[2] *and H. Abé*[1]

A Criterion Based on Crack-Tip Energy Dissipation in Plane-Strain Crack Growth Under Large-Scale Yielding

REFERENCE: Saka, M., Shoji, T., Takahashi, H., and Abé, H., **"A Criterion Based on Crack-Tip Energy Dissipation in Plane-Strain Crack Growth Under Large-Scale Yielding,"** *Elastic-Plastic Fracture: Second Symposium, Volume I—Inelastic Crack Analysis, ASTM STP 803*, C. F. Shih and J. P. Gudas, Eds., American Society for Testing and Materials, 1983, pp. I-130–I-158.

ABSTRACT: In order to assess the proper resistance of elastic-plastic materials for stable crack growth in Mode I under plane-strain conditions with large-scale yielding, a fracture criterion based on a rate of the crack-tip energy dissipation with crack extension was proposed. The criterion can be expressed as $T_W = (T_W)_c$, where T_W is a newly developed tearing parameter defined as $T_W = (1/R)(E/\sigma_0^2)(d\tilde{W}_p/da)$. The parameter T_W is a dimensionless representation of the increment of plastic work done, $d\tilde{W}_p/da$, in a circular region of characteristic radius R at the growing crack tip, and T_W is related directly as the amplitude to the singularity field at a growing crack tip.

A procedure of experimental determination of $(T_W)_c$ was described, where the crack-tip energy dissipation, W_p, was evaluated by means of the recrystallization-etch technique, and the width of recrystallized region observed parallel to the growing crack surface was used as the characteristic radius R. With special reference to a nonlinearity of the J-resistance curve for a wide range of crack extension, a conventional estimation procedure of the rate of crack-tip energy dissipation in terms of $\gamma[=(1/2)\beta^2\sigma_0^2R(T_W)_c]$ was also described.

Furthermore, the T_W criterion was verified experimentally and numerically in comparison with other candidate tearing parameters $T_J[=(E/\sigma_0^2)(dJ/da)]$ and $T_\delta[=(E/\sigma_0)(d\delta/da)]$. A significant superiority of T_W over T_J and T_δ [crack-opening angle (COA)] was emphasized as a characterizing parameter for crack growth. It was shown that the critical value of T_W during crack growth was not affected by specimen size, side groove depth, and an amount of crack extension.

KEY WORDS: ductile fracture, fracture criterion, stable crack growth, crack-tip energy dissipation, tearing parameter, intense strain region, recrystallization-etch technique, fi-

[1]Department of Mechanical Engineering, Tohoku University, Sendai, Japan.
[2]Research Institute for Strength and Fracture of Materials, Tohoku University, Sendai, Japan.

nite-element analysis, resistance curve, crack-opening angle, J-integral, elastic-plastic fracture

A large number of experimental and theoretical investigations on stable crack growth of ductile materials have been carried out, because the conventional assessment, based on the onset of crack growth, of the safety margin of flawed structures fabricated of high-toughness materials is conservative. The apparent toughness of these materials such as the J-integral, J, or the crack-opening displacement (COD), δ, increases with stable crack growth prior to crack instability as indicated by the J-resistance or the δ-resistance curves. Hence, the slopes of these curves dJ/da or $d\delta/da$ are being used as measures of resistance to ductile crack growth, where a is the crack length.

Shih et al [*1*][3] and Kanninen et al [*2*] have assessed numerically the Paris tearing modulus $T_J[=(E/\sigma_0^2)(dJ/da)]$ [*3*] as a crack extension parameter, where E is Young's modulus, and σ_0 is the flow stress. In their works, it has been shown that the critical value of T_J, $(T_J)_c$, decreases monotonically with a large amount of crack growth. Moreover, under the conditions of large amount of crack growth in both large-scale and small-scale yielding, many experimental results [*4–10*] have also demonstrated a significant nonlinearity of the J-resistance curve which results in the variation of $(T_J)_c$ with stable crack extension.

In contrast, the crack-tip opening angle (CTOA), $T_\delta[=(E/\sigma_0)(d\delta/da)]$, has also been examined numerically [*1,2*] as an alternative fracture parameter characterizing stable crack growth, and its critical value, $(T_\delta)_c$, as a measure of the material resistance, has been proven to be relatively constant over a wide range of crack growth. Rice and Sorensen [*11*] and Rice et al [*12*] took CTOA as a fracture parameter controlling stable crack growth in their theoretical analyses. Although the superiority of $(T_\delta)_c$ in comparison with $(T_J)_c$ has been demonstrated numerically, experimental measurement of CTOA is quite difficult due to the irregular profile of the zigzag fracture surface attributed to void growth and coalescence preceding crack growth. In fact, the value of crack opening angle (COA) has been measured experimentally instead of CTOA [*1,2*].

In order to assess the proper resistance of materials against stable crack growth, a viable fracture parameter is required, which must be a measure of the stress and deformation states in the vicinity of the growing crack tip, and whose critical value must be independent of specimen size and geometry, loading configuration, and an amount of crack extension.

In this paper, we propose a fracture criterion for stable crack growth based on a rate of the crack-tip energy dissipation associated with Mode I crack extension. An experimental procedure to determine the rate of crack-tip energy dissipation is described using compact tension specimens (A533B-1 steel)

[3]The italic numbers in brackets refer to the list of references appended to this paper.

under large-scale yielding, and a conventional estimation of this quantity is also discussed. Both the experimental and numerical verifications of the fracture criterion proposed here are given in comparison with $(T_J)_c$ and $(T_\delta)_c$.

Fracture Criterion for Stable Crack Growth

Fracture Parameter and Criterion Based on Crack-Tip Energy Dissipation

The rate of the crack tip energy dissipation with crack extension reflects directly the stress and deformation states in the vicinity of the growing crack tip. We introduce a new tearing parameter expressed in terms of the rate of crack-tip energy dissipation in the form

$$T_W = \frac{1}{R} \frac{E}{\sigma_0^2} \frac{d\tilde{W}_p}{da} \tag{1}$$

where $\tilde{W}_p$ is the plastic work done in the intense strain region with characteristic radius R at the growing crack tip (Fig. 1). The parameter T_W is a dimensionless quantity. By using this parameter, a fracture criterion for stable crack growth can be expressed as

$$T_W = (T_W)_c \tag{2}$$

where $(T_W)_c$ is the critical value of T_W which is considered as a material constant. The criterion (Eq 2) will be verified in the following sections.

Singularity Field at Growing Crack Tip in Terms of Tearing Parameter T_W

Consider an elastic-plastic material which is characterized by J_2 flow theory together with linear strain hardening. The relationship between the tearing

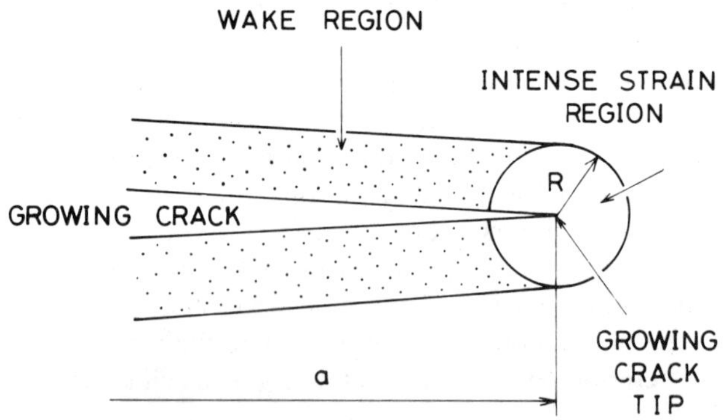

FIG. 1—*Intense strain region at growing crack tip.*

parameter T_W and the singular field near the steadily and quasi-statically growing crack tip can be obtained following the analysis by Amazigo and Hutchinson [13]. The stress and strain fields near the growing crack tip (Fig. 2) are expressed by

$$(\sigma_{ij}, s_{ij}, \overline{\sigma}) = K\sigma_0 r^s(\Sigma_{ij}(\theta), S_{ij}(\theta), \Sigma(\theta)) \tag{3}$$

$$(\dot{\sigma}_{ij}, \dot{\overline{\sigma}}) = K\sigma_0 r^{s-1}(t_{ij}(\theta), t(\theta)) \tag{4}$$

$$\alpha\dot{\epsilon}_{ij}{}^p = K\epsilon_0 r^{s-1} P_{ij}(\theta) \tag{5}$$

where

σ_{ij} = true stress,
s_{ij} = deviatoric true stress,
$\overline{\sigma} = \sqrt{(3/2)s_{ij}s_{ij}}$ = equivalent stress,
$\epsilon_{ij}{}^p$ = plastic strain,
$\alpha = E_t/E$,
E_t = tangent modulus,
$\epsilon_0 = \sigma_0/E$,
(r,θ) = polar coordinates defined in the vicinity of the crack tip,
$\Sigma_{ij}(\theta)$, $S_{ij}(\theta)$, $\Sigma(\theta)$,
$t_{ij}(\theta)$, $t(\theta)$, and $P_{ij}(\theta)$ = known functions of θ, and
$s(-1/2 < s < 0)$ = a material constant being a function of α.

($\dot{}$) denotes material differentiation with respect to any monotonically increasing quantity. Although the amplitude factor K provides a unique measure of the intensity of the singularity field in Eqs 3 to 5, it is still undetermined in the analysis by Amazigo and Hutchinson.

Now, we consider the determination of the amplitude of the singular field by using our new tearing parameter T_W. The rate of plastic work done in the circular region of radius R (Fig. 2) with crack extension is

$$\dot{\tilde{W}}_p = \lim_{\epsilon \to 0} \int_{-\pi}^{\pi} \int_{\epsilon}^{R} \sigma_{ij}\dot{\epsilon}_{ij}{}^p \, r dr d\theta$$
$$= \lim_{\epsilon \to 0} \int_{-\theta_p}^{\theta_p} \int_{\epsilon}^{R} \sigma_{ij}\dot{\epsilon}_{ij}{}^p \, r dr d\theta \tag{6}$$

since $\dot{\epsilon}_{ij}{}^p = 0$ for $\theta_p < |\theta| < \pi$. By substituting Eqs 3 and 5 into Eq 6 and by noting that $-1/2 < s < 0$, we obtain

$$\dot{\tilde{W}}_p = \frac{K^2}{\alpha} \frac{\sigma_0{}^2}{E} \frac{R^{2s+1}}{2s+1} \Psi \tag{7}$$

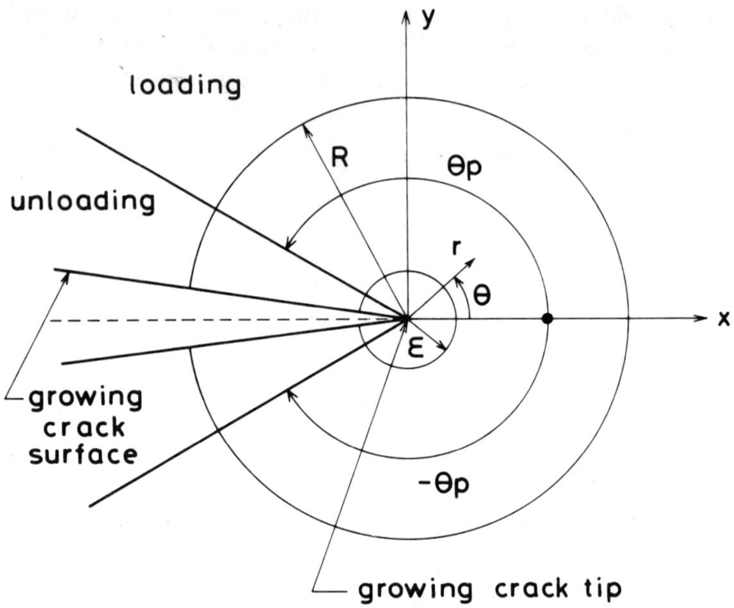

FIG. 2—*Crack-tip geometry and coordinates.*

where

$$\Psi = \int_{-\theta_p}^{\theta_p} \Sigma_{ij}(\theta) P_{ij}(\theta)\, d\theta \tag{8}$$

Then the parameter T_W is relatable to K by

$$T_W = \frac{1}{R}\, \frac{E}{\sigma_0{}^2}\, \frac{d\tilde{W}_p}{da} = K^2 \frac{\Psi}{\alpha(2s+1)} R^{2s} \tag{9}$$

By substituting Eq 9 into Eqs 3 to 5, we find that

$$(\sigma_{ij}, s_{ij}, \overline{\sigma}) = \sigma_0 \left[\frac{\alpha(2s+1)}{\Psi}\right]^{1/2} \left(\frac{r}{R}\right)^s T_W{}^{1/2} (\Sigma_{ij}(\theta), S_{ij}(\theta), \Sigma(\theta)) \tag{10}$$

$$(\dot{\sigma}_{ij}, \dot{\overline{\sigma}}) = \sigma_0 \left[\frac{\alpha(2s+1)}{\Psi}\right]^{1/2} \left(\frac{r}{R}\right)^s \frac{1}{r} T_W{}^{1/2} (t_{ij}(\theta), t(\theta)) \tag{11}$$

$$\dot{\epsilon}_{ij}{}^p = \frac{\epsilon_0}{\alpha} \left[\frac{\alpha(2s+1)}{\Psi}\right]^{1/2} \left(\frac{r}{R}\right)^s \frac{1}{r} T_W{}^{1/2} P_{ij}(\theta) \tag{12}$$

Equations 10 to 12 mean that the tearing parameter T_W is the amplitude of the singularity field near the tip of a growing crack, and these equations give the basis for the use of T_W as a key parameter for characterizing stable crack growth.

Experimental Determination of $(T_W)_c$

Experimental Procedure

Recently, Shoji [*14*] has investigated a significance of the crack-tip energy dissipation for the material resistance against stable crack growth by means of the recrystallization-etch technique [*15,16*]. In this paper, $(T_W)_c$ is measured by the same technique. By the use of this technique, a damage in the intense strain region at a crack tip which is attributed to the plastic strain concentration, or, in other words, the increase of dislocation density, can be quantitatively determined as the critical value of the energy dissipation rate T_W.

Figure 3 shows the dimensions of the compact tension specimens used. All

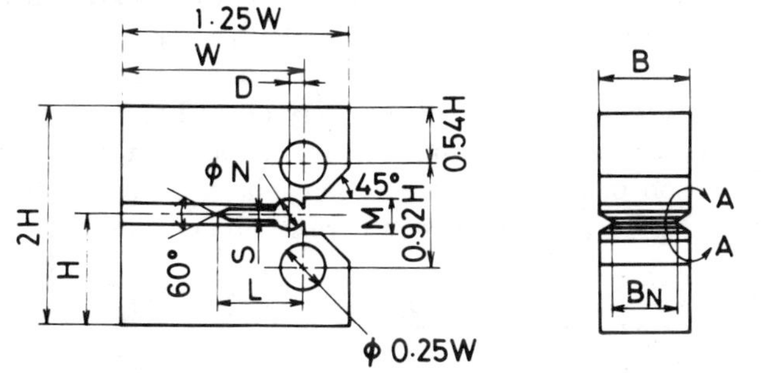

S.G.%	B_N 0.4TCT	1TCT	2TCT
0	10.0	25.0	
12.5	8.75	21.9	
25	7.50	19.5	37.7

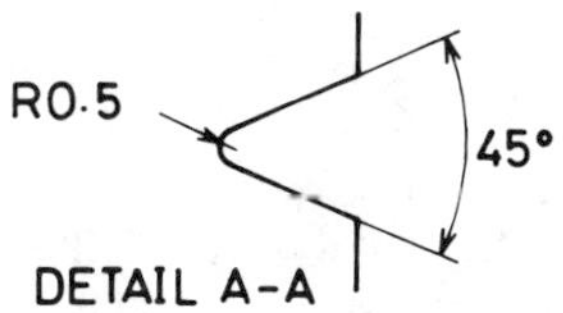

: mm

	W	H	B	L	M	S	N	D
0.4TCT	20	12	10	9.5	3.6	1.2	3.6	1.2
1 TCT	50	30	25	27.5	8	3	8	3
2 TCT	100	60	50	57.5	16	6	12	4

FIG. 3—*Dimensions of compact tension specimens used.*

specimens were extracted from an A533B-1 steel plate of 165 mm thickness, and were fatigue-precracked to a crack length of $a_0/W = 0.6$ in the L-T orientation, where a_0 is the original crack length and W the specimen width. The chemical composition and the mechanical properties of A533B-1 steel at room temperature are given in Table 1.

All tests were carried out at room temperature. The 0.4T and 1T compact tension specimens were loaded to predetermined levels along each curve of load versus load-point displacement and then unloaded. For tests of 2T compact tension specimens, the elastic unloading compliance method was employed to estimate an amount of crack growth. From compliance measurements, instantaneous values of crack length were determined by the formula of Saxena and Hudak [*17*].

After the toughness testings, many specimens were heat-treated at 700°C for 3 h in vacuum followed by furnace-cooling to recrystallize the highly strained region (intense strain region plus wake region in Fig. 1) [*15,16*]. This heat treatment can reveal the highly strained region by changing the original bainitic microstructure into a ferritic one, and the grain size of the ferrite nucleated and grown by recrystallization is a function of plastic strain given earlier. After the heat treatment, the specimens were cut into halves at midsections of thickness, and then the midsections of the specimens were polished. These were etched by 5 percent nital for metallographic observation.

Figure 4 shows a typical example of the recrystallized-etched region near the crack tip delineated by the technique, where an amount of crack extension at the midsection of a specimen was denoted by Δa_{mid}. Total plastic work, W_p, within the recrystallized region per unit thickness is [*16*]

$$W_p = \frac{4S}{4+\pi}\left\{\frac{\pi}{2}\left[\frac{\sigma_Y{}^*\bar{\epsilon}_0}{2-m} + \frac{H'\bar{\epsilon}_0{}^2}{4(1-m)}\right] + 2\left[\frac{\sigma_Y{}^*\bar{\epsilon}_0}{1-m} + \frac{H'\bar{\epsilon}_0{}^2}{2(1-2m)}\right]\right\} \quad (13)$$

where S is the total area of the recrystallized-etched region and obtained from a direct measurement. The critical strain above which the recrystallization phenomenon takes place, denoted by $\bar{\epsilon}_0$, is taken as 0.25 and $m = 0.7$ for A533B-1 steel [*16*]. The strain hardening rate, H', is 471 MPa and the pseudo-yield stress, $\sigma_Y{}^*$, obtained in the uniaxial true stress versus logarithmic tensile strain curve approximated by a line having linear strain hardening, is 630 MPa. Equation 13 has been derived from the plastic strain distribution near a tip of a stationary crack but extensively used for a growing crack as a first approximation, although Amazigo and Hutchinson [*13*] and Rice et al [*12*] analyzed the strain distribution around the growing crack tip and showed the different strain singularity from that for a stationary crack.

TABLE 1—*Chemical composition (% by weight) and mechanical properties of A533B-1 steel at room temperature.*

Chemical Composition												
C	Mn	Si	Ni	Cr	Mo	Cu	P	S	V	Sb	As	Sn
0.17	1.48	0.20	0.58	0.16	0.52	0.13	0.011	0.006	0.003	0.0041	0.015	0.008

Mechanical Properties			
Yield Strength, MPa	Ultimate Tensile Strength, MPa	Total Elongation, %	Reduction of Area, %
480	610	27.0	69.2

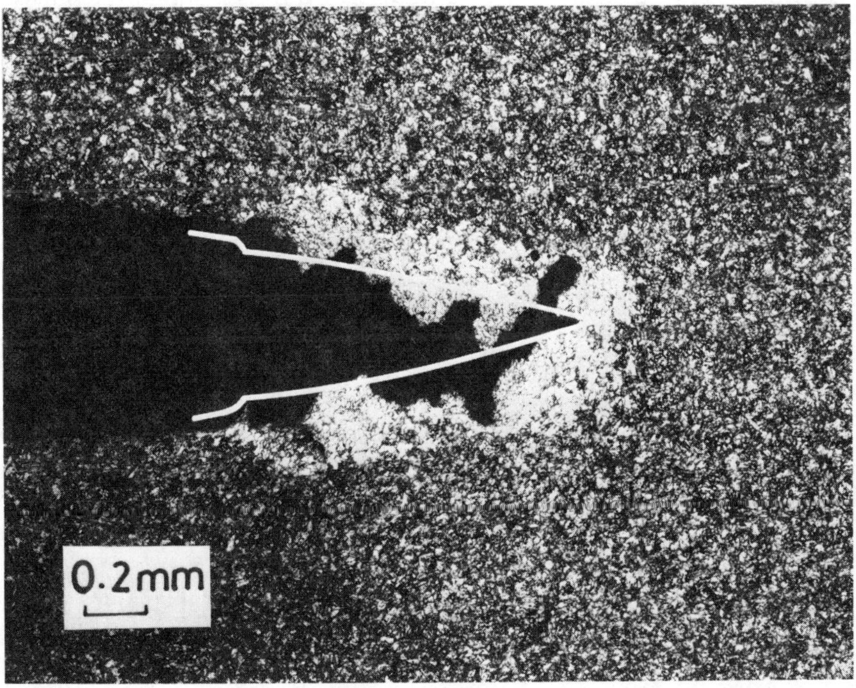

FIG. 4—*Typical example of recrystallized-etched region in 0.4T compact tension specimen (A533B-1) with 25 percent side groove at* Δa_{mid} *= 1.2 mm, tested at room temperature. Solid line shows crack-opening profile obtained in finite-element model simulation analysis of the same specimen.*

W_p-*Resistance Curve*

The W_p-resistance curve, W_p versus crack extension data, is given in Fig. 5. The W_p-resistance curve followed by every data point in this figure can be approximated by a line with $dW_p/da = 228$ kJm^{-2} for a wide range of stable crack growth except for the crack extension less than $\Delta a_{\text{mid}} = 0.3$ mm, where the effect of blunting of the original crack tip is significant on the resistance curve.

By using the value of dW_p/da which is independent of crack extension, the material proper resistance $(T_W)_c$ can be uniquely determined and this is 560 for the material treated. The characteristic radius R of the intense strain region was determined to be 0.28 mm as the width of the recrystallized-etched region delineated parallel to the crack surface. The effect of an amount of crack extension on $(T_W)_c$ will be discussed in detail later. Here, the value of $(T_W)_c$ was assessed by the rate of plastic work done within the total recrystallized region, dW_p/da, while the original definition of T_W given in Eq 1 is concerned with the rate of plastic work done $d\tilde{W}_p/da$ within the intense strain region having characteristic radius R at the growing crack tip (Fig. 1). Since restraining of the wake region (Fig. 1) is scarcely expected, the preced-

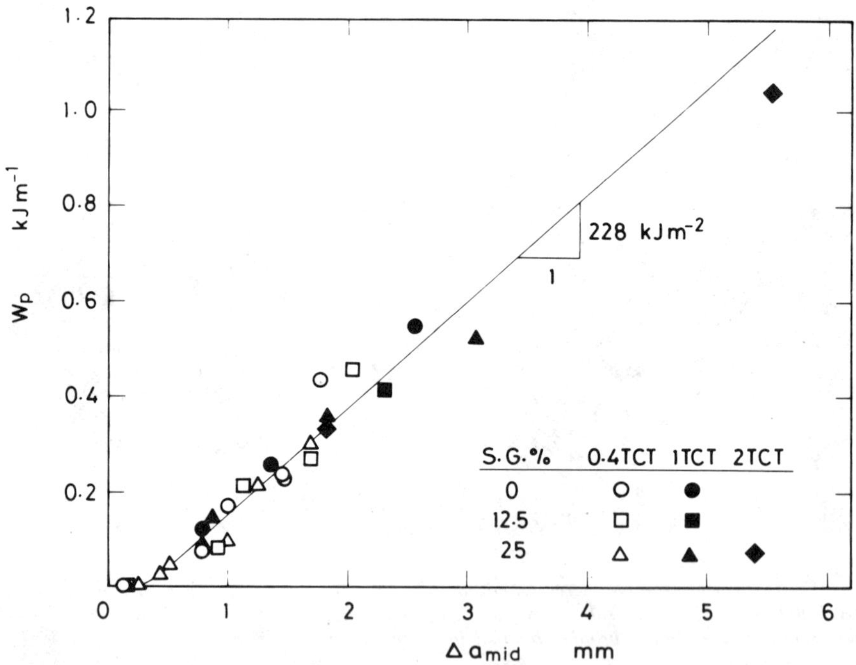

FIG. 5—*Relationship between plastic work done* (W_p) *within total recrystallized region and crack extension* (Δa_{mid}) *(A533B-1 steel tested at room temperature).*

ing assessment of $(T_W)_c$ by dW_p/da is reasonable. Details of this aspect will also be discussed later.

Description of *J*-Resistance Curve Based on T_W Criterion

As stated in previous sections, stable crack growth is controlled by the local resistance, that is, the rate of crack tip energy dissipation $(T_W)_c$. In this section, a relationship between the local resistance W_p and the global resistance *J* for crack growth is discussed and a conventional estimation procedure of the rate of crack-tip energy dissipation is also proposed by use of the *J*-resistance curve.

Figure 6 shows the J-integral versus W_p measurements in a wide range of crack growth, which was obtained experimentally in 1T and 2T compact ten-

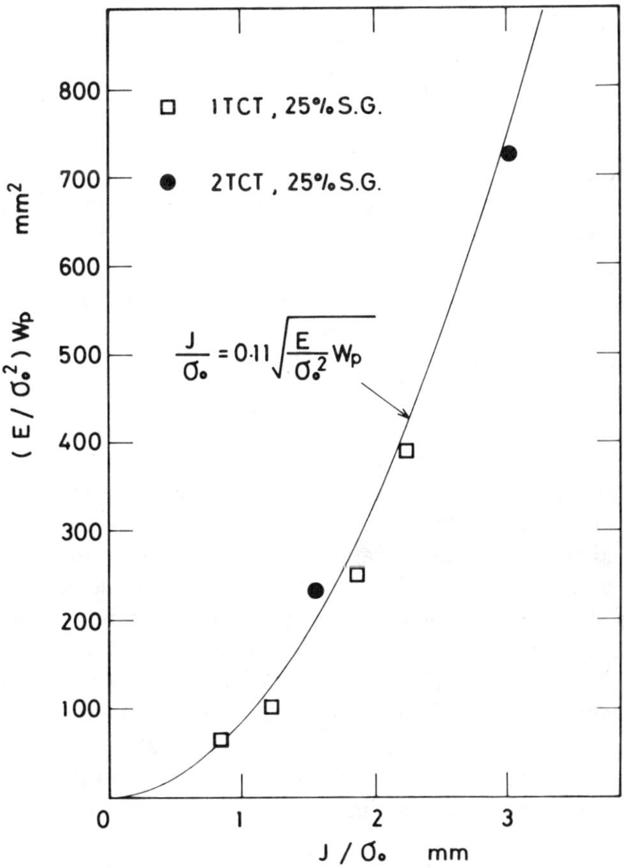

FIG. 6—*J-integral versus plastic work done,* W_p, *for 1T and 2T compact tension specimens with 25 percent side groove (A533B-1 steel tested at room temperature).*

sion specimens with 25 percent side groove regarded as in the plane-strain condition globally as well as locally. In the present investigation, the values of J were calculated using the Merkle-Corten relationship corrected for crack growth [*18*] to develop a more accurate resistance curve. The J-integral can be connected with W_p as

$$\frac{J}{\sigma_0} = \beta\sqrt{(E/\sigma_0^2)W_p} \tag{14}$$

where $\beta(=0.11)$ is a dimensionless material constant. Equation 14 was derived empirically, and the theoretical explanation of this expression has not been addressed here; however, for the deformation before the onset of crack growth only, an expression similar to Eq 14 was given by Saka et al [*19*].

Differentiating Eq 14 by a

$$J\frac{dJ}{da} = \gamma \tag{15}$$

where $\gamma[=(\frac{1}{2})\beta^2\sigma_0^2R(T_W)_c = 284 \text{ MJ}^2 \text{ m}^{-5}]$ is a material constant, since $(T_W)_c$ is 560 during stable crack growth. The newly introduced parameter γ is equivalent to $(T_W)_c$, since a characteristic radius R of the intense strain region is taken as a fixed value for crack growth. Equation 15 is the differential equation for J which governs the variation of J with increasing crack extension Δa during quasi-static crack growth. Solving Eq 15 for the initial condition of $J = J_i$ at $\Delta a = \Delta a_0$, the J-resistance curve based on the crack-tip energy dissipation is obtained as

$$J = \sqrt{2\gamma(\Delta a - \Delta a_0) + J_i^2} \tag{16}$$

where J_i and Δa_0 are critical values of J and Δa at the onset of crack growth. This equation indicates that the J-resistance curve is determined by two material constants, J_i for crack initiation which is equivalent to the critical value of W_p for crack initiation, $W_p^{\,i}$ [*14,19*], and γ for stable crack growth. Although, as shown in the following section, the J-resistance curve can be approximated by a straight line for a small amount of crack growth for the sake of convenience, Eq 16 implies the nonlinearity of the J-resistance curve. In order to evaluate the J-resistance curve for a wide range of crack extension, the nonlinear expression such as Eq 16 is more suitable and accurate than a straight-line approximation. It is worthwhile to mention that the nonlinear J-resistance curve can be approximated by a straight line only when the crack extension is small.

In turn, Eq 16 suggests that the crack growth parameter γ based on $(T_W)_c$ can be estimated from the nonlinear J-resistance curve. A conventional estimation procedure of γ and J_i in Eq 16 is based on a regression analysis for determination of two unknowns, where Δa_0 is related to J_i by the blunting line $J = 2\sigma_0 \Delta a$. More simply, J_i can be determined as the intersection point between the crack extension data and the blunting line. Figure 7 demonstrates a comparison of the J-resistance curves between measured data and an approximation by Eq 16, where γ is already given previously in terms of $(T_W)_c$. Good agreement in the preceding comparison indicates that the estimation procedure of γ might be feasible in general.

The J-resistance curves for different materials reported previously [*4, 7, 10*] can be also reevaluated by Eq 16; these are summarized in Fig. 8. For a wide range of crack extension of several materials cited here, it is verified that the J-resistance curves can be characterized by Eq 16, and this fact will give the evidence of stable crack growth in ductile materials controlled by the T_W criterion (Eq 2). In Table 2, the values of γ and J_i for several materials are summarized. It seems important to evaluate the proper resistance against ductile crack growth for different materials in terms of γ.

Fracture Parameter $(T_W)_c$ in Comparison with $(T_J)_c$ and $(T_\delta)_c$

A viable fracture parameter must be a measure of the stress and deformation states in the vicinity of the crack tip, and the critical value of this parameter must be independent of specimen size and geometry, loading configuration, and an amount of crack extension. In this section, the tearing parameter $(T_W)_c$ is verified from the viewpoint of the foregoing requirements necessary to a fracture criterion for stable crack growth in comparison with $(T_J)_c$ and $(T_\delta)_c$. For the foregoing verification, close-coupling results of experiment and computation are utilized. Details of the numerical procedure for finite-element analysis of stable crack growth are given in Appendix I.

As a limited case, in the case of the J-controlled crack growth, T_W has a direct relationship with T_J and T_δ as shown in Appendix II.

Invariance During Crack Growth

As shown in Fig. 5, the W_p-resistance curve was approximated by a line with $dW_p/da = 228$ kJ m^{-2} for a wide range of stable crack growth except for the crack extension less than $\Delta a_{\text{mid}} = 0.3$ mm. The computer simulation of plane-strain crack growth in Mode I provides a more precise understanding for the rate of energy dissipation near the crack tip. In the simulation analysis, the value of $(T_W)_c$ was calculated by

$$(T_W)_c = \frac{1}{R} \frac{E}{\sigma_0{}^2} \frac{\Delta \tilde{W}_p}{\overline{\Delta a}} \tag{17}$$

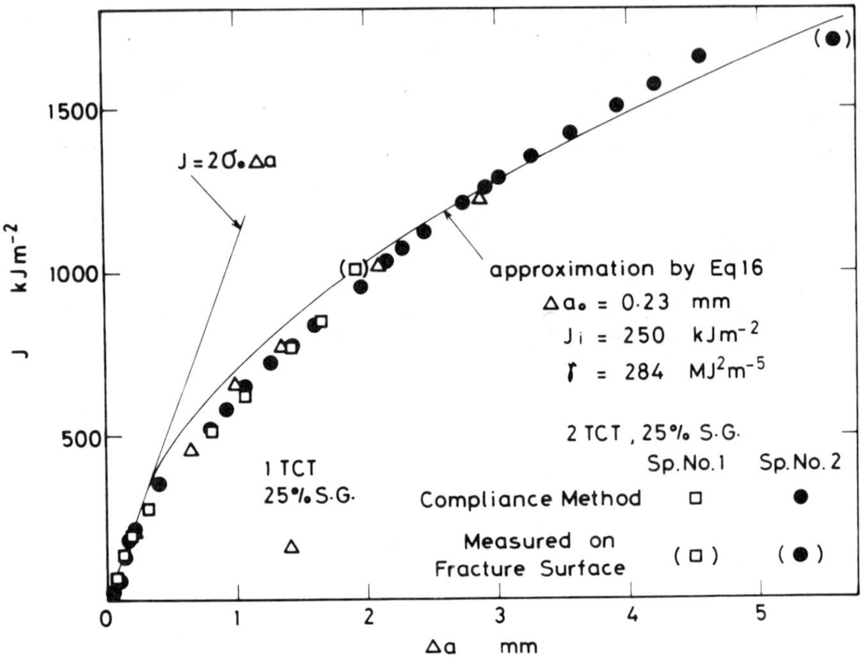

FIG. 7—*Nonlinear* J-*resistance curve for a wide range of crack extension (1T and 2T compact tension specimens with 25 percent side groove, A533B-1 steel tested at room temperature) and a comparison with an estimated curve by Eq 16.*

where $\Delta \tilde{W}_p$ is the increment of the plastic work occurring in the intense strain region of radius R (=0.35 mm) at the growing crack tip with instantaneous crack extension $\overline{\Delta a}$ (=0.125 mm). The characteristic radius R of the intense strain region was taken to be 0.35 mm, considering the finite-element grids, although it was 0.28 mm in the experiment as described previously. However, this difference in radius will have a negligible effect on the value of $(T_W)_c$ because s in Eq 9 is nearly zero for A533B-1 steel.

Figure 9 shows a computational result of $(T_W)_c$ during stable crack growth. Except for the first 0.3-mm crack extension, $(T_W)_c$ is relatively invariable during crack growth. In order to understand the invariance of the value of $(T_W)_c$ with respect to crack extension, next consider a distribution of an increment of plastic work density near the growing crack tip. Figure 10 shows the distribution of an increment of plastic work density, $\overline{\sigma}(\overline{\Delta \epsilon_p}/\overline{\Delta a})$, on one half of the specimen with instantaneous crack extension $\overline{\Delta a}$, observed at $\Delta a = 0.8$ mm in a 0.4T compact tension specimen, where $\overline{\Delta \epsilon_p} = \Delta\overline{\sigma}/H'$ is the equivalent plastic strain increment. In our analysis, the feature shown in Fig. 10 was maintained during stable crack growth except for crack extension less than 0.3 mm. Hence, such a fixed deformation state very near the growing crack tip resulted in the constant value of $(T_W)_c$ evaluated as

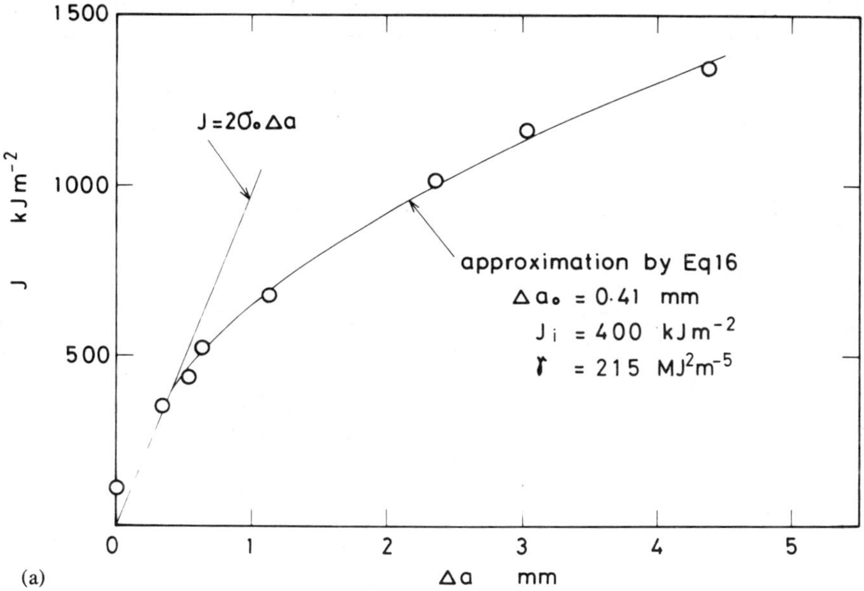

(a)

(b)

FIG. 8—J-*resistance curves for* (a) *A533B [4] (4T compact tension specimen with 12.5 percent side groove,* $a_0/W = 0.72$ *in L-T orientation, tested at 93°C),* (b) *A533B [10] (2T compact tension specimen with 20 percent side groove,* $a_0/W = 0.63$ *in T-L orientation, tested at 150°C), and* (c) *A533B, weld deposit [7] (1.6T compact tension specimen with 20 percent side groove,* $a_0/W = 0.52$ *in T-L orientation, tested at 200°C, irradiated with neutron (n) fluence:* 1.6×10^{19} $n/cm^2 > 1$ *MeV) and their comparison with estimated curves by Eq 16.*

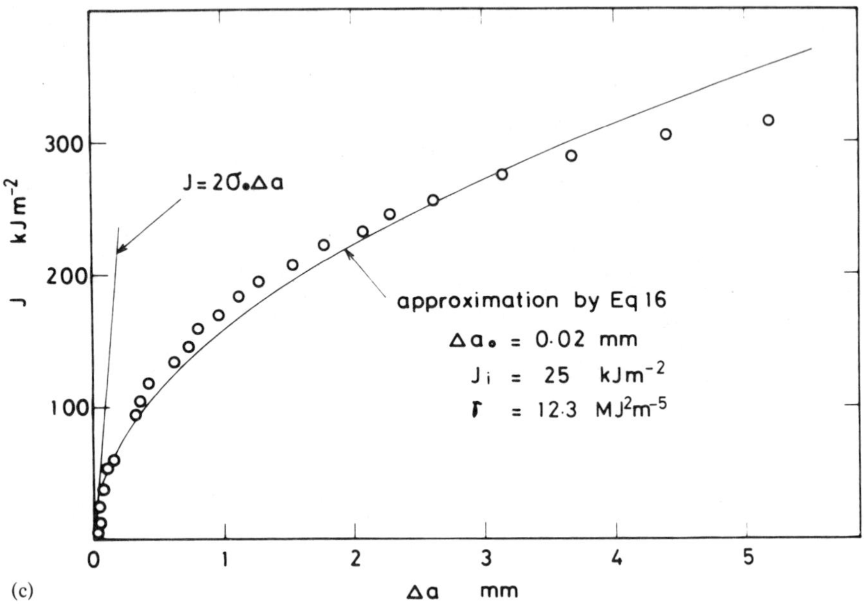

(c)

FIG. 8—*Continued.*

TABLE 2—*Values of* γ *and* J_i *for several materials.*

Materials	Specimens	Test Conditions	Δa_0, mm	J_i, kJ m^{-2}	γ, M J^2 m^{-5}
A533B-1 (Authors)	1T, 2T CT with 25% SG,[a] $a_0/W = 0.6$ in L-T orientation	room temperature	0.23	250	284[b]
A533B [*4*]	4T CT with 12.5% SG, $a_0/W = 0.72$ in L-T orientation	93°C	0.41	400	215
A533B [*10*]	2T CT with 20% SG, $a_0/W = 0.63$ in T-L orientation	150°C	0.14	160	27.9
A533B [*6*] Weld deposit	1T CT with 20% SG, $a_0/W = 0.62$ in T-L orientation	200°C	0.08	75	17.5
A533B [*6*] Weld deposit	1T CT with 20% SG, $a_0/W = 0.61$ in T-L orientation	200°C irradiated (neutron fluence = 1.2×10^{19} n/cm^2 > 1 MeV)	0.02	20	6.50

[a]SG = side groove.
[b]Estimated by $(T_W)_c$.

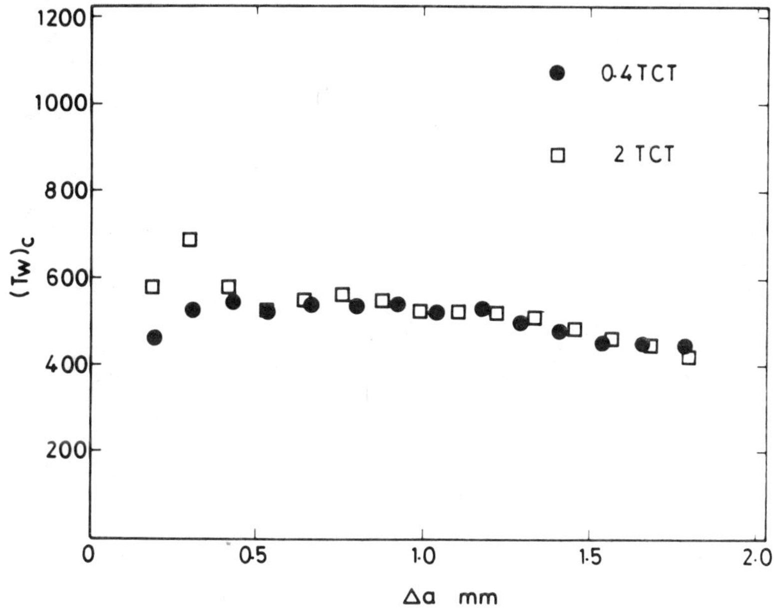

FIG. 9—*Calculated value of* $(T_W)_c$ *in plane strain simulation analysis of 0.4T and 2T compact tension specimens with* $a_0/W = 0.6$.

$$(T_W)_c = \frac{1}{R}\,\frac{E}{\sigma_0^2}\,\frac{d\tilde{W}_p}{da} = \frac{1}{R}\,\frac{E}{\sigma_0^2}\int_\Omega \bar{\sigma}\,\frac{d\bar{\epsilon}_p}{da}\,d\Omega \tag{18}$$

where Ω is the area of the intense strain region of radius R at the growing crack tip.

It can also be understood from Fig. 10 that the energy dissipation rate in the centered-fan zone in the well-known Prandtl slipline field [*11,12*] contributes dominantly to the value of $(T_W)_c$, and due to this fact experimental assessment of $(T_W)_c$ by the rate of plastic work done in the total recrystallized region agrees with the original definition of T_W given in Eq 1.

Another fracture parameter characterizing crack growth is $(T_\delta)_c$. The tearing parameter $(T_\delta)_c$ represents a crack-tip opening angle, in other words, a crack opening profile. An example of a crack opening profile obtained in the simulation analysis of a 0.4T compact tension specimen is given in Fig. 4 in comparison with the experimental result in the same specimen at the same amount of crack extension $\Delta a_{mid} = 1.2$ mm. Figure 4 shows good agreement in both results of experiment and computation. Figure 11 demonstrates a computational result of $(T_\delta)_c$ during stable crack growth, where the value of $(T_\delta)_c$, based on the opening displacement δ_t measured at the first node at a fixed distance, $\overline{\Delta a}$, behind the current crack tip, was calculated as

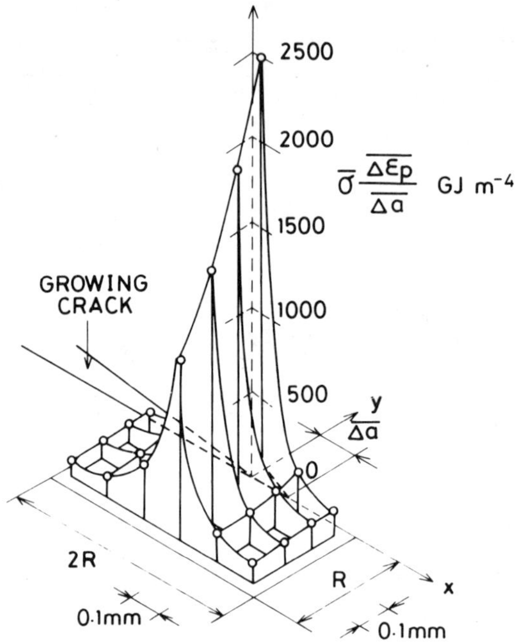

FIG. 10—*Distribution of an increment of plastic work density,* $\bar{\sigma}(\overline{\Delta\epsilon_p}/\overline{\Delta a})$*, with instantaneous crack extension* $\overline{\Delta a}$ *(=0.125 mm), which was obtained at* Δa = *0.8 mm in 0.4T compact tension specimen, and shown for one half of specimen.*

$$(T_\delta)_c = \frac{E}{\sigma_0}\,\frac{\delta_t}{\overline{\Delta a}} \tag{19}$$

Except for the first 0.3-mm crack extension, $(T_\delta)_c$ is relatively invariable during crack growth in the same manner as $(T_W)_c$.

In turn, Fig. 12 shows the *J*-resistance curves of each specimen measured experimentally up to crack extension of 2 mm. The resistance curves can be approximated by straight lines for a limited crack extension, and the resistance against crack growth is characterized by a single $(T_J)_c$-value irrespective of crack growth. It should be noted here, however, beyond this limitation, that the resistance curve cannot be described any more by a straight line or a unique value of $(T_J)_c$ as shown in Fig. 7 in the previous section.

Effect of Side Groove

As shown clearly also in Fig. 12, the *J*-resistance curves strongly depend on side groove depth. The side groove effect on resistance, measured globally as the J-integral which samples a large portion of the elastic-plastic strain field,

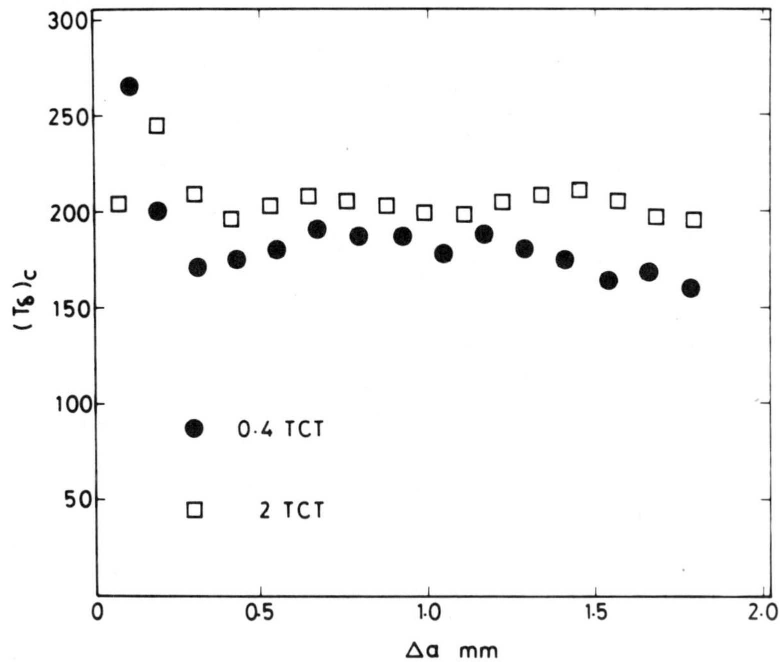

FIG. 11—*Calculated value of* $(T_\delta)_c$ *in plane-strain simulation analysis of 0.4T and 2T compact tension specimens with* $a_0/W = 0.6$.

arises from the degree of shear deformation and the resultant shear lip formation as shown in Fig. 13. Figure 13 indicates the recrystallized-etched region at the sections normal to the plate surface and the crack plane in 0.4T compact tension specimens with side groove of 0 and 25 percent. The same behavior of the shear lip formation was also observed in 1T and 2T compact tension specimens. The side groove suppresses the shear lip formation, and then promotes a practically uniform plane-strain constraint near the crack tip through the specimen net thickness. Hence, the side groove results in lowering the J-resistance curve and the resultant tearing modulus $(T_J)_c$ than non-side-grooved specimens.

In contrast, when we pay attention directly to the deformation in the vicinity of the crack tip at the midsection of the specimen, material resistance for plane-strain crack growth may be evaluated independently of side groove depth. In fact, it is understood from Fig. 5 that the W_p-resistance curve is independent of side groove depth, and hence the resultant $(T_W)_c$ is not affected by side groove depth.

Figure 14 shows the effect of side groove depth on $(T_\delta)_c$. Since CTOA cannot be measured experimentally by reason of the zigzag fracture surface, the value of $(T_\delta)_c$ was here evaluated in the form of, in 1.0 mm $< \Delta a_{mid} <$ 2.5 mm

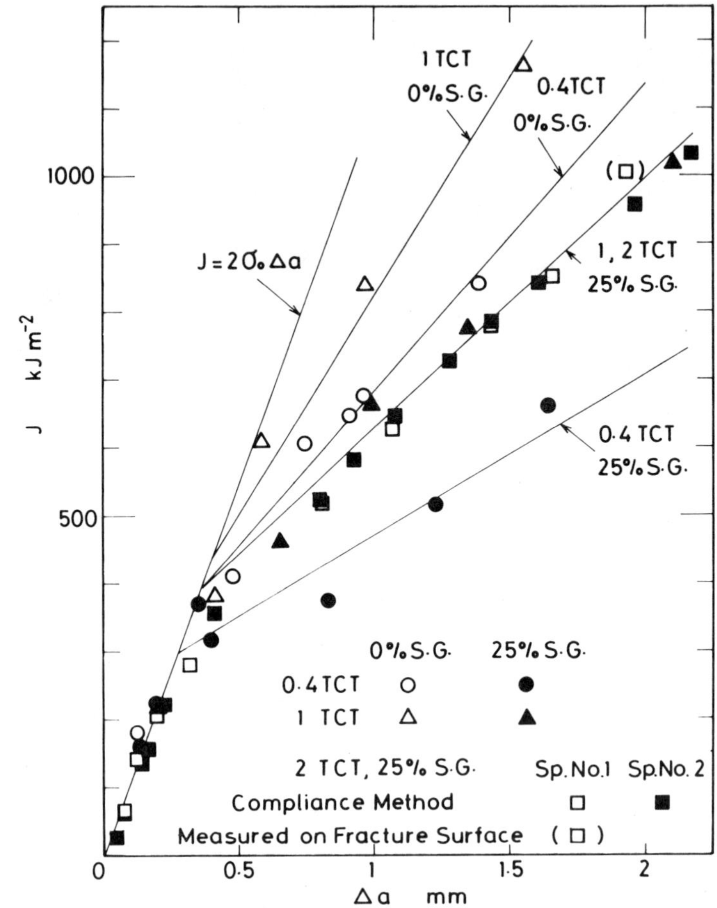

FIG. 12—*Effects of side groove depth and specimen size on* J*-resistance curve (A533B-1 steel tested at room temperature).*

$$(T_\delta)_c = 2\frac{E}{\sigma_0}\tan\left(\frac{\text{COA}}{2}\right) \tag{20}$$

where COA was obtained from the opening at the original crack tip at the midsection of specimen. It is clearly understood from Fig. 14 that $(T_\delta)_c$ is independent of side groove depth.

Effect of Specimen Size

Figure 12 demonstrates that 1T and 2T compact tension specimens with 25 percent side groove possess the same *J*-resistance curve. This curve can be

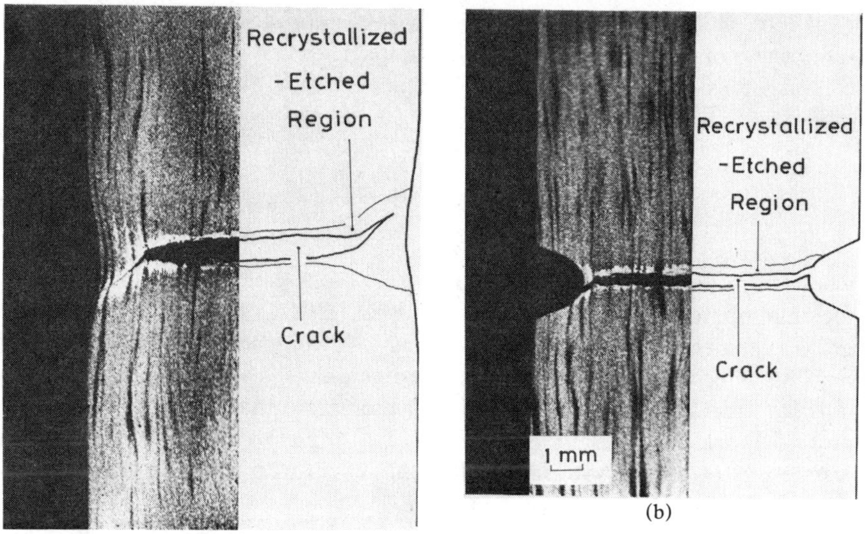

FIG. 13—*Through-thickness distribution of recrystallized-etched region at section normal to plate surface and crack plane in 0.4T compact tension specimens with* (a) *0 percent,* (b) *25 percent side groove (A533B-1 steel tested at room temperature).*

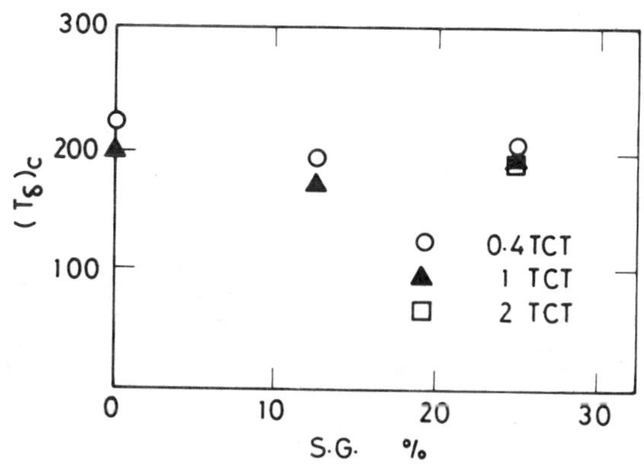

FIG. 14—*Effects of side groove depth and specimen size on* $(T_\delta)_c$, *where COA was measured within the crack extension (1.0 mm <* Δa_{mid} *< 2.5 mm) (A533B-1 steel tested at room temperature).*

regarded as the proper resistance of material for plane-strain crack growth. The resistance curve for a 0.4T compact tension specimen with 25 percent side groove, on the other hand, does not agree with this proper resistance curve of material. This result is associated with a fact that the stress state of a 0.4T compact tension specimen deviates from the plane-strain condition at the area far from the crack tip such as the compression region, and this deviation is caused by a low constraint in the thickness direction.

In fact, the plane-strain simulation analysis of crack growth resulted in a same value of $(T_J)_c$ even in the 0.4T compact tension specimen as in the 2T compact tension specimen (Fig. 15). Values of $(T_J)_c$ in Fig. 15 were determined by

$$(T_J)_c = \frac{E}{\sigma_0^2} \frac{\overline{\Delta J}}{\Delta a} \tag{21}$$

where ΔJ is an increment of J calculated far from the crack tip following the formulation for finite deformation analysis [*20*]. Consequently, the crack growth of specimens of small size such as 0.4T compact tension specimens is not characterized by the J-integral experimentally.

In contrast, Figs. 5, 9, 11, and 14 demonstrate that the values of $(T_W)_c$ and $(T_\delta)_c$ are independent of specimen size. Table 3 summarizes the numerical results of tearing parameters in comparison with those of experiments. The numerical results of $(T_W)_c$ and $(T_\delta)_c$, which are taken as the directly reflective parameter of the deformation state at the growing crack tip, show good agreements with experimental ones, respectively. These agreements of values obtained numerically and experimentally confirm the stress state, at

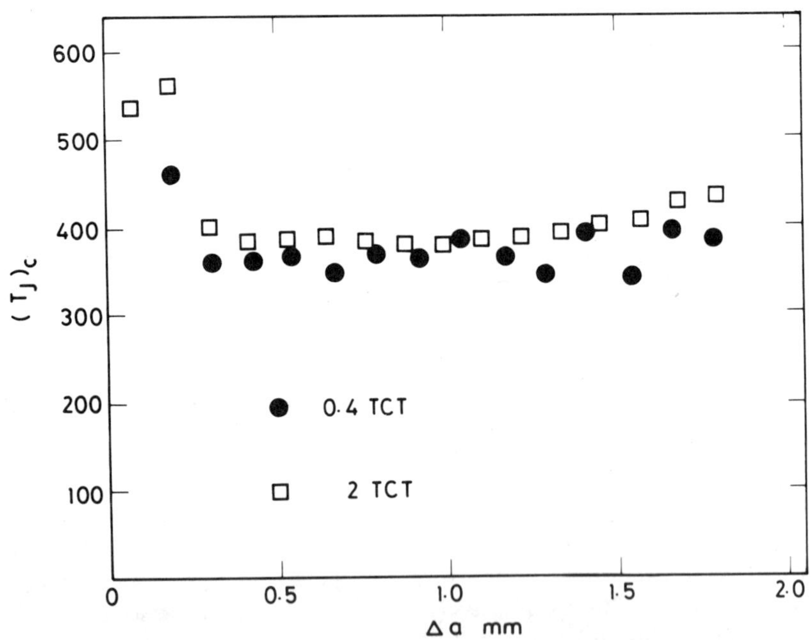

FIG. 15—*Calculated value of* $(T_J)_c$ *in plane-strain simulation analysis of 0.4T and 2T compact tension specimens with* $a_0/W = 0.6$.

TABLE 3—*Comparison of numerical results of tearing parameters with those of experiments in crack extension up to 2 mm, A533B-1 steel tested at room temperature.*

<table>
<tr><th rowspan="2"></th><th rowspan="2">$(T_W)_c$</th><th rowspan="2">$(T_\delta)_c$</th><th colspan="2">$(T_J)_c$</th></tr>
<tr><th>0.4T CT with 25% SG[a]</th><th>1T and 2T CT with 25% SG</th></tr>
<tr><td>Experiment (0.4T, 1T, and 2T CT)</td><td>560</td><td>200</td><td>170</td><td>260</td></tr>
<tr><td>Numerical analysis (0.4T and 2T CT in plane-strain)</td><td>530</td><td>190</td><td colspan="2">370</td></tr>
</table>

[a]SG = side groove.

least near the crack tip, to be of plane-strain type in experiments even in small specimen as assumed for numerical calculation.

Effect of Loading Configuration

Both experimental and numerical verifications for tearing parameters $(T_W)_c$, $(T_J)_c$, and $(T_\delta)_c$ suggest that the proper resistance of material for plane-strain crack growth seems to be assessed either by $(T_W)_c$ or $(T_\delta)_c$, which are the directly measurable parameters of deformation state in the vicinity of the growing crack tip. However, Shoji [*14*] reported recently that there is a large difference in COA between bend-type and compact tension-type specimens because of the effect of rigid-body rotation, in spite of the fact that dW_p/da can be regarded as still constant irrespective of loading configuration. Consequently, the tearing parameter $(T_W)_c$ based on the crack-tip energy dissipation is the most promising measure of the material resistance to plane-strain crack growth, and it was proven to be constant during crack growth and also to be independent of specimen size and geometry (side groove depth), and loading configuration.

Conclusions

1. We introduced a new tearing parameter expressed in terms of the rate of crack-tip energy dissipation in the form $T_W = (1/R)(E/\sigma_0^2)(d\tilde{W}_p/da)$, where $\tilde{W}_p$ is the plastic work done in the intense strain region with characteristic radius R at the growing crack tip. The parameter T_W is a dimensionless quantity. By using this parameter, a fracture criterion for stable crack growth was expressed as $T_W = (T_W)_c$, where $(T_W)_c$ is the critical value of T_W.
2. The value of $(T_W)_c$ can be determined experimentally by means of the recrystallization-etch technique as the slope of the W_p-resistance curve, where W_p is the plastic work done within the total recrystallized region and the char-

acteristic radius R is taken as the width of the recrystallized-etched region delineated parallel to the growing crack surface.

3. A conventional estimation procedure of the rate of crack-tip energy dissipation in terms of the parameter $\gamma[=(\frac{1}{2})\beta^2\sigma_0{}^2R(T_W)_c]$, equivalent to $(T_W)_c$, was also proposed with reference to the nonlinear J-resistance curve. Values of γ for several materials were evaluated by this procedure.

4. Finally, from a viewpoint of the several requirements necessary to a fracture criterion for stable crack growth, both experimental and numerical verifications of $(T_W)_c$ in comparison with $(T_J)_c$ and $(T_\delta)_c$ led to a conclusion that the tearing parameter $(T_W)_c$ is the most promising one for characterizing stable crack growth in ductile materials.

Acknowledgment

The authors wish to acknowledge Emeritus Professors M. Suzuki and O. Tamate, Tohoku University, for their kind encouragement and thankful advice during the present investigation. The authors must also express their gratitude to Messrs. H. Anzai, T. Hashida, H. Takeda, and N. Ebine, Department of Mechanical Engineering, Tohoku University, for their advice and helpful discussions. The numerical calculations were performed on the NEAC 2200 at the Computer Center, Tohoku University.

APPENDIX I

Numerical Procedure for Finite-Element Analysis of Stable Crack Growth

In order to analyze the large geometry change in crack-tip opening with crack growth, a Lagrangian incremental and piecewise linear finite-element method for elastic-plastic materials [*21*] was utilized under the plane-strain condition. In the analysis, geometrical nonlinearity was taken into consideration. A constitutive relation as a direct extension of the Prandtl-Reuss equation was adopted. Since the Jaumann stress increment is invariant with respect to rigid-body rotations, a linear relation between this stress increment and the increments of deformation tensors was employed. The uniaxial true stress versus logarithmic tensile strain curve of A533B-1 steel obtained from a uniaxial tension test at room temperature is shown in Fig. 16, where the curve is approximated with good accuracy by a line having bilinear work hardening.

A finite-element model of a 2T compact tension specimen ($a_0/W = 0.6$) is given in Fig. 17. For one half of the specimen, this model contains 506 constant-strain quadrilateral elements and 567 nodes. In a model of a 0.4T compact tension specimen ($a_0/W = 0.6$), finite-element grids around the crack tip were taken as equal in size to those of 2T compact tension specimen. A finite-element division of 398 constant-strain quadrilateral elements and 451 nodes was made for one half of the 0.4T compact tension specimen.

In all analyses, release of the nodal force at the node representing the original crack tip was started at $J = 100$ kJ m^{-2}, and then the load-line displacement, V_L, was increased simultaneously with the incremental release of the nodal forces in 20 steps following the experimentally measured V_L-versus-Δa records (Fig. 18). A finite step

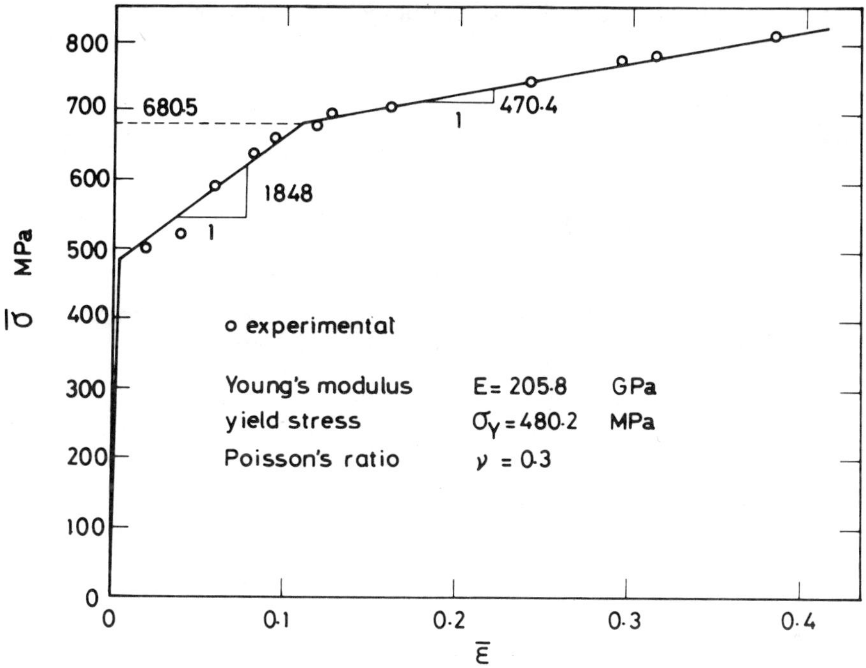

FIG. 16—*Uniaxial true stress ($\bar{\sigma}$) versus logarithmic tensile strain ($\bar{\epsilon}$) curve of A533B-1 steel at room temperature and a piecewise linear stress-strain curve having bilinear work-hardening for simulation analysis.*

size of crack extension was taken as 0.125 mm measured at the undeformed configuration, considering the microstructural quantity of A533B-1 steel, such as inclusion spacing being 0.1 ~ 0.2 mm [*22,23*].

APPENDIX II

Correlation of T_W with T_J and T_δ in *J*-Controlled Crack Growth

Shih et al [*24*] introduced the increment of crack-tip energy dissipation with stable crack extension analytically, applying the stresses and strains in the vicinity of the stationary crack tip in a power-law hardening material [*25,26*] to the growing crack. Their analysis revealed that the increment of plastic work in the material volume of radius *R* shown in Fig. 1 with increasing *J* and *a* is given by

$$\Delta \tilde{W}_p = \frac{n}{(n+1)I_n} \{S_n R \Delta J + t_n J \Delta a\} \tag{22}$$

where n is the hardening exponent and I_n, S_n, and t_n are dimensionless integration constants which are the functions only of n. The first term in Eq 22 corresponds to the

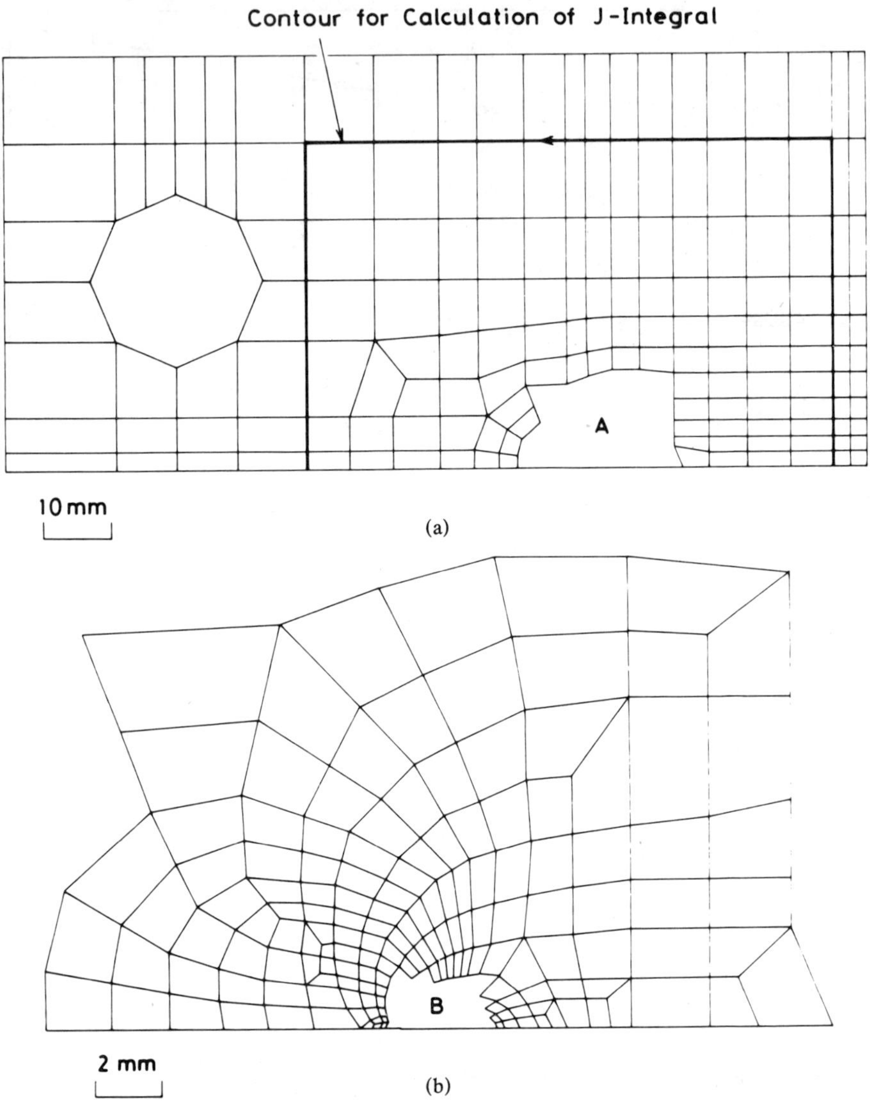

FIG. 17—*Undeformed configuration of finite-element model for one half of 2T compact tension specimen* ($a_0/W = 0.6$) *(506 constant-strain quadrilateral elements, 567 nodes):* (a) *general view,* (b) *detail of Section A, and* (c) *detail of Section B.*

increase of plastic work due to proportional loading, while the second term is nonproportional. If

$$\frac{R}{J}\frac{dJ}{da} \gg \frac{t_n}{S_n} \tag{23}$$

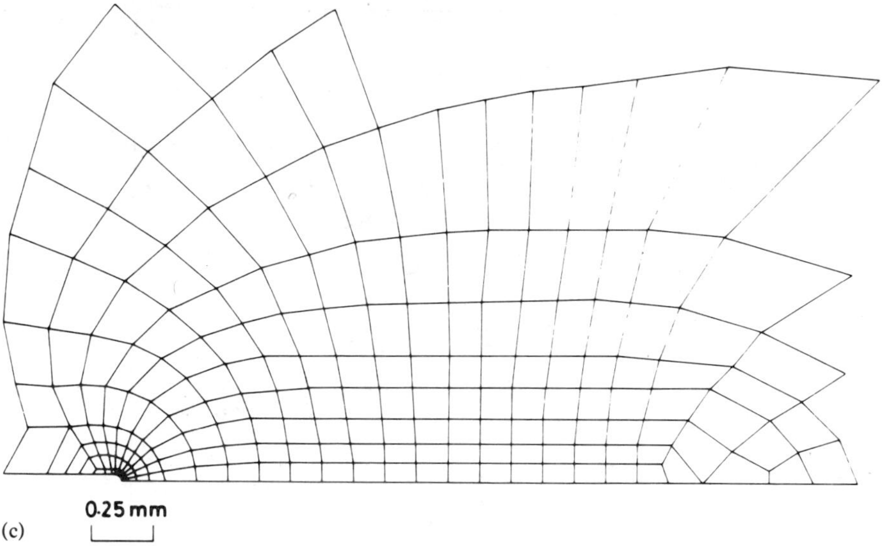

(c)

FIG. 17—*Continued.*

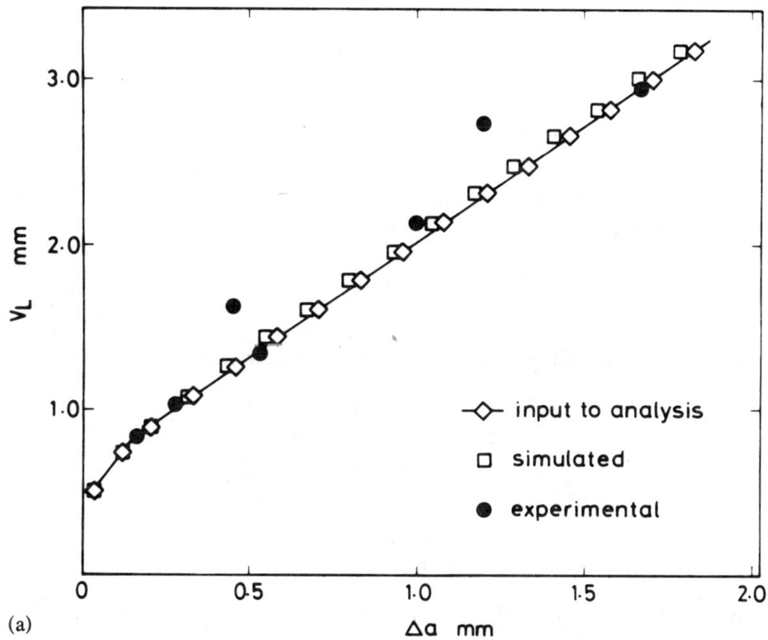

(a)

FIG. 18—*Load-line displacement* (V_L) *versus crack extension* (Δa) *for* (a) *0.4T,* (b) *2T compact tension specimens with 25 percent side groove.*

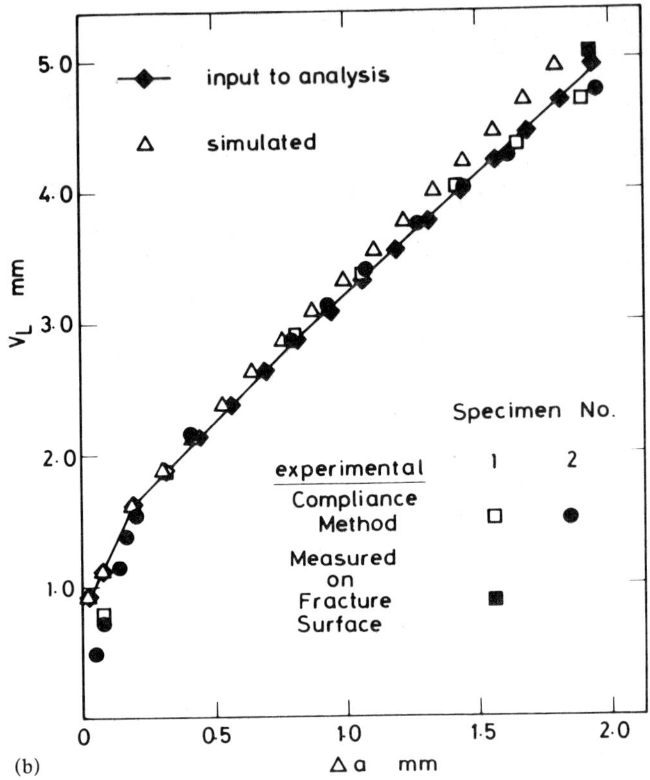

(b)

FIG. 18—*Continued.*

the increment of plastic work done associated with proportional loading is dominant, and then T_W defined in Eq 1 can be related to T_J as

$$T_W = \frac{n}{n+1} \frac{S_n}{I_n} T_J \tag{24}$$

When we use the relationship between T_J and T_δ [22], that is

$$T_J = \frac{1}{d_n} \frac{n+1}{n} T_\delta \tag{25}$$

for

$$\frac{r}{J} \frac{dJ}{da} \gg \frac{1}{n} \tag{26}$$

the relationship between T_W, T_J, and T_δ is obtained as

$$T_W = \frac{n}{n+1} \frac{S_n}{I_n} T_J = \frac{S_n}{I_n d_n} T_\delta \tag{27}$$

where d_n is a dimensionless constant which is a function of n and σ_0/E. Equation 27 is valid for the J-controlled crack growth in which crack growth is characterized by the J-integral based on the deformation theory of plasticity. Thus for the J-controlled crack growth with a limitation of a small crack extension, it was indicated that the tearing modulus T_W is equivalent to the other tearing parameters T_J and T_δ.

References

[*1*] Shih, C. F., deLorenzi, H. G., and Andrews, W. R. in *Elastic-Plastic Fracture, ASTM STP 668*, J. D. Landes, J. A. Begley, and G. A. Clarke, Eds., American Society for Testing and Materials, 1979, pp. 65–120.

[2] Kanninen, M. F., Rybicki, E. F., Stonesifer, R. B., Broek, D., Rosenfield, A. R., Marschall, C. W., and Hahn, G. T. in *Elastic-Plastic Fracture, ASTM STP 668*, J. D. Landes, J. A. Begley, and G. A. Clarke, Eds., American Society for Testing and Materials, 1979, pp. 121–150.

[*3*] Paris, P. C., Tada, H., Zahoor, A., and Ernst, H. in *Elastic-Plastic Fracture, ASTM STP 668*, J. D. Landes, J. A. Begley, and G. A. Clarke, Eds., American Society for Testing and Materials, 1979, pp. 5–36.

[*4*] Andrews, W. R. and Shih, C. F. in *Elastic-Plastic Fracture, ASTM STP 668*, J. D. Landes, J. A. Begley, and G. A. Clarke, Eds., American Society for Testing and Materials, 1979, pp. 426–450.

[*5*] Gudas, J. P., Joyce, J. A., and Davis, D. A. in *Fracture Mechanics, ASTM STP 677*, C. W. Smith, Ed., American Society for Testing and Materials, 1979, pp. 474–485.

[*6*] Loss, F. J., "Structural Integrity of Water Reactor Pressure Boundary Components," Quarterly Progress Report, Naval Research Laboratory, Washington, D. C., April–June 1979.

[*7*] Loss, F. J., "Structural Integrity of Water Reactor Pressure Boundary Components," Quarterly Progress Report, Naval Research Laboratory, Washington, D. C., April–June 1980.

[*8*] Takahashi, H., Khan, M. A., and Suzuki, M., *Journal of Testing and Evaluation*, Vol. 8, No. 2, March 1980, pp. 63–67.

[*9*] Hermann, L. and Rice, J. R., *Metal Science*, Vol. 14, Nos. 8 and 9, Aug.–Sept. 1980, pp. 285–291.

[*10*] Vassilaros, M. G., Joyce, J. A., and Gudas, J. P. in *Fracture Mechanics, ASTM STP 700*, American Society for Testing and Materials, 1980, pp. 251–270.

[*11*] Rice, J. R. and Sorensen, E. P., *Journal of the Mechanics and Physics of Solids*, Vol. 26, 1978, pp. 163–186.

[*12*] Rice, J. R., Drugan, W. J., and Sham, T-L. in *Fracture Mechanics, ASTM STP 700*, American Society for Testing and Materials, 1980, pp. 189–221.

[*13*] Amazigo, J. C. and Hutchinson, J. W., *Journal of the Mechanics and Physics of Solids*, Vol. 25, 1977, pp. 81–97.

[*14*] Shoji, T., *Journal of Testing and Evaluation*, Vol. 9, No. 6, Nov. 1981, pp. 324–334.

[*15*] Shoji, T., *Metal Science*, Vol. 10, No. 5, May 1976, pp. 165–169.

[*16*] Shoji, T., Takahashi, H., and Suzuki, M., *Metal Science*, Vol. 12, No. 12, Dec. 1978, pp. 579–586.

[*17*] Saxena, A. and Hudak, Jr., S. J., *International Journal of Fracture*, Vol. 14, No. 5, Oct. 1978, pp. 453–468.

[*18*] Ernst, H. and Paris, P. C., "Techniques of Analysis of Load-Displacement Records by J-Integral Methods," Report No. NUREG/CR-1222, U. S. Nuclear Regulatory Commission, Washington, D.C., Jan. 1980.

[*19*] Saka, M., Shoji, T., Takahashi, H., and Abé, H., *Journal of the Mechanics and Physics of Solids*, Vol. 30, 1982, pp. 209–224.

[*20*] McMeeking, R. M., *Journal of the Mechanics and Physics of Solids*, Vol. 25, 1977, pp. 357–381.

[*21*] Hibbitt, H. D., Marcal, P. V., and Rice, J. R., *International Journal of Solids and Structures*, Vol. 6, 1970, pp. 1069–1086.

[22] Shih, C. F., Andrews, W. R., German, M. D., VanStone, R. H., and Wilkinson, J. P. D., "Methodology for Plastic Fracture," 7th and 8th Quarterly Reports, General Electric Co., Schenectady, N. Y., Jan.–June 1978.

[23] Knott, J. F. in *Advances in Elasto-Plastic Fracture Mechanics*, L. H. Larsson, Ed., Applied Science Publishers, London, 1979, pp. 21–41.

[24] Shih, C. F., Andrews, W. R., German, M. D., Dean, R. H., and Wilkinson, J. P. D., "Methodology for Plastic Fracture," 9th and 10th Quarterly Reports, General Electric Co., Schenectady, N. Y., July–Dec. 1978.

[25] Hutchinson, J. W., *Journal of the Mechanics and Physics of Solids*, Vol. 16, 1968, pp. 13–31.

[26] Rice, J. R. and Rosengren, G. F., *Journal of the Mechanics and Physics of Solids*, Vol. 16, 1968, pp. 1–12.

Michael P. Wnuk[1]

Discontinuous Extension of Fracture in Elastic-Plastic Deformation Field

REFERENCE: Wnuk, M. P., **"Discontinuous Extension of Fracture in Elastic-Plastic Deformation Field,"** *Elastic-Plastic Fracture: Second Symposium, Volume I—Inelastic Crack Analysis, ASTM, STP 803*, C. F. Shih and J. P. Gudas, Eds., American Society for Testing and Materials, 1983, pp. I-159–I-175.

ABSTRACT: A procedure is developed whereby slow crack growth can be analyzed, and predictions are made regarding the transition from stable to spontaneous fracture in solids which are not susceptible to the linear elastic fracture mechanics formalism. Emphasis is placed on discerning the limit modes of fracture: cleavage versus tearing. These two extremes of material behavior, a very brittle one and a very ductile one, are described as certain limiting cases resulting from a single mathematical model. Certain numerical examples involving microstructural parameters are given to illustrate the use of the newly introduced quantities in analysis and in data reduction pertaining to the early stages of fracture.

KEY WORDS: ductile and brittle fractures, stable and unstable crack extension, essential work of fracture, final stretch, process zone, growth step, energy separation rate, stability analysis, elastic-plastic fracture

When fracture initiates in a solid capable of sustaining a stable crack growth over a certain range of the applied load, the apparent material resistance to cracking continues to rise steeply with crack advance. Such variations of fracture toughness cannot be explained solely on the basis of the linear elastic fracture mechanics (LEFM), and certain other mechanisms such as ductile void growth and coalescence, or creation of secondary microcracks some distance ahead of the front of a dominant crack, have to be admitted in order to explain the early stages of fracture development. It should be noted that even if the initiation parameters, such as K_{Ic} or J_{Ic}, are known from the observations made at the onset of crack growth, they alone are not sufficient means for predicting the instability which eventually terminates the

[1]Visiting professor, Technological Institute, Northwestern University, Evanston, Ill. 60201; presently at The University of Wisconsin, Milwaukee, Wis. 53211.

process of slow stable cracking. Therefore, it is necessary to devise a technique which would supply more complete information regarding material behavior in fracture occurring within the range in which the LEFM laws break down.

This study has been inspired by the recently proposed models for stable crack growth such as those described by Sih and Kiefer [*1*],[2] Wnuk and Sedmak [*2*], and Wnuk and Mura [*3*]. To describe a quasi-static crack extending in a stable manner prior to onset of spontaneous fracture propagation, a "fictitious crack" model is employed. The mathematics of the model is based on the "final stretch" concept which suggests that an incremental irreversible displacement is continually generated near the crack tip. As it turns out, in order to avoid certain physical inconsistencies (such as a vanishing energy release rate), it is necessary to discretize the process of crack growth and require that for each crack growth step an increment of displacement $\hat{\delta}$ is generated at a small distance Δ ahead of the crack front. Both $\hat{\delta}$ and Δ are considered to be microstructural parameters invariant to the geometry of fracture testing and reflecting the "true" or "local" material fracture resistance, as opposed by the "global" or "apparent" measures of toughness provided by such quantities as the critical level of a K-factor or a J-integral.

When this approach to quasi-static fracture is employed, the usual Griffith-Irwin criterion of fracture, $\mathcal{G} = \mathcal{G}_{Ic}$, is replaced by a certain differential equation

$$F(\mathcal{G}_R, a, d\mathcal{G}_R/da) = 0 \tag{1}$$

which describes variations of the apparent material resistance to cracking, denoted here by $\mathcal{G}_R$. The equality $\mathcal{G}_R = \mathcal{G}_{Ic}$, satisfied at the onset of stable crack growth, provides an initial condition for integration of Eq 1. Such a governing equation describing motion of a quasi-static crack results from the requirement $\mathcal{G}^\Delta = \hat{\mathcal{G}}$, in which the quantity on the left-hand side is a field characteristic representing the energy released from the strained solid during an incremental (but finite) crack extension, while the right-hand side, $\hat{\mathcal{G}} = \hat{\delta}\sigma_0$, represents the local or "true" material resistance to cracking. The symbol $\hat{\delta}$ is used to denote the final stretch and the "effective stress" σ_0 may be regarded as an average restraining stress $\langle S(x_1)\rangle$ which prevails within the end zone, $0 \leq x_1 \leq R$, while x_1 is the distance measured from the crack tip. When the apparent energy release rate $\mathcal{G}_R$ is replaced by the length of the end zone $R = R(a)$, and when the latter is used as a measure of material toughness, the energy released during a finite crack extension may be calculated as follows

$$\mathcal{G}^\Delta = 2\int_{t_1}^{t_2} \{S(x,t)\dot{u}_y(x,t)\}_{x=x_p}\,dt$$

[2]The italic numbers in brackets refer to the list of references appended to this paper.

$$= 2\int_{\Delta}^{0} S(x_1)\{du_y[x_1, R(x_1)]/dx_1\}\, dx_1 = F[R(a), dR/da, a] \quad (2)$$

Here $S(x_1)$ is the restraining force distribution prevailing over the end zone, while $\dot{u}_y(x_1, R)$ is the rate of change of the opening displacement within this zone as shown in Fig. 1. It is noted that the derivative dR/da serves as a measure of rate of redistribution of strains which occurs during subcritical crack propagation. We also note that in the classical limit of LEFM the rate dR/da vanishes. The coordinate of the control point P (see Fig. 1) is denoted by x_P and it should remain unchanged when the integral in Eq 2 is evaluated. Within the small-scale yielding range the opening displacement u_y for a Mode I crack is a known function provided that the restraining stress $S(x_1)$ is assumed constant throughout the nonlinear zone $0 \leq x_1 \leq R$. For such a simplified case the integral in Eq 2 reduces to

$$\mathcal{G}^{\Delta} = 2\sigma_0\,\{u_y[0, R(0)] - u_y[\Delta, R(\Delta)]\} \quad (2a)$$

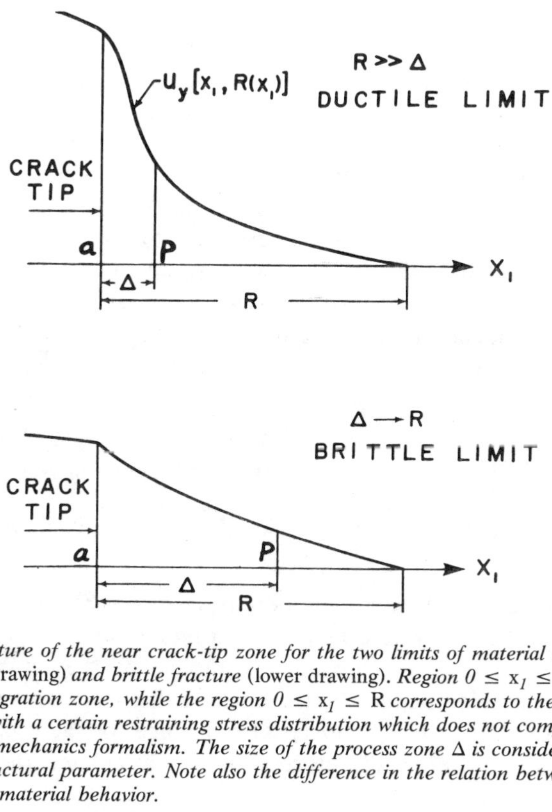

FIG. 1—*Structure of the near crack-tip zone for the two limits of material behavior: ductile fracture* (upper drawing) *and brittle fracture* (lower drawing). *Region* $0 \leq x_1 \leq \Delta$ *represents the process or disintegration zone, while the region* $0 \leq x_1 \leq R$ *corresponds to the entire nonlinear zone associated with a certain restraining stress distribution which does not comply with the continuum fracture mechanics formalism. The size of the process zone* Δ *is considered to be an important microstructural parameter. Note also the difference in the relation between* Δ *and* R *for the two limits of material behavior.*

Setting this quantity equal to the material property $\hat{\mathcal{G}} = \sigma_0\hat{\delta}$, the so-called "essential work of fracture," leads to the well-known final stretch criterion for the quasi-static tensile fracture, that is

$$u_y[0, R(0)] - u_y[\Delta, R(\Delta)] = \hat{\delta}/2 \tag{2b}$$

Function $F(dR/da, R, a)$ has a rather lengthy form, but it may be reduced (see Appendix) to certain asymptotic expressions corresponding to the two limits of material behavior, that is, brittle ($\Delta \rightarrow R$), and ductile ($\Delta \ll R$), namely

$$\mathcal{G}^{\Delta} = \frac{8\sigma_0{}^2}{\pi E_1} * \begin{cases} [\Delta\, dR/da + R - (2/3)\Delta(R/\Delta - 1)^{3/2}] & \text{brittle limit} \\ [\Delta\, dR/da + \Delta/2 + (\Delta/2)\log(4R/\Delta)] & \text{ductile limit} \end{cases} \tag{3}$$

We note that the first of the expressions written in the foregoing reduces to the classical Griffith-Irwin limit when dR/da is set zero and the size of the process zone is identified with the entire nonlinear zone, that is, when $\Delta = R$. The opposite extreme is observed when the process zone is deeply embedded within the end zone ($\Delta \ll R$); this situation corresponds to ductile fracture. In general, the degree of ductility associated with a fracture process will distinctly depend on the location of the control point P at which the irreversible deformation $\hat{\delta}$ is evaluated. This point lies within the end zone at the distance Δ from the crack front, coinciding with the outer edge of the process zone. Therefore, for $\Delta \rightarrow R$ we encounter brittle behavior, while for $\Delta \ll R$ the model describes the ductile fracture. As a suitable measure of the degree of ductility of a fracture process, we suggest the ratio of the toughness parameter R at the onset of stable crack growth (R_{ini}) to the size of the growth step, that is, $\rho_i = R_{\text{ini}}/\Delta$, the so-called ductility parameter.

Application of the criterion of fracture $\mathcal{G}^{\Delta} = \sigma_0\hat{\delta}$ leads to the following differential equations which define the resistance curve for each of the limiting cases considered here

$$\frac{dR}{da} = \begin{cases} M - R/\Delta + (2/3)(R/\Delta - 1)^{3/2} & \text{brittle limit} \\ M - 1/2 - (1/2)\log(4R/\Delta) & \text{ductile limit} \end{cases} \tag{4}$$

The symbol M denotes the tearing modulus, $M = (\pi/8)(E_1/\sigma_0)(\hat{\delta}/\Delta)$, which roughly equals 0.024 T_J, where T_J is the tearing modulus suggested by Paris.

Detailed discussion of the instabilities implied by these two equations is given in the companion paper [3]. To conclude this section we shall note that the maximum enhancement of material resistance to fracture attained during a quasi-static crack growth process may be calculated for both limiting cases as follows

$$m = \frac{R_{ss}}{R_{ini}} = \left(\frac{K_{ss}}{K_{ini}}\right)^2 = \begin{cases} f_1(\rho_i, \alpha), & \text{brittle case} \\ (4\rho_i)^{-1}\exp(2\alpha\rho_i - 1), & \text{ductile case} \end{cases} \tag{5}$$

The ratio m is a certain function of the ductility parameter, $\rho_i = R_{ini}/\Delta$, and of the toughness reduction index, $\alpha = \hat{\delta}/\delta_{ini}$. Introduction of the parameter α ($0 < \alpha \leq 1$) is necessitated by the experimentally observed drop in the resistance to crack extension immediately following termination of the crack blunting process, for example, Saka et al [4]. Reduction of material resistance to fracture which occurs at the onset of crack propagation is shown schematically in Fig. 2

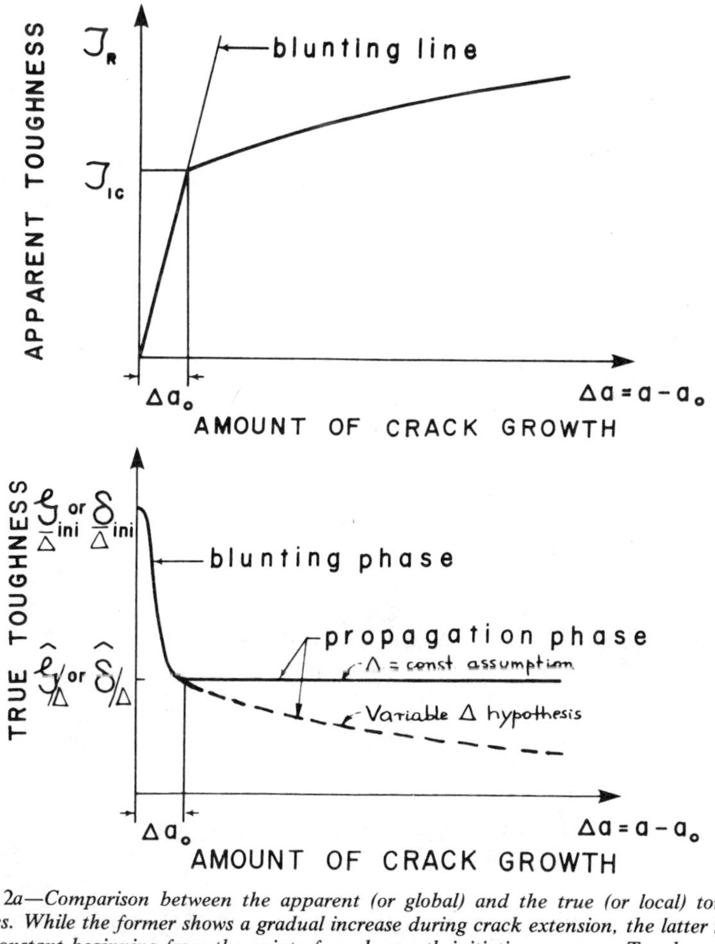

FIG. 2a—*Comparison between the apparent (or global) and the true (or local) toughness measures. While the former shows a gradual increase during crack extension, the latter remains nearly constant beginning from the point of crack growth initiation,* $a = a_0$. *Toughness reduction index* $\alpha = \hat{\mathcal{G}}/\mathcal{G}_{ini}$, *or* $\alpha = \hat{\delta}/\delta_{ini}$, *reflects the sudden drop of the local toughness parameter observed at the onset of crack growth. Symbol* Δa_0 *denotes the apparent crack extension determined at the termination of the blunting process.*

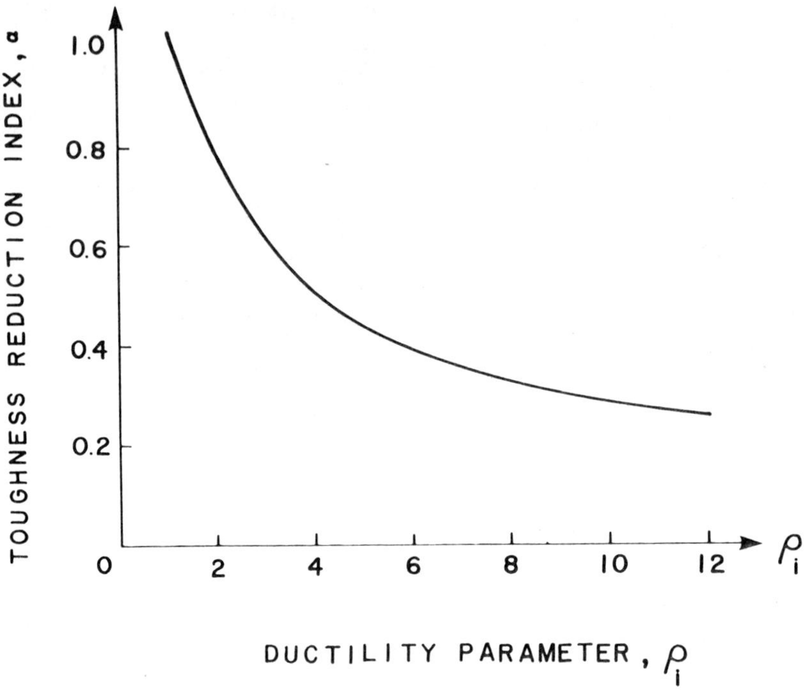

FIG. 2b—*Toughness reduction index as a function of the ductility parameter. Note that the roughness reduction index decreases as the ductility parameter* $\rho_i = R_{ini}/\Delta \simeq \epsilon^f/\epsilon_Y$ *increases. The function* $\alpha = \alpha(\rho_i)$ *plotted here is based on an analogy with Mode III fracture, namely,* $\alpha = (1/2\rho_i)(\ell n\, 4 + \sqrt{2\rho_i - 1})$.

Finally, the *lower* (K_{ini}) and the *upper* (K_{ss}) bounds for fracture toughness are predicted as follows

$$K_{ini} = 2\sqrt{\frac{2}{\pi}}\,\sigma_0\sqrt{\Delta}\,(\rho_i)^{1/2}, \quad \text{both modes}$$

$$K_{ss} = \begin{cases} 2\sqrt{\dfrac{2}{\pi}}\,\sigma_0\sqrt{\Delta}\,(\rho_i)^{1/2} f_1^{\,1/2}(\rho_i, \alpha), & \text{brittle limit} \\ \sqrt{\dfrac{2}{\pi}}\,\sigma_0\sqrt{\Delta}\,\exp(\alpha\rho_i - 1/2), & \text{ductile limit} \end{cases} \tag{6}$$

Applicability and interpretation of these fundamental relationships in view of microstructural considerations will be a subject of a follow-up paper.[3] Figure 3

[3]Wnuk, M. and Mura, T., *Mechanics of Materials*, 1983, Vol. 2, pp. 33–46.

shows enhancement of fracture toughness due to an increased ductility in both limits of the material behavior as predicted by Eqs 6.

Stability Analysis: Transition from Stable to Unstable Fracture

Extension of a quasi-static crack may be visualized as a sequence of local instability states which eventually lead to a catastrophic fracture, or to terminal instability, tantamount to the loss of global stability. Such transition from stable to unstable fracture propagation may be predicted from the considerations of energy exchange occurring at the tip of a moving crack. It is realized that while the load increases monotonically, the fracture criterion for a quasi-static extension of a crack has to be met continually. In terms of the energy release rate this criterion reads

$$\mathcal{G}^{\Delta} = \hat{\mathcal{G}} \tag{7}$$

The balance between the energy flowing into the crack tip and the energy absorbed within the process zone may also be expressed in terms of the field parameter R_A (index "A" is added here to emphasize the "applied" nature of the field intensity R), namely

$$R_A = R \tag{8}$$

In both of the foregoing equations, the left-hand side represents a certain measure of the energy supplied to the crack tip due to the work done by the external forces, while the right-hand side represents a demand for fracture energy, considered here to be a material property. Let us consider in a little more detail the second formulation of the fracture criterion, given by Eq 8. We note that during the stable crack growth phase both the quantities R_A and R are functions of the current crack length, while the intensity of the applied field, R_A, is also a function of the applied load Q. The material resistance to cracking, $R = R(a)$, is defined by one of the differential equations discussed in the preceding section. The applied R, on the other hand, depends on a in the following way

$$R_A = R_A[a, Q(a)] \tag{9}$$

or

$$R_A = R_A[a, \delta(a)] \tag{10}$$

in which Q denotes a nondimensional loading parameter, while δ is the displacement measured at the point of load application. The first of the preceding expressions applies to a load-controlled experiment, while the sec-

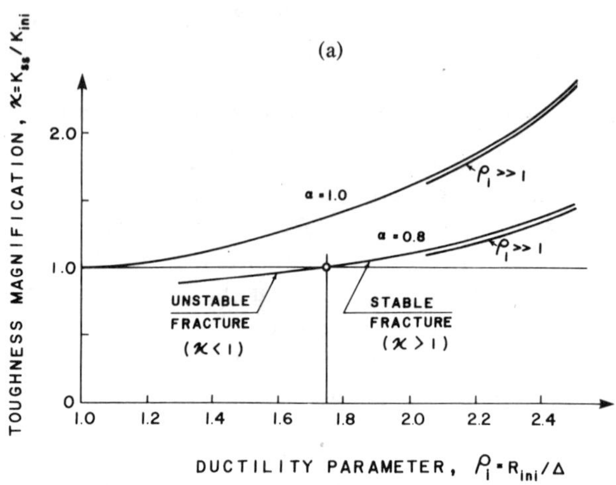

FIG. 3—*Toughness enhancement factor for Mode I fracture,* m = K_{ss}/K_{ini}, *plotted versus the ductility parameter* $\rho_i = R_{ini}/\Delta$. *The range* $1 \leq \rho_i < 2$ *corresponds to brittle behavior of material, that is, for the relatively low values of the ratio* $\rho_i \simeq \epsilon^f/\epsilon_Y$ = *strain at fracture/strain at yield. Beyond* $\rho_i = 2$ *the material response exhibits an increasing degree of ductility reflected by a growing span between the lower* (K_{ini}) *and the upper toughness bound* (K_{ss}).

(a) *Graphs* κ *versus* ρ_i *are obtained at a constant* α*-factor* ($\alpha = 1$ *and* 0.8), *assumed to be invariant to the variations in material ductility. It is noted that for the range shown there can be no stable fracture at* α *less than 0.7, since then* $\kappa < 1$. *The function* κ *is obtained by numerical solution of the equation*

$$\alpha\rho_i - \kappa^2\rho_i + \frac{2}{3}(\kappa^2\rho_i - 1)^{3/2} = 0$$

for the brittle limit ($\rho_i \to 1$), *or from the equation*

$$\alpha\rho_i - \kappa^2\rho_i + \kappa\sqrt{\rho_i(\kappa^2\rho_i - 1)} + \frac{1}{2}\log\frac{\kappa\sqrt{\rho_i} - \sqrt{\kappa^2\rho_i - 1}}{\kappa\sqrt{\rho_i} + \sqrt{\kappa^2\rho_i - 1}} = 0$$

for the intermediate range of the ductility parameter, and finally, from the asymptotic form valid for $\rho_i >> 1$ *(curves shown to the right of the figure):*

$$\kappa = \frac{1}{2\sqrt{\rho_i}}\exp\left(\alpha\rho_i - \frac{1}{2}\right)$$

(b) *Ductile material behavior. Semilogarithmic scale is used since the quantity* κ *attains rather large values if the* α*-factor is assumed to be independent of material ductility. Such an assumption does not appear to be supported by the experimental data. Curves shown here result from the equation*

$$\kappa = \frac{1}{2\sqrt{\rho_i}}\exp\left(\alpha\rho_i - \frac{1}{2}\right)$$

(c) *"Master curve"* κ *versus* ρ_i *represents the toughness enhancement factor as a single-valued function of the ductility parameter. Such unique "master curve" results from graphs shown in Fig.* 3b *if they are replotted with the toughness reduction index* α *assumed to be a function (see Fig.* 2b*) of material ductility. Equation defining the curve shown reads*

$$\kappa = \frac{1}{2\sqrt{\rho_i}}\exp\left\{\frac{1}{2}\left[\ln 4 - 1 + \sqrt{2\rho_i - 1}\,\right]\right\}$$

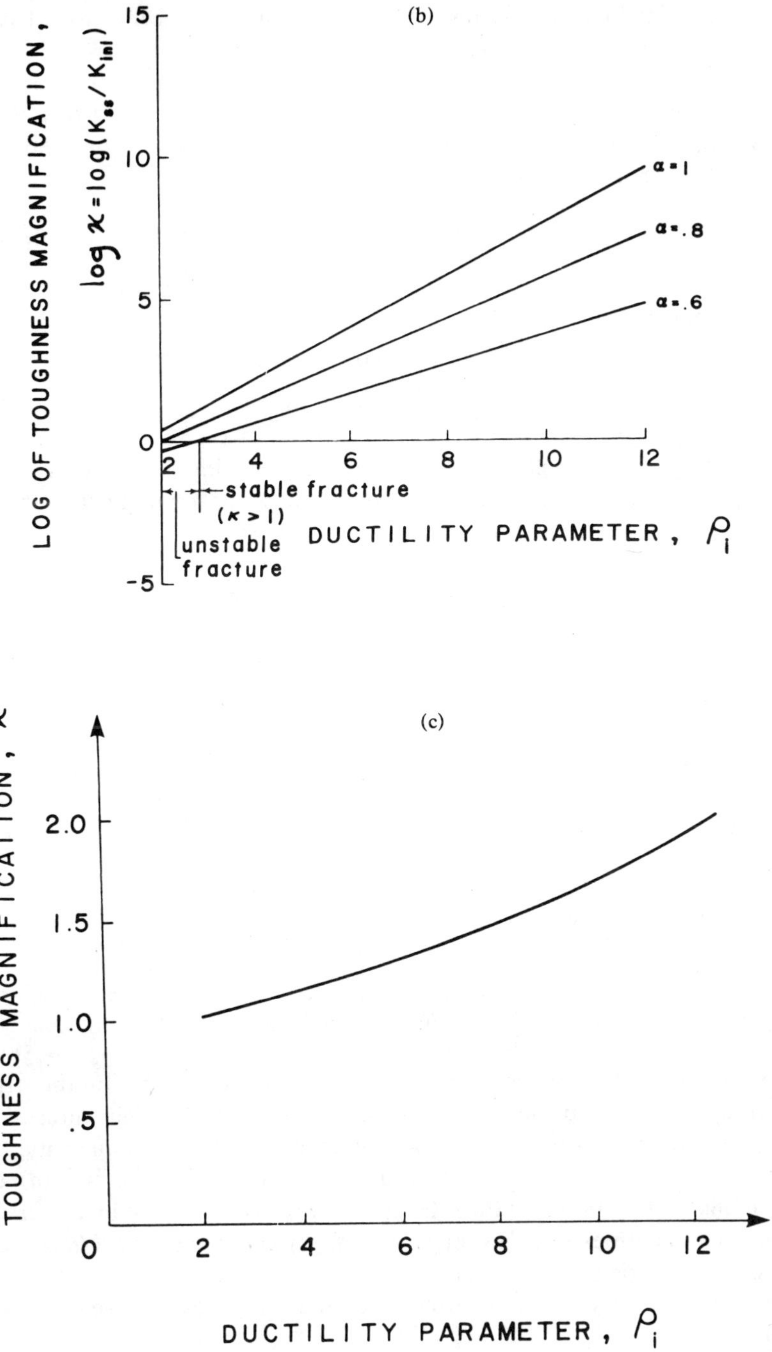

FIG. 3—*Continued.*

ond one pertains to a fracture test carried out under displacement-controlled condition.

The necessary condition for occurrence of the terminal instability may be formulated, compare Rice [9], by considering the second derivatives of the energy supplied and energy absorbed within the process zone. Considerations of this nature lead to the requirement

$$\frac{dR}{da} \geq \left.\frac{\partial R_A}{\partial a}\right)_Q \tag{11}$$

for a load-controlled test, and

$$\frac{dR}{da} \geq \left.\frac{\partial R_A}{\partial a}\right)_\delta \tag{12}$$

for a displacement-controlled test. The inequality sign implies stable crack extension, while the "equals" sign designates the onset of spontaneous fracture.

It is noted that vanishing of the difference

$$\frac{dR}{da} - \left(\frac{\partial R_A}{\partial a}\right)_Q \quad \text{or} \quad \frac{dR}{da} - \left(\frac{\partial R_A}{\partial a}\right)_\delta$$

corresponds to attainment of the terminal instability.

It is convenient to introduce a nondimensional quantity, referred to as a stability index [5] and defined as follows

$$\lambda_Q = \frac{dR}{da} - \left.\frac{\partial R_A}{\partial a}\right)_Q \tag{13}$$

for a prescribed loading parameter $Q = Q(t)$, and

$$\lambda_\delta = \frac{dR}{da} - \left.\frac{\partial R_A}{\partial a}\right)_\delta \tag{14}$$

for a situation in which a given displacement $\delta = \delta(t)$ is imposed on the structural component. The purpose of this study is to relate the stability indices, λ_Q and λ_δ, to the field parameters such as K-factor, the outer field intensity, R_A, and to the geometrical characteristics such as the shape factor $\phi = \phi(a)$ and the compliance function $Z = Z(a)$. Of course, the material resistance to cracking enters into both expressions for λ_Q and λ_δ through the term dR/da, which represents the slope of the resistance curve.

Let us consider first a load-controlled fracture test. Recalling the expression for the K-factor

$$K = \sqrt{a}\,\sigma\phi(a) \tag{15}$$

and the definition of the nondimensional loading parameter

$$Q = \pi\sigma/2\sigma_0 \tag{16}$$

we have

$$R_A = \frac{a}{2\pi} Q^2 \phi^2(a) \tag{17}$$

All that one needs to do now is to calculate the derivative

$$\left.\frac{\partial R_A}{\partial a}\right)_Q = \frac{Q^2}{2\pi}\left\{\phi^2 + 2a\phi\frac{d\phi}{da}\right\} = \frac{R_A}{a} + 2R_A\left(\frac{1}{\phi}\frac{d\phi}{da}\right) \tag{18}$$

Combining the latter expression with Eq 13 we obtain the desired form of the stability index

$$\lambda_Q = \frac{dR}{da} - \frac{R}{a}\left[1 + 2\frac{a}{\phi}\frac{d\phi}{da}\right] \tag{19}$$

Note that the index A has been dropped in Eq 19 since R_A and R must equal each other in order to satisfy the criterion for quasi-static fracture as given by Eq 8.

In a similar way, using Eq 10, we find the derivative $(\partial R_A/\partial a)_{\delta=\text{const.}}$. Before calculating this derivative, however, we must introduce the displacement at the point of load application δ into the expression for the K-factor, and subsequently into the formula for the R-parameter. This is done by replacing the loading parameter Q by the ratio δ/Z, in which the symbol Z denotes the compliance function, $Z = Z(a)$. Therefore, Eq 17 becomes now

$$R_A = \frac{a}{2\pi}\frac{\phi^2(a)}{Z^2(a)}\delta^2 \tag{20}$$

and

$$\left.\frac{\partial R_A}{\partial a}\right)_\delta = \frac{\delta^2}{2\pi}\left\{\frac{2a\phi}{Z^2}\frac{d\phi}{da} + \frac{\phi^2}{Z^2} - \frac{2a\phi^2}{Z^3}\frac{dZ}{da}\right\} = \frac{R_A}{a}\left\{1 + 2\frac{a}{\phi}\frac{d\phi}{da} - 2\frac{a}{Z}\frac{dZ}{da}\right\} \tag{21}$$

Inserting the latter expression into Eq 14 yields the desired form for the stability index

$$\lambda_\delta = \frac{dR}{da} - \frac{R}{a}\left\{1 + 2\frac{a}{\phi}\frac{d\phi}{da} - 2\frac{a}{Z}\frac{dZ}{da}\right\} \tag{22}$$

Again, the subscript A has been dropped.

It is of interest to note that the two stability indices, λ_Q and λ_δ, are related to each other by a simple formula

$$\lambda_\delta = \lambda_Q + 2\frac{R}{Z}\frac{dZ}{da} \tag{23}$$

On the other hand, differentiation of Eq 9 with respect to the current crack length gives

$$\frac{dR}{da} = \left.\frac{\partial R_A}{\partial a}\right)_Q + \frac{\partial R_A}{\partial Q}\frac{dQ}{da} \tag{24}$$

This relation implies the following useful identity

$$\frac{dR}{da} - \left.\frac{\partial R_A}{\partial a}\right)_Q = \lambda_Q = \frac{\partial R_A}{\partial Q}\frac{dQ}{da} \tag{25}$$

If we now substitute $(\partial R_A/\partial Q)(dQ/da)$ for λ_Q in Eq 23 and set λ_δ equal zero at the point of the terminal instability, we can demonstrate that the slope of the Q-curve at the terminal instability $(dQ/da)_f$ is a negative quantity which equals

$$\left(\frac{dQ}{da}\right)_f = \left(\frac{\partial R}{\partial Q}\right)^{-1}\left(-2\frac{R}{Z}\frac{dZ}{da}\right) \tag{26}$$

Since R is given by Eq 17, it is easily shown that

$$\frac{\partial R}{\partial Q} = 2R/Q \tag{27}$$

and thus the slope of the Q-curve at the critical point of transition to an unstable fracture, if the test is δ-controlled, is given by

$$\left(\frac{dQ}{da}\right)_f = -\frac{Q}{Z}\frac{dZ}{da} \tag{28}$$

This result was to be expected, since the terminal instability attained during a δ-controlled experiment is in fact determined by the requirement of vanishing slope $d\delta/da$. This slope, in turn, can be calculated from the simple relation $\delta = Z \cdot Q$, namely

$$\frac{d\delta}{da} = Z \frac{dQ}{da} + Q \frac{dZ}{da} \tag{29}$$

Hence, it follows

$$\left(\frac{dQ}{da}\right)_f = -\frac{Q}{Z}\frac{dZ}{da} \tag{30}$$

Which is identical to Eq 28. We shall conclude this section by noting that the pair of the critical parameters such as the load and the crack length at the point of transition to unstable fracture can be calculated by solving the set of equations

$$\begin{cases} R = R_A \\ \lambda_Q = 0 \end{cases} \tag{31}$$

for a load-controlled test, and

$$\begin{cases} R = R_A \\ \lambda_\delta = 0 \end{cases} \tag{32}$$

for a displacement-controlled test. Points marked by the little circles and by the stars shown in Figs. 4–6 were found numerically by seeking a solution to the set of Eqs 31, for the circles, and Eqs 32, for the stars. These three figures illustrate the rather strong effect of material ductility and the somewhat weaker effect of the specimen geometry on critical loads and crack sizes at which the transition between stable and unstable fracture takes place.

Conclusions

The model developed here is based on an assumption that the fracture zone formed during the early stages of fracture in inelastic solids is represented by a "fictitious" crack, or a "damage band" of finite width, preceding the dominant crack and capable to transfer stress. Such extension of a natural crack is referred to as a "nonlinear zone" in which the laws of the elastic-plastic fracture mechanics break down.

The stress-transferring ability of the damage zone depends on the separation distance created between the two opposite boundaries of a newly formed fracture zone. Although detailed information regarding the distribution of stress prevailing within the nonlinear zone is lacking, it is shown that certain plausible models may be constructed and employed to derive the differential equations which define a resistance curve.

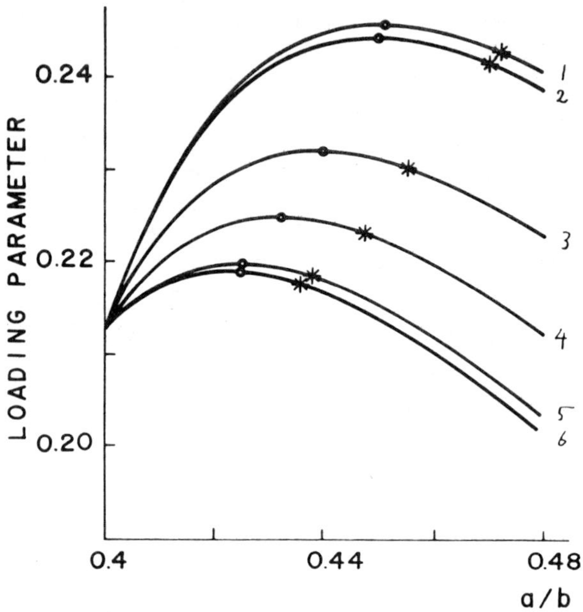

FIG. 4—*Dimensionless loading parameter plotted versus the ratio* a/b = *crack length/ width of the specimen for a single-edge-crack (SEN) configuration. Circles and stars denote the terminal instability points attained in a stress-controlled test (circles) and in a displacement-controlled test (stars). The uppermost curve represents ductile material behavior limit, while the bottom curve corresponds to brittle fracture. The asymptotic equations were used to plot these two opposite extremes, while for the intermediate range of the ductility parameter the complete R-curve equation has been used: namely, the equation* $\mathcal{G}^{\Delta} = \hat{\mathcal{G}}$ *in which the energy release rate* $\mathcal{G}^{\Delta}$ *is defined by Eq 35.* $\rho_i = 5$ *for Curves 1 and 2; 2.5 for Curve 3; 1.67 for Curve 4; and 1.1 for Curves 5 and 6.*

It is suggested that the "essential work of fracture" or so-called "final stretch," which serves as a measure of irreversible deformation, be calculated at each step of an advancing crack. Hence the variations of the toughness parameter with the extent of stable crack growth (an R-curve) may be predicted.

Such an approach requires that crack growth occur by a series of rapid but intermittent steps. Equations governing such discontinuous motion of a crack are provided, and the stability conditions for fracture developing either under a load-controlled situation or in a displacement-controlled test are briefly discussed.

Acknowledgments

Support of the Office of Naval Research under an ONR Grant No. N00014-81-K-0215 is gratefully acknowledged.

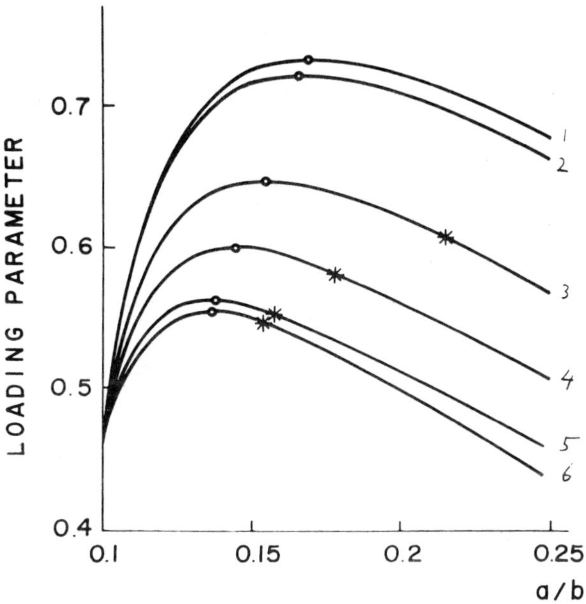

FIG. 5—*Dimensionless loading parameter plotted versus the ratio* a/b = *crack length/width of a three-point-bend specimen. Circles and stars denote the terminal instability points attained in a stress-controlled test (circles) and in a displacement-controlled test (stars). The uppermost curve represents ductile material behavior limit, while the bottom curve corresponds to brittle fracture. Note that the terminal instability points for a displacement-controlled test for curves* Q_1 *and* Q_2 *are beyond the range shown in the figure. Asymptotic equations were used to plot the two opposite extremes (curves* Q_1 *and* Q_6*), while for the intermediate range of the ductility parameter the complete R-curve equation has been used (*ρ_i *assumes the same values as in Fig. 4).*

APPENDIX

If the constant restraining stress $S(x_1)$ is assumed to prevail over the nonlinear zone, the model employed here is basically a modification of the Dugdale-Bilby-Cotterell-Swinden model which has been suggested for a *stationary* crack. The only important differences may be outlined as follows:

1. Time enters into the pertinent equations, since now the crack length and the extent of the plastic zone change while the crack propagates. A convenient way to represent time is to use either the crack length (a), or the distance (x_1) measured from a fixed point located on the perspective crack line to the approaching crack front.
2. The fracture criterion differs from the one used to describe the *initiation* of the crack extension. Either the work dissipated at the control point just prior to loss of cohesion of that point (the so-called essential fracture work, $\hat{\mathcal{G}}$) or the "final stretch" concept is employed to set up a condition for a continuing crack advancement. For the constant restraining stress, both criteria are equivalent as exemplified by Eqs 2*a* and 2*b* of the text.

In order to derive Eq 3, the opening displacement u_y is evaluated twice, first at the in-

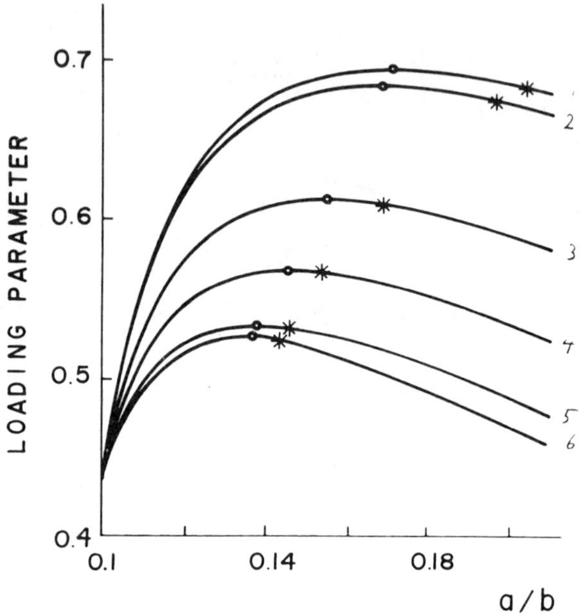

FIG. 6—*Dimensionless loading parameter plotted versus the ratio* a/b = *crack length/width of a four-point-bend specimen. Circles and stars denote the terminal instability points attained in a stress-controlled test (circles) and in a displacement-controlled test (stars). The uppermost curve represents ductile material behavior limit, while the bottom curve corresponds to brittle fracture. The asymptotic equations were used to plot these two opposite extremes, while for the intermediate range of the ductility parameter the complete R-curve equation has been used (ρ_i assumes the same values as in Fig. 4).*

stant t_1, that is, when the crack front is at the distance Δ from the control point P shown in Fig. 1, say $u_y(t_1)$, and second, at the instant t_2 when the point P coincides with the crack tip. Therefore, using Eq 9 we obtain

$$u_y(t_1) = \frac{4\sigma_0}{\pi E_1}\left\{\sqrt{R(R-x_1)} - \frac{x_1}{2}\log\frac{\sqrt{R}+\sqrt{R-x_1}}{\sqrt{R}-\sqrt{R-x_1}}\right\}_{\substack{x_1=\Delta \\ R=R(a)}} \tag{33}$$

and, when $x_1 = 0$

$$u_y(t_2) = \frac{4\sigma_0}{\pi E_1}\{R\}_{R=R(a+\Delta)} \simeq \frac{4\sigma_0}{\pi E_1}\left\{R + \Delta\frac{dR}{da}\right\}_{R=R(a)} \tag{34}$$

Substituting these expressions into Eq 2*a* yields the energy separation rate

$$\mathcal{G}^{\Delta} = \frac{8\sigma_0{}^2}{\pi E_1}\left\{R + \Delta\frac{dR}{da} - \sqrt{R(R-\Delta)} - \frac{\Delta}{2}\log\frac{\sqrt{R}+\sqrt{R-\Delta}}{\sqrt{R}-\sqrt{R-\Delta}}\right\}_{R=R(a)} \tag{35}$$

Finally, two asymptotic expansions performed for the case of $\Delta \rightarrow R$ and for $\Delta \ll R$ lead to Eq 3 given in the text as two representations of the extreme material behavior, the brittle one and the ductile one. It is noted that while for the former case the size of the Neuber zone (Δ) is on the same order of magnitude as the extent of the "plastic" or "cohesion" zone (R), in the latter case the Neuber zone is entirely embedded within the plastic zone. The crack growth step which is identified here with the process zone size may or may not be invariant to the amount of crack extension. The work presented here is based on the assumption of a constant growth step, but more recent experimental data seem to indicate a *variable* growth step (this case is indicated by a broken line in the lower half of Fig. 2*a*). Should the crack-tip opening angle (CTOA) be known, for example, through the direct measurements of the crack-tip configuration by an infiltration technique, then the second microstructural parameter, $\hat{\delta}$, may be estimated from the simple equality $\hat{\delta} = (\text{CTOA})\Delta$. The question which usually arises pertains to the determination of the length parameter Δ. This question has no easy answer, but a reasonable estimate of Δ may be given based on the studies of the plastic deformation field and on the considerations of the kinetics of microvoid formation [*8*]. Since Δ measures the extent of the intensive straining domain adjacent to the crack front it appears reasonable to equate it with the crack-tip opening displacement observed at the initiation of crack growth.

Future studies should address problems such as: (1) a more precise evaluation of the process zone size Δ, including its relation to the microstructure and the type of stress state generated around the crack tip. If only microstructure influences the process zone, then Δ will remain invariant to the crack extension. If, however, the near-tip state of stress has an effect on the process zone, then Δ will be a variable. This, of course, will alter the shape of the predicted R-curve. (2) Strain hardening may have an important effect on the shape of the resistance curve, too. To quantify this effect an extension of the present model is needed, so that some type of the strain hardening is incorporated into the stress-strain relations employed beyond the elastic limit.

References

[*1*] Sih, G. C. and Kiefer, B. V. in *Proceedings*, American Society of Mechanical Engineers Winter Annual Meeting, Vol. 35, Dec. 1979, pp. 135–156.

[*2*] Wnuk, M. P. and Sedmak, S. in *Fracture Mechanics: 13th Conference, ASTM STP 743*, American Society for Testing and Materials, 1981, pp. 236–249.

[*3*] Wnuk, M. P. and Mura, T. in *Fracture Mechanics: 14th Conference, ASTM STP 791*, American Society for Testing and Materials, 1982, pp. I-96–I-127.

[*4*] Saka, M., Shoji, T., Takahashi, H., and Abe, H., this publication, pp. I-130–I-158.

[*5*] Wnuk, M. P., *International Journal of Fracture Mechanics*, Vol. 15, No. 6, 1979, pp. 553–581.

[*6*] Wnuk, M. P., *Transactions*, American Society of Mechanical Engineers, *Journal of Applied Mechanics*, Vol. 41, 1974, pp. 234–242.

[*7*] Wnuk, M. P., *Transactions*, American Society of Mechanical Engineers, *Journal of Applied Mechanics*, Vol. 48, 1981, pp. 500–508.

[*8*] Curran, D. R., "Dependence of the J-R Curve on Microstructure," Poulter Laboratory Technical Report 003-81, SRI International, Menlo Park, Calif., 1981.

[*9*] Rice, J. R. in *Fracture: An Advanced Treatise*, H. Liebowitz, Ed., Vol. 2, Mathematical Fundamentals, Chapter 3, Academic Press, New York, 1968, pp. 191–311.

Yu-chen Gao[1]

Influence of Compressibility on the Elastic-Plastic Field of a Growing Crack

REFERENCE: Gao, Y.-C., **"Influence of Compressibility on the Elastic-Plastic Field of a Growing Crack,"** *Elastic-Plastic Fracture: Second Symposium, Volume I—Inelastic Crack Analysis, ASTM STP 803*, C. F. Shih and J. P. Gudas, Eds., American Society for Testing and Materials, 1983, pp. I-176–I-190.

ABSTRACT: The plastic singular field at the tip of a crack growing steadily in a perfectly plastic medium is solved in this paper by use of the basic equations and contiguity conditions given in a previous paper [*1*].[2] The fifth-order nonlinear differential equation has been integrated to a second-order equation. The five contiguity conditions are given in very simple form. By use of the general solution of the asymptotic equation and the contiguity conditions, the various possible cases of tip-field structures are discussed and the unique reasonable solution is obtained. For the angles of each sector, the numerical solution is given. In this paper, it is found that the strains also possess the singularity of $\ell n(A/r)$ in the sector ahead the crack tip, even if for the case of $\nu = 1/2$. Finally, a fracture criterion based on the analysis of singularity of strains is given.

KEY WORDS: elastic-plastic fracture, compatibility equation, contiguity conditions, unloading elastic sector, constant stress region, fan region

It is recognized more and more that the fracture criterion should be founded based on the analysis of the singularity of the stress and strain at the crack tip. But the singular field of a crack tip is a very difficult problem. Concerning the plasticity, creep, and dynamics, there is a lot of essential problems that have not been solved. For the most simple and important problem, namely, a crack growing steadily in a perfectly plastic medium, some results are obtained in special cases: Slepyan [*2*] adopted Tresca's yielding condition and Gao [*3*] adopted the Mises yielding condition and assumed $\nu = 1/2$. If the Mises yielding condition is adopted, when $\nu < 1/2$ the basic equations are

[1]Associate professor of mechanics, Harbin Shipbuilding Engineering Institute, Harbin, China.
[2]The italic numbers in brackets refer to the list of references appended to this paper.

elliptic, while when $\nu = 1/2$ the basic equations are hyperbolic. Therefore, in the two cases the tip fields essentially will be different from each other. Even when $\nu \rightarrow 1/2$, the tip field does not tend to that for the case of $\nu = 1/2$ obtained previously [*2,3*]. In Ref *1* it has been proved that the strains in the sector ahead the crack tip possess the singularity of $\ell n(A/r)$, but the whole crack tip field has not been obtained. In this paper, the elastic sectors and plastic sectors are assembled. The plastic sectors are composed of elliptic and hyperbolic regions. At the boundaries between the different regions all of the five contiguity conditions are satisfied.

Basic Equations

Let x, y denote the Cartesian coordinates: x^α = curvilinear coordinates; $g_{\alpha\beta}$ = fundamental metric tensor; $\sigma^{\alpha\beta}$ = stress, $\epsilon_{\alpha\beta}$ = strain; ϕ = stress function; and λ = flow factor. Then we have the following

Stress

$$\sigma_{\alpha\beta} = g_{\alpha\beta} \cdot \Delta\phi - \nabla_\alpha \nabla_\beta \phi \tag{1}$$

where

$$\Delta = \nabla_\alpha \nabla^\alpha.$$

Constitutive Relations

We consider the plane-strain problem only. From $\epsilon_z = 0$, we can obtain

$$\sigma_z = \nu\sigma_\alpha{}^\alpha + \epsilon\sigma_p, \qquad \epsilon = \frac{1}{2} - \nu \tag{2}$$

in which

$$\begin{cases} \sigma_p = \dfrac{2E}{3} \exp\left(-\dfrac{2E}{3}\Lambda\right) \displaystyle\int_{t_0}^{t} \lambda\sigma_\alpha{}^\alpha \cdot \exp\left(\dfrac{2E}{3}\Lambda\right) dt \\ \Lambda = \displaystyle\int_{t_0}^{t} \lambda dt \end{cases} \tag{3}$$

where t_0 denotes the initial yielding time.

Using Eq 2 and Reuss's flow relation, we have

$$\epsilon_{\alpha\beta} = \epsilon_{\alpha\beta}{}^e + \epsilon_{\alpha\beta}{}^p \tag{4}$$

$$\epsilon_{\alpha\beta}{}^e = \frac{1+\nu}{E}\sigma_{\alpha\beta} - \frac{\nu}{E}(1+\nu) g_{\alpha\beta}\sigma_\gamma{}^\gamma \tag{5}$$

$$\epsilon_{\alpha\beta}{}^{p} = \frac{\epsilon^2}{E} g_{\alpha\beta} \cdot \sigma_p + \int_{t_0}^{t} \lambda \left(\sigma_{\alpha\beta} - \frac{1}{2} g_{\alpha\beta} \cdot \sigma_\gamma{}^\gamma \right) dt \tag{6}$$

Yielding Condition

Using Eq 2, the Mises yielding condition can be written in the form

$$\frac{1}{2} \nabla^\alpha \nabla^\beta \phi \cdot \nabla_\alpha \nabla_\beta \phi - \frac{1}{4} (\Delta\phi)^2 + \frac{\epsilon^2}{3} (\Delta\phi - \sigma_p)^2 = k^2 \tag{7}$$

in which k is the shear yielding stress.

Compatibility Equation

$$\frac{1 - \nu^2}{E} \Delta\Delta\phi + \Delta\epsilon_\alpha{}^{\alpha p} - \nabla^\alpha \nabla^\beta \epsilon_{\alpha\beta}{}^{p} = 0 \tag{8}$$

or

$$\frac{1 - \nu^2}{E} \Delta\Delta\dot{\phi} + \Delta\dot{\epsilon}_\alpha{}^{\alpha p} - \nabla^\alpha \nabla^\beta \dot{\epsilon}_{\alpha\beta}{}^{p} = 0 \tag{9}$$

in which the dot denotes time derivative.

If $\nu < 1/2$, using Eqs 3 and 7, we obtain

$$\lambda = \frac{3}{2E\zeta} (\Delta\dot{\phi} - \dot{\zeta}) \tag{10}$$

$$\zeta = \Delta\phi - \sigma_p = \frac{\pm\sqrt{3}}{\epsilon} \left\{ k^2 + \frac{1}{4} (\Delta\phi)^2 - \frac{1}{2} \nabla^\alpha \nabla^\beta \phi \cdot \nabla_\alpha \nabla_\beta \phi \right\}^{1/2} \tag{11}$$

Finally, the compatibility equation can be written as

$$\frac{1 - \nu^2}{E} \Delta\Delta\dot{\phi} + \nabla^\alpha \nabla^\beta \phi \cdot \nabla_\alpha \nabla_\beta \lambda - \Delta\phi \cdot \Delta\lambda + \frac{1}{2} \Delta \left\{ \lambda \left(\Delta\phi + \frac{4}{3} \epsilon^2 \zeta \right) \right\} = 0 \tag{12}$$

Contiguity Conditions

Let $[\Phi]_r$ denote the jump of Φ across any curve Γ. Because of the continuity of two stresses and displacements, we can obtain the following conditions

$$[\phi]_r = \left[\frac{\partial \phi}{\partial n}\right]_r = \left[\frac{\partial^2 \phi}{\partial n^2}\right]_r = 0 \tag{13}$$

$$\frac{1-\nu^2}{E}\left[\frac{\partial^3 \phi}{\partial n^3}\right]_r - \frac{1}{V}[\lambda]_r\left\{\frac{1}{2}(\sigma_s - \sigma_n) + \frac{2}{3}\epsilon^2(\Delta\phi - \sigma_p)\right\}$$
$$- 2\frac{d}{ds}[\epsilon_{ns}{}^p]_r - \frac{d\vartheta}{ds}[\epsilon_n{}^p]_r = 0 \tag{14}$$

where ϑ denotes the angle from the x-axis to the normals of Γ and V is the normal component of the velocity of Γ. If t^* denotes the time when Γ sweeps the point under consideration, $t^* = t^*(x^\alpha)$, then

$$\frac{1}{V} = \frac{\partial t^*}{\partial n} \tag{15}$$

If $\dot{\phi}$ is taken as the fundamental unknown, then we have the fifth contiguity condition

$$\frac{1-\nu^2}{E}\left[\frac{\partial^3 \dot{\phi}}{\partial n^3}\right]_r + \left[\frac{\partial \dot{\epsilon}_s{}^p}{\partial n}\right]_r - 2\frac{d}{ds}[\dot{\epsilon}_{ns}{}^p]_r - \frac{d\vartheta}{ds}[\dot{\epsilon}_n{}^p]_r$$
$$+ 2V\frac{d^2\vartheta}{ds^2}[\epsilon_{ns}{}^p]_r = 0 \tag{16}$$

For the case of a crack growing steadily, we have

$$(\cdot) = \frac{\partial}{\partial t} = -\frac{\partial}{\partial x}, \qquad V = \cos\vartheta \tag{17}$$

therefore, Eqs 14 and 16 can be written as

$$\frac{1-\nu^2}{E}\left[\frac{\partial^3 \phi}{\partial n^3}\right]_r - \frac{[\lambda]_r}{\cos\vartheta}\left\{\frac{1}{2}(\sigma_s - \sigma_n) + \frac{2}{3}\epsilon^2(\Delta\phi - \sigma_p)\right\}$$
$$- 2\sigma_{ns}\left\{\frac{d}{ds}[\Lambda]_r - \tan\vartheta \cdot \frac{d\vartheta}{ds}[\Lambda]_r\right\} = 0 \tag{18}$$

$$\frac{1-\nu^2}{E}\left[\frac{\partial^4 \phi}{\partial n^4}\right]_r - \frac{1-\nu^2}{E}\tan\vartheta\,\frac{d}{ds}\left[\frac{\partial^3 \phi}{\partial n^3}\right]_r$$
$$- \frac{1}{\cos\vartheta}\left[\frac{\partial}{\partial n}\left\{\lambda\left(\frac{\sigma_s - \sigma_n}{2} + \frac{2\epsilon^2}{3}(\Delta\phi - \sigma_p)\right)\right\}\right]_r$$

$$+ \frac{1}{\cos \vartheta} \frac{d\vartheta}{ds} [\lambda]_r \left\{ \frac{1}{2} (\sigma_n - \sigma_s) + \frac{2}{3} \epsilon^2 (\Delta\phi - \sigma_p) \right\}$$

$$+ \frac{2}{\cos \vartheta} \frac{d}{ds} \{ \sigma_{ns} [\lambda]_r \} - 2 \frac{d^2\vartheta}{ds^2} \sigma_{ns} [\Lambda]_r = 0 \tag{19}$$

Asymptotic Equations

According to the results obtained in Ref *1*, for the first approximation, we have

$$\phi = r^2 \cdot f(\theta) \tag{20}$$

$$\lambda = \frac{1}{r} \lambda(\theta) \tag{21}$$

We assume that $\zeta \neq 0$, namely

$$f'^2 + \frac{1}{4} f''^2 < k^2 \tag{22}$$

then from Eqs 3 and 7 we can obtain

$$\begin{cases} \zeta = \pm \dfrac{\sqrt{3}}{\epsilon} \left(k^2 - f'^2 - \dfrac{1}{4} f''^2 \right)^{1/2} \\ \lambda(\theta) = \dfrac{3}{2E} \cdot \dfrac{\sin \theta}{\zeta} (f''' + 4f') \cdot \left(1 + \dfrac{3f''}{4\epsilon^2 \zeta} \right) \end{cases} \tag{23}$$

The compatibility equation can be written as

$$\left(\frac{d^2}{d\theta^2} + 1 \right) \left\{ \sin \theta (f''' + 4f') \left[1 - \nu^2 + \epsilon^2 \left(1 + \frac{3f''}{4\epsilon^2 \zeta} \right)^2 \right] \right\} = 0 \tag{24}$$

Let

$$F(\theta) = f'(\theta) \tag{25}$$

then we obtain

$$\frac{1}{E} \left[1 - \nu^2 + \epsilon^2 \left(1 + \frac{3F'}{4\epsilon^2 \zeta} \right)^2 \right] \cdot (F'' + 4F) = C_1 \cot \theta + C_2 \tag{26}$$

$$\zeta = \pm \frac{\sqrt{3}}{\epsilon}\left(k^2 - F^2 - \frac{1}{4}F'^2\right)^{1/2} \tag{27}$$

In Eq 26 if we take $C_1 = C_2 = 0$ then we obtain

$$\begin{cases} f(\theta) = a \pm \dfrac{k}{2}\cos 2(\theta - \theta_0) \quad \text{or} \\ f(\theta) = b \pm k\theta \end{cases} \tag{28}$$

The solutions shown in Eq 28 are just that obtained for the case of $\nu = 1/2$.
Using Eq 23, we can rewrite Eq 26 in the form

$$\frac{1-\nu^2}{E}(F'' + 4F) + \frac{\lambda}{\sin\theta}\left(\frac{F'}{2} + \frac{2}{3}\epsilon^2\zeta\right) = C_1\cot\theta + C_2 \tag{29}$$

Asymptotic Contiguity Conditions

For the discontinuity lines starting from the crack tip, we assume that

$$\begin{cases} \dfrac{d\vartheta}{ds} \ll \dfrac{1}{r}, \qquad [\epsilon_{r\theta}{}^{p}]_r = \Lambda_0 \ln\dfrac{A}{r} \\ \lambda = \dfrac{1}{r}\lambda(\theta) + \dfrac{1}{r}\ln\dfrac{A}{r}\cdot\lambda^*(\theta) \end{cases} \tag{30}$$

where A is a constant determined by the size of the whole plastic domain.
We consider the predominant singular terms only; using Eq 30, the contiguity condition Eqs 18 and 19 can be simplified as

$$\begin{cases} \dfrac{1-\nu^2}{E}[f''']_r + \dfrac{1}{\sin\theta}\left[\left\{\lambda(\theta) + \ln\dfrac{A}{r}\lambda^*(\theta)\right\}\left(\dfrac{f''}{2} + \dfrac{2\epsilon^2}{3}\zeta\right)\right]_r = 2\Lambda_0 f' \\ \dfrac{1-\nu^2}{E}\{[f'''']_r + \cot\theta[f''']_r\} + \dfrac{1}{\sin\theta}\left[\dfrac{\partial}{\partial\theta}\left\{\left(\lambda(\theta) + \ln\dfrac{A}{r}\cdot\lambda^*(\theta)\right)\right.\right. \\ \quad \left.\left.\cdot\left(\dfrac{f''}{2} + \dfrac{2}{3}\epsilon^2\zeta\right)\right\}\right]_r = \dfrac{2f'}{\sin\theta}[\lambda^*(\theta)]_r \end{cases} \tag{31}$$

Besides, Eq 13 becomes

$$[f]_r = [f']_r = [f'']_r = 0 \tag{32}$$

Using Eqs 26 and 29 we can rewrite Eq 31 as

$$\begin{cases} [C_1]_r \cot\theta + [C_2]_r = 2\Lambda_0 F(\theta) \\ -[C_1]_r + [C_2]_r \cot\theta = \dfrac{2F(\theta)}{\sin\theta}\,[\lambda^*(\theta)]_r \end{cases} \tag{33}$$

where C_1 and C_2 are constants in each sector. It should be noted that $\lambda^*(\theta)$ is constant in the fan sector and is zero in other sectors.

Various Assembly of Sectors

According to the proof given in Ref *1*, the sector ① in Figs. 1–3 ahead the crack tip must be a constant stress region, in which we have

$$f(\theta) = B + \frac{k}{2}\cos 2\theta \tag{34}$$

$$\lambda(\theta) = \frac{e\cdot\cos\theta}{\cos 2\theta}, \qquad \epsilon_y{}^p = k\cdot e\left(\ell\text{n}\,\frac{A}{r} + \frac{1}{2}\,\ell\text{n}\,\frac{1}{\cos 2\theta}\right) + 0(1) \tag{35}$$

From Eqs 34, 35, and 29, we obtain

$$C_1^{①} = -\,k\cdot e, \qquad C_2^{①} = 0 \tag{36}$$

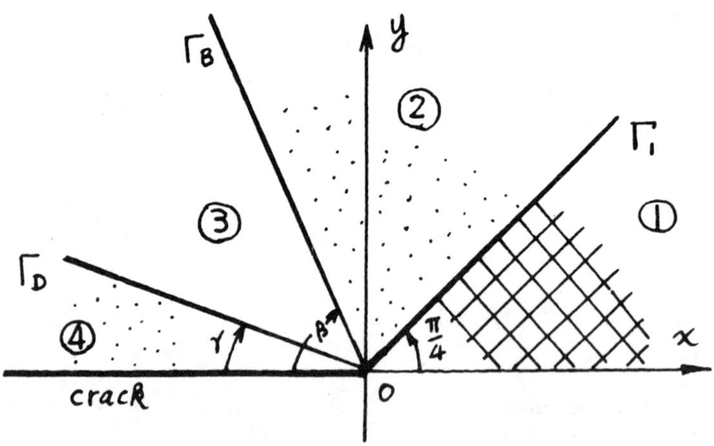

FIG. 1—*The first pattern.*

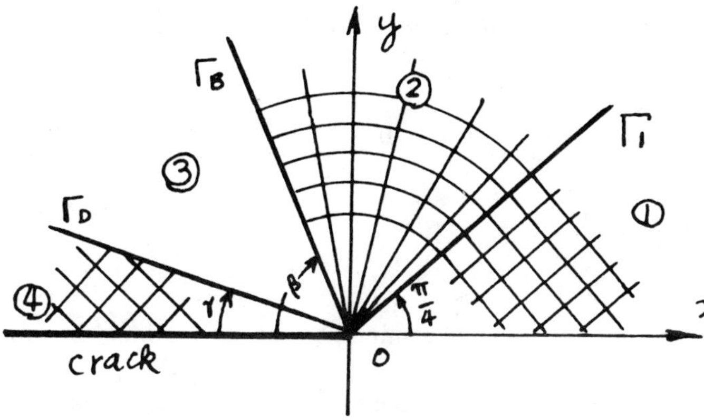

FIG. 2—*The second pattern.*

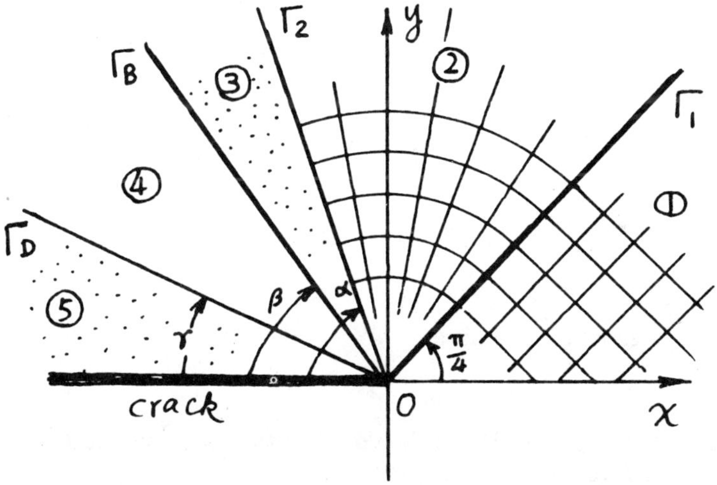

FIG. 3—*The third pattern.*

Evidently, when $\theta \to \pi/4$ Eq 35 is not valid; therefore, there will be a complex sector (inner boundary layer) at Γ_1.

In the following, we shall discuss all of the possible cases of other sectors.

The First Pattern

We assume that Sector ②, Fig. 1, is a general region, namely, that it is described by the general solution of Eq 26. Also, C_1 and C_2 are not both equal to zero. Then using Eq 32 we have

$$\begin{cases} f\left(\frac{\pi}{4}+0\right) = B, \qquad f'\left(\frac{\pi}{4}+0\right) = -k \\ f''\left(\frac{\pi}{4}+0\right) = 0 \end{cases} \tag{37}$$

Using Eq 33 we have

$$C_1^{②} = -k(e + \Lambda_0), \qquad C_2^{②} = -k\,\Lambda_0 \tag{38}$$

where Λ_0 is a constant to be determined by solving the inner boundary layer at Γ_1.

Further, from Eq 37 we have the following expansion near Γ_1 and in Sector ②

$$\begin{cases} f'''\left(\frac{\pi}{4}+\Delta\theta\right) = k \cdot R^* \\ f'\left(\frac{\pi}{4}+\Delta\theta\right) = -k + \frac{k}{2}(\Delta\theta)^2 R^* \\ f''\left(\frac{\pi}{4}+\Delta\theta\right) = \Delta\theta \cdot R^* k \\ \Delta\theta = \theta - \frac{\pi}{4} \end{cases} \tag{39}$$

From Eqs 22 and 39 we have

$$0 < R^* < 4 \tag{40}$$

Because $\lambda > 0$ (see Eq 23), we should select the minus sign in Eq 27. From Eqs 27 and 39 we obtain

$$\zeta = -\frac{\sqrt{3}}{2\epsilon} k \cdot \Delta\theta \sqrt{R^*(4 - R^*)} \tag{41}$$

Using Eqs 39, 41, and 31, and noting $\zeta \equiv 0$ in Sector ①, we have

$$e = \frac{\eta}{E}(4 - R^*) - 2\Lambda_0 \tag{42}$$

$$\eta = 1 - \nu^2 + \epsilon^2\left(1 + \frac{3f''}{4\epsilon^2\zeta}\right)^2 = 1 - \nu^2 + \epsilon^2\left(1 - \frac{\sqrt{3}}{2\epsilon} \cdot \sqrt{\frac{R^*}{4 - R^*}}\right)^2 \tag{43}$$

Sector ③ is assumed to be an unloading elastic sector described by

$$\frac{1-\nu^2}{E}(F''+4F)=C_1^{③}\cot\theta+C_2^{③} \tag{44}$$

The boundary Γ_B between Sectors ③ and ② is not an inner boundary layer. Using Eqs 33 at Γ_B we have

$$C_1^{③}=C_1^{②}, \qquad C_2^{③}=C_2^{②} \tag{45}$$

The position of Γ_B is determined by the unloading condition $\lambda|_{\Gamma_B}=0$, namely

$$\left(1+\frac{3f''}{4\epsilon^2\zeta}\right)_{\theta=\pi-\beta-0}=0 \tag{46}$$

Sector ④ is assumed to be a reloading (second) plastic sector in which $C_1^{④}=C_1^{③}$ and $C_2^{④}=C_2^{③}$. And the boundary Γ_D between Sectors ③ and ④ is determined by the reloading condition

$$\left(f'^2+\frac{1}{4}f''^2+\frac{\epsilon^2}{3}\zeta^2\right)_{\Gamma_D}=\left(f'^2+\frac{1}{4}f''^2+\frac{\epsilon^2}{3}\zeta^2\right)_{\Gamma_B} \tag{47}$$

It should be noted that σ_p = const. in Sector ③.

Because Λ_0 at Γ_1 is unknown before the inner boundary layer is solved, we take R^* and Λ_0 as both free parameters in the numerical calculation. The numerical results show that there are no proper values of R^* and Λ_0 that can satisfy the following boundary and plastic conditions

$$\begin{cases} f'(\pi)=0 \\ e\geq 0 \end{cases} \tag{48}$$

So, this assembly of sectors is denied.

The Second Pattern

We assume that the shape of the assembly of sectors is just like that for the case of $\nu = 1/2$ obtained before [3]; that is, in Fig. 2 Sector ② is a fan sector in which

$$\begin{cases} f(\theta)=B+k\cdot\left(\frac{\pi}{4}-\theta\right) \\ \lambda^*(\theta)=\text{const.}=\lambda^* \end{cases} \tag{49}$$

Equation 32 at Γ_1 is satisfied already.

Using Eqs 29, 10, and 49, we obtain

$$\begin{cases} C_1^{②} = 0 \\ C_2^{②} = -\dfrac{4k}{E}(1-\nu^2+\epsilon^2) \end{cases} \tag{50}$$

From Eqs 50, 33, and 31, we have

$$\begin{cases} \lambda^* = \dfrac{2\sqrt{2}}{E}(1-\nu^2+\epsilon^2) - \dfrac{1}{\sqrt{2}}\Lambda_0 \\ e = \dfrac{4}{E}(1-\nu^2+\epsilon^2) - 2\Lambda_0 \end{cases} \tag{51}$$

where Λ_0 is a constant as mentioned previously.

We assume that Sector ③ is an unloading elastic sector in which the governing equation is Eq 44. The boundary Γ_B between Sectors ③ and ② is an unloading boundary for which $\Lambda_0 = 0$. From Eq 33 at Γ_B, we obtain

$$\begin{cases} C_1^{③} = -2k\lambda^* \cdot \sin\beta \\ C_2^{③} = -\dfrac{4k}{E}(1-\nu^2+\epsilon^2) - 2k\lambda^* \cdot \cos\beta \end{cases} \tag{52}$$

Noting the continuity of f'' at Γ_B, we have

$$\{f''(f''' + 4f')\}_{\theta=\pi-\beta+0} = 0 \tag{53}$$

$$\left\{\frac{d^2}{d\theta^2}\left(f'^2 + \frac{1}{4}f''^2\right)\right\}_{\theta=\pi-\beta+0} = \frac{1}{2}f'''(f''' + 4f')_{\theta=\pi-\beta+0} \tag{54}$$

On the other hand, from Eq 44, in Sector ③ we have

$$f'''(f''' + 4f')|_{\theta=\pi-\beta+0} = \frac{E}{1-\nu^2}\left(C_2^{③} - C_1^{③}\cot\beta\right)$$

$$\cdot\left\{\frac{E}{1-\nu^2}(C_2^{③} - C_1^{③}\cdot\cot\beta) + 4k\right\}$$

$$= 16k^2\frac{\epsilon^2}{(1-\nu^2)^2}\cdot(1-\nu^2+\epsilon^2) > 0 \tag{55}$$

From Eqs 53–55, we obtain

$$\left(f'^2 + \frac{1}{4} f''^2\right)_{\theta > \pi - \beta} > k^2 \tag{56}$$

Therefore, when $\theta > \pi - \beta$, the yielding condition will be broken. So this assembly of sectors is denied, too.

The Third Pattern

We assume that Sector ① is a constant-stress region, Fig. 3, Sector ② is a fan-region, Sector ③ is a general region in which $\zeta < 0$, Sector ④ is an unloading elastic region, and Sector ⑤ is a general region in which $\zeta < 0$, too. The solutions in Sectors ① and ② are given by Eqs 34–36 and Eqs 49 and 50. The boundary Γ_2 between Sectors ② and ③ is an inner boundary layer, too. The position of Γ_2 is to be determined.

From Eq 32 at Γ_2, we have

$$\begin{cases} f'(\pi - \alpha + 0) = -k \\ f''(\pi - \alpha + 0) = 0 \end{cases} \tag{57}$$

In order that the yielding condition is not destroyed in Sector ③ we must take

$$\begin{cases} f'''(\pi - \alpha + 0) = 0 \\ \Lambda_0 = 0, \quad \text{at } \Gamma_2 \end{cases} \tag{58}$$

Further analysis shows that we must take

$$f''''(\pi - \alpha + 0) = 0 \tag{59}$$

Therefore, we have the following expansion near Γ_2

$$\begin{cases} f'''''(\pi - \alpha + 0) = R \cdot k \\ f' = -k + \dfrac{kR}{4!}(\Delta\theta)^4, \qquad f'' = \dfrac{kR}{3!}(\Delta\theta)^3 \\ \dfrac{\epsilon^2}{3}\zeta^2 = \dfrac{k^2}{12}\cdot R(\Delta\theta)^4, \qquad \dfrac{3f''}{4\epsilon^2\zeta} = -\dfrac{\Delta\theta}{4\epsilon}\sqrt{R} \\ \Delta\theta \mp \theta - (\pi - \alpha) \end{cases} \tag{60}$$

The fourth contiguity condition is satisfied already and from the fifth contiguity condition we obtain

$$R = \frac{9(1 - \nu^2 + \epsilon^2)^2}{2\epsilon^2 \sin^2 \alpha} \tag{61}$$

Besides, in Sector ③, $C_1^{③}$ and $C_2^{③}$ are given by

$$C_1^{③} = -2k\lambda^* \sin\alpha, \qquad C_2^{③} = -\frac{4k}{E}(1 - \nu^2 + \epsilon^2) - 2k\lambda^* \cos\alpha \tag{62}$$

$$C_1^{④} = C_1^{⑤} = C_1^{③}, \qquad C_2^{④} = C_2^{⑤} = C_2^{③} \tag{63}$$

In order to finish the numerical calculation, we must determine the unknown Λ_0 at Γ_1 included in Eq 51. But the work to analyze the inner boundary layer at Γ_1 has not yet been finished. According to the primary result obtained by the author, Λ_0 should be taken as the following

$$\Lambda_0 = \frac{1}{2}e, \quad \text{at } \Gamma_1 \tag{64}$$

then we have

$$\begin{cases} e = \dfrac{2}{E}(1 - \nu^2 + \epsilon^2) \\ C_1^{①} = -\dfrac{2k}{E}(1 - \nu^2 + \epsilon^2), \qquad C_2^{①} = 0 \\ C_1^{②} = 0, \qquad C_2^{②} = -\dfrac{4k}{E}(1 - \nu^2 + \epsilon^2) \\ C_1^{③} = -\dfrac{3}{E}\sqrt{2}(1 - \nu^2 + \epsilon^2)k \sin\alpha \\ C_2^{③} = -\dfrac{k}{E}(1 - \nu^2 + \epsilon^2)\cdot(4 + 3\sqrt{2}\cos\alpha) \\ C_1^{④} = C_1^{⑤} = C_1^{③}, \qquad C_2^{④} = C_2^{⑤} = C_2^{③} \end{cases} \tag{65}$$

Finally, there is only one unknown α to be determined with the help of the boundary condition

$$f'(\pi - 0) = 0 \tag{66}$$

It must be pointed out that the constant B in Eq 34 is an unknown, too, but it is not involved in the solving of Eq 26; therefore, B can be determined from the following boundary condition independently

$$f(\pi - 0) = 0 \tag{67}$$

Numerical Results

In Eqs 58–66, only α is unknown, so the numerical integration of Eq 26 begins from Γ_2, namely, from $\theta = \pi - \alpha$ up to $\theta = \pi$.

For the first step of integration from Γ_2, the expansion Eqs 60 and 61 are used, and $\Delta\theta$ is taken as $\Delta\theta = \pi/180$.

The results are given in Table 1 for various values of ν.

It should be pointed out that when $\nu \to 1/2$, the unloading boundary Γ_B will tend to Γ_2; this has been justified by the calculation. In fact, if we put $\alpha - \beta = \Delta\theta^*$, using Eqs 46, 60, and 61, we can obtain

$$\Delta\theta^* = \frac{4\sqrt{2}\epsilon^2}{3(1 - \nu^2 + \epsilon^2)} \cdot \sin\alpha \tag{68}$$

Evidently, $\Delta\theta^* \to 0$, as $\epsilon \to 0$.

On the Criterion of Fracture

It is well known that crack growth is controlled by the situation of the fracture-process zone. The size of the process zone is a microscopic size, and in this zone, the macroscopic concepts of mechanics are not valid. Therefore, the situation of the process zone cannot be obtained by means of macroscopic methods. On the other hand, however, no matter how complex the process zone is and no matter what the characteristic quantities of the process zone are, the situation of the process zone must be controlled by the macroscopic stress and strain field surrounding the process zone. Therefore, we can find a fracture criterion based on the asymptotic analysis of the macroscopic field. In Ref *4*, Rice discussed the fracture criterion based on the crack opening. Here, the fracture criterion will be discussed based on the analysis of strain singularity.

TABLE 1—*Numerical result.*

ν	B/k	α, deg	β, deg	γ, deg	$C_1^{③④⑤} \cdot E/k$	$C_2^{③④⑤} \cdot E/k$
0	0.8393	64.82	44.59	28.91	−4.796	−7.255
0.25	0.7903	61.97	53.57	24.57	−3.745	−5.994
0.45	0.7515	59.78	56.68	22.99	−2.939	−4.911
0.48	0.7447	59.43	59.04	22.03	−2.589	−4.473

From Eq 35 we can see that the asymptotic character of the strain ϵ_y ahead the crack tip is determined by e and A, namely

$$\epsilon_y = \frac{2k}{E}(1 - \nu^2 + \epsilon^2) \cdot \ell\text{n}\,\frac{A}{r} \tag{69}$$

where A is a constant determined by whole elastic-plastic field. A can be called the "strain singularity factor." For small-scale yielding, we have

$$A = \xi \cdot K_1{}^2/k^2 \tag{70}$$

in which K_1 is the stress-intensity factor and ξ is a coefficient which should be determined by the finite-element method. At any rate, the strain singularity factor A can be taken as a unique parameter of the tip field.

We assume that when the tensile strain ϵ_y at a reference point ahead the crack tip achieves a critical value ϵ_0, the crack will grow steadily, namely

$$\frac{2k}{E}(1 - \nu^2 + \epsilon^2) \cdot \ell\text{n}\,\frac{A_{\text{cr}}}{r_0} = \epsilon_0 \tag{71}$$

where ϵ_0 is called the tensile fracture-strain and r_0 is a microstructure size; without loss of generality, we can take the size of the process zone as r_0.

Equation 71 can be rewritten as

$$A_{\text{cr}} = r_0 \cdot \exp\left(\frac{\epsilon_0 E}{2k(1 - \nu^2 + \epsilon^2)}\right) \tag{72}$$

or

$$A_{\text{cr}} = r_0 \cdot e^T, \qquad T = \frac{\epsilon_0 E}{2k(1 - \nu^2 + \epsilon^2)} \tag{73}$$

where T denotes the toughness of the material.

It is natural that the plastic fracture criterion Eq 72 should be valid for the case of linear elastic fracture; therefore, in the special case, from Eq 70 we have

$$A_{\text{cr}} = \frac{\xi}{k^2} K_{\text{Ic}}{}^2 \tag{74}$$

References

[1] Gao, Y. C., "Elastic-Plastic Field at Crack-Tips in Perfectly Plastic Medium," reported at 15th ICTAM, Toronto, Ont., Canada, 1980; *Acta Mechanica Sinica*, 1981.

[2] Slepyan, L. I., *Izvestiia Akademii Nauk S.S.S.R., Mechanika Tverdogo Tela*, Vol. 9, 1974, pp. 57–67.

[3] Gao, Y. C., *Acta Mechanica Sinica*, No. 1, 1980, pp. 48–56.

[4] Rice, J. R., *Elastic-Plastic Crack Growth, Rodney Hill 60th Anniversary Volume*, Pergamon Press, New York, 1981.

Hugo A. Ernst[1]

Material Resistance and Instability Beyond J-Controlled Crack Growth

REFERENCE: Ernst, H. A., "**Material Resistance and Instability Beyond J-Controlled Crack Growth,**" *Elastic-Plastic Fracture: Second Symposium, Volume I—Inelastic Crack Analysis, ASTM STP 803,* C. F. Shih and J. P. Gudas, Eds., American Society for Testing and Materials, 1983, pp. I-191–I-213.

ABSTRACT: A modified version of the J-integral, J_M, is introduced extending previous results of Rice, Drugan, and Sham. It is proven that J_M always satisfies the Rice condition; that is, $\dot{J}_M$ must not depend on $\dot{a}$. The implications for instability are discussed in terms of the corresponding modified tearing moduli. Experimentally obtained data from compact specimens of a A508 Class 2A steel ranging from ½T to 10T were successfully correlated by using J_M even well beyond the J-controlled crack growth regime. Also, the correlation of data from bending and tension specimens is shown to improve significantly when J_M is used. Limitations for the use of the deformation theory J are set in terms of new parameters.

KEY WORDS: materials, fracture mechanics, J-R curves, plasticity, instability, modified J, tearing, crack growth, elastic-plastic fracture

The use of the J-integral [*1–3*][2] for characterizing the onset of crack extension has been accepted for some time, J being essentially the nonlinear counterpart of the linear G. Nevertheless, in the presence of significant amounts of crack extension, the adequacy of using J was seriously doubted until very recently. In 1977 Hutchinson and Paris [*4*] presented an analysis where they showed that there is a J-controlled crack growth regime if certain sufficient conditions are met. The first of these conditions simply says that the amount of crack extension Δa must be a small percentage of the remaining ligament b.

$$\Delta a \ll b \tag{1}$$

[1]Senior engineer, Materials Engineering Department, Westinghouse R&D Center, Pittsburgh, Pa. 15235.

[2]The italic numbers in brackets refer to the list of references appended to this paper.

The second condition refers to the ratio of proportional versus nonproportional strains in the presence of crack growth. Their logic was as follows. Close to the crack tip there is a zone of highly nonproportional strains, which is surrounded by another one where the stress-strains fields are correctly described by J. Thus, if the zone where J dominates is much bigger than the nonproportional one, it can be said that the latter is embedded in the former and then the crack growth process itself is controlled by J. The sufficient condition for J-controlled crack growth can be stated in terms of ω, being

$$\omega \equiv \frac{b}{J}\frac{dJ}{da} \gg 1 \tag{2}$$

where dJ/da is the slope of the J-resistance curve.

The aim of the Hutchinson-Paris analysis was to show that under certain conditions a J-controlled crack growth regime can exist. No attempt was made to answer the question of which is the minimum allowed value of ω, or how that minimum may change with geometry. Note that the expression for ω of Eq 2 is not specimen dependent, thus opening the possibility that the mentioned minimum value of ω may be different for different geometries. On the other hand, other investigations such as those by Wnuk [*5,6*], Rice et al [*7,9,10*], and Dean and Hutchinson [*8*] studied the actual process of crack growth in an elastic-plastic material. The basic conclusion is that if enough crack growth is allowed, the J-R curve will be dependent on size or type of specimen or both. Nevertheless, for small amounts of crack extension, the specimen geometry seems to have little effect in essential agreement with the Hutchinson and Paris work.

Specifically, the analysis of Rice, Drugan, and Sham [*7*] and Hermann and Rice [*9*] will be followed in some detail in the next section because of the significance to this work. In their analysis, among other things, they discussed the conditions that a J-like parameter, J_x, has to satisfy in order to be used in the presence of crack growth. And that is the rate at which J_x increases, $\dot{J}_x$, following the growth must not be a function of the rate of increase of crack length, $\dot{a}$

$$\dot{J}_x \neq \dot{J}_x(\dot{a}) \tag{3}$$

In the present work, the consequences of the condition of Eq 3 are explored and as a result a modified version of the J-integral, J_M, is defined. A general expression for J_M is given so that it can be calculated for any configuration. The properties of J_M as well as the implications for instability are discussed. In the second part of the paper, experimental data are used to show the adequacy of the approach. First, J-R curves of an A508 steel obtained from compact tension (CT) specimens ranging from ½T to 10T are plotted in the J-T frame. As expected, the curves are consistent at the early stages, but later on the branching of the smaller specimen curves starts to occur. When the plot is changed to a J_M-T_M one, a single curve results. That is, essentially, data obtained from ½T CT are in agreement with those from 10T CT even for situations where $\omega \approx 0$, and crack growth is up to 30 percent of the initial remaining ligament, when J_M is used. Thus, J_M seems to provide a way of greatly extending the spectrum of valid data. Second, resistance curves of an 2024 aluminum alloy obtained from CT specimens and center-cracked panels (CCP) with the "traditional" differences between bending and tension showed a very good correlation when plotted in terms of J_M.

Finally, two parameters of the ω-type are introduced to measure the error in both absolute value and slope in the R-curve when the deformation theory J is used.

The Rice, Drugan, and Sham Analysis

Rice, Drugan, and Sham [*7*] (RDS), in an extension of earlier studies by Rice and Sorensen [*10*], performed an analysis of the elastic-plastic stress and deformation fields at the tip of a crack which grows under plain-strain conditions. The analysis was performed for small-scale yielding, but it is argued that a solution of a similar form applies for large yielding with a possible variation of the constants involved if the triaxial constraint is maintained. They also reexamined the crack growth criterion of Rice and Sorensen [*10*], which requires a critical crack opening δ_t at some distance r from the tip during the growth process. Using this criterion, the differential equation involving J and dJ/da can be integrated giving as a result the J versus Δa curve. The results suggest that in principle the resistance curve will depend on the extent of yielding, that is, depend on the size or type of specimen or both used for its experimental determination.

In this section, attention will be devoted mainly to the parts of the RDS analysis, where they obtained the rate of crack opening displacement $\dot{\delta}_t$ and stated the property that the J-like parameter has to have.

The rate of crack opening obtained in Ref *7* is

$$\dot{\delta}_t = \beta \frac{\sigma_0}{E} \ell n \left(\frac{R}{r} \right) \dot{a} + \frac{\alpha}{\sigma_0} \dot{J}_x \tag{4}$$

where

α, β = parameters depending on amount of yielding,
σ_0 = yield strength,
E = Young's modulus,
R = length parameter associated with maximum radius of plastic zone
$\dot{J}_x$ = rate of increase of a parameter related to J-integral
$\dot{a}$ = rate of change of crack length.

As a consequence, they observed that the way R is defined will depend very much on what exactly J_x means. In particular, if J_x is the deformation theory value of the J-integral, J, then one can write

$$\dot{J} = \mu \dot{a} + \nu \dot{\delta} \tag{5}$$

where $\dot{\delta}$ is the rate of change of the externally applied displacement, and μ and ν are functions of a and δ. Substituting Eq 5 in Eq 4, gives

$$\dot{\delta}_t = \left(\beta \frac{\sigma_0}{E} \ell\text{n} \frac{R}{r} + \frac{\alpha}{\sigma_0} \mu \right) \dot{a} + \frac{\alpha}{\sigma_0} \nu \dot{\delta} \tag{6}$$

Now, it is a well-known fact that for a rigid perfectly plastic material, that is, $\sigma_0/E \rightarrow 0$, the crack opening displacement rate $\dot{\delta}_t$ and the applied displacement rate $\dot{\delta}$ are linearly related through a geometry-dependent factor for *both stationary and growing cracks*; that is

$$\dot{\delta}_t = s \, \dot{\delta} \quad \text{if } \sigma_0/E \rightarrow 0 \tag{7}$$

for stationary and growing cracks. But then, in order to meet this limit, the bracket multiplying $\dot{a}$ in Eq 6 has to vanish for $\sigma_0/E \rightarrow 0$. This means that either

$$\mu = 0$$

or (8)

$$\frac{\sigma\beta_0}{E} \ell\text{n} \frac{R}{r} = -\mu$$

If the second of Eq 8 is correct, this implies that

$$R \text{ proportional to } \exp\,(\text{constant}) \times E/\sigma_0 \tag{9}$$

which diverges for $\sigma_0/E \rightarrow 0$, thus yielding an unsatisfactory definition of R. As a consequence, $\mu = 0$ and this in its turn implies that the J-like parameter has to be such that its rate of change following the growth is independent of the rate of change of crack length for $\sigma/E \rightarrow 0$. That is substituting the condition $\mu = 0$ into Eq 5 gives

$$\dot{J}_x = \nu \dot{\delta} \tag{10}$$

or

$$\dot{J}_x \neq \dot{J}_x(\dot{a}) \quad \text{for } \sigma_0/E \rightarrow 0 \tag{11}$$

Once the condition of the foregoing equation was obtained, the authors studied the behavior of two J-like candidate parameters in different specimen configurations: the deformation theory J, and the so-called far-field J, J_f, which is the result of the line integral calculation taking the specimen boundary as the integration path. For the case of deeply cracked strip, leaving a remaining ligament b, subjected to bending in a rigid perfectly plastic material, the expressions of J and J_f in terms of the current angle of rotation θ are

$$J = 0.73\,\sigma_0\,\theta b$$
$$J_f = 0.73 \int_0^\theta b d\theta \tag{12}$$

As a result the rates are

$$\dot{J} = -\frac{J}{b}\dot{a} + 0.73\,\sigma_0\,b\dot{\theta}$$
$$\dot{J}_f = 0.73\,\sigma_0\,b\dot{\theta} \tag{13}$$

Thus, J_f has the required property of Eq 11, while J does not. Also the two can be connected as

$$\dot{J}_f = \dot{J} + \frac{J}{b}\dot{a} \tag{14}$$

If, instead, the specimen considered in the preceding represents one half of a double-edge-cracked tension specimen (assuming a Prandtl field exists over the remaining ligament), and δ is the displacement of one end of the specimen relative to the other, the value of J is

$$J = 2.97\,\sigma_0\delta$$
$$\dot{J} = 2.97\,\sigma_0\dot{\delta} \tag{15}$$

On the other hand, J_f does not seem to have a unique value due to the non-uniqueness of the stress field. Thus, in this case it is J which has the required property.

Finally, for the center-cracked specimen subjected to tension (CCP), J and J_f are given by

$$J = J_f = 1.15\,\sigma_0\delta$$

or (16)

$$\dot{J} = \dot{J}_f = 1.15\,\sigma_0\dot{\delta}$$

So for this case, although a Prandtl like stress state does not exist at the crack tip, both J-like parameters coincide and have the required property.

In conclusion, it can be seen that according to these examples, neither of the two candidate parameters examined exhibits always the required property. As a consequence, the use of either of them cannot be recommended in general.

The Modified J, J_M

As was seen in the last section, neither J nor J_f has always the required property; thus they cannot be used in general. Nevertheless, other J-like parameters can be defined in such a way as to guarantee that the Rice condition of Eq 11 is always satisfied.

Let us define the modified J, J_M, as

$$J_M = J - \int_{a_0}^{a} \left. \frac{\partial (J - G)}{\partial a} \right|_{\delta_{pl}} da \tag{17}$$

where

$J =$ J-integral, based on deformation theory,
$G =$ Griffith linear elastic energy release rate,
$a_0, a =$ initial and current crack length,
$J - G = J_{pl} =$ plastic part of the deformation theory J,
$\delta_{pl} = \delta - \delta_{el} =$ plastic part of the displacement or simply the difference between the total load-point displacement and the elastic one.

Correspondingly, the rates of increase of J_M, following crack extension, $\dot{J}_M$ and dJ_M/da, are

$$\begin{aligned} \dot{J}_M &= \dot{J} - \left. \frac{\partial J_{pl}}{\partial a} \right|_{\delta_{pl}} \dot{a} \\ \frac{dJ_M}{da} &= \frac{dJ}{da} - \left. \frac{\partial J_{pl}}{\partial a} \right|_{\delta_{pl}} \end{aligned} \tag{18}$$

In particular, for the case of a rigid perfectly plastic material, $\sigma_0/E \rightarrow 0$, $G \rightarrow 0$, and

$$\begin{aligned} J_{pl} &\approx J \\ \left. \frac{\partial J_{pl}}{\partial a} \right|_{\delta_{pl}} &\approx \left. \frac{\partial J}{\partial a} \right|_{\delta_{pl}} = \mu \end{aligned} \tag{19}$$

Substituting Eqs 5 and 19 in 18 gives

$$\dot{J}_M = \mu \dot{a} + \nu \dot{\delta} - \mu \dot{a} = \nu \dot{\delta} \tag{20}$$

which is not a function of $\dot{a}$. Thus, J_M is defined in such a way that it will always satisfy the Rice condition. Some examples follow.

For the case of the deeply cracked strip in bending [11]

$$\left.\frac{\partial J_{pl}}{\partial a}\right|_{\delta_{pl}} = -\frac{J_{pl}}{b} \tag{21}$$

Thus, J_M and the corresponding rates $\dot{J}_M$ and dJ_M/da are given by

$$J_M = J + \int_{a_0}^{a} \frac{J_{pl}}{b}\, da$$

$$\dot{J}_M = \dot{J} + \frac{J_{pl}}{b}\dot{a} \tag{22}$$

$$\frac{dJ_M}{da} = \frac{dJ}{da} + \frac{J_{pl}}{b}$$

As can be seen for this case comparing Eqs 14 and 22

$$J_M \equiv J_f \quad \text{for } \sigma_0/E \to 0 \tag{23}$$

For the case of a double-edge deeply notched specimen loaded in tension, following Rice, Paris, and Merkle [11] and assuming that load P and plastic displacement δ_{pl} are connected through a power law ($P \propto {\delta_{pl}}^{1/n}$)

$$\left.\frac{\partial J_{pl}}{\partial a}\right|_{\delta_{pl}} = \frac{J_{pl}}{bn} \tag{24}$$

Thus, J_M and the corresponding rates are given by

$$J_M = J - \int_{a_0}^{a} \frac{J_{pl}}{bn}\, da$$

$$\dot{J}_M = \dot{J} - \frac{J_{pl}}{bn}\dot{a} \tag{25}$$

$$\frac{dJ_M}{da} = \frac{dJ}{da} - \frac{J_{pl}}{bn}$$

In particular, for $\sigma_0/E \to 0$, $n \to \infty$, $J_{pl}/bn \to 0$ and thus

$$\frac{dJ_M}{da} \equiv \frac{dJ}{da}$$

or (26)

$$J_M \equiv J$$

The value of $\partial J_{pl}/\partial a$ can also be obtained for situations other than pure bending or pure tension, from plastic analysis. Following the Merkle-Corten analysis [*12*] of the CT specimen Ernst et al [*13*] obtained

$$\left.\frac{\partial J_{pl}}{\partial a}\right|_{\delta_{pl}} = -(1 + 0.76\,b/W)\frac{J_{pl}}{b} \tag{27}$$

Giving for J_M and the corresponding rates

$$J_M = J + \int_{a_0}^{a} (1 + 0.76\,b/W)\frac{J_{pl}}{b}\,da$$

$$\dot{J}_M = \dot{J} + (1 + 0.76\,b/W)\frac{J_{pl}}{b}\,\dot{a} \tag{28}$$

$$\frac{dJ_M}{da} = \frac{dJ}{da} + (1 + 0.76\,b/W)\frac{J_{pl}}{b}$$

More recently Ernst [*14*], generalizing the previously mentioned analysis of Merkle and Corten [*12*], obtained an expression for J_{pl} ranging continuously from pure bending to pure tension, in general; thus

$$\left.\frac{\partial J_{pl}}{\partial a}\right|_{\delta_{pl}} = -m\frac{J_{pl}}{b} \tag{29}$$

where m is a function of crack length a, and hardening exponent n. As a result J_M and the corresponding rates are given as

$$J_M = J + \int_{a_0}^{a} m\frac{J_{pl}}{b}\,da$$

$$\dot{J}_M = \dot{J} + m\frac{J_{pl}}{b}\,\dot{a} \tag{30}$$

$$\frac{dJ_M}{da} = \frac{dJ}{da} + m\frac{J_{pl}}{b}$$

In this section a new J-like parameter, the modified J, J_M, that always satisfies the Rice condition has been introduced. It is defined in a general way allowing its calculation for any configuration. Moreover, the different terms involved can be determined in different ways analytically, from finite elements, plastic analysis, etc., or experimentally. In fact, the value of the deformation theory J depends solely on current crack length and displacement, and it can be obtained from line integral calculations or experimentally following the different existing analyses, the multiple-specimen technique or P-δ related formulas for J. The other term, $\partial J_{pl}/\partial a$, can also be determined analytically or experimentally from calibration P-δ records.

The Modified Tearing Modulus—Implication of an Absolute Scale

In the previous section the modified J, J_M, was introduced and formulas for specific situations were discussed. In this section, the corresponding modified version of the Paris et al [*15,16*] material tearing modulus, $T_{M\text{mat}}$, is defined and the physical implications are explored.

First, consider the normalized version of Eq 18 as the definition of the $T_{M\text{mat}}$

$$T_{M\text{mat}} = \frac{E}{\sigma_0{}^2}\frac{dJ_M}{da} = \frac{E}{\sigma_0{}^2}\left[\frac{dJ}{da} - \left.\frac{\partial J_{pl}}{\partial a}\right|_{\delta_{pl}}\right] = T_{\text{mat}} - \frac{E}{\sigma_0{}^2}\left.\frac{\partial J_{pl}}{\partial a}\right|_{\delta_{pl}} \qquad (31)$$

Next, consider a load per unit thickness displacement (P-δ) test record as well as calibration functions, that is, curves of constant crack length, of a rigid plastic material as shown in Fig. 1. For this case

$$\delta_{pl} \approx \delta \qquad J_{pl} \approx J \qquad (32)$$

If J_A, J_B, and J_C represent the value of J_{pl} at Points A, B, and C, noting that Points A and C are at the same δ_{pl}, then J_C is

$$J_C = J_A + \left.\frac{\partial J_{pl}}{\partial a}\right|_{\delta_{pl}} \Delta a \qquad (33)$$

where Δa represents the change in crack length between Points A and C. At the same time T_{mat} and $T_{M\text{mat}}$ are given by

$$T_{\text{mat}} = \frac{E}{\sigma_0{}^2}\left(\frac{J_B - J_A}{\Delta a}\right) \qquad (34)$$

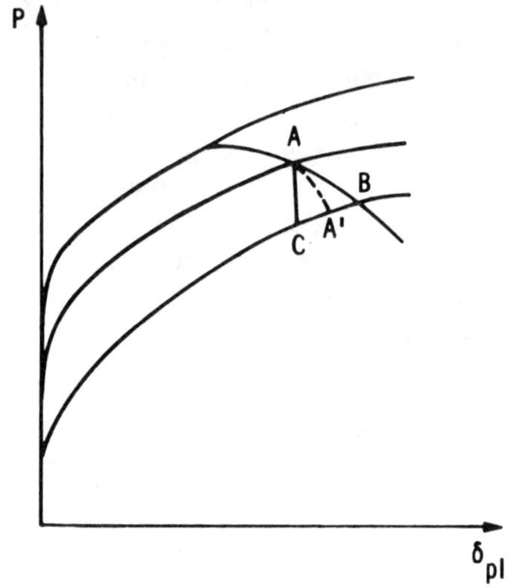

FIG. 1—*Load-displacement record.*

$$T_{M\text{mat}} = \frac{E}{\sigma_0{}^2}\left(\frac{J_B - J_A}{\Delta a} - \left.\frac{\partial J_{\text{pl}}}{\partial a}\right|_{\delta_{\text{pl}}}\right) \tag{35}$$

Introducing Eq 33 into Eq 35 gives

$$\begin{aligned} T_{M\text{mat}} &= \frac{E}{\sigma_o{}^2}\left(\frac{J_B - J_A}{\Delta a} - \frac{J_C - J_A}{\Delta a}\right) \\ &= \frac{E}{\sigma_0{}^2}\left(\frac{J_B - J_C}{\Delta a}\right) \end{aligned} \tag{36}$$

Thus, the effect of modifying T_{mat} by adding the term $-(E/\sigma_0{}^2)\ \partial J_{\text{pl}}/\partial a$, Eq 31, is simply to shift the zero or reference point. In other words the T_{mat}, Eq 34, is referenced to points of equal J, Points A-A′ in Fig. 1; that is, T_{mat} is zero when the P-δ record goes through Points A-A′, whereas $T_{M\text{mat}}$ is referenced to vertical drop, that is, $T_{M\text{mat}}$ is zero when P-δ is vertical. The saturation state that $d\delta_{\text{pl}} = 0$ represents in a given test corresponds to $T_{M\text{mat}} = 0$ (not to $T_{\text{mat}} = 0$).

Taking another viewpoint, it has been observed in R-curve testing that if enough crack extension is allowed the J-R curve may reach a plateau and eventually start decreasing ($dJ/da < 0$). It is obvious that this situation does not seem to make much sense from a physical point of view. A negative slope ($T_{\text{mat}} < 0$) would mean spontaneous instability and instead the test is very stable at that point. It is clear that by then the requirements to assure

J-controlled growth have been violated according to any standard, and thus J (and T_{mat}) does not seem to be the best parameter to use for describing the crack growth process. Instead, the use of J_M and the corresponding T_{Mmat} guarantees that $T_{Mmat} = 0$ only when $d\delta_{pl} = 0$. In this sense it seems possible that an "absolute scale" for T can be defined by the use of T_{Mmat}, that is, just shifting the zero by $(E/\sigma_0^2)\ \partial J_{pl}/\partial a$.

Instability Predictions—The Modified Applied Tearing Modulus

A word is in order here regarding the condition for instability. As is well known, instability results from a lack of balance between the rate of increase of the externally applied crack drive force and that of the material resistance to crack growth. According to the tearing instability theory based on the deformation theory J [*14,15*], instability will ensue when

$$T_{mat} < T_{app} \tag{37}$$

which represents a necessary and sufficient condition [*17*].

If the parameter describing the material resistance, T_{mat}, is changed to a modified version, T_{Mmat}, that one related to crack drive force, T_{app}, has to be changed consistently. Consider then the definition of the modified applied tearing modulus T_{Mapp} as

$$T_{Mapp} = T_{app} - \frac{E}{\sigma_0^2} \frac{\partial J_{pl}}{\partial a}\bigg|_{\delta_{pl}} \tag{38}$$

Combining Eqs 31 and 38, the modified tearing instability condition is obtained as

$$T_{Mmat} < T_{Mapp} \tag{39}$$

which obviously is completely consistent with Eq 37; that is, Eq 39 is just the result of adding the term $-(E/\sigma_0^2)\ \partial J_{pl}/\partial a$ to both sides of Eq 37. Thus

$$\Delta T = T_{mat} - T_{app} = T_{Mmat} - T_{Mapp} < 0 \quad \text{for instability} \tag{40}$$

In conclusion, the use of the modified tearing moduli T_{Mmat} and T_{Mapp} does not affect the instability condition, which is really dictated by the much more general principle of equilibrium and not by the way we choose to describe the problem [*17*].

Analysis of Experimental Results

In the previous sections the J_M was introduced and the implications of its use were explored in terms of the construction of an "absolute scale" for the material tearing modulus via the $T_{M\text{mat}}$, and the instability prediction with the use of $T_{M\text{app}}$. Still, the main justification for the use of the new parameter is its ability to correlate experimental data obtained from different size or type of specimens or both. This is the subject of the present section.

Experimental data obtained in a program for the Electrical Power Research Institute (EPRI) were used for this purpose. The program is described in detail in Refs *18* and *19*. Basically it consisted of the R-curve testing of CT specimens of different planar size, thickness B, and crack aspect ratio a/W, of an A508 Class 2 steel at 204°C (400°F). Results are presented in terms of J-Δa curves where Δa was measured using the unloading compliance technique and the deformation theory J was calculated using [*13*]

$$J_{i+1} = \left[J_i + \left(\frac{\eta}{b}\right)_i \frac{A_{i,i+1}}{B}\right]\left[1 - \frac{\gamma}{b}(a_{i+1} - A_i)\right] \tag{41}$$

where

η and γ = functions of a/W,
b = remaining ligament,
subscripts i (or $i+1$) = quantity evaluated at that stage, and
$A_{i,i+1}$ = area under the P-δ record between lines of constant displacement δ_i and δ_{i+1}.

Out of the large test matrix some specimens were selected for this study. These specimens were geometrically similar. All of them had an initial $a/W \approx$ 0.6 and the thickness B was half of the total width W in all cases. The overall size ranged from ½T; $W = 2.54$ cm (1 in.) to 10T; $W = 51$ cm (20 in.). The resulting J-R curves showed consistency at the early stages but, as expected, the different curves start deviating when the crack length increment Δa was significant compared with the ligament b.

The J-Δa curves were then converted to J-T plots [*19*]. The result is shown in Fig. 2. Here the differences between specimens are more obvious than in the regular J-R curve plot. It can be seen for lower T_{mat} (higher J) that "peeling off" of the smaller specimen curves occurs. In fact, as was mentioned before for some specimens, T_{mat} is close to zero (even negative for the ½T ones). The same data were replotted using the J_M in a J_M-$T_{M\text{mat}}$ frame. Using for J_M and T_M

$$J_M = J + \int_{a_0}^{a} (1 + 0.76\, b/W)\frac{J_{\text{pl}}}{b}\, da \tag{42}$$

$$T_{M\text{mat}} = \frac{E}{\sigma_0^2}\frac{dJ_M}{da} = T_{\text{mat}} + \frac{E}{\sigma_0^2}\left(\frac{1 + 0.76\, b/W}{b}\right)J_{\text{pl}} \tag{43}$$

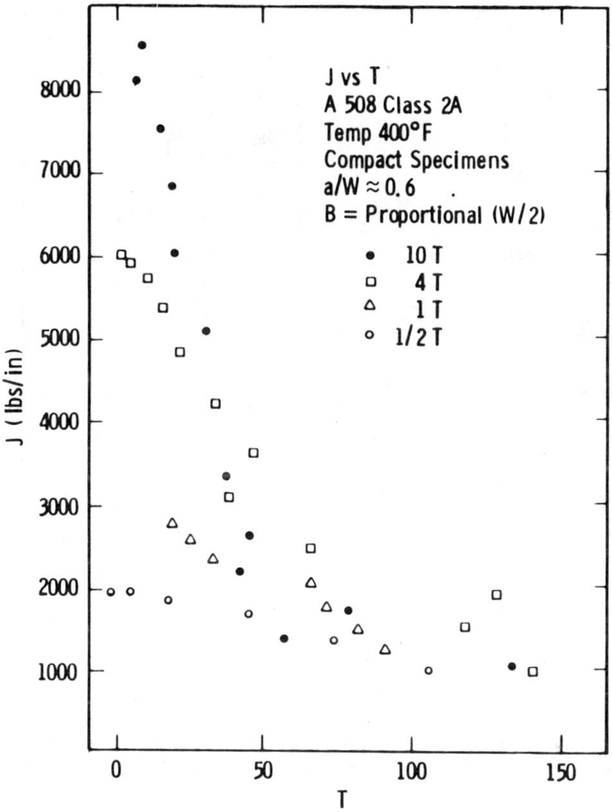

FIG. 2—J *versus* T *(deformation theory) (1 lb/in. = 0.1130 N·m).*

The first term of Eq 42 was obtained from Eq 41. The second term was evaluated by plotting the integrand versus Δa and measuring the area under the curve up to the corresponding value of a (see Appendix). The results are shown in Fig. 3. As can be seen, all the curves collapse into one, even for the last points of the smallest specimens. This basically means that J_M is correlating data for situations well beyond the so-called J-controlled crack growth, even for values of $\omega < 0$, and for crack growth up to 30 percent of the initial remaining ligament. This seems to indicate that the material resistance curve is specimen size independent if J_M is used.

Another test for J_M is obviously its ability to correlate data from different specimen geometries. In order to do so, experimental data previously obtained of CT and CCP specimens of a 2024 aluminum alloy were examined. First, the deformation theory J-resistance curves were compared, Fig. 4. Values of J were obtained using Eq 41 for the CT and for the CCP [*21*]

$$J = G + J_{pl}$$

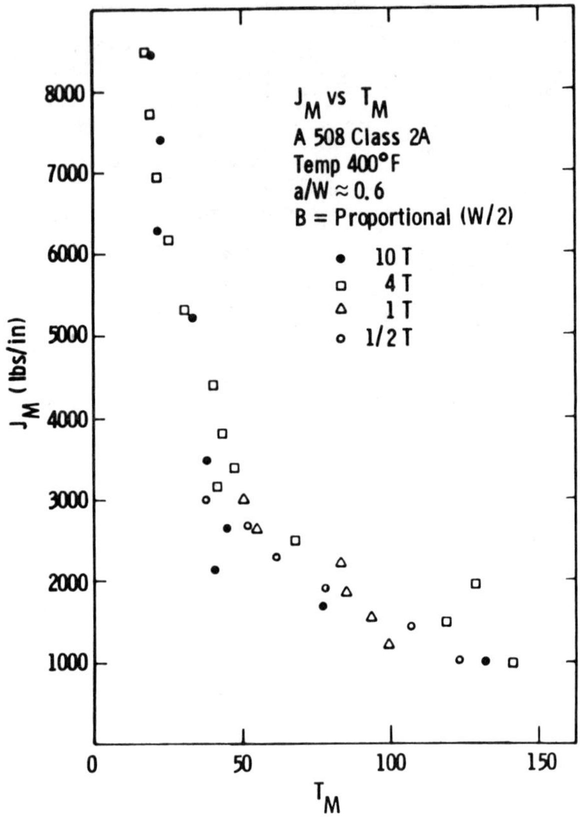

FIG. 3—J_M *versus* T_M *[°C = 5/9 (°F −32); 1 lb/in. = 0.1130 N·m].*

with

$$J_{pli+1} = \left[J_{pli} + \frac{J_{pli}\,\delta_{pli}}{b_i} + \frac{2}{b_i}A_{i,i+1}\right]\left[1 + \frac{\Delta a}{b_i}\left(\frac{P_i\,\delta_{pli} - J_{pli}\,b_i}{P_i\,\delta_{pli} + J_{pli}b_i}\right)\right] - \frac{P_{i+1}\,\delta_{pli+1}}{b_{i+1}} \tag{44}$$

where $\Delta a = b_i - b_{i+1}$ and b is the total remaining ligament ($W - 2a$).

Then the data were replotted using this time J_M given by Eqs 25 and 28 (see Appendix). The result is shown in Fig. 5. As can be seen, the correlation is significantly improved.

In conclusion, the use of J_M seems to greatly improve the correlation of data obtained from specimens of different size or type or both.

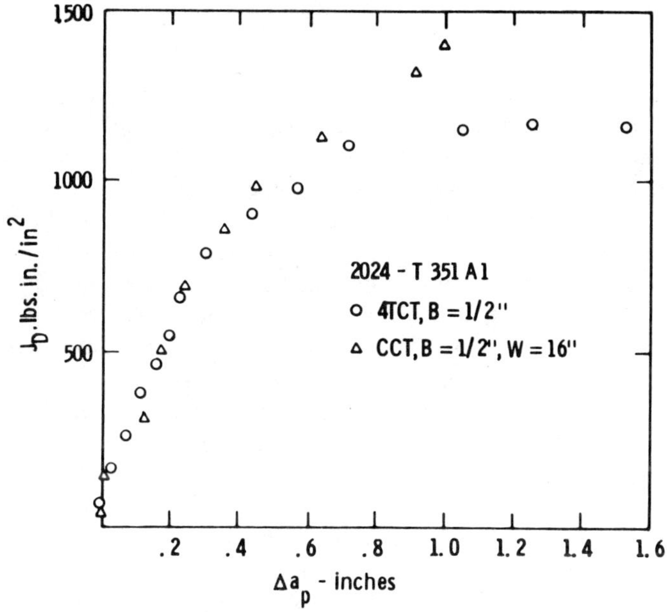

FIG. 4—*Compact specimen and center cracked panel J-R curve comparison using deformation theory* J *(1 in. = 2.54 cm; 1 lb-in./in.^{2} = 0.17512 kJ/m^2).*

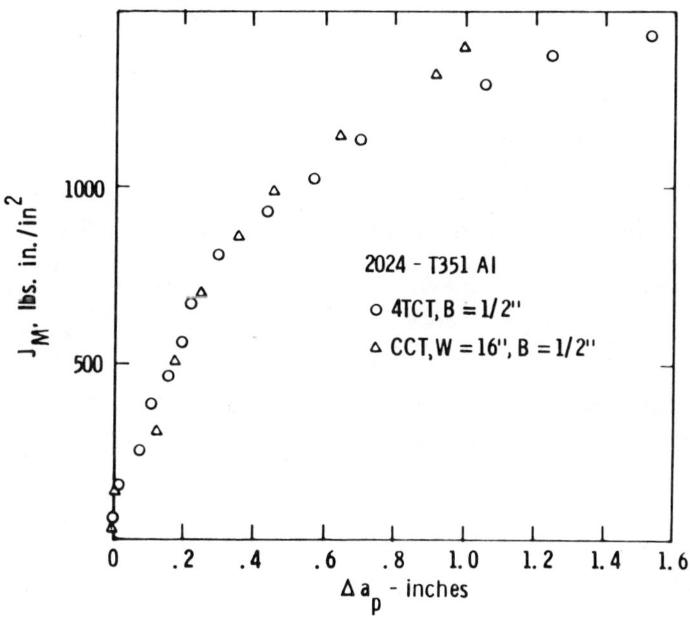

FIG. 5—*Compact specimen and center cracked panel J-R curve comparison using modified* J_M *(1 in. = 2.54 cm; 1 lb-in./in.^{2} = 0.17512 kJ/m^2).*

Discussion

Error Factors

In the previous sections a general definition was provided for J_M that without any ambiguity allows its evaluation for any configuration, experimentally, from finite elements or analytically using some plastic analysis. It was proven also that J_M always satisfies the Rice condition of Eq 11, that is, $\dot{J}_M \neq \dot{J}_M(\dot{a})$.

Next, the implications for stability analysis were discussed in terms of the modified tearing moduli. It was suggested that $T_{M\text{mat}}$ may provide an "absolute scale" for the material tearing modulus. Finally, experimental data for CT specimens ranging from ½T to 10T were successfully correlated when J_M was used even for situations where all requirements for J-controlled growth had been clearly violated. Also, the correlation of data from CCP and CT specimens of a 2024 aluminum alloy was significantly improved when J_M was used.

All of this seems to suggest that J_M is a better parameter to use for describing the material resistance to significant crack growth.

Following this line, parameters can be defined to measure the error both in slope and absolute value when J is used. Rearranging Eq 18 gives

$$\frac{dJ_M}{da} = \frac{dJ}{da}\left(1 - \frac{da}{dJ}\frac{\partial J_{\text{pl}}}{\partial a}\bigg|_{\delta_{\text{pl}}}\right) \tag{45}$$

Then the "slope error fraction" λ can be defined as

$$\lambda \equiv \left| \left(\frac{dJ_M}{da} - \frac{dJ}{da}\right) \bigg/ \frac{dJ}{da} \right| = \left| \left(\frac{1}{dJ/da}\frac{\partial J_{\text{pl}}}{\partial a}\right)_{\delta_{\text{pl}}} \right| \tag{46}$$

where the outside bars stand for absolute value; if Eq 29 applies

$$\lambda \equiv \left| \frac{m}{dJ/da}\frac{J_{\text{pl}}}{b} \right| \tag{47}$$

Note that this approach is completely consistent with that one of Hutchinson and Paris [*4*]. In fact, for a rigid-plastic material and pure bending

$$J \approx J_{\text{pl}} \qquad m = 1 \tag{48}$$

and then

$$\lambda \rightarrow \frac{1}{\omega} \tag{49}$$

Thus the modification to J proposed is of the order of $1/\omega$. Nevertheless, λ is defined in a more general way and provides a means of actually calculating the error associated with the use of J.

On the other hand, it can be argued that slopes are difficult to measure experimentally and thus a criterion regarding slopes cannot be very accurate. Thus an "absolute value error fraction," ϵ, can be defined based on Eq 17

$$J_M = J - \int_{a_0}^{a} \left.\frac{\partial J_{pl}}{\partial a}\right|_{\delta_{pl}} da$$

$$= J\left(1 - \frac{1}{J}\int_{a_0}^{a} \left.\frac{\partial J_{pl}}{\partial a}\right|_{\delta_{pl}} da\right) \tag{50}$$

Thus

$$\epsilon \equiv \left|\frac{1}{J}\int_{a_0}^{a} \left.\frac{\partial J_{pl}}{\partial a}\right|_{\delta_{pl}} da\right| \tag{51}$$

or if Eq 29 applies

$$\epsilon = |(J_M - J)/J| = \left|\left(\int_{a_0}^{a} m\,(J_{pl}/b)\,da\right)/J\right| \tag{52}$$

where again the bars stand for absolute value.

The two newly defined parameters λ and ϵ are clearly related to the error made if J is used. They are also geometry dependent through the factor m, which allows us to account for the fact that different configurations do not behave equally.

This represents a big advantage with respect to ω. In fact, if λ and ϵ are used, numbers can be set for the maximum error (0.30 and 0.20 respectively, for example) allowed in any configuration; that is, these numbers can be the same for all configurations.

Instead, if ω is used, a minimum value has to be found for each individual configuration: the fact that the expression for ω is the same for all configurations ($m = 1$) does not allow us to distinguish one configuration from the other.

Minimum Remaining Ligament

The newly introduced J_M allows us to extend greatly the spectrum of valid data. The obvious importance of this possibility will be emphasized further here.

Recently it has become common to use the J-T_{50} concept introduced by Paris [*20*]. It consists basically in finding the intersection of the J versus T_{mat} curve with the straight line described by the equation $J/T = 5.65$ N·m (50 lb/in.). The value of J of the intersection point, J_{50}, is the one recommended for design.

On the other hand, the ω-criterion can also be set in terms of a straight line in the J-T plane. Equation 2 can be rearranged to give

$$\frac{Jda}{dJ} = \frac{b}{\omega}$$

$$J/T = \frac{b}{\omega} \frac{\sigma_0^2}{E} \tag{53}$$

The intersection of this line with the J-T_{mat} curve gives the last valid point according to the particular ω selected (for the given σ_0, E, and b). Points to the left of that one are invalid points and those to the right are valid ones. Thus in order to have the J_{50} point in the region of valid points, the $J/T = 5.65$ N·m (50 lb/in.) line has to lie to the right of the limiting one based on ω, or

$$\frac{b}{\omega} \frac{\sigma_0^2}{E} \geq 5.65 \text{ N}\cdot\text{m (50 lb/in.)} \tag{54}$$

then

$$b \geq 5.65 \text{ N}\cdot\text{m (50 lb/in.)} \frac{\omega E}{\sigma_0^2} \tag{55}$$

This basically says that for a given material a certain minimum remaining ligament is required in order to get a valid J_{50} point. In particular, taking $\omega = 5$ as suggested by Paris [*20*], Eq 55 becomes

$$b \geq 28.25 \text{ N}\cdot\text{m (250 lb/in.)} \frac{E}{\sigma_0^2} \tag{56}$$

Moreover, it may be preferable to set the preceding condition in terms of the initial remaining ligament b_0 (rather than the current b), thus allowing 20 percent difference between b_0 and the current b; Eq 56 becomes

$$b_0 \geq 33.90 \text{ N}\cdot\text{m (300 lb/in.)} \frac{E}{\sigma_0^2} \tag{57}$$

This condition can be adopted as the criterion to select specimen size to assure a valid J_{50} point for a given material. A plot of Eq 57 setting $E = 207\,000$ MPa (30 000 ksi) is shown in Fig. 6.

For example, the yield stress of the A508 Class 2A steel reported previously was $\sigma_0 = 469$ MPa (68 ksi). According to Fig. 6 this implies that an initial re-

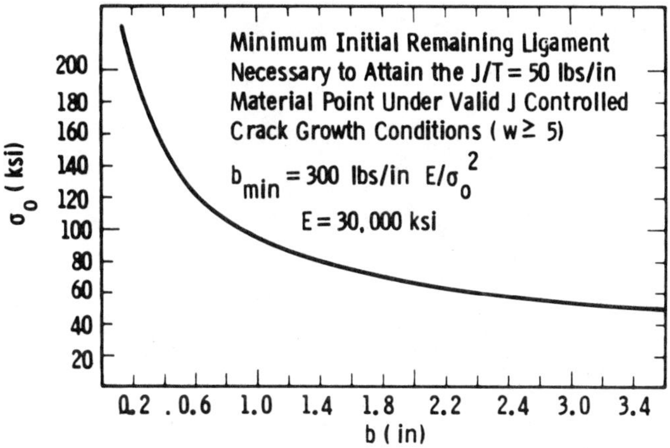

FIG. 6—*Yield strength versus minimum initial remaining ligament (1 lb/in. = 0.1130 N·m; 1 ksi = 6.90 MPa; 1 in. = 2.54 cm).*

maining ligament of at least 4.8 cm (1.9 in.) is required, that is, a 2T specimen. Now, if the only available specimens were the ½T ones (typical case of surveillance specimens), it is obvious that the J_{50} point cannot be obtained under valid conditions. In fact, setting for this case $b = 1.27$ cm (0.5 in.), $\omega = 5$, $\sigma_0 = 469$ MPa (68 ksi), and $E = 207\,000$ MPa (30 000 ksi), the ω-based limiting line of Eq 53 can be obtained as

$$J/T = \frac{b}{\omega}\,\frac{\sigma_0{}^2}{E} = 1.695\,\text{N}\cdot\text{m}\,(15\,\text{lb/in.}) \tag{58}$$

which is smaller than the required $J/T = 5.65$ N·m (50 lb/in.). This is illustrated in Fig. 7, where best-fitted lines of the ½T and 10T experimental data are plotted as well as the J/T lines of 5.65 and 1.695 N·m (50 and 15 lb/in.). As was shown, the former intersects the J-T_{mat} line of the ½T specimens well into the invalid region, yielding an intersection point of $J_{50} = 192$ N·m (1700 lb/in.), whereas the intersection point for the 10T specimens (believed to be closer to the structure) is $J_{50} = 305$ N·m (2700 lb/in.).

Thus, if the only available specimens were the ½T ones, the only resulting information would be a J_{50} point known to be invalid and too conservative.

Instead, by considering again Fig. 3, it is clear that using J_M and T_M the material curves of the different specimen size converge into one and thus the intersection point, J_{50}, is essentially the same for all specimen sizes. This basically means that *the use of* J_M *allows us to make maximum meaningful use of the information obtained from small specimens, well beyond the* J*-controlled crack growth regime*; in particular it allows us to get to the J_{50} level with no unnecessary reduction in the predicted toughness capability.

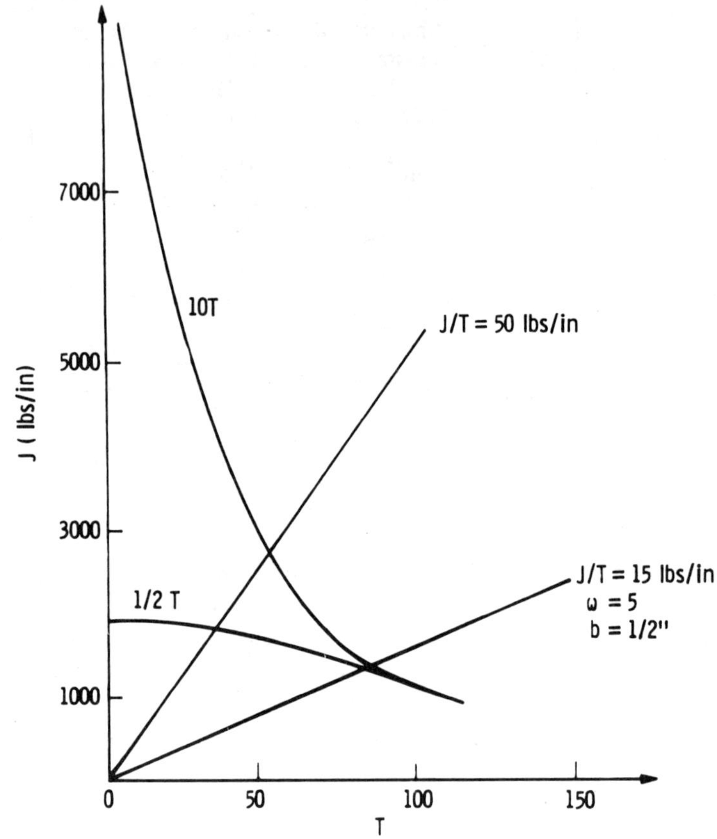

FIG. 7—J *versus* T_{mat} *and* J/T *lines A508 Class 2A steel (1 lb/in. = 0.1130 N·m; 1 in. = 2.54 cm).*

Summary and Conclusions

In this work a modified version of the J-integral, J_M, has been introduced. First, its properties and the consequences of its definition were analyzed. Then the advantages of its use were clearly demonstrated by showing the correlation of experimental data. The salient features can be summarized as follows:

1. A general definition was provided for J_M that allows its evaluation, without ambiguity, for any configuration.
2. It was shown that J_M always satisfies the Rice condition, that is, $\dot{J}_M \neq \dot{J}_M(\dot{a})$.
3. The resulting modified tearing moduli, $T_{M\text{mat}}$ and $T_{M\text{app}}$, were introduced. It was shown that the instability condition they provide is completely consistent with that obtained from tearing instability theory based on deformation J, which in its turn is a necessary and sufficient condition for instability.

4. It was noted that the use of $T_{M\text{mat}}$ provides an "absolute (or thermodynamic) scale" for the material tearing modulus, that is, $T_{M\text{mat}} \geq 0$ always.

On the other hand, the advantages of actually using J_M can be summarized as follows:

1. The correlation of resistance curves of a 2024 aluminum alloy obtained from CT and CCT specimens was significantly improved when J_M was used.
2. Experimental data of A508 steel CT specimens ranging from ½T to 10T were successfully correlated using J_M even for situations where $\omega < 0$ and $\Delta a \approx 0.3\, b_0$; that is, all conditions for deformation J-controlled crack growth had been grossly violated.
3. Error factors were defined that allow the evaluation of the error made when J is used.
4. Examples were provided to emphasize the advantages of using J_M: First, the J_{50} obtained from ½T specimens was compared with that one from 10T. The significant difference, 192 versus 305 N·m (1700 versus 2700 lb/in.), shows the restrictive nature of using deformation J for small specimens or large amounts of crack extension or both. On the other hand, when J_M is used the J_{50} point is basically the same for all specimen sizes, thus avoiding unnecessary loss in toughness capability. Second, the relation between the yield strength of the material and the minimum initial remaining ligament necessary to attain the J_{50} point under valid conditions was derived.

In conclusion, J_M provides a way of characterizing the material resistance to crack growth in a way that is specimen-size independent. It allows a large relaxation of the restrictions on the amount of crack extension or initial remaining ligament or both needed to produce valid data. Moreover, J_M permits the full use of the material toughness capability without the unnecessary reduction that previously resulted when specimens of insufficient size were used.

Acknowledgments

This study was partially supported by EPRI under Contract No. RP 1238-2, Dr. Douglas M. Norris, project manager. The author wants to thank his colleagues from Westinghouse R&D Center, D. E. McCabe and J. D. Landes, for their invaluable help.

APPENDIX

The method to obtain J_M can be summarized here as follows. In general

$$J_M = J + \int_{a_0}^{a} \frac{m}{b} J_{\text{pl}}\, da \tag{59}$$

$$\frac{dJ_M}{da} = \frac{dJ}{da} + \frac{m}{b} J_{pl} \tag{60}$$

Thus, the second term of Eq 59 can be evaluated by plotting mJ_{pl}/b versus Δa and measuring the area under the curve from the origin up to the point of interest (in Δa); alternatively, the following incremental form of Eq 59 can be used

$$J_{Mi+1} = J_{i+1} + \Delta J_{i+1} \tag{61}$$

where J_{i+1} represents the value of the deformation theory J at that level and ΔJ_{i+1} is given by

$$\Delta J_{i+1} = \Delta J_i + \left(\frac{m}{b} J_{pl}\right)_i (a_{i+1} - a_i) \tag{62}$$

where $\Delta J_i = 0$ for $i = 0$ and the symbol $(\ )_i$ indicates that the inside quantities are to be evaluated at stage i.

In particular for the CT specimen

$$m = (1 + 0.76\, b/W) \tag{63}$$

and for the CCP

$$m = -\frac{1}{n} = -N \tag{64}$$

where N is the strain-hardening index ($0 \leq N \leq 1$).

References

[1] Rice, J. R., *Journal of Applied Mechanics*, Vol. 35, 1968, pp. 379–386.
[2] Hutchinson, J. W., *Journal of the Mechanics and Physics of Solids*, Vol. 16, No. 1, 1968, pp. 13–31.
[3] Rice, J. R. and Rosengren, G. F., *Journal of the Mechanics and Physics of Solids*, Vol. 16, No. 1, 1968, pp. 1–12.
[4] Hutchinson, J. W. and Paris, P. C. in *Elastic-Plastic Fracture, ASTM STP 668*, J. D. Landes, J. A. Begley, and G. A. Clarke, Eds., American Society for Testing and Materials, 1979, pp. 37–64.
[5] Wnuk, M. P., *Transactions*, American Society of Mechanical Engineers, *Journal of Applied Mechanics*, Vol. 41, March 1974, pp. 234–242.
[6] Wnuk, M. P. and Sedmak, S. in *Fracture Mechanics: 13th Conference, ASTM STP 743*, American Society for Testing and Materials, 1981, pp. 236–249.
[7] Rice, J. R., Drugan, W. J., and Sham, T. L. in *Fracture Mechanics: 12th Conference, ASTM STP 700*, American Society for Testing and Materials, 1980, pp. 189–221.
[8] Dean, R. H. and Hutchinson, J. W. in *Fracture Mechanics: 12th Conference, ASTM STP 700*, American Society for Testing and Materials, 1980, pp. 383–405.
[9] Herman, L. and Rice, J. R., "Comparison of Experiment and Theory for Elastic Plastic Plane Strain Crack Growth," Brown University Report No. 76, Providence, R.I., Feb. 1980.
[10] Rice, J. R. and Sorensen, E. P., *Journal of the Mechanics and Physics of Solids*, Vol. 26, 1978, pp. 163–186.
[11] Rice, J. R., Paris, P. C., and Merkle, J. G. in *Progress in Flaw Growth and Fracture Toughness Testing, ASTM STP 536*, American Society for Testing and Materials, 1973, pp. 231–245.

[*12*] Merkle, J. and Corten, H., *Transactions*, American Society of Mechanical Engineers, *Journal of Pressure Vessel Technology*, 1974, pp. 286–292.
[*13*] Ernst, H. A., Paris, P. C., and Landes, J. D. in *Fracture Mechanics: 13th Conference, ASTM STP 743*, American Society for Testing and Materials, 1981, pp. 476–502.
[*14*] Ernst, H. A. in *Fracture Mechanics: 14th Conference, ASTM STP 791*, American Society for Testing and Materials, 1982, pp. 499–519.
[*15*] Paris, P. C., Tada, H., Zahoor, A., and Ernst, H. A. in *Elastic-Plastic Fracture, ASTM STP 668*, American Society for Testing and Materials, 1979, pp. 5–36.
[*16*] Paris, P. C., Tada, H., Ernst, H., and Zahoor, A. in *Elastic-Plastic Fracture, ASTM STP 668*, American Society for Testing and Materials, 1979, pp. 251–265.
[*17*] Ernst, H. A., this publication, pp. II-133–II-155.
[*18*] McCabe, D. E. and Landes, J. D., this publication, pp. II-353–II-371.
[*19*] McCabe, D. E., Landes, J. D., and Ernst, H. A., this publication, pp. II-560–II-579.
[*20*] Paris, P. C. and Johnson, R. E., this publication, pp. II-5–II-40.
[*21*] Ernst, H. A., work in progress.

Jalees Ahmad,[1] *Clarence R. Barnes,*[1] *and Melvin F. Kanninen*[1]

An Elastoplastic Finite-Element Investigation of Crack Initiation Under Mixed-Mode Static and Dynamic Loading

REFERENCE: Ahmad, J., Barnes, C. R., and Kanninen, M. F., **"An Elastoplastic Finite-Element Investigation of Crack Initiation Under Mixed-Mode Static and Dynamic Loading,"** *Elastic-Plastic Fracture: Second Symposium, Volume I—Inelastic Crack Analysis, ASTM STP 803*, C. F. Shih and J. P. Gudas, Eds., American Society for Testing and Materials, 1983, pp. I-214–I-239.

ABSTRACT: The fracture mechanics techniques needed for materials that fracture in a highly ductile manner must give explicit attention to the extensive plastic deformation surrounding the crack tip. Large-scale crack-tip plasticity has two effects. First, because of crack-tip blunting, crack growth initiation tends to occur under conditions for which conventional linear elastic fracture mechanics analyses are invalid. Second, significant amounts of stable crack growth can occur prior to fracture instability. To cope with these essentially inelastic processes, elastic-plastic fracture mechanics techniques are now being widely pursued [*1*].[2] The research reported in this paper builds on these developments to provide the basis for fracture mechanics treatments under the more general conditions involved in dynamic mixed-mode loading.

Current work in elastic-plastic fracture mechanics is largely confined to quasi-static loading conditions and to crack growth in the opening mode (Mode I). Yet, in impact-loaded structures, crack initiation could occur under dynamic conditions with a combination of both the opening mode and the sliding mode (Mode II). Thus, it is necessary to obtain a quantitative understanding of crack growth initiation, propagation, and arrest under initial combined-mode dynamic loading conditions. The research described here is aimed at developing this understanding through a program of integrated experimental and finite-element analysis work.

KEY WORDS: elastic-plastic fracture, mixed mode, impact loading

In a critical survey of progress in elastic-plastic fracture mechanics, Kanninen et al [*1*] found that most efforts are focused on the *J*-resistance curve

[1]Research scientists and project manager, respectively, Battelle Columbus Laboratories, Columbus, Ohio 43201.

[2]The italic numbers in brackets refer to the list of references appended to this paper.

approach. Despite the fact that such approaches are inherently limited to small amounts of crack growth prior to fracture instability, they do provide a significant improvement over the conventional linear-elastic fracture mechanics techniques. However, the current approaches are restricted to Mode I conditions under quasi-static loading. While a generalization of the J-integral valid for mixed-mode and dynamic loading conditions exists, it has not previously been critically examined. Nevertheless, in view of the success that has been achieved with J and the complete lack of a viable alternative, this parameter, known as the $\hat{J}$-integral, was selected for use in this study. An integrated program of finite-element analysis and experimentation was pursued to assess its feasibility for the objectives of this research.

The experimental work employed HY-100 steel three-point-bend specimens with slant edge cracks as shown in Fig. 1. This configuration has a number of advantages for this research. First, crack initiation and propagation can be obtained for both quasi-static and dynamic loading in the same specimen geometry. Second, by varying the crack angle and the crack length, a range of Mode I and Mode II combinations can be achieved. Third, this specimen can be economically analyzed by a finite-element method. Fourth, because of the constraint that arises in bending, the J-resistance curves obtained may be lower-bound values that will provide conservative predictions when utilized for structural integrity assessments.

A dynamic elastic-plastic finite-element analysis was used in the analysis portion of the research. The objective was to learn if crack growth initiation in HY-100 steel is governed by critical values of the $\hat{J}$-parameter. The analyses were of two main types. First, for quasi-static loading, critical $\hat{J}$-values were determined by coupling the finite-element analysis to the experiment. In essence, by calculating the vectorial components of $\hat{J}$ (that is, $\hat{J}_1$ and $\hat{J}_2$) as a function of load and using the load level at initiation, critical values for

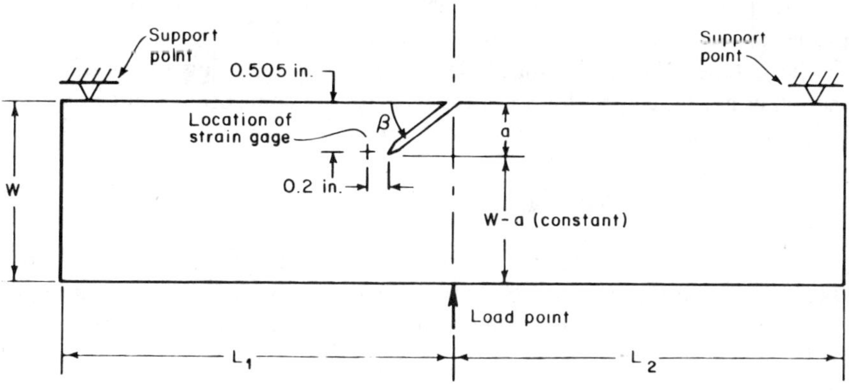

FIG. 1—*Three-point-bend specimen with a slant edge crack (1 in. = 25.4 mm).*

each crack angle could be determined. This provided a tentative elastic-plastic mixed-mode fracture criterion in terms of the parameter $\hat{J}_{1c}$.

The second type of analysis was performed for impact loading. Here, the fracture criterion deduced under quasi-static loading was used as input to a finite-element simulation of the impact-loaded experiment. In this way a prediction of the initiation time was made for comparison with the experiments. An assessment of the accuracy of this prediction then gives an indication of the usefulness of the approach. Unfortunately, the experimental techniques applied (mainly developed for application to materials less ductile than HY-100 steel) were not sufficiently refined to give an unequivocal comparison with the analysis predictions.

Fracture Criterion Development

The $\hat{J}$-Integral

The path-independent integral J as originally proposed by Rice [*2*] is valid only for Mode I loading under plane-stress or plane-strain conditions. Also, it is based on the assumptions that all inertia and body forces are zero, thermal stresses are zero, and the material behavior is (linear or nonlinear) elastic. Several attempts have been made in recent years to arrive at a more general definition of J which would minimize these assumptions while still retaining its desirable features—the energy release rate interpretation of J and the rather straightforward way in which it can be computed using the contour integral definition. Among others, these attempts are due to Hellen and Blackburn [*3*], Blackburn [*4*], Neale [*5*], Bergez [*6*], Bui [*7*], Freund [*8*], Wilson and Yu [*9*], Mall [*10*], Kishimoto et al [*11*], and Miyamoto et al [*12*]. While some of the proposed formulations extend the definition of J to axisymmetric and three-dimensional problems [*3–6,12*], others consider elastic-plastic material behavior [*4*] and more general loading conditions [*7–11*]. Kishimoto et al [*11*] have presented a unified expression for J (called $\hat{J}$) which encompasses almost all the other proposed generalizations.

The detailed derivation for the $\hat{J}$-integral by integrating the equation of motion of a solid can be found in Ref *11*. From a conceptual point of view, a convenient expression for $\hat{J}$ can be written as

$$\hat{J}_k = J_{k_e} + J_{k_d} + J_{k_t} + J_{k_p} + J_{k_b} \tag{1}$$

where the lower-case letter subscripts stand for the elastic (e), dynamic (d), thermal (t), plastic (p), and body force (b) contributions to the $\hat{J}_k$-integrals and k (=1,2) indicates that each term in Eq 1 is a vector. Kishimoto et al [*11*] define $\hat{J}$ as follows

$$\hat{J} = \hat{J}_1 \cos\theta + \hat{J}_2 \sin\theta \tag{2}$$

where θ is the direction of virtual crack extension measured anticlockwise from the crack line (Fig. 2). The integrals J_{k_e} through J_{k_b} in Eq 2 may be expressed as

$$J_{ke} = \int_{\Gamma_w} \left(W_e n_k - T_i \frac{\partial u_i}{\partial x_k} \right) d\Gamma \tag{3}$$

$$J_{kd} = \iint_A \rho \ddot{u}_i \frac{\partial u_i}{\partial x_k} dA \tag{4}$$

$$J_{kp} = \iint_A \sigma_{ij} \frac{\partial \epsilon_{ij}{}^*}{\partial x_k} dA \tag{5}$$

$$J_{k_t} = \iint_A \alpha_1 \epsilon_{ii} \frac{\partial T}{\partial X_k} dA - \int_{\Gamma_w} {}^1\!/\!_2 \alpha_1 T \epsilon_{ii} n_k d\Gamma \tag{6}$$

$$J_{kb} = - \iint_A F_i \frac{\partial u_i}{\partial X_k} dA \tag{7}$$

$$\Gamma_w = \Gamma + \Gamma_s - \Gamma_{\text{end}}$$

where

W_e = elastic strain energy density,
T_i = traction vector,
σ_{ij} = stress tensor,
ϵ_{ij} = strain tensor,
$\epsilon_{ij}{}^*$ = plastic strain tensor,
u_i = displacement vector,
$\ddot{u}_i$ = acceleration vector,
T = temperature increment,
F_i = body force,
n_k = components of unit outward normal to contour,
x_k = crack-tip coordinates (Fig. 1),
A = area enclosed between contours Γ and Γ_{end},
θ = direction of virtual crack extension,
α = coefficient of thermal expansion,
ρ = mass density, and
α_1 = $\alpha E/(1 - 2\nu)$—plane strain = $\alpha E/(1 - \nu)$—plane stress.

It can be shown [*11,13*] that $\hat{J}$ represents the energy flow rate into the region A_{end} (Fig. 2) when the externally applied forces on the body (including the body force) remain constant. In general, all of the energy flowing into A_{end} is not available for crack advance since a part of it must be spent in such

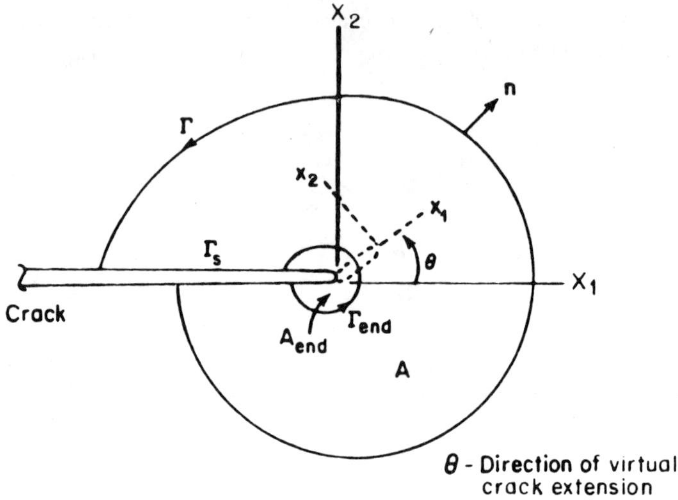

FIG. 2—*Crack-tip coordinates used in the definition of* $\hat{J}_k$.

microprocesses as void nucleation and coalescence occurring around the crack tip. However, under small-scale yielding conditions, it can be assumed that the energy spent in micromechanical processes is very small and can be ignored. With this assumption it becomes possible to treat even the region A_{end} using a continuum mechanics approach, and the limit as $A_{\text{end}} \to 0$ (Γ_{end} coincides with the crack boundary) can be considered.

For the mixed-mode problem addressed in this research, it can be considered that temperature changes and body forces are zero. Then, with the small-scale yielding assumption ($A_{\text{end}} \to 0$) it follows that

$$\hat{J} = \int_{\Gamma+\Gamma_s} \left(W_e n_k - T_i \frac{\partial u_i}{\partial x_k} \right) d\Gamma + \iint_A \sigma_{ij} \frac{\partial \epsilon_{ij}^*}{\partial x_k} dA + \iint_A \rho \ddot{u}_i \frac{\partial u_i}{\partial x_k} dA \tag{8}$$

Note that W_e is the elastic strain energy density which includes any nonlinear elastic (deformation plasticity) contributions. Therefore, where the material behavior can be described using the deformation theory of plasticity, the area integral term containing ϵ_{ij}^* would be zero.

Invoking the assumptions involved in the derivation of the J-integral [2], both the area integral terms in Eq 8 are identically zero. Also, since only Mode I loading is assumed ($\theta = 0$, $k = 1$) and the crack faces are traction free, it is found that

$$\hat{J} = \hat{J}_1 = \int_{\Gamma} \left(W_e n_1 - T_i \frac{\partial u_i}{\partial x_1} \right) d\Gamma \tag{9}$$

Starting from Eq 9, for linear elastic material behavior and mixed mode loading it can be shown that

$$\hat{J}_1 = \frac{\kappa + 1}{8\mu}\left[K_I^2(t) + K_{II}^2(t)\right] + \frac{1}{2\mu}K_{III}^2(t) \tag{10}$$

$$\hat{J}_2 = -\frac{\kappa + 1}{4\mu}K_I(t)\,K_{II}(t) \tag{11}$$

where μ is the shear modulus, and $\kappa = 3 - 4\nu$ for plane strain and $\kappa = (3 - \nu)/(1 + \nu)$ for plane stress.

The validity of the $\hat{J}$ approach was verified by solving a number of special cases for which solutions could be found in the literature. These include (1) an inclined edge-cracked plate under uniform tension, (2) a center-cracked plate under suddenly applied tension, (3) constant-velocity crack propagation, and (4) an unrestrained impact-loaded bend specimen under elastic and elastic-plastic conditions.

In all cases it was found that the use of the Ĵ-integral provided results which were in good agreement with the reference solutions. Details of these analyses can be found in Refs *14* and *15*.

Experimental Procedure

The specimen type used in this research is shown in Fig. 1. Specimen blanks were cut from a HY-100 steel plate. The specimen blanks were oriented such that the crack propagation direction for the specimens with $\beta = 90$ deg (see Fig. 1) was parallel to the plate rolling direction. Crack direction in specimens with $\beta = 30$ and $\beta = 45$ deg was at 60 and 45 deg, respectively, to the rolling direction.

The general specimen configuration conformed with the American National Standards Institute/American Society for Testing and Materials (ANSI/ ASTM) E6077-77 Standard for Dynamic Tear Specimens except for the thickness and notch treatment. While the ANSI/ASTM standard requires a thickness of 15.875 mm, 29.21-mm-thick specimens (full plate thickness) were used in this investigation to promote plane-strain behavior at the notch tip. The standard also requires that the notch root be sharpened by pressing a preformed tool into the notch root.

In this work, however, the notches for the quasi-static tests were sharpened using a 127-μm jewelers saw. This resulted in a notch root radius of about 82.55 μm. This "blunted" notch was adopted to conform with the notches used on the first three impact-loaded specimens described later.

The dynamic specimens were prepared in accord with the procedure used for the static specimens except that three notch sharpening procedures were used. The first of these, the jeweler's saw sharpening procedure, was adopted because past experience at Battelle has demonstrated that many otherwise ductile materials fail either partially or totally by cleavage when tested in specimens with a pressed notch as specified by the ANSI/ASTM standard. (One such material is HY-130 steel.) In addition, the specimen configuration precluded the use of fatigue (as used in the ASTM standards for K_{Ic} specimens) to sharpen the notch root. It was anticipated that the blunt notches used would store enough energy to promote rapid unstable crack growth which would broaden the scope of the experimental results.

The quasi-static tests were performed using a Baldwin-Tate Emery Test Machine having a load capacity of 27 000 kg. The loading rate was adjusted such that the load-point displacement (LPD) was 254 μm min^{-1}. The loading was interrupted at preselected points on the load-displacement curve to fill the notch with a silicone rubber compound.[3] This procedure was used to determine the load at crack initiation. It was determined that the cracks initiated internally and tunneled for a distance roughly equal to the shear lip width (5.0 to 8 mm) before reaching the surface.

A Battelle-designed and built machine, similar in concept to a Charpy machine, was used to conduct the impact tests. It is a multipurpose pendulum-type machine with a total impact energy of approximately 21693 J. A variety of specimen geometries can be accommodated by changing the bolt-on supports.

To assess the practicality of determining crack initiation and crack propagation speeds in HY-100 steels, a crack monitoring gage was deposited upon the specimen surface. This technique involves the vapor deposition of an electrically conducting grid upon a thin epoxy layer bonded to the test specimen surface. In addition, the impact machine was electrically wired so that the tup striking the test specimen would provide a closing switch, producing a voltage indicating tup-specimen contact. The tup contact switch and output from the crack monitoring gage were recorded simultaneously by a Biomation 8100 transient recorder and a frequency-modulated (FM) tape recorder. The strain gage outputs were recorded on two adjacent tape channels.

In the revised procedure, each specimen surface was roughened by grit-blasting to promote epoxy adhesion. Following this, care was taken to obtain a small (0.05 mm) notch root radius using electric discharge machining (EDM). Experience has generally shown that a sharper crack produces a smaller plastic zone. Smaller zones minimize epoxy debonding. Each specimen was also strain-gaged to more accurately determine crack initiation. Results from the first three tests indicated that the crack initiated at a time of 300 to 400 μs after tup contact.

Even with these precautions, some of the revised test procedure data were less definitive than the data from the initial series. The crack monitoring output was particularly bad. Crack speeds could not be deduced from the grid

[3]Kerr Citrocone, mfg. by Kerr/Sybron and available at most dental supply companies.

output signal. On the other hand, better notch preparation reduced the energy required for specimen fracture by 17 to 24 percent.

These inconsistent initiation times prompted another effort to more accurately predict crack initiation. The last three impact tests (112D, 113D, and 114D) contained semicircular crack starter notches sharpened by EDM. This notch configuration was adopted so that a strain gage could be applied in the notch center and thus indicate crack initiation more precisely than would a gage placed upon the specimen surface. Another gage was positioned at the side of the sharpened notch as in previous specimens. This gage was used to detect the time at which the internally initiated crack surfaced.

Analysis of Crack Initiation in HY-100 Steel Under Mixed-Mode Loading

The specimen geometries analyzed in this study are shown in Fig. 3. The different types of analyses performed on each specimen type are given in Table 1. The results for the static and dynamic elastic-plastic analyses of the midpoint cracked specimen and the static elastic analyses of the quarter-point cracked specimen are presented in the following.[4]

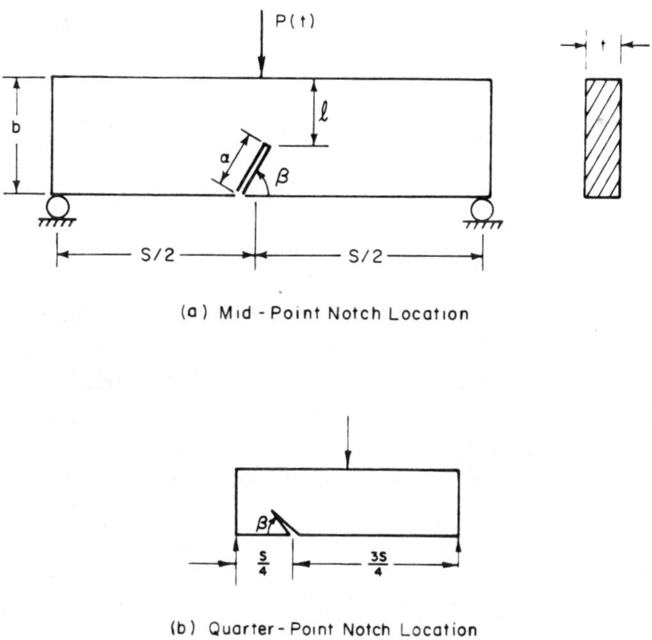

FIG. 3—*Three-point-bend specimen with a slant edge crack:* s = *165.1 mm,* b = *40.64 mm,* l = *28.58 mm,* t = *29.21 mm;* β = *30, 45, and 90 deg (1 mm = 0.04 in.).*

[4]The stress-strain curve used in the analysis is shown in Fig. 4.

TABLE 1—*List of finite-element analyses performed on mixed-mode test specimen geometries of Fig. 3.*

	Analyses Performed		
Analysis No.	Crack Position	β, deg	Type of Analysis
1	midpoint	30	static elastic-plastic
2	midpoint	45	static elastic-plastic
3	midpoint	90	static elastic-plastic
4	midpoint	30	dynamic elastic-plastic
5	midpoint	45	dynamic elastic-plastic
6	midpoint	90	dynamic elastic-plastic
7	quarter-point	30	static elastic
8	quarter-point	45	static elastic
9	quarter-point	90	static elastic

Static Elastic-Plastic Analyses of Midpoint Cracked Specimens with β = 30, 45, and 90 Deg

The finite-element meshes used in these analyses are shown in Fig. 5. In the case of $\beta = 90$ deg, symmetry about the x-axis was exploited so that only one half of the specimen geometry had to be analyzed. For $\beta = 30$-deg and $\beta = 45$-deg specimen geometries, 74 eight-noded isoparametric elements and 255 nodes (510 deg of freedom) were used. For $\beta = 90$-deg specimen, 52 elements and 195 nodes (390 degrees of freedom) were employed. In all cases, the smallest element dimension used around the crack tip was 0.04 times the crack length. For each analysis $\hat{J}$ was computed using three integration paths as indicated in Fig. 5.

The results of these analyses are shown in Figs. 6, 7, and 8. It is clear that path independence of the $\hat{J}_1$ component of the $\hat{J}$-integral is more consistent than of the $\hat{J}_2$-integral. Also, there is a deterioration in path independence with decreasing notch angle β. While there is no obvious reason for this behavior, it is consistent with the findings of Kishimoto et al [*11*] and Hellen [*3*].

In general, all three contours yielded very similar results at lower load levels. The path independence deteriorated slowly as more load was applied. Specifically, the values from the outermost contour seemed to diverge from the values of the two inner contours. At first it was thought that this is due to interference with the plastic field around the load point. Further scrutiny of the results for the 90-deg notch angle revealed that lack of adequate convergence in the incremental plasticity solution may also have contributed to the deterioration in path independence. Although such convergence problems cannot be entirely eliminated (especially for materials exhibiting only small amounts of strain hardening), reducing the incremental load step size and imposing a more strict convergence criterion will surely improve the results.

At the experimentally measured loads at initiation the corresponding

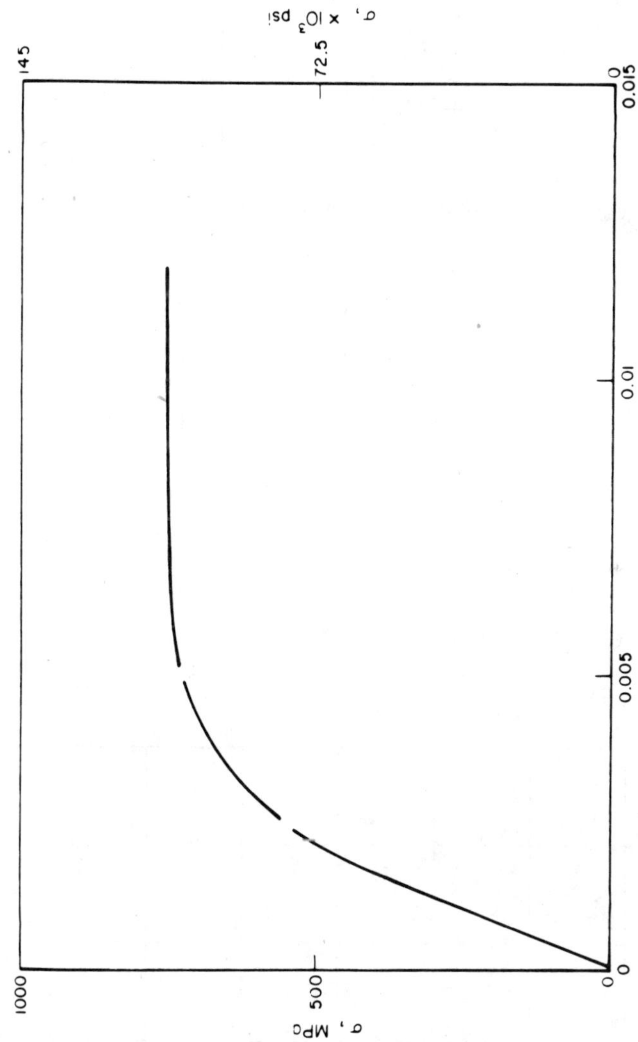

FIG. 4—*Stress-strain curve for HY-100 obtained by static uniaxial tension test.*

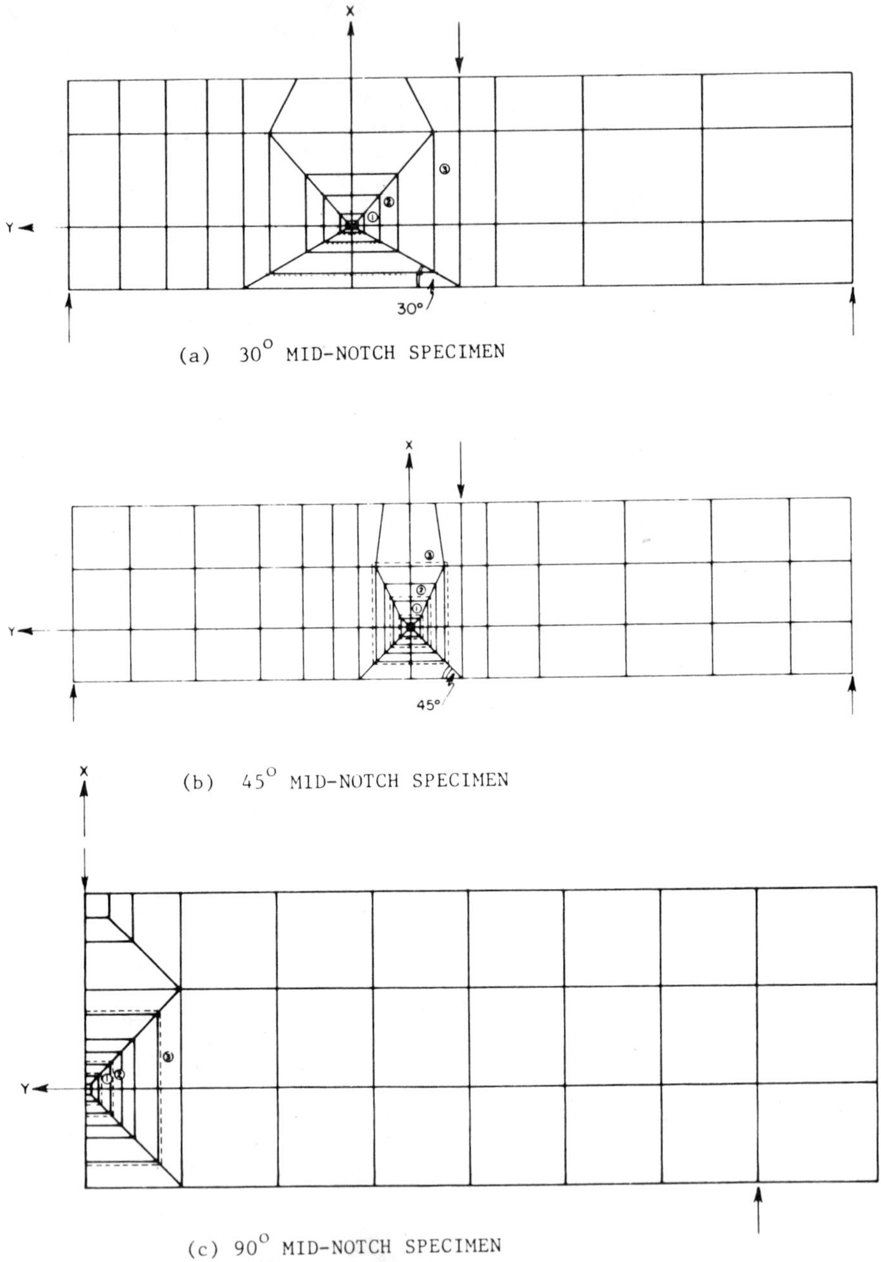

FIG. 5—*Finite-element mesh for 30, 45, and 90-deg mid-notch specimen.*

values for $\hat{J}_{1c}$ and $\hat{J}_{2c}$ were obtained using Figs. 6–8. These values are given in Table 2.

Dynamic Elastic-Plastic Analyses of Midpoint Cracked Specimens with $\beta = 30$, 45, and 90 Deg

The finite-element meshes used in these analyses were the same as those used in the corresponding static analyses except that the length of the specimens was changed to 152.4 mm (as used in the impact experiments). In all analyses a time step of 2 μs was used. The impact load was simulated by applying a constant-displacement-rate boundary condition at the load point. The load-point displacement was obtained from an estimate of 6.88-m/s tup speed in the actual experiments with the assumption being made that the tup is sufficiently massive that its progress is unimpeded by the specimen. The results of these analyses are shown in Figs. 9, 10, and 11.

Regarding the path independence of $\hat{J}_1$ and $\hat{J}_2$, the trends were found to be similar to those in the static analyses; that is, the path independence deteriorates with decreasing angle β and with increasing time.

Static Elastic Analysis of Quarter-Point Cracked Specimens with $\beta = 30$, 45, and 90 Deg

The finite-element meshes used in these analyses are shown in Figs. 12*a* to 12*c*. Results of the analyses are summarized in Table 3. Once again, the results from the three contours are more consistent for $\hat{J}_1$ and for larger β-values.

For $\beta = 90$ deg, ignoring the relatively smaller $\hat{J}_2$-value and using a $\hat{J}_{1c}$ of 254 000.0 N/m (from Table 2), the load at initiation can be roughly estimated as 421 011.3 ($\sqrt{\hat{J}_{1c}/1.433 \times 10^{-6}}$). This load is high enough (as was experimentally found) to cause gross yielding of the specimen under the load point before initiation occurs. This was consistent with the experimental observations.

In view of this it was decided not to perform elastic-plastic static and dynamic analyses of these specimens.

Discussion of Analyses

While a fairly detailed description of the $\hat{J}$-integral was given in the preceding, one computational detail which possibly needs some further discussion is the treatment of the terms containing elastic strain energy density (W_e) and the plastic strain ϵ_{ij}^* in Eqs 1 through 5. It was mentioned earlier that W_e in J_{k_e} contains nonlinear elastic (deformation plasticity) contribution. The term J_{k_p}, on the other hand, contains plastic strain ϵ_{ij}^* in the incremental plasticity sense. Now, in a nonlinear finite-element program, the separation of non-

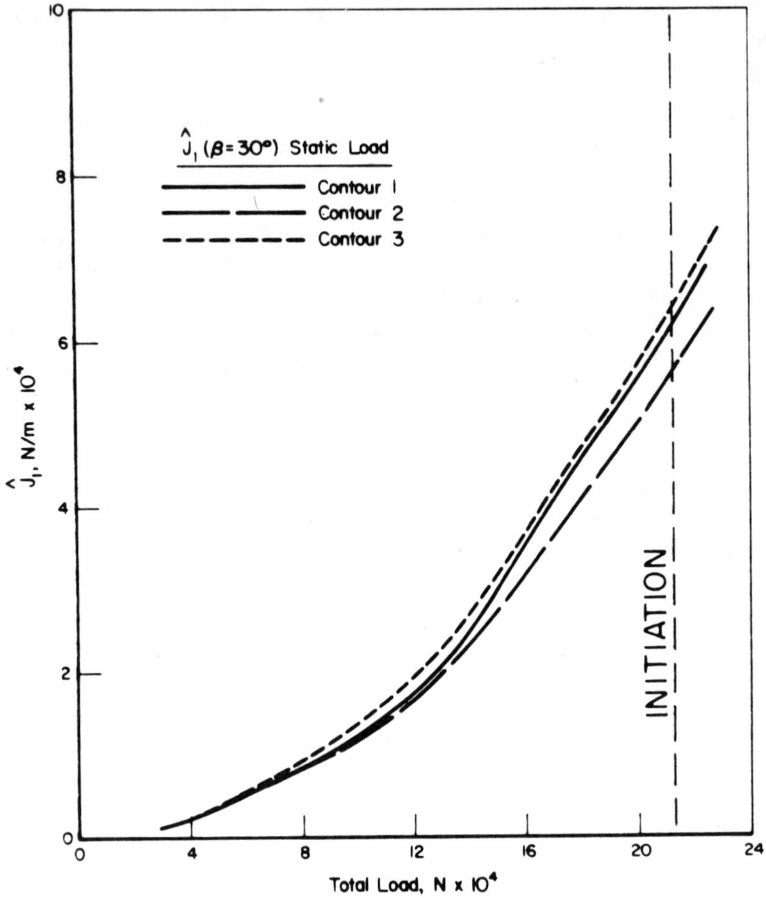

(a) VARIATION OF $\hat{J}_1$, WITH LOAD

FIG. 6—*Static elastic-plastic analysis of 30-deg mid-notch (1 N/m = 0.74 lb/ft).*

linear elastic strain from plastic strain can be accomplished only by invoking an empirical criterion. To overcome this difficulty the following equations were utilized

$$\iint_A \sigma_{ij} \frac{\epsilon_{ij}}{\partial x_k} dA = \iint_A \sigma_{ij} \frac{\sigma\epsilon_{ij}^e}{\partial x_k} dA + \iint_A \sigma_{ij} \frac{\epsilon_{ij}^*}{\partial x_k} dA \quad (12)$$

Equation 12 is a result of dividing the total strain into its elastic and plastic components. Now, recognizing that

$$\sigma_{ij} = \frac{\partial W_e(\epsilon_{ij}^e)}{\partial \epsilon_{ij}^e} \quad (13)$$

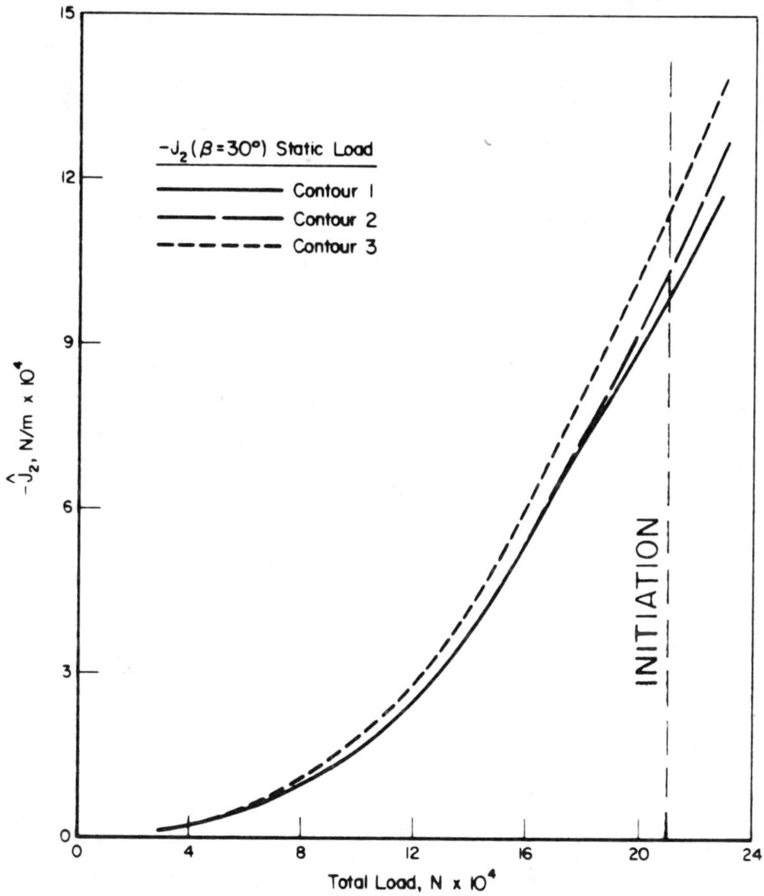

(b) VARIATION OF $\hat{J}_2$ WITH LOAD

FIG. 6—*Continued.*

and using the Green's theorem in Region A of Fig. 2 (where it is certainly applicable) the first area integral on the right-hand side of Eq 12 can be reduced to a contour integral. In fact, It can be easily shown that even in the limit as $A_{end} \rightarrow 0$ this term becomes identical to the first term in the expression for J_{k_e} in Eq 3. Therefore, the left-hand side of Eq 12 simply represents the sum of J_{k_p} and the first term of J_{k_e}. Further, it avoids having to separate the elastic and plastic strain components.

An interesting fact about the Ĵ-integrals is that, long before Kishimoto et al's formulation [*11*] appeared, Hellen and Blackburn [*3*] and Bergez [*6*] had arrived at identical results in their attempts to find energy release rate expressions for linear elastic mixed-mode problems. The work of Bergez [*6*] clearly suggests that $\hat{J}_1$ is the energy release rate for mixed-mode problems. The use of

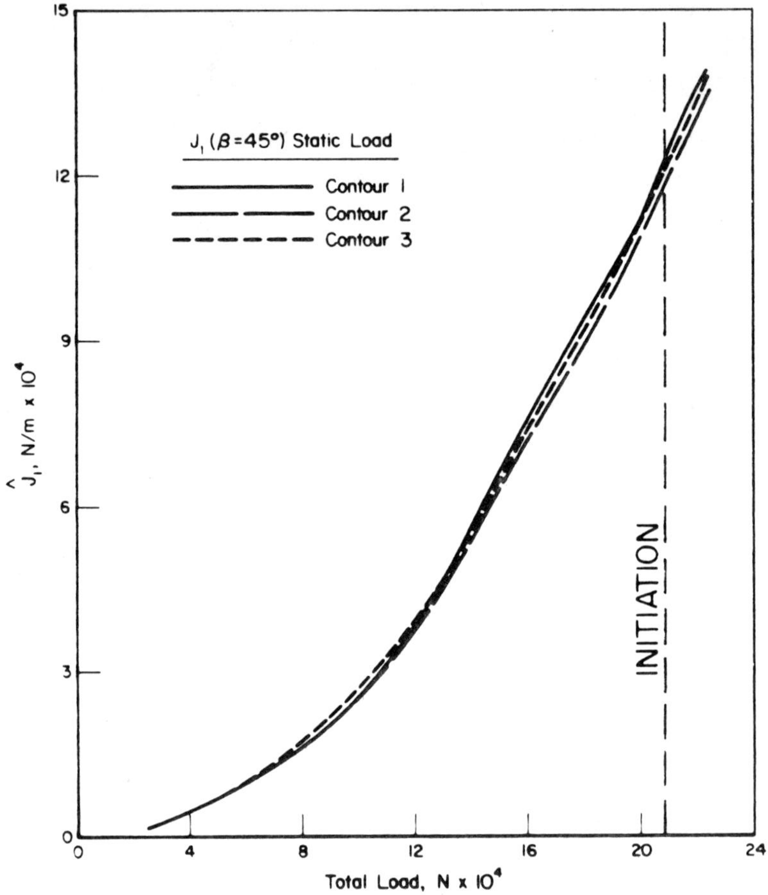

(a) VARIATION OF $\hat{J}_1$ WITH LOAD

FIG. 7—*Static elastic-plastic analysis of 45-deg mid-notch (1 N/m = 0.74 lb/ft).*

$\hat{J}_2$ in his work (along with several other path-independent integrals) serves only as a means of separately evaluating the Mode I and Mode II stress-intensity factors. For example, in order to evaluate K_I and K_{II} from Eqs 10 and 11 at least two equations are necessary. The suggestion that $\hat{J}_1$ is (in linear elastic case) the energy release rate is also supported by the fact that under the Mode I condition, Eq 10 reduces exactly to Griffith's energy release rate G.

Results and Discussion

The key results of the quasi-static experiments and the finite-element analyses of the inclined midpoint-notch-bend specimen for the three notch angles considered are contained in Table 2. Using these results a plot of

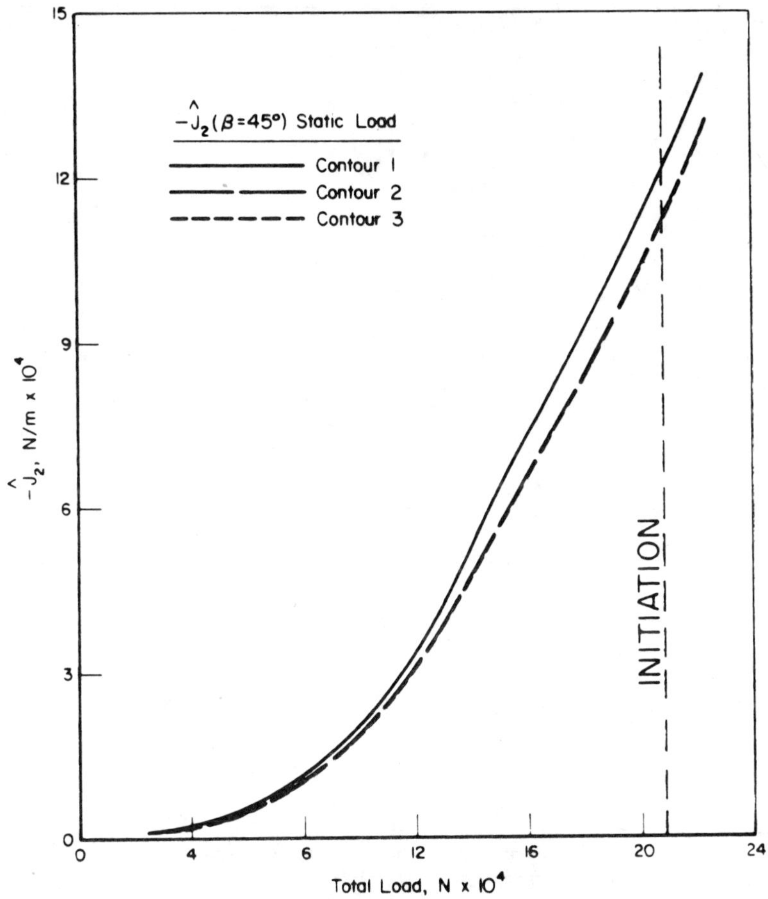

(b) VARIATION OF $\hat{J}_2$ WITH LOAD

FIG. 7—*Continued.*

notch angle (β) versus the $\hat{J}_1$ at crack initiation ($\hat{J}_{1c}$) can be made. This is shown in Fig. 13. An empirical fit of these data can be made which suggests that initiation occurs when

$$\hat{J}_1 = J_c \sin^2\beta \tag{14}$$

where J_c is the value of J_{1c} at $\beta = 90$ deg. Note that within the deformation plasticity assumption (no unloading), J_c coincides exactly with the usual J-integral value at crack initiation. As shown in Fig. 13 and Table 3, a value of $J_c = 0.254$ MN/m was found in this work.[5]

[5]A brief literature search for J_{1c}-values for HY-100 steel, made for comparison, was not successful.

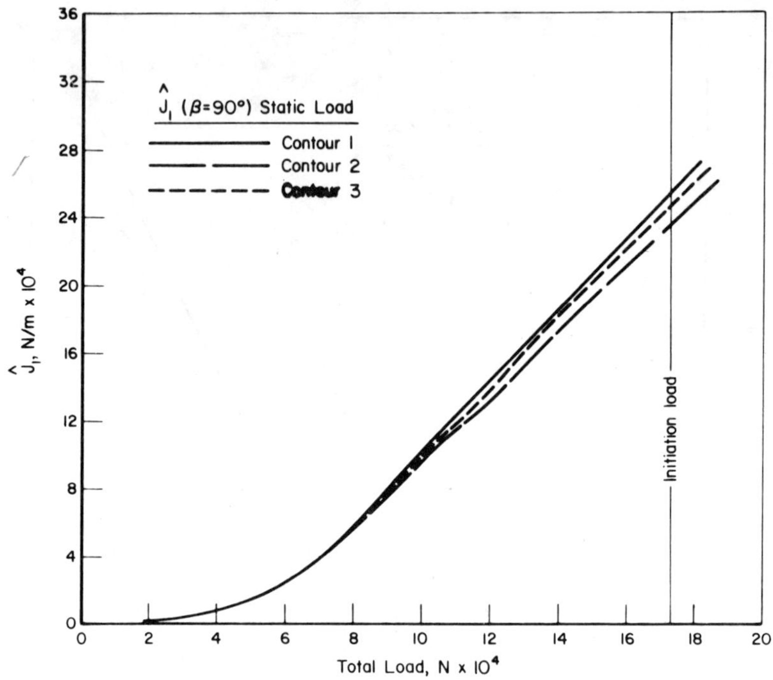

FIG. 8—*Static elastic-plastic analysis of 90-deg specimen (1 N/m = 0.74 lb/ft).*

TABLE 2—$\hat{J}_1$ *and* $\hat{J}_2$ *at initiation for static loading.*

β, deg	$\hat{J}_{1c}$, N/m	$-\hat{J}_{2c}$, N/m
30	6.26×10^4	10.99×10^4
45	12.84×10^4	12.5×10^4
90	25.4×10^4	...

The crack initiation criterion suggested by Eq 14 is based on fitting a curve through only four data points (three experimental results together with the origin). Could some other function of β provide just as good a fit? To explore this possibility, consider the problem of an infinite plate containing a crack of length $2a$ subjected to remote uniaxial applied stress σ whose direction makes an angle β with the crack line (Fig. 14). The linear elastic solution of this problem gives [*16*]

$$K_I = \sigma(\pi a)^{1/2} \sin^2\beta \tag{15}$$

$$K_{II} = \sigma(\pi a)^{1/2} \sin\beta \cos\beta \tag{16}$$

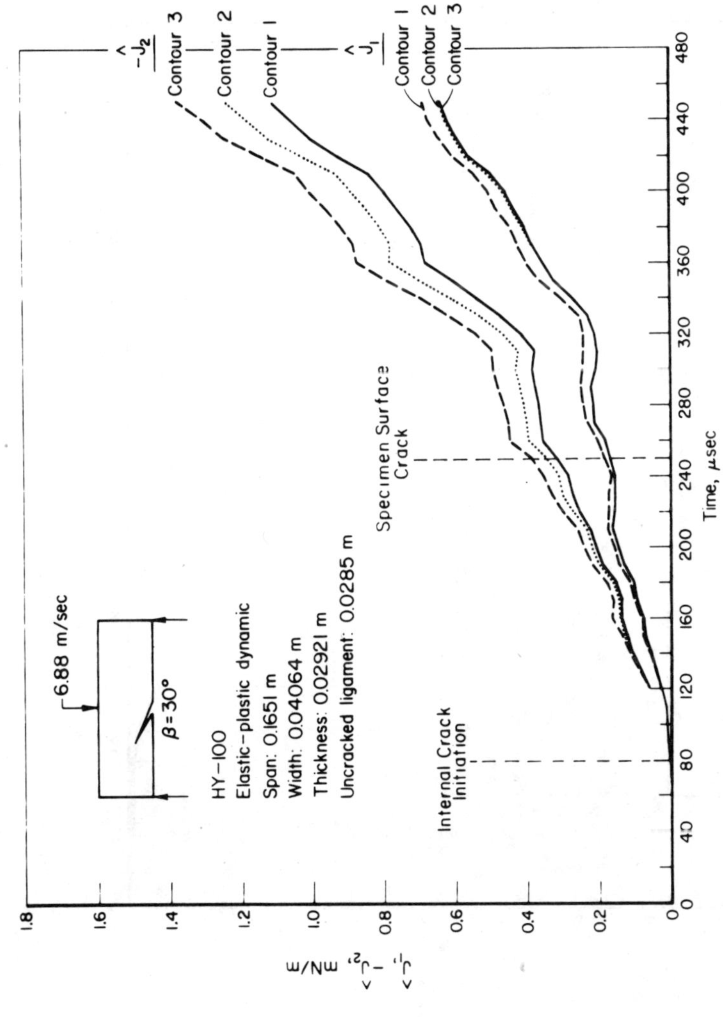

FIG. 9—*Dynamic elastic-plastic analysis of 30-deg mid-notch specimen (1 N/m = 0.74 lb/ft).*

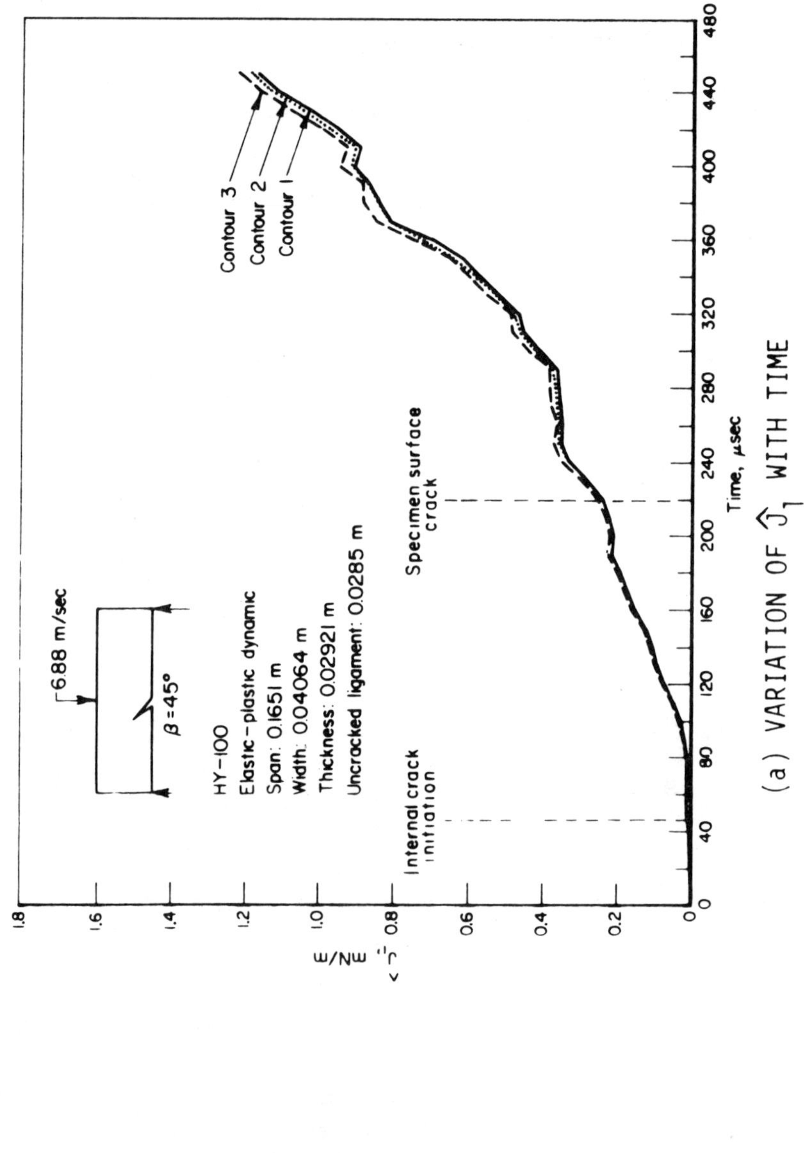

FIG. 10—*Dynamic elastic-plastic analysis of 45-deg mid-notch (1 N/m = 0.74 lb/ft).*

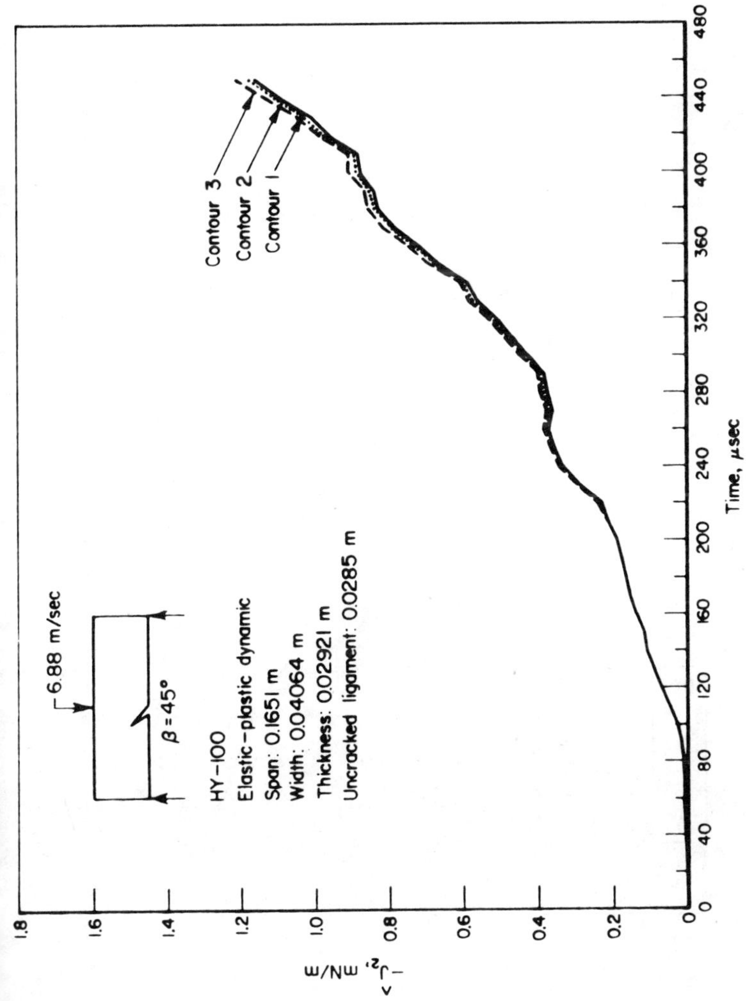

(b) VARIATION OF $\hat{J}_2$ WITH TIME

FIG. 10—*Continued.*

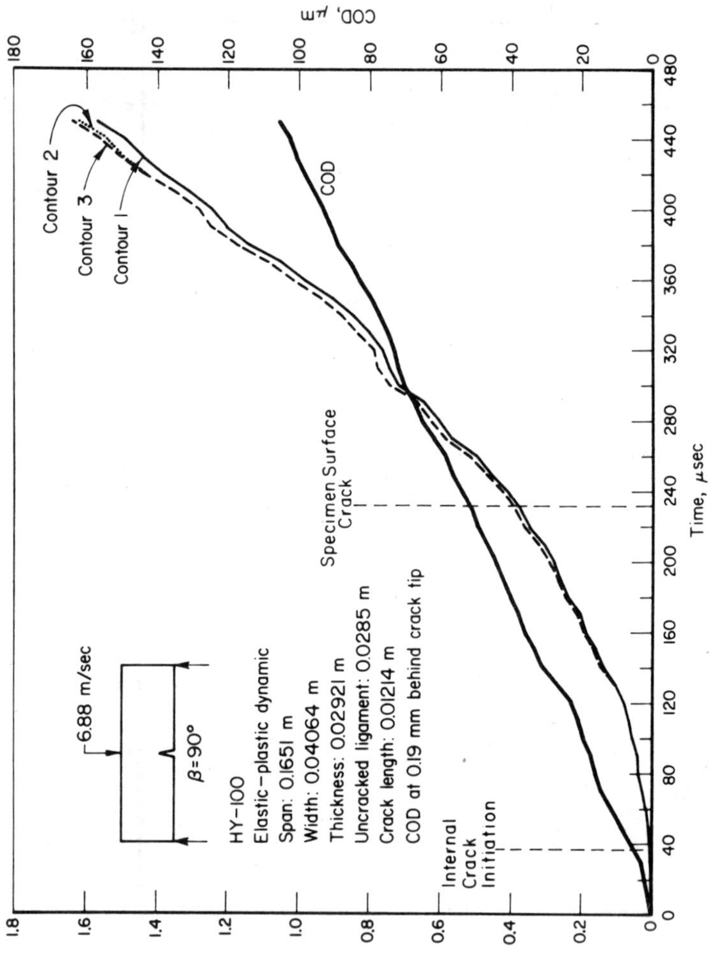

FIG. 11—*Dynamic elastic-plastic analysis of 90-deg mid-notch specimen (1 N/m = 0.74 lb/ft).*

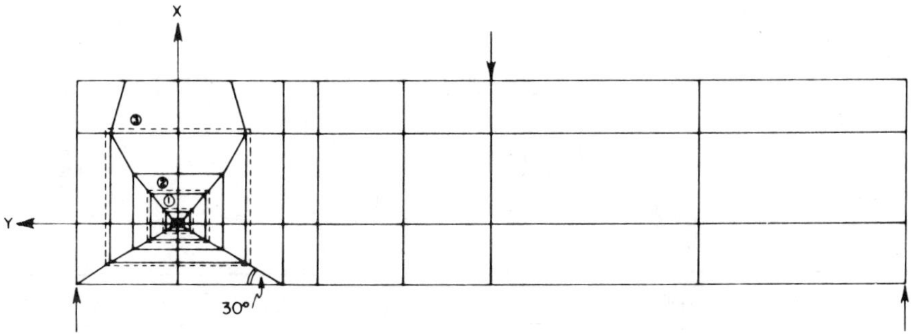

(a) 30° QUARTER-POINT SPECIMEN

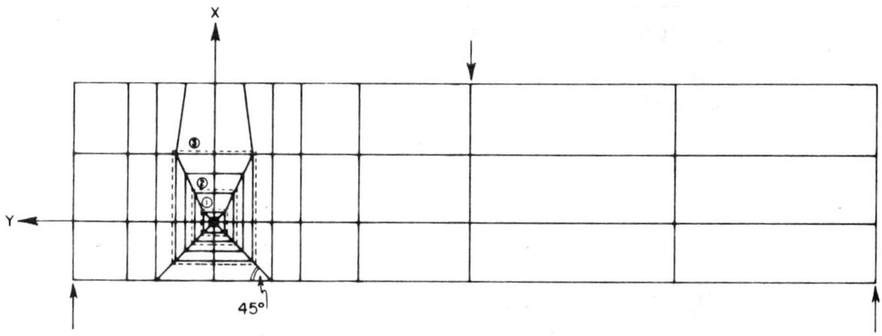

(b) 45° QUARTER-POINT SPECIMEN

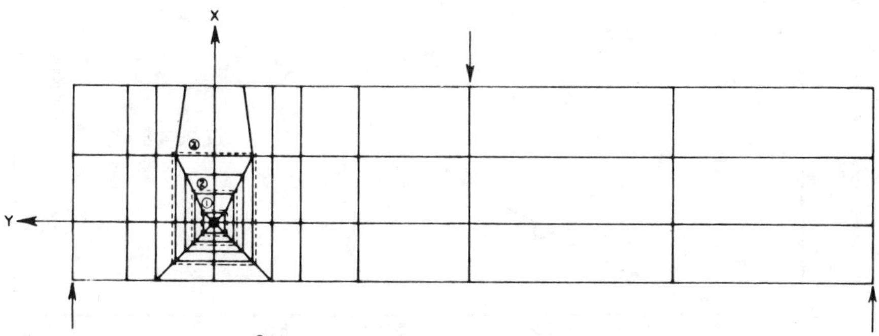

(c) 90° QUARTER-POINT SPECIMEN

FIG. 12—*Finite element mesh for 30, 45, and 90-deg quarter-point specimens.*

TABLE 3—*Results of static elastic analyses of quarter-point notch specimens.*

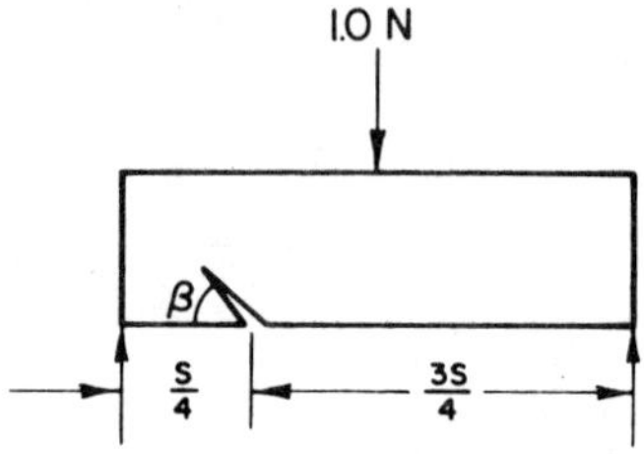

	$\hat{J}_1$, N/m × 10^{-6}			$-\hat{J}_2$, N/m × 10^{-6}		
β, deg	Contour 1	Contour 2	Contour 3	Contour 1	Contour 2	Contour 3
30	0.238	0.265	0.209	0.353	0.403	0.398
45	0.490	0.498	0.499	0.615	0.615	0.600
90	1.440	1.433	1.428	0.268	0.245	0.227

Quarter-point notch specimens (static analysis):
span = 0.1651 m
width = 0.04064 m
thickness = 0.2921 m
uncracked ligament = 0.02858 m

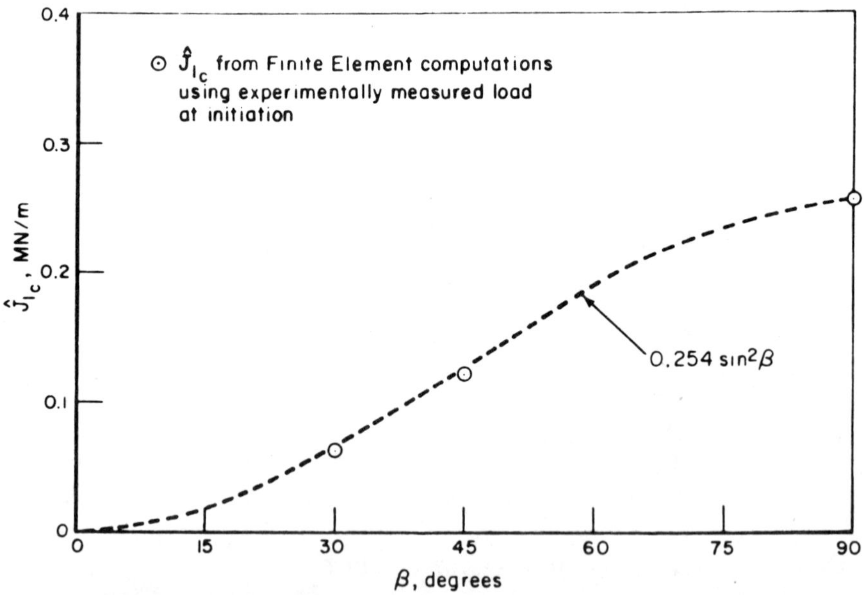

FIG. 13—*Variation of* J_{Ic} *steel with notch angle in the quasi-static loaded inclined midpoint notch bend specimen of HY-100 steel (1 N/m = 0.74 lb/ft).*

The relation between $\hat{J}_1$ and the stress-intensity factors K_I and K_{II} given by Eq 10, together with Eqs 15 and 16, will give

$$\hat{J}_1 = \left\{\frac{\kappa + 1}{8\mu}\,\sigma(\pi a)^{1/2}\right\}\sin^2\beta \tag{17}$$

The expression within the brackets is the energy release rate G which, in a linear elastic problem, is just J. Therefore, Eq 17 can be written as

$$\hat{J}_1 = J\sin^2\beta \tag{18}$$

Now, if fracture occurs when J achieves a critical value that is independent of β, then Eq 18 will be identical to Eq 14.

Despite their similarity, there is a significant difference between the origins of Eqs 14 and 18. Specifically, Eq 18 contains $\hat{J}_1$ found by taking the material to be linear elastic and *assuming* a fracture criterion; that is, that $J = J_c$ for all β-values. In contrast, Eq 14 was obtained by computing $\hat{J}_1$ by elastic-plastic finite-element computations and using actual fracture data. Two conclusions can tentatively be drawn. First, the functional relationship between $\hat{J}_1$ and J in the linear elastic case may remain unaffected by plasticity. Second, this criterion may be geometry-independent. If indeed this is the case, the problem of predicting crack initiation reduces down to knowing J_{1c} for the material and performing a linear elastic analysis of the problem to deter-

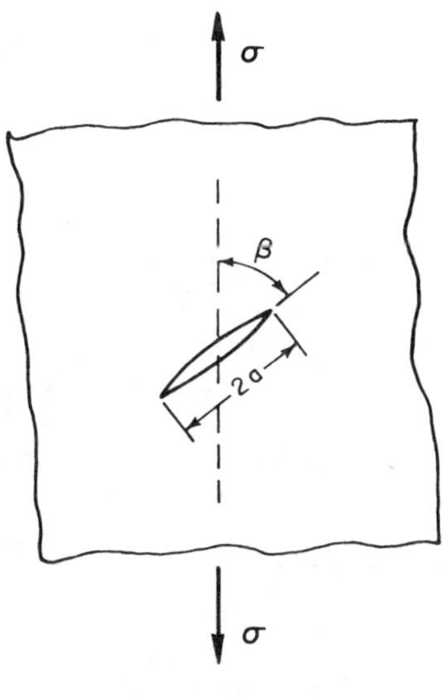

FIG. 14—*Infinite plate with inclined crack.*

mine the relationship between J and $\hat{J}_1$. These implications present intriguing possibilities which must be further explored.

Turning now to impact loading conditions, a dynamic generalization of Eq 14 would be given by

$$\hat{J}_1 = DJ_c \sin^2\beta \tag{19}$$

where $\hat{J}_1$ now contains inertia effects while D is a rate-dependent material constant. If the loading rate effect is small, then $D \simeq 1$. This assumption will be made here on the basis that it is consistent with the use of the static stress-strain curve in finite-element calculations. Therefore, strain-rate effects are omitted from both sides of Eq 19. Now, using Eq 19 and the results of dynamic finite-element computations (Figs. 9, 10, and 11), a prediction of the time at which crack growth will initiate can be made. A comparison of these predictions with the experimentally measured time at initiation is contained in Fig. 15.

Note that the differences in the results of the three series of experiments shown in Fig. 15 should not be viewed as scatter in the data. Rather, in each

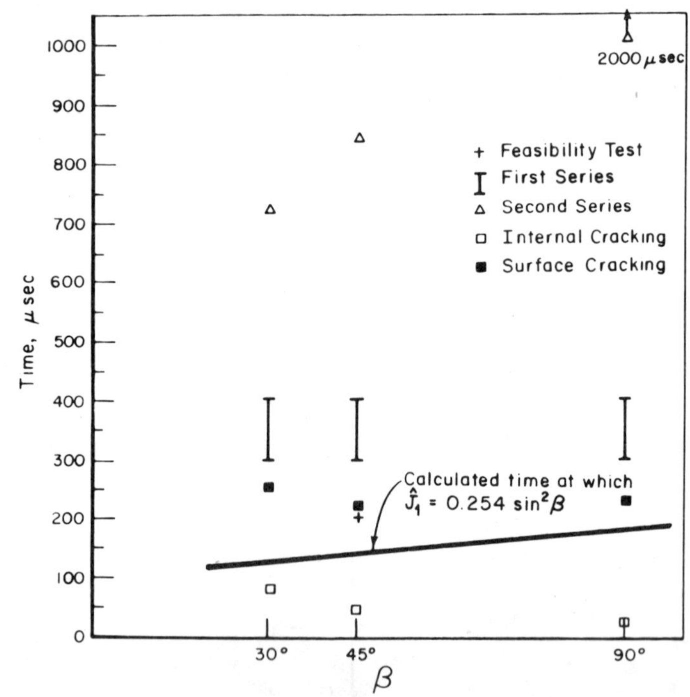

FIG. 15—*Comparison of predicted and experimentally measured time at initiation in impact-loaded inclined midpoint notch bend specimens of HY-100 steel.*

series a different experimental technique was used to detect the instant of crack initiation. The last series of experiments, represented by the two sets of square symbols, should be viewed as the most accurate that could be obtained. It can readily be seen that the average of the internal cracking initiation time (the open squares) and the surface cracking initiation time (the solid squares) is in quite good agreement with the analysis prediction. The significance of this observation is not certain at this time. However, the fact that the predicted time at initiation in all three cases lies within the measured time interval from the instant when the crack initiation was detected below the surface of the specimen to the instant when it appeared at the surface is very encouraging. That is, since the two-dimensional finite-element analyses used in the prediction do not model crack tunneling, the predicted time should be comparable to some instant inside the measured interval. While it can therefore be claimed that the last series of experiments does provide data which are quite comparable to the analysis predictions, more experiments are needed to draw any decisive conclusions.

References

[*1*] Kanninen, M. F., Popelar, C. H., and Broek, D., *Nuclear Engineering and Design*, Vol. 68, 1981; also, NUREG/CR-2110, Nuclear Regulatory Commission, May 1981.
[*2*] Rice, J. R., *Transactions*, American Society of Mechanical Engineers, *Journal of Applied Mechanics*, Vol. 35, June 1968.
[*3*] Hellen, T. K. and Blackburn, W. S., *International Journal of Fracture*, Vol. 11, No. 4, 1975.
[*4*] Blackburn, W. S., *International Journal of Fracture Mechanics*, No. 8, 1972.
[*5*] Neale, B. K., "Elastic Plastic Analysis of Cracked Bodies Using the J-Integral Method," CEGB, Rep. No. RD/B/N3253, Central Electricity Generating Board, Berkely, U.K., 1975.
[*6*] Bergez, D., *International Journal of Fracture*, Vol. 12, No. 4, Aug. 1976.
[*7*] Bui, H. E. in *Proceedings*, Fourth International Conference on Fracture, Waterloo, Ont., Canada, Vol. 3, 1977.
[*8*] Freund, L. B., *Journal of Elasticity*, Vol. 2, 1972, pp. 341–349.
[*9*] Wilson, W. K. and Yu, I. W., *International Journal of Fracture*, Vol. 15, No. 4, Aug. 1979.
[*10*] Mall, S. in *Proceedings*, 5th International Conference on Fracture, Cannes, France, April 1981.
[*11*] Kishimoto, K., Aoki, S., and Sakata, M., *Engineering Fracture Mechanics*, Vol. 13, 1980, pp. 841–850.
[*12*] Miyamoto, H., Kikuchi, M., Sakaguchi, Y., and Takenori, S. in *Proceedings*, U.S.-Japan Conference on Fracture Mechanics, Tokyo, 1980.
[*13*] Kanninen, M. F. and Ahmad, J., "Generalized Forms of J-Integral," manuscript under preparation.
[*14*] Jung, J., Ahmad, J., Kanninen, M. F., and Popelar, C. H. in *Failure Prevention and Reliability—1981*, F. T. C. Loo, Ed., American Society of Mechanical Engineers, 1981.
[*15*] Ahmad, J., Jung, J., Barnes, C. R., and Kanninen, M. F., "The Development of a Dynamic Elastoplastic Finite Element Analysis for Fast Fracture Under Impact Loading," submitted for publication in *Engineering Fracture Mechanics*.
[*16*] Paris, P. C. and Sih, G. C. in *Fracture Toughness Testing and Its Applications, ASTM STP 381*, American Society for Testing and Materials, 1965, p. 30.

Wolfgang Brocks,[1] *Heinz-Hermann Erbe,*[2] *Hans-Dieter Noack,*[1] *and Helmut Veith*[1]

Elastic-Plastic Analysis of a Nozzle Corner Crack by the Finite-Element Method

REFERENCE: Brocks, W., Erbe, H.-H., Noack, H.-D., and Veith, H., **"Elastic-Plastic Analysis of a Nozzle Corner Crack by the Finite-Element Method,"** *Elastic-Plastic Fracture: Second Symposium, Volume I—Inelastic Crack Analysis, ASTM STP 803,* C. F. Shih and J. P. Gudas, Eds., American Society for Testing and Materials, 1983, pp. I-240–I-255.

ABSTRACT: The paper presents the results of an elastic-plastic three-dimensional finite-element analysis for a nozzle corner crack in a pressurized reactor test vessel. The calculations were performed by the finite-element program ADINA using the von Mises yield condition and isotropic hardening. The crack plane was taken parallel to the axis of the vessel and the crack front straight and perpendicular to the symmetry line of the nozzle corner in order to obtain the worst position for a nozzle corner crack. The calculations were performed up to that pressure level where general yield of the ligament in the nozzle corner section takes place.

For the chosen crack length, which is relatively small in comparison with the ligament, the stress normal to the ligament plane and the crack-opening displacement have a maximum in the center of the crack. The differences between the maximum and minimum values along the crack front increase rapidly in the plastic range when the pressure load is raised above half of the general yield pressure. The results of the finite-element analysis are compared with figures obtained from analytical procedures of elastic-plastic fracture mechanics.

KEY WORDS: elastic-plastic fracture, finite-element analysis, fractures (materials), reactor vessels, stress-intensity factor, crack-opening displacement, nozzle corner crack

The inner corner of the junction between nozzle and vessel of a pressure vessel is a region of high stress concentration. Therefore, this region seems to be sensitive to the initiation of cracks and, in the further course of operation, also to stable or unstable crack growth. Various evaluations have already been performed on nozzle corner crack problems in the elastic regime by

[1]Bundesanstalt für Materialprüfung (BAM), Berlin, West Germany.
[2]University of Bremen, Bremen, West Germany.

analytical methods as well as by three-dimensional finite-element procedures [*1–3*].[3] These results were confirmed widely by experimental examinations on the basis of photoelasticity [*2,4*].

As many pressure vessels operate at temperatures where major plastification occurs before stable or unstable crack growth takes place, an elastic-plastic analysis is necessary in order to get a realistic estimate of the structural behavior. Studies in this regime have been conducted with a two-dimensional model based on the tearing-modulus concept [*5*].

The paper presents a three-dimensional elastic-plastic finite-element analysis on a straight nozzle corner crack.

Finite-Element Analysis

The pressure vessel under consideration, Fig. 1, has a ratio of wall thickness of 130 mm to outer diameter of 2024 mm, which is similar to that of real reactor pressure vessels. Thus the results of the analysis should also be relevant to full-scale pressure vessels. A straight crack with a maximum crack length of 30 mm is assumed in the worst position of the nozzle corner, which is the xz-plane where the hoop stresses of nozzle and vessel act as crack-opening stresses.

By reason of symmetry only a quarter section of the nozzle vessel junction has to be modeled (see Fig. 2). The finite-element mesh contains 496 isoparametric three-dimensional elements with 8 to 20 nodes. It has 1293 nodal points, which means 3620 degrees of freedom. Around the crack front, specially distorted elements were used by collapsing the brick elements to wedge-shaped elements and moving the midpoint nodes to the quarter-point position, as proposed by Barsoum [*6*] for getting a singular strain field solution.

The calculations were performed by the finite-element (FE) program ADINA from the Massachusetts Institute of Technology (MIT) [*7*]. The results are based on an incremental theory of plasticity using the von Mises yield condition and isotropic strain hardening. The material properties for room temperature were taken from the German standard steel 22Ni-Mo-Cr37, which is comparable to the American standard steel A508 Class 2. So a Young's modulus of 210 000 N/mm^2, a Poisson's ratio of 0.3, a yield strength of 439 N/mm^2, and a strain hardening modulus of 1500 N/mm^2, which is 1/140 of the Young's modulus, were used.

Results

Figure 3 shows the intersection of the nozzle corner with the xz-plane and the evaluation paths used in the diagrams herewith. These paths are the con-

[3]The italic numbers in brackets refer to the list oi references appended to this paper.

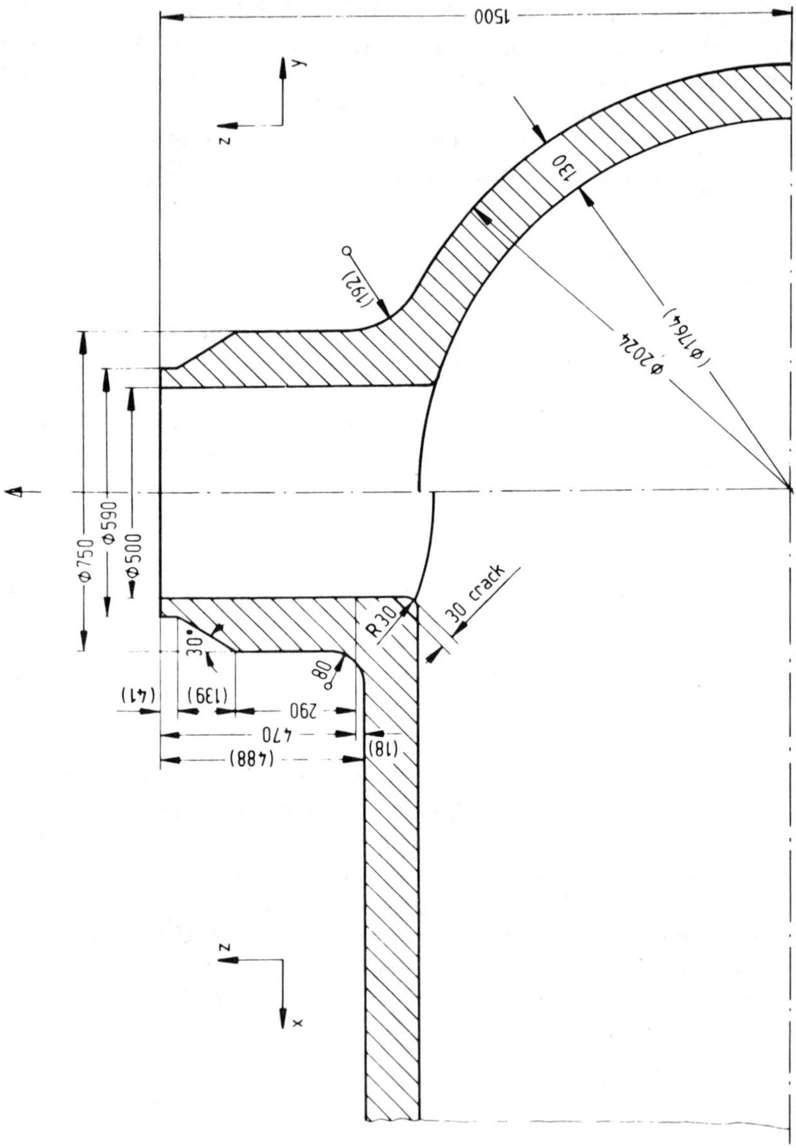

FIG. 1—*Dimensions of nozzle and vessel.*

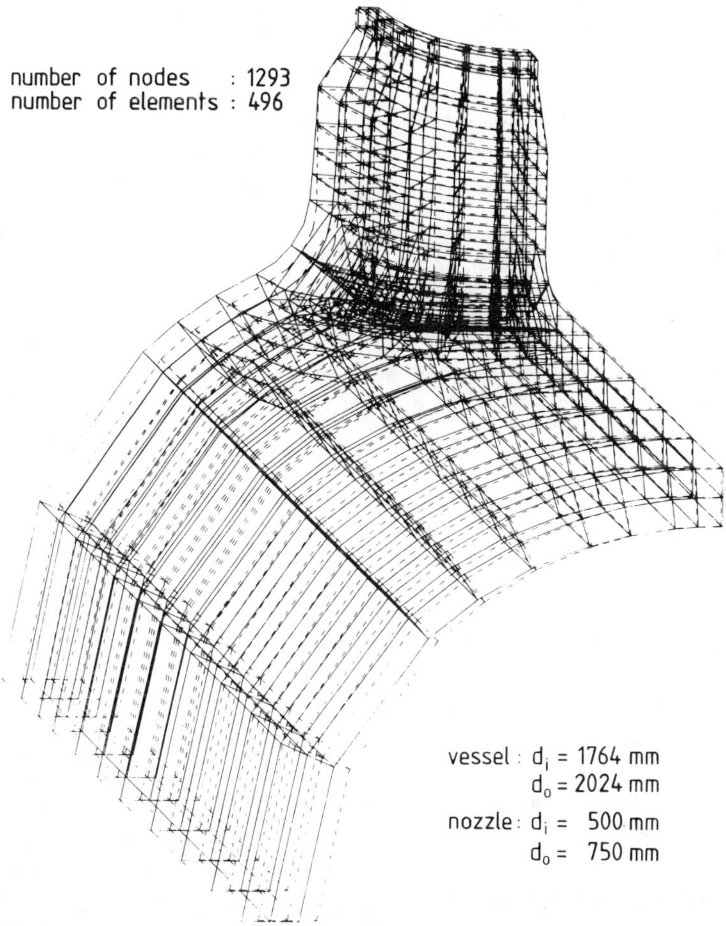

FIG. 2—*Finite-element mesh and displacement plot of the whole structure at 550 bar, displacement scaling factor 8.*

necting lines of the integration points where the FE program evaluates stresses and strains. The integration points of the crack-tip elements lie a distance of about 0.4 mm ahead of the crack front (Path 2).

The first diagram, Fig. 4, allows the comparison of the hoop stresses of nozzle and vessel in the crack plane near the inner surfaces with the stresses σ_{yy} normal to the ligament plane near the crack front. As the integration points are a finite distance from the crack front, the stresses remain finite, too. Here and in the other figures, all the stresses are normalized by the uniaxial yield strength σ_Y.

At a pressure of 100 bar, where the hoop stresses are still far below the yield strength σ_Y, σ_{yy} already exceeds σ_Y by a factor of more than 2. At 550

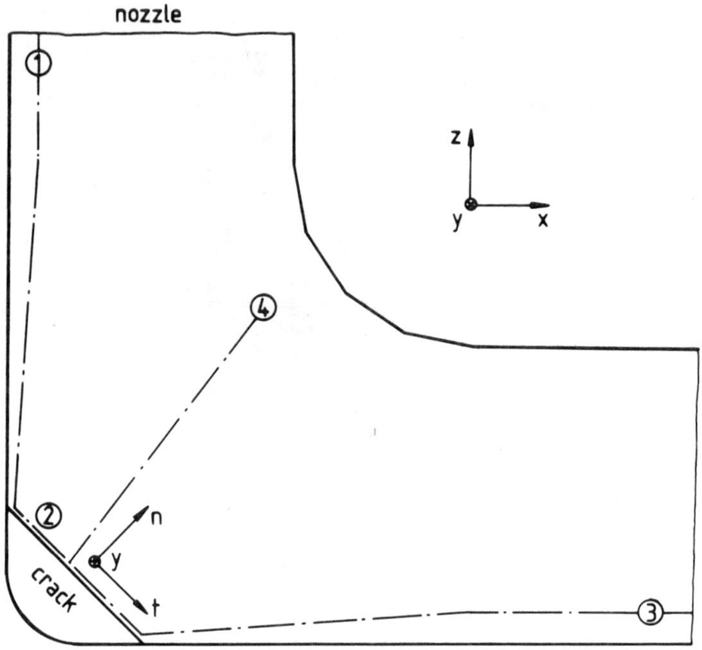

FIG. 3—*Nozzle corner with crack and evaluation paths of stresses:* (1) *path near inner nozzle surface;* (2) *path near crack front;* (3) *path near inner vessel surface;* (4) *ligament path near crack center.*

bar, where general yield of the ligament occurs, maximum values of σ_{yy} of more than five times σ_Y are observed. Major oscillations of the stress results near the crack front, especially at higher pressure levels, are caused by numerical discontinuities due to the applied element shape functions and the condition of isochoric plastic deformation. A larger number of values along the crack front with a better possibility for curve fitting would be obtained by refining the element mesh in this region. But with respect to the restricted central memory of the computer, there were still enough degrees of freedom in the structure as it was.

The stress σ_{yy} as well as the stresses σ_{nn} normal and σ_{tt} parallel to the crack front have their maxima near the crack front (see Figs. 5 and 6). All three stresses decrease rapidly within a few millimetres of the ligament. At the center of the crack front (Fig. 6) higher maximum stresses are observed than those near the inner surfaces of nozzle or vessel (Fig. 5). Near the inner surfaces of nozzle or vessel σ_{nn} and σ_{tt} tend to zero so that a nearly uniaxial stress state will be reached there. For this reason, σ_{yy} just equals the yield strength σ_Y at the general yield pressure of 550 bar as Fig. 6 demonstrates.

A gradient of the maximum principal stress σ_{yy} at a distance of about 0.4 mm along the crack front is already observed at a pressure of 50 bar as shown

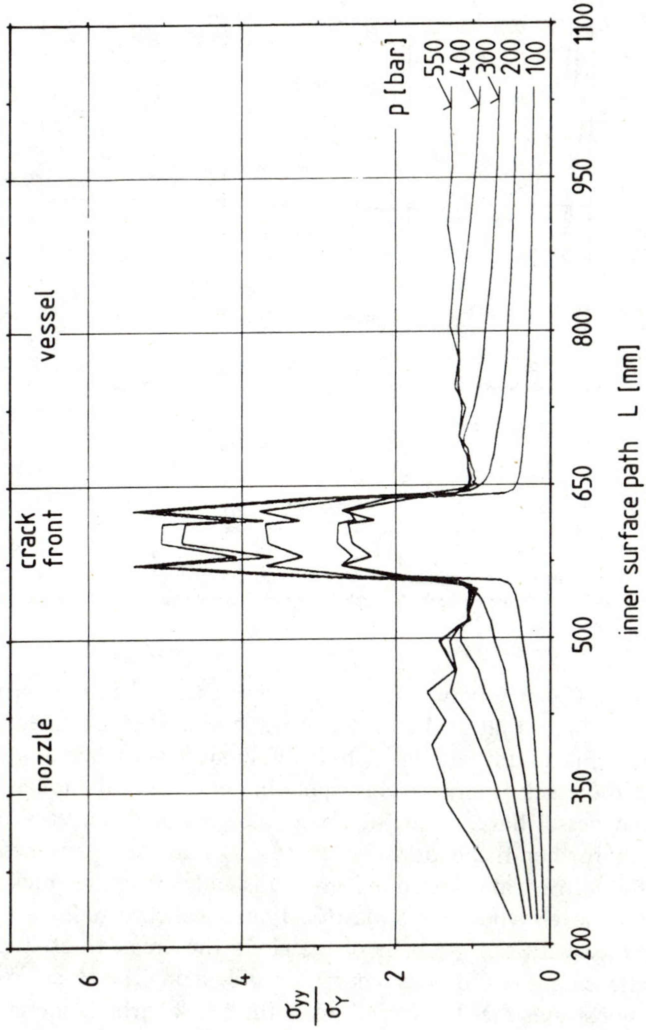

FIG. 4—*Stress σ_{yy} near the inner surfaces of nozzle and vessel and near the crack front at different pressure loadings.*

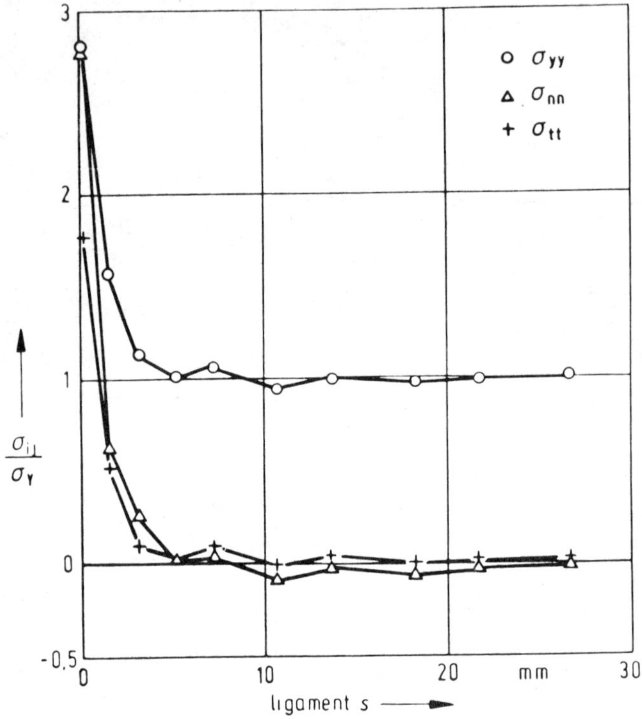

FIG. 5—*Stress normal to the ligament plane* σ_{yy}, *stress normal to the crack front* σ_{nn}, *and stress parallel to the crack front* σ_{tt} *along a path near the inner surface of the nozzle at 550 bar.*

in Fig. 7. The difference between the stress levels near the inner surfaces of nozzle and vessel and that at the center of the crack front increases with rising pressure. A maximum ratio of σ_{yy} to σ_Y of about 5 is reached at 400 bar in the center of the crack front; the corresponding minima at the inner surfaces of nozzle and vessel have values of about 3.3. A small decrease of σ_{yy} is observed with further rising pressure up to 550 bar. As mentioned in the preceding, the numerical values of σ_{yy} grow unsteady along the crack front at higher pressure levels where large plastifications have occurred.

In accordance with the gradient of σ_{yy} along the crack front, the crack-opening displacement (COD) has a curved profile along the crack front, too. Figure 8 demonstrates that the curvature of the crack surfaces increases with rising pressure. An essential growth of COD takes place above 400 bar. At general yield (550 bar), COD values of about 0.6 mm in the center of the crack front and of about 0.2 mm near the inner surfaces of nozzle and vessel are observed, if COD is taken to be twice the displacement of the quarter-point nodes in the crack-opening direction.

The curvature of the crack surfaces along the crack front can also be

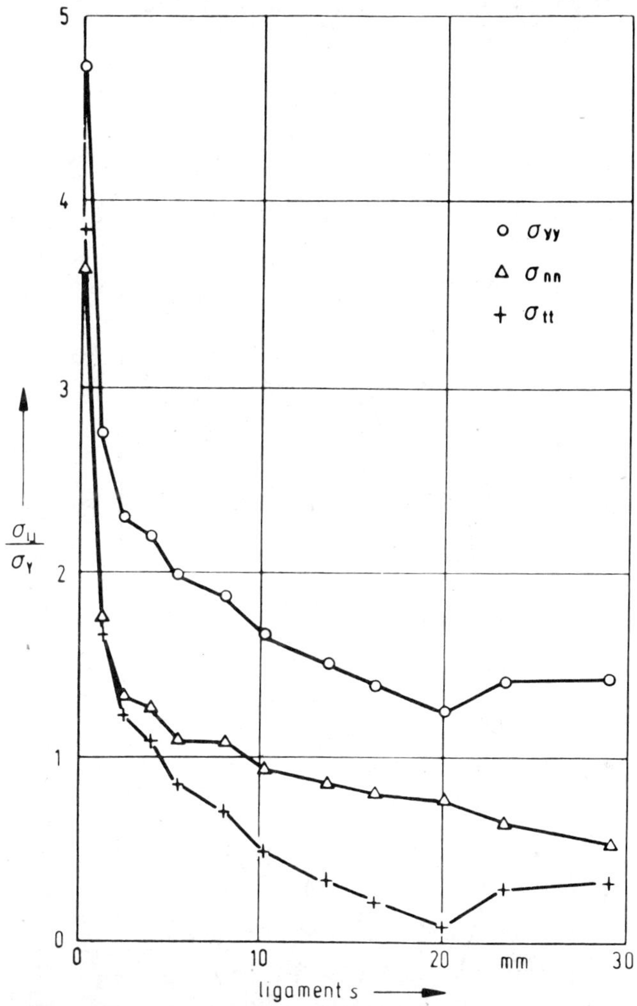

FIG. 6—*Stress normal to the ligament plane σ_{yy}, stress normal to the crack front σ_{nn}, and stress parallel to the crack front σ_{tt} along a path in the center of the crack at 550 bar.*

recognized by comparing the crack opening profiles in the middle of the crack front with those at the inner surfaces of nozzle and vessel. Figure 9 shows again that in the vicinity of the crack front the differences between COD in the center and COD on the nozzle surface increase with rising pressure. The crack surfaces are less crooked with increasing distance *R* from the crack front.

Figure 10 demonstrates how under rising pressure the plastic zones extend through the nozzle corner cross section in the crack plane as well as on the inner circumferences of nozzle and vessel near the crack front. Up to a pressure

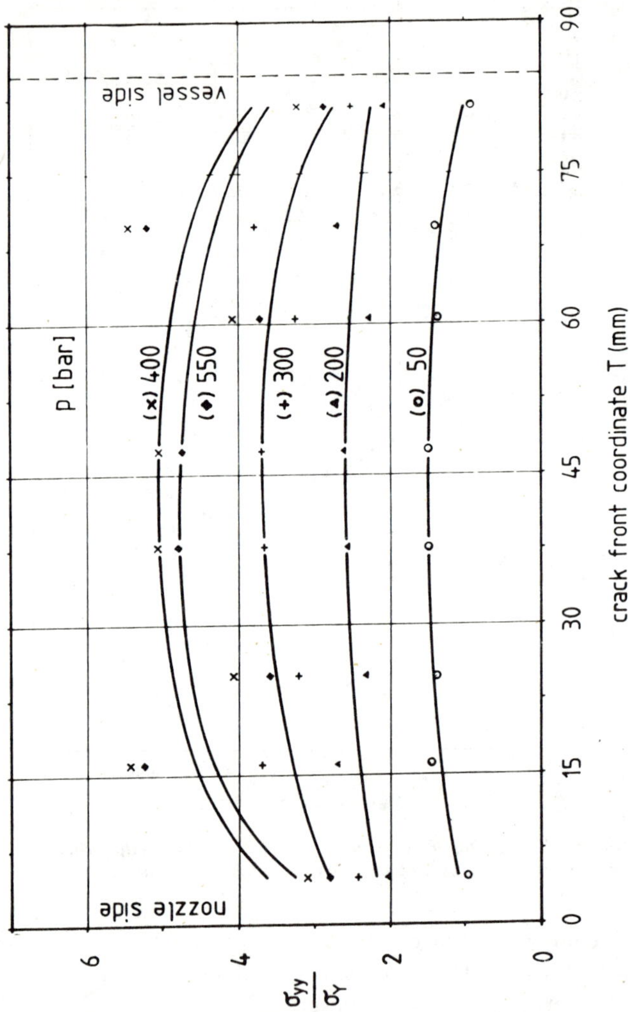

FIG. 7—*Stress normal to the ligament plane* σ_{yy} *along the crack front at different pressure loadings.*

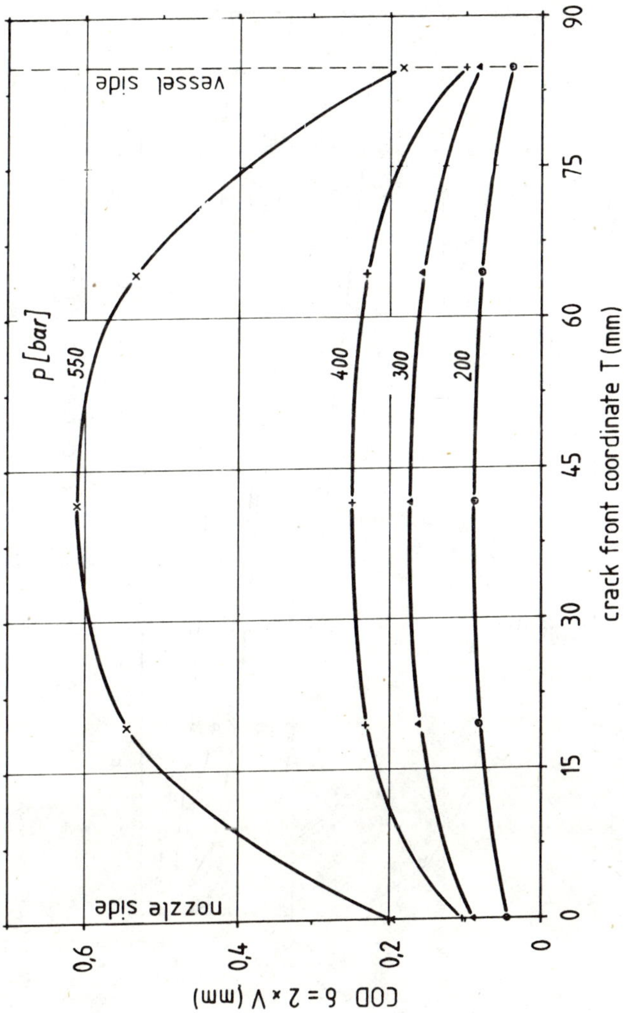

FIG. 8—*Crack-opening displacement along the crack front.*

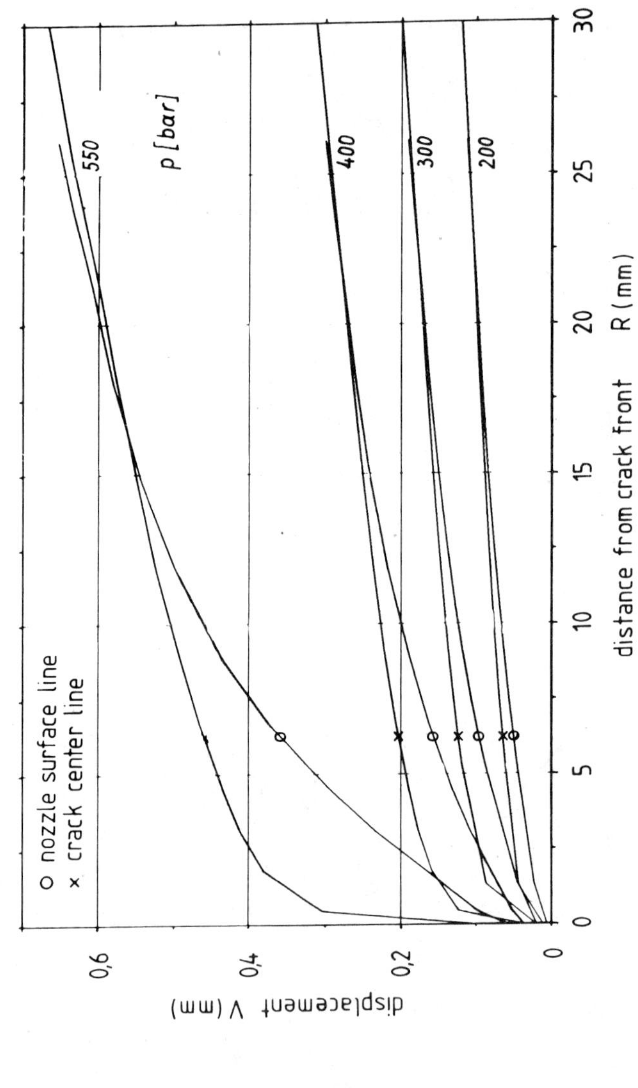

FIG. 9—*Crack-opening profiles along the centerline of the crack and along the inner surface of the nozzle.*

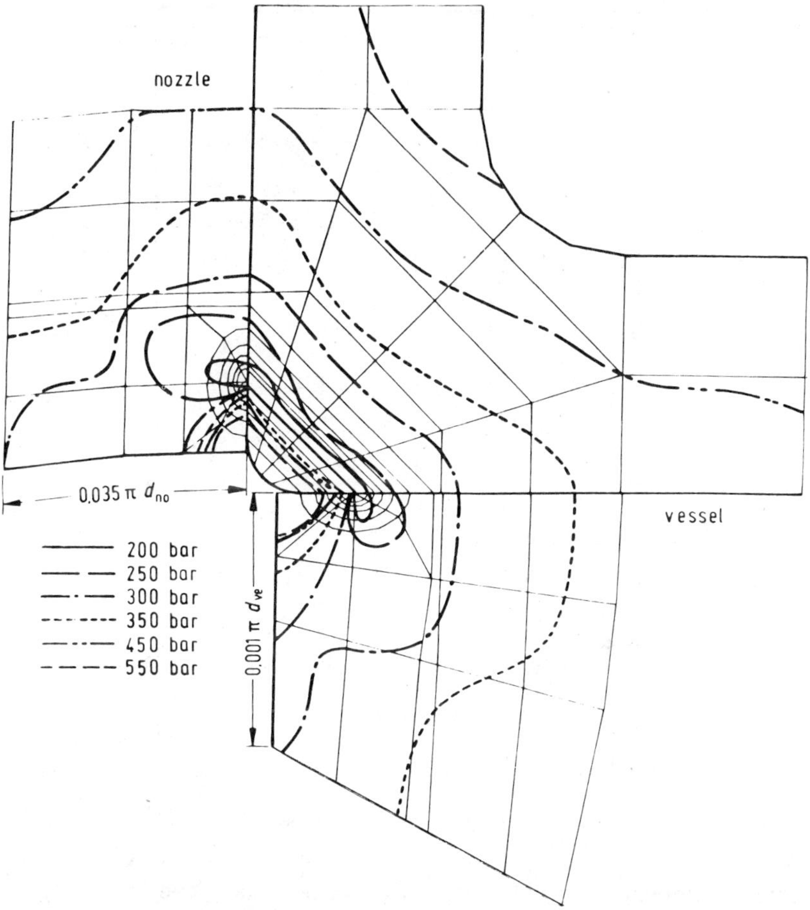

FIG. 10—*Spreading of the plastic zone through the nozzle corner with the crack.*

of about 250 bar the plastic zone spreads just slowly into the ligament. Circular-shaped regions of plastification are observed on the inner surfaces of nozzle and vessel. At about 300 bar the plastic zone reaches the edge of the nozzle vessel junction. A further increasing pressure causes a rapid spread of the plastic zone through the cross section, where at 550 bar general yield of the ligament takes place. In addition, Fig. 11 shows the spread of the plastic zone when looking into the nozzle. Above 400 bar the plastic zone extends rapidly into the vessel region.

Figure 2 shows a displacement plot of the whole structure. Comparatively small displacements are found in that region where the vessel stiffness is increased by the nozzle, whereas larger displacements are observed in the plane perpendicular to the nozzle axis, so that the cross section of the vessel

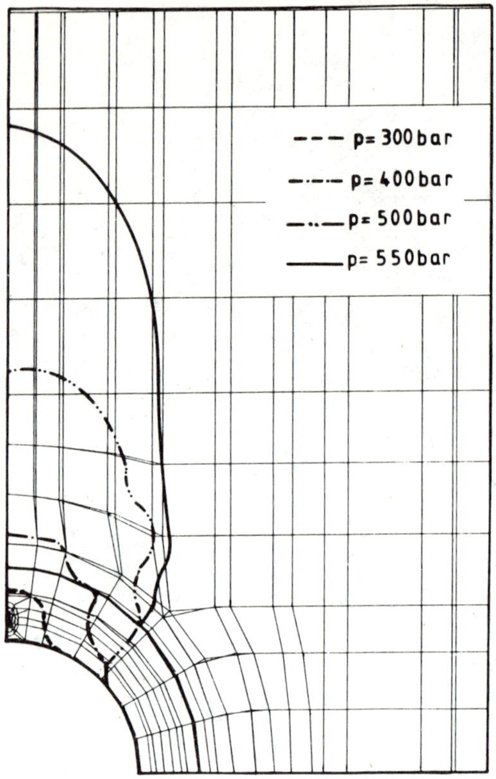

FIG. 11—*Spreading of the plastic zones in the vessel.*

becomes elliptical. Likewise, large displacements are concentrated in the region of the smooth nozzle vessel junction where they result in an increase of the junction radius.

Discussion

The calculated variables necessary for a comparison of different fracture mechanics concepts, as for instance the stress-intensity factor K, the radius r_y of the plastic zone in the ligament, and the crack-opening displacement δ, vary along the crack front. In the following discussion, the values in the middle of the crack front are used where K and δ have their maxima.

An elastic FE analysis yields the stress-intensity factor K depending linearly on pressure, which is presented by a straight line called "K_{elast}" in Fig. 12. Taking a critical value K_{Ic} of about 6000 N/mm$^{3/2}$ from the ASME Boiler and Pressure Vessel Code (Section XI, Appendix A) for the considered steel, a critical pressure of about 360 bar results for the purely elastic case. However, the elastic-plastic analysis shows that a major plastic zone will

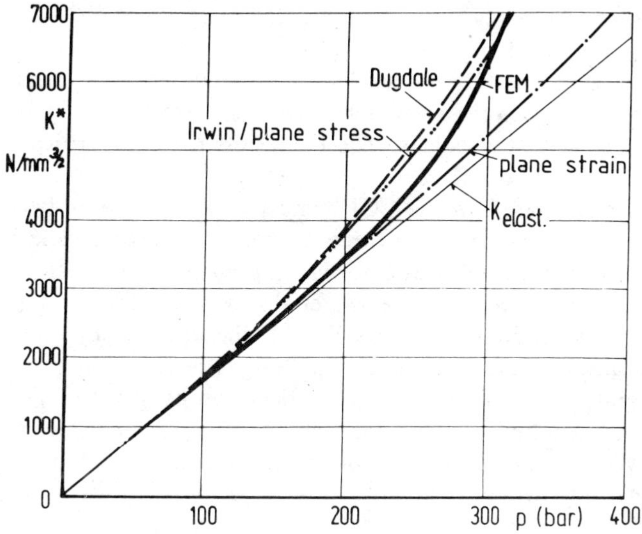

FIG. 12—*Comparison of different analytical approaches for a corrected stress-intensity factor* K* *with results of finite-element computations.*

already exist in the ligament at this pressure. Hence, a plastic zone correction of K should be performed by using Irwin's approach [*10*] of a "notional crack tip" at r_y

$$K^* = K\sqrt{(a + r_y)/a} \tag{1}$$

Taking the results of the FE calculations for r_y (see Fig. 10), the curve called "FEM" in Fig. 12 will be found. For comparison, various two-dimensional analytical approaches for the estimation of the plastic zone size under small-scale yielding conditions were considered. These approaches by Irwin [*10*] for plane strain

$$r_{y_1} = \frac{(1 - 2\nu)^2}{2\pi} \cdot \frac{K^2}{{\sigma_y}^2} \tag{2}$$

and plane stress

$$r_{y_2} = \frac{1}{2\pi} \cdot \frac{K^2}{{\sigma_y}^2} \tag{3}$$

and by Dugdale [*11*]

$$r_{y_3} = \frac{\pi}{16} \cdot \frac{K^2}{{\sigma_y}^2} \tag{4}$$

are also presented in Fig. 12. Up to a pressure of about 150 bar there is no substantial difference between the various approaches. Above 150 bar the plane-strain solution of Irwin results in a nonconservative estimate for the critical pressure, while the approaches for plane stress of Irwin and Dugdale give slightly conservative values. The FE results predict a critical pressure of about 300 bar.

An assessment based on the COD concept is possible, too, because COD-values have been calculated by three-dimensional elastic-plastic FE analysis (see Fig. 8), and critical values δ_i for crack growth initiation are known from tests on CT50 specimens at the Materialprüfanstalt Stuttgart [*12*] for the 22Ni-Mo-Cr37 steel considered. In Fig. 13, the numerically evaluated COD in the middle of the crack front is plotted depending on pressure (curve "FEM") and is compared with analytical results gained from the two-dimensional Irwin and Dugdale models. The analytical δ-values are deduced from the Westergaard equations for the displacement field, inserting Eqs 2 to 4, respectively, which yields to

$$\delta_1 = \frac{4}{\pi} \cdot \frac{K^2}{E\sigma_y}(1 - \nu^2)(1 - 2\nu) \tag{5}$$

for the plane-strain approach of Irwin

$$\delta_2 = \frac{4}{\pi} \cdot \frac{K^2}{E\sigma_y} \tag{6}$$

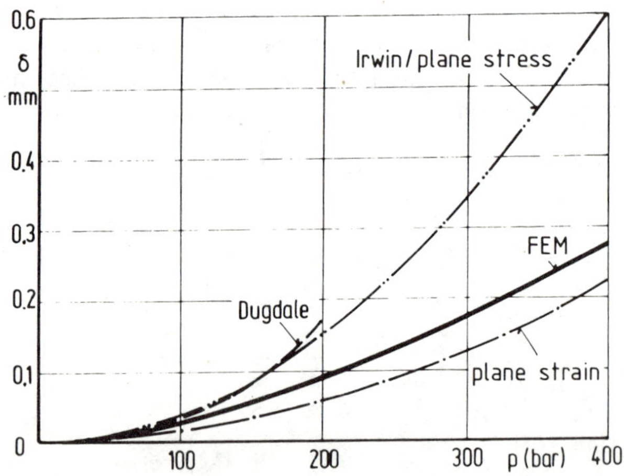

FIG. 13—*Crack-tip opening displacement δ gained by finite-element computation and compared with results of different analytical approaches.*

for the plane-stress approach of Irwin, and

$$\delta_3 = \frac{8\sigma_y}{\pi E} a \,\ell n \left[\sec\left(\sqrt{\frac{\pi}{a}} \frac{K}{2\sigma_y} \right) \right] \tag{7}$$

for the Dugdale model.

Using the test result of $\delta_i = 0.18$ mm for crack growth initiation as a yardstick, once more a critical pressure of about 300 bar results from the FE calculations. The differences from the various two-dimensional models are remarkable. Again, the plane-strain approach of Irwin is nonconservative. On the other side, the plane-stress solution gives a much too conservative pressure limit, which is about 100 bar below the value gained from the FEM curve.

References

[1] Broekhoven, M. J. G., "Computation of Stress Intensity Factors for Nozzle Corner Cracks by Various Finite Element Procedure," 3rd International Conference on Structural Mechanics in Reactor Technology, G4/6, 1975.

[2] Robert, A., Bourdon, C., Baylac, G., D'Escatha, Y., and Vagner, J., "Pressure Induced Stresses in the Upper Part of a PWR Vessel, Using Photoelastic and 3 D Finite Element Methods," 4th International Conference on Structural Mechanics in Reactor Technology, G 7/3, 1977.

[3] Atluri, S. N. and Kathiresan, K., "Stress Intensity Factor Solutions for Arbitrarily Shaped Surface Flaws in Reactor Pressure Vessel Nozzle Corners," 5th International Conference on Structural Mechanics in Reactor Technology, G 4/3, 1979.

[4] Smith, C. W., Peters, W. H., Hardrath, W. T., and Fleischmann, T. S., "Stress Intensity Distributions in Nozzle Corner Cracks of Complex Geometry," 5th International Conference on Structural Mechanics in Reactor Technology, G 4/4, 1979.

[5] Tada, H., Musicco, G., and Paris, P. in *Proceedings*, Specialists Meeting on Elastoplastic Fracture Mechanics, Daresbury, U.K., Paper No. 16, May 1978.

[6] Barsoum, R. S., *International Journal of Numerical Methods in Engineering*, Vol. 11, 1977, pp. 85–98.

[7] Bathe, K. J., "ADINA—A Finite Element Program," Report 82448-1, Department of Mechanical Engineering, Massachusetts Institute of Technology, Cambridge, Mass., 1978.

[8] Boddenberg, R., Herrmann, H., Hoffmann, J., and Matzkows, J., "FEZZIK—FEM-Postprocessor," Institut für Luft- und Raumfahrt, Technical University, Berlin, 1979.

[9] Knott, J. J., *Fundamentals of Fracture Mechanics*, Butterworths, London, 1973.

[10] Irwin, G. R., *Applied Materials Research*, Vol. 3, 1964, pp. 65–81.

[11] Dugdale, D. S., *Journal of the Mechanics and Physics of Solids*, Vol. 8, 1960, pp. 100–104.

[12] Krägeloh, E., Issler, L., and Zirn, R., "Analyse von Großprobenversuchen mit Zähbruchkonzepten," 6th Seminar, Staatliche Materialprüfanstalt Stuttgart, Oct. 1980.

Peter D. Hilton[1] *and L. Nash Gifford*[2]

Elastic-Plastic Finite-Element Analysis for Two-Dimensional Crack Problems

REFERENCE: Hilton, P. D. and Gifford, L. N., **"Elastic-Plastic Finite-Element Analysis for Two-Dimensional Crack Problems,"** *Elastic-Plastic Fracture: Second Symposium, Volume I—Inelastic Crack Analysis, ASTM STP 803*, C. F. Shih and J. P. Gudas, Eds., American Society for Testing and Materials, 1983, pp. I-256–I-273.

ABSTRACT: The specialized finite-element program, PAPST (plastic axisymmetric and planar structures), has been developed by the Navy for analysis of elastic-plastic crack problems. After a number of years of PAPST development, internal usage [*1,2*][3] and refinement, preliminary documentation [*3*] has now been issued for public release. PAPST includes a finite strain treatment [*4*] (with a small strain option) for incremental (J_2) plasticity. Both power-hardening and multilinear models for uniaxial response are available. The singular core element [*5*] can be employed to enforce the Hutchinson-Rice-Rosengren crack-tip singularity for power-hardening materials [*6, 7*] while the enriched element [*8*] may be used to model crack-tip response for the multilinear material model. For crack problems, both path values for the J-integral and values based on the amplitude of the singular solution can be automatically computed.

PAPST has been used within the Navy research and development community to evaluate response of specimens of varying size, geometries, and material properties. Results for planar and axisymmetric configurations are presented to demonstrate the applications of PAPST to elastic-plastic crack problems.

KEY WORDS: crack, enriched elements, finite element, fracture mechanics, J-integral, nodal release, plasticity, tearing modulus, elastic-plastic fracture

Nomenclature

a_{ij} Coordinates of center of yield surface
D_{ijkl} Tensor relating stress matrix to strain matrix
E Young's modulus

[1]Consultant, Arthur D. Little, Inc., Cambridge, Mass. 02140.
[2]David W. Taylor Naval Ship R&D Center, Bethesda, Md. 20034.
[3]The italic numbers in brackets refer to the list of references appended to this paper.

e_{ij}	Deviatoric strain matrix
F	Correction load vector
J	J-integral
K	Stiffness matrix
K_p	Plastic intensity factor
n	Hardening exponent
$N(s, t)$	Element shape function
(r, θ)	Polar coordinates centered at crack tip
$\dot{R}$	Loading rate vector
s_{ij}	Deviatoric stress vector
s_{ij}'	Deviatoric stress vector measured from center of yield surface
T_i	ith component of the traction vector
T	Tearing modulus
u_i	Displacement component
(u,v)	Displacement components
$\dot{W}_{ij}$	Rotation rate tensor
W	Strain energy
(x,y,z)	Geometric coordinates
x	Hardening coefficient
β	Parameter; 0 implies isotropic loading and 1 implies kinematic hardening
$\dot{\epsilon}\epsilon_{ij}$	Strain rate tensor
ϕ	Airey stress function
ν	Poisson's ratio
σ_e	Effective stress
σ_e'	Effective stress measured from center of yield surface
$\dot{\sigma}_{ij}$	Stress rate tensor
σ_{yd}	Yield stress

The computer program PAPST (plastic axisymmetric and planar structures) has been developed for the Navy with the aim of addressing the elastic-plastic crack problems which arise from materials and structures considerations associated with the use of high-toughness materials. The program has been used in support of laboratory programs to develop material property data for J_{Ic} and other pertinent parameters for fracture characterization. Precracked test articles have been developed and tested to improve our understanding of expected component behavior and some of these have also been analyzed using PAPST. More recently, the capabilities of PAPST have been extended to include simulation of quasi-static crack growth. This procedure is of interest in determination of the tearing modulus and in assessing the stability of crack growth.

The present paper describes the theoretical foundation for PAPST and gives example applications.

Finite Strain Treatment

PAPST incorporates finite strain effects through the use of an updated Lagrangian formulation [4]; that is, the coordinate system is convected with the deformation. In this coordinate system the relationship between true strain rate and the deformation rate (or velocity) is unchanged from the small-strain-theory equations. Also the true (or Cauchy) stress rate tensor relates to the true strain rate in the classical sense, that is

$$\dot{\sigma}_{ij}{}^{c} = D_{ijkl}\dot{\epsilon}_{kl} \tag{1}$$

where subscripts take values of one to three and repeated subscripts imply summation.

The Cauchy stress tensor is not invariant under finite rotations, but rather at zero strain rate it changes at the rate [9]

$$\dot{\sigma}_{ij}{}^{c} = \dot{W}_{ip}\,\sigma_{pj}{}^{c} - \dot{W}_{pj}\sigma_{ip}{}^{c} \tag{2}$$

where $\dot{W}_{ij}$ is the rotation rate tensor. Equations 1 and 2 are combined to express the Cauchy stress rate in terms of the true strain and rotation rates as

$$\dot{\sigma}_{ij}{}^{c} = D_{ijkl}\dot{\epsilon}_{kl} + \dot{W}_{ik}\,\sigma_{kj}{}^{c} - \dot{W}_{kj}\sigma_{ik}{}^{c} \tag{3}$$

Plasticity Theory

PAPST incorporates incremental J_2-flow theory plasticity (von Mises yield criterion). The components of the deviatoric strain rate are expressed in terms of the current deviatoric stress components (s_{ij}) and the components of the deviatoric stress rate ($\dot{s}_{ij}$) as

$$\dot{e}_{ij} = \begin{cases} \dfrac{1+\nu}{E}\,\dot{s}_{ij} + \dfrac{3}{2}f(\sigma_e)s'_{ij}\dot{\sigma}'_e & (\text{for } \sigma'_e = \sigma_{\text{yd (current)}} \text{ and } \dot{\sigma}'_e > 0) \\ \dfrac{1+\nu}{E}\,\dot{s}_{ij} & (\text{otherwise}) \end{cases} \tag{4}$$

where

$\dot{e}_{ij}$ = deviatoric strain rate components ($\dot{e}_{ij} = \dot{\epsilon}_{ij} - \frac{1}{3}\dot{\delta}_{pp}\delta_{ij}$),
s_{ij} = current deviatoric stress components ($s_{ij} = \sigma_{ij} - \frac{1}{3}\sigma_{pp}\delta_{ij}$),
s'_{ij} = deviatoric stress components measured from center of current yield surface ($s'_{ij} = s_{ij} - a_{ij}$),
a_{ij} = coordinates in stress space of center of yield surface,
σ_e = von Mises effective stress ($\sigma_e = \sqrt{(3/2)s_{ij}s_{ij}}$),
$\sigma'_e = \sqrt{3s'_{ij}s'_{ij}/2}$, and
σ_{yd} = yield stress.

The hydrostatic strain rate is given in terms of the mean stress rate as

$$\dot{\epsilon}_{pp} = \frac{1 - 2\nu}{E} \dot{\sigma}_{pp}$$

The center of the yield surface can move under plastic deformation at a rate proportional to the projection of the stress rate vector onto the local normal to the yield surface. The motion of the yield surface center is directed along the vector s'_{ij} describing the deviatoric stress relative to the yield surface center, that is,

$$\dot{\alpha}_{ij} = \begin{cases} \frac{3}{2} \beta s'_{kl} \dot{s}_{kl} s'_{ij} / \sigma_e'^2 & \text{when } \sigma'_e = \sigma_{yd}, \quad \dot{\sigma}'_e > 0 \\ 0 & \text{otherwise} \end{cases} \tag{5}$$

where $0 \leq \beta \leq 1$. Isotropic hardening is achieved by setting β equal to zero and kinematic hardening results when β is set equal to one.

The function $f(\sigma_e)$ contained in the first of Eqs 4 is determined from the uniaxial stress-strain curve of material response. For a uniaxial applied stress $\sigma = \sigma_e$, the first of Eqs 4 reduces to

$$\frac{2}{3}(\dot{\epsilon}_{\text{axial}} - \dot{\epsilon}_{\text{transverse}}) = \frac{2}{3} \frac{1 + \nu}{E} \dot{\sigma}_e + f(\sigma_e)\sigma'_e \dot{\sigma}_e \tag{6}$$

Dividing the strain rate into elastic and plastic parts

$$\dot{\epsilon} = \dot{\epsilon}^E + \dot{\epsilon}^p$$

with $\dot{\epsilon}^E$ transverse $= - \dot{\nu}^E_{\text{axial}}$ and $\dot{\epsilon}^p_{\text{transverse}} = -0.5\ \dot{\epsilon}^p_{\text{axial}}$. Inserting these results into Eq 6 and solving for $f(\sigma_e)$ yields

$$f(\sigma_e) = \dot{\epsilon}^p_{\text{axial}} / \sigma'_e \dot{\sigma}_e \tag{7}$$

This result is used to obtain specific forms for $f(\sigma_e)$ which depend on the mathematical model used to represent the uniaxial stress-strain curve.

Material Models

In PAPST an experimentally determined true stress-true strain curve can be approximated using either a power-hardening model or a multilinear model to describe the uniaxial response.

In the case of the power hardening simulation (Ramberg-Osgood), the uniaxial response is given as

$$\epsilon = \begin{cases} \sigma/E & \text{for} \quad \sigma \leq \sigma_{yd} \\ \sigma/E + \alpha\left[\dfrac{\sigma}{E}\left(\dfrac{\sigma}{\sigma_{yd}}\right)^{n-1} - \dfrac{\sigma_{yd}}{E}\right] & \text{for} \quad \sigma > \sigma_{yd} \end{cases} \tag{8}$$

where n, α, and σ_{yd} (the initial yield stress) are chosen to best model the observed behavior.

Equation 8 corresponds to

$$f(\sigma_e) = \frac{\alpha n}{E\sigma_{yd}}\left(\frac{\sigma_e}{\sigma_{yd}}\right)^{n-2} \tag{9}$$

The mathematical form for the multilinear model of material behavior is

$$\epsilon = \frac{\sigma}{E} + \frac{\alpha_1}{E}(\sigma_1 - \sigma_{yd}) + \frac{\alpha_2}{E}(\sigma_2 - \sigma_1) + \cdots + \frac{\alpha_m}{E}(\sigma - \sigma_{m-1}) \tag{10}$$

where $\sigma_{m-1} < \sigma \leq \sigma_m$ and α_m is defined as $(E\Delta\epsilon_m - \Delta\sigma_m)/\Delta\sigma_m$. For this material model, the plastic strain rate is given by

$$\dot{\epsilon}_{\text{plastic}} = \alpha_m \dot{\sigma}_e/E \tag{11}$$

and

$$f(\sigma_e) = \frac{\alpha_m}{E\sigma_e}$$

Iterative Solution Procedure

The governing equations for the finite-element approximation of the incremental plasticity problem are of the form

$$[K(u)]\dot{u} - \dot{R} = 0 \tag{12}$$

where $\dot{R}$ is the equivalent nodal loading rate vector and $\dot{u}$ the corresponding nodal velocity vector. The stiffness matrix $[K]$ depends on the current nodal displacement vector through the matrix $[D(u)]$ relating stress and strain rates and through the matrix $[B(u)]$ relating the strain rate to the nodal displacement rates, that is,

$$[K(u)] = \Sigma_{\text{elements}} \int_{\text{area}} [B(u)][D(u)][B(u)]dA \tag{13}$$

The set of rate equations is integrated over a load increment (ΔR) to determine Δu. The Newton-Raphson solution procedure is used to solve this nonlinear problem

$$[K(u_L)]\Delta u^1_{L+1} - \Delta R_{L+1} = 0 \tag{14}$$

at increment $L + 1$ yields the first estimate Δu^1_{L+1} of Δu_{L+1}. The correction to Δu^i_{L+1} which is Δu^{i+1}_{L+1} is then obtained from

$$\left[K\left(u_L + \sum_{j=1}^{i} \Delta u^j_{L+1}\right)\right]\Delta u^{i+1}_{L+1} = R_{L+1} - \int_{U_L}^{u_L + \sum_{j=1}^{i} \Delta u^j_{L+1}} [K(u)]du = F^{i+1}_{L+1} \tag{15}$$

Here the subscript indicates the increment number while the superscript refers to the iteration number. Equation 15 is to be solved for the $(i+1)$th correction to the $(L+1)$th nodal displacement vector increment Δu^{i+1}_{L+1}. The implied integration over the displacement increment for the right-hand side of Eq 15 is approximated by a Simpson's rule summation procedure.

At each iteration of each load increment, the set of governing equations given in the preceding is solved for changes in the nodal displacement vector, for example, Δu^{i+1}_{L+1}. Then the iteration step is divided into m subincrements with nodal displacement change $(\Delta u^{i+1}_{L+1})/m$. In each of these subincrements the grid pattern is convected with the deformation $(X_{\text{new}} = X + (\Delta u^{i+1}_{L+1})/m)$ and the element quadrature point strain component changes are found in terms of the nodal displacement changes using the current element geometry to calculate the matrix $[B]$ which relates the strain vector to the nodal displacement vector $\{\Delta\epsilon = [B]\ (\Delta u^{i+1}_{L+1})/m$ and $\epsilon_{\text{new}} = \epsilon + \Delta\epsilon\}$. Next, the stress state and the coordinates of the center of the yield surface are updated at each subincrement. A radial return procedure [*10*] is employed to maintain the current stress state on (or within, if no yielding) the yield surface for each quadrature point of each element.

The center of the yield surface is moved in the direction of the stress deviator vector as measured from the current yield surface center $(s' = s - a)$ an amount proportional to the stress deviator subincrement $\Delta s'$ multiplied by the factor β during the portion of the subincrement in which yielding is occurring, that is,

$$\Delta a = \frac{\beta s'(s' \cdot \Delta s'_{\text{plastic}})}{\sigma_e'^2} \tag{16}$$

If there has been yielding, the change in stress state during the subincrement is then adjusted so that the current stress vector lies on the yield surface, that is

$$\Delta s_{\text{adjusted}} = \Delta s - s'(\sigma_e'^2 - \sigma_{\text{yd}}^2)/2\sigma_e'^2 \tag{17}$$

Equation 17 must be altered for the special case of plane stress such that the adjustment to the increment in the deviatoric stress vector satisfies the condition $\sigma_z = 0$. This is accomplished by adjusting the hydrostatic component of stress accordingly as

$$\Delta\sigma_{pp(\text{adjusted})} = \Delta\sigma_{pp} + 3s_z'\left(\frac{\sigma_e'^2 - \sigma_{yd}^2}{2\sigma_e'^2}\right) \tag{18}$$

Fracture Mechanics

The J-integral [7] for elastic-plastic crack problems is given in terms of the strain energy density

$$W = \int_0^{\epsilon_{ij}} \sigma_{ij} d\epsilon_{ij} \tag{19}$$

as

$$J = \int_\tau \left(W dy - T_i \frac{\partial u_i}{\partial x} ds\right)$$

where the crack lies parallel to the local x-axis and the contour τ is traversed counterclockwise from one crack face to the other.

For elastic-plastic crack problems, J can be related to the amplitude of the plastic singular solution K_p[5]. The asymptotic solution [6] for the case of a multilinear model of material response has an Airy stress function

$$\phi = \frac{K_p r^{3/2}}{\sqrt{2\pi}}\left(\cos\frac{\theta}{2} + \frac{1}{3}\cos\frac{3\theta}{2}\right) \tag{20}$$

applicable to both plane stress and plane strain, where r and θ are polar coordinates centered at the crack tip. Thus the plastic singular solution for the near-tip stress components has the same (r, θ) dependence as in the elastic case.

The corresponding results for the displacement components are

$$\begin{aligned} u &= u_0 + \frac{K_p}{4E}\sqrt{\frac{2r}{\pi}}\left[\left(5 - 3\nu + \frac{7}{2}\alpha_N\right)\cos\frac{\theta}{2} \right. \\ &\quad \left. - \left(1 + \nu + \frac{3}{2}\alpha_N\right)\cos\frac{3\theta}{2}\right] + 0(r) \\ v &= \frac{K_p}{4E}\sqrt{\frac{2r}{\pi}}\left[\left(7 - \nu + \frac{13}{2}\alpha_N\right)\sin\frac{\theta}{2} \right. \\ &\quad \left. - \left(1 + \nu + \frac{3}{2}\alpha_N\right)\sin\frac{3\theta}{2}\right] + 0(r) \end{aligned} \tag{21}$$

for plane stress and

$$u = u_0 + \frac{K_p}{4E}\sqrt{\frac{2r}{\pi}}\left[\left(5 - 3\nu + \frac{7}{2}\alpha_N - 8\beta\right)\cos\frac{\theta}{2} - \left(1 + \nu + \frac{3}{2}\alpha_N\right)\cos\frac{3\theta}{2}\right] + 0(r)$$

$$v = \frac{K_p}{4E}\sqrt{\frac{2r}{\pi}}\left[\left(7 - \nu + \frac{13}{2}\alpha_N - 8\beta\right)\sin\frac{\theta}{2} - \left(1 + \nu + \frac{3}{2}\alpha_N\right)\sin\frac{3\theta}{2}\right] + 0(r) \tag{22}$$

for the plane-strain case with $\beta = (\nu + \alpha_N/2)^2/(1 + \alpha_N)$. In the foregoing, α_N is the slope of the last segment defining the multilinear stress-strain curve.

The J-integral is expressed in terms of the plastic intensity factor K_p by carrying out the integration on a path in the region governed by the plastic singularity solution. The results are

$$J = (1 + \alpha_N)K_p{}^2/E \quad \text{(plane stress)}$$

$$J = \left[1 + \frac{\gamma^2}{2} - \frac{3}{2}\nu\gamma + \alpha_N\left(1 - \frac{3\gamma}{4} + \frac{\gamma^2}{2}\right)\right]\frac{K_p{}^2}{E} \quad \text{(plane strain)} \tag{23}$$

with $\gamma = (\nu + \alpha_N/2)(1 + \alpha_N)$.

The near-field behavior described in the preceding may be included explicitly in the finite-element approximation to the solution [2]. Two distinct "special" elements to accomplish this are included in PAPST:

1. The enriched element, in which K_p is carried as an additional unknown, thereby allowing direct calculation of J according to Eq 23.
2. The 1/9th–4/9th element which permits imposition of the correct singularity in the strain energy density in the near-tip region.

Enriched Element

The concept of enriching the conventional element displacement assumption with the asymptotic displacement field appropriate for fracture analysis originated with Benzley [8] and has been extensively applied by the present authors for the elastic ease [11]. For the elastoplastic Mode I crack problem, the multilinear material model is used and the leading term of the asymptotic expansion for the near-tip crack displacement, Eq 24 or Eq 25, is added to the usual polynomial displacement field within the 12-node element. The result is

$$
\begin{aligned}
u &= \sum_{i=1}^{12} N_i(s,t)u_i + K_p\left[\tilde{u}(s,t) - \sum_{i=1}^{12} N_i(s,t)\tilde{u}_i\right] \\
v &= \sum_{i=1}^{12} N_i(s,t)v_i + K_p\left[\tilde{v}(s,t) - \sum_{i=1}^{12} N_i(s,t)\tilde{v}_i\right]
\end{aligned}
\tag{24}
$$

where

(s, t) = local coordinates associated with the isoparametric element,
N_i = usual isoparametric shape functions, and
$\tilde{u}(s, t)$ = term (transformed to local element coordinates) in Eqs 21 or 22 in the expression for u whose coefficient is K_p.

The $\tilde{v}$ component of displacement is defined similarly.

Equations 24 are the enriched element displacement approximation, giving the displacement components within the element in terms of their nodal values (u_i and v_i) and the plastic intensity factor K_p. The enriched element stiffness matrix is developed from these equations in the usual manner, with the exception that higher-order Gauss quadrature (8×8) is employed in the area integration to properly handle the steep gradients near the crack tip.

The J-integral is calculated from K_p using Eqs 23 as described previously. In addition, the J-integral may be evaluated on paths.

The value of J obtained directly from the enriched elements using Eqs 24 is generally too low and its inaccuracy increases with decreasing slope of the strain-hardening portion of the stress-strain curve [*2*]. Path values for J, on the other hand, do not possess this inaccuracy, and the use of the enriched element actually improves them.

1/9th–4/9th Element

Barsoum [*12*] has shown that when mid-side nodes of 8-node quadrilateral isoparametric elements are moved to the quarter-point nearest the crack tip, a $1/\sqrt{r}$ singularity in the strain field is developed. Following this work, Pu and Hussain [*13*] found that the same effect could be achieved with 12-node quadrilateral isoparametric elements by moving intermediate nodes to the 1/9th and 4/9th positions nearest the crack tip. The authors have performed elastic calculations with 1/9th–4/9th quadrilateral elements (rather than corresponding collapsed triangular elements currently preferred by most investigators) and have obtained highly accurate results. Benzley [*14*] has shown that the use of quarter-point 8-node elements in elastic-plastic computations leads to improved path values of the J-integral. At the same time, the plastic singular solution associated with a multilinear material model contains the square-root singularity associated with the 1/9th–4/9th element. Thus the 1/9th–4/9th element may be employed with a multilinear material model as an additional plastic singularity element, and improved values of J may be obtained by evaluation over paths remote from the crack tip. The authors [*2*] have observed comparable accuracy of J (path) values when either enriched or 1/9th–4/9th crack tips are employed.

Tearing Modulus

The applied tearing modulus [*15*] is defined by

$$T_{\text{APPL}} = \frac{E}{\sigma_0^2} \frac{dJ}{da} \tag{25}$$

where

E = Young's modulus,
σ_0 = flow stress (average of yield and ultimate stresses), and
dJ/da = rate of change of the J-integral with respect to crack length.

Ductile crack growth stability is assessed by the tearing modulus as follows. Fracture is stable upon reaching J_{Ic} if $T_{\text{APPL}} < T_{\text{MATL}}$, or unstable if $T_{\text{APPL}} \geq T_{\text{MATL}}$, where T_{MATL} is the tearing modulus of the material taken from test records (resistance curves) of J versus Δa in standard J_{Ic} tests.

In calculating T_{APPL}, the crack tip is advanced as shown in Fig. 1 using the approach recommended by Shih et al [*16*]. The crack-tip node is first moved incrementally until the crack is advanced over half the width of the element originally ahead of the crack. The crack plane nodes of this element are then gradually released, again over a number of user-specified increments, until the crack tip corresponds with the next corner node in the plane of the crack. The procedure can be repeated so that the crack grows across a number of elements.

Thick-Walled Cylinder

The plane-strain idealization of an internally pressurized thick-walled cylinder (an axisymmetric problem) is shown in Fig. 2. The method of load incrementation, convergence characteristics, displacements at the inside and outside surfaces, and extent of the plastic zone are all indicated in Table 1 and comparison is made with the results of Ref *17*. Slightly higher PAPST displacement values at higher load levels are caused by the large-strain, large-displacement capabilities of PAPST which are not contained in Ref *17*. The faster convergence characteristics of PAPST are apparent, but do not reflect actual computational efficiency since the solution techniques differ.

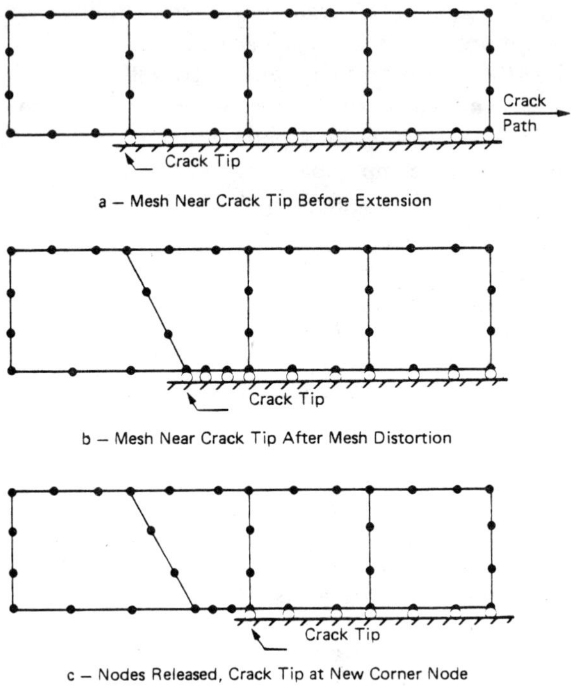

FIG. 1—*Procedure for advancing crack to calculate applied tearing modulus.*

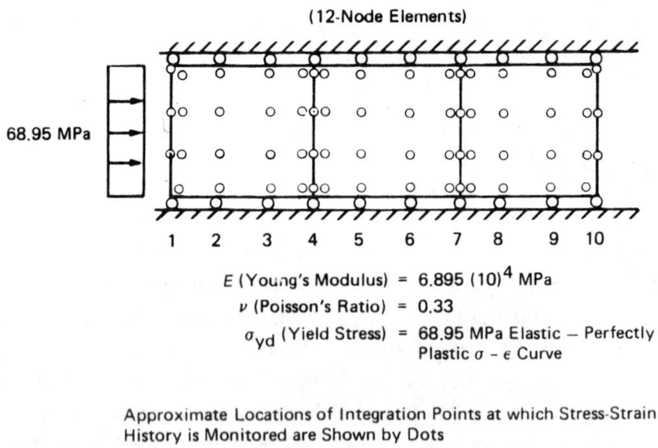

FIG. 2—*Idealization of thick wall cylinder for plastic analysis.*

TABLE 1—*Results for three-element idealization of thick-walled cylinder.*

Increment Number I	Internal Stress / Yield Stress	Initial Stress Method (Ref. 16)	Tangent Stiffness Method (This Study)	Yielded Area (This Study)
1	0.453	Not Available	2 (1) 0.0665 mm (2) 0.0411 mm (3)	
2	0.5	3 0.0747 mm 0.0460 mm	2 0.0747 mm 0.0460 mm	
3	0.55	3 0.0243 mm 0.0515 mm	2 0.0859 mm 0.0523 mm	
4	0.6	4 0.0993 mm 0.0599 mm	2 0.0985 mm 0.0600 mm	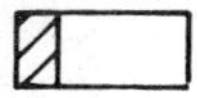
5	0.65	4 0.116 mm 0.0091 mm	2 0.1162 mm 0.0093 mm	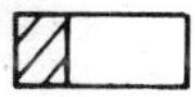
6	0.7	5 0.139 mm 0.0815 mm	3 0.1398 mm 0.0820 mm	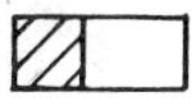
7	0.75	6 0.175 mm 0.100 mm	3 0.177 mm 0.102 mm	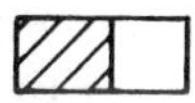
8	0.77	7 0.196 mm 0.111 mm	3 0.200 mm 0.114 mm	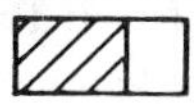
9	0.79	8 0.228 mm 0.128 mm	3 0.241 mm 0.135 mm	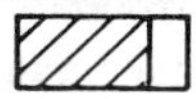

(1) Number of iterations to converge
(2) Radial displacement at inside surface, mm
(3) Radial displacement at outside surface, mm

(Earlier experience with PAPST indicated that the presently used tangent stiffness method is about twice as efficient as the initial stress approach coupled with the alpha acceleration technique [*18*].)

Compact J_{Ic} Specimen

Although the geometry of the standard precracked compact-tension J_{Ic} test specimen is simple, it serves as a good example of the use of the nonlinear fracture analysis capabilities of PAPST.

An idealization typical for such a specimen (modeling only the symmetric upper half of the specimen) is shown in Fig. 3. The two small elements surrounding the crack tip are enriched elements. The actual pin loading (or displacement) applied to the specimen is simulated by applying tractions (or displacements) along the left edge of the specimen, which corresponds to the

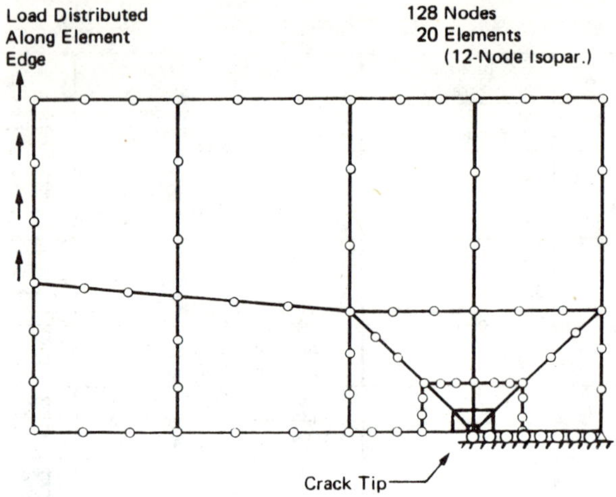

FIG. 3—*Finite-element model of compact* J_{Ic} *specimen.*

centerline of the loading pins. A bilinear material model was used to approximate the experimental stress-strain curve with Young's modulus taken as $199(10)^3$ MPa, the yield stress as 965 MPa, and the slope of the hardening portion of the curve as 1207 MPa.

The purpose of this example is to calculate the J-integral as a function of applied displacement and to indicate the variation of *J* from path to path. Complete computational details for this example have been documented informally by Gifford [*19*]. The six paths selected for evaluation of the J-integral are shown in Fig. 4.

Results of the computations and a comparison with experimental results are given in Table 2. (Experimental values of *J* are obtained by application of the Merkle-Corten formula [*20*] for *J* to the actual experimental load-deflection curve.) The variation of *J* from path to path is somewhat higher than one would hope, on the order of 17 percent (maximum) at the highest applied displacement.[4]

The computations were performed under a plane-strain assumption. This is valid near the crack tip, but a plane-stress assumption would be more appropriate over most of the specimen. Thus the model is too stiff; under displacement control, calculated values of *J* at a given load level tend to be higher than experimental values, while under load control the opposite occurs. In fact, it was found that calculated values of *J* (under load control) for conditions of plane strain and plane stress actually did bracket the experimental results.

[4]The computations were done before strain smoothing was introduced into PAPST, an improvement that diminishes somewhat the differences in *J* among the various paths.

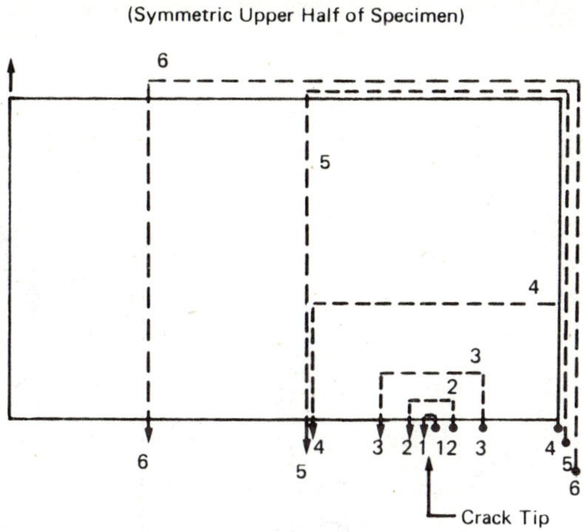

FIG. 4—*Paths for evaluation of the J-integral.*

TABLE 2—*Typical path results for the J-integral.*

Applied Measured Displacement, mm	Corresponding Load, kN	J-Integral, kJ/m^2						
		Path 1	Path 2	Path 3	Path 4	Path 5	Path 6	Experimental
0.358	8.896	12.1	11.7	11.9	13.6	15.0	14.4	12.6
0.564	13.344	29.8	28.9	29.0	33.4	36.9	35.5	29.4
0.790	17.792	59.4	56.0	56.4	64.8	71.8	69	55.2
0.914	20.016	79.3	73	74.8	85.6	94.9	91.4	72.7
1.054	22.24	102.6	94.7	98.1	110	122	117	95.1
1.143	23.352	118.6	110.5	114.5	127	138.7	133.6	109
1.245	24.464	137.6	129.1	133.8	147	158	152	128
1.397	25.576	168.3	159.7	162.5	177	187.6	181	155
1.588	26.2432	209.4	198.6	196.7	219.1	226.6	218.7	192.6

NOTE: σ_{yd} = 965.3 MPa; strain hardening $\Delta\sigma/\Delta\epsilon$ = 1206.6 MPa.

Crack Growth Simulation

A plane-stress, elastic-plastic analysis for crack growth in a compact-tension specimen is presented to demonstrate this use for PAPST. The specimen geometry and grid pattern are shown in Fig. 5. The uniaxial material behavior is approximated as bilinear with σ_{yd} = 462 MPa and $\sigma_{ultimate}$ = 620 MPa at 22 percent strain.

For purposes of demonstration, the specimen has been loaded to approximately 11.96 kN with a corresponding load-point displacement of 0.7 mm prior to crack growth. Crack extension has been simulated separately for constant-load-point displacement.

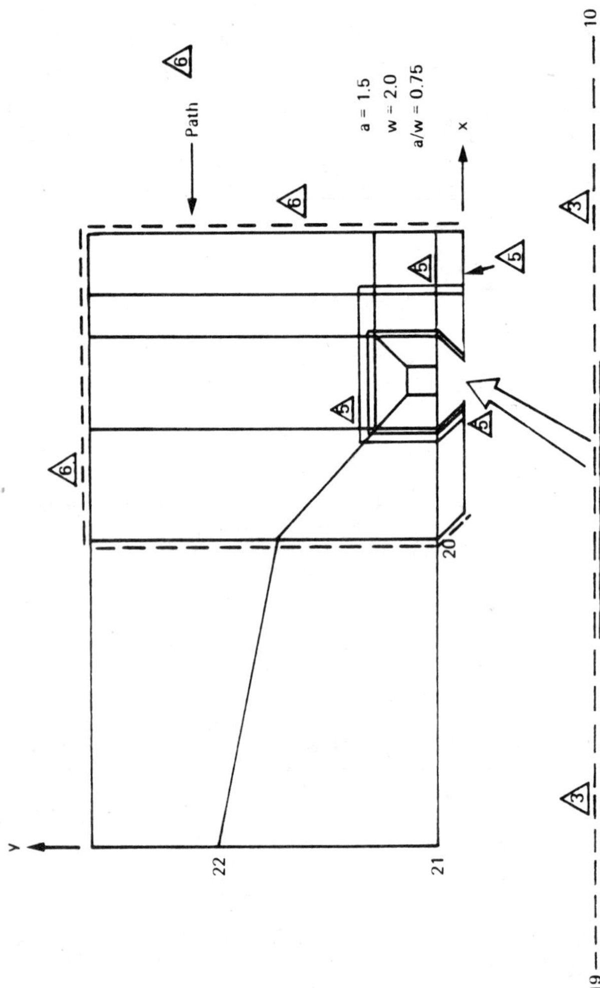

FIG. 5—*Crack growth compact tension specimen: geometry, grid pattern, J-integral paths.*

The element over which the crack extension occurs has dimensions of 0.635 mm. As the growth is simulated in two parts by first moving the crack-tip node an amount equal to half the element dimension and then releasing the nodes along this element ahead of the crack, two equal increments of crack extension $\Delta a = 0.03175$ mm are simulated. Increment-by-increment results for load, load-point displacement, and the corresponding path values of J-integral are given in Table 3. Note that for the crack advance under displacement control, the reaction load decreases. Quadrature points which unloaded during this crack growth increment were modeled as having elastic incremental stiffness properties.

Figures 6, 7, and 8 contain the load/load-point displacement, J (from Path 5) versus crack extension, and the plastic zone boundaries at the end of the

TABLE 3—*Crack growth simulation results.*

Increment No.	Load, kN	Load-Point Displacement, mm	J-Path Values, kJ/m^2 2	3	4	5	6	J (LEFM)[a] k_1^2/E
1	2.30	0.085	0.6	0.7	0.6	0.7	0.7	0.74
2	8.56	0.356	12.0	11.6	9.5	10.3	11.6	10.3
3	10.73	0.510	20.3	20.5	18.0	19.8	21.5	16.2
4	11.55	0.610	25.7	26.4	23.8	27.0	28.7	18.7
5	11.97	0.710	33.8	33.4	29.9	33.8	36.4	20.1
7	11.39	0.710	27.7	26.6	23.1	32.9	22.2	...
9	10.39	0.710	29.2	26.4	23.1	30.8	20.8	...

[a]LEFM = linear elastic fracture mechanics.

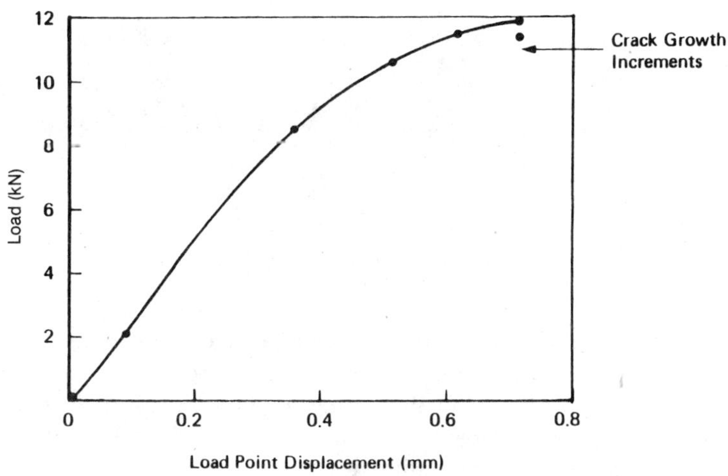

FIG. 6—*Load versus load-point displacement.*

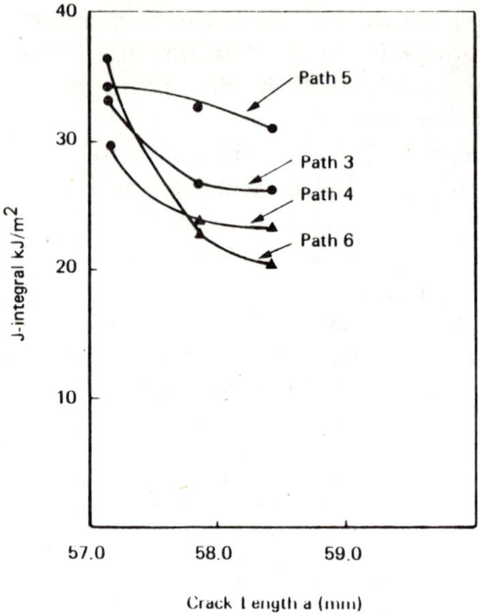

FIG. 7—*Path value—predictions for* J *during crack extension.*

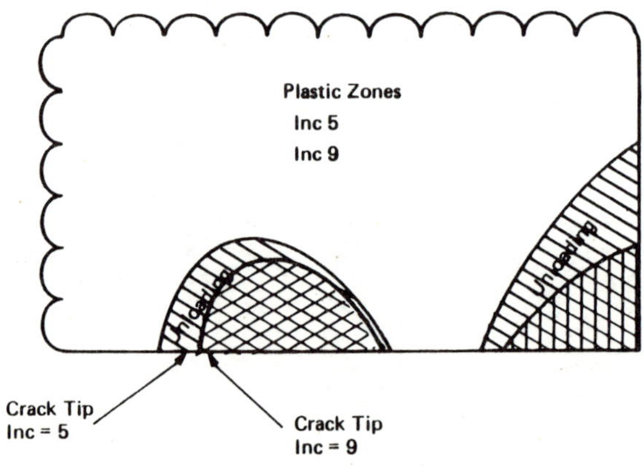

FIG. 8—*Plastic zones changes during crack growth.*

fifth and ninth increments, respectively. Note that there exists substantial variations in the J-integral predictions during crack growth from path to path. This is clearly related to the highly nonproportional behavior of the stress field during crack growth as reflected in the change in plastic zone geometry. The authors are continuing their research in an attempt to understand the source of these variations and to develop procedures which minimize them.

References

[1] Hilton, P. D., *Journal of Pressure Vessel Technology*, Feb. 1976.

[2] Hilton, P. D. and Gifford, L. N., "Evaluation of Some Crack Tip Finite Elements for Elastoplastic Fracture Analysis," DTNSRDC Report 79/052, David W. Taylor Naval Ship R&D Center, Bethesda, Md., July 1979.

[3] Hilton, P. D. and Gifford, L. N., "Preliminary Documentation of PAPST Nonlinear Fracture and Stress Analysis by Finite Elements," David Taylor Naval Ship R&D Center, Bethesda, Md., Jan. 1981.

[4] Key, S. W., *Computer Methods in Applied Mechanics and Engineering*, Vol. 4, 1974, pp. 195–218.

[5] Hilton, P. D. and Hutchinson, J. W., *Engineering Fracture Mechanics*, Vol. 3, 1971, pp. 435–451.

[6] Hutchinson, J. W., *Journal of the Mechanics and Physics of Solids*, Vol. 16, 1968, pp. 13–31.

[7] Rice, J. R. and Rosengren, G. F., *Journal of the Mechanics and Physics of Solids*, Vol. 16, No. 1, 1968, pp. 1–12.

[8] Benzley, S. E., *International Journal for Numerical Methods in Engineering*, Vol. 8, 1974, pp. 537–545.

[9] Praeger, W., *Introduction to Mechanics of Continua*, Dover Press, New York, 1971, pp. 154–156.

[10] Krieg, R. D. and Krieg, D. B., *Transactions*, American Society of Mechanical Engineers, *Journal of Pressure Vessel Technology*, Series J, Vol. 99, No. 4, Nov. 1977, pp. 510–515.

[11] Gifford, L. N. and Hilton, P. D., *Engineering Fracture Mechanics*, Vol. 10, No. 3, 1978, pp. 485–496.

[12] Barsoum, R. S., *International Journal for Numerical Methods in Engineering*, Vol. 10, 1976, pp. 25–37.

[13] Pu, S. L. and Hussain, M. A., *International Journal for Numerical Methods in Engineering*, Vol. 12, No. 11, 1978, pp. 1727–1742.

[14] Benzley, S. E., *International Journal of Fracture*, Vol. 12, No. 1, June 1976, pp. 475–477.

[15] Paris, P. C. et al, "A Treatment of the Subject of Tearing Instability," Nuclear Regulatory Commission Report NUREG 0311, Washington, D.C., July 1977.

[16] Shih, C. F., deLorenzi, H. G., and Andrews, W. R. in *Elastic-Plastic Fracture, ASTM STP 668*, J. D. Landes, J. A. Begley, and G. A. Clarke, Eds., American Society for Testing and Materials, 1979, pp. 65–120.

[17] Nayak, G. C. and Zienkiewicz, O. C., *International Journal for Numerical Methods in Engineering*, Vol. 5, No. 1, 1972, pp. 113–135.

[18] Nayak, G. C. and Zienkiewicz, O. C., *International Journal for Numerical Methods in Engineering*, Vol. 4, 1972, pp. 579–582.

[19] Gifford, L. N., "J-Integral Analysis of a Compact J_{Ic} Specimen," DTNSRDC Structures Department Technical Note m^4," David W. Taylor Naval Ship R&D Center, Bethesda, Md., Sept. 1978.

[20] Merkle, J. G. and Corten, H. T., *Transactions*, American Society of Mechanical Engineers, *Journal of Pressure Vessel Technology*, Series J, Vol. 98, No. 4, Nov. 1974, pp. 286–292.

Fully Elastic Crack and Surface Flaw Analysis

Ming Yuan He[1] *and John W. Hutchinson*[2]

Bounds for Fully Plastic Crack Problems for Infinite Bodies

REFERENCE: He, M. Y. and Hutchinson, J. W., **"Bounds for Fully Plastic Crack Problems for Infinite Bodies,"** *Elastic-Plastic Fracture, Volume I—Inelastic Crack Analysis, ASTM STP 803*, C. F. Shih and J. P. Gudas, Eds., American Society for Testing and Materials, 1983, pp. I-277–I-290.

ABSTRACT: For cracks in infinite bodies it is shown that modified principles of complementary potential energy and potential energy can be used to generate upper and lower bounds to the J-integral of the deformation theory of plasticity. These principles are used to obtain relatively tight numerical bounds on J for two basic plane strain problems: the finite crack in an infinite plane and the edge-crack in a semi-infinite plane. In both problems the material is incompressible with a pure power relation between stress and strain. Upper bounds for the plane stress problems are also given.

KEY WORDS: nonlinear fracture mechanics, fully plastic crack problems, J-integral, bounds, elastic-plastic fracture

In the first section of this paper we derive principles for obtaining upper and lower bounds for the J-integral in certain infinite or semi-infinite crack problems in which the only length quantity involved is the crack length itself. Under these circumstances J, the energy release rate per unit of crack extension in a deformation theory material, is simply related to the minimum of the potential energy functional and the minimum of the complementary energy functional, each of which is modified appropriately for an unbounded region. Any statically admissible stress field in conjunction with the modified complementary energy functional generates an upper bound to J, while any kinematically admissible displacement field used with the modified potential energy functional gives a lower bound.

The principles are then implemented to obtain numerical bounds for two basic fully plastic problems in plane strain: the finite crack of length $2a$ in the

[1]Researcher, Institute of Mechanics, Chinese Academy of Sciences, Beijing, China.

[2]Professor of applied mechanics, Division of Applied Sciences, Harvard University, Cambridge, Mass. 02138.

infinite plane and the crack of length a normal to the edge of a semi-infinite plane. Numerical results for the upper bound to J are also given for the corresponding problems under plane-stress conditions.

The numerical results are obtained for the case of a small strain incompressible solid characterized in simple tension by

$$\epsilon/\epsilon_0 = \alpha(\sigma/\sigma_0)^n \tag{1}$$

where ϵ_0 and σ_0 are a reference strain and stress and α is an extra constant introduced for convenience of application. The tensile relation is generalized to multi-axial states by J_2 deformation theory according to

$$\frac{\epsilon_{ij}}{\epsilon_0} = \frac{3}{2}\alpha\left(\frac{\sigma_e}{\sigma_o}\right)^{n-1}\frac{s_{ij}}{\sigma_0} \tag{2}$$

where s_{ij} is the stress deviator and σ_e is the effective stress defined by

$$\sigma_e = (3/2\, s_{ij}\, s_{ij})^{1/2} \tag{3}$$

With an effective strain defined by

$$\epsilon_e = (2/3\, \epsilon_{ij}\, \epsilon_{ij})^{1/2} \tag{4}$$

σ_e and ϵ_e satisfy the tensile relation, Eq 1.

The numerical results are obtained using a Rayleigh-Ritz procedure to minimize the respective functionals with respect to a family of kinematically or statically admissible fields with free amplitude factors. This same numerical procedure was employed in an earlier paper [*1*][3] to bound J from below for the finite crack in the infinite plane and for a penny-shaped crack in an infinite body. In still earlier work, Ranaweera and Leckie [*2*] used finite-element procedures based on both displacements and stresses together with the corresponding two unmodified minimum principles to evaluate J for several configurations of finite extent. Their results do appear to have a bound-like character even though the bounding principles do not strictly apply to finite bodies.

Upper and Lower Bounds to *J*

Attention is restricted to a small strain deformation theory of plasticity for homogeneous bodies in which the strain energy density and complementary stress energy density are

$$W(\epsilon) = \int_0^{\epsilon} \sigma_{ij} d\epsilon_{ij} \quad \text{and} \quad U(\sigma) = \int_0^{\sigma} \epsilon_{ij} d\sigma_{ij} \tag{5}$$

[3]The italic numbers in brackets refer to the list of references appended to this paper.

To construct the minimum principles for an infinite or semi-infinite body loaded at infinity, let σ^∞, ϵ^∞, and $\mathbf{u}^\infty$ denote the stresses, strains, and displacements associated with the uniform field of the loaded body in the absence of the crack. Denote additional quantities by a tilde so that the total quantities in the presence of a crack are given by

$$\sigma = \sigma^\infty + \tilde{\sigma}, \qquad \epsilon = \epsilon^\infty + \tilde{\epsilon}, \qquad \mathbf{u} = \mathbf{u}^\infty + \tilde{\mathbf{u}} \tag{6}$$

Consider a family of finite bodies with an outer boundary S_R of radius R on which tractions $T_i = \sigma_{ij}{}^\infty n_j$ are prescribed as shown in Fig. 1. The traction-free crack has length (or half-length) a and for the edge-crack problem the edge ($x_1 = 0$) is also traction-free. The remote stress σ^∞ must be consistent with the traction-free conditions on boundaries other than the crack. For example, $\sigma_{ij}{}^\infty n_j$ must vanish on the edge $x_1 = 0$ for the edge-crack problem.

The potential and complementary potential energy functionals for the finite body are

$$PE = \int_{V_R} W(\epsilon)dV - \int_{S_R} \sigma_{ij}{}^\infty n_j u_i dS \tag{7}$$

and

$$CE = \int_{V_R} U(\sigma)dV \tag{8}$$

where V_R denotes the volume per unit thickness exterior to the crack and interior to S_R. The stress field in Eq 8 must satisfy equilibrium and the traction-free boundary conditions on the crack and on any other traction-free

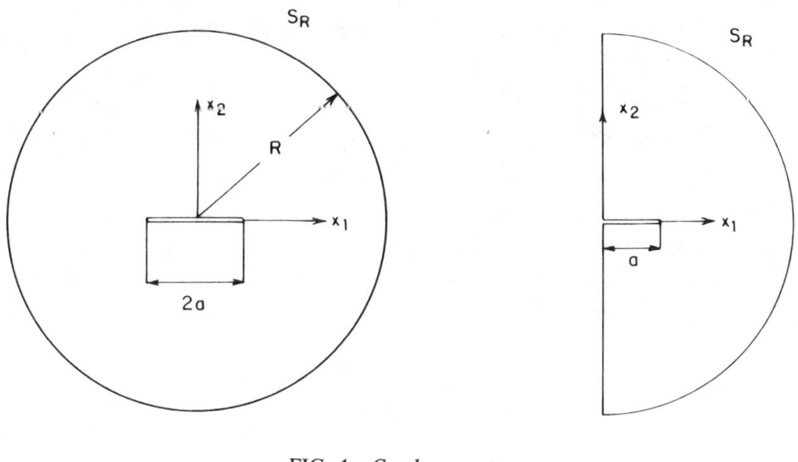

FIG. 1—*Crack geometry.*

boundary. It must satisfy the traction condition $\sigma_{ij}n_j = \sigma_{ij}{}^{\infty}n_j$ on S_R. The strain in Eq 7 is derived from the displacement field using the linear strain-displacement equations. Subject to standard convexity conditions on the energy density functions, which will be assumed, the two functionals are minimized by the respective exact fields. For the exact solution it is easily shown that

$$CE = -PE \tag{9}$$

and J, defined as the energy release rate with $\sigma_{ij}{}^{\infty}n_j$ held constant on S_R, is

$$J = -\frac{1}{m}\frac{dPE}{da} = \frac{1}{m}\frac{dCE}{da} \tag{10}$$

Here $m = 1$ when one crack tip is present, as in the edge-crack problem, and $m = 2$ when the crack has two tips.

We now manipulate Eqs 7 and 8 into forms such that the contributions which depend on a remain bounded as R becomes infinite. First we deal with Eq 7. By the principle of virtual work

$$\int_{S_R} \sigma_{ij}{}^{\infty}n_j\tilde{u}_i dS = \int_{V_R} \sigma_{ij}{}^{\infty}\tilde{\epsilon}_{ij} dV_R + \int_{S} \sigma_{ij}{}^{\infty}n_j\tilde{u}_i dS \tag{11}$$

where S is the surface of the crack and $\mathbf{n}$ is the unit normal to S pointing into V_R. Equation 7 can be written as

$$PE = \Phi + C \tag{12}$$

where

$$\Phi(\tilde{\mathbf{u}}) = \int_{V_R} \{W(\epsilon) - W(\epsilon^{\infty}) - \sigma_{ij}{}^{\infty}\tilde{\epsilon}_{ij}\} dV - \int_{S} \sigma_{ij}{}^{\infty}n_j\tilde{u}_i dS \tag{13}$$

and

$$C = \int_{V_R} W(\epsilon^{\infty}) dV - \int_{S_R} \sigma_{ij}{}^{\infty}n_j u_i{}^{\infty} dS \tag{14}$$

Now, as long as $\tilde{\mathbf{u}}$ decays faster than $r^{-1/2}$ as $r \to \infty$, where $r = (x_i x_i)^{1/2}$, it can be shown that Φ remains bounded as $R \to \infty$. Furthermore, among all such admissible $\tilde{\mathbf{u}}$, the exact solution $\tilde{\mathbf{u}}$ minimizes Φ for the infinite problem [*3*]. Since C is independent of a, it follows that

$$J = -\frac{1}{m}\frac{d\Phi}{da} \tag{15}$$

and this relation is valid in the limit as R becomes unbounded.

For either the infinite plane with crack of length $2a$ or the semi-infinite plane with the edge-crack of length a, the crack length is the only length quantity and thus Φ must have the form

$$\Phi = a^2 f(\sigma^\infty) \tag{16}$$

where f depends implicitly on the material properties and on σ^∞ but not on a. Thus, from Eqs 15 and 16

$$J = -\frac{2}{ma}\,\Phi_{\min} \tag{17}$$

where the notation $\Phi_{\min}$ is used to emphasize that Φ is evaluated using the exact solution. Any estimate of J obtained from

$$J = -\frac{2}{ma}\,\Phi(\tilde{u}) \tag{18}$$

using an admissible additional field $\tilde{\mathbf{u}}$ necessarily leads to a *lower bound* to J since $\Phi(\tilde{\mathbf{u}}) \geq \Phi_{\min}$.

Next, we rewrite Eq 8 as

$$CE = F + C' \tag{19}$$

where

$$F(\tilde{\sigma}) = \int_{V_R} \{U(\sigma) - U(\sigma^\infty) - \tilde{\sigma}_{ij}\epsilon_{ij}{}^\infty\}\,dV \tag{20}$$

and

$$C' = \int_{V_R} U(\sigma^\infty)dV \tag{21}$$

Here we have made use of the fact that

$$\int_{V_R} \tilde{\sigma}_{ij}\epsilon_{ij}{}^\infty dV = 0$$

since $\tilde{\sigma}_{ij}n_j = 0$ on S_R and $\tilde{\sigma}_{ij}n_j u_i{}^\infty = -\sigma_{ij}{}^\infty n_j u_i{}^\infty$ on the crack faces, which integrates to zero. If the additional stresses $\tilde{\sigma}$ decay faster than $r^{-3/2}$ as $r \to \infty$, then Eq 20 is bounded in the infinite problem. Since C' is independent of a

$$J = \frac{1}{m}\frac{dF}{da} \tag{22}$$

Furthermore, among all statically admissible $\tilde{\sigma}$ the exact solution minimizes F, as can be shown using an approach similar to that employed in Ref *3* for the potential energy formulation. From dimensional considerations it follows that F is proportional to a^2 for either of the plane problems so that

$$J = \frac{2}{ma} F_{\min} \tag{23}$$

By virtue of the minimizing role of the exact solution, any estimate of J obtained from

$$J = \frac{2}{ma} F(\tilde{\sigma}) \tag{24}$$

using a statically admissible additional stress field gives an *upper bound*.

Numerical Method

For the upper bounds a stress function was used to generate equilibrium additional stress fields according to

$$\tilde{\sigma}_{11} = \chi_{,22}, \qquad \tilde{\sigma}_{22} = \chi_{,11}, \qquad \text{and} \quad \tilde{\sigma}_{12} = -\chi_{,12} \tag{25}$$

The power-law material, Eq 2, is inherently incompressible and thus in the lower-bound calculations a stream function could be used to give the additional displacements as

$$\tilde{u}_1 = \psi_{,2} \quad \text{and} \quad \tilde{u}_2 = -\psi_{,1} \tag{26}$$

In each case, the function was represented as a linear sum of admissible functions in the form

$$\chi = \sum_{i=1}^{K} A_i \chi^{(i)} \quad \text{or} \quad \psi = \sum_{i=1}^{K} A_i \psi^{(i)} \tag{27}$$

where the A_i's are free variables chosen to minimize F or Φ. This minimization was achieved by a Newton-Raphson method. With the upper-bound formulation to illustrate the method, the minimum condition is

$$\frac{\partial F}{\partial A_k} = 0 \qquad k = 1, K \tag{28}$$

where

$$\frac{\partial F}{\partial A_k} = \int_V (\epsilon_{ij} - \epsilon_{ij}^{\infty}) s_{ij}^{(k)} dV \tag{29}$$

and $\mathbf{s}^{(k)}$ is derived from $\chi^{(k)}$ so that

$$s_{ij} = \sum_{k=1}^{K} A_k s_{ij}^{(k)} \tag{30}$$

If $\{A_j\}$ is an estimate of the solution to Eq 28, the improved estimate $\{A_j + \Delta A_j\}$ from the next iteration is obtained from

$$\sum_{m=1}^{K} \Delta A_m \frac{\partial^2 F}{\partial A_m \partial A_p} = -\frac{\partial F}{\partial A_p} \tag{31}$$

where

$$\frac{\partial^2 F}{\partial A_m \partial A_p} = \int_V \frac{3}{2} \sigma_e^{n-1} \left[s_{ij}^{(m)} s_{ij}^{(p)} + \frac{3}{2} \frac{(n-1)}{\sigma_e^2} (s_{ij} s_{ij}^{(m)})(s_{kl} s_{kl}^{(p)}) \right] dV \tag{32}$$

Corresponding expressions for the lower bound based on Φ and the first of Eq 27 are readily derived.

The integrals over the body in Eqs 20, 29, and 32 were evaluated using a mapping technique together with numerical integration in the mapping plane. With reference to Fig. 2, the physical plane in both plane problems investigated here was mapped into the unit circle in the ζ-plane using the conformal map

$$z = \omega(\zeta) \equiv \frac{a}{2}(\zeta + \zeta^{-1}) \tag{33}$$

where $z = x_1 + ix_2$ and $\zeta = \xi + i\eta = \mu i e^{-i\phi}$. The planar-polar coordinates μ and ϕ in the mapping plane were taken as the independent variables in the functional representation of χ and ψ, as described in the following. Integrations over the body were performed using iterated 10-point Gaussian quadrature formulas for μ ranging from 0 to 1 and ϕ ranging from θ to $\pi/2$. This same technique was used in Refs *1* and *4*, where it is described in more detail. Test cases based on known linear solutions and on special trial functions for the nonlinear material indicated that the numerical integration of F or Φ is accurate to at least three or four significant figures.

The numerical method in the present study differed from that in our previous paper [*1*] in that here explicit expressions for the partial derivatives of Φ and F with respect to the amplitude factors (for example, Eqs 29 and 32) were used. In the earlier work these derivatives were evaluated directly from the functional by numerical differentiation. The present procedure resulted in a considerable reduction in computation time and led to better numerical conditioning in some instances.

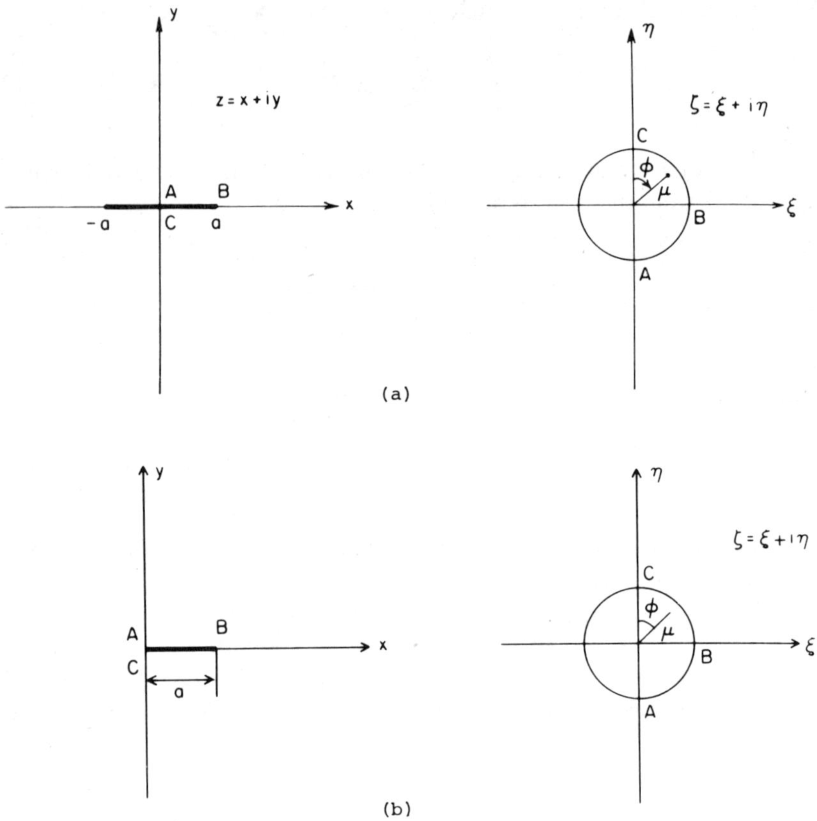

FIG. 2—*Mapping from physical plane to mapping plane:* (a) *central crack;* (b) *edge-crack.*

Bounds for the Crack in the Infinite Plane

As indicated in Fig. 1, the crack of length $2a$ lies on the x_1-axis, the one nonzero remote in-plane stress component is $\sigma_{22}{}^{\infty} = \sigma^{\infty}$, and plane-strain conditions ($\epsilon_{33} = 0$) are assumed. The form of the solution for J for the pure power-hardening material, Eq 2, is [*1*]

$$J = a\sigma_e{}^{\infty}\,\epsilon_e{}^{\infty}h(n) \tag{34}$$

where, from Eqs 3 and 4

$$\sigma_e{}^{\infty} = \frac{\sqrt{3}}{2}|\sigma^{\infty}| \quad \text{and} \quad \epsilon_e{}^{\infty} = \frac{2}{\sqrt{3}}|\epsilon_{22}{}^{\infty}| = \alpha\epsilon_0\left(\frac{\sigma_e{}^{\infty}}{\sigma_0}\right)^n \tag{35}$$

The bounds given in the following are for $h(n)$.

For the *lower bound* the stream function, Eq 26, was taken as

$$\psi = \sum_{k=1}^{N} \sum_{j=1}^{M} A_{kj} \, [\mu^{j-1} \sin 2k\phi + (-1)^{k+1} \, 2k(\phi - \pi/2)] \qquad (36)$$

which is the same representation used in Ref *1*, except that more terms were used to obtain the present results. This choice is consistent with the symmetry of the solution with respect to the two Cartesian axes and, as discussed in Ref *1*, gives rise to a crack opening of the approximate form $f(x_1)\sqrt{a^2 - x_1{}^2}$ where $f(a)$ is bounded. Expressions for partial derivatives of ψ (and χ) with respect to Cartesian coordinates, which are needed in the evaluation of Φ, are obtained in terms of the derivatives with respect to μ and ϕ with the aid of standard change of variable formulas as illustrated in Refs *1* and *4*.

The stress function for the additional stresses Eq 25 for the *upper bound* was taken to be

$$\chi = \frac{\sigma^{\infty}}{2} \, (\mu + \mu^{-1})^2 [(1 - \mu^2)^s - 1] \sin^2\phi +$$

$$- (1 + b_i) f_p(\phi)] + \sum_{k=1}^{N} \sum_{j=1}^{M} A_{kj} \, (1 - \mu)^{3-[2/(n+1)]+j} \cos 2k\phi \qquad (37)$$

This choice is also consistent with the twofold symmetry and each individual contribution gives rise to additional stress components which decay faster than $r^{-3/2}$ for large r. The first term in Eq 37 gives $\tilde{\sigma}_{12} = 0$ and $\tilde{\sigma}_{22} = -\sigma^{\infty}$ on the crack faces, while each of the other terms makes no traction contribution. Thus, from Eq 6, the total traction vanishes on the crack faces. By numerical experimentation it was found that a relatively large value of the exponent s in the first term was best and the choice $s = 7$ was made. The individual terms in the double sum in Eq 37 with $j = 1$ each has a stress singularity at the tip of the order $r^{-1/(n+1)}$, which is in agreement with the order of the actual singularity. The terms in the double sum make a contribution to σ_{11} which is zero on the crack faces. The second set of terms in Eq 37 is capable of representing an arbitrary distribution of σ_{11} along the crack faces. The functions $f_i(\phi)$ form a complete set with the required symmetry conditions $f = f' = 0$ at $\phi = 0$ and $\pi/2$. They are given by

$$f_i(\phi) = \cos \lambda_i(\phi - \pi/2) + b_i \cosh \lambda_i(\phi - \pi/2) \qquad (38)$$

where λ_i is the root of (ordered in ascending magnitude with $\lambda_1 = 1.5056187$)

$$\cos(\lambda_i \pi/2) \sinh(\lambda_i \pi/2) + \sin(\lambda_i \pi/2) \cosh(\lambda_i \pi/2) = 0 \qquad (39)$$

and

$$b_i = \sin(\lambda_i \pi/2)/\sinh(\lambda_i \pi/2) \qquad (40)$$

The computed bounds to $h(n)$ as defined in Eq 34 are given in Table 1 for values of n between 1 and 7. For $n \leq 3$ the spread between the bounds is less than 2 percent. For $n = 5$ it is 4 percent and for $n = 7$ it is 7 percent. The lower bound was computed with $N = 5$ and $M = 6$ for a total of 30 free amplitudes.[4] The upper bound was determined using 34 free amplitudes with $N = 5$, $M = 6$, and $P = 5$.

Included in Table 1 and plotted in Fig. 3 are numerical values based on the

TABLE 1—*Upper- and lower-bound estimates of* h *for plane-strain cracks in an infinite plane.*

	$n = 1.0$	$n = 1.5$	$n = 2.0$	$n = 3.0$	$n = 4.0$	$n = 5.0$	$n = 7.0$
Upper bound	3.156	3.884	4.517	5.616	6.579	7.456	9.052
Lower bound	3.141	3.856	4.470	5.511	6.390	7.152	8.421
$\pi\sqrt{n}$	3.141	3.848	4.443	5.441	6.282	7.024	8.312

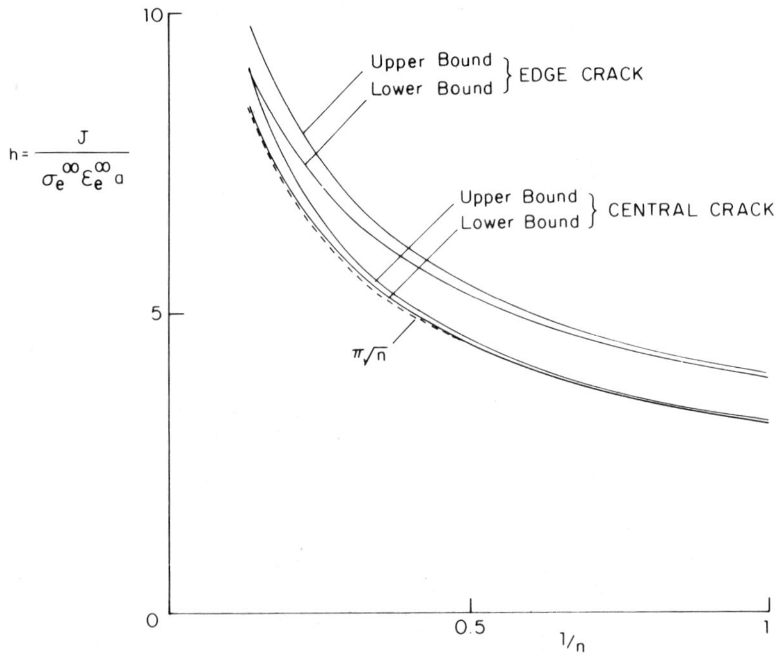

FIG. 3—*Bounds for the central crack in the infinite body and the edge crack in the semi-infinite plane under plane-strain conditions.*

[4]The results reported in Ref *1* were computed with $N = 3$ and $M = 4$; for $n = 5$ the present results are about 5 percent higher. This reflects the convergence rate with an increasing number of terms.

approximate result $h(n) = \pi\sqrt{n}$ obtained in Ref *1*. It is seen that this simple approximation is only slightly below the lower bound in the range $1 < n \leq 7$ and is within 3 percent of the upper bound for $n \leq 3$ and within 8 percent for $n = 7$. As described in Ref *1* this result is a special case of a formula derived under more general remote stress conditions where

$$\sigma_{22}{}^{\infty} = S, \qquad \sigma_{11}{}^{\infty} = T, \qquad \text{and} \quad \sigma_e{}^{\infty} = \frac{\sqrt{3}}{2}|S - T| \tag{41}$$

That formula is

$$J = \frac{3\pi\sqrt{n}}{4} a\sigma_e{}^{\infty}\epsilon_e{}^{\infty}\left(\frac{S}{\sigma_e{}^{\infty}}\right)^2 \tag{42}$$

For $n = 1$ the formula is exact for all S and T, while for $n \neq 1$ the formula has an error of order $(S/\sigma_e{}^{\infty})^4$ when $S/\sigma_e{}^{\infty}$ is regarded as small. The present bounds indicate that the formula retains its accuracy (for most purposes) for $S/\sigma_e{}^{\infty}$ at least as large as $2/\sqrt{3}$, corresponding to remote plane-strain tension. The reader is referred to the discussion in Ref *1* for an explanation of the unbounded character of the results in the rigid-perfectly plastic limit as $n \rightarrow \infty$.

Bounds for the Edge-Crack in the Semi-Infinite Plane

Here the crack of length a is again under plane-strain conditions with a remote stress $\sigma_{22}{}^{\infty} = \sigma^{\infty}$ so that Eqs 34 and 35 still hold.

For the *lower bound* the stream function was taken as

$$\psi = (\phi - \pi/2)\left[A_0(1 - \mu) + \sum_{i=1}^{P} A_i(\mu^{i+1} - \mu)\right] + \sum_{k=1}^{N}\sum_{j=1}^{M} A_{kj}\left[\mu^{j-1}\sin 2k\phi + 2k(-1)^{k+1}(\phi - \pi/2)\mu\right] \tag{43}$$

This choice, like the previous one for the crack in the infinite plane, has a crack opening of the form $f(x_1)\sqrt{a - x_1}$ with associated strains and stresses which are less singular than the actual solution when $n > 1$. The representation for the stress function χ used to calculate the *upper bound* is the same as Eq 37, except that the double sum is replaced by

$$\sum_{k=1}^{N}\sum_{j=1}^{M} A_{kj}(1 - \mu)^{3-[2/(n+1)]+j} f_k(\phi) \tag{44}$$

where $f_k(\phi)$ is defined by Eq 38. Each term in the representation for χ satisfies the zero traction condition on the edge of the plane $x_1 = 0$. The total stress distribution meets the zero traction condition on the crack faces. Thus, χ produces a statically admissible stress distribution for any set of amplitudes.

Computed bounds to $h(n)$ in Eq 34 are presented in Table 2 and plotted in Fig. 3. The lower bound was determined using 34 free parameters ($N = 6$, $M = 5$, $P = 3$), and the upper bound was computed with a total of 29 free parameters ($N = 5$, $M = 5$, $P = 5$). Included in Table 2 are values of the normalized crack mouth opening, $\delta = 2u_2(0, 0^+)$, at the edge of the plane where

$$g(n) = \frac{\delta}{a\epsilon_e^{\infty}} \tag{45}$$

These values were obtained from the stream function computation.

No comparably simple formula to Eq 42 is available for the edge crack. For $n = 1$ at the same remote state of uniaxial tension σ^∞, $h = (1.1215)^2\pi$ for the edge crack while $h = \pi$ for the crack in the infinite plane [5]. In Table 2 we have included the values from $(1.1215)^2\pi\sqrt{n}$, which would be a good approxi-

TABLE 2—*Upper- and lower-bound estimates of* h *and* g *for plane-strain edge cracks.*

		$n = 1.0$	$n = 1.5$	$n = 2.0$	$n = 3.0$	$n = 4.0$	$n = 5.0$	$n = 7.0$
h	Upper bound	3.952	4.771	5.448	6.568	7.507	8.336	9.792
	Lower bound	3.871	4.639	5.264	6.281	7.108	7.812	8.972
	$(1.1215)^2\pi\sqrt{n}$	3.951	4.844	5.593	6.853	7.909	8.843	10.46
g		4.726	5.271	5.731	6.467	7.080	7.613	8.486

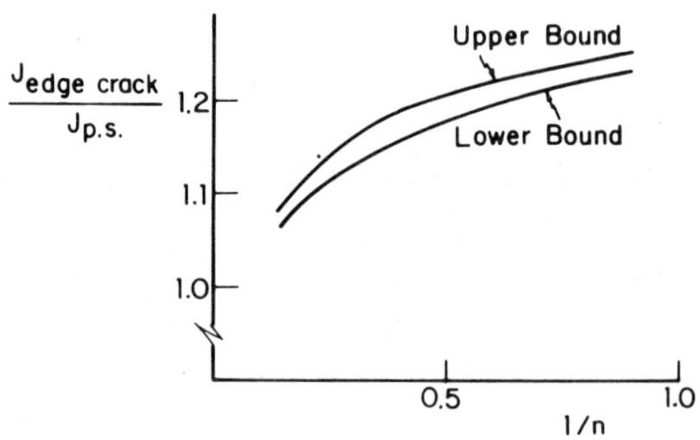

FIG. 4—*Ratio of bounds to* J *for edge crack to estimate of* J *for central crack as a function of* 1/*n*.

TABLE 3—*Upper-bound estimates of* h *for plane-stress cracks.*

	$n = 1.0$	$n = 1.5$	$n = 2.0$	$n = 3.0$	$n = 4.0$	$n = 5.0$	$n = 7.0$
Edge crack	3.955	4.874	5.656	6.985	8.119	9.124	10.88
Central crack	3.190	3.914	4.561	5.723	6.774	7.751	9.560

mation if the ratio of J for the edge crack to J for the crack in the infinite plane were independent of n. This is clearly not the case as can also be seen from the plot of this ratio in Fig. 4. The curves of Fig. 4 suggest that the ratio of the two J-values may approach unity for large n.

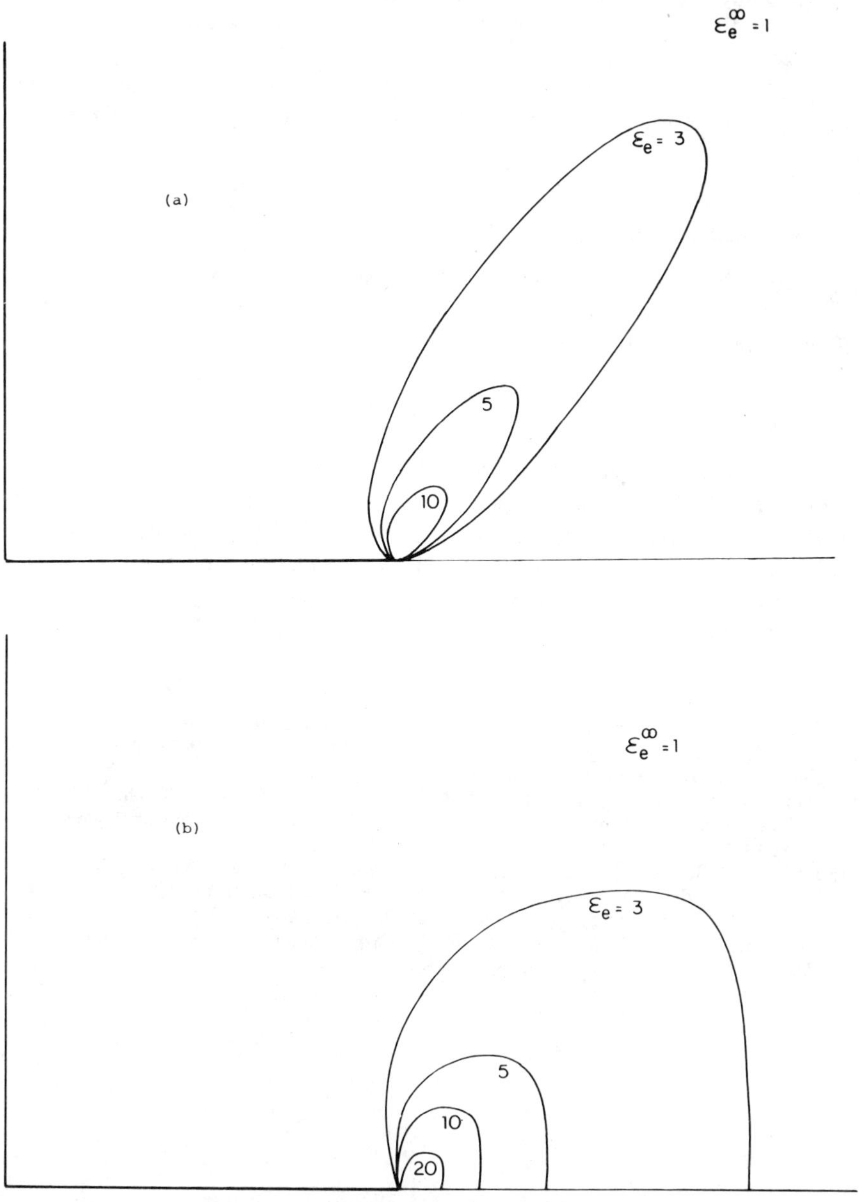

FIG. 5—*Contours of constant effective strain with the remote effective strain as unity and* n = *7:* (a) *edge crack in plane strain;* (b) *edge crack in plane stress.*

Upper Bounds for the Plane-Stress Problems

The method outlined in the preceding for determining the upper bound to J in plane strain applies to plane stress with the minor modification that $\sigma_{33} = 0$ and ϵ_{33} is not required to be zero. The representations for the stress functions are still applicable and $h(n)$ continues to be defined by Eq 34. The upper bounds to $h(n)$ for the two problems are given in Table 3. In each case 19 free amplitudes are used in the calculation ($N = M = P = 4$). It is noted that the numerical values for $h(n)$ are fairly close to the respective values for the plane-strain problems.

Constant-Strain Contours

Contours of constant effective strain ϵ_e are shown in Fig. 5 for the edge-crack problem in plane strain and in plane stress. In each plot $n = 7$ and the remote effective strain is unity.

Acknowledgments

The work of M.Y.H. was supported in part by the Chinese Academy of Sciences and by the National Science Foundation under Grant CME-78-10756. The work of J.W.H. was supported in part by the National Science Foundation under Grants DMR-80-20247 and CME-78-10756, and by the Division of Applied Sciences, Harvard University.

References

[1] He, M. Y. and Hutchinson, J. W., *Journal of Applied Mechanics*, Vol. 48, 1981, pp. 830–840.
[2] Ranaweera, M. P. and Leckie, F. A., *Computer Methods in Applied Mechanics and Engineering*, Vol. 19, 1979, pp. 367–389.
[3] Budiansky, B., Hutchinson, J. W., and Slutsky, S. in *Mechanics of Solids*, The Rodney Hill 60th Anniversary Volume, H. G. Hopkins and M. J. Sewell, Eds., Pergamon Press, New York, 1981.
[4] Budiansky, B. and Hutchinson, J. W. in *Proceedings*, 15th International Congress on Theoretical and Applied Mechanics (Postprints), F. P. J. Rimrott and B. Tabarrok, Eds., North-Holland Publishing Co., Amsterdam, 1980, pp. 243–249.
[5] Tada, H., Paris, P. C., and Irwin, G. R., *The Stress Analysis of Cracks Handbook*, Del. Research Corp., Hellertown, Pa., 1973, p. 8.1.

Ming Yuan He[1] *and John W. Hutchinson*[2]

Penny-Shaped Crack in a Round Bar of Power-Law Hardening Material

REFERENCE: He, M. Y. and Hutchinson, J. W., **"Penny-Shaped Crack in a Round Bar of Power-Law Hardening Material,"** *Elastic-Plastic Fracture: Second Symposium, Volume I—Inelastic Crack Analysis, ASTM STP 803*, C. F. Shih and J. P. Gudas, Eds., American Society for Testing and Materials, 1983, pp. I-291–I-305.

ABSTRACT: A study is made of the deformation of a round bar of pure power-law hardening material under tension containing a central penny-shaped crack. Several important issues related to fracture mechanics are examined, including the near-tip stress and strain field and its relation to the analogous problem in plane strain.

KEY WORDS: nonlinear fracture mechanics, penny-shaped crack, J-integral, fully plastic crack problems, elastic-plastic fracture

This paper continues a study initiated in a previous paper [*1*][3] in which the problem of a penny-shaped crack in an infinite body subject to general remote axisymmetric stressing conditions was studied. Now the objective is the finite problem of a round bar containing a centered penny-shaped crack.

The material is an incompressible, isotropic solid. A pure power hardening relation between stress and strain is assumed so that in simple tension

$$\epsilon/\epsilon_0 = \alpha(\sigma/\sigma_0)^n \tag{1}$$

where ϵ_0 and σ_0 are a reference strain and stress and α is an extra constant introduced for convenience of application. For J_2 deformation theory, Eq 1 generalizes to

$$\frac{\epsilon_{ij}}{\epsilon_0} = \frac{3}{2}\alpha\left[\frac{\sigma_e}{\sigma_0}\right]^{n-1}\frac{S_{ij}}{\sigma_0} \tag{2}$$

[1]Researcher, Institute of Mechanics, Chinese Academy of Sciences, Beijing, China.

[2]Professor of applied mechanics, Division of Applied Sciences, Harvard University, Cambridge, Mass. 02138.

[3]The italic numbers in brackets refer to the list of references appended to this paper.

where S_{ij} is the stress deviator and σ_e is the effective stress defined by

$$\sigma_e = \left(\frac{3}{2} S_{ij} S_{ij}\right)^{1/2} \tag{3}$$

It is also convenient to define an effective strain

$$\epsilon_e = \left(\frac{2}{3} \epsilon_{ij} \epsilon_{ij}\right)^{1/2} \tag{4}$$

which coincides with the tensile strain in uniaxial tension so that σ_e and ϵ_e satisfy Eq 1. A catalog of fully plastic solutions for this class of materials is now available [*2*].

The incompressible finite-element program (INFEM) developed by Needleman and Shih [*3*] will be used. In this program the incompressibility constraints are imposed directly on the nodal degrees of freedom. We recapitulate briefly the method which has been explained in detail in Ref *3*.

The basic finite element employed is a quadrilateral composed of four constant-strain triangles arranged so that the diagonals and edges of the quadrilateral form the sides of the triangles as depicted in Fig. 1. For each triangle in the quadrilateral there is one constraint equation arising from the incompressibility condition. For each quadrilateral there are four such equations, but only three of these are independent. Two of these three constraint equations can be used to express the nodal displacements of the central node of the quadrilateral in terms of the nodal displacements of the corners. Therefore there is only one independent constraint to be satisfied in each quadrilateral. The quadrilaterals are arranged in strips ranging from one boundary of the grid to another, as shown in Fig. 2. Two such strips are considered as one substructure, and the centerline of nodes is employed to satisfy the remaining constraints.

Problem and Method

Axisymmetric fully plastic analyses were carried out for a round bar of radius b under uniform tension containing a central penny-shaped crack of radius a, shown in Fig. 3, for a/b = 0.25, 0.5, 0.75 and n = 1, 3, 5, 10. As a comparison, analyses were also carried out for a plane-strain center-cracked panel (CCP) under remote uniform tension for a/b = 0.25, n = 1, 3, 5, 10

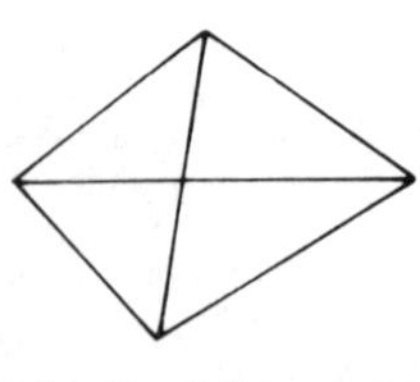

FIG. 1—*Quadrilateral element.*

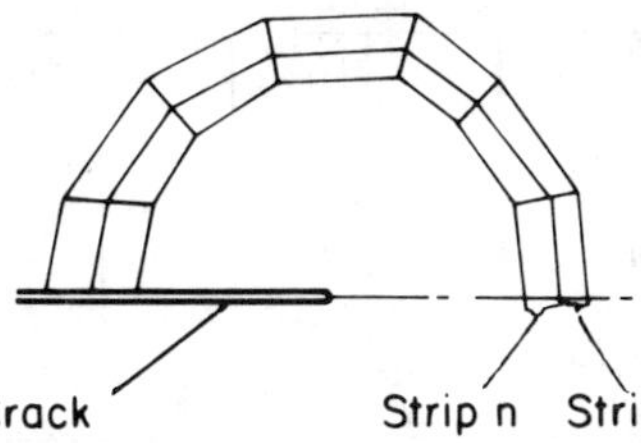

FIG. 2—*Strip arrangement of elements.*

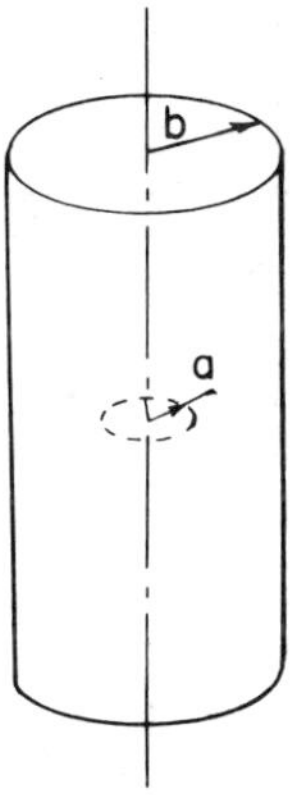

FIG. 3—*Geometry of round bar with penny-shaped crack.*

and $a/b = 0.75$, $n = 1, 3, 5$. Because of the symmetry with respect to the crack, only the upper half of the specimen was modeled. A typical finite-element mesh for the case $a/b = 0.25$ is shown in Fig. 4; the mesh has 12 quadrilateral elements in circumferential direction about the crack tip and 27 elements in the radial direction from the tip to the external boundaries. The meshes used for the other ratios a/b were similar to the one shown here. The meshes have a small hole at the crack tip with radius equal to 1 to 2 percent of the crack length or the remaining ligament, whichever is the smaller dimension. According to Ref *4*, this hole has virtually no effect on the J-integral and mouth opening displacement for $n = 1 \sim 20$ for plane-strain fully plastic analysis.

The INFEM program employs linear or modified Newton-Raphson iteration to obtain the desired solution. The parameter tracking technique was used by which the solution corresponding to $n = 1$ was taken as the initial guess to obtain the solution for $n = 3$ and which in turn is used as the initial guess to obtain the solution for $n = 5$ and so on. In our calculation, for $a/b = 0.25$ and 0.5, Newton's iterations were used. It typically took three or four iterations to reach the desired accuracy. For $a/b = 0.75$, several linear iterations were used at the start and then Newton's procedure was used to obtain convergence.

For convenience, uniform displacement boundary conditions were imposed

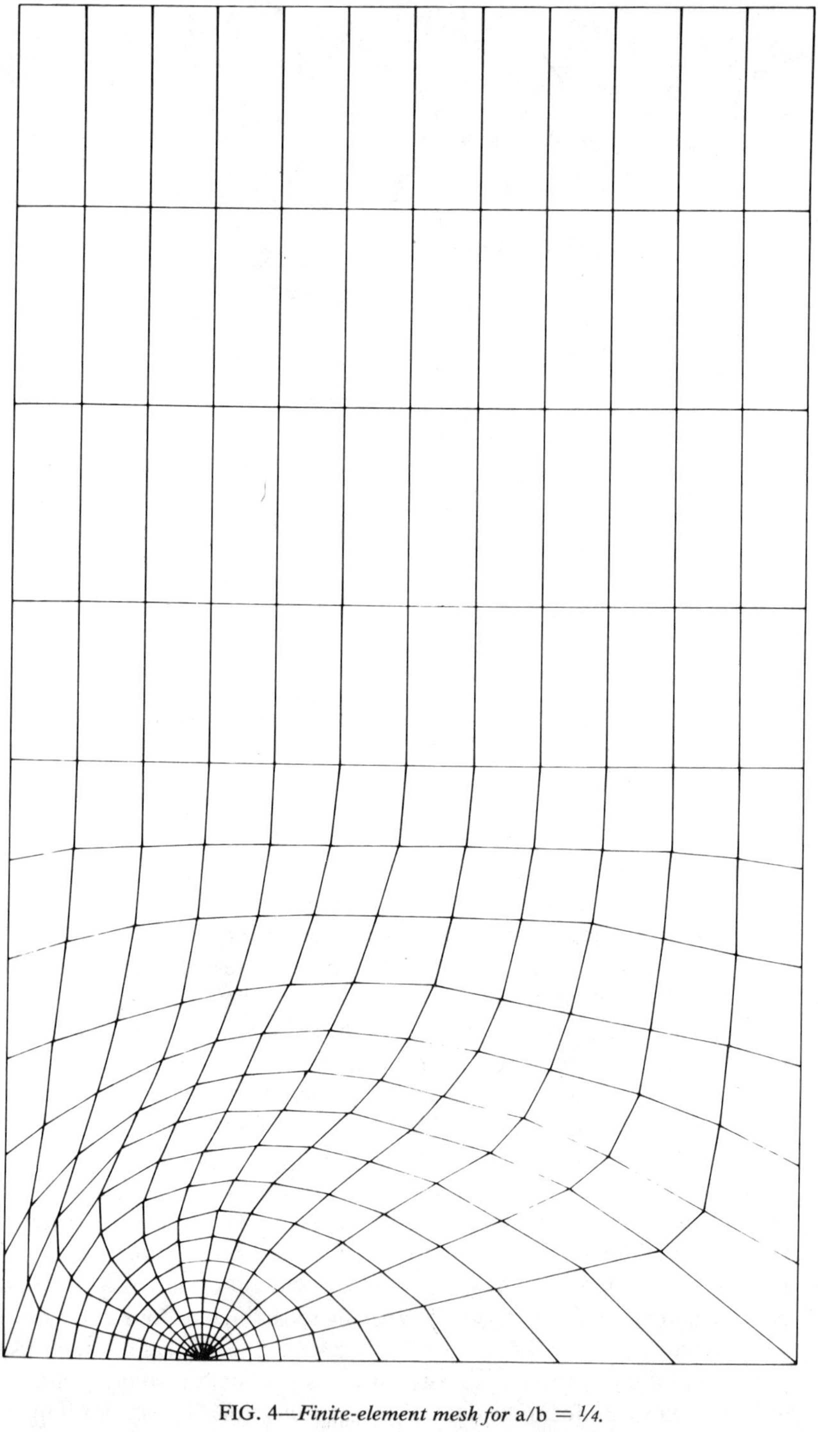

FIG. 4—*Finite-element mesh for* a/b = 1/4.

on the ends. ($L/b = 1.7$ for $a/b = 0.25$ and $L/b = 2.3$ for $a/b = 0.75$, where $2L$ is the length of the bar.) The results show that the stresses on the end boundary are very close to uniform tension.

For the axisymmetric problem the M-integral was evaluated. For the power-law material, the M-integral is given by [5]

$$M = \int_S \left[Wx_i n_i - \sigma_{ij} n_j u_{i,k} x_k - \frac{(2n-1)}{(n+1)} \sigma_{ij} n_j u_i \right] ds \tag{5}$$

where

$$w(\epsilon) = \int_0^{\epsilon} \sigma_{ij} d\,\epsilon_{ij} = \frac{n}{n+1} \alpha \sigma_0 \epsilon_0 \left[\frac{\epsilon_e}{\alpha \epsilon_0} \right]^{(n+1)/n} \tag{6}$$

and S is any closed surface which encloses the crack and n is the unit outward normal to S at each point. If J is defined as the energy release rate per unit length of crack edge for a penny-shaped crack in a round bar as in the plane case, that is

$$J = -\frac{1}{2\pi a} \frac{dPE}{da} \tag{7}$$

where PE is the potential energy of the cracked body, then it follows that

$$J = \frac{M}{2\pi a^2} \tag{8}$$

As discussed in Ref *1*, J also plays the role of the amplitude of the plane-strain crack-tip singularity fields which dominate as the crack is approached. As a check on the numerical accuracy, the M-integral was evaluated along a number of torus-like surfaces ringing the edge of the penny-shaped crack. In all cases the computed values of M were found to be within about 1 percent of the mean value (over all the surfaces).

Results and Discussion

Results for J-Integral and Crack-Opening Displacement

The normalizations of J and δ_0 are given by

$$\begin{aligned} h_1 &= \frac{J}{\sigma_e^\infty \, \epsilon_e^\infty \, a} [1 - (a/b)^2]^n \\ h_2 &= \frac{\delta_0}{a \epsilon_0^\infty} [1 - (a/b)^2]^n \end{aligned} \tag{9}$$

where σ_e^∞ and ϵ_e^∞ are the remote effective stress and effective strain which satisfy the power-law relation, Eq 1, and δ_0 is the opening of the crack at its center. Numerical results for h_1 and h_2 for $a/b = 0.25, 0.5, 0.75$ and $n = 1, 3, 5, 10$ are given in Table 1 and Figs. 5 and 6. The results for h_1 and h_2 for $a/b = 0$ from Ref *1* are also included for comparison purposes.

TABLE 1—h_1 *and* h_2 *for penny-shaped crack in a round bar.*

$$h_1 = \frac{J(1-(a/b)^2)^n}{\sigma_e^\infty \epsilon_e^\infty a}$$

a/b	$n = 1$	$n = 3$	$n = 5$	$n = 10$
0	0.9549	1.331	1.484	1.639
0.25	0.9369	1.267	1.335	1.264
0.5	0.8350	0.9163	0.8379	0.5520
0.75	0.8010	0.5959	0.3926	0.1334

$$h_2 = \frac{\delta_0(1-(a/b)^2)^n}{\epsilon_e^\infty a}$$

a/b	$n = 1$	$n = 3$	$n = 5$	$n = 10$
0	1.909	2.375	2.525	2.637
0.25	1.858	2.131	2.107	1.819
0.5	1.626	1.451	1.183	0.6631
1.0	0.7957	0.4369	0.2640	0.08439

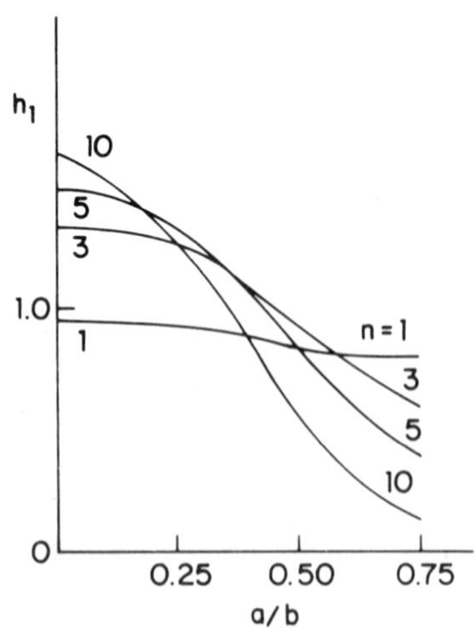

FIG. 5—h_1(a/b, n) *defined in Eq 9.*

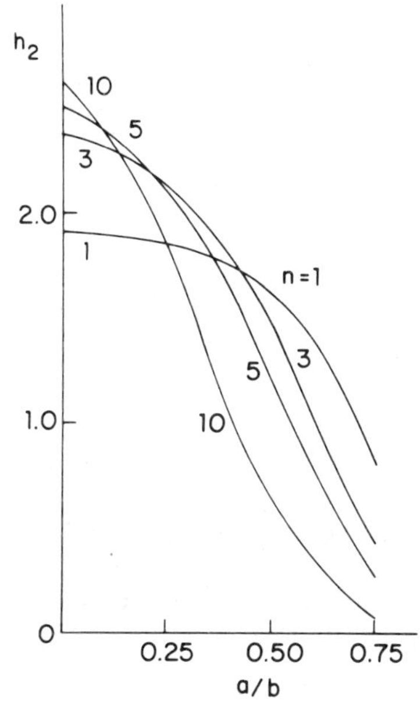

FIG. 6—$h_2(a/b, n)$ *defined in Eq 9.*

The present results for the linear elastic case corresponding to $n = 1$ were compared with the results of Benthem [*6*]. It was found that for all a/b, the averages of the J-integral over all the valid contours enclosing the crack tip were within 3 percent of the values given in Ref *6*.

Stress Field Near the Crack Tip

According to Hutchinson-Rice-Rosengren (HRR) theory [*7,8*] for a plane-strain crack the stress field of the dominant singularity at the crack tip has the form

$$\sigma_{ij} = \sigma_0 \left[\frac{J}{\alpha \sigma_0 \epsilon_0 I_n r} \right]^{1/(n+1)} \tilde{\sigma}_{ij}(\theta, n) \qquad (10)$$

where r is the distance from the crack tip and θ is the angle measured from directly ahead of the crack. The plane-strain singular fields, Eq 10, also hold at the edge of a penny-shaped crack, where r and θ are then local coordinates in a plane perpendicular to the edge of the crack. At issue is the size of the crack-tip region over which the singular field, Eq 10, dominates. We compared the calculated stress distributions from the finite-element method with the HRR field. The normal stress ahead of the crack σ_{yy} is normalized in the following way

$$\frac{\sigma_{yy}/\sigma_e{}^\infty}{\left(\dfrac{J}{\sigma_e{}^\infty \epsilon_e{}^\infty a}\right)^{1/(n+1)}} = \frac{\sigma_{yy}/\sigma_0}{\left(\dfrac{J}{\alpha\sigma_0\epsilon_0 a}\right)^{1/(n+1)}} \tag{11}$$

which is necessarily independent of J (and of load). According to HRR theory, from Eq 10

$$\frac{\sigma_{yy}/\sigma_0}{\left(\dfrac{J}{\alpha\sigma_0\epsilon_0 a}\right)^{1/(n+1)}} = \left(\frac{a}{rI_n}\right)^{1/(n+1)} \tilde{\sigma}_{yy}(0, n) \tag{12}$$

where I_n is an integration constant. I_n and $\tilde{\sigma}_{yy}$ can be found in Ref 7.

The plots of this ratio are given in Fig. 7 for $n = 1$, 3, and 10 and $a/b = 0.25$ and 0.75. The corresponding curves for the plane-strain centered crack are also included. It is seen that for the penny-shaped crack the normalized normal stress ahead of the crack falls well below that for the HRR singular field where $r/a > 0.02$ when $n \geq 3$. For $r/a < 0.02$ our finite-element resolution of

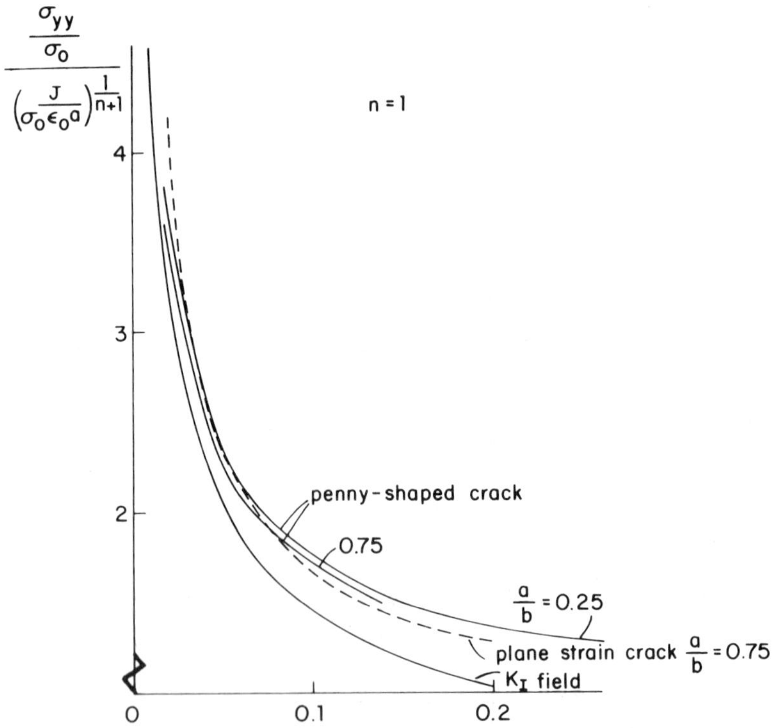

FIG. 7a—*Normalized stress ahead of crack edge for* n = *1.*

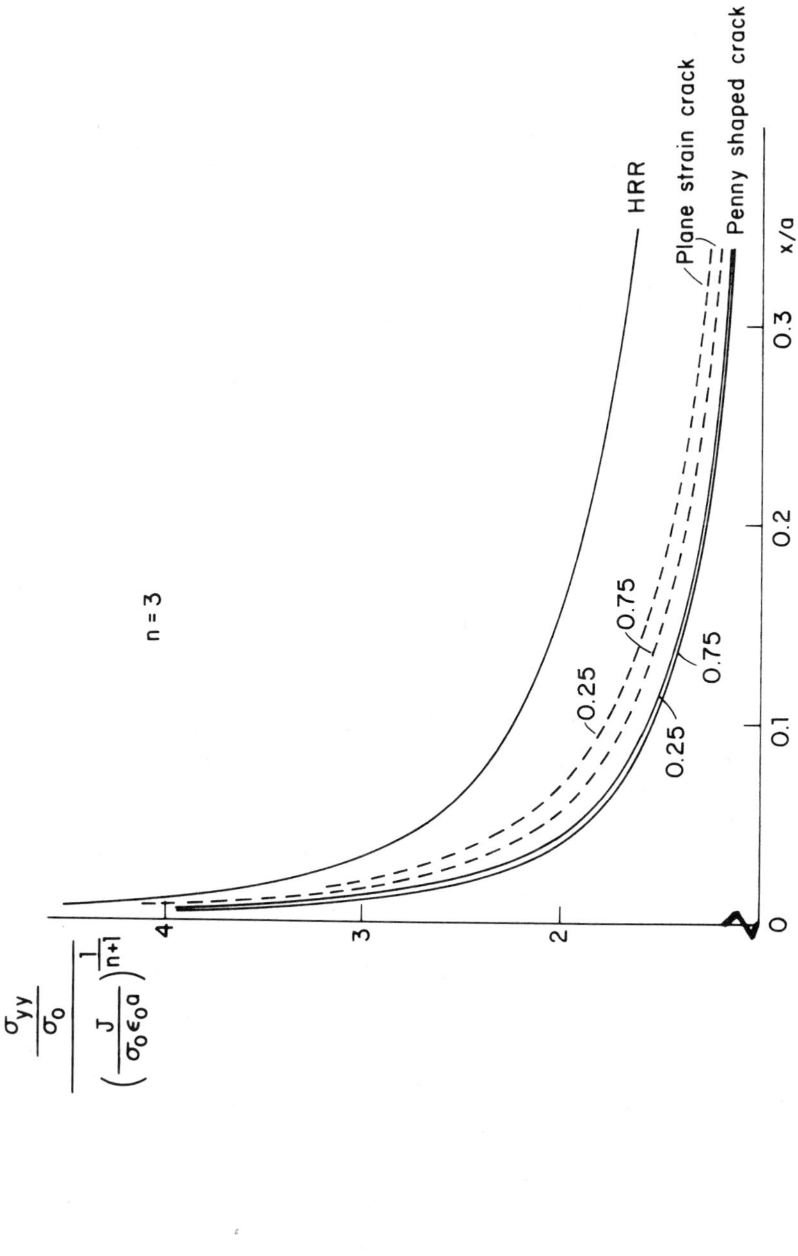

FIG. 7b—*Normalized stress ahead of crack edge for* n = 3.

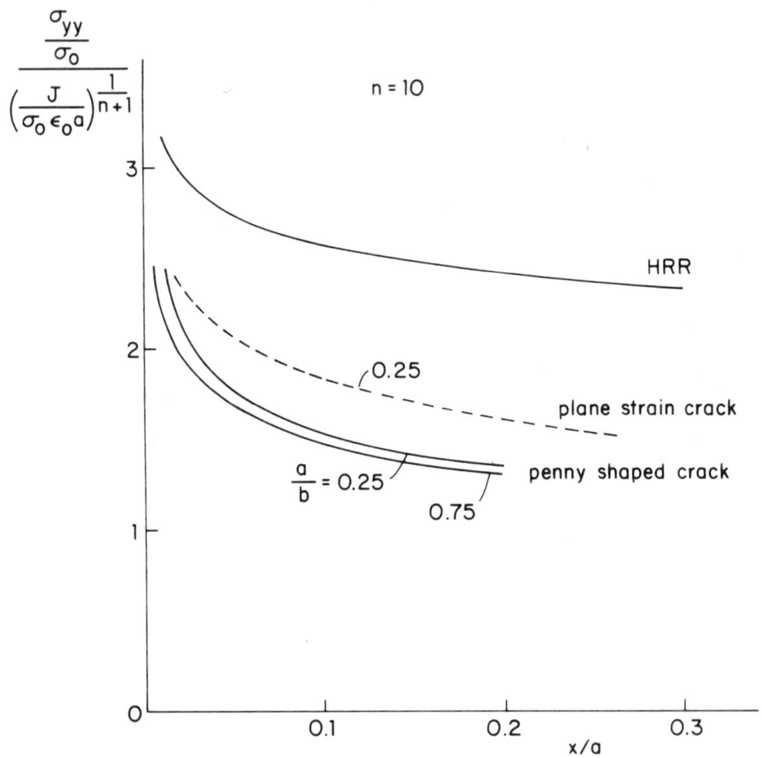

FIG. 7c—*Normalized stress ahead of crack edge for* n = *10.*

the stress field is not sufficient to make a more accurate comparison. What is clear, however, is that the size of the region in which the stresses are well approximated by the singular field is less than about $0.02a$. It should be noted that for a given n these curves depend only slightly on a/b and the differences between the results of the penny-shaped crack and the plane-strain crack are relatively small.

Displacement and Near-Tip Strain of the Bar

A significant difference between the axisymmetric problem and the plane-strain problem is brought out by the plots of Fig. 8, which show the displacement of the boundaries for the case $a/b = 0.25$ and $n = 10$. Included in the figure for the plane-strain specimen is the rigid-perfectly plastic slipline mode of deformation. It is seen that the solution for $n = 10$ reflects the limiting discontinuous behavior of the perfectly plastic solution. By contrast, no slip discontinuities are possible in the perfectly plastic axisymmetric problem and, in fact, it can be seen that the presence of the crack does not significantly per-

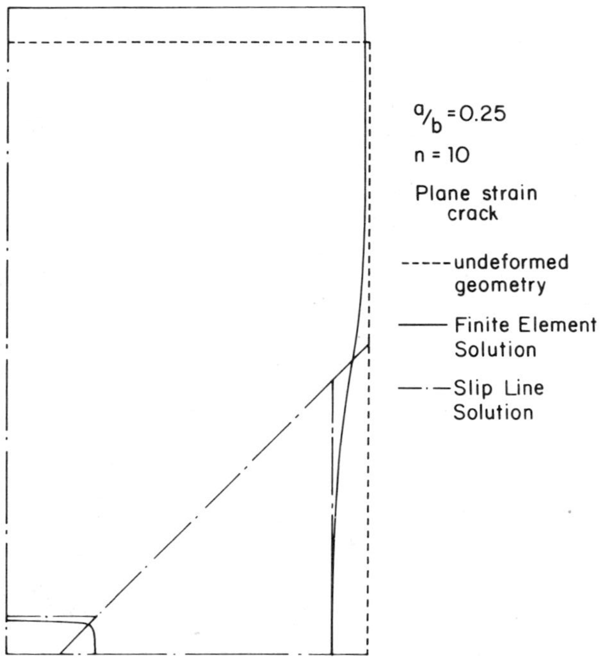

FIG. 8*a*—*Displacement of the boundary for the plane-strain problem and comparison with the slipline solution for the rigid-perfectly plastic limit.*

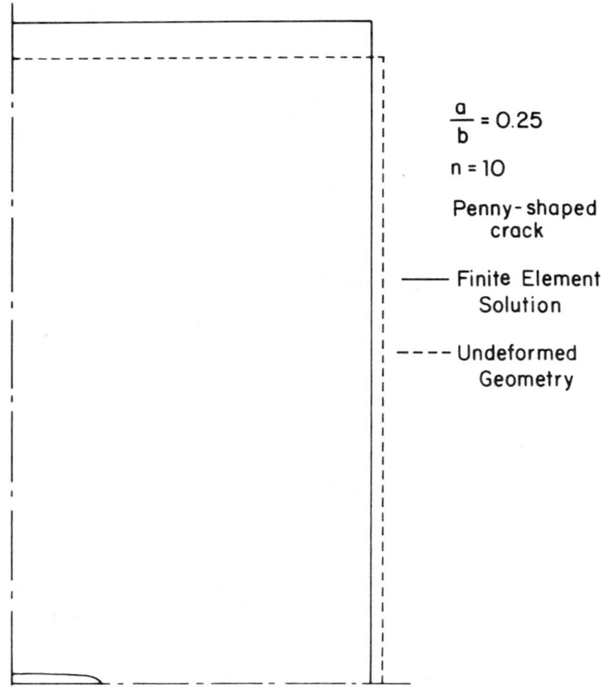

FIG. 8*b*—*Displacement of boundary for round bar with penny-shaped crack.*

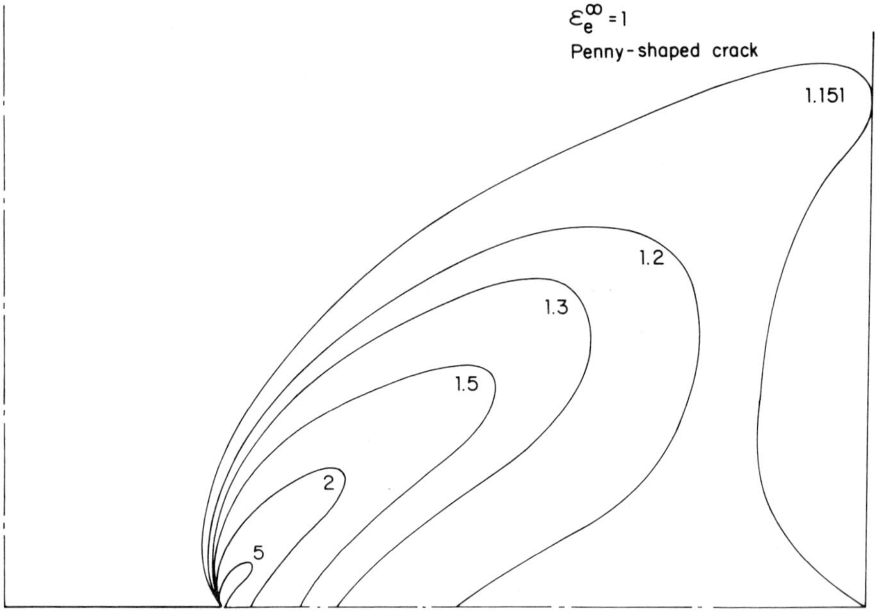

FIG. 9—*Contours of constant effective strain* ϵ_e *for penny-shaped crack in round bar with* n = *10,* a/b = *1/4, and* $\epsilon_e^{\infty} = 1$.

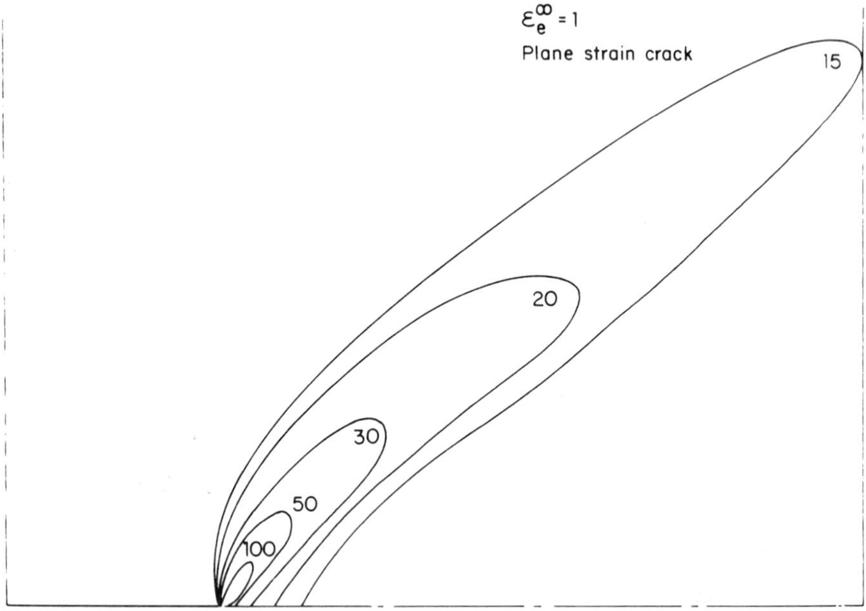

FIG. 10—*Contours of constant effective strain* ϵ_e *for plane-strain crack-problem with* n = *10,* a/b = *1/4, and* $\epsilon_e^{\infty} = 1$.

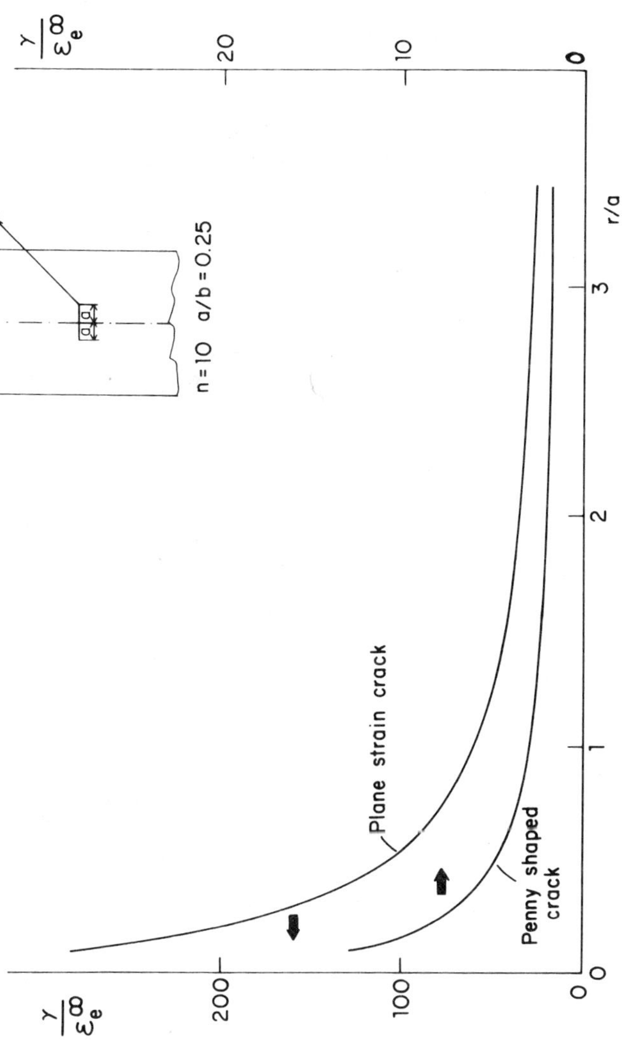

FIG. 11—*Comparison of shear strain levels on plane at 45 deg emananting from crack tip.*

turb the boundary of this specimen. Contours on which the effective strain is a constant are shown in Fig. 9 for the penny-shaped crack and in Fig. 10 for the plane-strain crack, both for $n = 10$, $a/b = 0.25$, and loaded such that $\epsilon_e^\infty = 1$. Obviously, there are significant differences between these two cases. For the plane-strain case the strain concentrates in rather narrow bands emanating from the tips. On the other hand, for the penny-shaped crack, the strain field is much more diffuse. A comparison of Fig. 10 with Fig. 9 reveals the substantially larger strain concentration in the plane-strain problem at a given distance from the curve tip.

The shear strains ($\gamma \equiv 2\epsilon_{r\theta}$) on the 45-deg line emanating from the tip are shown in Fig. 11. It is again seen that for the plane-strain crack the shear strain is 15 to 20 times higher than the shear strain for the penny-shaped crack.

Finally, to aid in obtaining the limit load for the rigid-perfectly plastic penny-shaped cracked round bar, we define as σ_L^∞ the remote stress at which ϵ_e first attains the value ϵ_0 at the outer cylindrical boundary. Computed values of $\sigma_L^\infty/\{\sigma_0[1 - (a/b)^2]\}$ are shown as solid lines in Fig. 12 for $n = 1, 3, 5$, and 10. The dashed curve for $n = \infty$ is the estimate of the limit load and this was obtained by cross-plotting against $1/n$ and extrapolating to the limit $n \rightarrow \infty$ for given values of a/b.

Acknowledgment

We are indebted to A. Needleman and C. F. Shih for supplying us with a copy of their finite-element program INFEM and for helpful discussions. This

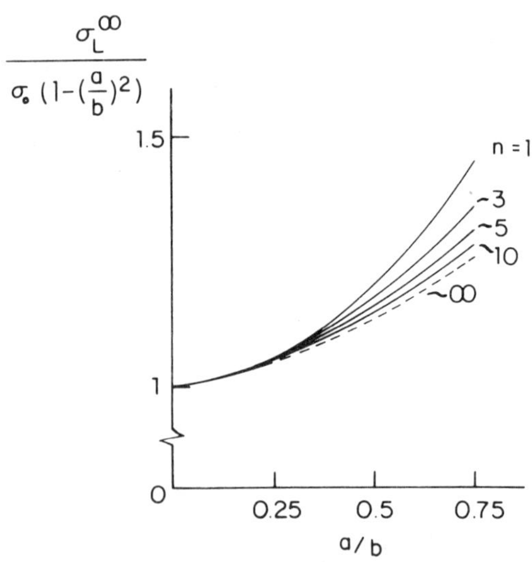

FIG. 12—*Remote stress σ_L^∞ at which ϵ_e/ϵ_0 first attains unity at the external surface of a round bar, and extrapolation to limit load corresponding to* n $= \infty$.

work was supported in part by the National Science Foundation under Grants DMR-80-20247 and CME-78-10756, and by the Division of Applied Sciences, Harvard University.

References

[1] He, M. Y. and Hutchinson, J. W., *Journal of Applied Mechanics*, Vol. 48, 1981, pp. 830–840.
[2] Kumar, V., German, M. D., and Shih, C. F., "An Engineering Approach for Elastic-Plastic Fracture Analysis," NP-1931, Research Project 1237-1. Prepared for Electric Power Research Institute, Palo Alto, Calif., 1981.
[3] Needleman, A. and Shih, C. F., *Computer Methods in Applied Mechanics and Engineering*, Vol. 15, 1978, pp. 223–240.
[4] Shih, C. F. and Kumar, V., "Estimation Technique for the Prediction of Elastic-Plastic Fracture of Structural Components of Nuclear Systems." First Semiannual Report, General Electric Co., Schenectady, N.Y., 1 June 1979.
[5] Budiansky, B. and Rice, J. R., *Journal of Applied Mechanics*, Vol. 40, 1973, pp. 201–203.
[6] Tada, H., Paris, P. C., and Irwin, G. R., *The Stress Analysis of Cracks Handbook*, Del Research Corp., Hellertown, Pa., 1973, p. 28.1.
[7] Hutchinson, J. W., *Journal of the Mechanics and Physics of Solids*, Vol. 16, 1968, pp. 13–31.
[8] Rice, J. R. and Rosengren, G. K., *Journal of the Mechanics and Physics of Solids*, Vol. 16, 1968, pp. 1–12.

Virendra Kumar,[1] *Marjorie D. German,*[1] *and C. Fong Shih*[1]

Elastic-Plastic and Fully Plastic Analysis of Crack Initiation, Stable Growth, and Instability in Flawed Cylinders

REFERENCE: Kumar, Virendra, German, M. D., and Shih, C. F., **"Elastic-Plastic and Fully Plastic Analysis of Crack Initiation, Stable Growth, and Instability in Flawed Cylinders,"** *Elastic-Plastic Fracture: Second Symposium, Volume I—Inelastic Crack Analysis, ASTM STP 803*, C. F. Shih and J. P. Gudas, Eds., American Society for Testing and Materials, 1983, pp. I-306–I-353.

ABSTRACT: In this paper a procedure has been developed for examining crack initiation, stable growth, and instability in flawed cylinders. Elastic and fully plastic solutions are presented for cylinders containing axial and circumferential cracks located at the inner surface. In the case of an axial crack the loading is due to internal pressure and a plane-strain analysis is carried out. The circumferentially cracked cylinder is loaded by remote uniform tension and is treated as an axisymmetric problem. The solutions are based on the deformation theory of plasticity and employ a special incompressible finite-element method. Solutions cover a broad range of values for the wall thickness to radius ratio, crack length to wall thickness ratio a/b, and strain-hardening exponent n. The relevant crack parameters such as the J-integral, crack mouth opening displacement, and load-point displacement are suitably normalized and tabulated. Accuracy of the elastic solutions is assessed by a comparison with available literature solutions and some other consistency checks.

By appropriately combining the elastic and fully plastic solutions developed, simple estimation formulas are derived through which the relevant crack parameters can be readily computed in the entire range of elastic-plastic deformation. Corresponding results for some illustrative problems are compared with detailed finite-element calculations based on both flow and deformation theories of plasticity. Example J-integral crack driving force diagrams are presented for materials like A533B and 304 stainless steel. Since equilibrium of crack growth requires that applied J equal the material resistance, a superposition of the material J_R-curve on such diagrams yields complete information regarding crack initiation, amount of stable growth, instability, and the overall deformation behavior.

[1]Mechanical engineers, General Electric Co., Corporate Research and Development Schenectady, N.Y. 12301; coauthor C. F. Shih is presently at Division of Engineering, Brown University, Providence, R.I. 02912.

Finally, the instability of crack growth under J-controlled growth conditions is examined along the lines of the tearing modulus, T_J, approach. J versus T_J stability diagrams are presented for both axial and circumferential cracks covering a broad range of practical problems of interest. Instability develops when the applied tearing modulus exceeds the material resistance to crack growth. Effects of material properties, flaw size, and cylinder dimensions, including system compliance, on the crack growth and instability phenomenon are investigated in some detail.

KEY WORDS: cylinders, axial cracks, circumferential cracks, elastic solutions, fully plastic solutions, elastic-plastic estimation procedure, incompressible deformation, J-integral, crack opening displacement, J-resistance curve, crack initiation, stable growth, instability, tearing modulus, ductile fracture, limit load, creep crack growth, elastic-plastic fracture

In ductile materials such as pressure vessel and piping steels, large plastic strains are required for crack initiation and propagation. In addition, crack initiation is usually followed by stable growth prior to fast unstable fracture. Being inherently nonlinear, the problem cannot be resolved in any satisfactory manner by methods based on linear elastic fracture mechanics (LEFM) and must be addressed within the framework of nonlinear fracture mechanics. Recent analytical, experimental, and computational studies suggest that the J-integral and the crack opening displacement (COD), denoted by δ, are the most viable parameters for characterizing crack initiation and stable growth in ductile materials. The rates of change of J and δ with respect to crack length (dJ/da and $d\delta/da$) appear to be the controlling parameters from the viewpoint of crack growth instability. The parameters J and δ as well as their rates are also found to be related through a simple relationship. Furthermore, resistance curves based on J and δ can be regarded as the material property for limited range of crack growth provided certain requirements are satisfied. These findings, discussed in detail for example in Refs *1–20*, constitute the concept of J- or COD-controlled growth. An important consequence of J-controlled crack growth is that the characterizing parameters J and δ can be obtained from analyses employing the deformation theory of plasticity. Although comparisons between J- and COD-based fracture criteria have not been studied extensively in the literature, presently J is considered more attractive than COD because of the relative ease of computing it in the laboratory as well as for fracture analysis. However, it has some limitations concerning size requirement and the amount of crack growth [*9–12*].[2] COD has a more fundamental appeal and, therefore, may have fewer limitations also, but it is relatively more difficult from a measurement and computational viewpoint.

With J and COD as the characterizing crack-tip parameters, an engineering methodology has been developed by Kumar et al [*21*] for examining crack initiation, stable growth, and instability in flawed structural components in the elastic-plastic and fully plastic regimes of material deforma-

[2]The italic numbers in brackets refer to the list of references appended to this paper.

tions. The key elements of the approach are: (1) a catalog of fully plastic solutions for J, COD, and other relevant fracture parameters for a variety of crack configurations; (2) an estimation procedure which combines elastic and fully plastic solutions to obtain relatively simple formulas for the elastic-plastic solutions; and (3) a method in which the J-integral (or COD) crack driving force estimates are compared against the corresponding material resistance curve, determined from standard specimen tests, to predict crack initiation, stable growth, and instability behavior, thus providing all important details on the load-carrying capacity and safety margin of a flawed structure. The elastic solutions can be found in elastic fracture handbooks such as the ones by Tada et al [*22*] and Rooke and Cartright [*23*]. With regard to fully plastic solutions, the emphasis so far has been primarily on the planar test specimen configurations [*24–34*]. The present study is concerned with cylinders containing two-dimensional axial and circumferential cracks.

This paper presents elastic and fully plastic analyses of axially and circumferentially cracked cylinders covering a wide range of values for flaw size, cylinder dimensions, and material strain-hardening exponent. Solutions developed are appropriately normalized and tabulated in a convenient format. Accuracy of the elastic solutions is assessed by comparisons with available literature solutions [*35,36*] and alternate finite-element calculations which were also carried out as a part of this investigation. Estimation formulas are given for obtaining solutions to various fracture parameters in the entire range of elastic-plastic deformation from small-scale yielding to fully plastic conditions including contained plasticity. The corresponding results are compared with detailed finite-element crack computations based on flow and deformation theories of plasticity. The J-integral crack driving force diagrams for materials like A533B and 304 stainless steels are presented. By superposing the material J-resistance curve on such diagrams, predictions are made regarding crack initiation, amount of stable crack extension, the maximum load, and overall deformation behavior. Finally, the instability of crack growth under J-controlled conditions is examined using the tearing modulus, T_J, approach [*8*]. J versus T_J stability diagrams are presented for both axial and circumferential cracks covering a broad range of practical problems of interest. Effects of strain hardening, crack depth, cylinder dimensions, and compliance of the system on the extent of stable growth and instability load are reported in some detail. Before proceeding to the analysis details, a brief outline of the underlying approach is given in the following section.

Brief Outline of the Engineering Approach

The engineering approach is described in detail in Ref *21*. To stimulate the forthcoming discussion, a brief summary is provided here with respect to central ideas, namely, the fully plastic solutions, the estimation procedure, and the resistance-curve approach.

Fully Plastic Solutions

Fully plastic solutions are associated with the analysis of a boundary-value problem, based on the deformation theory of plasticity, for a crack in an incompressible power-law material which in uniaxial tension can be characterized by a stress-strain law of the type

$$\epsilon/\epsilon_0 = \alpha(\sigma/\sigma_0)^n \tag{1}$$

where

σ_0 = some reference stress, usually taken to be the yield stress,
ϵ_0 = σ_0/E,
n = strain hardening exponent, and
α = a material constant.

At one limit given by $n = 1$, Eq 1 represents linear elastic behavior while at the other limit of $n \to \infty$, it gives rigid/perfectly plastic behavior. Generalizing this uniaxial relationship to three-dimensional stress state by the J_2 deformation theory of plasticity leads to

$$\frac{\epsilon_{ij}}{\epsilon_0} = \frac{3}{2}\alpha\left(\frac{\sigma_e}{\sigma_0}\right)^{n-1}\frac{S_{ij}}{\sigma_0} \tag{2}$$

where

E = Young's modulus,
S_{ij}, ϵ_{ij} = stress and strain deviators, and
σ_e = effective stress defined by $\sigma_e{}^2 = 3S_{ij}S_{ij}/2$.

As discussed by Ilyushin [*37*], a solution to the boundary-value problem involving a single load or displacement parameter has the property that all stress quantities are linearly proportional to the applied load P, and all strain and displacement quantities are proportional to P^n. One implication of this result is that, if P is increased monotonically, the solution based on the foregoing formulation is also an exact solution to the same problem for the J_2 flow theory of plasticity. Another consequence as noted by Goldman and Hutchinson [*24*] is that similar to linear elastic crack problems there exist scaling laws for various fracture parameters, such as the J-integral and crack opening displacement, associated with the foregoing nonlinear problem also. In other words, dependence of fracture parameters on load can be separated from its dependence on geometric and material variables. Consequently, it is feasible to tabulate the solutions for J, δ, and other parameters of interest as functions of crack depth and strain-hardening exponent n. Such solutions have been developed for a variety of crack configurations in Refs *1* and *24–34*. Other similar tabulations are reviewed in Ref *38*. Thus, in a given application the relevant crack parameters can be readily calculated by using relevant formulas and tables without having to perform detailed finite-element computations.

Estimation Procedure

Linear elastic solutions obtainable from handbooks such as Refs *22* and *23* and fully plastic solutions as discussed in the preceding are applicable to small-scale yielding and fully yielded situations, respectively. Most crack problems of practical interest lie in the elastic-plastic regime and it is primarily for this range that estimation procedures have been developed [*21,25,26,31–34*]. The estimation procedure provides an interpolation over the entire range of elastic-plastic deformations from small-scale yielding to fully yielded conditions, including contained plasticity. In this, the elastic and fully plastic contributions are appropriately superposed to arrive at the elastic-plastic estimates. For a material governed by the Ramberg-Osgood stress-strain law

$$\epsilon/\epsilon_0 = \sigma/\sigma_0 + \alpha(\sigma/\sigma_0)^n \tag{3}$$

estimation formulas are of the form

$$J(a, P) = J^e(a_e, P) + J^p(a, P) \tag{4}$$

$$\delta(a, P) = \delta^e(a_e, P) + \delta^p(a, P) \tag{5}$$

$$\Delta(a, P) = \Delta^e(a_e, P) + \Delta^p(a, P) \tag{6}$$

where a is the crack length and P the applied load. δ and Δ are mouth opening displacement (MOD) and load-point displacement (LPD), respectively. The first term with superscript e denotes the adjusted elastic contributions while the second term with superscript p represents the fully plastic part. a_e is the effective crack length adjusted to include Irwin's small-scale yielding correction and is given by Refs *21,25, 26, 31–34*

$$a_e = a + \phi r_y \tag{7}$$

where

$$r_y = \frac{1}{\beta\pi}\left[\frac{n-1}{n+1}\right]\left(\frac{K_I}{\sigma_0}\right)^2 \tag{8}$$

and

$$\phi = 1/[1 + (P/P_0)^2] \tag{9}$$

$\beta = 2$ for plane stress and 6 for plane strain. P_0 is the limit load corresponding to a perfectly plastic material ($n = \infty$). r_y in Eq 8 is taken as an *explicit* function of the elastic stress-intensity factor K_I.

Elastic-plastic results obtained via the estimation method have been found to be in good correspondence with experimental data and detailed finite-element crack analyses for several example problems [*21,25,26,31–34*]. It should be noted that the approach yields accurate predictions for parameters of interest which would otherwise require elaborate finite-element computations. An important feature is that the tearing modulus [*8*], used for crack growth instability predictions, can be readily calculated with the aid of the estimation method [*21*].

Resistance-Curve Approach for Crack Growth and Instability Analyses

The applied crack driving force in terms of J or COD, computed as described in the foregoing, can be used in conjunction with the experimentally determined material J- (or COD-) resistance curve to predict crack initiation, amount of stable growth, instability load, and other quantities relevant to overall fracture behavior. The resistance-curve approach, based on K, for treating crack growth and instability under LEFM conditions and small-scale plasticity is fairly well-known and is discussed in detail in Ref *39*. An extension of this approach to deal with large-scale yielding situations has been described by Paris et al [*8*], Hutchinson and Paris [*9*], and Shih et al [*12*]. Under J-controlled crack growth conditions, the equilibrium condition for stable growth is stated as

$$J(a, P) = J_R(a - a_0) \tag{10}$$

The left-hand term represents the applied crack driving force and the right-hand term the material resistance to crack extension. For a given geometry and material properties the crack driving force is a function of only the current crack length a and applied load P. The material resistance is a function of the amount of crack growth ($a - a_0$) only. Crack growth is unstable if

$$(\partial J/\partial a)_{\Delta_T} \geq dJ_R/da \tag{11}$$

The subscript in Eq 11 denotes a partial derivative with the total displacement Δ_T held fixed; Δ_T is defined by

$$\Delta_T = \Delta + C_M P \tag{12}$$

where Δ is the load-point displacement through which P works and C_M the system compliance (in effect, a linear spring placed in series with the cracked body). The crack driving force can be expressed to reveal the role of the system compliance through the relationship [9]

$$\left(\frac{\partial J}{\partial a}\right)_{\Delta_T} = \left(\frac{\partial J}{\partial a}\right)_P - \left(\frac{\partial J}{\partial P}\right)_a \left(\frac{\partial \Delta}{\partial a}\right)_P \left[C_M + \left(\frac{\partial \Delta}{\partial P}\right)_a\right]^{-1} \tag{13}$$

Paris et al [8] introduced a nondimensional quantity, termed tearing modulus, defined by

$$T_J = \frac{E}{\sigma_0{}^2}\left(\frac{\partial J}{\partial a}\right)_{\Delta_T} \quad \text{and} \quad T_{JR} = \frac{E}{\sigma_0{}^2}\frac{dJ_R}{da} \tag{14}$$

The instability criterion, Eq 11, can be restated in terms of tearing modulus as

$$\begin{aligned} T_J &< T_{JR}; \quad \text{stable} \\ T_J &\geq T_{JR}; \quad \text{unstable} \end{aligned} \tag{15}$$

The stability of crack growth can also be analyzed through the use of COD as outlined in Refs *21* and *31–34*. As discussed in Ref *21*, the partial derivatives appearing in Eq 13 can be evaluated by analytical or numerical differentiation of estimation formulas for J and Δ; these can be utilized via Eqs 13 and 14 to yield T_J. The material resistance to tearing instability T_{JR} is obtained by differentiating the material J_R-curve as in Eq 14.

J and T_J calculated as described in the preceding paragraphs can be plotted in a desired format to construct the so-called "stability diagrams." For example, one can plot J versus T_J curves with crack depth and C_M as parameters; by superposing on these the material resistance (J_R versus T_{JR}) curve, the regimes of stable and unstable behavior can be identified through the instability criterion, Eq 15. Such information used in conjunction with the crack driving force diagram yields the flaw size and load associated with instability. Further details are provided in Ref *21*.

The approach summarized in the foregoing is used in this paper to examine crack initiation, stable growth, and instability behavior in axially and circumferentially flawed cylinders. We next present a brief discussion on the incompressible finite-element method employed in the present study.

Method of Finite-Element Analysis

Fully plastic crack problems described in the previous section correspond to a nonlinear and incompressible material. In plane-stress situations the incompressible deformation does not introduce any additional complexity in the finite-element method. Under plane-strain and axisymmetric conditions, however, the enforcement of incompressible constraints within the framework of the displacement finite-element method is quite complicated and can be accommodated in a variety of ways as briefly discussed by Needleman and Shih [*40*]. The finite-element method developed in Refs *32, 33, 40,* and *41* has been employed in the present investigation, which involves plane-strain and axisymmetric deformation. The method is formulated from the virtual work equation based on deviatoric quantities. The incompressibility constraint is imposed directly on the admissible displacement field by direct elimination of nodal displacements. This leads to a stiffness matrix which is symmetric and positive definite. Modified Newton-Raphson and linear iterative schemes coupled with a parameter tracking technique are employed to solve the nonlinear system of equations. Once a convergent solution for the displacement has been obtained, the hydrostatic stress is determined through the principle of virtual work.

Internally Pressurized Cylinder with an Internal Axial Crack

Figure 1 describes a cylinder containing an internal radial axial crack of length a, with internal radius R_i and outer radius R_0. The crack front is parallel to the axis of the cylinder. A uniform pressure p acts on the inside of the cylinder and on the crack face. The wall thickness is denoted by $b = R_0 - R_i$, and $c = b - a$ is the uncracked ligament. This crack geometry is analyzed as a plane-strain problem. Elastic and fully plastic analyses were performed using the incompressible finite-element method discussed earlier. It is noted that the J-integral formulation for such crack problems, which involve crack face loading, is different from the conventional two-dimensional

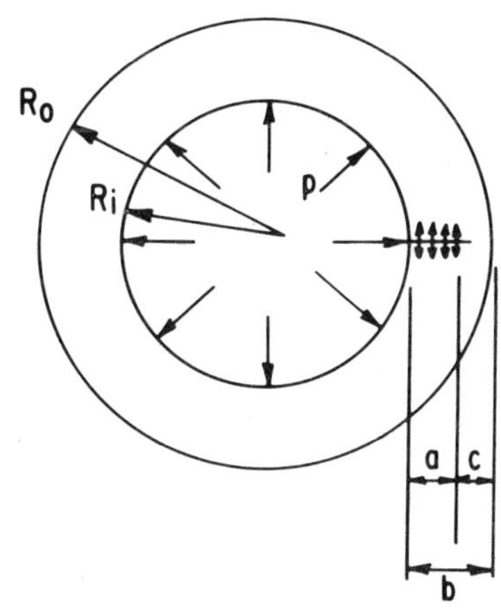

FIG. 1—*Axially cracked cylinder under internal pressure.*

expression [*1*]. The modification accounting for this effect is summarized in the Appendix and discussed in greater detail in Ref *42*. The solutions reported in this paper are calculated using these modified forms of the J-integral; this point should be kept in mind when applying the material given here to specific problems.

Three values of wall thickness to inside radius ratio, b/R_i, are considered, namely $1/5$, $1/10$, and $1/20$. For each b/R_i, the crack length to wall thickness ratio, a/b, is taken to be $1/8$, $1/4$, $1/2$, and $3/4$. The strain hardening exponent n is assigned values of 1, 2, 3, 5, 7, and 10. For symmetry considerations it was sufficient to model only half the cross section. A typical finite-element mesh is depicted in Fig. 2 for the case of $a/b = 1/2$ and $b/R_i = 1/10$. The quadrilateral elements consist of four constant-strain triangles; the former is grouped in substructures which are employed as the basic building blocks of the finite-element grid [*40*]. The model has 14 quadrilaterals in the circumferential direction around the crack tip and ten elements across the thickness away from the crack-tip region. The mesh has about 600 elements in total. The meshes used for other cases were similar and had approximately the same number of elements. The main consideration in mesh generation was to produce acceptable accuracy in the elastic cases. The singular crack-

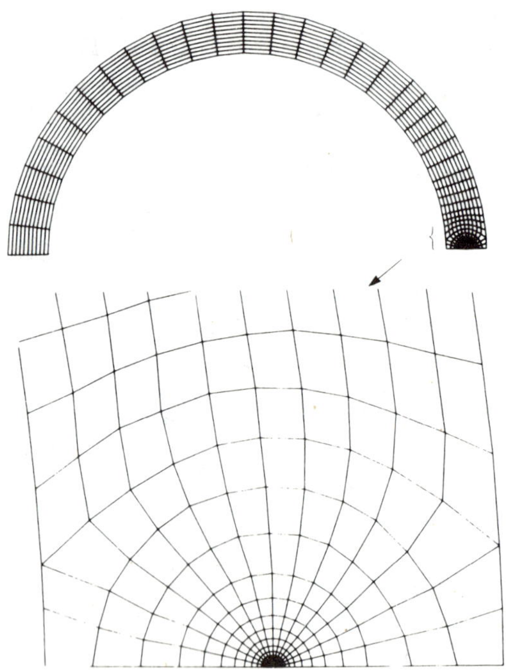

FIG. 2—*Finite-element mesh for the upper half of an axially cracked cylinder with* a/b = ½.

tip element was not used in these analyses because of the associated convergence problems for large n-values. Instead, the mesh had a small hole at the crack tip with a radius of less than 1 percent of the crack length or the remaining ligament, whichever was the smaller dimension. This circumvented some of the numerical difficulties at the crack tip and gave satisfactory convergence properties. This hole was very small in size, and numerical experimentation showed that it had virtually no effect on the J-integral and mouth opening displacement for the range of n examined. In most cases, the difference amounted to less than 0.5 percent. For high n-values, the effect of this hole is expected to be even smaller.

Following the procedure in Ref *40*, linear or modified Newton-Raphson iterations were employed to obtain the fully plastic solution. The parameter tracking technique was used in which the appropriately scaled solution corresponding to $n = 1$ was taken as the initial guess to obtain the $n = 2$ solution, and which in turn is used as the starting point for $n = 3$ and so on. For $n > 5$, the Newton iterations did not always converge and it seemed necessary to employ several linear iterations before using the Newton method. For $n > 7$, direct application of Newton's iterations could lead to numerical instability in some cases and, therefore, mostly linear iterations were employed. It typically took three Newton's iterations for $n < 5$ and five linear iterations plus three Newton's iterations for $5 \leq n \leq 10$. The computations were performed in single precision (about 14 digits) on the CDC 176. Since elastic solutions are not available in the literature for all the values of R_i/b and a/b examined, the elastic results obtained are also tabulated. Accuracy of these solutions was assessed through comparisons against available literature solutions and other consistency checks which will be discussed in the following paragraph.

The stress-intensity factor, K_I, solutions for some combinations of a/b and b/R_i have been reported by Bowie and Freese [*43*], Rooke and Cartright [*23*], Buchalet and Bamford [*35*], and Labbens et al [*36*]. For $b/R_i = 1/10$, the calculated results were compared against those obtained by Buchalet and Bamford. Close agreement between the two results is observed, with a discrepancy of less than 4 percent in J. Elastic solutions for $b/R_i = 1/5$ and $1/20$ are not available for all a/b-values examined, so the accuracy of computed results was assessed by consistency checks at limits. A normalized stress-intensity factor was plotted versus R_i/R_0 with a/b as the parameter as shown in Fig. 3. As $R_i/R_0 \rightarrow 1$ (R_i and $R_0 \rightarrow \infty$), the solution should approach the single-edge crack solution which is obtainable from Ref *22*. Bowie and Freese [*43*] have cataloged the solutions for $R_i/R_0 \leq 0.8$ and these are plotted here. We note that the present solutions do interpolate between the Bowie and Freese solution and the single-edge-cracked panel (SECP) solution except for the Bowie and Freese solution at $R_i/R_0 = 0.8$. We believe that Bowie and Freese may be in error for $R_i/R_0 = 0.8$, because of the excellent agreement with Buchalet and Bamford for $R_i/R_0 = 0.91$. This consistency

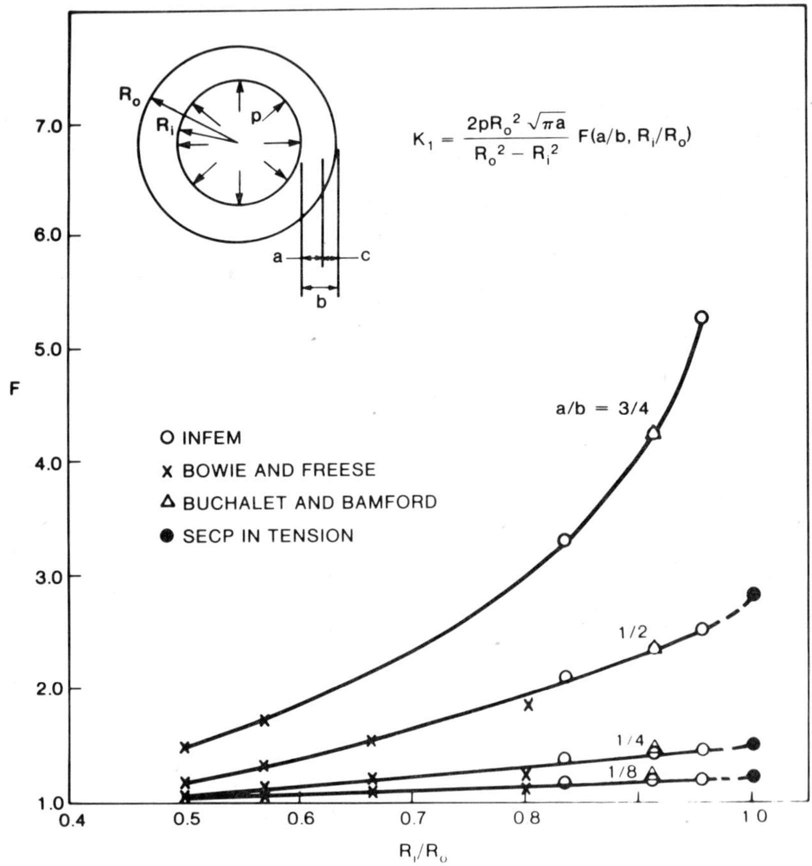

FIG. 3—*Elastic solutions for axially cracked cylinders under internal pressure.*

study suggests that the solutions developed are accurate to within 4 percent in J in the elastic case ($n = 1$). In the linear elastic range, the stress-intensity factor K_I, the J-integral, and the mouth opening displacement δ can be written as

$$K_\text{I} = \frac{2pR_0^2\sqrt{\pi a}}{R_0^2 - R_i^2} F\left(\frac{a}{b}, \frac{R_i}{R_0}\right) \tag{16}$$

$$\delta = \frac{8pR_0^2 a}{(R_0^2 - R_i^2)E'} V_1\left(\frac{a}{b}, \frac{R_i}{R_0}\right) \tag{17}$$

$$J = K_\text{I}^2/E' \tag{18}$$

where $E' = E/(1 - \nu^2)$, ν being Poisson's ratio. It is noted that the dimensionless functions F and V_1 depend upon two ratios, a/b and R_i/R_0. The developed solutions for F and V_1 are cataloged in Table 1.

As discussed in an earlier section, an important characteristic feature of fully plastic solutions is that they are scalable. The fully plastic solutions for J and δ are expressed as[3]

$$J = \alpha\sigma_0\epsilon_0 a \frac{c}{b} h_1\left(\frac{a}{b}, n; \frac{R_i}{R_0}\right)\left(\frac{p}{p_0}\right)^{n+1} \tag{19}$$

$$\delta = \alpha\epsilon_0 a h_2\left(\frac{a}{b}, n; \frac{R_i}{R_0}\right)\left(\frac{p}{p_0}\right)^{n} \tag{20}$$

where p_0 is the limit pressure for this geometry in the perfectly plastic case ($n = \infty$). A lower-bound value for this is given by

$$p_0 = \frac{2}{\sqrt{3}} \frac{c\sigma_0}{R_c} \tag{21}$$

where $R_c = R_i + a$ is the radial distance from the centerline to the crack tip. The dimensionless functions h_1 and h_2 in Eqs 19 and 20 are functions of a/b and n with R_i/R_0 as the parameter. The computed h-functions are cataloged for a wide range of a/b and n-values in Tables 2–4 for $b/R_i = 1/5$, $1/10$, and $1/20$, respectively. The results for $b/R_i = 1/10$ are also plotted in the form of h versus $1/n$ in Fig. 4 and h versus a/b in Fig. 5. Such curves are useful for interpolating and extrapolating the h's corresponding to desired values of a/b and n. Similar plots of h_1 and h_2 versus R_i/R_0 with a/b and n as the parameters can also be readily constructed.

TABLE 1—*F and V_1 for an axially cracked cylinder under internal pressure.*

		$a/b = 1/8$	$a/b = 1/4$	$a/b = 1/2$	$a/b = 3/4$
$b/R_i = 1/5$	F	1.19	1.38	2.10	3.30
	V_1	1.51	1.83	3.44	7.50
$b/R_i = 1/10$	F	1.20	1.44	2.36	4.23
	V_1	1.54	1.91	3.96	10.4
$b/R_i = 1/20$	F	1.20	1.45	2.51	5.25
	V_1	1.54	1.92	4.23	13.5

[3]Solutions for the CTOD δ_t are not cataloged here since the relationship between J and δ_t is not completely established in the presence of crack face loading.

TABLE 2—h_1 *and* h_2 *for an axially cracked cylinder under internal pressure with* $b/R_i = 1/5$.

		$n = 1$	$n = 2$	$n = 3$	$n = 5$	$n = 7$	$n = 10$
$a/b = 1/8$	h_1	6.32	7.93	9.32	11.5	13.12	14.94
	h_2	5.83	7.01	7.96	9.49	10.67	11.96
$a/b = 1/4$	h_1	7.00	8.34	9.03	9.59	9.71	9.45
	h_2	5.92	8.72	7.07	7.26	7.14	6.71
$a/b = 1/2$	h_1	9.79	10.37	9.07	5.61	3.52	2.11
	h_2	7.05	6.97	6.01	3.70	2.28	1.25
$a/b = 3/4$	h_1	11.00	5.54	2.84	1.24	0.83	0.493
	h_2	7.35	3.86	1.86	0.556	0.261	0.129

TABLE 3—h_1 *and* h_2 *for an axially cracked cylinder under internal pressure with* $b/R_i = 1/10$.

		$n = 1$	$n = 2$	$n = 3$	$n = 5$	$n = 7$	$n = 10$
$a/b = 1/8$	h_1	5.22	6.64	7.59	8.76	9.34	9.55
	h_2	5.31	6.25	6.88	7.65	8.02	8.09
$a/b = 1/4$	h_1	6.16	7.49	7.96	8.08	7.78	6.98
	h_2	5.56	6.31	6.52	6.40	6.01	5.27
$a/b = 1/2$	h_1	10.5	11.6	10.7	6.47	3.95	2.27
	h_2	7.48	7.72	7.01	4.29	2.58	1.37
$a/b = 3/4$	h_1	16.1	8.19	3.87	1.46	1.05	0.787
	h_2	9.57	5.40	2.57	0.706	0.370	0.232

TABLE 4—h_1 *and* h_2 *for an axially cracked clyinder under internal pressure with* $b/R_i = 1/20$.

		$n = 1$	$n = 2$	$n = 3$	$n = 5$	$n = 7$	$n = 10$
$a/b = 1/8$	h_1	4.50	5.79	6.62	7.65	8.07	7.75
	h_2	4.96	5.71	6.20	6.82	7.02	6.66
$a/b = 1/4$	h_1	5.57	6.91	7.37	7.47	7.21	6.53
	h_2	5.29	5.98	6.16	6.01	5.63	4.93
$a/b = 1/2$	h_1	10.8	12.8	12.8	8.16	4.88	2.62
	h_2	7.66	8.33	8.13	5.33	3.20	1.65
$a/b = 3/4$	h_1	23.1	13.1	5.87	1.90	1.23	0.883
	h_2	12.1	7.88	3.84	1.01	0.454	0.240

Elastic and fully plastic solutions presented in the foregoing can be combined through Eqs 4–6 to yield elastic-plastic estimation formulas. For a material governed by the Ramberg-Osgood stress-strain representation in uniaxial tension, Eq 3, this procedure gives the following expressions for J and δ

$$J = f_1\left(a_e, \frac{R_i}{R_0}\right) \frac{p^2}{E'} + \alpha\sigma_0\epsilon_0 a \frac{c}{b} h_1\left(\frac{a}{b}, n; \frac{R_i}{R_0}\right)\left(\frac{p}{p_0}\right)^{n+1} \tag{22}$$

$$\delta = f_2\left(a_e, \frac{R_i}{R_0}\right) \frac{p}{E'} + \alpha\epsilon_0 a h_2\left(\frac{a}{b}, n; \frac{R_i}{R_0}\right)\left(\frac{p}{p_0}\right)^{n} \tag{23}$$

a_e is the effective crack length as defined in Eqs 7–9. The functions f_1 and f_2 in Eqs 22 and 23 are related to F and V_1, defined in Eqs 16 and 17, in the following manner

$$f_1\left(a, \frac{R_i}{R_0}\right) = 4\pi a\left[\frac{R_0^2}{R_0^2 - R_i^2}\right]^2 F^2\left(\frac{a}{b}, \frac{R_i}{R_0}\right) \tag{24}$$

$$f_2\left(a, \frac{R_i}{R_0}\right) = 8a\left[\frac{R_0^2}{R_0^2 - R_i^2}\right]^2 V_1\left(\frac{a}{b}, \frac{R_i}{R_0}\right) \qquad (24)$$

Thus, given the material properties (E, ν, α, σ_0, and n), crack depth (a), cylinder dimensions, and internal pressure (p), the parameters J and δ can be calculated for the complete range of elastic-plastic deformation by using Eqs 22–25 and Tables 1–4. Results thus obtained compare favorably with detailed finite-element elastic-plastic crack computations. This will be elaborated upon in a later section.

Cylinder with an Internal Circumferential Crack Under Tension

Consider a cylinder with inner radius R_i, outer radius R_0, wall thickness $b = R_0 - R_i$, and length $2L$, containing an internal axisymmetric part-through crack of depth a as illustrated in Fig. 6. The wall thickness is denoted by $b = R_0 - R_i$ and the uncracked ligament by $c = b - a$. The cylinder is subjected to a uniform tensile stress field at its ends given by

$$\sigma_{zz}(r, \pm L) = \sigma^\infty = P/\pi(R_0^2 - R_i^2) \tag{26}$$

where P is the total tensile load carried by the cylinder. This configuration is treated as an axisymmetric problem. We note that the associated J-integral must be modified to account for axisymmetric deformation; this is briefly discussed in the Appendix and details in Refs *44* and *45*.

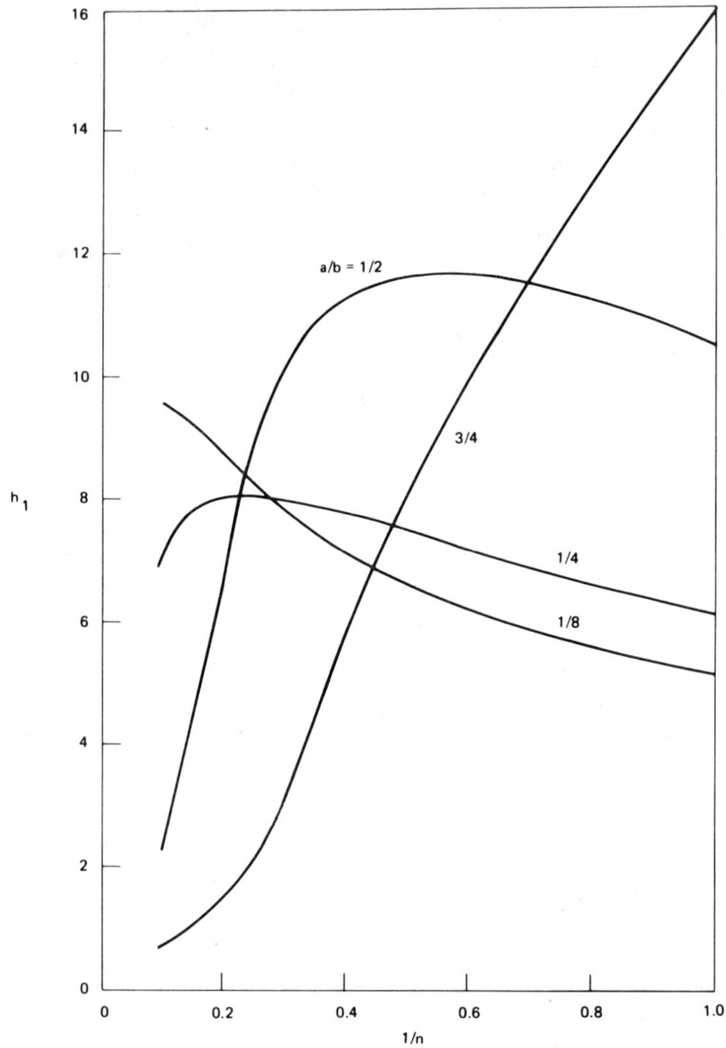

FIG. 4—h *versus* 1/n *for an axially cracked cylinder with* $b/R_i = 1/10$.

Similar to the axially cracked cylinder of the previous section, elastic and fully plastic analyses are carried out using the incompressible finite-element method described earlier. Solutions obtained correspond to b/R_i of ⅕, ⅒, and 1⁄20. For each b/R_i, the crack length to wall thickness ratio is taken to be ⅛, ¼, ½, and ¾, and the strain-hardening exponent n of 1, 2, 3, 5, 7, and 10. Because of the symmetry, only half the cylinder was modeled. A typical finite-element mesh corresponding to $b/R_i = 1/10$ and $a/b = 1/2$ is diagrammed in Fig. 7. The model has 16 quadrilateral elements around the crack tip and

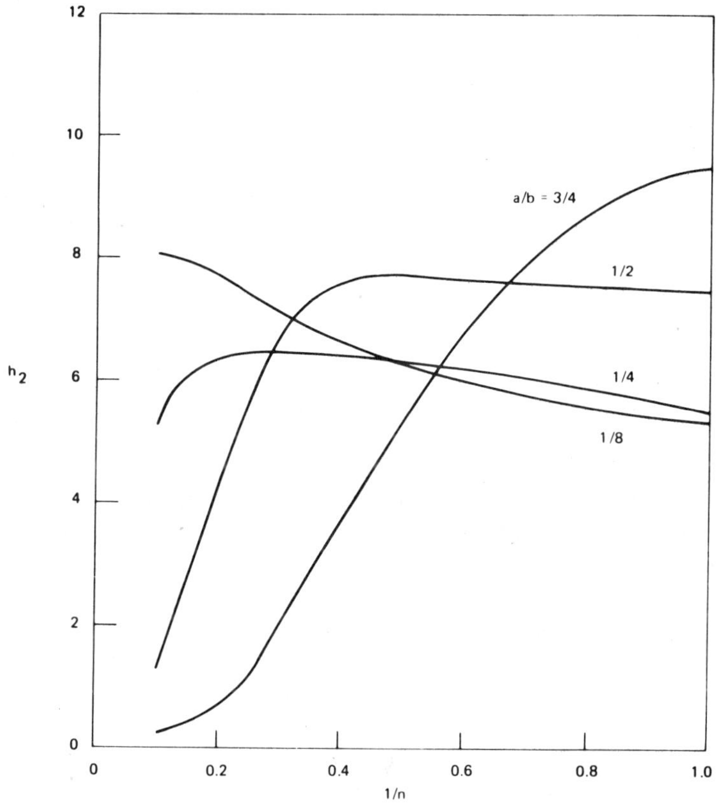

FIG. 4—*Continued.*

12 elements across the thickness as shown. The cylinder length is taken to be sufficiently large so that the cylinder can be treated as infinitely long. Specifically, the cylinder length ($2L$) to inside radius ratio, $2L/R_i$, is taken to be 4, 2, and 1 relative to R_i/b of 5, 10, and 20, respectively. Other computational details are similar to those for the axially cracked cylinder and, therefore, will not be repeated here.

The assessment of K_I results for this problem was made in a manner similar to that for axial cracks. The solutions developed corresponding to $b/R_i = 1/10$ agree within 1 to 4 percent with those of Buchalet and Bamford [*35*]. To provide some comparison for the present solutions for cylinders with $b/R_i = 1/5$ and $1/20$, the elastic solutions were also obtained using Automatic Dynamic Incremental Nonlinear Analysis (ADINA) [*46*] for a/b of $1/4$ and $3/4$. A typical mesh employed in the ADINA analysis for $a/b = 3/4$ is shown in Fig. 8; it consists of approximately 70 eight-noded isoparametric elements. A 2×2 integration procedure was used in the analysis and the J-integral was calculated by the virtual crack extension method [*46,47*]. The

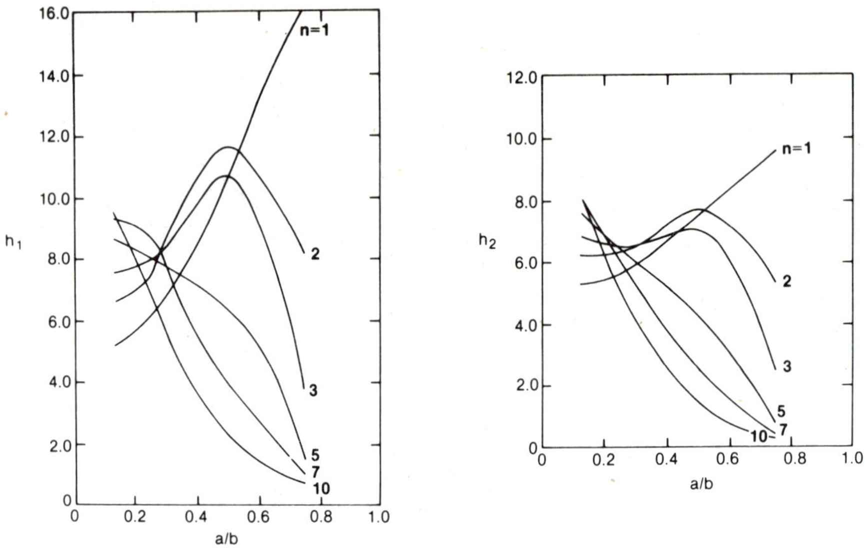

FIG. 5—h *versus* a/b *for an axially cracked cylinder with* $b/R_i = 1/10$.

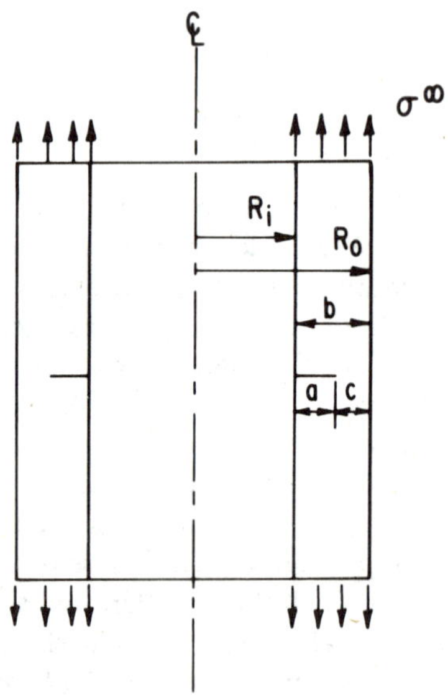

FIG. 6—*Circumferentially cracked cylinder subjected to remote uniform tension.*

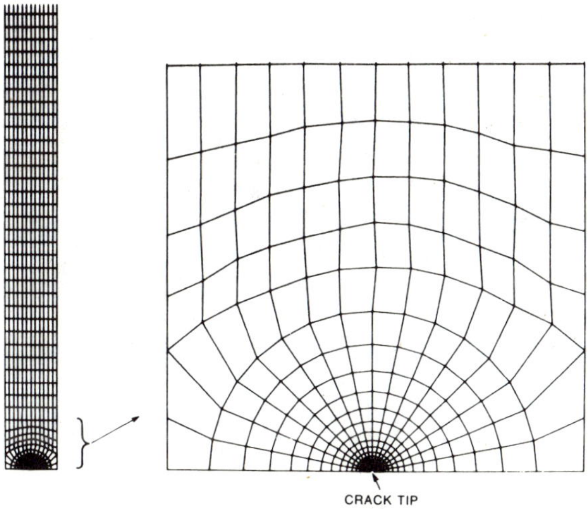

FIG. 7—*Finite-element mesh employed in the analysis of circumferentially cracked cylinders with* a/b = ½.

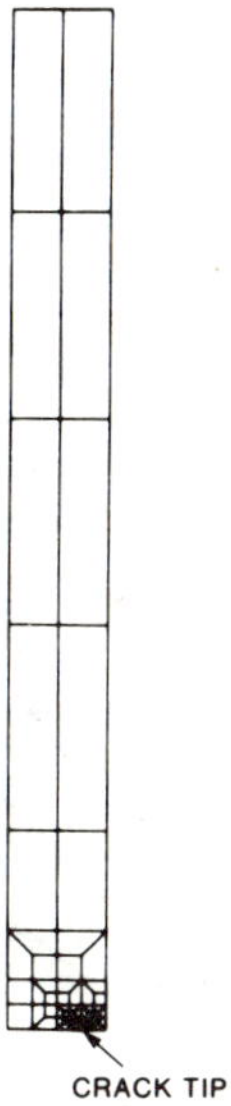

FIG. 8—*Finite-element mesh employed in the ADINA analysis of circumferentially cracked cylinders with* a/b = ¾.

present solutions, ADINA results, and Buchalet and Bamford solutions are plotted in different formats in Figs. 9 and 10. For $R_i/R_0 \rightarrow 1$ (R_i and $R_0 \rightarrow \infty$) the solution should approach the SECP solution given in Ref *22*. We observe several points. First, for both $b/R_i = 1/5$ and $1/20$, the present solutions for $a/b = 1/4$ as well as $3/4$ are within 4 percent of the ADINA results for K. Second, in cases where direct comparisons cannot be made, the present solutions do appear to interpolate between the available reference points. These consistency checks suggest that the solutions obtained here are accurate to within 1 to 4 percent in K_I.

In the linear elastic case, the fracture parameters K_I, δ, and Δ_c are given by

$$K_I = \sigma^\infty \sqrt{\pi a}\ F(a/b, R_i/R_0) \tag{27}$$

$$\delta = \frac{4\sigma^\infty a}{E'} V_1(a/b,\ R_i/R_0) \tag{28}$$

$$\Delta_c = \frac{4\sigma^\infty a}{E'} V_2(a/b,\ R_i/R_0) \tag{29}$$

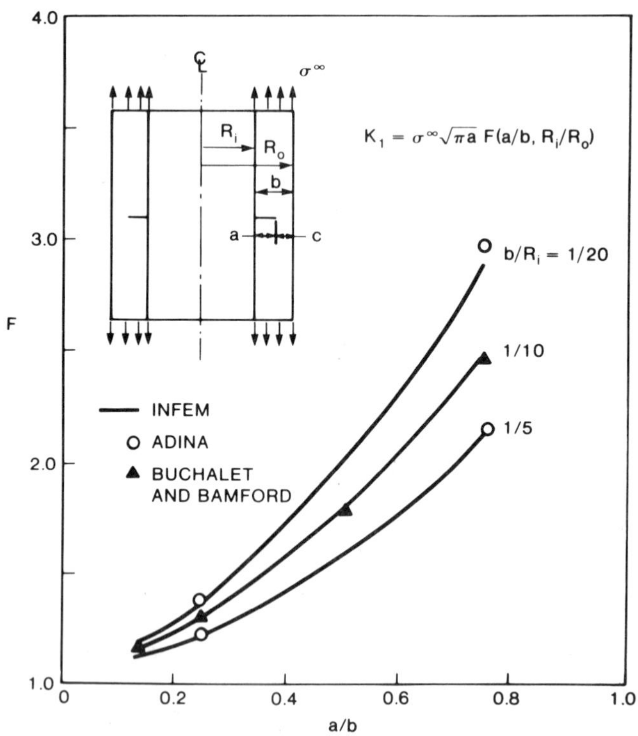

FIG. 9—*Plot of elastic solutions versus crack depth for circumferentially cracked cylinders.*

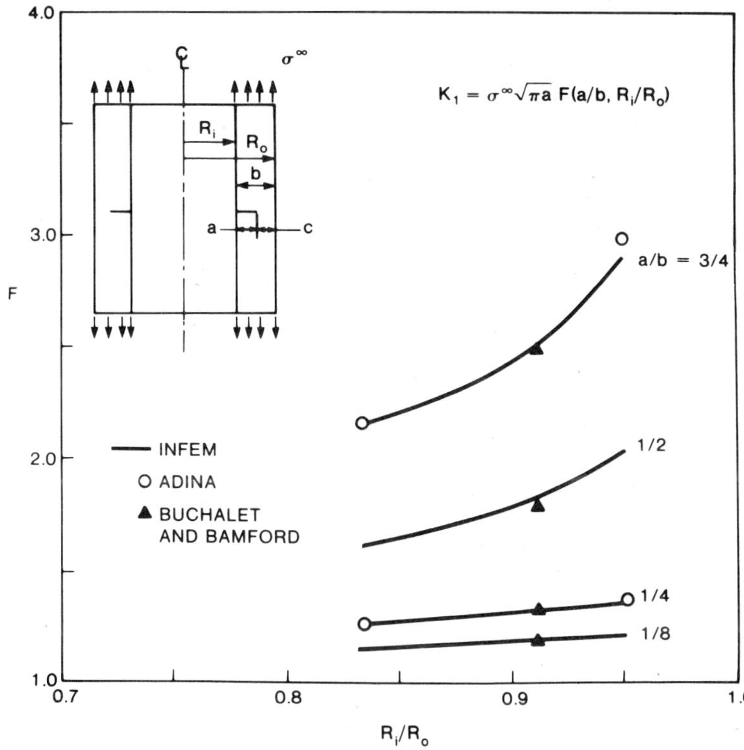

FIG. 10—*Plot of elastic solutions versus* R_i/R_0 *for circumferentially cracked cylinders.*

where $E' = E/(1 - \nu^2)$ and F, V_1, V_2 are dimensionless functions of a/b and R_i/R_0. Sneddon and Lowengrub [*49*] and Rice [*2*] have shown that the singular stress and strain fields in the crack-tip vicinity of an axisymmetric crack correspond to those in plane strain. Consequently, J is related to K_I via the plane-strain relationship, namely

$$J = K_I^2/E' \tag{30}$$

The calculated values of F, V_1, and V_2 as functions of a/b and R_i/R_0 are presented in Table 5.

The fully plastic solutions for this crack geometry can be expressed as

$$J = \alpha\sigma_0\epsilon_0 a \frac{c}{b} h_1\left(\frac{a}{b}, n; \frac{R_i}{R_0}\right)\left(\frac{P}{P_0}\right)^{n+1} \tag{31}$$

$$\delta = \alpha\epsilon_0 a h_2\left(\frac{\alpha}{b}, n; \frac{R_i}{R_0}\right)\left(\frac{P}{P_0}\right)^{n} \tag{32}$$

TABLE 5—F, V_1, and V_2 *for a circumferentially cracked cylinder in tension.*

		$a/b = 1/8$	$a/b = 1/4$	$a/b = 1/2$	$a/b = 3/4$
	F	1.16	1.26	1.61	2.15
$b/R_i = 1/5$	V_1	1.49	1.67	2.43	3.76
	V_2	0.117	0.255	0.743	1.67
	F	1.19	1.32	1.82	2.49
$b/R_i = 1/10$	V_1	1.55	1.76	2.84	4.72
	V_2	0.180	0.290	0.885	2.09
	F	1.22	1.36	2.03	2.89
$b/R_i = 1/20$	V_1	1.59	1.81	3.26	5.99
	V_2	0.220	0.315	1.04	2.74

$$\Delta_c = \alpha\epsilon_0 a h_3\left(\frac{a}{b}, n; \frac{R_i}{R_0}\right)\left(\frac{P}{P_0}\right)^n \tag{33}$$

$$\delta_t = \alpha\epsilon_0 a \frac{c}{b} h_4\left(\frac{a}{b}, n; \frac{R_i}{R_0}\right)\left(\frac{P}{P_0}\right)^{n+1} \tag{34}$$

where P_0 is the limit load for perfectly plastic material ($n = \infty$). A lower-bound expression for this is

$$P_0 = \frac{2}{\sqrt{3}}\sigma_0\pi(R_0^2 - R_c^2) \tag{35}$$

Here $R_c = R_i + a$ is the radial distance from the centerline to the crack tip. In Eq 32, $\delta = U_z(R_i, 0^+) - U_z(R_i, 0^-)$ is the mouth opening displacement. δ_t in Eq 34 is the crack-tip opening displacement as defined in Ref *15*. Δ_c in Eq 33 is the load-point displacement due to the crack and is given by

$$\Delta_c = \Delta - \Delta_{nc} \tag{36}$$

where Δ is the remote load-point displacement through which P works and Δ_{nc} the load-point displacement in the absence of crack. Under fully plastic conditions it can be written as

$$\Delta_{nc} = 2\alpha\epsilon_0 L\left[\frac{P}{\pi\sigma_0(R_0^2 - R_i^2)}\right]^n \tag{37}$$

The nondimensional quantities h_1, h_2, h_3, and h_4 in Eqs 31–34 are functions of a/b, n, and R_i/R_0. The h-functions computed in the present analysis are presented in Tables 6-8. h_4 is obtained from h_1 through the plane-strain relationship between J and δ_t [*15*]

$$h_4 = d_n h_1 \tag{38}$$

where d_n is a function of σ_0/E and $1/n$ can be obtained from Ref *15*. The graphs of h's versus a/b and $1/n$ are given in Figs. 11 and 12; similar plots can also be readily drawn with R_i/R_0 as the variable. These graphs are convenient in obtaining the h-functions for a given set of a/b, n, and R_i/R_0-values by an interpolation or extrapolation procedure.

The elastic and fully plastic solutions can then be employed according to the estimation procedure, Eqs 4-6, to obtain expressions for the elastic-plastic solutions. In the case of a Ramberg-Osgood stress-strain law, Eq 3, such formulas are given by

$$J = f_1\left(a_e, \frac{R_i}{R_0}\right)\frac{P^2}{E'} + \alpha\sigma_0\epsilon_0 a\frac{c}{b}h_1\left(\frac{a}{b}, n; \frac{R_i}{R_0}\right)\left(\frac{P}{P_0}\right)^{n+1} \quad (39)$$

TABLE 6—h_1, h_2, *and* h_3 *for a circumferentially cracked cylinder in tension with* $b/R_i = 1/5$.

		$n = 1$	$n = 2$	$n = 3$	$n = 5$	$n = 7$	$n = 10$
$a/b = 1/8$	h_1	3.78	5.00	5.94	7.54	8.99	11.1
	h_2	4.56	5.55	6.37	7.79	9.10	11.0
	h_3	0.369	0.700	1.07	1.96	3.04	4.94
$a/b = 1/4$	h_1	3.88	4.95	5.64	6.49	6.94	7.22
	h_2	4.40	5.12	5.57	6.07	6.28	6.30
	h_3	0.673	1.25	1.79	2.79	3.61	4.52
$a/b = 1/2$	h_1	4.40	4.78	4.59	3.79	3.07	2.34
	h_2	4.36	4.30	3.91	3.00	2.26	1.55
	h_3	1.33	1.93	2.21	2.23	1.94	1.46
$a/b = 3/4$	h_1	4.12	3.03	2.23	1.546	1.30	1.11
	h_2	3.46	2.19	1.36	0.638	0.436	0.325
	h_3	1.54	1.39	1.04	0.686	0.508	0.366

TABLE 7—h_1, h_2, *and* h_3 *for a circumferentially cracked cylinder in tension with* $b/R_i = 1/10$.

		$n = 1$	$n = 2$	$n = 3$	$n = 5$	$n = 7$	$n = 10$
$a/b = 1/8$	h_1	4.00	5.13	6.09	7.69	9.09	11.1
	h_2	4.71	5.63	6.45	7.85	9.09	10.9
	h_3	0.548	0.733	1.13	2.07	3.16	5.07
$a/b = 1/4$	h_1	4.17	5.35	6.09	6.93	7.30	7.41
	h_2	4.58	5.36	5.84	6.31	6.44	6.31
	h_3	0.757	1.35	1.93	2.96	3.78	4.60
$a/b = 1/2$	h_1	5.40	5.90	5.63	4.51	3.49	2.47
	h_2	4.99	5.01	4.59	3.48	2.56	1.67
	h_3	1.555	2.26	2.59	2.57	2.18	1.56
$a/b = 3/4$	h_1	5.18	3.78	2.57	1.59	1.31	1.10
	h_2	4.22	2.79	1.67	0.725	0.48	0.300
	h_3	1.86	1.73	1.26	0.775	0.561	0.360

TABLE 8—h_1, h_2, and h_3 for a circumferentially cracked cylinder in tension with $b/R_i = 1/20$.

		$n = 1$	$n = 2$	$n = 3$	$n = 5$	$n = 7$	$n = 10$
$a/b = 1/8$	h_1	4.04	5.23	6.22	7.82	9.19	11.1
	h_2	4.82	5.69	6.52	7.90	9.11	10.8
	h_3	0.680	0.759	1.17	2.13	3.23	5.12
$a/b = 1/4$	h_1	4.38	5.68	6.45	7.29	7.62	7.65
	h_2	4.71	5.56	6.05	6.51	6.59	6.39
	h_3	0.818	1.43	2.03	3.10	3.91	4.69
$a/b = 1/2$	h_1	6.55	7.17	6.89	5.46	4.13	2.77
	h_2	5.67	5.77	5.36	4.08	2.97	1.88
	h_3	1.80	2.59	2.99	2.98	2.50	1.74
$a/b = 3/4$	h_1	6.64	4.87	3.08	1.68	1.30	1.07
	h_2	5.18	3.57	2.07	0.808	0.472	0.316
	h_3	2.36	2.18	1.53	0.772	0.494	0.330

$$\delta = f_2\left(a_e, \frac{R_i}{R_0}\right)\frac{P}{E'} + \alpha\epsilon_0 a h_2\left(\frac{a}{b}, n; \frac{R_i}{R_0}\right)\left(\frac{P}{P_0}\right)^n \tag{40}$$

$$\Delta_c = f_3\left(a_e, \frac{R_i}{R_0}\right)\frac{P}{E'} + \alpha\epsilon_0 a h_3\left(\frac{a}{b}, n; \frac{R_i}{R_0}\right)\left(\frac{P}{P_0}\right)^n \tag{41}$$

In the foregoing, a_e is the effective crack length as defined by Eqs 7–9. The functions f_1, f_2, and f_3 are obtained from F, V_1, and V_2 as follows

$$f_1\left(a, \frac{R_i}{R_0}\right) = \frac{aF^2\left(\frac{a}{b}, R_i/R_0\right)}{\pi(R_0^2 - R_i^2)^2} \tag{42}$$

$$f_2\left(a, \frac{R_i}{R_0}\right) = \frac{4aV_1\left(\frac{a}{b}, R_i/R_0\right)}{\pi(R_0^2 - R_i^2)} \tag{43}$$

$$f_3\left(a, \frac{R_i}{R_0}\right) = \frac{4aV_2\left(\frac{a}{b}, R_i/R_0\right)}{\pi(R_0^2 - R_i^2)} \tag{44}$$

The load-point displacement is evaluated by adding to Δ_c, Eq 41, the Δ_{nc} contribution which in the elastic-plastic range is given by

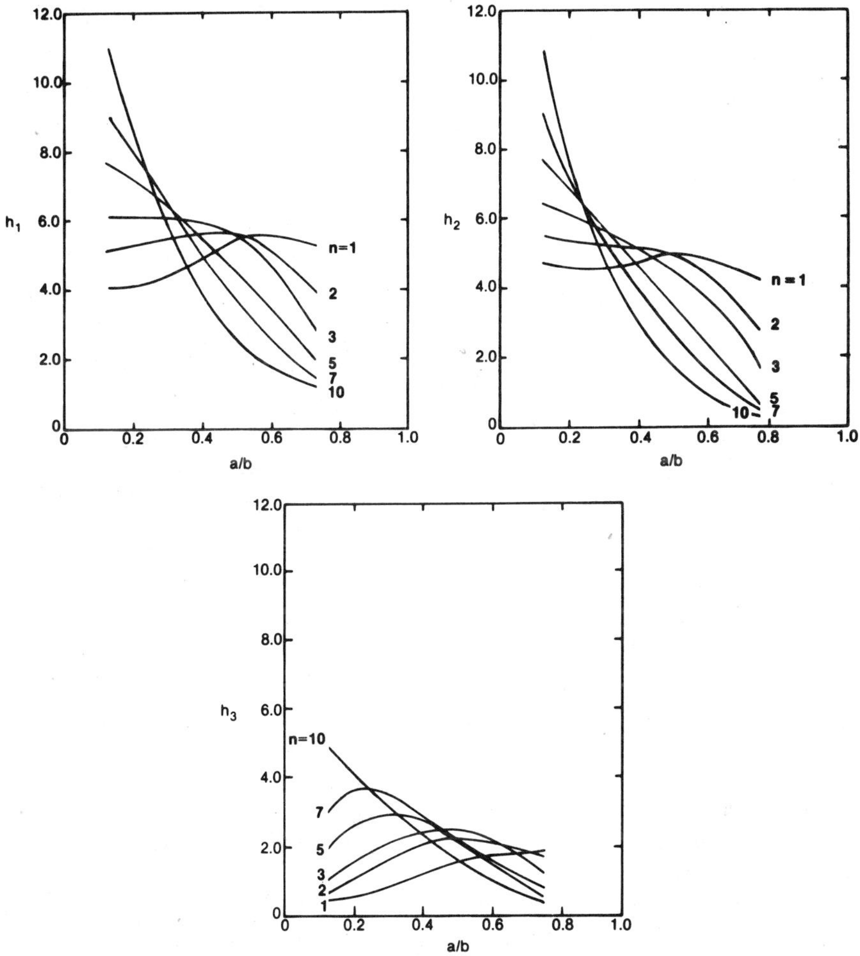

FIG. 11—*Plot of* h *versus* a/b *for a circumferentially cracked cylinder with* b/R$_i$ = 1/10.

$$\Delta_{nc} = 2L\epsilon_0 \left[\frac{P}{\pi\sigma_0 (R_0^2 - R_i^2)} + \alpha \left[\frac{P}{\pi\sigma_0 (R_0^2 - R_i^2)} \right]^n \right] \tag{45}$$

Illustrative application and verification examples are described in the following section.

Estimation Method Comparisons with Elastic-Plastic Finite-Element Calculations

To provide some verification of the elastic and fully plastic solutions and the associated elastic-plastic estimation formulas as presented in the previous sec-

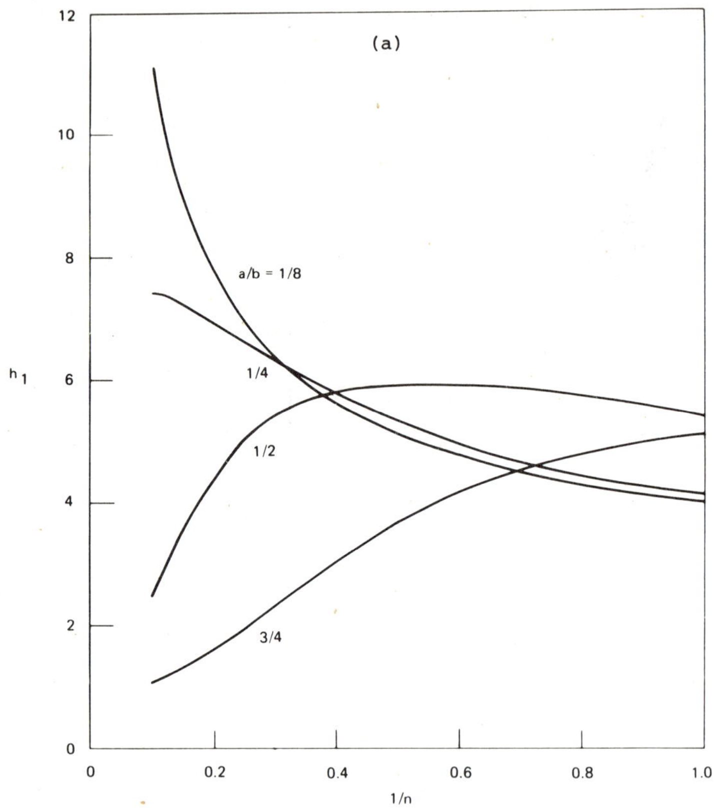

FIG. 12—*Plot of* h *versus* 1/n *for a circumferentially cracked cylinder with* $b/R_i = 1/10$.

tions, the corresponding results are next compared with elastic-plastic finite element crack computations for a few illustrative examples

First, consider an axially cracked cylinder with $R_i = 50.8$ cm (20 in.), $R_0 = 55.88$ cm (22 in.), $b = 5.08$ cm (2 in.), and $a/b = 1/4$. The material is modeled by the Ramberg-Osgood stress-strain representation with the following parameters: $E = 2.068 \times 10^5$ MPa (30×10^3 ksi), $\nu = 0.3$, $\sigma_0 = 4.14 \times 10^2$ MPa (60 ksi), $\alpha = 1$, and $n = 2$, 5, and 10. An elastic-plastic plane-strain analysis based on the deformation theory of plasticity is carried out using the finite-element code ADINA [*46*]. The mesh employed is depicted in Fig. 13; it consists of eight-noded isoparametric elements. The crack-tip blunting associated with elastic-plastic deformation is modeled by the collapse-node technique [*12*]. The computed results for J and δ are plotted in normalized forms in Figs. 14*a* and 14*b*. Also shown in these figures are the corresponding estimation method predictions. A close correspondence between the two solu-

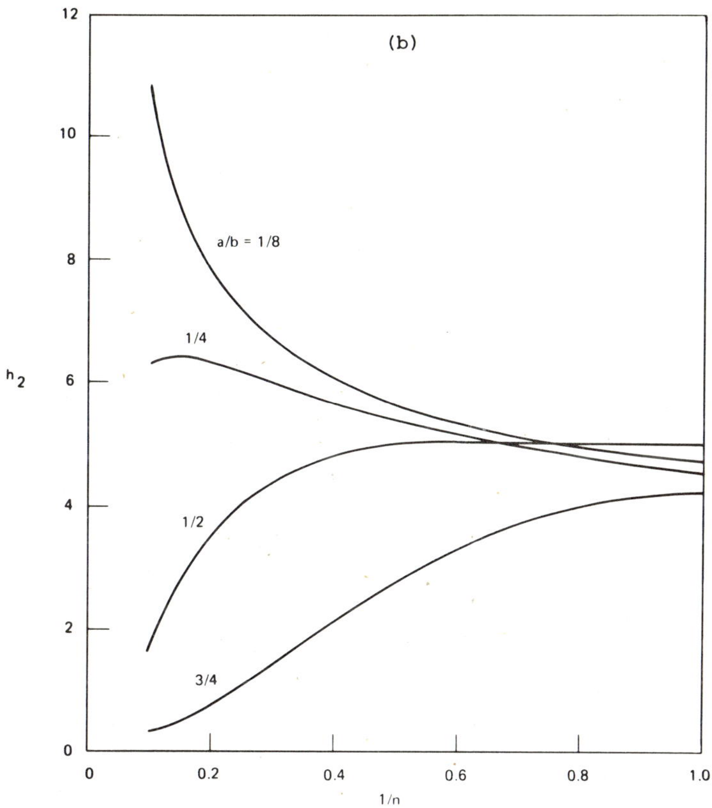

FIG. 12—*Continued.*

tions is noted, with the difference becoming smaller as n decreases. This is consistent with similar observations made in Refs *25* and *26*.

The second example is concerned with a circumferentially cracked cylinder of 304 stainless steel (Fig. 6). Following Ref *21*, the Ramberg-Osgood parameters for 304 stainless steel (SS) are taken to be: $E = 2.068 \times 10^5$ MPa (30×10^3 ksi), $\nu = 0.3$, σ_0 (yield stress) $= 2.07 \times 10^2$ MPa (30 ksi), $\alpha = 1.69$, and $n = 5.42$. Geometrical dimensions of the cylinder are taken to be as follows: $R_i = 2.286$ m (90 in.), $R_0 = 2.515$ m (99 in.), $b = 22.86$ cm (9 in.), and $2L = 4.572$ m (180 in.). The crack is assumed to be 5.715 cm (2.25 in.) deep, which corresponds to $a/b = \frac{1}{4}$ (these somewhat unrealistic dimensions for a material like 304 SS should not cause any concern since the objective here is only to provide a comparison for the estimation method). An elastic-plastic axisymmetric analysis based on the flow as well as the deformation theory of plasticity is performed using ADINA. The finite-element mesh employed is similar to the one shown in Fig. 13 except that it is straight and has

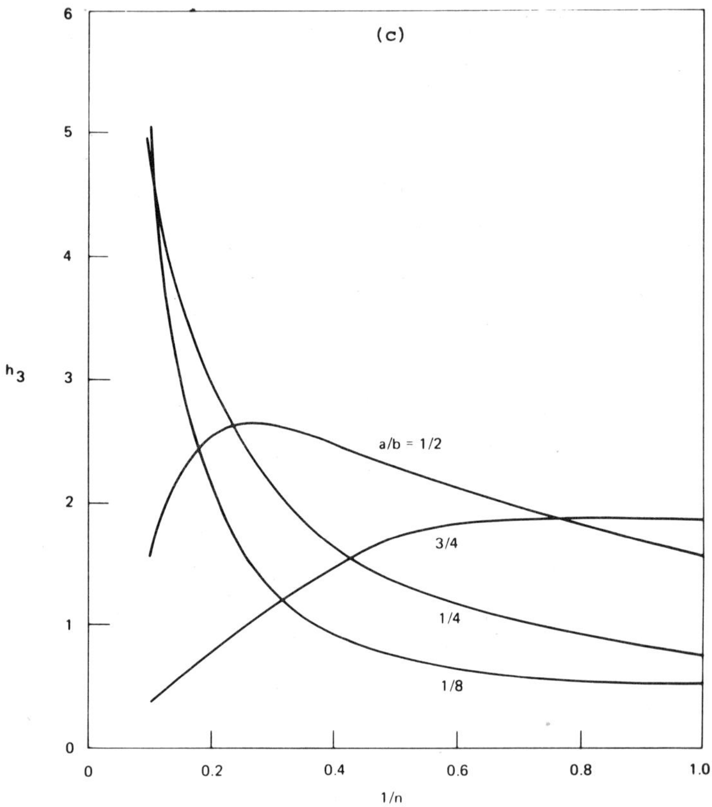

FIG. 12—*Continued.*

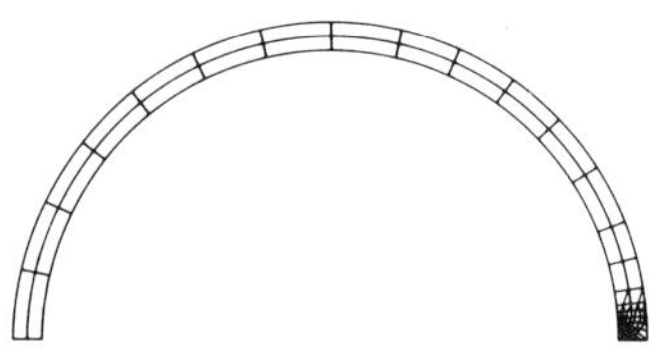

FIG. 13—*Finite-element mesh employed in elastic-plastic ADINA analysis of an axially cracked cylinder.*

about 30 percent fewer elements. The solutions for J, δ, and Δ thus obtained are compared against the respective estimation method results as shown in Figs. 15*a*–15*c*. We observe an excellent agreement between the two calculations. These limited verification studies support the predictive capability of the engineering approach developed.

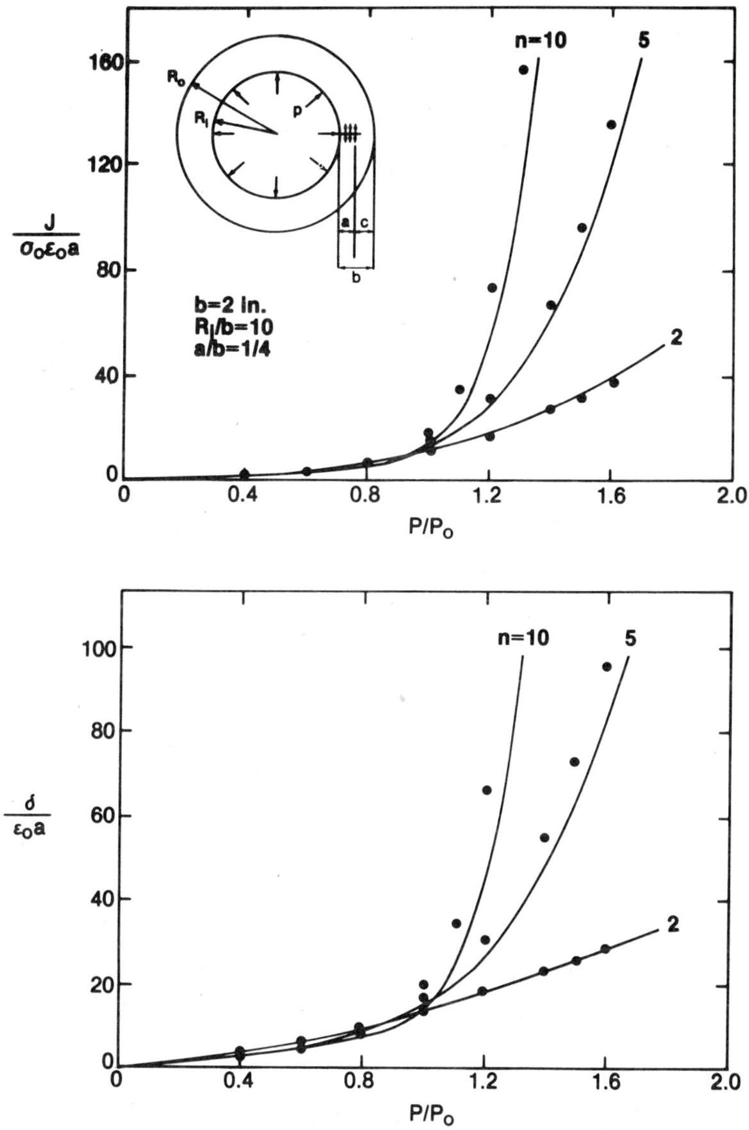

FIG. 14—*Comparison of the estimation method and finite-element results in the case of axially cracked cylinders for* (a) *the J-integral and* (b) *the mouth-opening displacement (1 in. = 2.54 cm).*

Analysis of Crack Initiation, Stable Growth, and Instability

Given the material properties (E, ν, α, σ_0, n), crack length (a), cylinder dimensions (R_i, b, L), and applied load (P) in a specific application, the crack driving force in terms of "applied" J (or COD) can be obtained by employing

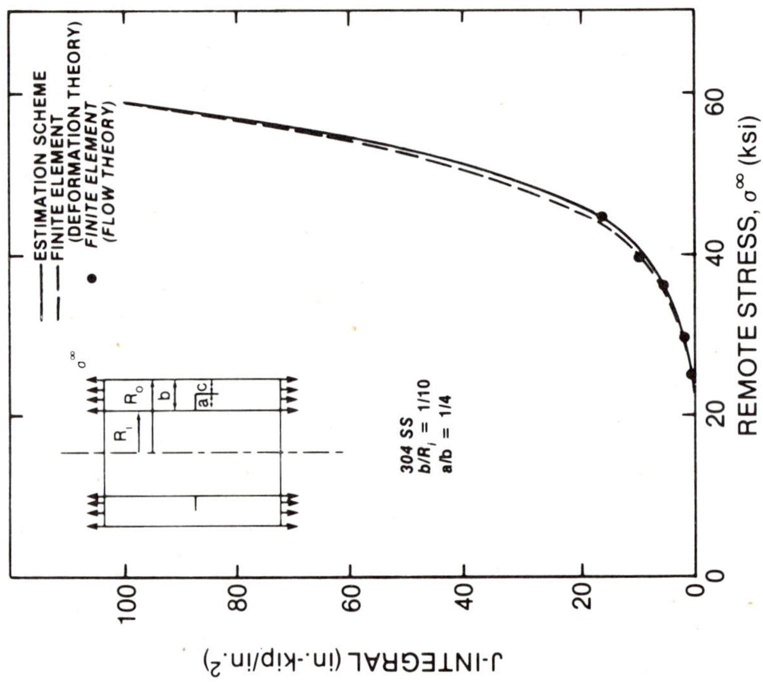

ESTIMATION SCHEME
FINITE ELEMENT (DEFORMATION THEORY)
FINITE ELEMENT (FLOW THEORY)
304 SS
b/R_i = 1/10
a/b = 1/4
R_i
R_o
b
a
c
σ^∞
J-INTEGRAL (in.-kip/in.²)
REMOTE STRESS, σ^∞ (ksi)
0
20
40
60
80
100

(a)

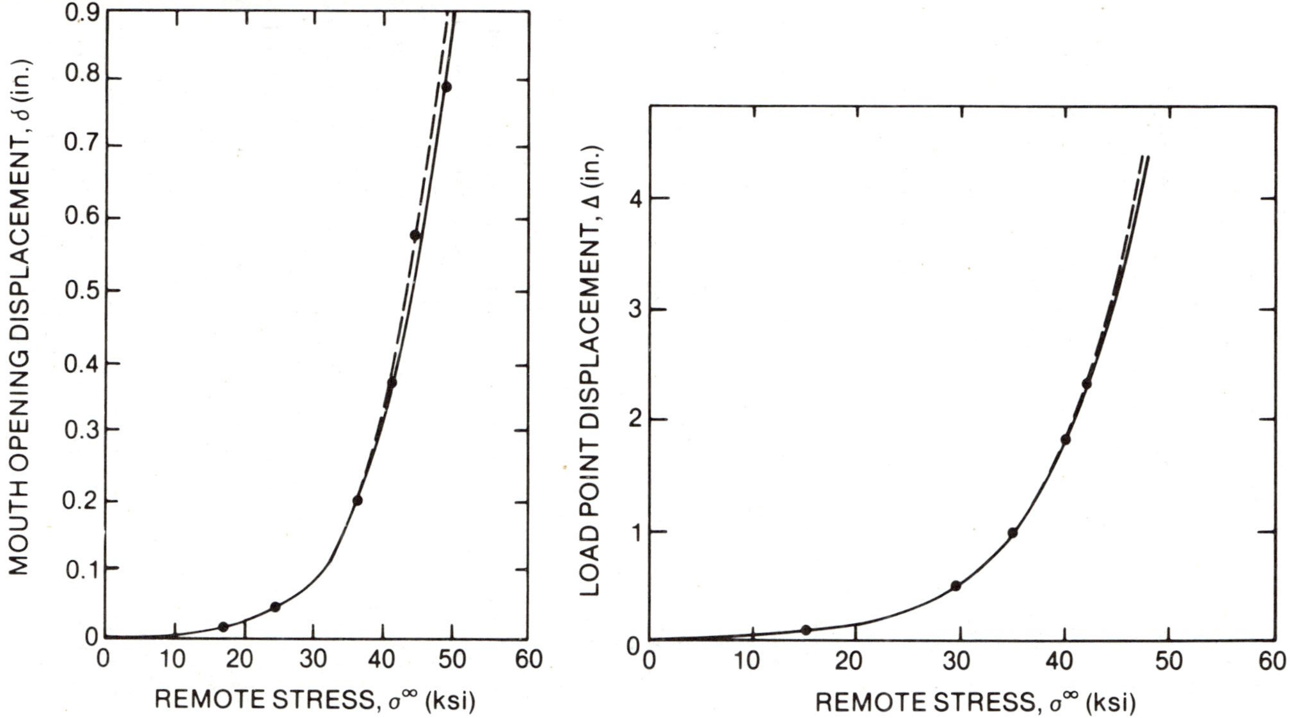

(b)

(c)

FIG. 15—*Comparison of the estimation method and finite-element results for a 304 stainless steel circumferentially cracked cylinder for* (a) *the J-integral,* (b) *the mouth-opening displacement, and* (c) *the load-point displacement (1 in.-kip/in.^{2} = 175 kNm/m^2, 1 ksi = 6.895 MPa, 1 in. = 0.0254 m).*

the relevant elastic-plastic estimation formulas and tables as discussed in the earlier sections. Following the crack growth and stability conditions as posed by Eqs 10 and 11, this is compared with the material resistance to crack growth (the J_R-curve), determined experimentally, to obtain predictions for crack initiation, stable growth, instability, and other quantities relevant to the overall fracture behavior. Depending upon the application, several different analysis diagrams can be generated. These are customarily referred to as the crack driving force diagram, the stability assessment diagram, and the failure assessment diagram. In this paper the attention is focused on crack driving force and stability diagrams; the topic of failure assessment diagrams is elaborated upon in Refs *21*, *34*, and *50*.

A crack driving force diagram is basically a plot of the J-integral (or COD) versus crack length and applied load. The superposition of the material *J*- (or COD-) resistance curve on this diagram enables one to perform a comprehensive analysis of the crack growth process, from initiation to instability. This diagram is actually the graphical solution to crack growth stability as posed by Eqs 10 and 11. It is important to note that the condition for instability of crack growth, Eq 11, and the effect of system compliance are directly accounted for in the crack driving force diagram. Stability assessment diagrams, on the other hand, consists of T_J-curves plotted against variables like the crack length, applied loading, and system compliance. This diagram is basically a compression of the information contained in the crack driving force diagram via Eqs 14 and 15. The customary plot is made on coordinates of J-integral versus the tearing modulus T_J. Such diagrams are particularly useful if only limited information on the fracture behavior is needed, that is, whether a flawed structure is stable or unstable. As pointed earlier with further details in Ref *21*, the applied T_J can be obtained from the elastic-plastic estimation formulas for J and Δ by analytical or numerical differentiation. The latter method was employed in the results presented here.

We next present some example crack driving force and stability diagrams for axially and circumferentially flawed cylinders. Material properties are assumed to correspond to A533B and 304 stainless steels which are commonly employed in pressure vessel and piping applications. Following Ref *21*, the Ramberg-Osgood parameters for A533B steel at 200°F are taken to be as follows: $E = 2.068 \times 10^5$ MPa (30×10^3 ksi), $\nu = 0.3$, σ_0 (yield stress) $= 4.14 \times 10^2$ MPa (60 ksi), $\alpha = 1.12$, and $n = 9.7$. The material constants for 304 stainless steel as given in the previous section are employed here also. The material *J*-resistance curves associated with these two steels is taken from Ref *21*. In the case of A533B steel the cylinder wall thickness, b, is assumed to be 22.86 cm (9 in.), and the inside radius, R_i, is assigned three different values 1.143, 2.286, and 4.572 m (45, 90, and 180 in.) so as to give R_i/b of 5, 10, and 20. The value of b for 304 stainless steel is assumed to be 1.78 cm (0.7 in.), and R_i is chosen to give the R_i/b range from 5 to 20. These dimensions are selected so that they fall in the range of real-life applications.

Using elastic-plastic estimation formulas, Eqs 22–23, in conjunction with elastic and fully plastic solutions cataloged in Tables 1–4, a J-integral driving force diagram is generated as depicted in Fig. 16 for an A533B steel axially cracked cylinder with R_i/b of 10 and initial crack length to wall thickness ratio, a_0/b of ¼. The diagram plots J versus crack length with internal pressure p as the parameter. The J_R-curve is superimposed on this diagram at the initial crack length of 5.715 cm (2.25 in.). The instability pressure is determined from the constant pressure line which is tangent to the J_R-curve. The extent of stable growth is calculated by subtracting the initial crack length from the crack length associated with the instability point. Since internal pressure in the cylinder acts as a dead load (that is, $C_M = \infty$), the instability pressure also gives the maximum load-carrying capacity. Sliding the J_R-curve along the crack length axis allows straightforward predictions of structural response corresponding to different initial crack sizes.

Figure 17 shows a recasting of the crack driving force diagram for A533B axially cracked cylinders on $J/(\sigma_0\epsilon_0 a)$ versus p/p_0 coordinates, with a/b varying. The effect of cylinder diameter is clearly demonstrated. Such diagrams are particularly useful in obtaining the load corresponding to crack initiation for a given J_{Ic}. The related stability assessment diagram is given in Fig. 18 in the form of J versus T_J curves for several values of a/b and R_i/b. The material J_R-T_{JR} curve, obtained from the material J_R-curve via Eq 14, is shown by the heavy solid line. This curve separates the stable and unstable regions. In the stable region, T_J is less than T_{JR}, and in the unstable regime T_J exceeds T_{JR}.

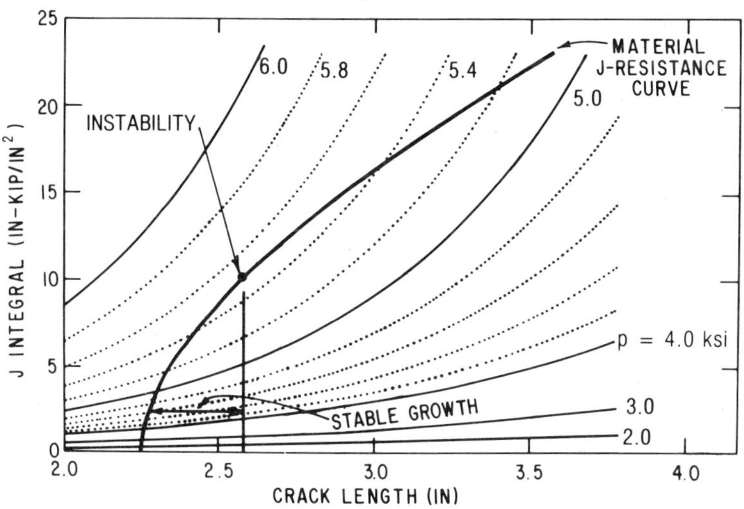

FIG. 16—*Crack driving force diagram for an axially cracked cylinder of A533B steel under internal pressure loading and with* b = *22.86 cm (9 in.). The material J_R-curve is shown by the heavy solid line and the instability point is also clearly indicated (1 in.-kip/in.² = 175) kNm/m², 1 ksi = 6.895 MPa, 1 in. = 0.0254 m).*

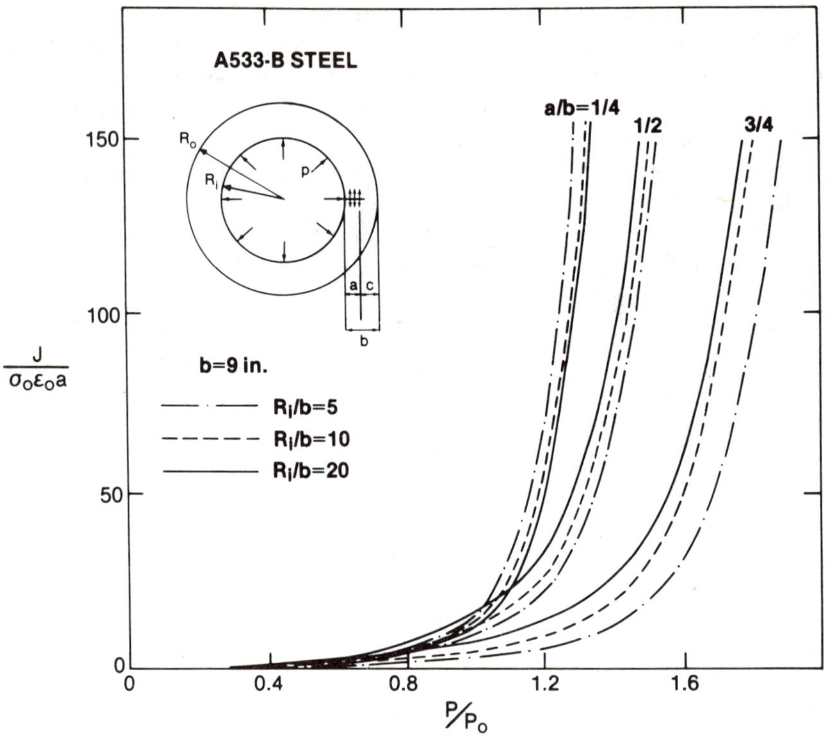

FIG. 17—*Plots of* $J/\sigma_0\epsilon_0 a$ *versus* p/p_0 *for A533B steel axially cracked cylinders (1 in. = 0.0254 m).*

For a given *a/b*, the critical value of *J* corresponding to instability is given by the point of intersection of the J_R-T_{JR} curve with a J-T_J curve. The critical value of *J* thus obtained can be used to enter the crack driving force diagram of Fig. 17 to predict the instability pressure. Several observations can be made on Fig. 18. First, the effect of cylinder diameter on the stability behavior is virtually none for a small crack ($a/b = 1/8$) and becomes more pronounced as *a/b* increases. This is not surprising since a small crack does not see the influence of cylinder curvature. Similar observations were made with respect to the elastic *K* solutions in Figs. 3, 9, and 10. Secondly, we note that for $a/b > 1/8$ a larger R_i/b-value gives higher T_J, meaning that large-diameter cylinders will promote unstable fracture. Finally, T_J decreases with increase in crack length, indicating that deeper cracks are more stable.

Similar results for circumferentially cracked cylinders of A533B steel are diagrammed in Figs. 19 and 20. Figure 19 consists of *J* versus the applied stress, σ^∞, curves for several cases of *a/b* and R_i/b. Again, we notice that (1) the influence of cylinder diameter is more for deeper cracks and (2) for a given *a/b* the effect of increase in R_i is to increase *J* and, therefore, to enhance the

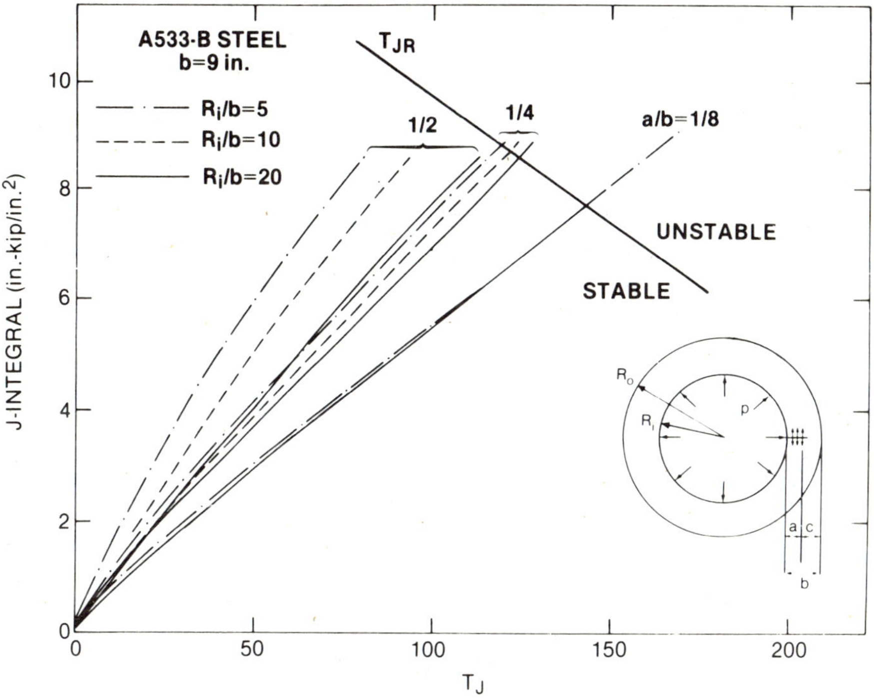

FIG. 18—J *versus* T_J *stability diagram for internally pressurized axially cracked cylinders of A533B steel. The material* J_R-T_{JR} *curve is shown by the heavy solid line (1 in.-kip/in.*^{2} = *175 kNm/m*^{2}*, 1 in. = 0.0254 m).*

likelihood of the crack initiation event. By looking at the J versus T_J curves for load-controlled ($L/R = \infty$) and displacement-controlled situations ($L/R = 1$) as plotted in Figs. 20*a* and 20*b*, the same trend as in the case of axial cracks is seen with respect to the effect of crack depth and cylinder diameter (for $L/R = \infty$, $a/b = 1/4$ results are almost the same as $a/b = 1/2$ and therefore are not presented here). The exception is the $a/b = 3/4$ ratio which gives higher T_J, at least under load-controlled conditions, than smaller a/b values. Furthermore, an increase in R_i now decreases the tearing modulus, which is in contrast to earlier observations. From the results presented it can be concluded that moderate-size cracks ($a/b = 1/4$ and $1/2$) will be more stable than the extreme-size flaws such as $a/b = 1/8$ and $3/4$.

The load-displacement and crack growth behavior of a circumferentially cracked cylinder can be determined in somewhat more detail from a crack driving force diagram of the type shown in Fig. 21 for the case $R_i/b = 5$, $L/R = 1$, and $a_0/b = 1/2$. The solid lines indicate the variation of J with crack length for applied stress, σ^∞, held fixed, and the dashed lines for load-point displacement, Δ, held constant. The material J_R-curve is superposed at the in-

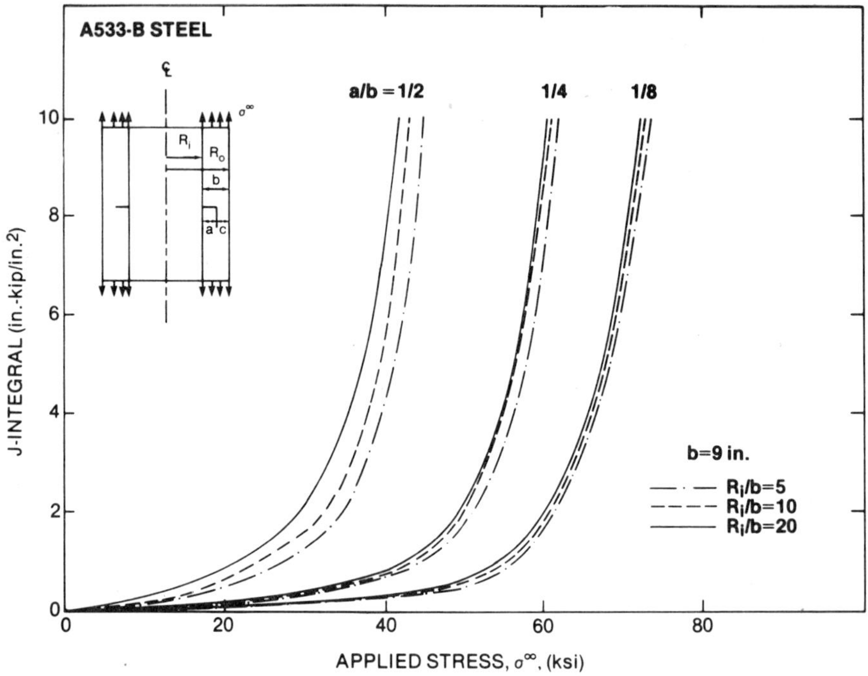

FIG. 19—*J-integral versus applied stress, σ^{∞}, curves for A533B steel circumferentially cracked cylinders (1 in.-kip/in.^{2} = 175 kNm/m^2, 1 ksi = 6.895 MPa, 1 in. = 0.0254 m).*

itial crack length of 4.5 in. Values of load and displacement associated with various points on the J_R-curve yield the load displacement behavior during crack growth. Other quantities of interest such as maximum load, values of load at crack initiation and instability, and the extent of stable growth are readily obtained from this diagram. Also, sliding the J_R-curve along the crack length axis will yield the crack growth and deformation behavior corresponding to different initial crack sizes.

The predicted results for the load-displacement behavior in the form of a σ^{∞} versus Δ_c curve are shown in Fig. 22 (note that $\Delta_c = \Delta - \Delta_{nc}$ where Δ is obtained from Fig. 21 and Δ_{nc} from Eq 45). Load at the onset of crack propagation is clearly indicated and the effect of cylinder length on the instability behavior is also demonstrated. Since Δ_c does not depend upon L, the σ^{∞} versus Δ_c curve would also be independent of L. However, L does control the instability load and the extent of stable growth; the effect of L would, therefore, be to shift the instability point on this curve as illustrated on the figure. The $L/R = 1$ case would give only stable crack extension, the effect of increase in L would be to decrease the amount of stable growth and consequently shift the instability point from right to left as shown. We further

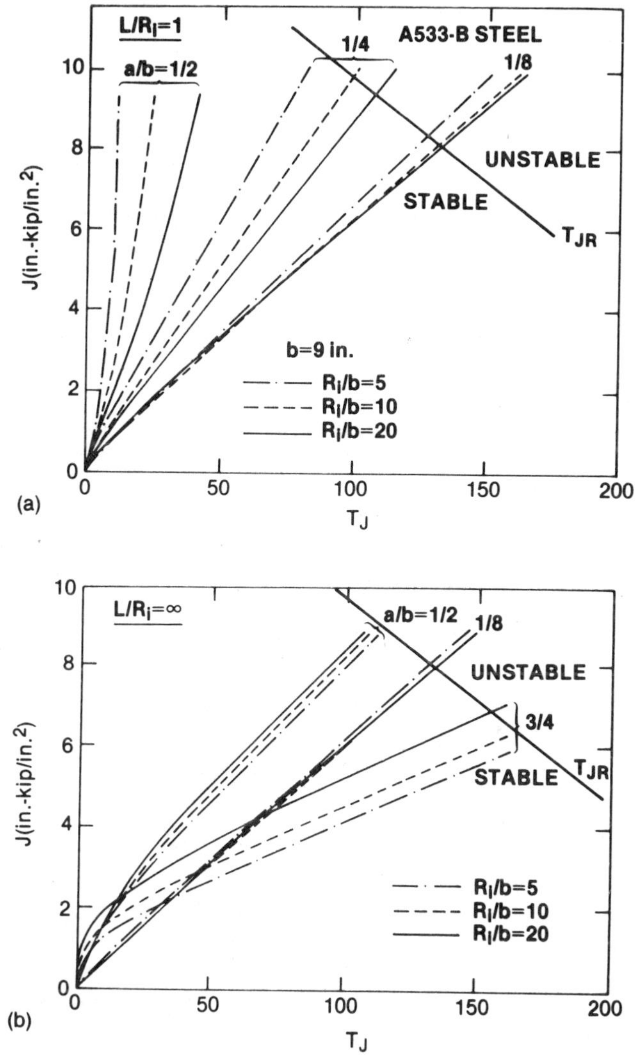

FIG. 20—J *versus* T_J *stability diagrams for A533B steel circumferentially cracked cylinders:* (a) L/R = *1 and* (b) L/R = ∞ *(1 in.-kip/in.*^{2} *= 175 kNm/m*^{2}*, 1 in. = 0.0254 m).*

remark that an L/R of 20 can be treated as being infinity for all practical purposes.

By constructing the crack driving force diagrams as described in the foregoing, the results for crack initiation and instability loads and the amount of stable growth were calculated for various cases of axially and circumferentially cracked cylinders of A533B steel. These are shown in Figs. 23 and 24. Effects of cylinder diameter and crack depth are consistent with our

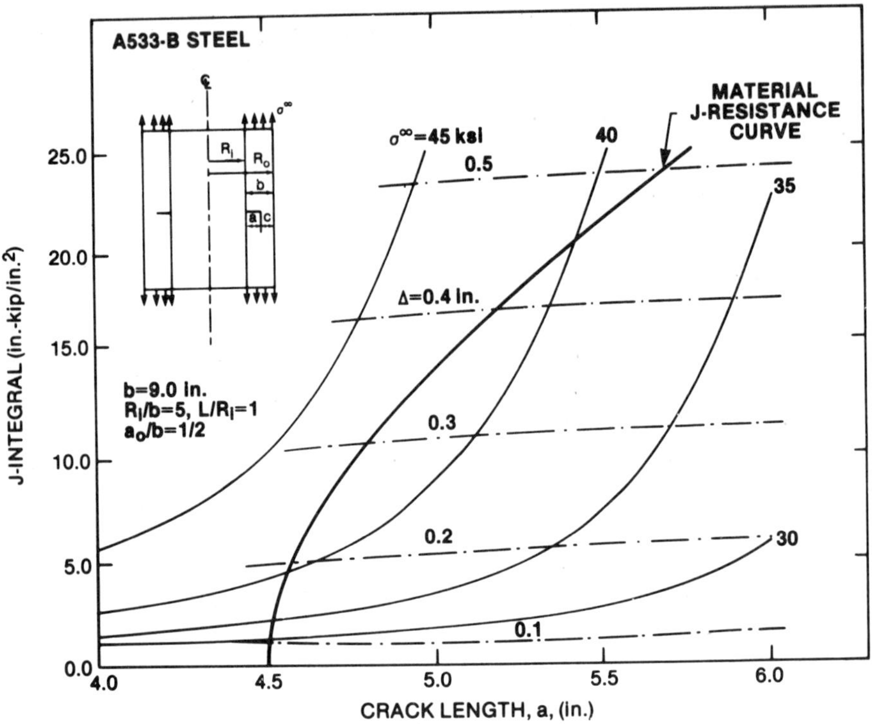

FIG. 21—*J-integral crack driving force diagram for an A533B steel circumferentially cracked cylinder with* $R_i/b = 5$ *and* $a_0/b = 1/2$ *(1 in.-kip/in.*^{2} *= 175 kNm/m*^{2}*, 1 ksi = 6.895 MPa, 1 in. = 0.0254 m).*

earlier observations on the related crack driving force and stability assessment diagrams. The circumferential crack results correspond to the worst-case situation of $L/R = \infty$. In the case of axial cracks the crack initiation and instability pressures are plotted in terms of an average hoop stress, pR/b, where $R = (R_i + R_0)/2$, mainly because this represents the crack opening stress. It is interesting to remark that the quantity pR/b is virtually insensitive to the cylinder diameter for the crack initiation and instability phenomena.

The crack growth and instability behavior of part-through circumferential cracks in 304 SS piping has been investigated in sufficient detail in Refs *51* and *52*. Some additional results for this case are briefly reported here. Using the material properties, cylinder and flaw dimensions, and the relevant elastic-plastic estimation formulas and tables discussed earlier, crack driving force and stability diagrams have been generated in several different formats. Figure 25 shows the curves of J-integral versus applied stress with a/b and R/b as parameters. Similar to A533B steel, we observe that (1) for a given

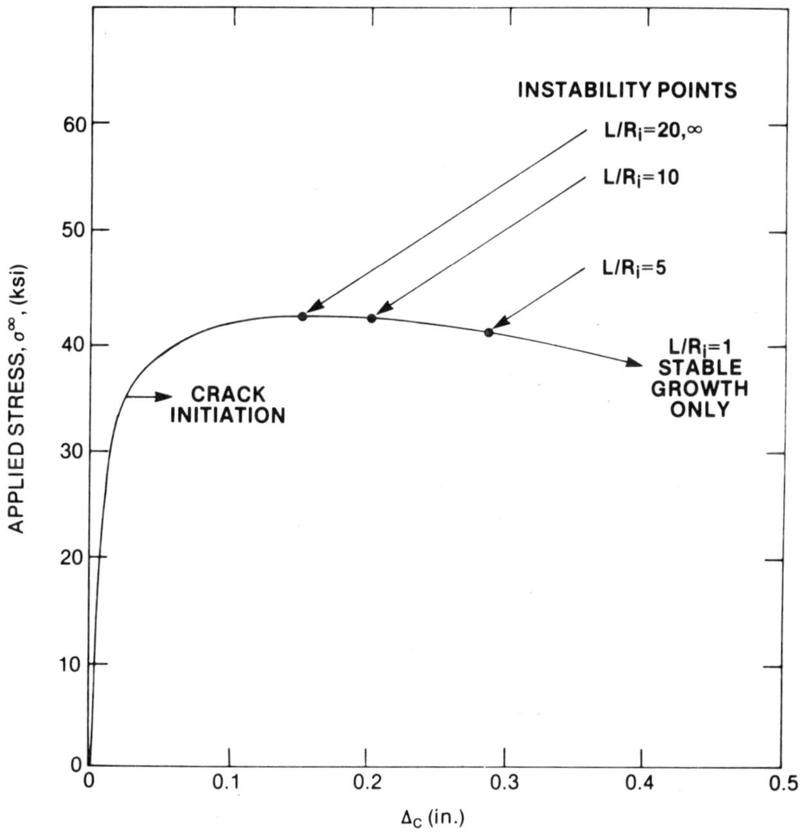

FIG. 22—*Load-displacement behavior, in the form of* σ^{∞} *versus* Δ *plot, for A533B steel circumferentially cracked cylinder with* $R_i/b = 5$ *and* $a_0/b = 1/2$. *The effect of* L/R *is demonstrated (1 ksi = 6.895 MPa, 1 in. = 0.0254 m).*

a/b, higher R/b yields larger J, and (2) the effect of R/b enhances as a/b increases. Figure 26 illustrates the J versus T_J stability diagrams for $a/b = 1/8$, $1/4$, and $1/2$, and for extreme values of the cylinder length-to-radius ratio, L/R (which can be assumed to represent limiting values, 0 and ∞, of the cylinder compliance C_M). Several points can be noted. First, as anticipated the variable R/b has virtually no influence on the stability behavior for shallow cracks; the effect, however, increases as cracks become deeper. Secondly, small cracks are rather insensitive to cylinder length as compared with longer cracks. Finally, for a given J the tearing modulus T_J increases as a/b decreases, meaning that smaller cracks are more susceptible to tearing instability than longer cracks. The J versus T_J curves for $a/b = 3/4$ are drawn in Fig. 27 for several values of L/R. Clearly, the cylinder length has a significant effect on the crack growth stability for such a large flaw size.

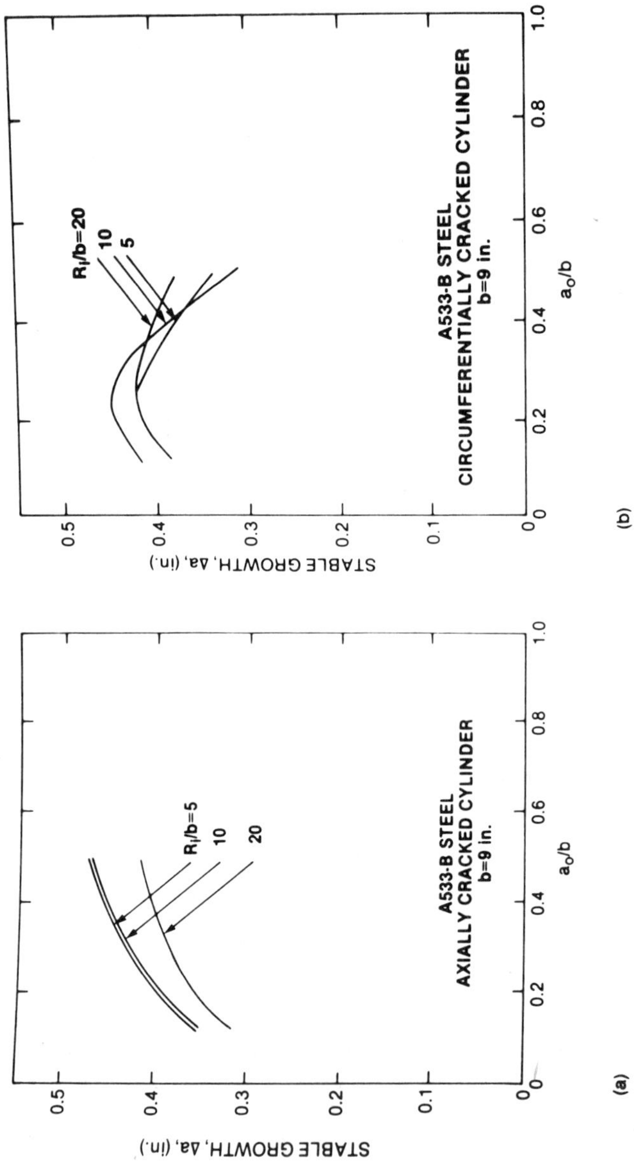

FIG. 23—*Plots of amount of stable crack growth* (Δa) *versus* a/b *for A533B steel cylinders:* (a) *axial cracks and* (b) *circumferential cracks (1 in. = 0.0254 m).*

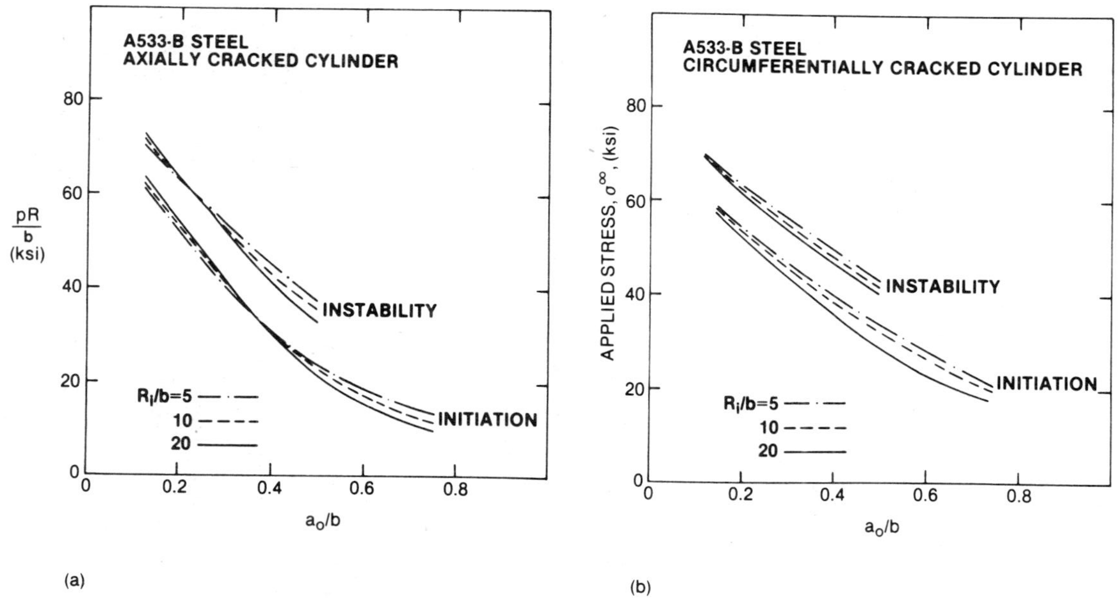

FIG. 24—*Plots of initiation and instability loads versus* a/b *for A533B steel cylinders:* (a) *axial flaws and* (b) *circumferential flaws (1 ksi = 6.895 Pa).*

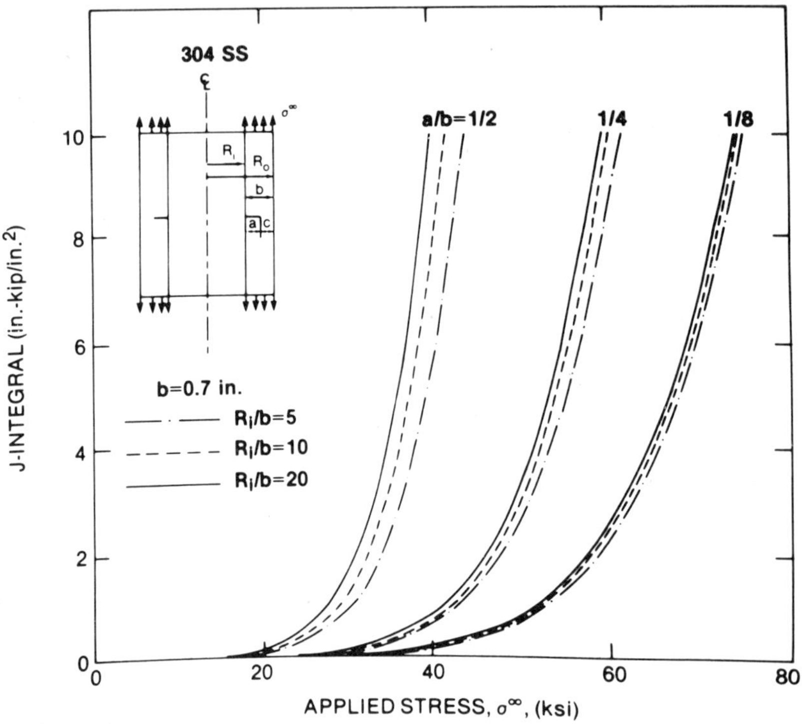

FIG. 25—*J-integral versus applied stress crack driving force diagram for 304 stainless steel circumferentially cracked cylinders (1 in.-kip/in.^{2} = 175 kNm/m^2, 1 ksi = 6.895 MPa, 1 in. = 0.0254 m).*

Discussion

In the present study the applied J-integral obtained by the elastic-plastic estimation procedure has been employed together with the material J_R-curve to predict crack initiation, stable growth, and instability in two-dimensional flawed cylinders. This approach tacitly assumes that J is a valid description of stress and strain fields near the crack tip during the crack growth process. In other words the so-called Hutchinson-Rice-Rosengren (HRR) singularity [*3,4*] with J as the amplitude is assumed to prevail over a microstructurally meaningful distance at the crack tip for stationary as well as growing cracks. Conditions and requirements essential to the validity of this assumption have been addressed in Refs *9*, *10*, *12*, and *14*. These studies have, however, been restricted to plane-strain analysis of cracks in flat plates in tension or in bending. A limited investigation of J-dominance in two-dimensional flawed cylinders and in round bars with penny-shaped cracks has recently been carried out by Shih [*41*] and Hutchinson [*53*] under stationary crack conditions. The remaining ligament associated with flawed cylinders examined in this

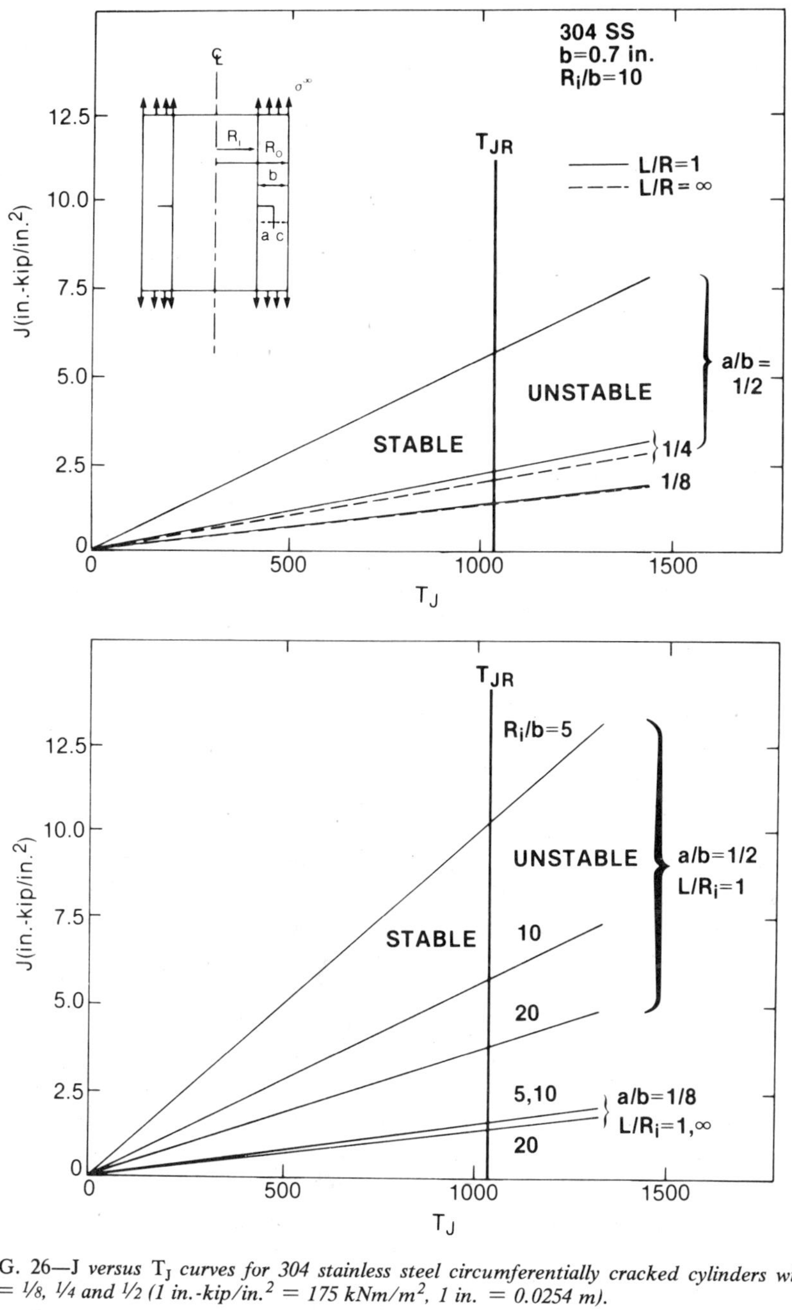

FIG. 26—J *versus* T_J *curves for 304 stainless steel circumferentially cracked cylinders with* a/b = 1/8, 1/4 and 1/2 *(1 in.-kip/in.^{2} = 175 kNm/m^2, 1 in. = 0.0254 m).*

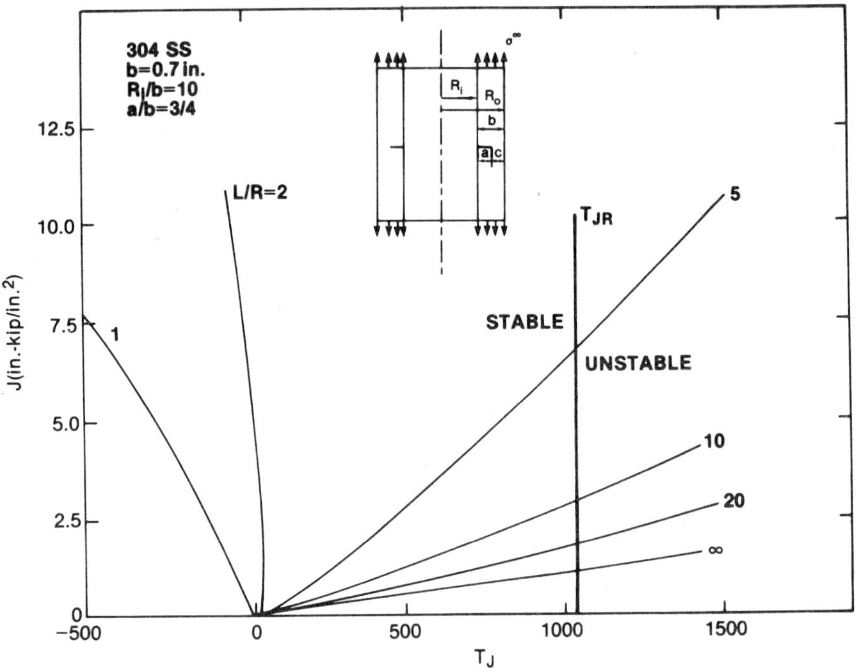

FIG. 27—*Effect of* L/R *on the stability behavior of a 304 stainless steel circumferentially cracked cylinder with* a/b = ¾ *(1 in.-kip/in.*^{2} *= 175 kNm/m*^{2}*, 1 in. = 0.0254 m).*

paper is subjected to tensile as well as bending loads and further studies are required to establish conditions for *J*-controlled growth in such crack configurations. Nevertheless, following Refs *31* and *34* it can be argued that predictions for the load-carrying capacity of flawed cylinders as studied here will be conservative even under non-*J*-controlled crack growth conditions so long as the J_R-curve employed in the analysis corresponds to plane-strain pure bend configurations.

It is also noted that by interpreting strains and displacements as their rate counterparts and *J* as C^*, the fully plastic solutions presented here can be directly applied to examine crack propagation under time-dependent creep conditions. Recently, attempts have been made to correlate the observed creep crack growth rate with various parameters such as K_I [*54*], the net section stress σ_{net} [*55*], the crack opening displacement rate $\dot{\delta}$ [*56,57*], and the path-independent energy rate line integral C^* [*58,59*]. All the experimental investigations [*54–60*] suggest that although none of these parameters correlates with creep crack data in a unified manner, C^* appears to be the most viable parameter for characterizing creep crack growth under steady-state conditions. This observation is also supported by recent analytical studies carried out in Ref *61*. Modified forms of C^* can also be used to characterize crack growth in materials governed by strain-hardening or time-hardening theories of creep as discussed in Ref *62*. Extension of modified C^* to somewhat more general situations has recently been discussed in Ref *63*. C^* and other modified expressions for it can be directly evaluated by employing the solutions reported.

Summary and Conclusions

Crack growth and instability in flawed cylinders under large-scale yielding have been addressed in this paper within the framework of an engineering approach developed by the present authors. Results of this study can be summarized as follows:

1. Elastic and fully plastic solutions have been obtained for axially and circumferentially cracked cylinders covering a wide range of values for the crack length, cylinder dimensions, and strain-hardening exponent. The solutions have been appropriately normalized, tabulated, and plotted in a convenient format.
2. Elastic-plastic estimation formulas have been derived for obtaining the relevant crack parameter at all levels of plastic deformation. Corresponding results for some illustrative examples are compared with detailed finite-element computations. A reasonable agreement between the two solutions is observed.
3. Example crack driving force and stability diagrams have been generated for problems of practical interest. These diagrams are used in conjunction with the material J_R-curve to predict crack initiation, extent of stable growth, instability load, and overall deformation behavior.
4. Effects of material properties, cylinder dimensions, and flaw type and size on the crack growth, instability, and deformation behavior have been examined in a somewhat detailed manner.

Acknowledgments

This work was sponsored by the Electric Power Research Institute (EPRI), Palo Alto, Calif., under Project RP1237-1 administered by Drs. D. M. Norris, Jr., R. L. Jones, and T. U. Marston. Encouragement and discussions provided by Dr. D. F. Mowbray of the General Electric Co. and Dr. D. M. Norris, Jr., of EPRI are gratefully acknowledged. Collaboration with Professor Alan Needleman of Brown University, Providence, R.I., on the development of the incompressible finite-element program employed in this study is also sincerely appreciated.

APPENDIX

Modified J-Integral Expressions

J-Integral Expression for Loaded Crack Surfaces

A modified expression for the J-integral to account for crack face loading has been derived by Karlsson and Backlund [*42*]. For the case of a constant pressure p on the crack surface, it is given by

$$J = \int_{\Gamma}\left(W\,dy - \mathbf{T}\cdot\frac{\partial \mathbf{U}}{\partial x}\,ds\right) + pU_y(-d, 0) \tag{46}$$

where the symbols are illustrated in Fig. 28. For further details on the derivation, the reader is referred to Ref *42*.

J-Integral Expression for Axisymmetric Bodies

Astiz et al [*44*] and Bergquist and Lan Huong [*45*] obtained modified expressions for energy release rate or the J-integral in problems of axial symmetry. In this case the J-integral, interpreted as energy release rate per unit crack advance in the radial direction and per unit length along the circumferential direction, is given by

$$J = \frac{1}{R_i + a}\left\{\int_{\Gamma} r\left[W\,dz - \left(T_r\frac{\partial U_r}{\partial r} + T_z\frac{\partial U_z}{\partial r}\right)ds\right] - \int_{S}\left(W - T_\Theta\frac{U_r}{r}\right)dS\right\} \tag{47}$$

where the coordinate system and other notations in the foregoing equation are described in Fig. 29.

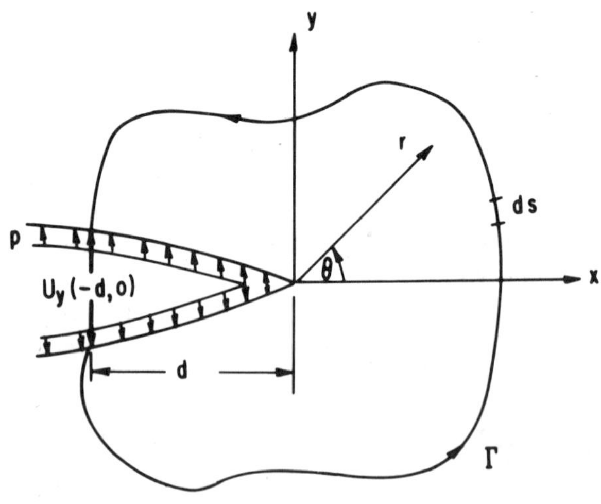

FIG. 28—*Schematic of a cracked body with uniform pressure on the crack surface.*

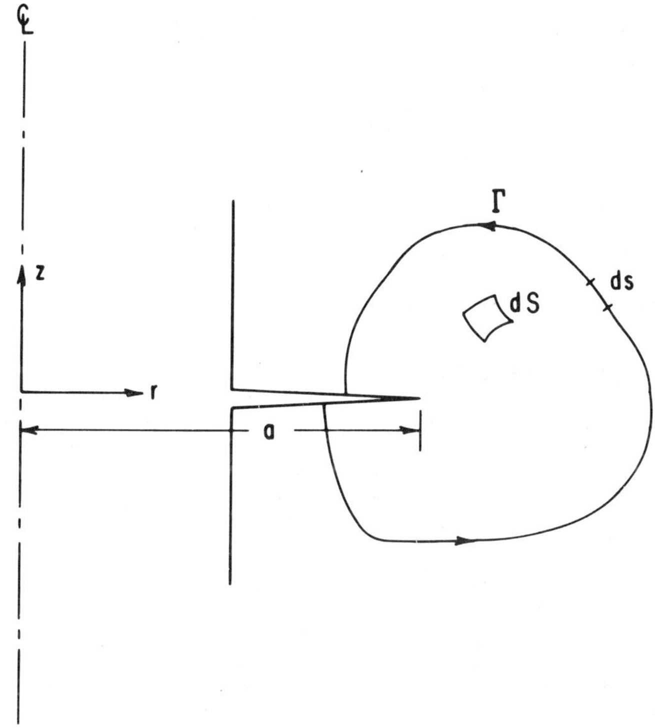

FIG. 29—*Coordinate system and nomenclature employed in definition of J-integral for an axisymmetric crack. S is the area enclosed by contour* Γ.

References

[*1*] Rice, J. R., *Journal of Applied Mechanics*, Vol. 35, 1968, pp. 379–386.
[*2*] Rice, J. R. in *Fracture*, H. Liebowitz, Ed., Vol. 2, Academic Press, New York, 1968, pp. 191–311.
[*3*] Hutchinson, J. W., *Journal of the Mechanics and Physics of Solids*, Vol. 16, No. 1, 1968, pp. 13–31.
[*4*] Rice, J. R. and Rosengren, G. F., *Journal of the Mechanics and Physics of Solids*, Vol. 16, No. 1, 1968, pp. 1–12.
[*5*] Begley, J. A. and Landes, J. D. in *Fracture Toughness, ASTM STP 514*, American Society for Testing and Materials, 1972, pp. 1–23 and pp. 24–39.
[*6*] Bucci, R. J., Paris, P. C., Landes, J. D., and Rice, J. R. in *Fracture Toughness, ASTM STP 514*, American Society for Testing and Materials, 1972, pp. 40–69.
[*7*] Rice, J. R., Paris, P. C., and Merkle, J. G. in *Progress in Flaw Growth and Fracture Toughness Testing, ASTM STP 536*, American Society for Testing and Materials, 1973, pp. 231–245.
[*8*] Paris, P. C., Tada, H., Zahoor, A., and Ernst, H. in *Elastic-Plastic Fracture, ASTM STP 668*, American Society for Testing and Materials, J. D. Landes, J. A. Begley, and G. A. Clarke, Eds., 1979, pp. 5–36 and pp. 251–265.
[*9*] Hutchinson, J. W. and Paris, P. C. in *Elastic-Plastic Fracture, ASTM STP 668*, American Society for Testing and Materials, J. D. Landes, J. A. Begley, and G. A. Clarke, Eds., 1979, pp. 37–64.

[10] McMeeking, R. M. and Parks, D. M. in *Elastic-Plastic Fracture, ASTM STP 668*, American Society for Testing and Materials, J. D. Landes, J. A. Begley, and G. A. Clarke, Eds., 1979, pp. 175–194.
[11] Shih, C. F., deLorenzi, H. G., and Andrews, W. R. in *Elastic-Plastic Fracture, ASTM STP 668*, American Society for Testing and Materials, J. D. Landes, J. A. Begley, and G. A. Clarke, Eds., 1979, pp. 65–120.
[12] Shih, C. F., Andrews, W. R., deLorenzi, H. G., German, M. D., Van Stone, R. H., and Mowbray, D. F., "Methodology for Plastic Fracture," Final Report No. EPRI-NP-1735, Research Project 601-2, General Electric Co., Schenectady, N.Y., March 1981.
[13] Kanninen, M. F. et al, "Development of a Plastic Fracture Methodology," Final Report No. EPRI NP-1734, Research Project 601-1, Battelle Columbus Laboratories, Columbus, Ohio, March 1981.
[14] Shih, C. F. and German, M. D., *International Journal of Fracture*, Vol. 17, No. 1, 1981, pp. 27–43.
[15] Shih, C. F., *Journal of the Mechanics and Physics of Solids*, to appear.
[16] Chitaley, A. D. and McClintock, F. A., *Journal of the Mechanics and Physics of Solids*, Vol. 19, 1971, pp. 147–163.
[17] Amazigo, J. C. and Hutchinson, J. W., *Journal of the Mechanics and Physics of Solids*, Vol. 25, 1977, pp. 81–97.
[18] Rice, J. R., Drugan, W. J., and Sham, T. L. in *Fracture Mechanics, ASTM STP 700*, American Society for Testing and Materials, 1980, pp. 189–221.
[19] Dean, R. H. and Hutchinson, J. R. in *Fracture Mechanics, ASTM STP 700*, American Society for Testing and Materials, 1980, pp. 383–405.
[20] Wells, A. A. in *Crack Propagation Symposium Proceedings*, Cranfield College of Aeronautics, Vol. 1, 1961, p. 210.
[21] Kumar, V., German, M. D., and Shih, C. F., "An Engineering Approach for Elastic-Plastic Fracture Analysis," Topical Report No. EPRI-NP-1931, Research Project 1237-1, General Electric Co., Schenectady, N.Y., July 1981.
[22] Tada, H., Paris, P. C., and Irwin, G. R., *The Stress Analysis of Cracks Handbook*, Del Research Corp., Hellertown, Pa., 1973.
[23] Rooke, D. P. and Cartwright, J. C., *Compendium of Stress Intensity Factors*, Her Majesty's Stationery Office, London, 1976.
[24] Goldman, N. L. and Hutchinson, J. W., *International Journal of Solids and Structures*, Vol. 11, 1975, pp. 575–591.
[25] Shih, C. F. in *Mechanics of Crack Growth, ASTM STP 590*, American Society for Testing and Materials, 1976, pp. 3–22.
[26] Shih, C. F. and Hutchinson, J. R., *Transactions*, American Society of Mechanical Engineers, *Journal of Engineering Materials and Technology*, Vol. 98, No. 4, Oct. 1976, pp. 289–295.
[27] Hutchinson, J. W., Needleman, A., and Shih, C. F. in *Fracture Mechanics*, N. Perrone et al, Eds., University Press of Virginia, Charlottesville, Va., 1978, pp. 515–528.
[28] Yagawa, G., Aizawa, T., and Ando, Y. in *Fracture Mechanics, ASTM STP 700*, American Society for Testing and Materials, 1980, pp. 439–452.
[29] Ranaweera, M. P. and Leckie, F. A. in *Numerical Methods in Fracture Mechanics*, A. R. Luxmoore and D. R. J. Owen, Eds., University College of Wales, Swansea, U.K., 1978, pp. 450–463.
[30] Parks, D. M. in *Numerical Methods in Fracture Mechanics*, A. R. Luxmoore and D. R. J. Owen, Eds., University College of Wales, Swansea, U.K., 1978, pp. 464–478.
[31] Kumar, V. and Shih, C. F. in *Fracture Mechanics, ASTM STP 700*, American Society for Testing and Materials, 1980, pp. 406–438.
[32] Shih, C. F. and Kumar, V., "Estimation Technique for the Prediction of Elastic-Plastic Fracture of Structural Components of Nuclear Systems," 1st Semiannual Report to EPRI, Contract No. RP1237-1, General Electric Co., Schenectady, N.Y., 1 July 1978–31 Jan. 1979.
[33] Kumar, V., German, M. D., and Shih, C. F., "Estimation Technique for the Prediction of Elastic-Plastic Fracture of Structural Components of Nuclear Systems," Combined 2nd and 3rd Semiannual Reports to EPRI, Contract No. RP1237-1, General Electric Co., Schenectady, N.Y., 1 Feb. 1979–31 Jan. 1980.
[34] Shih, C. F., German, M. D., and Kumar, V., *International Journal of Pressure Vessels and Piping*, Vol. 9, 1981, pp. 159–196.

[35] Buchalet, C. B. and Bamford, W. H. in *Mechanics of Crack Growth, ASTM STP 590*, American Society for Testing and Materials, 1976, pp. 385–402.
[36] Labbens, R., Pellissier-Tannon, A., and Heliot, J. in *Mechanics of Crack Growth, ASTM STP 590*, American Society for Testing and Materials, 1976, pp. 368–384.
[37] Ilyushin, A. A. in *Prikadnaia Matematika i Mekhanika*, Vol. 10, 1946, p. 347.
[38] Parks, D. M., Kumar, V., and Shih, C. F., this publication, pp. I-370–I-383.
[39] *Fracture Toughness Evaluation by R-Curve Methods, ASTM STP 527*, American Society for Testing and Materials, 1973.
[40] Needleman, A. and Shih, C. F., *Computer Methods in Applied Mechanical Engineering*, 1978, pp. 223–240.
[41] Kumar, V., deLorenzi, H. G., Andrews, W. R., Shih, C. F., German, M. D., and Mowbray, D. F., "Estimation Technique for the Prediction of Elastic-Plastic Fracture of Structural Components of Nuclear Systems," 4th Semiannual Report to EPRI, Contract No. RP1237-1, General Electric Co., Schenectady, N.Y., 1 July 1980–31 Jan. 1981.
[42] Karlsson, A. and Backlund, J., *International Journal of Fracture*, Vol. 14, 1978, pp. R311–R314.
[43] Bowie, O. L. and Freese, C. E., *Engineering Fracture Mechanics*, Vol. 4, 1972, pp. 315–321.
[44] Astiz, M. A., Elices, M., and Sanchez Galvez, V. in *Fracture 1977, Proceedings*, Fourth International Conference on Fracture, D. M. R. Taplin, Ed., Waterloo, Ont., Canada, 19–24 June 1977, pp. 395–400.
[45] Bergkvist, H. and Lan Huang, G.-L., *International Journal of Fracture*, Vol. 13, 1977, pp. 556–558.
[46] deLorenzi, H. G., *Journal of Computer Structures*, Special Issue, June 1981.
[47] Parks, D. M., *Computer Methods in Applied Mechanical Engineering*, Vol. 12, 1977, p. 353.
[48] deLorenzi, H. G., "Finite Element Implementation of the Virtual Crack Extension Method," General Electric Co. TIS Report, Schenectady, N.Y., to be published.
[49] Sneddon, I. N. and Lowengrub, M., *Crack Problems in Classical Theory of Elasticity*, Wiley, New York, 1969.
[50] Shih, C. F., Kumar, V., and German, M. D., this publication, pp. II-239–II-261.
[51] German, M. D., Andrews, W. R., Kumar, V., Shih, C. F., deLorenzi, H. G., and Mowbray, D. F., "Elastic-Plastic Fracture Analysis of Flawed Stainless Steel Pipes," Final Report to EPRI, Research Project T118-8, General Electric Co., Schenectady, N.Y., Nov. 1981.
[52] German, M. D. and Kumar, V. in *Aspects of Fracture Mechanics in Pressure Vessels and Piping*, ASME Publication No. PVP-Vol. 58, American Society of Mechanical Engineers, 1982, pp. 109–142.
[53] Hutchinson, J. W., private communication, 1982.
[54] Silverns, M. J. and Price, A. T., *International Journal of Fracture*, No. 2, 1973, pp. 193–207.
[55] Harrison, C. B. and Sandor, G. N., *Engineering Fracture Mechanics*, Vol. 3, No. 4, 1971, pp. 403–420.
[56] Haigh, J. R., *Materials Sciences and Engineering*, Vol. 20, 1975, pp. 213–223.
[57] McEvilly, A. J. and Wells, C. H. in *Proceedings*, International Conference on Creep and Fatigue in Elevated Temperature Applications, Institution of Mechanical Engineers, U.K., 1975.
[58] Landes, J. D. and Begley, J. A. in *Mechanics of Crack Growth, ASTM STP 590*, American Society for Testing and Materials, 1976, pp. 128–148.
[59] Nikbin, K. M. and Webster, G. A. in *Cracks and Fracture, ASTM STP 601*, American Society for Testing and Materials, 1976, pp. 47–62.
[60] Sandonanda, K. and Shahinian, P., *Engineering Fracture Mechanics*, Vol. 15, Nos. 3–4, 1981, pp. 327–342.
[61] Riedel, H. and Rice, J. R. in *Fracture Mechanics, ASTM STP 700*, American Society for Testing and Materials, 1980, pp. 112–130.
[62] Riedel, H., *Journal of the Mechanics and Physics of Solids*, to appear.
[63] Riedel, H., this publication, pp. I-505–I-520.

Genki Yagawa[1] *and Tatsuhiko Aizawa*[2]

A Superposition Method for Nonlinear Crack Problems

REFERENCE: Yagawa, G. and Aizawa, T., **"A Superposition Method for Nonlinear Crack Problems,"** *Elastic-Plastic Fracture: Second Symposium, Volume I—Inelastic Crack Analysis, ASTM STP 803*, C. F. Shih and J. P. Gudas, Eds., American Society for Testing and Materials, 1983, pp. I-354–I-369.

ABSTRACT: The finite-element evaluation of nonlinear cracks in power-hardening materials involves some numerical difficulties: the incompressibility in the plane-strain condition and the representation of the nonlinear crack singularity. In the present paper, the eigenmodes of Hutchinson-Rice-Rosengren (HRR) singularity at the nonlinear crack tip are obtained numerically by the finite-element eigenfunction calculation. Superposing the singularity functions thus obtained on the finite-element displacement, the J-integral values for various structures are determined directly from a coefficient of the superposed function. As for the numerical examples, we take the center-cracked plate, the double-edge-cracked plate, and the single-edge-cracked plate. Several comparison studies are made between the present and the other relevant results.

KEY WORDS: crack analysis, finite-element method, power hardening material, Hutchinson-Rice-Rosengren (HRR) singularity, penalty function, superposition method, incompressibility, numerical eigenfunction calculation, J-integral value, crack-opening displacement, residual load-point displacement, elastic-plastic fracture

The progress of linear fracture mechanics and finite elements [*1,2*][3] in the last decade has enabled many engineers to estimate potential crack sensitivity, calculate the stress-intensity factor of existing cracks, and evaluate the subcritical crack growth in various structural components. The recent requirement for more intimate safety analysis about structural integrity consists in the evaluation of cracks with consideration of material nonlinearity, because ductile materials, such as A533B and some types of stainless steels, are used in the structural members [*3*]. Among several approaches for

[1]Associate professor, Department of Nuclear Engineering, Faculty of Engineering, University of Tokyo, Hongo, Bunkyo-ku, Tokyo 113, Japan.
[2]Research associate, Department of Fundamental Science, University of Tokyo, Komaba, Meguro-ku, Tokyo 153, Japan.
[3]The italic numbers in brackets refer to the list of references appended to this paper.

nonlinear crack analysis, the assumption of proportional loading and deformation theory of plasticity is known to play the important role [4]. According to this assumption, a specific dominant singularity of nonlinear cracks is obtained: Hutchinson-Rice-Rosengren (HRR) singularity [5–7]. Corresponding to this type of singularity, some nonlinear fracture parameters, including the J-integral value, can be defined. Though these parameters may be obtained by evaluating nonlinear cracks, analysts face some numerical difficulties: the consideration of incompressibility of plastic or creep strain and the representation of nonlinear crack singularity.

It is well recognized that the penalty function method is effective in solving various boundary-value problems [8]. The present authors [9] applied the method with the finite elements and the superposition techniques to analysis of the nonlinear fracture parameters for the plane-stress problem. Also, they extended the method to solve the incompressible nonlinear cracked plates in the plane-strain condition [10], where the new penalty term was added to the original functional to take account of the incompressibility. In the present paper, we further the development of the method by selecting the approximate function which is obtained through the numerical eigenfunction analysis. As for the numerical examples, we take the center-cracked plate, double-edge-cracked plate and single-edge-cracked plate. Some comparison studies are made between the present and other relevant results [11–14].

Theoretical Background

Nonlinear analysis of cracked plates of power-hardening material under plane strain and incompressibility requires the special formulation of the finite-element method. For simplicity, the discussion in this section is limited to the linear crack problems. If Poisson's ratio ν is less than 0.5, the usual elastic constitutive equations and the potential functional are given as follows

$$\epsilon_{ij} = \frac{1+\nu}{E}\sigma'_{ij} + \frac{1}{E}(1-2\nu)\,\sigma_m \tag{1}$$

$$\sigma_m = \frac{E}{3(1-2\nu)}(\epsilon_{11}+\epsilon_{22}) \tag{2}$$

$$\Pi = \frac{1}{2}\int_V \sigma'_{ij}\,\epsilon_{ij} dV - \int_{S_\sigma} \bar{p}_i u_i dS + \int_V \sigma_m(\epsilon_{11}+\epsilon_{22})dV \tag{3}$$

where

E = Young's modulus,

σ'_{ij}, ϵ'_{ij}, u_i = deviatoric stress, strain tensor, and the displacement vector, respectively,

σ_m = mean stress, and
$\overline{p}_i$ = prescribed load vector.

As Poisson's ratio approaches 0.5, Eqs 1, 2, and 3 become, respectively:

$$\epsilon_{ij} = \frac{3}{2E} \sigma'_{ij} \tag{4}$$

$$\epsilon_{11} + \epsilon_{22} = 0 \tag{5}$$

$$\Pi(u_i, \sigma_m) = \frac{E}{2} \int_V \overline{\epsilon}^2 (u_i) dV - \int_{S_\sigma} \overline{p}_i u_i dS + \int_V \sigma_m(\epsilon_{11} + \epsilon_{22}) dV \tag{6}$$

where $\overline{\epsilon}$ is the equivalent strain, Eq 4 becomes the multiaxial stress-strain law due to the J_2 deformation theory in linear elasticity, and Eq 5 is the so-called incompressibility condition.

In view of these two groups of equations, we notice two important features:

1. Mean stress should keep proportional to the dilatation, $\epsilon_{11} + \epsilon_{22}$, even when the Poisson's ratio is near 0.5.
2. The solutions (u_i, σ_m), minimizing Eq 3, coincide with those in Eq 6 as Poisson's ratio approaches 0.5.

Suppose we formulate mixed finite elements in a conventional way, using Eqs 4, 5, and 6, and assuming that u_i and σ_m are approximated by the same interpolation functions. In this case, we might encounter with the numerical difficulty of approximation because of lack of consideration of the No. 1 requirement of the foregoing [*15*]. According to the theory of mixed variational principle [*15*], there are two ways to avoid this difficulty:

1. Match the degrees of u_i and σ_m interpolations; for example, u_i is quadratic and σ_m is linear.
2. Satisfy the incompressibility condition, $\epsilon_{11} + \epsilon_{22} = 0$, at selected points in an element, instead of pointwise satisfaction; for example

$$\int_{V_e} (\epsilon_{11} + \epsilon_{22}) dV = 0 \tag{7}$$

$$\epsilon_{11} + \epsilon_{22} = 0 \quad \text{at Point } C \tag{8}$$

where $\int_{V_e}$ is the integral in an element and Point C denotes the central point of the element.

The second assumption, No. 2, leads to the direct representation of σ_m by u_i as follows

$$\sigma_m = \frac{E}{3(1-2\nu)} \int_{V_e} \left(\frac{\partial u_1}{\partial x_1} + \frac{\partial u_2}{\partial x_2}\right) dV \tag{9}$$

$$\sigma_m = \frac{E}{3(1-2\nu)}\left(\frac{\partial u_1}{\partial x_1} + \frac{\partial u_2}{\partial x_2}\right) \quad \text{at Point } C \tag{10}$$

where (x_1, x_2) is the cartesian coordinate. Further, substituting Eq 9 into Eq 6, we have the following functional to be minimized

$$\Pi(u_i) = \sum_{n=1}^{N_e}\left\{\frac{E}{2}\int_{V_{e(n)}} \bar{\epsilon}^2(u_i)dV + \frac{\beta}{2}\left[\int_{V_{e(n)}}\left(\frac{\partial u_1}{\partial x_1} + \frac{\partial u_2}{\partial x_2}\right)dV\right]^2\right\} - \int_{S_\sigma} \bar{p}_i u_i dS \tag{11}$$

where N_e is the number of elements and $\beta = 2E/[3(1-2\nu)]$ is the penalty number.

When the crack analysis of an incompressible plate is made with use of the aforementioned finite elements, the analytical trial functions are superposed at the crack tip to represent the singularity, in the following manner

$$u_i = u_i^F + u_i^A \tag{12}$$

Here, u_i^F and u_i^A denote the finite element and analytical displacement vector, respectively. The general form of analytical functions to be employed is given by

$$u_1^A = \sqrt{r}\sum_{m=1}^{N} a_m F_u^m(\theta), \qquad u_2^A = \sqrt{r}\sum_{m=1}^{N} a_m F_v^m(\theta) \tag{13}$$

where

N = number of terms,
$a_m\,(m = 1, \ldots, N)$ = unknown coefficients, and
(r, θ) = polar coordinates with the origin at the crack tip.

Considering the constraints, $u_2^A = 0$, on OC and

$$\frac{\partial u_1^A}{\partial x_1} + \frac{\partial u_2^A}{\partial x_2} = 0, \quad \text{in } V$$

we can select the following analytical functions for the cracked plates, which are shown in Fig. 1

$$N = 1; \qquad F_u^1 = \sum_{m=1}^{M} \tilde{a}_m \left\{\left(\frac{1}{2} - \frac{m}{6}\right)\cos\left(\frac{m}{2}+1\right)\theta\right.$$

$$
\begin{aligned}
&\qquad -\left(\frac{m}{6}+\frac{1}{2}\right)\cos\left(\frac{m}{2}-1\right)\theta\Bigg\} \\
&F_v{}^1 = \sum_{m=1}^{M} \tilde{a}_m \left\{\left(\frac{1}{2}-\frac{m}{6}\right)\sin\left(\frac{m}{2}+1\right)\theta\right. \\
&\qquad \left. +\left(\frac{m}{6}+\frac{1}{2}\right)\sin\left(\frac{m}{2}-1\right)\theta\right\}
\end{aligned}
\tag{14}
$$

In order to determine the coefficients $\tilde{a}_m (m = 1, \ldots, M)$, we have made a finite-element eigenfunction analysis [17], and after that, using the obtained numerical results, have fitted the curve, Eq 14, through the method of least squares. The employment of Eq 14 in the present method has such a good feature that the stress-intensity factor, K, can be obtained from the unknown amplitude a_1 by

$$K = \frac{E}{3}\sqrt{2\pi}\, a_1 \tag{15}$$

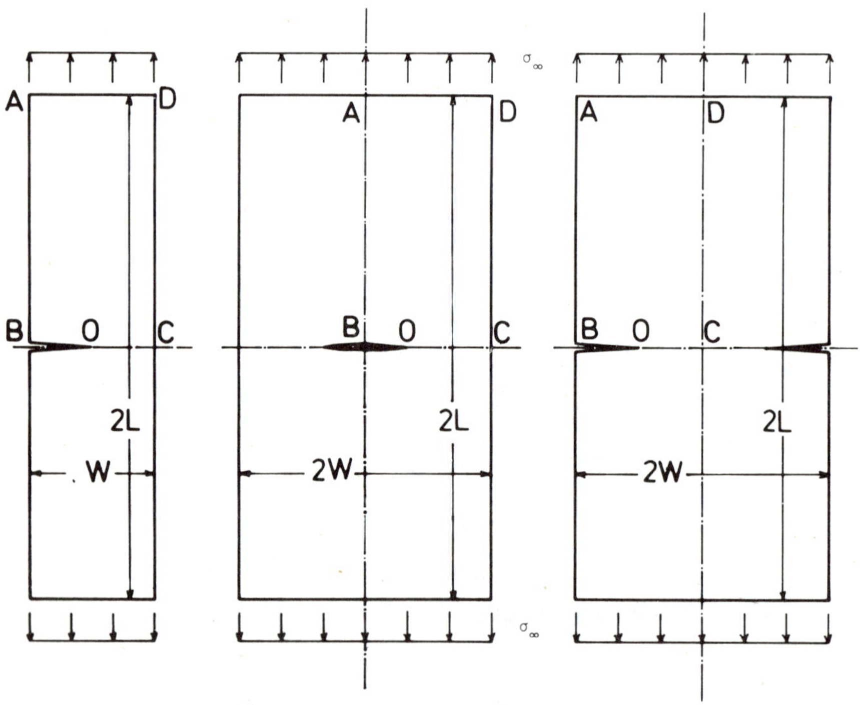

FIG. 1—*Three types of typical cracked specimens: single-edge-cracked, center-cracked, and double-edge-cracked plates.*

Once the finite elements for incompressibility and the analytical trial functions are determined, we can perform the crack analysis using the following functional

$$\Pi(u_i^F, a_1) = \sum_{n=1}^{N_e} \left\{ \frac{E}{2} \int_{V_{e(n)}} \bar{\epsilon}^2 (u_i) dV + \frac{\beta}{2} \left[\int_{V_{e(n)}} \left(\frac{\partial u_1^F}{\partial x_1} + \frac{\partial u_2^F}{\partial x_2} \right) dV \right]^2 \right\}$$

$$+ \frac{\alpha}{2} \int_{S_u} (u_1^F + u_1^A)^2 dS - \int_{S_\sigma} \bar{p}_i u_i dS \qquad (16)$$

where α and β are the penalty numbers, and the third term on the right-hand side is necessary to consider the essential boundary conditions on S_u: $u_1 = 0$ on AB for a center-cracked plate or $u_1 = 0$ on CD for a double-edge-cracked plate, as depicted in Fig. 1.

The extension of the present method to nonlinear crack analysis needs no more complicated formulations except the analytical representation of nonlinear singularity and the iterative procedure to consider the material nonlinearity. Here, we assume that the multi-axial constitutive equation can be written through the J_2 deformation theory as follows

$$\epsilon_{ij} = \frac{3}{2} \kappa \bar{\sigma}^{n-1} \sigma'_{ij} \qquad (17)$$

where κ is the proportional coefficient and n is the strain hardening index. For the power-hardening law material, the displacement at the near-crack tip is dominated by the following HRR singularity

$$u_i^A = \left(\frac{J}{I_n} \right)^{n/(n+1)} r^{1/(n+1)} \tilde{u}_i(\theta) \qquad (18)$$

where

I_n = constant value, which is dependent only on n,
$\tilde{u}_i(\theta)$ = eigenfunction at the crack tip, and
J = the J-integral value.

The approximate analytical functions for Eq 18 are given by

$$u_1^A = r^{1/(n+1)} \sum_{m=1}^{N} a_m F_u^m(\theta), \qquad u_2^A = r^{1/(n+1)} \sum_{m=1}^{N} a_m F_v^m(\theta) \qquad (19)$$

where $F_u^m(\theta)$ and $F_v^m(\theta)$ are assumed as follows

$$N = 1; \qquad F_u^1 = \sum_{m=1}^{M} \tilde{a}_m \left\{ \left(\frac{1}{2} - \frac{m(n+1)}{4(n+2)} \right) \cos \left(\frac{m}{2} + 1 \right) \theta \right.$$

$$
\begin{aligned}
&\quad - \left(\frac{1}{2} + \frac{m(n+1)}{4(n+2)}\right) \cos\left(\frac{m}{2} - 1\right)\theta\Bigg\} \\
F_v{}^1 &= \sum_{m=1}^{M} \tilde{a}_m \left\{\left(\frac{1}{2} - \frac{m(n+1)}{4(n+2)}\right) \sin\left(\frac{m}{2} + 1\right)\theta \right. \\
&\quad \left. + \left(\frac{1}{2} + \frac{m(n+1)}{4(n+2)}\right) \sin\left(\frac{m}{2} - 1\right)\theta\right\}
\end{aligned} \tag{20}
$$

Here, the coefficients $\tilde{a}_m$ ($m = 1, \ldots, M$) have also been determined by the finite-element eigenfunction analysis [*17*] and curve-fitting technique. When Equation 19 is employed, the J-integral value can be obtained directly, without any contour integral calculations [*18*], from the unknown coefficient a_1 by

$$
J = I_n(a_1)^{(n+1)/n} \tag{21}
$$

With the use of the previously stated incompressible finite-element formulations and the trial functions, Eq 19, the nonlinear crack problems are solved by minimizing the functional

$$
\begin{aligned}
\Pi(u_i{}^F, a_1) = \sum_{m=1}^{N_e} &\left\{\frac{n}{n+1}\kappa^{-1/n} \int_{V_{e(m)}} \bar{\epsilon}^{(n+1)/n}(u_i) dV \right. \\
&\left. + \frac{\beta}{2}\left[\int_{V_{e(m)}} \left(\frac{\partial u_1{}^F}{\partial x_1} + \frac{\partial u_2{}^F}{\partial x_2}\right) dV\right]^2\right\} \\
&+ \frac{\alpha}{2}\int_{S_u} (u_1{}^F + u_1{}^A)^2 dS - \int_{S_\sigma} \bar{p}_i u_i dS
\end{aligned} \tag{22}
$$

Using the strain nodal displacement relation and the stress-strain law, Eq 17, Eq 22 can be changed to be a nonlinear function of the discretized displacement d^F and the analytical coefficient, a_1.

The resulting functional is minimized with respect to d^F and a_1, using the Ritz method, which yields the following stiffness equation

$$
([K] + [K_P{}^\alpha] + [K_P{}^\beta]) \begin{Bmatrix} d^F \\ a_1 \end{Bmatrix} = \{f\} \tag{23}
$$

where

$[K_P{}^\beta]$, $[K_P{}^\alpha]$ = pseudo-stiffness matrices resulting from the second and third terms on the right-hand side of Eq 22, respectively,

$\{f\}$ = force vector from the fourth term, and

$[K]$ = nonlinear stiffness matrix with respect to d^F and a_1; that is

$$[K] = \frac{2}{3}\kappa^{-1/n} \sum_{m=1}^{N_e} \int_{V_{e(m)}} \bar{\epsilon}(d^F, a_1)^{(1-n)/n}[B]^t[B]dV \quad (24)$$

where $[B]$ is the strain-displacement matrix. Though several methods are studied to solve the preceding kinds of nonlinear equations, we adopt the simple iteration method. As the difficulty comes from the variable factor $\bar{\epsilon}(d^F, a_1)^{(1-n)/n}$, we replace this factor by a unit in the first step. In the following steps, we calculate $\bar{\epsilon}(d^F, a_1)$ using the solutions in the previous step, and substitute these into Eq 24. The newly built stiffness matrix $[K]$ being constant, the linearized stiffness equations can be solved in the conventional way. The convergence rate of this method is not so fast as the Newton method, but the numerical stability might be improved for the various power exponents.

Some Preliminary Tests on Accuracy and Convergence

Some tests are performed in the linear or nonlinear analysis of cracked plates as shown in Fig. 1. In the first place, we examine the influences of the mesh arrangement and the penalty number β in Eq 16 on the accuracy of the nondimensional stress-intensity factor f $(=K/\sigma_\infty\sqrt{\pi a})$ through the linear crack analysis of a single-edge-cracked plate, as follows

$$L/W = 1.0, \qquad a/W = 0.5, \qquad \alpha = 0.0, \qquad \kappa = \sigma_\infty = 1.0$$

As can be seen from Table 1, even with such a fine mesh arrangement as 10 × 10, the value of f cannot converge to the exact one [*19*] if β is less than 10^2. On the other hand, with β kept 10^5 and varying the mesh arrangement from 2 × 2 to 10 × 10, the discrepancy between the numerical and the exact values diminishes monotonically to less than 1.0 percent. In the case of

TABLE 1—*Effect of the number of degrees of freedom and penalty number* β *on the nondimensional stress-intensity factor* f.

Nondimensional Stress-Intensity Factor					
Mesh Arrangement	Number of Degrees of Freedom	Penalty Number β			
		10^0	10^2	10^5	10^7
2 × 2	22	3.003	2.492	2.481	2.481
4 × 4	54	3.768	2.830	2.795	2.795
6 × 6	102	4.077	2.880	2.800	2.800
8 × 8	140	4.264	2.951	2.800	2.800
10 × 10	246	4.384	3.042	2.800	2.800
Ref *19*		2.825			

center-cracked or double-edge-cracked plates, the essential condition on AB or CD is considered by the introduction of another penalty number, α. The effect of α on the accuracy is examined through the linear analysis of center-cracked plate as follows

$$L/W = 1.0, \quad a/W = 0.5, \quad \beta = 10^5, \quad \kappa = \sigma_\infty = 1.0, \quad 10 \times 10 \text{ mesh}$$

With increasing α in Eq 16, Fig. 2 shows that the ratio, $R(\alpha)$, of the difference of the nondimensional stress-intensity factor between the present result and the exact solution [*20*] to the exact one can be settled within the discretized error; that is, 1 percent for the mesh arrangement as 10×10.

In the foregoing, one half of a single-edge-cracked plate and one quarter of center-cracked plate are analyzed by using 4-node quadrilateral elements as the finite-element displacement u_i^F superposed on u_i^A.

Next, the convergence in the simple iteration and the accuracy of the J-integral value are studied for the crack analysis of the work-hardening material of $n = 3$. The geometry of a center-cracked plate and other input data for calculation are given as

$$L/W = 2.0, \qquad a/W = 0.5, \qquad \alpha = \beta = 10^5, \qquad \kappa = \sigma_\infty = 1.0$$

As shown in Table 2, to have the ratio of the difference of the functionals between two consecutive iteration steps to that of the previous step settled

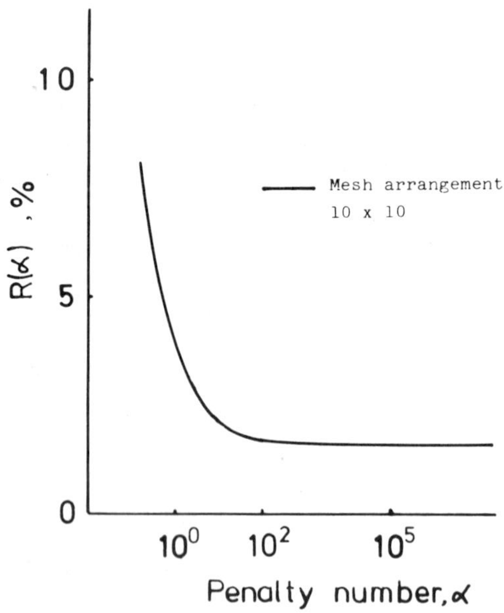

FIG. 2—*Convergence behavior of* R(α) *with increase of the penalty number*, α.

TABLE 2—*Effect of number of degrees of freedom on the nonlinear fracture parameters and comparison with other results.*

Mesh Arrangement	Number of Degrees of Freedom	Singular Amplitude	J-Integral Value	Number of Iterations
6 × 12	186	0.991	5.440	10
8 × 16	310	0.981	5.368	10
12 × 24	654	0.969	5.281	11
Yagawa, Aizawa, and Ando [*10*]			5.218	...
Goldman and Hutchinson [*11*]			5.625	...
Hutchinson, Needleman, and Shih [*13*]			4.905	...

under 1 percent, that is, $|\Pi_i - \Pi_{i-1}|/\Pi_i < 0.01$ with Π_i and Π_{i-1} being the functionals at the ith and $(i - 1)$th steps, respectively, about ten iterations were required. It is also noticed that a comparatively fine-mesh arrangement is necessary to get enough accurate results, and that the J-integral value which is obtained by Eq 21 makes good agreement with that calculated by the contour integral [*18*], while other two results [*11,13*] are higher or lower than the present one.

Application of the Superposition Method to Nonlinear Cracks

The J-integral value, J, the crack opening displacement, δ, and the residual load-point displacement, Δ, are known to be important parameters of fracture mechanics in the cracks of work-hardening materials, and can be determined as the function of geometry of structure, applied load, and material constants. Assuming that the material obeys the power law as defined in Eq 17, we take into account the HRR singularity by forming the trial analytical functions as Eq 19 in the calculation, in order to obtain the fracture parameters with enough accuracy.

According to the dimensional analysis [*12*], we define the nondimensional forms of J, δ, and Δ as

$$J = a\kappa(1 - a/W)h_1(a/W, n)(P/P_0)^{n+1} \tag{25a}$$

$$\delta = a\kappa h_2(a/W, n)(P/P_0)^n \tag{25b}$$

$$\Delta = a\kappa h_3(a/W, n)(P/P_0)^n \tag{25c}$$

where

h_1, h_2, h_3 = nondimensional values of J-integral value, crack-opening displacement, and residual load point displacement, respectively,

P = applied load per thickness, and
P_0 = limit load per thickness which is calculated for each type of specimen as shown in Fig. 1; for example

$$P_0 = 4(W - a)\sigma_0/\sqrt{3} \tag{26}$$

for a center-cracked plate under uniform tension. Here, σ_0 is the yield stress. In the present analysis, the J-integral is calculated directly from the analytical coefficient, a_1, by Eq 21. The crack-opening displacement is obtained from the vertical displacement at Point B in Fig. 1, and the load-point displacement is determined by

$$\Delta = \Delta_{\text{crack}} - \Delta_{\text{no crack}} \tag{27}$$

where Δ_{crack} and $\Delta_{\text{no crack}}$ are the load-point displacements of cracked plate and uncracked one, respectively. Δ_{crack} is calculated commonly for these three types of specimens in Fig. 1 by

$$\Delta_{\text{crack}} = \frac{1}{2W}\int_{-W}^{W}[u_2(x, L) - u_2(x, -L)]dx = \frac{2}{W}\int_{0}^{W} u_2(x, L)dx \tag{28}$$

$\Delta_{\text{no crack}}$ is defined, for example, in case of a center-cracked plate analysis by

$$\Delta_{\text{no crack}} = \sqrt{3}\kappa L\left(\frac{\sqrt{3}P}{4W\sigma_0}\right)^n \tag{29}$$

At the beginning, in order to show the effectiveness of the method, we have performed some calculations with various values of power index, n, on the center-cracked plate as follows

$$a/W = 0.5, \quad L/2 = W = 1.0, \quad \kappa = \sigma_\infty = 1.0, \quad M = 10, \quad \alpha = \beta = 10^5$$

The mesh arrangement is 12 × 24 for one quarter of a center-cracked plate.

Figure 3 shows the comparison of the nondimensional J-integral value, h_1, which is defined by Eq 25*a*, between the present results and the others [*10–12,14*]. As can be seen, the present solutions agree well with those in Refs *10* and *12* when n is lower than 5, but they seem to have some differences with other results, in case n is higher than 5. In Fig. 4 is shown the nondimensional crack-opening displacement, h_2, which is defined by Eq 25*b*. The comparatively good agreement is acknowledged between the present results and the other ones, unless n is too high. In the case of full plasticity, the ratio of residual load-point displacement to crack-opening displacement Δ/δ, becomes 1.0 for any value of a/W, as the power index, n, ap-

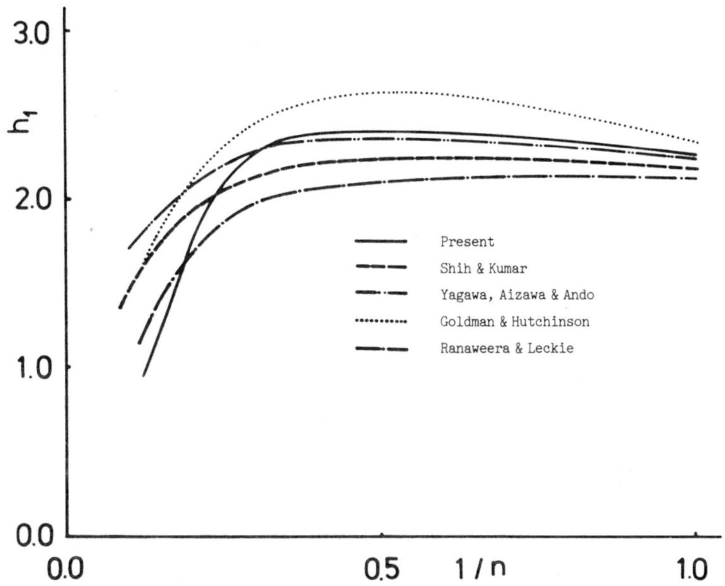

FIG. 3—*Comparison of nondimensional J-integral value,* h_1, *between several numerical methods in the case of a center-cracked plate analysis.*

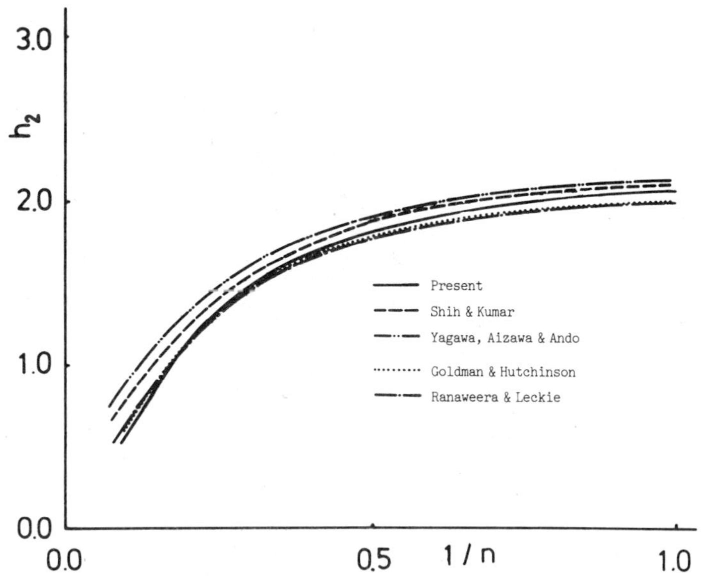

FIG. 4—*Comparison of nondimensional crack-opening displacement,* h_2, *between several numerical methods in the case of a center-cracked plate analysis.*

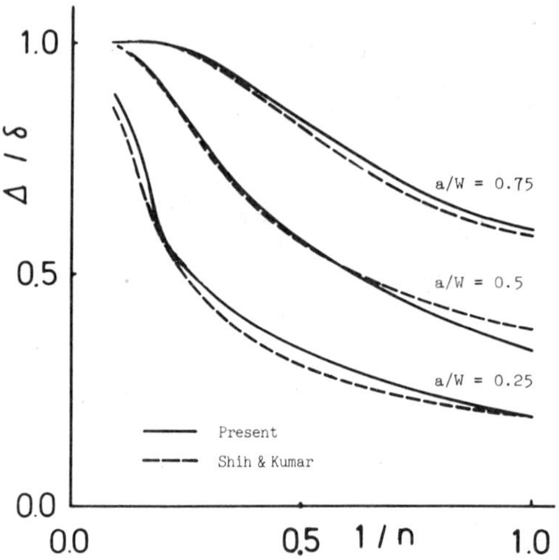

FIG. 5—*Variation of the ratio of residual load-point displacement to crack-opening displacement,* Δ/δ, *as* n *increases.*

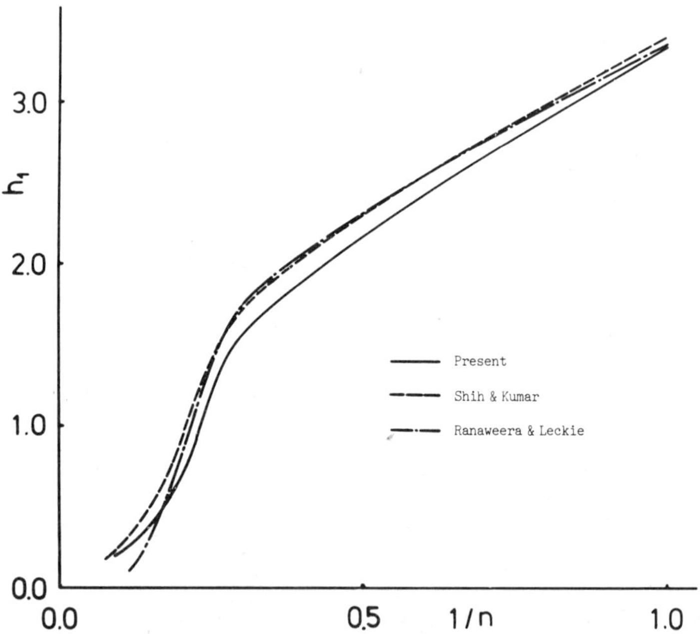

FIG. 6—*Comparison of nondimensional J-integral value,* h_1, *between several numerical methods in the case of a single-edge-cracked plate analysis.*

proaches infinity. As shown in Fig. 5, both the present results and those in Ref *12* reach 1.0 when $1/n$ approaches zero.

Next, we analyze a single-edge-cracked plate under uniform tension as follows:

$$a/W = 0.5, \quad L/2 = W = 1.0, \quad \kappa = \sigma_\infty = 1.0, \quad M = 10, \quad \beta = 10^5$$

The mesh arrangement is 12 × 24 for one half of a single-edge-cracked plate.

Figures 6 and 7 show the comparison between the present results and the others [*12,14*] for h_1 and h_2, respectively. Three solutions are in good agreement when n is less than 5, but, as n increases, the results in Ref *14* seem to become lower than those in Ref *12* and in this paper.

Finally, the following double-edge-cracked plate under uniform tension is analyzed:

$$L/2 = W = 1.0, \quad \kappa = \sigma_\infty = 1.0, \quad M = 10, \quad \alpha = \beta = 10^5$$

The mesh arrangement is 12 × 24 for one quarter of a double-edge-cracked plate.

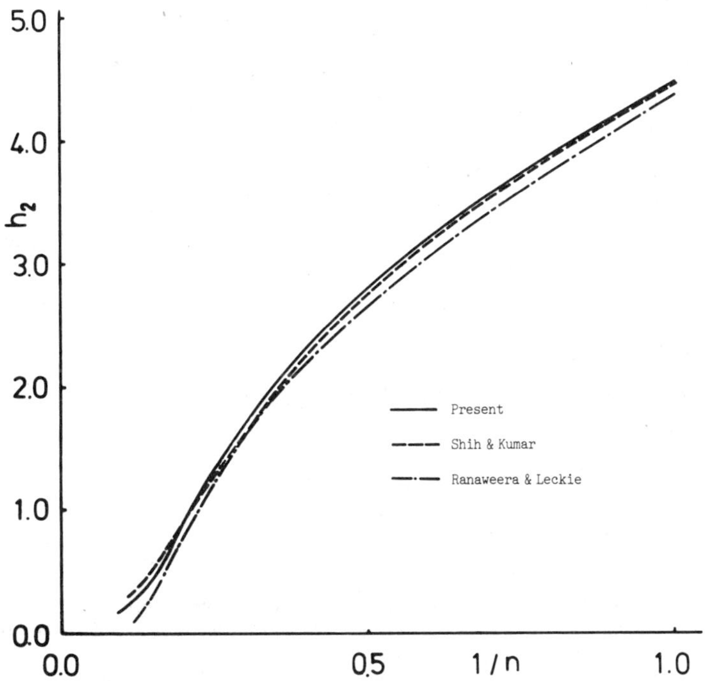

FIG. 7—*Comparison of nondimensional crack-opening displacement,* h_2, *between several numerical methods in the case of a single-edge-cracked plate analysis.*

TABLE 3—*Nondimensional nonlinear fracture parameters* h_1 *and* h_2 *with the values of the strain hardening exponent* n *and the crack length to plate width ratio* a/W.

		n							
		1.0		3.0		5.0		7.0	
a/W		Present	Ref *14*	Present	Ref *14*	Present	Ref *14*	Present	Ref *14*
1/4	h_1	19.33	...	129.0	...	674.1	...	3314.0	...
	h_2	9.600	...	52.45	...	268.3	...	1337.0	...
1/2	h_1	14.50	14.22	53.34	54.45	171.6	172.0	513.7	517.2
	h_2	6.476	6.374	19.66	19.21	58.89	55.27	178.4	156.4
3/4	h_1	11.53	...	23.25	...	40.74	...	66.52	...
	h_2	3.648	...	4.918	...	7.437	...	11.59	...

The nondimensional parameters, h_1 and h_2, are shown in Table 3. Comparing the present results and those in Ref *14* for $a/W = 0.5$, it is found that the present results have enough accuracy even for this type of specimen.

Conclusions

In this paper, a new technique combining the superposition method and the penalty function method is proposed to deal with the incompressible cracked plate in plane strain and full plasticity. With this approach, we could solve various nonlinear crack problems in a straightforward manner, avoiding the numerical difficulties associated with incompressibility as well as singularity at the crack tip. Also, using the approximate eigenfunctions of singularity, the J-integral value is obtained directly from the amplitude of the superposed analytical functions.

Acknowledgment

This work was supported in part by a Grant-in-Aid for Scientific Research from the Ministry of Education.

References

[*1*] Gallagher, R. H. in *Proceedings*, First International Conference on Structural Mechanics in Reactor Technology, Vol. 5, L6/9, West Berlin, 1972.

[*2*] Apostal, M. C., Jordan, S., and Marcal, P. V., "Finite Element Techniques for Posturated Flaws in Shell Structures," EPRI SR-22 Special Report, Electric Power Research Institute, Palo Alto, Calif., 1975.

[*3*] Shih, C. F. et al, "Methodology for Plastic Fracture," 1st through 10th Quarterly Reports to Electric Power Research Institute, Contract No. RP601-2, General Electric Co., Schenectady, N.Y., 1976–1979.

[*4*] Hutchinson, J. W. and Paris, P. C. in *Elastic-Plastic Fracture, ASTM STP 668*, J. D.

Landes, J. A. Begley, and G. A. Clarke, Eds., American Society for Testing and Materials, 1979, pp. 37–64.

[5] Hutchinson, J. W., *Journal of the Mechanics and Physics of Solids*, Vol. 16, No. 4, 1968, pp. 13–31.

[6] Hutchinson, J. W., *Journal of the Mechanics and Physics of Solids*, Vol. 16, No. 4, 1968, pp. 337–347.

[7] Rice, J. R. and Rosengren, G. F., *Journal of the Mechanics and Physics of Solids*, Vol. 16, No. 1, 1968, pp. 1–12.

[8] Zienkiewicz, O. C., *The Finite Element Method in Engineering Science*, 3rd ed., McGraw-Hill, New York, 1977.

[9] Yagawa, G., Aizawa, T., and Ando, Y. in *Fracture Mechanics, ASTM STP 700*, American Society for Testing and Materials, 1980, pp. 439–452.

[10] Yagawa, G., Aizawa, T., and Ando, Y., *International Journal for Numerical Methods in Engineering*, Vol. 17, 1981, pp. 719–733.

[11] Goldman, N. L. and Hutchinson, J. W., *International Journal of Solids and Structures*, Vol. 11, 1975, pp. 575–591.

[12] Shih, C. F. and Kumar, V., "Estimation Technique for the Prediction of Elastic Plastic Fracture of Structural Components of Nuclear Systems," Contract No. RP1237-1, General Electric Co., Schenectady, N.Y., 1978–1979.

[13] Hutchinson, J. W., Needleman, A., and Shih, C. F., "Fully Plastic Crack Problems in Bending and Tension," Report MECH-6, Harvard University, Cambridge, Mass., 1978.

[14] Ranaweera, M. P. and Leckie, F. A., *Computer Methods in Applied Mechanics and Engineering*, Vol. 19, 1979, pp. 367–389.

[15] Malkus, D. S. and Hughes, T. J. R., *Computer Methods in Applied Mechanics and Engineering*, Vol. 15, 1978, pp. 63–81.

[16] Bercovier, M., *Revue Française D'Automatique D'Informatique et Recherche Operationnelle*, Vol. 12, 1978, pp. 211–236.

[17] Aizawa, T., Yagawa, G., and Ando, Y. in *Proceedings*, 15th Conference, Japan Society of Steel Construction, 1981, pp. 131–136 (in Japanese).

[18] Rice, J. R., *Transactions*, American Society of Mechanical Engineers, *Journal of Applied Mechanics*, Vol. 35, 1968, pp. 379–386.

[19] Tada, H., Paris, P., and Irwin, G., *The Stress Analysis of Cracks Handbook*, Del Research Corp., St. Louis, Mo., 1973.

[20] Ishida, M. in *Methods of Analysis and Solution of Crack Problems*, G. C. Shih, Ed., Noordhoff, Leiden, The Netherlands, 1973, pp. 56–130.

David M. Parks,[1] *Virendra Kumar,*[2] *and C. Fong Shih*[3]

Consistency Checks for Power-Law Calibration Functions

REFERENCE: Parks, D. M., Kumar, V., and Shih, C. F., **"Consistency Checks for Power-Law Calibration Functions,"** *Elastic-Plastic Fracture: Second Symposium, Volume I—Inelastic Crack Analysis, ASTM STP 803*, C. F. Shih and J. P. Gudas, Eds., American Society for Testing and Materials, 1983, pp. I-370–I-383.

ABSTRACT: Recently, considerable numerical effort has been devoted to obtaining solutions to crack problems in incompressible J_2 deformation theory materials of pure power-law type ($\epsilon \propto \sigma^n$). Following Il'yushin's observation of the scalability of such solutions, many investigators have conducted broad parametric studies on crack length and material exponent, and have tabulated normalized J and deformation magnitudes. Due to numerical difficulties such as slow convergence, incompressibility, or inadequate finite-element mesh refinement, the solutions do not always agree.

We outline the basis for a very stringent general test of the "quality" of proposed solutions by applying the well-known equivalence of J and energy difference rate with respect to crack length.

The test is applied to data available in the literature, and conclusions are drawn as to their accuracy.

KEY WORDS: J-integral, power-law calibration functions, *J*-estimation procedures, consistency checks, elastic-plastic fracture

Recently, a considerable amount of analytical and computational effort [*1–13*][4] has been directed toward obtaining solutions for cracked bodies of incompressible power-law constitutive behavior. The uniaxial tensile stress-strain response of such materials is taken to be

$$\epsilon/\epsilon_0 = \alpha(\sigma/\sigma_0)^n \tag{1}$$

[1]Associate professor of mechanical engineering, Department of Mechanical Engineering, Massachusetts Institute of Technology, Cambridge, Mass. 02139.

[2]Mechanical engineer, Corporate Research and Development, General Electric Co., Schenectady, N.Y. 12301.

[3]Associate professor, Division of Engineering, Brown University, Providence, R.I. 02912.

[4]The italic numbers in brackets refer to the list of references appended to this paper.

where α, ϵ_0, σ_0, and n are material constants. The generalization of Eq 1 to multiaxial stress states using J_2 deformation theory of plasticity gives

$$\epsilon_{ij}/\epsilon_0 = \frac{3}{2}\alpha\left(\frac{\sigma_e}{\sigma_0}\right)^{n-1}\frac{s_{ij}}{\sigma_0} \tag{2}$$

where s_{ij} is the stress deviator

$$s_{ij} = \sigma_{ij} - \tfrac{1}{3}\,\delta_{ij}\sigma_{kk} \tag{3}$$

and the "effective stress" σ_e is given as

$$\sigma_e = (\tfrac{3}{2}\,s_{ij}s_{ij})^{1/2}$$

For materials characterized by Eq 2, the principal feature of solutions to traction boundary problems involving a single load parameter Q is that they are scalable. This result, due to Il'yushin, can be rationalized by noting that the equilibrium equations, being linear, lead to stress fields proportional to the generalized load Q. From Eq 2, strain is proportional to stress raised to the power n and hence, to Q^n. Finally, with strain as a linear differential operator on displacement field, the displacements are also proportional to Q^n.

In solutions to crack problems in these materials, the asymptotic near-tip fields are given by the Hutchinson-Rice-Rosengren (HRR) singularities [*14,15*], with an amplitude which can be characterized by the value of the J-integral. An examination of the well-known J-integrand (proportional to the product of stress and strain), in conjunction with Il'yushin's results, gives a scalable J, proportional to Q^{n+1}.

These observations have led many workers to undertake broad parametric studies of the material exponent n, crack length, and structural or specimen geometry, cataloguing normalized J-solutions in handbook format [*1*] for subsequent use in J-estimation procedures.

Unfortunately, the normalized solutions of different workers, using different numerical methods, have shown considerable disagreement at times. Until a numerical consensus is established (uniqueness is applicable!), there remains some uncertainty as to which normalized solution is correct.

In this paper we outline a general procedure which can serve as a stringent test of the "quality" of a parametric numerical study of differing crack lengths in any one load configuration and material exponent. The method is based on the well-known equivalence of J and energy differences with respect to crack length.

Formulation

Consider a planar body subject to a generalized force Q per unit thickness with crack length a and remaining ligament c as shown in Fig. 1. Also, let $b =$

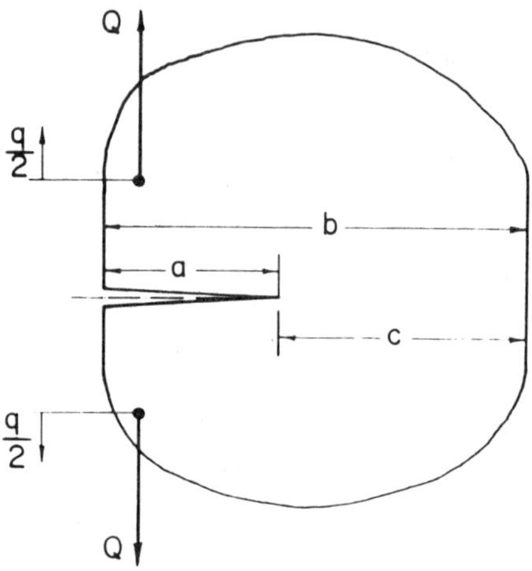

FIG. 1—*Schematic illustration of a planar cracked body subject to generalized force per unit thickness* Q, *with work-conjugate displacement* q. *Crack length is* "a," *remaining ligament is* "c," *and* b = a + c *is the unchanging length of the body in the plane of the crack.*

$a + c$ be the (unchanging) dimension on the plane of the crack. Let the generalized displacement work-conjugate to Q be q. We will let $q = q_c + q_{nc}$ be the sum of "cracked" and "no-crack" parts. Also, we will introduce a generalized reference load for the structure as

$$Q_0 = \sigma_0 L(a,b) \tag{4}$$

where L is a given function of the geometry. In order that the J-normalizations remain reasonably bounded in the near nonhardening limit of large n, it is appropriate to choose L so that Q_0 is the limit load (if known) of the perfectly plastic ($n \to \infty$) limit of Eq 2. Finally, let the relative crack length be $a/b \equiv \xi$.

With these preliminaries, we can generalize the standard normalizations for this specimen as

$$J = \alpha\epsilon_0\sigma_0 c g_1(\xi) h_1(\xi, n)(Q/Q_0)^{n+1} \tag{5}$$

$$q_c = \alpha\epsilon_0 g_3(a, b) h_3(\xi, n)(Q/Q_0)^n \tag{6}$$

Here g_1 and g_3 are some "given" functions of their arguments (g_1 is dimensionless) and h_1 and h_3 are dimensionless functions of their dimensionless arguments which are tabulated at a discrete number of points (ξ, n). If the functions L, g_1, and g_3 have been judiciously chosen, the h_1 and h_3 functions should not vary too rapidly and should reasonably permit interpolation or extrapolation with respect to the tabulated values.

Now consider the energy difference relation indicated in Fig. 2 by the generalized load-displacement curves of neighboring crack lengths a and $a + \Delta a$. The area between the two curves, up to load Q, is $J\Delta a$ so that

$$J = \int_0^Q \left.\frac{\partial q}{\partial a}\right|_Q dQ = \int_0^Q \left.\frac{\partial q_c}{\partial a}\right|_Q dQ \tag{7}$$

where the second integral follows from the additive decomposition of q and the fact that q_{nc} is independent of a.

When the partial derivative of Eq 6 is inserted into Eq 7, there results, after algebraic manipulation

$$J = \alpha\epsilon_0\sigma_0\left(\frac{Q}{Q_0}\right)^{n+1}\left(\frac{L}{n+1}\right)\left\{h_3\frac{\partial g_3}{\partial a} + \frac{g_3}{b}\frac{\partial h_3}{\partial \xi} - \frac{ng_3h_3}{L}\frac{\partial L}{\partial a}\right\} \tag{8}$$

But this is the same J as described by Eq 5, so that common terms can be cancelled, leading to the "consistency relation"

$$cg_1h_1 = \frac{L}{n+1}\left\{h_3\frac{\partial g_3}{\partial a} + \frac{g_3}{b}\frac{\partial h_3}{\partial \xi} - \frac{ng_3h_3}{L}\frac{\partial L}{\partial a}\right\} \tag{9}$$

Equation 9 can be rearranged to solve for the rate of change of h_3 with respect to relative crack length, $\partial h_3/\partial\xi$, as an analytically known linear combination of h_1 and h_3. Now suppose that we plot $h_3(\xi, n)$ versus ξ for fixed n and a discrete

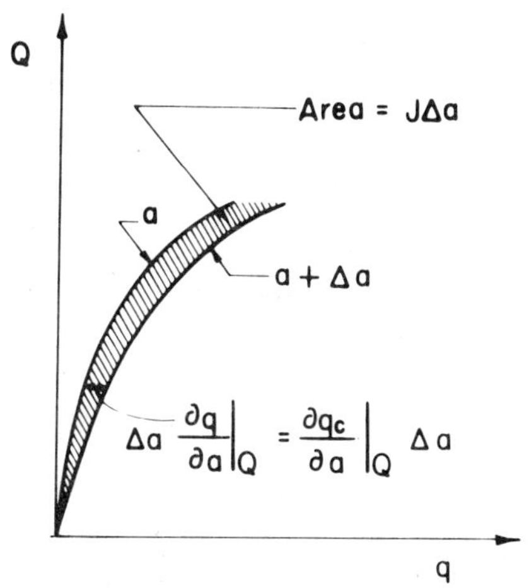

FIG. 2—*Schematic illustration of two generalized load/displacement curves in a power-law material for two bodies containing slightly different crack lengths* a *and* a + Δa. *To first order, the area between the curves is* JΔa.

number of ξ-values, then fit a smooth curve through the data points. We see that Eq 9 permits us to calculate the analytical tangent to this curve in terms of our estimates of h_1 and h_3. If the calculated tangent is indeed close to the slope of our fitted curve, we may term the calculations "consistent." On the other hand, a disagreement of these two tangent measures may be an indication of an inconsistency, involving either errors in h_1 or h_3 or both at the (ξ, n) point in question or errors in h_3 at neighboring relative crack lengths, ξ.

This test, when applied, can indicate that some aspect of the solutions is wrong, but apparently cannot pinpoint the source. Still, it can be a useful check of solutions since it compares a linear combination of h_1 and h_3 from *one mesh geometry* with the variation in h_3 as inferred from *two or more meshes* of varying crack lengths. Furthermore, we will see that a useful application of this consistency check can lead to smoother interpolations (with respect to ξ) of the normalized displacements.

Applications

As a specific application of the test, we examine the plane-strain solutions of the compact specimen reported by Kumar et al [*1*,*5*]. In this example, $Q = P$, the pin load/unit thickness; $q_c = \Delta_L$ is taken as the load line-displacement; and $Q_0 = P_0 = 1.455\sigma_0 c\eta(2a/c)$ is the reference load, with the dimensionless function η given by

$$\eta\left(\frac{2a}{c}\right) = \left[\left(\frac{2a}{c}\right)^2 + 2\left(\frac{2a}{c}\right) + 2\right]^{1/2} - \left[\left(\frac{2a}{c}\right) + 1\right] \tag{10}$$

Kumar and Shih normalize J and Δ_L according to

$$J = \alpha\epsilon_0\sigma_0 c h_1(P/P_0)^{n+1} \tag{11}$$

$$\Delta_L = \alpha\epsilon_0 a h_3(P/P_0)^n \tag{12}$$

giving $g_1 = 1$, $g_3 = a$, and $L = 1.455\,c\eta$. When these terms are inserted in the consistency relation Eq 9, there results

$$\frac{\partial h_3}{\partial \xi} = \frac{1}{\xi}\left\{\left[\frac{n+1}{1.455\eta}\right]h_1 - \left[1 + \frac{\xi n}{1-\xi}\left(1 + \frac{2}{(1-\xi)\eta + 1 + \xi}\right)\right]h_3\right\} \tag{13}$$

In Fig. 3, we plot h_3 versus ξ for n-values of 1, 3, 5, and 10 at the relative crack depths $\xi = 1/4, 3/8, 1/2, 5/8$, and $3/4$. We also calculate the tangents $\partial h_3/\partial\xi$ according to Eq 13 for several of the data points. It is noted that the agreement is generally quite good, and the computed tangents drawn through the

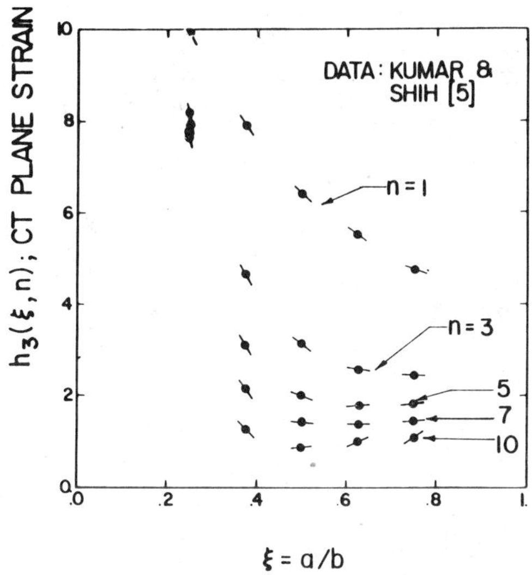

FIG. 3—*Graph of* h_3 *data for plane-strain compact specimen [5] versus relative crack depth* a/b = ξ, *including analytically calculated slope* $\partial h_3/\partial \xi$ *obtained from consistency check.*

h_3 data points are strongly suggestive of the propriety of "smooth" curves connecting neighboring ξ-values. We may note that the current normalization of h_3 gives rise to steep gradients in cases of relatively short cracks and high exponents, and this is borne out both in the relative values of h_3 (0.25, n) and h_3 (0.375, n) and also in the large negative gradients $\partial h_3/\partial \xi$ inferred from Eq 13 at $\xi = 0.25$ for $n = 3$, 5, 7, and 10. These large gradients make interpolation of h_3 somewhat difficult for relative crack depths between 0.25 and 0.375 and suggest that more suitable normalizations could be applied in this regime. However, such short cracks are not of great significance in the compact specimen, and no further consideration of this topic is presented.

A new phenomenon can be observed in Fig. 4, which shows the results of the consistency check when applied to solutions of the plane-strain single-edge-cracked plate (SECP) in tension, reported by Kumar et al [1]. We note that, in this case, $Q = P = \sigma^{\infty} b$ is the tensile load per unit thickness, $q_c = \Delta_c$ is the total "cracked" extension across the ends of the long tensile strip, and $Q_0 = P_0 = 1.455\sigma_0 c\eta$ is the reference load, giving $L = 1.455c\eta$. Here the dimensionless function η is given by

$$\eta(a/c) = [1 + (a/c)^2]^{1/2} - (a/c) \tag{14}$$

The displacement Δ_c and J are given by [1]

$$J = \alpha\epsilon_0\sigma_0 c \cdot \xi \cdot h_1(P/P_0)^{n+1} \tag{15}$$

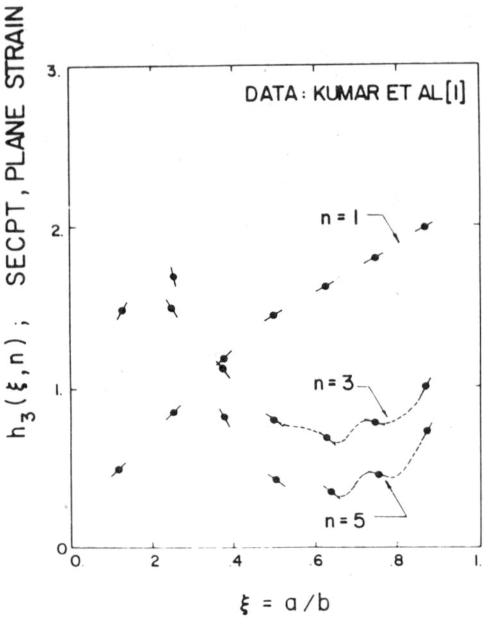

FIG. 4—*Graph of* h_3 *data for plane-strain single-edge-cracked plate (SECP) in tension [1] versus relative crack depth* $a/b = \xi$. *Analytically calculated tangents* $\partial h_3/\partial \xi$ *obtained from consistency check are also shown.*

$$\Delta_c = \alpha \epsilon_0 a h_3 (P/P_0)^n \tag{16}$$

so that $g_1 = \xi$ and $g_3 = a$. When these results are inserted into the general consistency Eq 9, one obtains

$$\frac{\partial h_3}{\partial \xi} = \frac{1}{\xi}\left[\frac{(n+1)}{1.455\eta} h_1 - \left(1 + \frac{n\xi}{(1-\xi)^2}\left[\frac{(1-\xi)(\eta+\gamma)+1}{\eta+\gamma}\right]\right)h_3\right] \tag{17}$$

where $\gamma = a/c = \xi/(1-\xi)$. Figure 4 shows h_3 data for this specimen, as well as the tangent calculated according to Eq 17. For purposes of clarity, only n-values of 1, 3, and 5 are included in Fig. 4. (Note: The h_3 data given in Tables 3–7 of Ref *1* contain some numerical errors. The corrected values are presented here in Table 1.)

The calculated tangents are in good agreement with a consensus smoothed curve through the discrete (h_3, ξ) points for the linear material. By "consensus smooth curve" is meant nothing more than a simple curve which a draftsman might pass through the data points. However, the calculated slopes are rather

TABLE 1—h_3 *functions for plane-strain single-edge notched plate in tension.*

	$n = 1$	$n = 2$	$n = 3$	$n = 5$	$n = 7$	$n = 10$
$a/b = 1/4$	0.833	1.207	1.495	1.664	1.541	1.198
$a/b = 3/8$	1.150	1.248	1.141	0.777	0.487	0.238
$a/b = 1/2$	1.409	1.092	0.799	0.410	0.215	0.0851
$a/b = 5/8$	1.614	0.995	0.670	0.329	0.171	0.0667
$a/b = 3/4$	1.794	1.094	0.765	0.435	0.262	0.125
$a/b = 7/8$	1.968	1.272	0.988	0.713	0.564	0.424

different from "smooth" curves connecting h_3 data points at large ξ-values for the nonlinear materials with exponents $n = 3$ and 5. In fact, the calculated tangents seem to be suggesting an oscillatory behavior in the h_3-function, as indicated schematically by the dotted lines passing through the h_3 data points with a slope calculated from Eq 17.

At this point, however, it should be noted that another, more serious, interpretation of the disagreement in calculated and smoothed/consensus tangents in Fig. 4 could equally well be made. That is, the disagreement in analytically calculated and smoothed/consensus tangents *could* indicate that the original h_1, h_3 data contain numerical errors which lead either to incorrect tangents through Eq 17 or to incorrect placement of the raw h_3 data points or to both, hence affecting our "smooth curve" passing through the data.

Fortunately, we have reason to believe that the data are essentially correct, and that the "true" h_3 functions defined implicitly by Eq 13 may indeed contain the oscillations indicated by the dotted lines in Fig. 4. The reason is shown in Fig. 5 for the $n = 3$ data. Here we show the original h_3 data and calculated tangents $\partial h_3/\partial\xi$ of Fig. 4 as well as the results of consistency checks when new normalizing functions $g_{3j}(a, b)$ and $h_{3j}(\xi, n)$ are introduced into Eq 16 according to

$$\Delta_c = \alpha\epsilon_0 a h_3 (P/P_0)^n \tag{16}$$

$$= \alpha\epsilon_0 a(1 - \xi)^j h_{3j}(P/P_0)^n \tag{16a}$$

where, for this problem

$$g_{3j} = a/(1 - \xi)^j \tag{18}$$

for integer $j = 0, 1, 2 \ldots$. Elimination of common terms between Eq 16 and Eq 16*a* gives

$$h_{3j}(\xi, n) = (1 - \xi)^j h_3(\xi, n) \tag{19}$$

Now, the general consistency check Eq 9 can be applied to the data in terms of the same g_1, h_1, and L-functions as before, but at each point replacing g_3 by

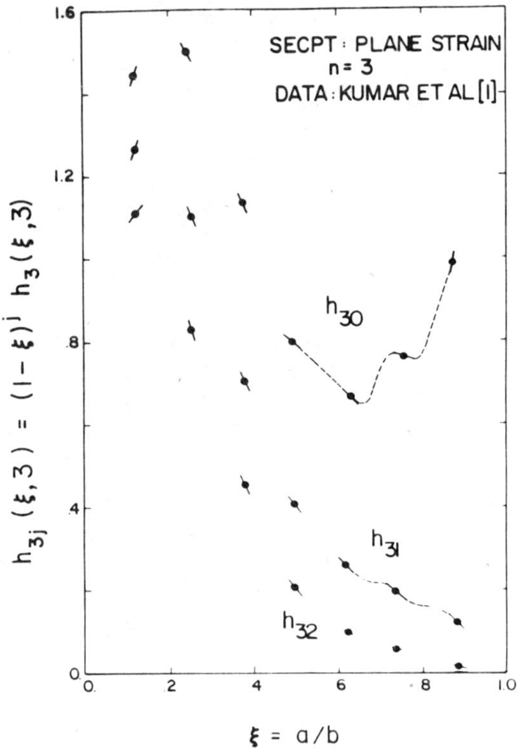

FIG. 5—*Graph of various normalizations* h_{3j} *of crack displacement in the plane-strain SECP tension specimen [1] for material exponent* n = 3, *versus relative crack depth* ξ. *Also shown are calculated tangents* $\partial h_{3j}/\partial\xi$ *obtained from consistency check.*

g_{3j} and h_3 by h_{3j}, as well as in expressions for the derivatives. The curves shown in Fig. 5 thus represent h_{30}, h_{31}, and h_{32} data points and their respective tangents calculated by the consistency condition. As can be seen, this renormalization of the data substantially damps the magnitude of the oscillations, and for $j = 2$ there is quite good agreement between calculated tangents $\partial h_{32}/\partial\xi$ and the slope inferred from a "smooth curve" connecting neighboring values of h_{32}.

It is important to realize that the three curves shown in Fig. 5 are simply three alternative ways of processing the same raw data. If the data from one of the processings, say, for $j = j^*$, look consistent, then all of the data, say, for $j = j^{**}$, are similarly consistent, provided that the smoothed curved $h_{3j^{**}}$ is calculated at *all* ξ-values (not just at the discrete ξ-values of tabulated relative crack lengths) from the smooth (and consistent) curve h_{3j^*}. When this is done, of course, the curves $h_{3j^{**}}$ may exhibit oscillations, but these oscillations will be consistent with the computed oscillations in the slope at the discrete data points.

Indeed, when this procedure was applied to other data which, at first examination, seemed "inconsistent," it was often found that relatively simple renormalizations led to smoother and more consistent curves. Due to constraints on manuscript size, we will not present a complete consistency check analysis of all solutions available, nor even of all solutions contained in Ref *1*. We have processed several of these solutions, however, and found generally that the existing normalizations either give directly a good sense of consistency, as do the plane-strain compact specimen results of Fig. 3, or indicate oscillations which can be reduced to consistency by renormalization of the sort indicated in Figs. 4 and 5.

As a final point on this matter, we consider the possible role of the reference load $Q_0 = \sigma_0 L$ in giving rise to oscillations in straightforwardly normalized h_3 data, by considering two plane-strain geometries for which the limit load of the corresponding rigid/perfectly plastic material is known. Note that since L enters expressions for J and q_c to the powers $n + 1$ and n, respectively, there is some sensitivity to computed h_1 and h_3 values, even if J and q_c are precisely computed. In the case of the plane-strain tension of the center-cracked plate of total width $2b$ and total crack length $2a$, the limit load is [*1,4*]

$$Q_0 = \sigma_0 L(a, b) = \frac{4}{\sqrt{3}} \sigma_0 (b - a) \tag{20}$$

For application of the normalization, the generalized force $Q = P = 2\sigma^{\infty} b$ is the tensile load per unit thickness, $q_c = \Delta_c$ is the cracked relative separation of the strip, and the normalizations used [*1*] are

$$J = \alpha \epsilon_0 \sigma_0 a \cdot (c/b) \cdot h_1 (P/P_0)^{n+1} \tag{21}$$

$$\Delta_c = \alpha \epsilon_0 a h_3 (P/P_0)^n \tag{22}$$

giving $g_3 = a$ and $g_1 = a/b = \xi$. When the consistency check is applied, there results good agreement, as shown in Fig. 6. A tentative conclusion might be that a good (that is, exact) limit load function L could reduce oscillations and seeming inconsistencies.

However, this hypothesis is not borne out by the results shown in Fig. 7 for pure plane-strain bending of an edge-crack strip. In this case, the generalized load is $Q = M$, the bending moment/thickness, $q_c = \theta_c$ is the cracked relative rotation of the ends of the strip, and the limit load, due to Green and Hundy [*16*], is

$$Q_0 = M_0 = 0.364 \sigma_0 (b - a)^2 \tag{23}$$

Normalizations for this specimen are [*6*]

$$J = \alpha \epsilon_0 \sigma_0 c h_1 (M/M_0)^{n+1} \tag{24}$$

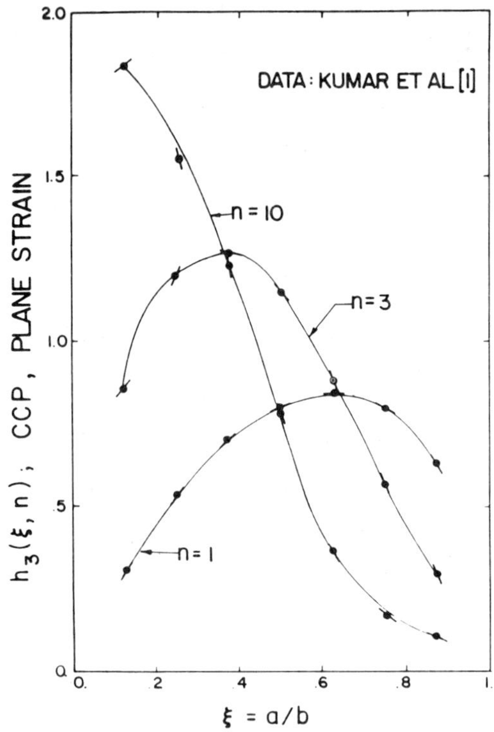

FIG. 6—*Graph of* h_3 *data for plane-strain tension of center-cracked plate [1] versus relative crack depth, including tangents* $\partial h_3/\partial \xi$ *calculated from consistency check.*

$$\theta_c = \alpha \epsilon_0 h_3 (M/M_0)^n \tag{25}$$

giving $g_1 = g_3 = 1$. Plotted in Fig. 7 are the consistency check results

$$\frac{\partial h_3}{\partial \xi} = \frac{1}{1-\xi}\left[\frac{n+1}{0.364}\,h_1 - 2nh_3\right] \tag{26}$$

applied to the (spatially-limited) data of Hutchinson et al [*6*] and to recent work of Sauter and Kumar [*17*] for material exponents $n = 1, 3, 5,$ and 10. For clarity, only h_3 data points and their tangents are drawn. It appears that oscillations persist in the large n, large ξ data even though the limit function L was essentially exact.

We note also that the form of Eq 26 clearly shows how stringent a criterion the consistency check is, even when oscillations persist. Note that both h_1 and h_3, when normalized according to Eqs 24 and 25, are typically of order unity, but for $n = 10$ and $\xi = 0.75$ the coefficients of h_1 and h_3 in Eq 26 are 120.88 and -80, respectively. Hence, even small inaccuracies in h_1 or h_3 or both will

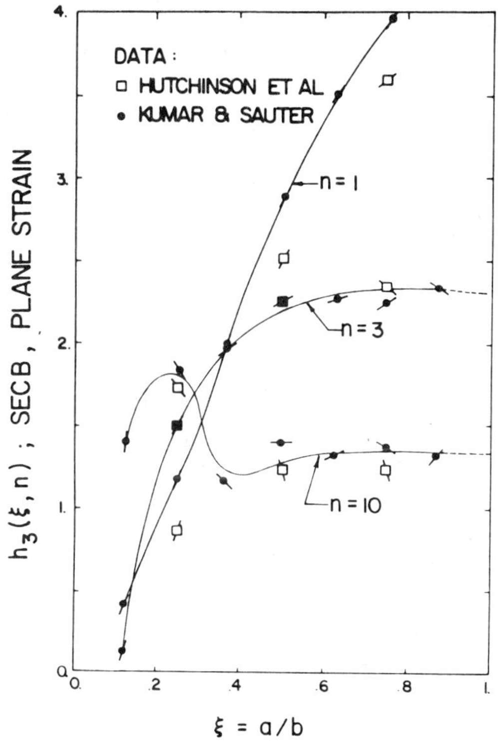

FIG. 7—*Graph of* h_3 *data for plane-strain pure bending of the single-edge cracked specimen [3,17] versus relative crack depth* ξ, *including tangents* $\partial h_3/\partial \xi$ *calculated from consistency checks.*

manifest themselves as "large" contributions to the slope. As a contrast in degree of stringency, we note that in the deep-crack limit ($\xi \rightarrow 1$), where it is known that $(1 - \xi)\partial h_3/\partial \xi \rightarrow 0$, we can rearrange Eq 26 to give

$$1 = \frac{0.728\,n}{(n + 1)} \frac{h_3}{h_1} \tag{27}$$

which is a special form of the consistency check noted by Kumar et al [*18*]. They also noted that the data of Ref *6* for $\xi = 0.75$ and $n = 10$ failed to satisfy Eq 27 by some 10 percent, suggesting errors of up to this amount in either h_1 or h_3. On the other hand, Eq 26 serves to greatly amplify errors of this magnitude, resulting in the very steep calculated slopes shown in Fig. 7.

Summary

We have used the equivalence of energy differencing and J to develop a consistency check for numerical data normalizing J and work-conjugate displacement in planar crack problems of power-law deformation theory materials subject to a single generalized load. The method may be straightforwardly generalized for axisymmetric configurations. The test is a very sensitive indicator of possible numerical errors in the solutions which may be caused by slow convergence, inadequate mesh refinement, computational difficulties associated with incompressibility, or other sources.

In certain of the present applications of the method, the test at first suggested that the numerical data failed to satisfy the analytically necessary conditions. On closer examination, however, it was discovered that at least part of the observed disagreement could be due to an impulse to attribute more smoothness than warranted to a curve passing through a discrete set of data points. When the data were renormalized, the tendency toward inferred oscillation was reduced, resulting in a more "consistent" interpretation. New normalizations of solutions to this class of problems need not be restricted to additional simple polynomial factors in ξ as was done here with g_3. Also, we expect that new normalizations of g_1 and L may prove useful in applying the test as well. One important point to consider, however, is that *any new normalizations must not overwhelm the signal contained in existing tabulations.* For example, if h_3 is bounded as $\xi \rightarrow 1$, then $h_{3j} = (1 - \xi)^j h_3$ will very rapidly go to zero for large j, and so will the slope $\partial h_{3j}/\partial \xi$. "Consistency" is bound to result, regardless of the quality of the data.

As more analytical information becomes available, including asymptotic forms for short [*10,11*] or deep cracks, it is likely that new normalizations of existing tabulated data [*1*] will be suggested. The aim of such new normalizations should be to provide h_1, h_3-values which are everywhere of order unity over their range of exponents and relative crack depths, so that accurate interpolation becomes more likely.

If a given set of numerical solutions is deemed consistent, then the present formulas provide higher-order accuracy in interpolation of compliance data; that is, with both h_3 and $\partial h_3/\partial \xi$ known at two neighboring ξ values, we need not resort to linear interpolation of h_3 for intermediate ξ-values, but can use, for instance, local Hermitian cubic interpolation as in beam finite elements [*19*] to infer work-conjugate deformations with higher accuracy. Such applications may be of use in experimental evaluation of the C^*-parameter for steady-state creep crack growth.

It is suggested that the consistency test be applied to any existing or proposed solutions to problems of this type to serve as a qualitative indicator of their accuracy, and perhaps as a suggestion of more readily useful normalizations of such solutions.

Acknowledgment

The work of D. M. Parks was sponsored by the Electric Power Research Institute (EPRI) under Contract #RP1237-3 with the Massachusetts Institute of

Technology. The work of V. Kumar and C. F. Shih was supported by the General Electric Corp. Research and Development Center. Dr. D. M. Norris, Jr. was the EPRI project administrator.

References

[*1*] Kumar, V., German, M. D., and Shih, C. F., "An Engineering Approach for Elastic-Plastic Fracture Analysis," EPRI Topical Report NP-1931, Electric Power Research Institute, Palo Alto, Calif., July 1981.

[*2*] Shih, C. F. in *Mechanics of Crack Growth, ASTM STP 590*, American Society for Testing and Materials, 1976, pp. 3–22.

[*3*] Shih, C. F. and Hutchinson, J. W., *Transactions*, American Society of Mechanical Engineers, *Journal of Engineering Materials and Technology*, Series H, Vol. 98, 1976, pp. 289–295.

[*4*] Goldman, N. L. and Hutchinson, J. W., *International Journal of Solids and Structures*, Vol. 11, 1975, pp. 575–591.

[*5*] Kumar, V. and Shih, C. F. in *Fracture Mechanics, ASTM STP 700*, American Society for Testing and Materials, 1980, pp. 406–438.

[*6*] Hutchinson, J. W., Needleman, A., and Shih, C. F. in *Fracture Mechanics*, N. Perrone et al., Eds., University Press of Virginia, Charlottesville, 1978, pp. 515–528.

[*7*] Yagawa, G., Aizawa, T., and Ando, Y. in *Fracture Mechanics, ASTM STP 700*, American Society for Testing and Materials, 1980, pp. 439–452.

[*8*] Ranaweera, M. P. and Leckie, F. A. in *Numerical Methods in Fracture Mechanics*, A. R. Luxmoore and D. R. J. Owen, Eds., University College of Wales, Swansea, U.K., 1978, pp. 450–463.

[*9*] Parks, D. M. in *Numerical Methods in Fracture Mechanics*, A. R. Luxmoore and D. R. J. Owen, Eds., University College of Wales, Swansea, U.K., 1978, pp. 464–478.

[*10*] He, M. Y. and Hutchinson, J. W., *Transactions*, American Society of Mechanical Engineers, *Journal of Applied Mechanics*, Vol. 48, 1981, pp. 830–840.

[*11*] He, M. Y. and Hutchinson, J. W., this publication, pp. I-277–I-290.

[*12*] Amazigo, J. C., *International Journal of Solids and Structures*, Vol. 10, 1974, pp. 1003–1015.

[*13*] Amazigo, J. C., *International Journal of Solids and Structures*, Vol. 11, 1975, pp. 1291–1299.

[*14*] Hutchinson, J. W., *Journal of the Mechanics and Physics of Solids*, Vol. 16, 1968, pp. 13–31.

[*15*] Rice, J. R. and Rosengren, G. F., *Journal of the Mechanics and Physics of Solids*, Vol. 16, 1968, pp. 1–12.

[*16*] Green, A. P. and Hundy, B. B., *Journal of the Mechanics and Physics of Solids*, Vol. 4, 1956, pp. 128–144.

[*17*] Sauter, A. and Kumar, V., "Elastic-Plastic and Fully Plastic Analysis of Single-Edge Cracked Plate in Pure Bending," General Electric Co. Report T 15, Schenectady, N.Y., to appear.

[*18*] Kumar, V., German, M. D., and Shih, C. F., "Estimation Technique for the Prediction of Elastic-Plastic Fracture of Structural Components of Nuclear Systems," General Electric Co. Report SRD-80-094 to EPRI, Contract RP1237-1, Combined Second and Third Semi-Annual Report, Schenectady, N.Y., 1 June 1980.

[*19*] Tong, P. and Rossettos, J. N., *Finite Element Method Basic Technique and Implementation*, M.I.T. Press, Cambridge, Mass., 1977.

Charles S. White,[1] Robert O. Ritchie,[1] and David M. Parks[1]

Ductile Growth of Part-Through Surface Cracks: Experiment and Analysis

REFERENCE: White, C. S., Ritchie, R. O., and Parks, D. M., **"Ductile Growth of Part-Through Surface Cracks: Experiment and Analysis,"** *Elastic-Plastic Fracture: Second Symposium, Volume I—Inelastic Crack Analysis, ASTM STP 803*, C. F. Shih and J. P. Gudas, Eds., American Society for Testing and Materials, 1983, pp. I-384–I-409.

ABSTRACT: It is an essentially open question as to how well characterizations of through-thickness crack growth based on parameters such as J, crack-tip opening displacement, and tearing modulus describe ductile cracking in important structural configurations such as the part-through surface-cracked plate.

In this work we compare experimental observations and predictions of ductile crack extension in surface-cracked plates subject to bending. Predictions are made by performing finite-element analysis of this three-dimensional geometry using the line-spring model, in conjunction with experimental plane-strain crack growth data from R-curves obtained from through-thickness-cracked compact-tension specimens.

Analysis of the surface flaw is based upon the incremental plasticity theory line-spring element, which has been incorporated into the ABAQUS© general-purpose finite-element program.

The material under study is a low-alloy bainitic pressure vessel steel; namely, fossil energy ASTM A542, Class 3.

Results show that for the crack configuration considered here the amount of stable tearing is quite uniform along the crack front as is the local value of the J-integral. Crack growth initiation occurs at similar J's for both through-crack and surface-crack specimens, but the surface-crack specimens tested exhibit a higher apparent tearing modulus.

KEY WORDS: stable crack growth, fracture, plastic properties, surface crack, line spring, J-integral, resistance curves, elastic-plastic fracture

Elastic-plastic fracture mechanics has developed to the point where there exists an American Society for Testing and Materials (ASTM) standard test

[1]Graduate fellow and associate professors of mechanical engineering, respectively, Department of Mechanical Engineering, Massachusetts Institute of Technology, Cambridge, Mass. 02139; coauthor Ritchie is currently professor in the Department of Materials Science and Mineral Engineering, University of California, Berkeley, Calif., and coauthor White is currently 1st lieutenant, U.S. Army, Materials and Mechanics Research Center, Watertown, Mass.

[Standard Test for J_{Ic}, a Measure of Fracture Toughness (E 813-81)] for fracture initiation toughness as measured by J_{Ic}, and considerable activity in testing methods, analysis, and computation is evidenced in the recent literature. To date, most attention has been focussed on characterizing material ductile behavior in well-defined laboratory specimens such as the face-grooved compact-tension (CT) geometry, and rather less work has been directed to elastic-plastic applications in structural crack geometries.

The great utility of linear elastic fracture mechanics (LEFM) in engineering design is the capability it provides for making quantitative assessments of the crack growth behavior of flawed engineering structures or components. Experiments are performed on laboratory specimens, and material "properties" such as K_{Ic} or da/dN versus ΔK_I fatigue crack growth curves are obtained. We note in passing that the conversion of experimental raw data, in the form of load level, crack length, etc., into the fracture mechanics parameter K_I requires elastic stress analysis of the test specimen such as in Tada et al [*1*].[2] Further laboratory testing on different specimen types and sizes can support the self-consistency of the LEFM approach, but the "proof of the pudding" for the designer is the confirmation that these material characterizations, combined with appropriate stress analyses of crack configurations encountered in practice, indeed adequately describe the crack growth behavior of the structure.

The transferability of LEFM from the laboratory to the designer (subject to the usual limitations on plastic zone size, etc.) has been satisfactorily established many times over the past 30 years. At this point, however, elastic-plastic fracture mechanics (EPFM) has essentially reached the point where the requisite limitations for obtaining consistent results from differing laboratory specimens are known. The demonstration that the EPFM laboratory characterizations in terms of J_{Ic}, J_R-curves, etc. adequately predict the elastic-plastic cracking of structural flaw geometries, especially the part-through surface crack in a plate or shell, is largely lacking.

In this paper, we present preliminary work at establishing the correlation of ductile extension of surface-cracked plates with a J_R-curve obtained from CT specimens.

As a schematic of the intent of the paper, we note the two block diagrams shown in Fig. 1. In Fig. 1*a* we note aspects of the testing and analysis procedures for obtaining EPFM parameters from through-thickness CT specimens. Initial geometry (crack-length) is noted and records are taken of load, load-point displacement, and of the evolution of crack length $a - a_0 \equiv \Delta a$ during the test. A stress analysis converts the raw data into parameters such as $J_{applied}$, crack-tip opening displacement (CTOD), or δ_t. The sophistication of the analysis may vary from detailed finite-element analysis to the simple $J = 2A/Bb$ relation for deep cracks in predominant bending, where b is re-

[2]The italic numbers in brackets refer to the list of references appended to this paper.

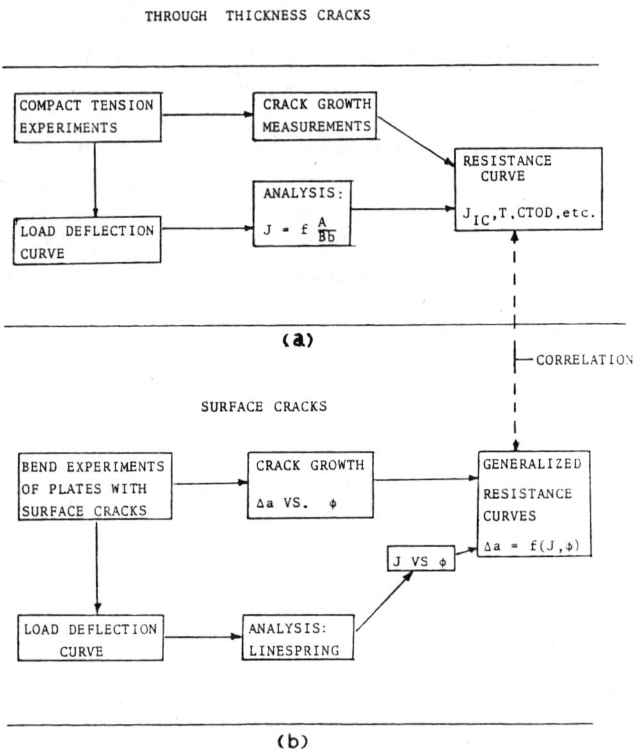

FIG. 1—(a) *Schematic test, observation, and analysis procedures required to characterize ductile crack growth in a material by using laboratory compact-tension specimens. Results may be reduced to materials properties like* J_{Ic} *and a* J_R*-curve.* (b) *Schematic test, observation, and analysis procedures required to characterize ductile crack growth in a structural configuration such as a surface-cracked plate. Can the behavior of such cracks be correlated with the results of the "laboratory" tests of* (a) *above?*

maining ligament length, B is crack-plane specimen thickness, and A is the area under the load/load-point-displacement curve. Correlation of the evolution of inferred J or δ_t with that of observed crack length, a, provides, for example, the J_R-curve.

Fig. 1*b* illustrates an analogous sequence of testing and analysis procedures for the part-through surface-cracked plate. Applied load and various displacements are monitored during the test, as is the crack growth *function* $\Delta a(\phi)$. Thus with ϕ serving as a parameter specifying position along the crack front (see Fig. 3 for the definition of ϕ for a semi-elliptical surface flaw), we may expect at any moment to observe different amounts of crack extension at different locations along the crack front, presumably indicative of varying intensity of crack-tip deformation along the crack front. We must also perform a stress analysis of the surface-flawed geometry in order to relate the history of applied loading to the evolution of $J(\phi)$, the crack-front distribution of J. If the history of both $\Delta a(\phi)$ and $J(\phi)$ are cross-plotted at various crack front locations, the obvious question of consistency is, "How well does the J_R-curve from the CT specimen correlate with the $J(\phi)/\Delta a(\phi)$ relations?"

There are numerous difficulties in actually carrying out the steps in the loop of Fig. 1*b*. For example, *local* measurements of crack advance along a crack front are quite difficult, as unloading compliance and potential drop techniques provide only *global* information on crack growth. Relatively sophisticated acoustic or ultrasonic techniques or both could perhaps be used for continuous monitoring of a growing surface crack front, but such methods were not explored here. Rather, we relied on "snapshot" crack front marking strategies, including the "patient experimenter" technique of loading several specimens to differing amounts of imposed deformation, resulting in microscopically ductile crack advance, then cooling each to liquid nitrogen temperature, and reloading to obtain microscopic cleavage crack extension.

In addition, the analysis of load and deformation information required to obtain $J(\phi)$ is made formidable by the decidedly three-dimensional nature of the surface flaw. In principle, detailed three-dimensional elastic-plastic finite-element methods [*2,3*] are available for this task. However, the computer costs of such analyses are large, and a good deal of manpower is required for mesh generation, data reduction, etc. Although models of the surface crack can be formulated which are sufficiently idealized so as to be (algebraically) tractable, it is difficult to assess the accuracy of such (necessarily) highly simplified models. In the present work, we perform the analysis of the surface flaw using the line-spring model of Rice [*4,5*] which has recently been further developed and applied by Parks and co-workers [*6–10*].

In the following sections, we describe the experiments performed, including J_R-tests on CT specimens and four-point bending of surface-cracked (SC) plates. Subsequently, the line-spring model is briefly reviewed, including certain details of the elastic-plastic spring constitutive law. The calculated load/crack face displacement data are found to be in good agreement with the SC experiments. Further, the point of initiation of crack extension in the SC specimens correlates well with a calculated value of $J \sim J_{Ic}$. However, subsequent stable crack growth in the SC specimen seems to occur with a much steeper $\Delta J/\Delta a$ slope than obtained from the face-grooved CT specimen.

Experiment

The material chosen for testing is a low-alloy 2¼Cr-1Mo pressure vessel steel, ASTM A542, Class 3. Table 1 gives the chemical composition. The material was machined and tested in the as-received condition with a heat-

TABLE 1—*Chemistry of SA542 Class 3 material (from Lukens Steel Co.).*

Heat	Lukens #3707
C	0.12
Mn	0.45
P	0.014
S	0.015
Cu	0.12
Si	0.21
Ni	0.11
Cr	2.28
Mo	1.05

treatment consisting of austenitizing for 5½ h at 954°C followed by a water quench, and tempering for 1½ h at 663°C, followed by stress relief of 15 h at 593°C, 22 h at 649°C, and 18 h at 663°C. The resulting microstructure was entirely bainitic, primarily lower bainite, with molybdenum and chromium carbide particles. The grain size was 70 μm. Room temperature mechanical properties are given in Table 2 [*11*].

All specimens were machined for crack growth in the TL orientation. This notation is sufficient for the CT specimens, but in the surface-flawed specimens the TL orientation of crack extension was for through-thickness penetration of the deepest part of the surface flaw.

The CT specimens were standard 1-T specimens with certain modifications as allowed in Method E 813-81. The pinhole spacing was increased from 0.275 W to 0.375 W to centerline to allow measurement of load-line displacement with a clip gage. Additionally, the CT specimens were face-grooved 10 percent on each side of the centerline with a 90-deg included angle. The initial fatigue crack length was grown to a relative crack length of $a/W = 0.6$.

Loading of the CT specimens was performed with manual crosshead displacement control in a closed-loop servohydraulic testing machine. Specimens were loaded to varying levels of deformation, then heat-tinted, at

TABLE 2—*Room temperature mechanical properties of as-received steel* [11].

Yield Strength, MPa	496 ±7
Ultimate Tensile Strength, MPa	610 ±17
Hardness, R_B	93
% Elongation (45-mm Gage Length)	25
% Reduction in Area	77
Charpy V-Notch, J	197 ±18

fixed displacement, to record crack extension. One data point was obtained for each specimen.

The value of the J-integral was inferred from the load/displacement record by

$$J = f\,A/Bb \tag{1}$$

where

A = area under load/displacement curve,
b = remaining ligament length,
B = specimen reduced width under face-grooves, and
f = scaling factor, depending on a/W, obtained from E 813.

The fracture surfaces exhibited fully flat ductile texture and the face grooves suppressed shear lip formation. A stretch zone was observed on each specimen, and was included in the crack growth measurement. Crack extension, measured with an optical microscope on a calibrated moving stage, was found to be quite uniform along the crack front.

The resistance curve obtained from the CT specimens is shown in Fig. 2. The intersection of the blunting and tearing lines gives a value of J_{Ic} of 350 kN/m. The material tearing modulus T_{mat} is given by

$$T_{mat} = \frac{E}{\sigma_{flow}^2}\,\frac{dJ_R}{da} \tag{2}$$

and, with σ_{flow} = 550 MPa, the tearing line of Fig. 2 gives $T_{mat} \sim 60$. These values are representative of this class of tough steels [*11*].

The SC specimens were 159 mm long, with width b = 102 mm, and thickness t = 12.7 mm. A surface defect in the shape of an arc of a circle was

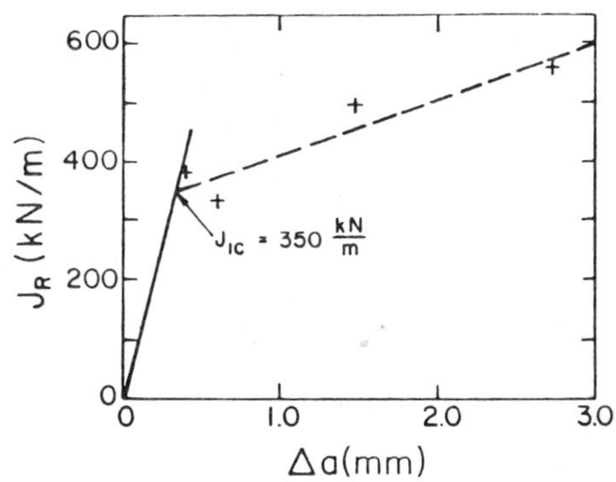

FIG. 2—*J_R-curve obtained from multispecimen compact-tension tests of SA542 Class 3 steel.*

introduced using a 70-mm-diameter by 0.36-mm-thick slotting saw. The maximum depth of the saw cut was 1.9 mm.

The plate was then fatigued in four-point bending under constant displacement limits until obtaining a final nominal crack depth (at centerline) of $a_0/t = 0.35$ and aspect ratio $a_0/c = 0.4$, where $2c$ is the projected total surface length.

The shape of resulting fatigue-cracked surface flaws was quite reproducible. Table 3 gives the dimensions of the crack on the different specimens. For all practical purposes, the surface crack shape was semi-elliptical, and a parameter ϕ locating points on a semi-elliptical surface crack is shown in Fig. 3.

The four-point-bending apparatus used in fatigue crack growth was also used in the ductile fracture experiments. The loading is schematically illustrated in Fig. 4. The large-diameter rollers were rigidly attached to the hydraulic frame 132 mm apart and the crosshead displacement was applied to the smaller-diameter rollers. The smaller-diameter rollers, made of hardened steel, were spaced 50.8 mm apart and were free to rotate.

TABLE 3—*Geometry of surface-cracked specimens, imposed gage displacement, and observed centerline crack extension.*

Specimen No.	a_0/t	a_0/c	Total Δa Centerline, mm	δ_g, mm
S11	0.403	0.404	0.23	1.49
S12	0.406	0.490	0.49	3.16
S13	0.352	0.374	0.63	2.21
S18	0.346	0.346	0.50	3.34
S19	0.348	0.350	0.34	2.02

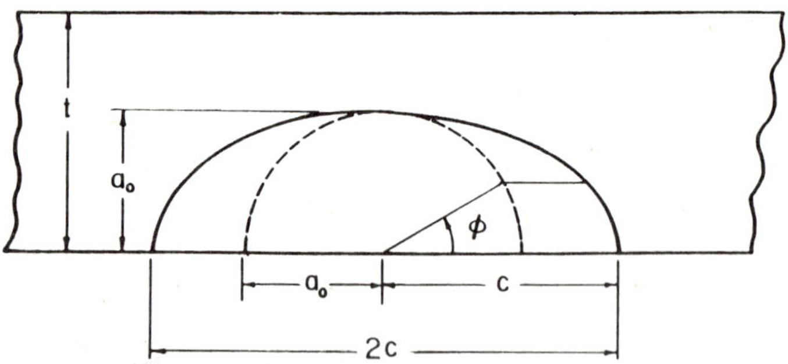

FIG. 3—*Schematic illustration of a semi-elliptical surface flaw of maximum depth* a_0 *and length* 2c *in a plate of thickness* t. *The parameter* ϕ *locates points on the semi-elliptical crack front, with* $\phi = 0$ *at the free surface and* $\phi = \pi/2$ *at specimen centerlines.*

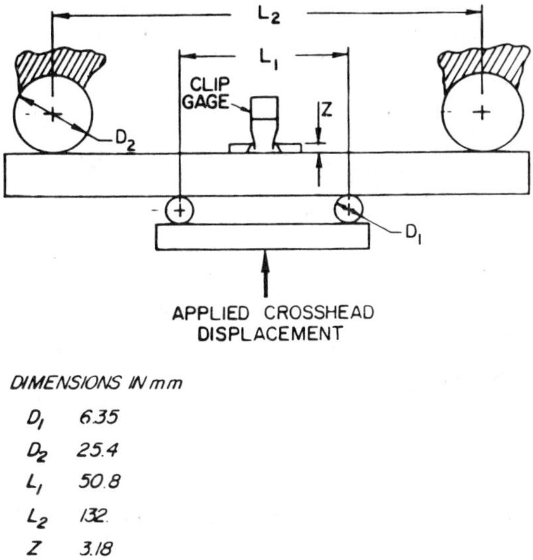

FIG. 4—*Schematic illustration of the four-point bending imposed on the surface-cracked specimens indicating the measurement point for crack gage opening displacement,* δ_g.

As indicated in Fig. 4, knife edges were attached to the cracked side of the plate on either side of surface crack at centerplane. The knife edges were 3.2 mm above the plate surface. As the specimen was loaded, a clip gage was used to monitor the relative separation of the knife edges. This displacement was termed δ_g.

The specimens were deformed by imposing crosshead displacement and monitoring load cell reaction and δ_g. Each plate was loaded to a different displacement level, but all were within the rising portion of reaction force versus clip-gage displacement. Specimens were held at the maximum imposed crosshead displacement and heat-tinted with a propane torch. Subsequently, large portions of the specimen were cut away, the remaining piece cooled to liquid nitrogen temperature, and fractured.

The fracture surfaces of the SC specimens were also examined for stable crack advance by using a calibrated traveling microscope. Stable cracking was again fully flat ductile tearing, and the amount of local normal crack advance was quite uniform along the crack front, as is indicated in Fig. 5. In Fig. 5, Δa is plotted versus parametric angle ϕ for different specimens which were loaded to the δ_g-values indicated in Table 3. Note that for semi-ellipses of aspect ratio $a/c = 0.4$, the parameter ϕ is quite nonlinear in its correlation with arc length along the crack front. Although data are shown in Fig. 5 only for $\phi > 30$ deg or so, this includes almost all of the crack front. Essentially no crack advance was observed right *at* the free surfaces, $\phi = 0$.

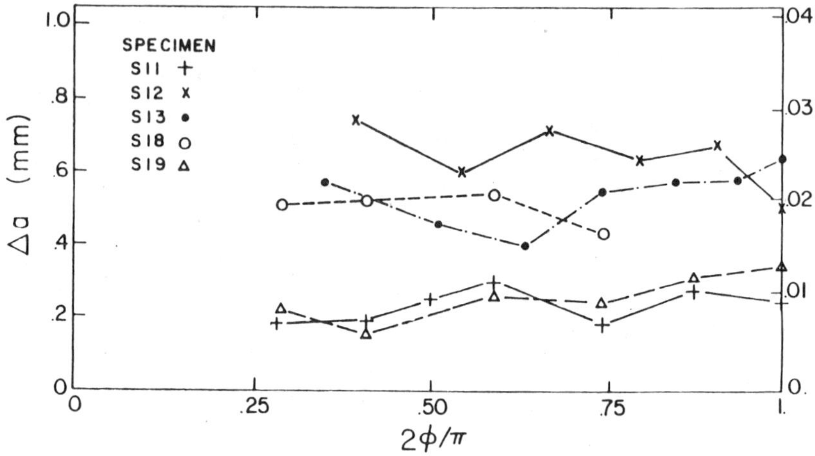

FIG. 5—*Observed distribution of crack extension* Δa *versus parametric loaction* ϕ *in surface-cracked specimens loaded to varying imposed displacements.*

Analysis

In this section, we describe our implementation of the line-spring model in the analysis of the SC specimen tests described in the preceding. The formulation is essentially that of incorporating the incremental elastic-plastic strain-hardening spring constitutive model proposed by Parks [*6*] into the line-spring finite element which Parks, Lockett, and Brockenbrough [*9*] put into the ABAQUS© finite-element program.

The first notion of the line-spring model is that the additional compliance introduced into a plate or shell structure by a part-through surface crack can be accounted for by introducing a conceptual through-crack of total length $2c$ equal to the maximum length of the surface flaw in the shell/plate mid-surface. Then, a generalized foundation is placed across the two sides of this cut, so that the traction and moment resultants carried by the uncracked portion of the ligament are transmitted across the sides of the opening through-crack. This foundation, schematically illustrated in Fig. 6, has a stiffness which varies according to the local depth $a(x)$ of the surface flaw.

For symmetric loading, the foundation can transmit a shell normal force $N(x)$ and bending moment $M(x)$ per unit length. These local forces have work-conjugate deformations $(\delta(x), \theta(x))$ consisting of the relative separation and rotation of the sides of the model through-crack.

In an incremental plasticity formulation the foundation is described by symmetric rate moduli $S_{ij}(x)$ such that

$$\begin{bmatrix} \dot{N} \\ \dot{M} \end{bmatrix} = \begin{bmatrix} S_{11} & S_{12} \\ S_{21} & S_{22} \end{bmatrix} \begin{bmatrix} \dot{\delta} \\ \dot{\theta} \end{bmatrix} \tag{3}$$

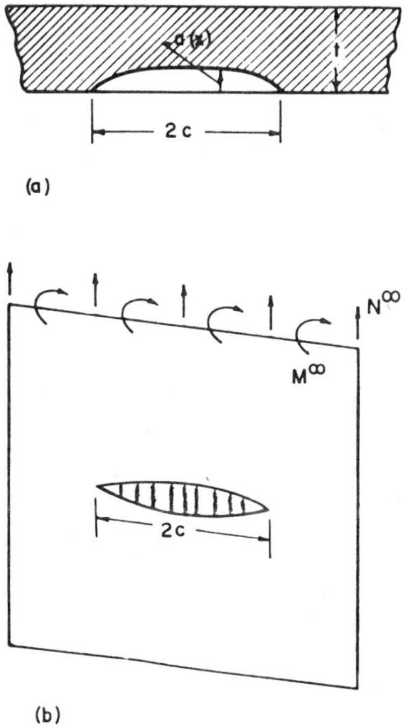

FIG. 6—*Schematic illustration of the line-spring model, indicating that the part-through surface crack is modeled by a plate containing a through-crack which has a generalized foundation across its flanks.*

The model is completed by basing the choice of S_{ij} on the behavior of the plane-strain edge-cracked strip subject to tension and bending (rates) shown in Fig. 7. The increments in total separation and rotation of the load points in Fig. 7 can be additively decomposed into the sum of a "cracked" part, $(\dot{\delta}_c, \dot{\theta}_c)$, and an uncracked part, proportional to the length of the strip. The line-spring model effectively takes S_{ij} as those moduli connecting $(\dot{\delta}_c, \dot{\theta}_c)$ and $(\dot{N}, \dot{M})$ in Fig. 7. As discussed elsewhere [*4–10*], the linear elastic connection $S_{ij}^{(e)}$ is well in hand. Parks [*6*] presented the theory for combining the elastic stiffnesses with the description of a nonhardening yield-surface, $\Phi(N, M; a, t; \sigma_0) = 0$, to obtain elastic-plastic rate moduli: $S_{ij}^{(ep)}$. Here $\sigma_0 \equiv \sqrt{3}\,\tau_0$ is meant to be a cross-sectional "average" tensile yield stress of the material, and τ_0 is the corresponding shear yield stress, according to a Mises criterion. Some representative normalized yield surfaces are shown in Fig. 8. These are based on slipline theory and upper-bound analysis as given by Rice [*5*].

The steps in formulating the elastic-plastic rate moduli are given in Ref *6*, and are briefly repeated here. Rename the generalized stresses and displacements as $Q_1 = N$, $Q_2 = M$; $q_1 = \delta$, and $q_2 = \theta$. Then in incremental form $\dot{q}_i$ is additively decomposed into elastic and plastic parts:

$$\dot{q}_i = \dot{q}_i^{(e)} + \dot{q}_i^{(p)} \tag{4}$$

The plastic generalized displacement increments are obtained from normality to the yield surface to within a nonnegative multiplicative factor Λ

$$\dot{q}_i^{(p)} = \Lambda \frac{\partial \Phi}{\partial Q_i} \tag{5}$$

The scalar Λ is determined from the consistency condition that, during plastic response, $\dot{\Phi} = 0$, giving

$$0 = \dot{\Phi} = \frac{\partial \Phi}{\partial Q_i} \dot{Q}_i + \frac{\partial \Phi}{\partial \sigma_0} \dot{\sigma}_0 \tag{6}$$

with understood summation on subscript "*i*."

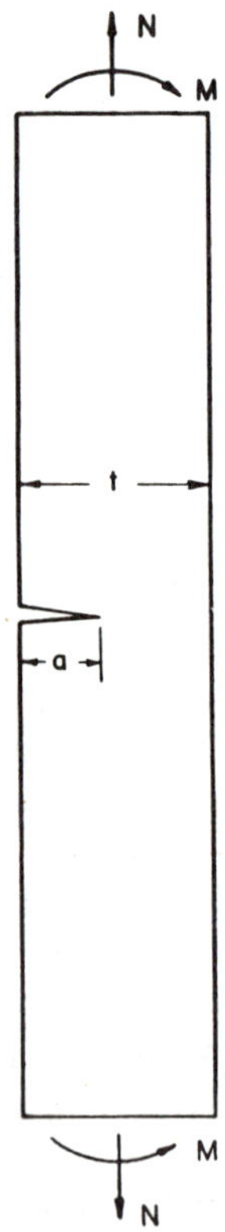

FIG. 7—*Schematic illustration of the plane-strain edge-cracked specimen subject to combined tension and bending. This specimen provides the line-spring foundation stiffness as well as the J-calibration.*

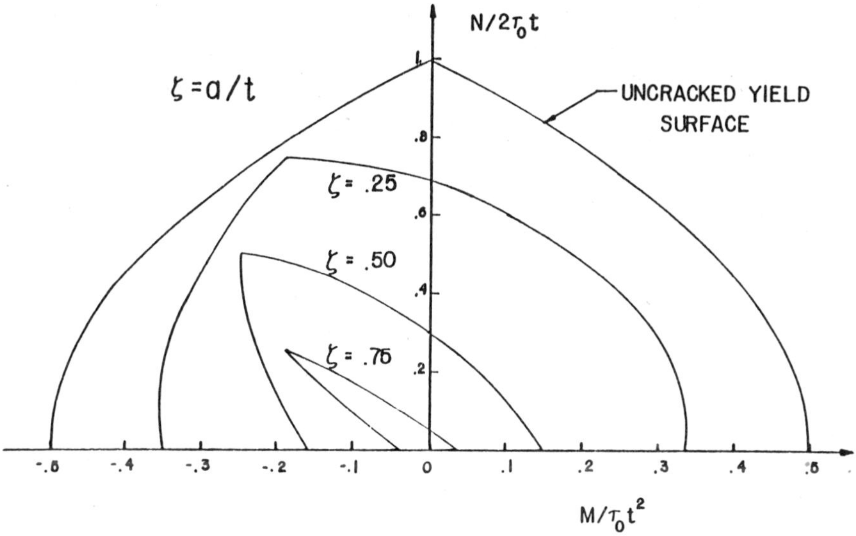

FIG. 8—*Yield surfaces used in elastic-plastic line-spring calculations.*

We can replace $\dot{Q}_i$ by

$$\begin{aligned}\dot{Q}_i &= S_{ij}^{(e)}\,\dot{q}_j^{(e)} = S_{ij}^{(e)}\,(\dot{q}_j - \dot{q}_j^{(p)})\\ &= S_{ij}^{(e)}\,\dot{q}_j - \Lambda\, S_{ij}^{(e)}\,\frac{\partial\Phi}{\partial Q_j}\end{aligned} \tag{7}$$

Furthermore, the plastic work rate per unit thickness $\dot{W}_p$ can be expressed both in terms of the generalized stress and strain measures as

$$\dot{W}_p = Q_i\,\dot{q}_i^{(p)} = \Lambda\, Q_i\,\frac{\partial\Phi}{\partial Q_i} \tag{8}$$

and as the integral over the cross section A^* of the edge-cracked specimen in Fig. 7 of continuum plastic work rate

$$\dot{W}_p = \int_{A^*} \sigma_{ij}\,\dot{\epsilon}_{ij}^{(p)}\,dA^* = \int_{A^*} \overline{\sigma}\,\dot{\overline{\epsilon}}^{(p)}\,dA^* \tag{9}$$

Here the quantities $\overline{\sigma}$ and $\dot{\overline{\epsilon}}^{(p)}$ are local values of tensile equivalent stress and plastic strain rate, respectively. It is inferred from the slipline field shown schematically in Fig. 9 (and used to generate the yield surfaces of Fig. 8) that the area of cross section where plastic deformation is occurring will be proportional to the square of the remaining ligament, $t - a$. Thus, one can make an approximate evaluation of the second integral, in terms of σ_0 and $\dot{\bar{\epsilon}}_0^{(p)}$, a section-average plastic strain rate connected to σ_0 by a plastic hardening modulus h

$$\dot{\sigma}_0 = h\,\dot{\bar{\epsilon}}_0^{(p)} \tag{10}$$

The approximate evaluation of $\dot{W}_p$ then becomes

$$\dot{W}_p = f_A \cdot \sigma_0 \cdot \dot{\bar{\epsilon}}_0^{(p)} \cdot (t - a)^2 \tag{11}$$

where the dimensionless f_A is expected to be of order unity.

When Eqs 6–11 are combined, the scalar Λ is obtained as

$$\Lambda = \frac{\dfrac{\partial \Phi}{\partial Q_i} S_{ij}^{(e)}\,\dot{q}_j}{\dfrac{\partial \Phi}{\partial Q_m} S_{mn}^{(e)} \dfrac{\partial \Phi}{\partial Q_n} - \dfrac{\partial \Phi}{\partial \sigma_0} \cdot \dfrac{hQ_k \dfrac{\partial \Phi}{\partial Q_k}}{f_A \sigma_0 (t - a)^2}} \tag{12a}$$

$$\equiv \frac{\eta_j \dot{q}_j}{D} \tag{12b}$$

where the scalar D represents the denominator of Eq 12*a* and η_j is the plastic normal multiplied by elastic stiffness.

With Λ thus determined, the plastic rate moduli simply become

$$S_{ij}^{(ep)} = S_{ij}^{(e)} - \frac{\eta_i \eta_j}{D} \tag{13}$$

In the present work, we defined the plastic hardening of Eq 10, starting at an initial value of $\sigma_0 = 496$ MPa, with a constant $h = 1780$ MPa up to an equivalent plastic strain of 0.10 above which the response is considered nonhardening ($h = 0$). This is simply a coarse, trilinear approximation to the stress-strain response which we inferred from Ref *11*.

The value of the dimensionless scalar f_A was obtained by conducting plane-strain finite-element analyses of the edge-crack specimen with $a/t = 0.5$ subject to pure bending and predominant bending with a small proportional amount of axial tension or compression. The "uncracked" components were subtracted from the total extension and rotation of the continuum finite-element solution, to obtain (δ_c, θ_c) as a function of imposed load.

We then integrated the response of a single spring element of the cor-

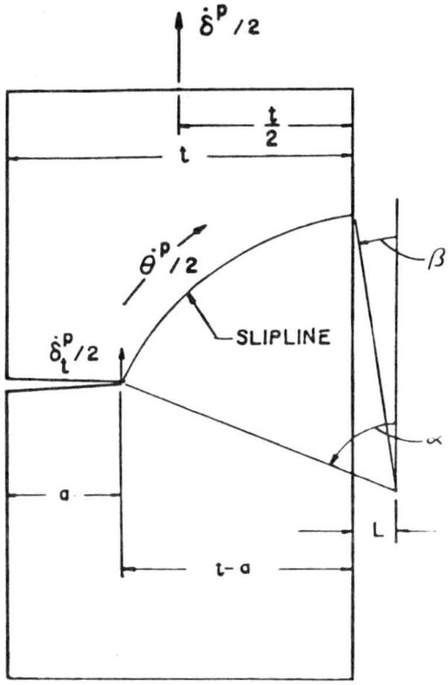

$$\dot{\delta}_t^P = \dot{\delta}^P + (t/2 - a)\dot{\theta}^P$$

FIG. 9—*Schematic illustration of the upper-bound/slipline fields used to generate yield surfaces shown in Fig. 8, indicating the kinematics of crack-tip opening in terms of load-point generalized displacements.*

responding dimensions, material properties, and load histories for different values of the parameter f_A. It was found that good agreement between the cracked displacements as computed by the spring and in the continuum was obtained with $f_A = 0.4$.

This value gave a good match to the overall load/displacement history for each of the three load cases considered.

With the compliance of the line-spring thus specified, the second, and most important, feature of the line spring comes into play. If the surface-cracked plate were sectioned at some position x by a plane normal to the surface flaw and to the plate surface, the section would look very like the single edge-cracked specimen of Fig. 7. As an estimate of the J-value of the surface crack at the point cut by the section, we take the J-value which would result in the corresponding plane strain edge-cracked specimen when subject to the local load history of (N,M) which the foundation has transmitted at that point.

In order to do this [*6*], divide J into the sum of elastic and plastic parts

$$J = J^{(e)} + J^{(p)} \tag{14}$$

where $J^{(e)}$ is obtained from a stress intensity factor K_I in the standard way as

$$J^{(e)} = (1 - \nu^2)K_I^2/E \tag{15}$$

The stress-intensity factor K_I is computed by superposition of the K_I-values which would result from applying the current values of (N,M) to the specimen in Fig. 7 using the K_I-calibrations of Tada et al [*1*].

The increment of $J^{(p)}$ is related to the increment of plastic crack-tip opening displacement $\dot{\delta}_t^{(p)}$ by

$$\dot{J}^{(p)} = m\,\sigma_0\,\dot{\delta}_t^{(p)} \tag{16}$$

where m is a scalar which depends on the geometry of the active slipline field [*12,13*]. More will be said later regarding the value of m. The plastic part of the CTOD increment can be straightforwardly related to the plastic parts of overall displacement increments through the kinematics of the assumed slipline field by

$$\dot{\delta}_t^{(p)} = \dot{\delta}^{(p)} + (t/2 - a)\,\dot{\theta}^{(p)} \tag{17}$$

This relation is schematically illustrated in Fig. 9. When Eqs 14–17 are combined and integrated over load history, local J-values along the surface crack may be inferred.

We note that the resulting form is quite approximate. For example, in the case of fully plastic response of a given spring, the elastic K_I used in defining $J^{(e)}$ is not particularly meaningful. On the other hand, once plasticity is fully developed it is anticipated (and, indeed, observed) that the magnitude of $J^{(p)}$ totally dominates that of any conceivable $J^{(e)}$.

The value of the scalar m is intimately coupled with the active slipline field, which in turn depends on the ratio of tension to bending, for example, cN/M, where c is the remaining ligament. Since the current tests imposed pure bending in the far field, it is to be expected that the line-spring elements will be in a load state of predominant bending, with "small" superposed axial force. The appropriate slipline field for this spring load state is a modification [*5*] of the Green and Hundy (GH) field [*14*] for pure bending. The modified field is shown in Fig. 10, where the total remaining ligament c is the sum of two parts of lengths c_1 and c_2

$$c = c_1 + c_2 \tag{18}$$

If c_2 were zero, the remaining field would be the complete GH field. The stresses of this field, scaled to length c_1, are statically equivalent to a pure moment of magnitude

$$M_1 = 0.63\,\tau_0\,c_1^2$$

$$= 0.63\ \tau_0\ (c - c_2)^2 \tag{19}$$

as is indicated in Fig. 10. Furthermore, the extension of the bending slipline field into the region of width c_2 means that this material must be in a state of compression, and the stresses of this portion of the slipline field are statically equivalent to a compression force of magnitude $2\tau_0 c_2$, centered a distance $c_2/2$ from the free surface, as is also shown in Fig. 10. By taking a free body of the upper-half specimen, including the static resultants of the slipline fields, one obtains the equation of yielding states as

$$M/\tau_0 c^2 = 0.63 + (N/2\tau_0 c)\cdot(0.26 - a/c) - 0.37\,(N/2\tau_0 c)^2 \tag{20}$$

An equivalent definition of this portion of the yield surface was noted by Shiratori and Miyoshi [*13*]. Note that for $c_2 > 0$

$$c_2 = -N/2\tau_0 \tag{21}$$

we find a state of net compression. The maximum value of such compression is, of course, $-2\tau_0 c = N_{min}$ and corresponds to midligament compression with no GH field: $c_1 = 0$. The modified field of Fig. 10 also can accommodate a limited amount of tension, corresponding to a negative value of c_2.

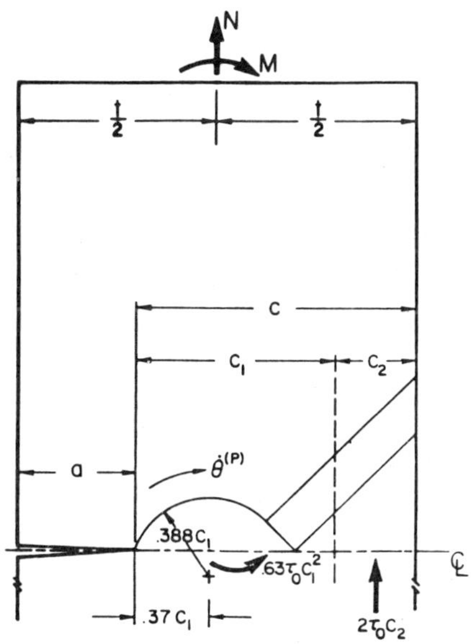

FIG. 10—*Modified Green and Hundy slipline fields used for cases of predominant bending. Static equivalent resultants of the stress fields in Regions* c₁ *and* c₂ *of the remaining ligament are indicated.*

This simply means that part of the GH field which was in compression is "shaved off," and the circular arc part is scaled up in size to occupy a larger fraction of the total remaining ligament. The maximum value of the tension which can be accommodated in this way occurs when the circular arc reaches the orientation of -45 deg just at the free boundary, with $N_{max} = 2\tau_0 c \cdot (0.355)$. The generalized stress states obtained on the yield surface of Eq 20 always had $N_{max} > N > N_{min}$.

The J-integral can be defined as the negative of the difference in work done on two bodies differing slightly in crack length a in bringing each body to the same prescribed deformation. With the generalized loads Q_i and conjugate deformations q_i noted in the preceding

$$J = -\int_0^{q_j} \left.\frac{\partial Q_i}{\partial a}\right|_{q_k} dq_i \tag{22a}$$

or, explicitly, using the fact that $-\partial(\ )/\partial a = \partial(\ )/\partial c$

$$J = \int_0^{q_j} \left.\frac{\partial N}{\partial c}\right|_{\theta,\delta} d\delta + \int_0^{q_j} \left.\frac{\partial M}{\partial c}\right|_{\theta,\delta} d\theta \tag{22b}$$

As Rice [*12*] noted, one may formally compute J from Eq 22*a* in the rigid/perfectly plastic case as the product of generalized displacement and rate of change of limit load with respect to crack length. This operation is straightforward in cases of a single generalized force/displacement pair. In the case of Eq 22*b* the expression becomes

$$J = \left.\frac{\partial N}{\partial c}\right|_{\theta,\delta} \delta + \left.\frac{\partial M}{\partial c}\right|_{\theta,\delta} \theta \tag{22c}$$

In order to evaluate the indicated partial derivatives, we note from the kinematics of the modified GH field that

$$\delta/\theta = c\{0.37\,(1 + N/2\tau_0 c) - 1\} + t/2 \equiv K \tag{23}$$

Further, we may take the partial derivative of Eq (23)

$$0 = \left.\frac{\partial(\delta/\theta)}{\partial c}\right|_{\delta,\theta} \tag{24}$$

to obtain

$$\left.\frac{\partial N}{\partial c}\right|_{\delta,\theta} = \frac{2\tau_0(1 - 0.37)}{0.37} = 2\tau_0 \cdot \left(\frac{0.63}{0.37}\right) \tag{25}$$

We may take the partial derivative with respect to c of Eq 20 defining the yield surface, insert Eq 25, and obtain

$$\left.\frac{\partial M}{\partial c}\right|_{\delta,\theta} = \frac{0.63}{0.37}\,\tau_0 t - a\left.\frac{\partial N}{\partial c}\right|_{\delta,\theta}$$

$$= \frac{0.63}{0.37}\,\tau_0(t - 2a) \tag{26}$$

We may now combine Eqs 22c, 23, 25, and 26 to obtain one expression for J, proportional to the θ-component, using $\delta = K\theta$ from Eq 23

$$J = \tau_0\theta\cdot\left(\frac{0.63}{0.37}\right)\cdot[(t - 2a) + 2K] \tag{27}$$

But from the (total) form of Eq 16 with $\sigma_0 = \sqrt{3}\,\tau_0$, and using Eqs 17 and 23 to express δ_t in terms of θ, we find

$$J = \tau_0\theta\cdot\frac{m\sqrt{3}}{2}\cdot[(t - 2a) + 2K] \tag{28}$$

so that a comparison of Eqs 27 and 28 provides

$$m = \frac{2}{\sqrt{3}}\cdot\left(\frac{0.63}{0.37}\right) \doteq 1.97 \tag{29}$$

for all points on this yield surface portion. Rice [*12*] obtained this value for pure bending, and Shiratori and Miyoshi [*13*] noted the constancy of this parameter in the analysis of pin-loaded edge-cracked specimens in tension for those crack lengths deep enough to admit the modified GH field.

When this value for m was used in the one-spring test problems, the inferred J versus load of the single spring was in good agreement with the J-values obtained from the continuum solutions.

Results

The mesh used for the line-spring calculations is shown in Fig. 11. There were four line-spring elements and 28 plate elements in one quarter of the specimen.

Seven Simpson's rule integration points were located through the thickness of the plate with 2 × 2 Gaussian integration in the plane. The line-spring elements were integrated using the three-point Simpson's rule along their

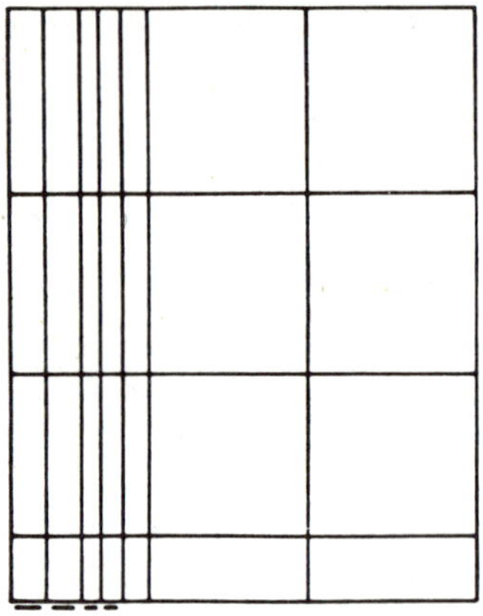

FIG. 11—*Finite-element mesh of plate and line-spring elements used to model one fourth of the surface-cracked specimen.*

length. This was found to essentially eliminate the spurious oscillation in stress-intensity factor, and generalized resultants which Parks et al [*9*] reported when two- or three-point Gaussian integration was used on the line springs. Appropriate symmetry boundary conditions were imposed on the bottom row and left column of nodes. The top row of nodes was constrained to have no displacement normal to the plane, in accord with the reaction at the large-diameter rollers. The row of nodes on the interelement boundary between the next-to-top and next-to-bottom row of elements was given uniform imposed vertical displacements consistent with the crosshead and inner-diameter roller's action. The reaction force was taken as the sum of nodal forces on either row.

Although the specimens were deformed far into the plastic range, so that $J^{(p)} \gg J^{(e)}$, we indicate the elastic accuracy of the line-spring model in Fig. 12 by comparing normalized K_I-variations along the semi-elliptical

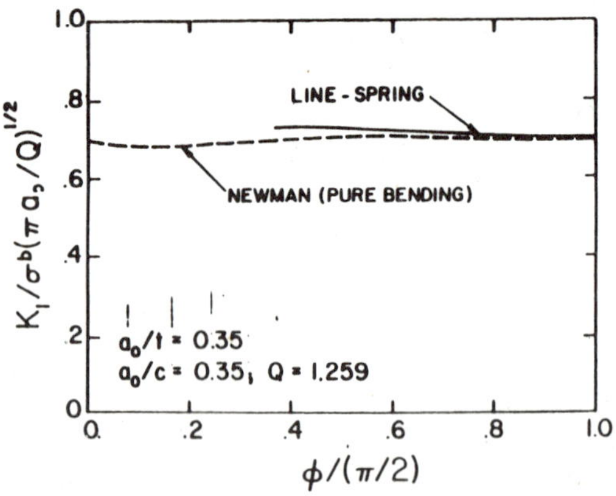

FIG. 12—*Elastic* K_I *distribution for pure bending of the surface-cracked plate as computed by the line-spring and by Newman and Raju [15].*

crack as obtained from the line spring and from interpolation of the three-dimensional finite-element solutions of Newman and Raju [*15*]. The reference stress σ_b is the tensile stress at the free surface of a beam of width t subject to a pure bending moment M per unit thickness: $\sigma_b = 6M/t^2$. As can be seen, the solutions are in good agreement, and both indicate little variation in K_I along the crack front for this aspect ratio and relative depth.

Figure 13 compares an experimental reaction load, P, versus clip gage opening displacement δ_g for Specimen No. SC18 with the value inferred from the line-spring finite-element solution. The plastic part of δ_g is related to $(\delta^{(p)}, \theta^{(p)})$ in a manner similar to that used to obtain $\delta_t^{(p)}$ in Eq 17 and Fig. 9, but noting that $\delta_g^{(p)}$ is measured at a location 3.2 mm above the free surface. In order to estimate the elastic part $\delta_g^{(e)}$, we made use of the calibrations of Tada et al [*1*] for mouth opening displacements of the edge-cracked specimen due to tension and bending. The contribution of N to the part of $\delta_g^{(e)}$ was taken as the mouth opening displacement due to tension. The contribution of M to the part of $\delta_g^{(e)}$ was taken as the mouth opening displacement due to bending, but magnified by a ratio of lengths corresponding to an assumed rotation about the middle of the remaining ligaments. These estimates are not precise since the elastic rigid-body rotation of the crack mouth point is not calibrated, but they seem reasonable. Further, the initial slopes

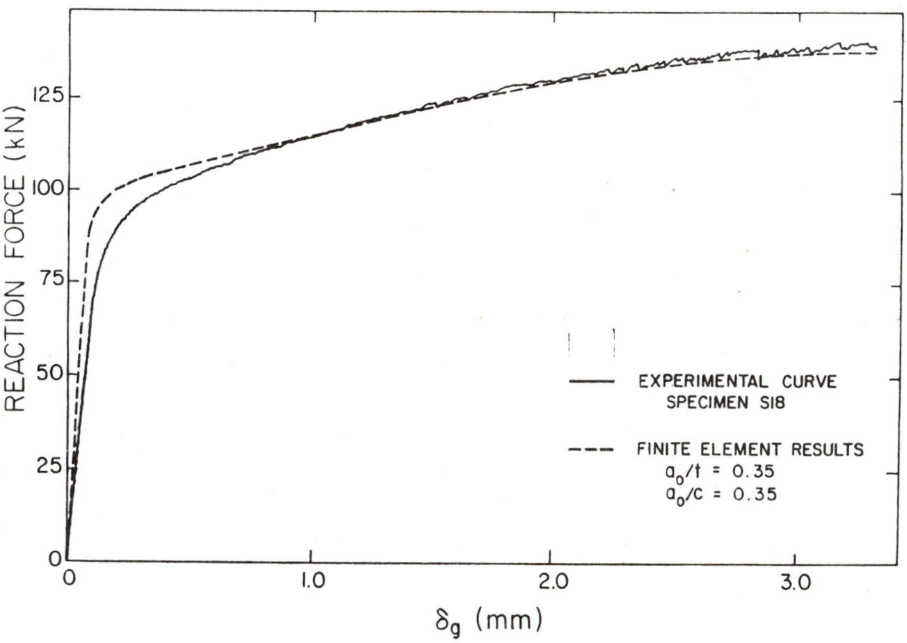

FIG. 13—*Experimental and calculated curves of axial load cell resultant versus crack displacement gage opening* δ_g.

measured experimentally are in good agreement with this estimate, as seen in Fig. 13. In the fully plastic range, when δ_g is dominated by $\delta_g^{(p)}$, the agreement is remarkable. This agreement would seem to suggest that the line-spring kinematics is developing in a quite plausible manner.

Figure 14 shows the computed distribution of normalized moment and force transmitted across the cut at two values of imposed deformation. Near the specimen centerline the state is predominant bending, with a compressive resultant which is becoming more negative with increasing deformation. The axial force goes into tension at ϕ-values near the free surface. However, the line-spring model has no basis near the intersection of crack front and free surface, so the computed distribution has been suppressed in that region.

Figure 15 shows the distribution of computed J-values at two imposed

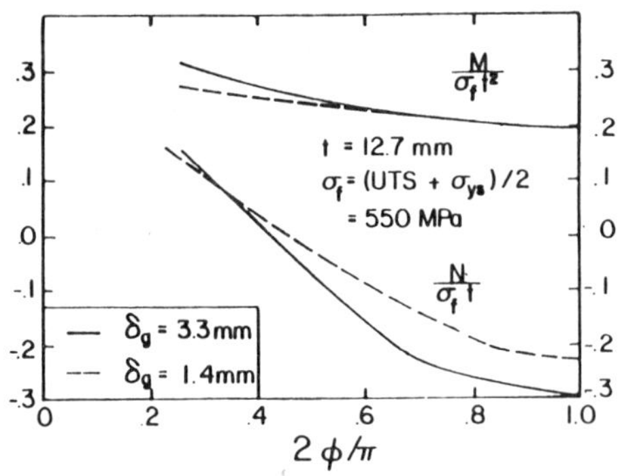

FIG. 14—*Distribution of calculated normalized force and bending moment transmitted across the line-spring cut versus parametric position* ϕ *at two levels of imposed deformation.*

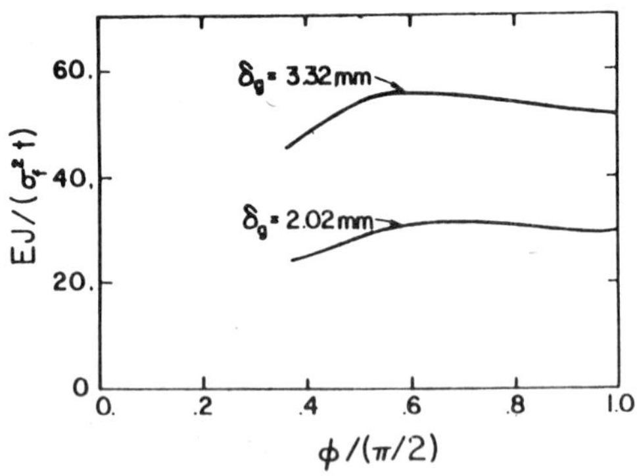

FIG. 15—*Calculated distribution of* J *versus position* ϕ *along crack front at two different levels of imposed deformation. Note that the* J*-distribution is quite uniform, as was the observed crack extension in Fig. 5.*

levels of deformation. Again, there is little variation in inferred J over the portion of crack front in which the line spring could be applicable. It is noteworthy that this inferred uniformity of imposed J is in some sense consistent with the uniformity of observed ductile crack growth shown in Fig. 5. Since in the present case both $J(\phi)$ and $\Delta a(\phi)$ are nearly uniform along the crack front, in comparing a J_R-curve from the SC specimens with that of the CT geometry, it suffices to compare only at a single location of the crack front, and we have accordingly used the specimen centerplane.

Figure 16 compares the J_R-curves of SC and CT geometries. In this figure we have subtracted from Δa the contribution due to blunting, and plot J versus amount of observed *tearing* crack extension, $\Delta a_{\text{tearing}}$. Two points are clearly evident in Fig. 16. First, the J-value at which tearing is observed to commence is, within a rather larger window of uncertainty in back-extrapolation, apparently quite similar in the CT and SC-bend specimens tested here. The initiation value, as noted earlier, is near $J_{Ic} \simeq 350$ kN/m. Secondly, the slope of the R-curve for the SC bend specimen is much steeper than that obtained on the side-grooved CT specimen.

Discussion

The purpose of this paper has been to arrive at the results shown in Fig. 16. From an engineering point of view, it may be considered a fortunate circumstance that the R-curve of the cracked-structure is steeper than that of the laboratory (CT) specimen. From a methodological point of view,

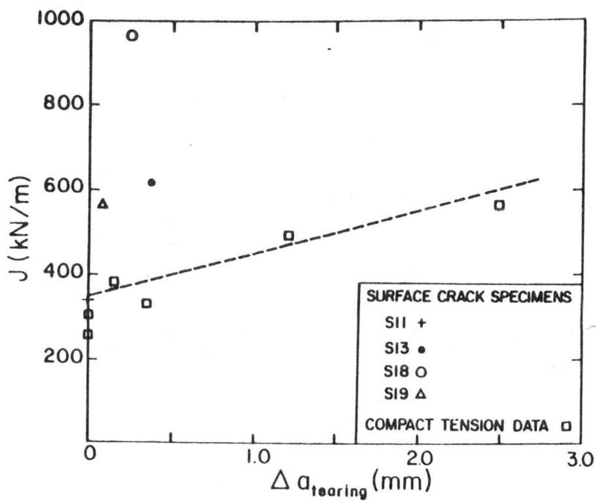

FIG. 16—J *versus amount of tearing crack extension,* $\Delta a_{tearing}$, *for the compact-tension specimen (same data as Fig. 2) and that inferred for the centerline location of the surface-cracked plates in bending.*

however, one might have hoped for a closer agreement. If closer agreement between SC and CT crack behavior is to be obtained in other cases, what conditions will be required?

It is likely that the two R-curves shown in Fig. 16, like the two differing R-curves for CT and center-cracked plate (CCP) geometries provided by Begley and Landes [*16*], are indicative in the steeper case of a loss of Hutchinson-Rice-Rosengren (HRR) dominance of the crack-tip fields. This subject has been analyzed for differing plane-strain geometries by McMeeking and Parks [*17*] and Shih and German [*18*]. In order that J-controlled HRR fields truly dominate the crack-tip deformations over physical size scales, it is necessary that geometrical dimensions be sufficiently large by comparison to the crack-tip opening displacement, given to order of magnitude by the length J/σ_f, where σ_f is a representative flow stress such as half the sum of the ultimate tensile strength and tensile yield strength. In the material tested, $\sigma_f \sim 550$ MPa. In high-constraint fields it is recommended that ligament c satisfy $c\sigma_f/J > 25$. For the CT specimens, with $c = 20.3$ mm, this means $J < 447$ kN/m. McMeeking and Parks [*17*], found very close agreement with J-dominance in pure bending for values of $c\sigma_f/J$ as low as 12, however, so that it is likely that the initiation and at least some stable extension was J-controlled in the CT specimen.

The SC specimens were in full section only $t = 12.7$ mm thick, and with $a_0/t = 0.35$, the remaining ligament at deepest penetration of the surface crack was $c = 8.3$ mm. If the permissible ratio of ligament to J/σ_f is as small as 12 for SC specimens in bending, then J-dominance would be obtained up to $J = 378$ KN/m in the SC specimen at centerline. However, the deformations imposed in order to obtain crack growth of order 1 mm in the SC specimens likely correspond to conditions beyond J-controlled HRR dominance of the crack front fields.

Another way in which crack-tip constraint could be lost pertains to short crack lengths. Using slipline theory, Ewing [*19*] has analyzed conditions in pure plane-strain bending under which the crack flank can have a negative rotation rate relative to the $\dot{\theta}$ of the half of the specimen to which it is attached. This corresponds to a breaking back of plastic deformation to the crack-side free suface, as opposed to remaining confined to the uncracked ligament. Rephrasing Ewing's critical conditions, we can find a statically admissible slipline extension of the modified GH field for cases when the ratio of crack length "a" to radius of the circular arc "ρ" of the GH field becomes less than the critical value

$$a/\rho \leq 1.09 \tag{30}$$

This was not encountered in the present study except near the free surface where a was small and, due to the extra tension, $\rho = 0.388\, c_1 = 0.388\, c\,(1 + N/2\tau_0 c)$ was largest. For values of ϕ greater than approximately 0.5, the axial force data of Fig. 14 and the crack depth profile were sufficient to ensure that plane-strain break-back would not occur.

In all of the aforementioned possible limits of HRR-dominance, the conditions under consideration were plane strain, both close to the crack front (as it must be at positions which are finitely removed from the intersection of the crack front and a free surface) but also *throughout the body*. Crack-tip uniqueness is likely most easily lost in the near-hyperbolic conditions of the equations of continuing equilibrium in low-hardening materials under plane strain. For a surface-cracked body, plane-strain conditions rigorously apply only *very* close to the crack front, and it is not at all clear under what conditions HRR-dominance is either assured or lost in general. This topic would seem to merit further careful examination, both experimentally and computationally.

No obvious manifestations of a loss of HRR-dominance such as off-plane cracking or shear band localization of deformation were observed in the SC specimens.

It is known that the presence or absence of side grooves on CT specimens can have effects qualitatively similar to the varying $\Delta J/\Delta a$ slopes found here. The SC specimen has a crack front which does intersect a plane-stress free surface which is not grooved. It is perhaps possible that this region exerted a "force" which tended to reduce the crack growth at other locations on the SC crack front in a manner similar to the effects of shear lips in CT specimens. It seems unlikely that this effect could be very strong at the specimen centerline, however.

No account of SC crack growth was taken in the computations, although such effects on evolving crack shape could, in principle, be performed. In view of the rather small remaining ligament in the first place, it is possible that "small" SC extensions, of order 1 mm, could have pronounced effects on the local force and moment distributions, as well as on the accompanying kinematics. Note, for example, that as centerline N becomes more negative with increasing deformation, the radius ρ of the GH field becomes progressively smaller. Indeed ρ vanishes when the load on the model edge crack is statically equivalent to midligament compression. The center of rotation tends to the crack tip and, although an increasing gage displacement δ_g would be observed, the deformation of the crack-tip region would essentially cease. Perhaps some aspect of this "self-limiting" cracking or load redistribution or both is taking place in plastic SC plate bending experiments.

In the experiments, imposed crosshead displacements exceeded 10 mm at the largest δ_g-values reported. It is possible, then, that some errors were introduced in the analysis by not updating the plate/spring geometry to account for large rotations, but still small strains. If significant additional compression were to have occurred on the surface crack cut, due to large geometry change effects, the problem of small GH field radius noted in the foregoing would be increased. Although these displacements are not truly "large," the observed Brinelling of the small rollers and the fixity of the large rollers could still introduce important membrane effects into the problem. This topic is under investigation.

Summary

Compact-tension and surface-cracked plate specimens of SA542 Class 3 steel were tested to obtain load/deformation/crack extension behavior. The data for the CT specimens was processed into a J_R-curve in a manner consistent with that of ASTM Method E 813-81. Using the elastic-plastic line-spring model, as implemented in the ABAQUS finite-element program, J-estimates were made for the SC specimen as well.

The computed force versus surface crack opening was in excellent agreement with experiments. The inferred J-variation along the crack front, like the observed crack extension, was extremely uniform for this choice of specimen and crack dimensions. In the elastic regime, this uniformity of J also matches another detailed three-dimensional analysis of the problem.

The J-values at initiation of crack tearing were inferred to be similar in the CT and SC specimens, but the tearing slopes $\Delta J/\Delta a$ were found to differ considerably.

A number of possible reasons for the observed differences were considered, but no definite conclusions could be made. Further careful specimen design, testing, and analysis are recommended. Nonetheless, it would seem that the elastic-plastic line spring shows promise of providing useful, economical information regarding the behavior or surface-cracked plates in ductile materials.

Acknowledgments

We are pleased to acknowledge the sources of support which have made this work possible. One of us (C. S. White) was supported by an International Harvester Co. fellowship. The participation of David M. Parks was supported by the U.S. Office of Naval Research Structural Mechanics Program under Grant N000014-80-C-0706 to the Massachusetts Institute of Technology. Robert O. Ritchie was supported in part by the U.S. Department of Energy, under contracts with the Office of Basic Energy Sciences and the Division of Fossil Energy Research.

We are also pleased to acknowledge the support of the ABAQUS finite-element program by Hibbitt, Karlsson and Sorensen, Inc., and by Drs. H. D. Hibbitt and E. P. Sorensen, in particular.

We are grateful to Dr. John Landes for his assistance in making the SA542 steel available to us.

The manuscript was carefully prepared by Mary Toscano.

References

[1] Tada, H., Paris, P. C., and Irwin, G. R., *The Stress Analysis of Cracks Handbook*, Del Research Corp., Hellertown, Pa., 1973.

[2] Parks, D. M., *Computer Methods in Applied Mechanics and Engineering*, Vol. 12, 1977, pp. 353–364.

[3] *ABAQUS Theory Manual*, Hibbitt, Karlsson and Sorensen, Inc., Providence, R.I., Jan. 1981.

[4] Rice, J. R. and Levy, N., *Transactions*, American Society of Mechanical Engineers, *Journal of Applied Mechanics*, Vol. 39, 1972, pp. 185–194.

[5] Rice, J. R. in *The Surface Crack: Physical Problems and Computational Solutions*, J. L. Swedlow, Ed., American Society of Mechanical Engineers, 1972.

[6] Parks, D. M., *Transactions*, American Society of Mechanical Engineers, *Journal of Pressure Vessel Technology*, Vol. 103, 1981, pp. 246–254.

[7] Parks, D. M. in *Advances in Fracture (Fracture 1981)*, D. Francois, Ed. Pergamon Press, Oxford, U.K., Vol. 5, 1981, pp. 2589–2598.

[8] Parks, D. M. and Lockett, R. R., *EPRI Ductile Fracture Review Document*, EPRI NP-80-10-LD, Electric Power Research Institute, Palo Alto, Calif., 1980, pp. 9.1–9.12.

[9] Parks, D. M., Lockett, R. R., and Brockenbrough, J. R. in *1981 Advances in Aerospace Structures and Materials*, S. S. Wang and W. J. Renton, Eds. American Society of Mechanical Engineers, 1981, pp. 279–285.

[10] Parks, D. M., Rodin, G. J., and Lockett, R. R., "The Line-Spring Model for *J*-Analysis of Surface-Cracked Plates and Shells: Calibration of the Power-Law Spring," presented at the Second ASTM Symposium on Elastic-Plastic Fracture, Philadelphia, Pa., 6–9 Oct. 1981.

[11] McCabe, D. E. and Landes, J. D., "Design Properties of Steels for Coal Conversion Vessels: Mechanical Properties of Materials," EPRI Report AP-1338, Research Project 627-1 (Westinghouse R&D Center, Pittsburgh, Pa.), Electric Power Research Institute, Palo Alto, Calif., Jan. 1980.

[12] Rice, J. R. in *Mechanics and Mechanisms of Crack Growth*, M. J. May, Ed. British Steel Corp. Physical Metallurgy Centre Publication; *Proceedings*, Conference at Churchill College, Cambridge, U.K., 1973.

[13] Shiratori, M. and Miyoshi, T. in *Proceedings*, Third International Conference on Materials, Vol. 3, Cambridge, U.K., 1980, pp. 425–434.

[14] Green, A. P. and Hundy, B. B., *Journal of the Mechanics and Physics of Solids*, Vol. 4, 1956, pp. 128–144.

[15] Newman, J. C., Jr., and Raju, I. S., "Analysis of Surface Cracks in Finite Plates Under Tension or Bending Loads," NASA Technical Paper 1578, National Aeronautics and Space Administration, Washington, D.C., Dec. 1979.

[16] Begley, J. A. and Landes, J. D., *International Journal of Fracture*, Vol. 12, 1976, pp. 764–766.

[17] McMeeking, R. M. and Parks, D. M. in *Elastic-Plastic Fracture, ASTM STP 668*, J. D. Landes, J. A. Begley, and G. A. Clarke, Eds., American Society for Testing and Materials, 1979, pp. 175–194.

[18] Shih, C. F. and German, M. D., *International Journal of Fracture*, Vol. 17, 1981, pp. 27–43.

[19] Ewing, D. J. F., *Journal of the Mechanics and Physics of Solids*, Vol. 16, 1968, pp. 205–213.

Masaki Shiratori[1] *and Toshiro Miyoshi*[2]

Evaluation of J-Integral for Surface Cracks

REFERENCE: Shiratori, M. and Miyoshi, T., **"Evaluation of J-Integral for Surface Cracks,"** *Elastic-Plastic Fracture: Second Symposium, Volume I—Inelastic Crack Analysis, ASTM STP 803*, C. F. Shih and J. P. Gudas, Eds., American Society for Testing and Materials, 1983, pp. I-410–I-424.

ABSTRACT: A new finite-element program has been developed in order to analyze the stress-intensity factor K and J-integral for surface cracked plates and shells. The program is composed of the flat shell elements and the line spring elements which are based upon the line spring model proposed by Rice. Summary of results analyzed by this program is as follows:

1. In the analysis of K for surface-cracked rectangular plates, the results have been in good agreement with the three-dimensional finite-element results obtained by Raju and Newman. The computing time is one order less than the three-dimensional finite-element analysis.
2. Nearly the same results as those by Parks have been obtained in the analysis of the J-integral for surface-cracked rectangular plates.
3. As a practical application of this program, K and J have been analyzed for surface-cracked part-circular cylindrical shells. The effect of curvature has been found to lower the values of K and J slightly. Therefore, estimates of K and J by those of flat plates for the equivalent crack configuration give conservative results.

KEY WORDS: J-integral, stress-intensity factor, surface crack, line spring model, finite-element method, tearing instability, fracture, elastic-plastic fracture

The J-integral is effective in determining the plane-strain fracture toughness K_{Ic} with small laboratory specimens. This J_{Ic} testing method is now under study for standardization [*1*].[3] In this case the J-integral is adequate to be evaluated only for three-point bend bars and compact specimens. Therefore, simple and reliable equations to evaluate J are proposed for such specimens

[1]Associate professor, Department of Mechanical Engineering, Faculty of Engineering, Yokohama National University, 156 Tokiwadai, Hodogaya-ku, Yokohama, 240 Japan.

[2]Associate professor, Department of Precision Machinery Engineering, Faculty of Engineering, University of Tokyo, 7-3-1 Hongo, Bunkyo-ku, Tokyo, 113 Japan.

[3]The italic numbers in brackets refer to the list of references appended to this paper.

[*2,3*]. In recent years, however, as the toughness of the structural materials is highly increased, it is sometimes necessary to apply the fracture criterion based upon J-integral, instead of the K_{Ic} criterion, directly to the structural members. The tearing instability of the pipeline of a nuclear reactor may be the most important problem in the field. In this case J-integral has to be evaluated for cracks embedded in the general structural members such as pipes and pressure vessels.

Surface cracks embedded in the plates or shells are extremely important from the engineering point of view. As for the surface cracks, there are few studies about J-integral though a lot about the stress-intensity factor [*4–9*]. The reason comes from a big barrier that the ordinary three-dimensional finite-element analysis is too time-consuming to evaluate J-integral for surface cracks. To remove the barrier the authors have developed a new finite-element program which utilizes the ordinary flat shell elements and the newly developed line spring elements. It is successfully applied to the analysis of K and J for the tension of rectangular plates and part-circular cylindrical shells with surface cracks.

Method of Analysis

Figure 1 shows the line spring model proposed by Rice [*10,11*]. The problem of a three-dimensional surface crack (Fig. 1*a*) is reduced to that of stretching and bending of a plate with a line spring at $x_2 = 0$ and $|x_1| \leq c$ (Fig. 1*b*). Later, Parks [*12*] extended the model to the analysis of J-integral. He analyzed K and J for infinite plates with semi-elliptical surface cracks where the boundary conditions are simple enough to lead to the simple dominating integral equations. But application of the approach is limited to such simple problems only.[4] On the other hand, the finite element method (FEM) is more useful for more complex problems. Therefore, the authors have developed a new FEM program which is based upon the line spring model. Flat shell elements and line spring elements are used in the program. The flat shell element is composed of the plane stress element with constant strains and the bending element with constant curvatures (Fig. 2). It is one of the most fundamental elements for the analysis of plates and shells [*14*]. On the other hand, the line spring element is newly developed in order to carry out the analysis of the line spring model by FEM. The formulation of this element is as follows:

1. The crack border in Fig. 1*b*, $x_2 = 0$ and $|x_1| \leq c$, is discretized into a finite number of line spring elements as shown in Fig. 3*a*.
2. The line spring element is composed of two straight sides $\overline{14}$ and $\overline{23}$ with four nodes as shown in Fig. 3*b*. Both sides lie upon another on the ξ-axis

[4] Recently, Parks has applied his method to shallow cylindrical shells with longitudinal surface flaws as well [*13*].

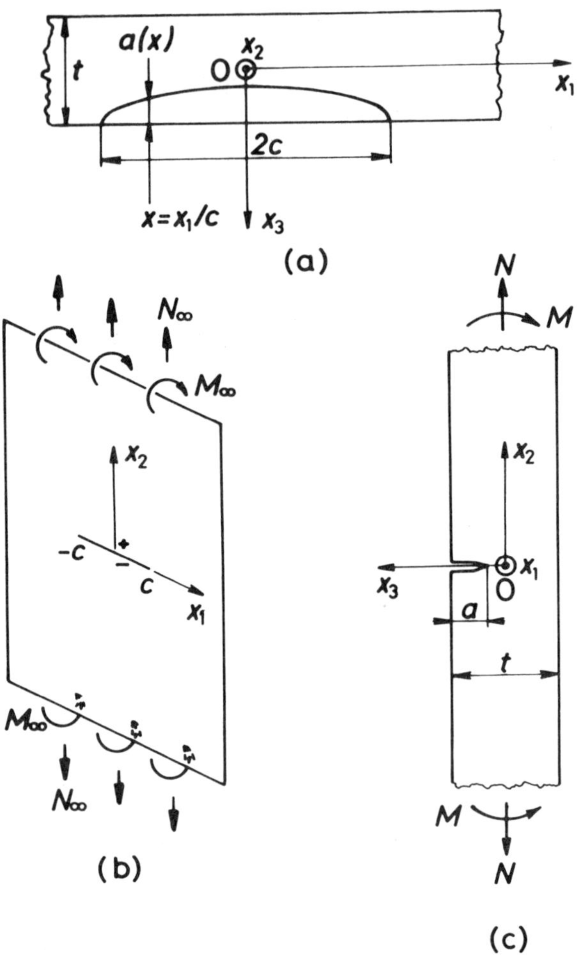

FIG. 1—*Schematic illustrations of the line spring model.*

before deformation. After deformation, each side displaces upward and downward, respectively, but these are connected to each other by the springs $\overline{12}$ and $\overline{43}$.

3. If (δ_A, θ_A) and (δ_B, θ_B) are denoted to be the generalized displacements of the line spring at $\xi = -1$ and 1, respectively, the relations between these and the nodal displacements (v_i, ϕ_i) $(i = 1\text{-}4)$ are

$$\delta_A = v_1 - v_2, \qquad \theta_A = \phi_1 - \phi_2 \qquad \delta_B = v_4 - v_3, \qquad \theta_B = \phi_4 - \phi_3 \quad (1)$$

If the generalized displacements (δ, θ) at an arbitrary point ξ are assumed to vary linearly within the element, the displacement function is given by

$$\begin{Bmatrix} \delta \\ \theta \end{Bmatrix} = \begin{bmatrix} N_1' & 0 & N_2' & 0 \\ 0 & N_1' & 0 & N_2' \end{bmatrix} \begin{Bmatrix} \delta_A \\ \theta_A \\ \delta_B \\ \theta_B \end{Bmatrix} \tag{2}$$

where

$$N_1' = \frac{1-\xi}{2}, \qquad N_2' = \frac{1+\xi}{2} \tag{3}$$

4. The equivalent nodal forces for this element (N_i, M_i) $(i = 1\text{-}4)$ can be calculated by the same method as the distributed load is evaluated, that is

$$\begin{Bmatrix} N_1 \\ M_1 \\ N_4 \\ M_4 \end{Bmatrix} = -\begin{Bmatrix} N_2 \\ M_2 \\ N_3 \\ M_3 \end{Bmatrix} = \int_{L_e} \begin{bmatrix} N_1' & 0 \\ 0 & N_1' \\ N_2' & 0 \\ 0 & N_2' \end{bmatrix} \begin{Bmatrix} N \\ M \end{Bmatrix} ds \tag{4}$$

where the integral is taken over the length L_e of the element, that is, for $|\xi| \leq 1$.

5. The relation between the generalized force (N, M) and the corresponding generalized displacement (δ, θ) at the point ξ within the element has the form of

$$\begin{Bmatrix} \delta \\ \theta \end{Bmatrix} = \begin{bmatrix} c_{11} & c_{12} \\ c_{21} & c_{22} \end{bmatrix} \begin{Bmatrix} N \\ M \end{Bmatrix} \quad \text{or conversely} \quad \begin{Bmatrix} N \\ M \end{Bmatrix} = \begin{bmatrix} s_{11} & s_{12} \\ s_{21} & s_{22} \end{bmatrix} \begin{Bmatrix} \delta \\ \theta \end{Bmatrix} \tag{5}$$

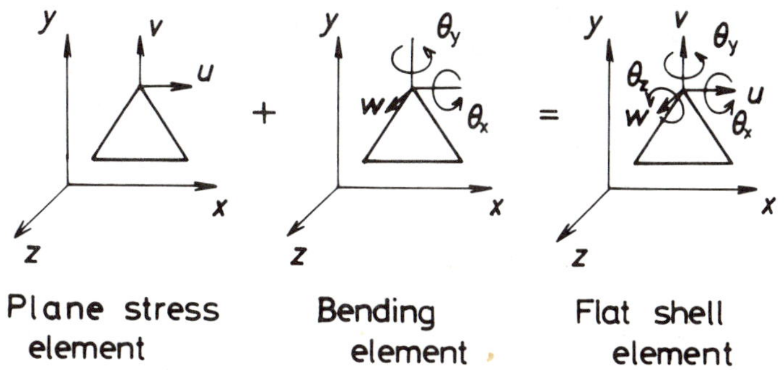

FIG. 2—*Flat shell element:* u, v, w = *components of displacements parallel to the indicated axis, and* θ_x, θ_y, θ_z = *components of rotations about the indicated axis.*

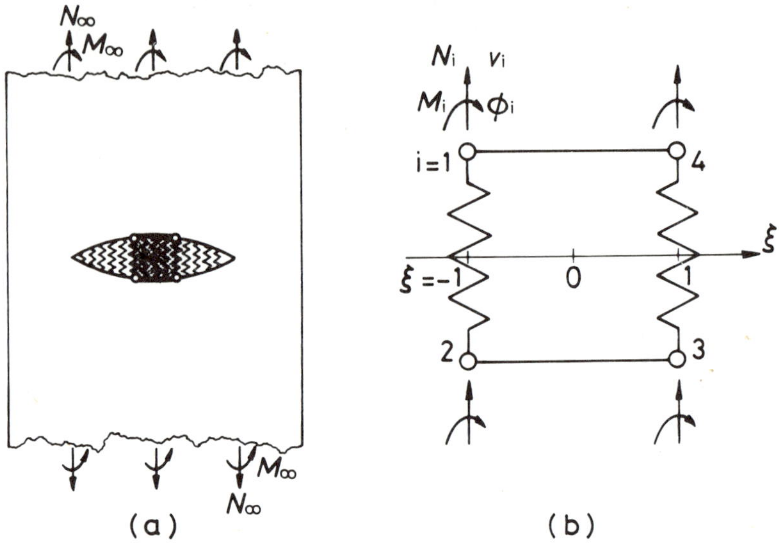

FIG. 3—*Line spring element.*

where c_{ij} and s_{ij} are the compliance and the stiffness of the line spring at ξ, respectively. These are evaluated by those of the single-edge-cracked strip with the same crack length versus thickness ratio a/t (Fig. 1*c*) in the same way indicated in Refs *10–12*. That is to say, the stress magnification factors F_N for tension and F_M for bending from Ref *15* are utilized to evaluate the elastic compliance while the upper-bound solution from Ref *11* is used to obtain the plastic stiffness. The constitutive equation for plasticity is expressed in an incremental form instead of Eq 5.

6. Substituting Eq 2 into Eq 5 and then into Eq 4

$$\begin{Bmatrix} N_1 \\ M_1 \\ N_4 \\ M_4 \end{Bmatrix} = - \begin{Bmatrix} N_2 \\ M_2 \\ N_3 \\ M_3 \end{Bmatrix}$$

$$= \int_{L_e} \begin{bmatrix} N_1' & 0 \\ 0 & N_1' \\ N_2' & 0 \\ 0 & N_2' \end{bmatrix} \begin{bmatrix} s_{11} & s_{12} \\ s_{21} & s_{22} \end{bmatrix} \begin{bmatrix} N_1' & 0 & N_2' & 0 \\ 0 & N_1' & 0 & N_2' \end{bmatrix} ds \begin{Bmatrix} \delta_A \\ \theta_A \\ \delta_B \\ \theta_B \end{Bmatrix} \tag{6}$$

Now define the nodal displacement vector $\{\delta\}_e$ and the corresponding equivalent nodal force vector $\{F\}_e$ of the line spring element by

$$\begin{aligned}\{\delta\}_e^T &= \lfloor v_1\,\phi_1\,v_2\,\phi_2\,v_3\,\phi_3\,v_4\,\phi_4 \rfloor \\ \{F\}_e^T &= \lfloor N_1\,M_1\,N_2\,M_2\,N_3\,M_3\,N_4\,M_4 \rfloor\end{aligned} \tag{7}$$

Then the stiffness equation for the line spring element

$$\{F\}_e = [K]_e\,\{\delta\}_e \tag{8}$$

can be obtained easily by integrating and rearranging Eq 6. Since $\phi = -\theta_x$ in Figs. 2 and 3, it is easy to see that the element is compatible with the flat shell element.

The generalized displacement (δ, θ) and then the generalized force (N, M) along the crack border can be obtained by applying this FEM program to the problem to be solved. Then, K and J can be evaluated in the manner indicated in Refs *10–12*, that is

$$K_I = N\cdot F_N + M\cdot F_M \tag{9}$$

$$J = J^e + J^p \tag{10}$$

$$J^e = \frac{1-\nu^2}{E}K_I^2, \qquad J^p = m\sigma_0\delta_t^p, \qquad \dot{\delta}_t^p = \dot{\delta}^p + \left(\frac{t}{2} - a\right)\dot{\theta}^p$$

where

E = Young's modulus,
ν = Poisson's ratio,
σ_0 = flow stress,
$m = 2/\sqrt{3}$ = parameter affected by the plastic constraint at the notch root, and
δ_t^p = plastic crack tip opening displacement for the edge-cracked strip in Fig. 1*c*.

Superscripts e and p denote the elastic and plastic components, respectively, and "·" means the increment of the indicated variables.

Tension of Flat Plate with a Semi-Elliptical Crack

Analysis of the Stress-Intensity Factor

At first the stress-intensity factors for semi-elliptical cracks embedded in the rectangular flat plates subject to tension have been calculated by this program in order to examine the accuracy and the computing time of the proposed method. Figure 4 shows the geometry and the element breakdown of one quarter of the specimen. Due to symmetry, one quarter is enough to be analyzed. The dimensions of the specimen have been taken as

$$2c/W = 0.2, \qquad 2c/L = 0.2, \qquad a/c = 0.2$$

$$a/t = 0.2, 0.4, 0.6, \quad \text{and} \quad 0.8$$

Raju and Newman [7] calculated K-values for the same geometry by three-dimensional FEM. Therefore, results of the present analysis can be compared with those in Ref 7. Figure 5 shows the stress-intensity factors along the leading edges of the cracks. The vertical axis shows the stress-intensity factor K_I normalized by $\sigma_\infty\sqrt{\pi a/Q}$ where σ_∞ is applied tensile stress and Q is the square of the elliptical integral of the second kind, while the horizontal axis shows the portion of the leading edge of the crack defined in Fig. 4.

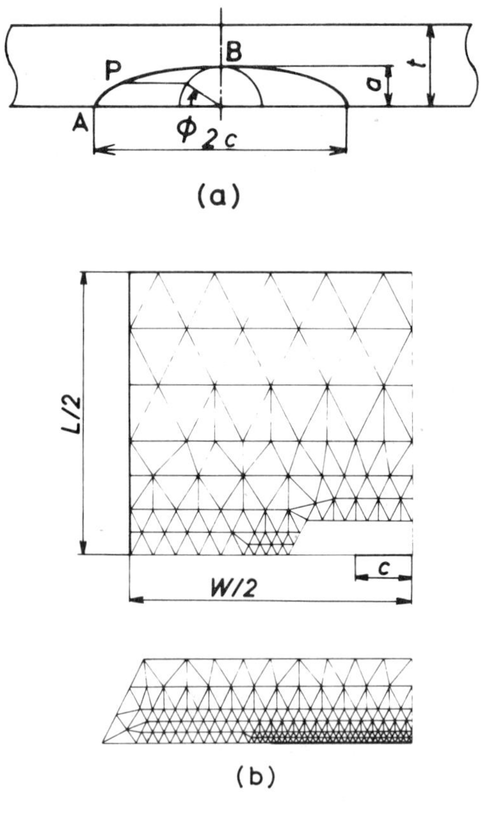

FIG. 4—*Shape of the specimen and element breakdown for elastic analysis.*

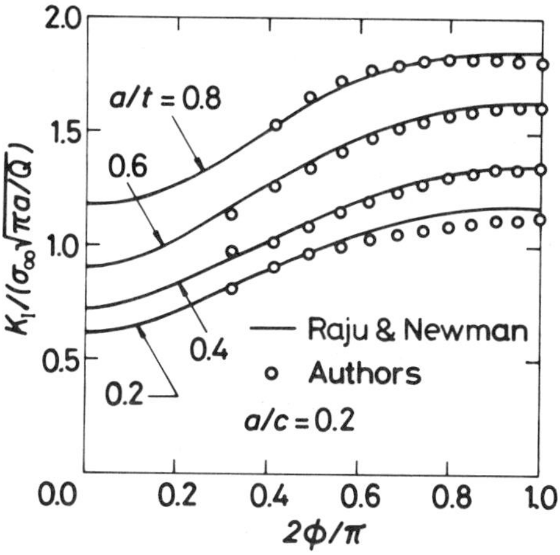

FIG. 5—*Normalized stress-intensity factor along the leading edge of the crack in a rectangular plate subject to tension.*

Figure 5 shows that the present results are in good agreement with those by Raju and Newman except for $2\phi/\pi < 0.4$, that is, near the surface. Generally, the solution based upon the line spring model is inaccurate near the surface as shown by the dotted line in Fig. 6*b*. The following fact may be one of the reasons for the error. Consider the stress-intensity factor at point P in Fig. 6*a*. The K-value obtained by Raju and Newman is the one normal to the leading edge of the crack. Let it be K_1. On the other hand, the K-value obtained by the line spring model is the one parallel to the x_3-axis. Let it be K_2. For Points P, Q, and R in Fig. 6*a*, K_1 and K_2 have the following relation

$$\sigma_{x_2}(Q) = K_1(R)/\sqrt{r_1} = K_2(P)/\sqrt{r_2}$$

Therefore, in the limit of $R \rightarrow P$

$$\frac{K_1}{K_2} = \lim_{R \rightarrow P} \frac{K_1(R)}{K_2(P)} = \lim_{r_1, r_2 \rightarrow 0} \frac{\sqrt{r_1}}{\sqrt{r_2}} = \sqrt{\sin \theta_0} \tag{11}$$

The open circles in Fig. 6*b* show the K_1-values modified through Eq 11. The modification improves the accuracy in the range $0.2 < 2\phi/\pi < 0.4$.

Analysis of J-Integral

The general elastic-plastic analysis including the domain around the line spring is possible by FEM while the method in Ref *12* has to confine the plastic region to the line spring only. But in this paper the authors have carried out the elastic-plastic analysis by confining the plastic zone to the line spring only in order to compare the results with those in Ref *12*. Therefore, it is as-

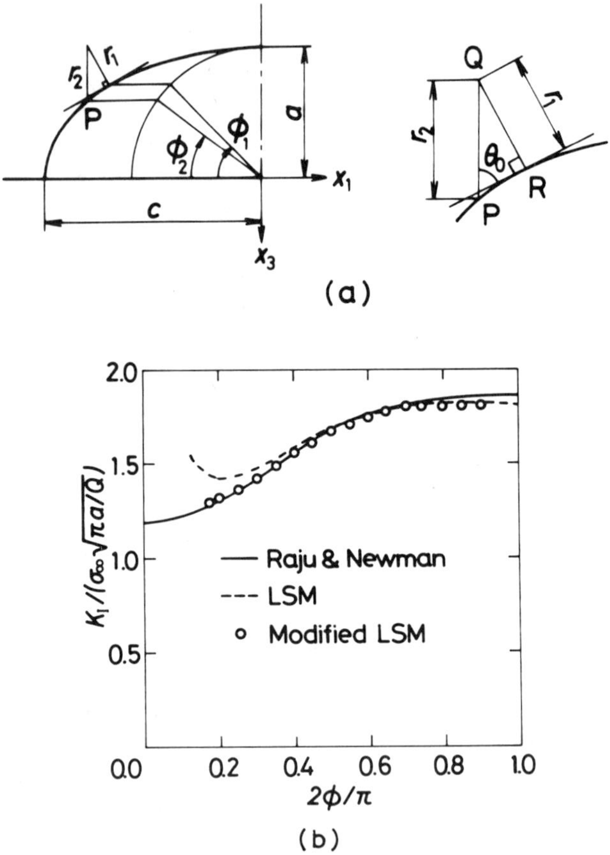

FIG. 6—*Modification of the results calculated by the line spring model due to a geometrical consideration.*

sumed that plasticity occurs in the line spring elements while the constitutive equations for the flat shell elements are held elastic throughout the analysis. Incremental procedures are adopted by changing the stiffness of the line spring step by step from elastic to plastic. The judgment of yielding condition is made at integrating points for numerical integral.

J-integral is analyzed for the same configurations and the same load conditions as the analysis of K. But a slightly coarser mesh is used to save computing time (Fig. 7). Work-hardening is neglected; therefore, the flow stress σ_0 in Eq 10 is taken to be the initial yielding stress σ_Y. Figure 8 shows the normalized J-integral versus the normalized applied tensile stress at the center of the crack, that is, at Point B in Fig. 4a. The open circles show the results of the present FEM analysis while the solid curves show those by Parks, who

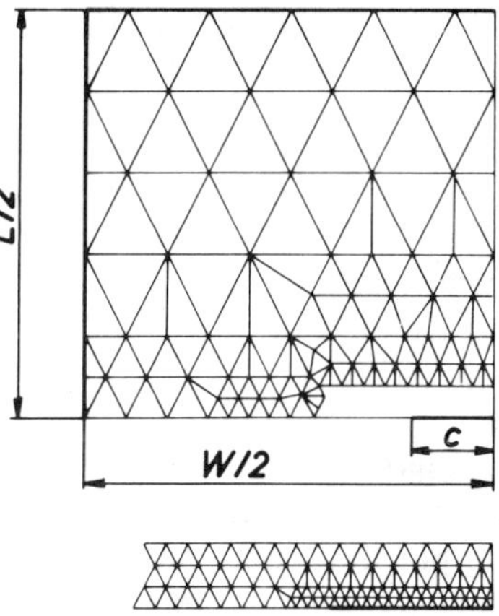

FIG. 7—*Element breakdown for elastic-plastic analysis.*

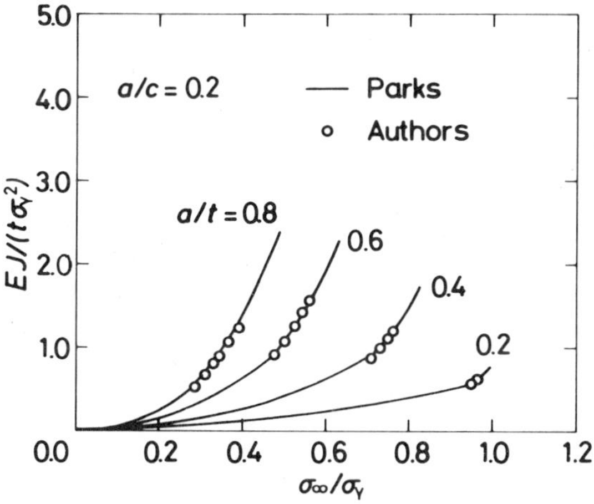

FIG. 8—*Normalized J-integral versus normalized applied stress at the center of the semi-elliptical crack in a rectangular plate subject to tension.*

solved the dominating integral equations by a point matching technique. As the coincidence is remarkable, errors caused by the difference in methodology for solving the dominating equations are negligible as far as the line spring model is concerned. Although Parks has applied his method not only to infinite plates but also to shallow shells [*13*], it is obvious that the present finite-element method has more potential for wide application than his method, as he himself recognizes [*16*]. On the other hand, three-dimensional FEM may be hopelessly time-consuming at the moment.

Tension of Part-Circular Cylindrical Shell with a Surface Crack

In this section the program is applied to the problem of the tension of part-circular cylindrical shells with surface cracks, and the effect of the curvature of the cylindrical shell on K and J has been considered. The problem is important because the solution may give a good approximation for a line pipe with a surface crack subject to bending. Especially, the solution of J-integral may be essential to the leak-before-break (LBB) analysis of pipelines.

Figure 9*b* shows the shape of the half of the flat plate with a semi-elliptical surface crack which has been analyzed in the previous section. The plate is bent to a part-circular cylinder shown in Fig. 9*a*, which is the object of the present analysis. Therefore, the configuration of the surface crack is not semi-elliptical any more, but the corresponding configuration of the crack in the flat plate is semi-elliptical. From geometrical consideration in Fig. 9

$$\Theta = \frac{W}{2R}, \qquad \theta_c = \frac{c}{R}, \qquad \theta = \frac{x}{R} \tag{12}$$

where R is the radius of the cylindrical shell, and other parameters are defined in the figure. Uniform tensile stress is applied to the end of the speci-

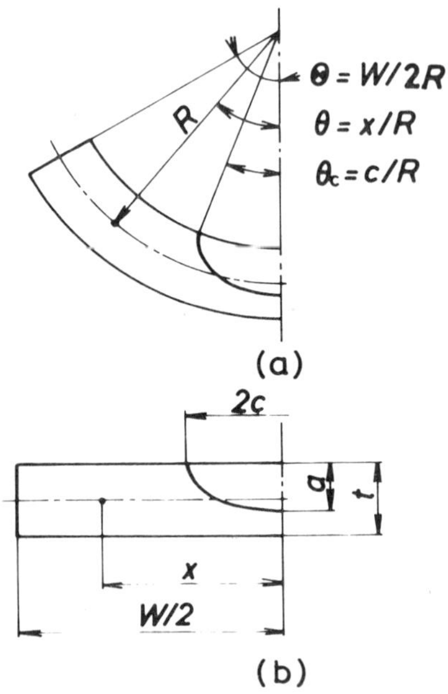

FIG. 9—*Configuration of a part-circular cylindrical shell containing a surface crack formed by bending a rectangular plate with a semi-elliptical crack.*

men perpendicular to the paper surface in the same way as in the previous section. For the parametric study of K, a/t and Θ are taken to be

$$\frac{a}{t} = 0.2, \qquad 0.4 \quad 0.6 \quad \text{and} \quad 0.8$$

$$\Theta = 0, \qquad \frac{\pi}{12}, \qquad \frac{\pi}{4}, \qquad \frac{\pi}{3}, \quad \text{and} \quad \frac{\pi}{2}$$

while a/c and c/W are held constant ($a/c \quad 2c/W = 0.2$) through the analysis. The case of $\Theta = 0$ corresponds to the flat plate analyzed in the previous section. In the analysis of J-integral, $a/t = 0.5$ while all other parameters are the same as in the analysis of K. The same finite-element meshes as those in the previous section are used in the elastic (Fig. 4) and elastic-plastic (Fig. 7) analyses.

Figures 10*a*–10*d* show the distributions of the stress-intensity factor along the leading edge of the surface cracks for $a/t = 0.2$, 0.4, 0.6, and 0.8, respectively. For simplicity the stress-intensity factor is normalized by $\sigma_\infty\sqrt{\pi a}$ because the configuration of the surface crack is not semi-elliptical any more. These figures show that the values of K decrease with increase of Θ for all cases analyzed. This fact means that the K-values for part-circular cylindrical shells with surface cracks can be approximated with a little conservative safety factor by those for flat plates with the corresponding semi-elliptical cracks.

Figure 11 shows the normalized J-integral versus the normalized applied stress at the center of the crack ($\theta = 0$) for $a/t = 0.5$. The values of J decrease with increase of Θ; therefore, a conservative approximation is possible in the same manner as in the case of K-value.

Further work is necessary, however, to apply these results to the problems of bending of a pipe with a circumferential surface flaw. For example, the stress distribution due to pure bending is different from the uniform tensile stress, and therefore the foregoing results for the flat plates might be extremely conservative for the pipe with large curvature and large surface flaw subject to bending.

Discussion

The computing time by the proposed program is about a tenth part in the analysis of K and far less in the analysis of J when compared with three-dimensional FEM. Although the method in Ref *12* is less time-consuming than the proposed method, it should be remembered that the method is applicable to such simple problems only as, for example, infinite plate. Therefore, the newly developed finite element program is reliable and effective

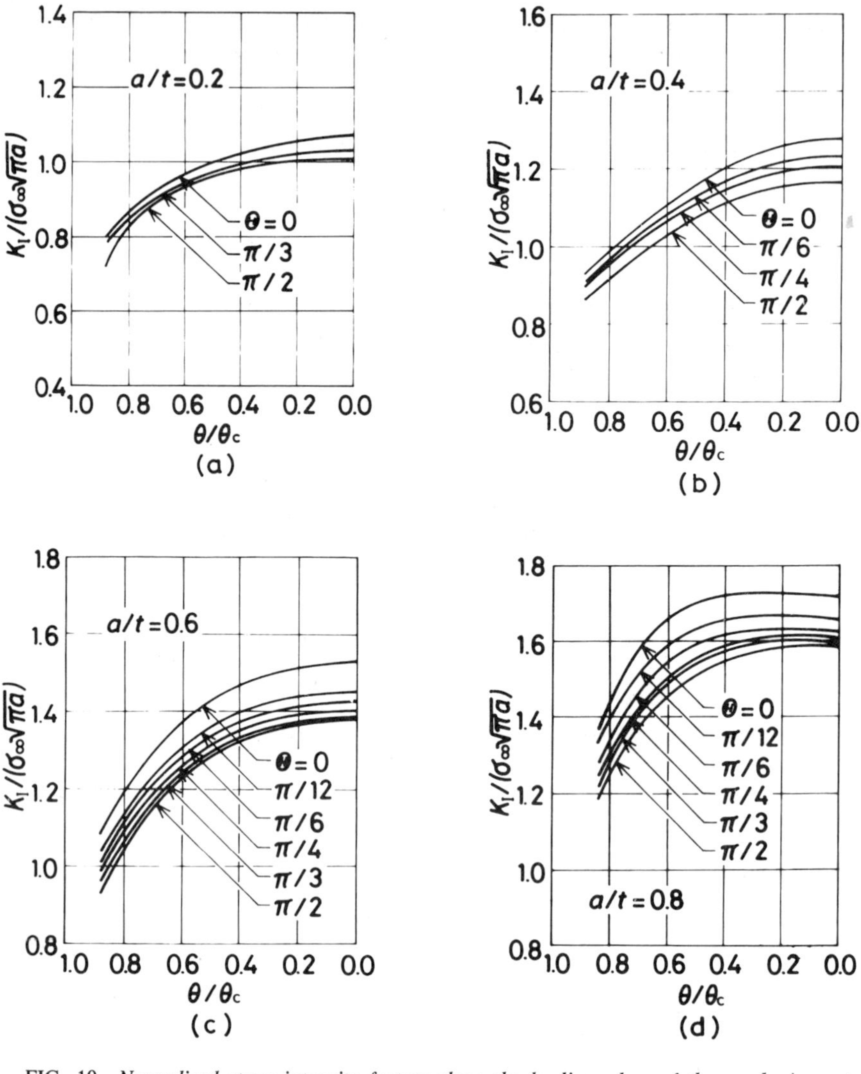

FIG. 10—*Normalized stress-intensity factors along the leading edges of the cracks in part-circular cylindrical shells subject to tension.*

from the viewpoint of accuracy, economical computing time, and wide range of applications.

A short-coming of the method is that accuracy is not assured when the leading edge of the crack approaches nearly normal to the surface of the plate, as in the case already discussed in connection with Figs. 5 and 6. It may be a serious problem for relatively short cracks with large a/c ratios.

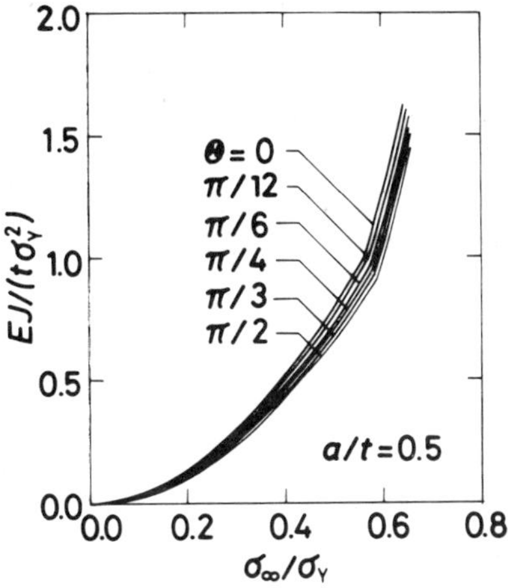

FIG. 11—*Normalized J-integral at the center of the surface cracks in part-circular cylindrical shells subject to tension.*

Therefore, further improvements are necessary to overcome the difficulty. The method based upon Eq 11 is one such improvement. The flat shell element used in this paper is based upon Kirchhoff plate theory, while Parks [*16*] and Delale and Erdogan [*17*] have recently obtained the line spring solutions based upon Reissner plate theory. Their results show that the accuracy is improved for the cases of relatively large a/c ratios.

As an important application of the proposed method, there is the simulation of the tearing instability of line pipes with surface cracks. The problem is important from the viewpoint of the leak-before-break analysis. The simulation is possible through computing the crack propagation process controlled by the tearing modulus criterion [*18*]. It is also possible to consider the effects of the compliance of the pipeline systems in the same way as proposed by Hutchinson and Paris [*19*]. This is the topic of our further work.

Conclusions

The newly developed finite-element program, which is based upon the line spring model, has been found reliable and effective in the analysis of the stress-intensity factor and the J-integral for plates and shells with surface cracks from the viewpoints of accuracy, economical computing time, and wide range of applications. In the analysis of K and J for part-circular cylindrical shells with circumferential inner-surface cracks, it has been found that they can be approximated conservatively by those for flat plates with the corresponding crack configurations.

Acknowledgment

The authors would like to acknowledge the help of Mr. Y. Yoshida and Mr. M. Nemoto in the present finite-element work.

References

[*1*] Clarke, G. A., Andrews, W. R., Begley, J. A. Donald, J. K., Embley, G. T., Landes, J. D., McCabe, D. E., and Underwood, J. H., *Journal of Testing and Evaluation*, Vol. 7, No. 1, 1979, pp. 49–56.

[*2*] Rice, J. R., Paris, P. C., and Merkle, J. G. in *Progress in Flaw Growth and Fracture Toughness Testing, ASTM STP 536*, American Society for Testing and Materials, 1973, pp. 231–245.

[*3*] Merkle, J. G., and Corten, H. T., *Journal of Pressure Vessel Technology, Transactions*, American Society of Mechanical Engineers, Series J, Vol. 94, No. 4, 1974, pp. 286–292.

[*4*] Miyamoto, H. and Miyoshi, T. in *Proceedings*, Symposium on High Speed Computing of Elastic Structures, Liege, Belgium, Aug. 1970.

[*5*] Yagawa, G., Ichimiya, M., and Ando, Y. in *Proceedings*, First International Conference on Numerical Methods in Fracture Mechanics, A. R. Luxmoore and D. R. J. Owen, Eds., Swansea, Wales, 1978, pp. 249–267.

[*6*] Newman, J. C., Jr., in *Part-Through Crack Fatigue Life Prediction, ASTM STP 687*, J. B. Chang, Ed., American Society for Testing and Materials, 1979, pp. 16–42.

[*7*] Raju, I. S., and Newman, J. C., Jr., *Journal of Engineering Fracture Mechanics*, Vol. 11, No. 4, 1979, pp. 817–829.

[*8*] Shah, R. C. and Kobayashi, A. S. in *The Surface Crack: Physical Problems and Computational Solutions*, J. L. Swedlow, Ed., American Society of Mechanical Engineers, 1972, pp. 79–124.

[*9*] Kobayashi, A. S. in *Proceedings*, 2nd International Conference on Mechanical Behavior of Materials, American Society of Mechanical Engineers, 1976, p. 1073.

[*10*] Rice, J. R. and Levy, N., *Journal of Applied Mechanics, Transactions*, American Society of Mechanical Engineers, Vol. 39, 1972, pp. 185–194.

[*11*] Rice, J. R. in *The Surface Crack: Physical Problems and Computational Solutions*, J. L. Swedlow, Ed., American Society of Mechanical Engineers, New York, 1972, pp. 171–185.

[*12*] Parks, D. M., *Journal of Pressure Vessel Technology, Transactions*, American Society of Mechanical Engineers, Vol. 103, 1981, pp. 246–254.

[*13*] Parks, D. M. in *Fracture 1981*, D. Francois, Ed., Pergamon Press, Oxford, England, 1981, pp. 2589–2598.

[*14*] Zienkiewicz, O. C., *The Finite Element Method*, 3rd ed., McGraw-Hill, London, 1977, p. 329.

[*15*] Tada, H., Paris, P. C., and Irwin, G. R., *The Stress Analysis of Cracks Handbook*, Del Research Corp., Hellertown, Pa., 1973, pp. 2.10–2.16.

[*16*] Parks, D. M. in *EPRI Ductile Fracture Review Document*, EPRI NP-80-10-LD, WS 80-912, Electric Power Research Institute, Palo Alto, Calif., 1980.

[*17*] Delale, F. and Erdogan, F., *International Journal of Engineering Science*, Vol. 19, 1981, pp. 1331–1340.

[*18*] Paris, P. C., Tada, H., Zahoor, A., and Ernst, H. in *Elastic-Plastic Fracture*, ASTM STP 668, J. D. Landes, J. A. Begley, and G. A. Clarke, Eds., American Society for Testing and Materials, 1979, pp. 5–36.

[*19*] Hutchinson, J. W., and Paris, P. C. in *Elastic-Plastic Fracture*, J. D. Landes, J. A. Begley, and G. A. Clarke, Eds., American Society for Testing and Materials, 1979, pp. 37–64.

Masaru Sakata,[1] Shigeru Aoki,[1] Kikuo Kishimoto,[1] Morihito Kanzawa,[2] and Naoaki Ogure[2]

Effects of Thickness on J-Integral in Structures

REFERENCE: Sakata, M., Aoki, S., Kishimoto, K., Kanzawa, M., and Ogure, N., **"Effects of Thickness on J-Integral in Structures,"** *Elastic-Plastic Fracture: Second Symposium, Volume I—Inelastic Crack Analysis, ASTM STP 803*, C. F. Shih and J. P. Gudas, Eds., American Society for Testing and Materials, 1983, pp. I-425–I-443.

ABSTRACT: The distribution of the $\hat{J}$-integral, an extension of the J-integral, is calculated along the crack front of compact specimens of varying thicknesses by a three-dimensional elastic-plastic finite-element method. The $\hat{J}$-integral takes the maximum value at the midthickness and smaller values near the plate surfaces, and the $\hat{J}$-value at the midthickness coincides with the J-value determined from the computed load versus load-line displacement curve by using the Merkle-Corten formula. These results explain why the plane-strain fracture toughness value J_{Ic} can be experimentally determined by using specimens of moderate thickness.

To avoid the expense of three-dimensional computations, an empirical formula based on a two-dimensional finite-element method is proposed to determine the J-value which coincides with the $\hat{J}$-value at midthickness in a cracked structure. Experimental results on compact specimens of 304 and 4340 steels and also on rotating disks of 4340 steel indicate that the onset of crack growth in a structure with finite thickness can be estimated by using the empirical formula and the J_{Ic}-value of the compact specimen.

KEY WORDS: elastic-plastic fracture, structure, compact specimen, rotating disk, J-integral, finite-element method

It is pertinent to apply the J_{Ic} concept [*1*][3] to evaluation of the integrity of structures made of high-toughness materials. The J-values in structures have to be estimated from the line integral by numerical calculation. Although a two-dimensional finite-element method (2DFEM) is usually used by assuming plane-strain conditions, it is thought that the plane-stress state is

[1]Professor, associate professor, and research associate, respectively, Department of Physical Engineering, Tokyo Institute of Technology, Meguro-ku, Tokyo 152, Japan.

[2]Research engineers, Research Laboratory, Ebara Corp., Fujisawa, Kanagawa 251, Japan.

[3]The italic numbers in brackets refer to the list of references appended to this paper.

predominant in the far field even when the plane-strain state arises in the near field of a crack tip. Andrews and Shih [2] and Shiratori and Miyoshi [3] have reported that the J versus load-line-displacement curves span a range for varying thicknesses. The effects of thickness have to be considered in order to apply the J_{Ic} concept to fracture of structures with finite thicknesses.

In the present paper, the distribution of the Ĵ-integral [4], an extension of the J-integral, is calculated along the crack front of a compact specimen by a three-dimensional finite-element method (3DFEM). It is shown that the Ĵ-integral takes the maximum value at the midthickness and smaller values near the plate surfaces. It is also shown that the $\hat{J}$-value at the midthickness coincides with the J-value determined from the computed load versus load-line displacement curve by using the Merkle-Corten formula [5]. These results explain why the plane-strain fracture toughness value J_{Ic} can be experimentally determined by using specimens of moderate thickness.

To avoid the expense of three-dimensional computations, an empirical formula is proposed to determine from 2DFEM computation the J-integral which coincides with the $\hat{J}$-value at midthickness in a cracked structure with a finite thickness. In order to demonstrate its validity, a test program was conducted on compact specimens of ASTM A276 Type 304 and AISI 4340 steels, and its usefulness in predicting fracture initiation is shown by comparison with the experimental results on rotating disks of 4340 steel [6].

Analysis of the Effects of Thickness on J-integral in Compact Specimens

An analysis is carried out to study the effects of thickness on the J-integral in structures so that the distribution of the Ĵ-integral along the crack front of compact specimens of varying thicknesses is computed by using 3DFEM. The Ĵ-integral is an extension of the J-integral and it has been defined in a previous paper [4]. The three-dimensional version of the Ĵ-integral is written as

$$\left.\begin{aligned} \hat{J} &= \int_{\Gamma} W^e\, dX_2 - \int_{\Gamma} T_i u_{i,1} d\Gamma + \iint_A \{\sigma_{ij}\epsilon_{ij,1}{}^p - (\sigma_{i3}u_{i,1}),_3\}dA \\ W^e &= W^e(\epsilon_{ij}{}^e) = \int_0^{\epsilon_{ij}{}^e} \sigma_{ij} d\epsilon_{ij}{}^e \end{aligned}\right\} \quad (1)$$

where

σ_{ij} = stress tensor,
ϵ_{ij}^e = elastic component of strain tensor,
ϵ_{ij}^p = plastic component of strain tensor,
T_i = surface traction,
u_i = displacement,

W^e = elastic strain energy density function,
Γ = any curve surrounding the crack tip, and
A = area surrounded by both curve Γ and crack surfaces as shown in Fig. 1.

It is noted that the $\hat{J}$-integral given by Eq 1 does not lose the property of path independence and its physical significance as the energy release rate even for material described by the Prandtl-Reuss constitutive relation [*4*]. It was reported that the $\hat{J}$-value almost coincided with the J-value under monotonic loading conditions of two-dimensional problems [*7*]. In the present paper, the $\hat{J}$-integral defined by Eq 1 is computed by employing a 3DFEM. In the FEM, 20-node isoparametric brick elements are used and the $3 \times 3 \times 3$ Gaussian integration rule [*8*] is adopted in evaluating the components of the stiffness matrices and also the $\hat{J}$-integral. The procedure for computation was fully described elsewhere [*7*].

The effects of specimen thickness on the $\hat{J}$-integral are studied by computing elastic-plastic deformation and the $\hat{J}$-integral of a compact specimen as shown in Fig. 2*a*. The specimen geometry has been modified for ease of computation, and the finite-element mesh pattern of a quarter of the specimen and typical integration paths for $\hat{J}$-integral are shown in Fig. 2*b*. Computations are carried out for specimens of varying thicknesses with otherwise the same dimensions, and with the number of elements in the thickness direction being maintained. The material behavior is modeled as a linear strain-hardening material with Young's modulus E = 206 GPa, Poisson's ratio $\nu = 0.3$, tensile yield stress σ_y = 480 MPa, and the tangent modulus for plasticity $H' = d\bar{\sigma}/d\bar{\epsilon}^p = E/100$, where $\bar{\sigma}$ and $d\bar{\epsilon}^p$ are the von Mises effective stress and increment of the effective plastic strain, respectively.

The results of 3DFEM for the load per unit thickness versus load-line-displacement of the specimen with various thicknesses are summarized in

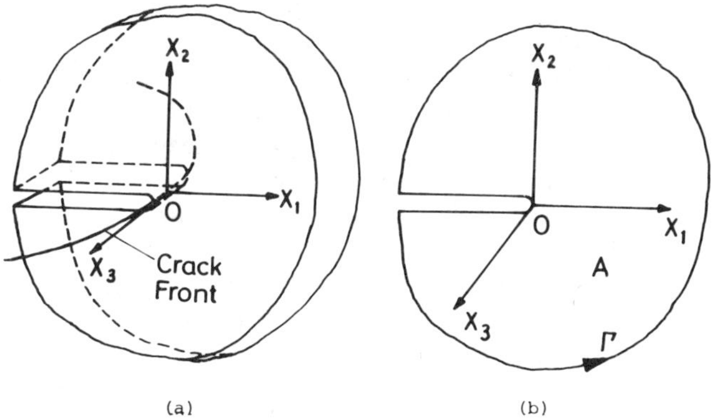

FIG. 1—*Three-dimensional crack in elastic-plastic solid.*

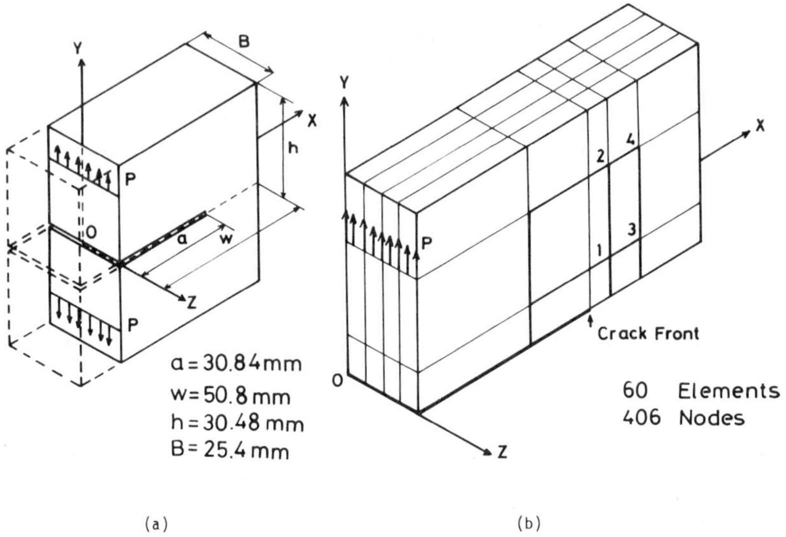

FIG. 2—*Modified compact specimen and 3D finite-element breakdown.*

Fig. 3, where the dashed curve represents the result obtained by the conventional 2DFEM using constant strain triangular elements. It is thought that the accuracy of the present 3DFEM is not satisfactory compared with the 2DFEM and a finer mesh would be required for more precise computation. However, the tendency of the effects of specimen thickness on the distribution of the $\hat{J}$-integral may be studied by the results of the present 3DFEM since the general tendency of the effects of thickness conforms with the experimental evidence reported [*2*].

The distribution of the $\hat{J}$-integral along the crack front is shown in Fig. 4, where open circles indicate the average values of $\hat{J}$-integral computed along four paths in a plane as illustrated in Fig. 2*b*. It is observed that with increase in load, the $\hat{J}$-value at the midthickness increases more remarkably than near the surface. This tendency may explain the fact that fracture intiation in real material occurs at the midthickness of the specimen. The plateau in the $\hat{J}$-distribution at the midthickness may indicate that the plane-strain condition is more predominant for thicker specimens.

The values of the $\hat{J}$-integral at the midthickness are compared with those of the J-integral determined from the computed load versus load-line-displacement curves as shown in Fig. 3 by using the Merkle-Corten formula [*5*]; the results are shown in Fig. 5. It is observed that the $\hat{J}$-values at the midthickness (represented by the symbols) agree well with the J-values determined from the Merkle-Corten formula (solid curves). This may explain the fact that the plane-strain fracture toughness value J_{Ic} can be experimentally determined by using moderate thickness specimens which satisfy the recom-

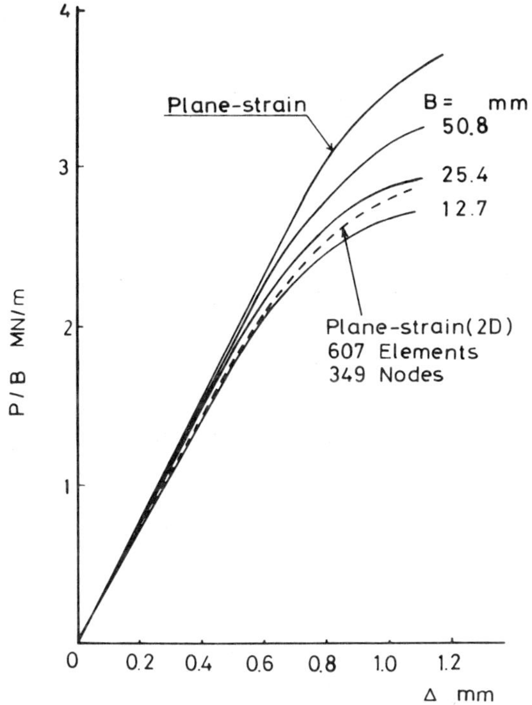

FIG. 3—*Computed results for load per unit thickness versus load-line displacement of modified compact specimens with various thicknesses.*

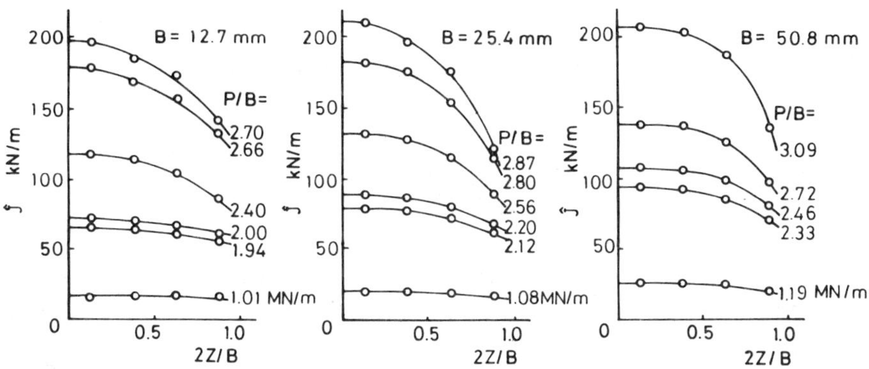

FIG. 4—*Distribution of Ĵ across the thickness of modified compact specimens with various thicknesses.*

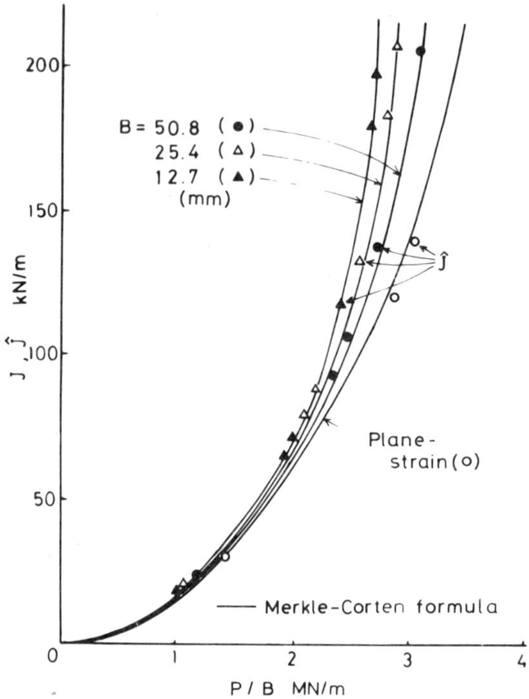

FIG. 5—*Comparison of* Ĵ *at the midthickness with the* J *obtained by the Merkle-Corten formula.*

mended size criterion for J_{Ic} testing [9], since in such a specimen the plane-strain fracture initiates at the midthickness even when the plane-stress state is predominant in the rest.

Empirical Formula for Estimating J-integral in Structures of Finite Thickness

The preceding computation suggests that the $\hat{J}$-values at midthickness have to be calculated in order to apply the J_{Ic} concept to fracture of structures with finite thickness, and that the $\hat{J}$-values at the midthickness coincide with the J-values determined from load-displacement curves. Analyses considering the effects of thickness by 3DFEM require a fine mesh and a lot of computation time and expense, so that an empirical formula based on 2DFEM is proposed for estimating the J-integral which coincides with the $\hat{J}$-values at midthickness.

It is assumed that the load of a structure is represented by the generalized force P and its associated generalized displacement by Δ [10]. The load per unit thickness, P/B, versus displacement, Δ, determined by 2DFEM is assumed to be represented by

$$(P/B)_\epsilon = k_\epsilon(a, \Delta_\epsilon)\,\Delta_\epsilon \quad \text{for plane-strain} \tag{2}$$

$$(P/B)_\sigma = k_\sigma(a, \Delta_\sigma)\,\Delta_\sigma \quad \text{for plane-stress} \tag{3}$$

where

a = crack-length,
$k_\epsilon(a, \Delta_\epsilon)$ or $k_\sigma(a, \Delta_\sigma)$ = nonlinear spring constant, and
ϵ and σ = subscripts designating, respectively, plane-strain and plane-stress condition.

We introduce an empirical formula for the load-displacement curve of a finite thickness-structure as

$$\frac{P}{B} = \{(1 - \exp[-\alpha(b/B)^{1/2}])k_\sigma(a, \Delta) + \exp[-\alpha(b/B)^{1/2}]k_\epsilon(a, \Delta)\}\,\Delta \tag{4}$$

or

$$\frac{P}{B} = k(a, \Delta)\,\Delta \tag{5}$$

where

$$k(a, \Delta) = \{1 + (m - 1)\exp[-\alpha(b/B)^{1/2}]\}\,k_\sigma(a, \Delta) \tag{6}$$

$$m = k_\epsilon(a, \Delta)/k_\sigma(a, \Delta) \tag{7}$$

where b is the ligament length and α a constant to be determined experimentally. We note that the value of α determined in the experimental stage of the present investigation was 1.0 for both 304 and 4340 steels.

The J-integral is determined from the $P - \Delta$ curve by [*11*]

$$J = -\int_0^\Delta \left.\frac{\partial(P/B)}{\partial a}\right|_\Delta d\Delta \tag{8}$$

Substitution of Eq 5 into Eq 8 leads to

$$J = -\{1 + (m - 1)\exp[-\alpha(b/B)^{1/2}]\}\int_0^\Delta \frac{\partial k_\sigma(a, \Delta)}{\partial a}\,\Delta d\Delta \tag{9}$$

By using the theorem of average in integration, this equation is written as

$$J = -\frac{1}{2}\{1 + (m - 1)\exp[-\alpha(b/B)^{1/2}]\}\frac{\partial k_\sigma(a, \Delta_R)}{\partial a}\,\Delta^2 \tag{10}$$

where Δ_R is an appropriate value satisfying the condition $0 < \Delta_R < \Delta$. Substitution of Eq 5 leads to

$$J = -\frac{1}{2}\{1 + (m-1)\exp[-\alpha(b/B)^{1/2}]\}\frac{\partial k_\sigma(a, \Delta_R)}{\partial a}\frac{1}{k^2(a, \Delta)}\left(\frac{P}{B}\right)^2 \tag{11}$$

The plane-stress state is represented in the limiting case $b/B \rightarrow \infty$ so that we have from Eqs 10 and 11

$$J_\sigma = -\frac{1}{2}\frac{\partial k_\sigma(a, \Delta_R)}{\partial a}\Delta_\sigma^2 \tag{12}$$

$$J_\sigma = -\frac{1}{2}\frac{\partial k_\sigma(a, \Delta_R)}{\partial a}\frac{1}{k_\sigma^2(a, \Delta)}\left(\frac{P}{B}\right)_\sigma^2 \tag{13}$$

For a given value of J $(=J_\sigma)$, we obtain the following relations by equating Eq 10 to Eq 12 and also Eq 11 to Eq 13

$$\Delta = \{1 + (m-1)\exp[-\alpha(b/B)^{1/2}]\}^{-1/2}\Delta_\sigma \qquad \text{for } J = J_\sigma \tag{14}$$

$$\frac{P}{B} = \{1 + (m-1)\exp[-\alpha(b/B)^{1/2}]\}^{1/2}\left(\frac{P}{B}\right)_\sigma \qquad \text{for } J = J_\sigma \tag{15}$$

The fact that the limiting case $b/B \rightarrow 0$ represents the plane-strain condition leads to

$$\frac{\Delta_\epsilon}{\Delta_\sigma} = m^{-1/2} \quad \text{for} J = J_\sigma = J_\epsilon \tag{16}$$

$$\frac{(P/B)_\epsilon}{(P/B)_\sigma} = m^{1/2} \quad \text{for} J = J_\sigma = J_\epsilon \tag{17}$$

Now that we can obtain the $J_\epsilon - \Delta_\epsilon$ or $J_\sigma - \Delta_\sigma$ relation as well as $J_\epsilon - (P/B)_\epsilon$ or $J_\sigma - (P/B)_\sigma$ relations by 2DFEM, the value of m for a specified $J(=J_\epsilon = J_\sigma)$ can be determined by Eq 16 or 17 even when it is difficult to determine the value of m from Eq 7. Thus, for a given value of J we can obtain the value of Δ or P/B for a finite-thickness structure from Eqs 14 and 15 so that the J versus Δ or P/B curve which coincides with the $\hat{J}$ at midthickness versus Δ or P/B curves can be determined.

For a rotating disk, the centrifugal force field is proportional to the square of the rotating speed, N, so that $P \propto N^2$. For a given value of J, Eqs 15 and 17 are written as

$$N = \{1 + (m - 1)\exp[-\alpha(b/B)^{1/2}]\}^{1/4} N_\sigma \quad \text{for } J = J_\sigma \tag{18}$$

$$N_\epsilon / N_\sigma = m^{1/4} \quad \text{for } J = J_\sigma = J_\epsilon \tag{19}$$

by using the preceding equations, the J-values of a rotating disk with a finite thickness can be determined from the results of 2DFEM.

The accuracy of the empirical formula proposed will be discussed by comparison with the experimental results.

Two-Dimensional Elastic-Plastic Finite-Element Analysis

In order to determine the numerical values of the empirical formula, stress analysis was performed by 2DFEM with a constant-strain triangular element employing the incremental theory of plasticity. The material behavior was modeled by

$$\begin{aligned} \bar{\epsilon} &= \bar{\epsilon}^e = \bar{\sigma}/E && \text{for } \bar{\sigma} < \sigma_y \\ \bar{\epsilon} &= \bar{\epsilon}^e + \bar{\epsilon}^p = \bar{\sigma}/E + \kappa(\bar{\sigma}^n - \sigma_y{}^n) && \text{for } \bar{\sigma} > \sigma_y \end{aligned} \tag{20}$$

Where σ_y denotes yield stress. The von Mises effective plastic strain $\bar{\epsilon}^p$ is assumed to be related to the effective stress $\bar{\sigma}$ by

$$\bar{\epsilon}^p = \kappa(\bar{\sigma}^n - \sigma_y{}^n) \tag{21}$$

where κ and n are material constants referred to as the plastic constant and the plastic exponent, respectively.

Figure 6 shows the finite-element representation of a half of 1T or ½T compact specimen and also the paths over which J is estimated. Stress analysis of rotating disks was also performed by 2DFEM with 768 elements and 425 nodes in a similar procedure as reported previously [*6*]. The J versus rotating speed of a disk with a crack of 20 or 40 mm length is shown in Fig. 7.

Experimental

Materials and Specimens

Two materials were tested in the program, a stainless steel ASTM A276 Type 304 and a rotor forging steel AISI 4340. The chemical composition of the test material is given in Table 1. A long bar of 304 steel of 230-mm

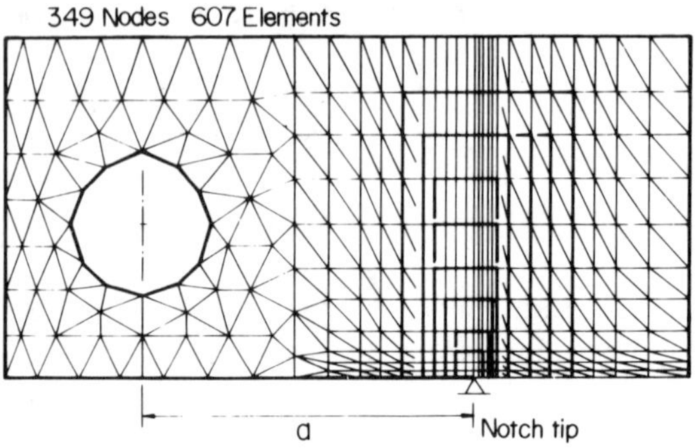

FIG. 6—*2D finite-element breakdown for compact specimens.*

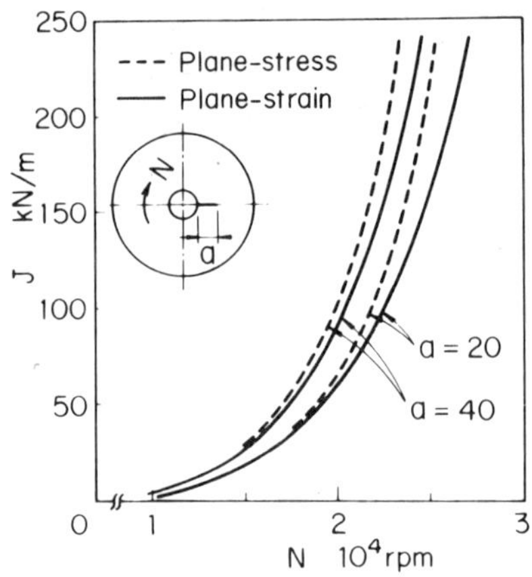

FIG. 7—*Computed results for the* J *versus rotating speed of a disk with a crack of 20- or 40-mm length (2DFEM).*

TABLE 1—*Chemical composition, weight percent.*

Alloy	C	Si	Mn	P	S	Ni	Cr	Mo
ASTM A276 Type 304	0.056	0.50	1.26	0.028	0.020	8.52	18.76	0.13
AISI 4340	0.40	0.27	0.78	0.011	0.003	1.65	0.92	0.23

diameter was received from the manufacturer. As to 4340 steel, a long bar of 220-mm diameter was cut into cylinders of 225-mm length. They were forged by upsetting into disks of 380 mm diameter and oil-quenched at 870°C and tempered at 600°C [*6*].

Tension Specimens—Tension specimens were taken from the location as shown in Fig. 8. The room temperature mechanical properties and material constants obtained by tension specimens are given in Table 2. The stress-strain relations at room temperature are shown as the solid curves in Fig. 9, where the fine lines represent the best fit curves.

Compact Specimens—Compact specimens of 304 steel were machined with notches oriented in the radial direction with the notch tip at the radius of 85 mm as shown in Fig. 8*a*. As to 4340 steel, compact specimens were machined from the disks. The specimen orientation within the disk placed the specimen notches in the radial direction with the crack tip at the radius of 52 mm as shown in Fig. 8*b*. The specimen geometry was fundamentally based on the standard compact specimen [ASTM Test for Plane-Strain Fracture Toughness of Metallic Materials (E 399-74)] as shown in Fig. 10. The varied

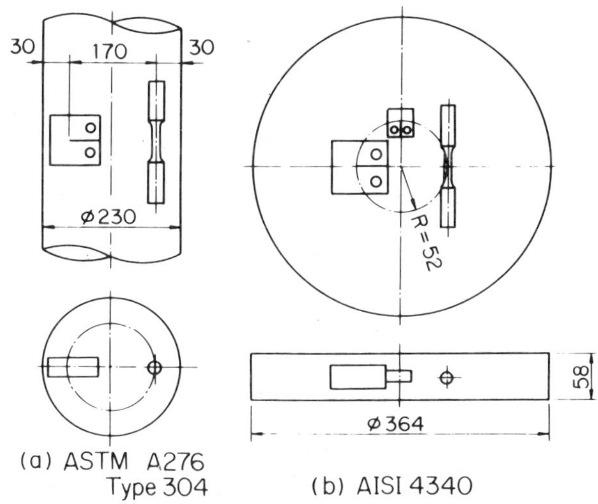

FIG. 8—*Location of test specimen taken from bar or disk material.*

TABLE 2—*Material constants and mechanical properties at 20°C.*

Identification	ASTM A276 Type 304	AISI 4340
Density, Mg/m^3	...	7.86
Young's modulus, GPa	196	206
Poisson's ratio	0.3	0.3
Plastic constant, MPa unit	1.52×10^{-5}	4.08×10^{-4}
Plastic exponent	1.5	1.0
Tensile strength, MPa	693	801
Yield strength, MPa	284	637
Elongation, %	82.4	25.7
Reduction of area, %	68.5	57.8

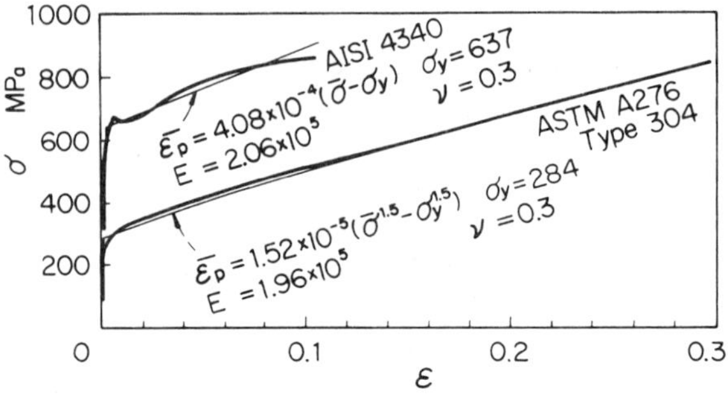

FIG. 9—*Stress-strain diagram of test materials.*

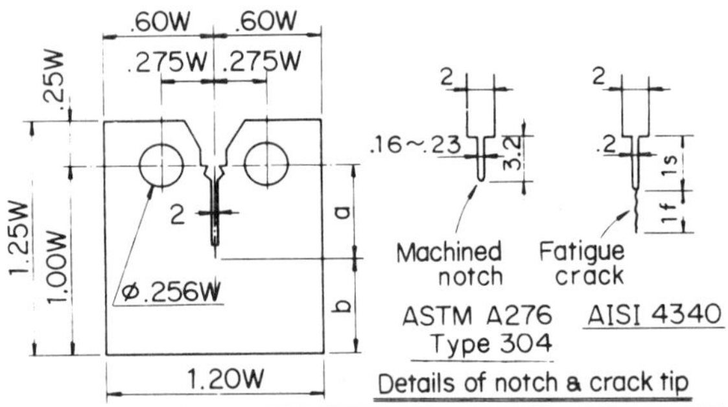

	W	Thickness B	a	1s	1f	a/W
ASTM A276 Type 304	50.8	6.35 12.7 25.4 50.8	29.9	—	—	0.588
AISI 4340	50.8	25.4	30.8	7	3	0.605
	25.4	12.7	15.5	5	3	0.610

FIG. 10—*Configuration of compact specimen used in the experiment.*

specimen thickness is summarized in the figure. The specimens of 304 steel were machined with a blunt notch of radius 0.1 mm to eliminate the possible scatter of the crack length and also any crack growth during the test. In order to check the difference in deformation between a specimen with blunt notch and that with fatigue crack, additional specimens were precracked in air at room temperature following the guidelines of Method E 399-74.

Rotating Disks—Details of the rotating disks were reported previously [*6*], and a brief description is given here. Rotating disks of outer diameter 300 mm, inner diameter 60 mm, and thickness 25.4 or 12.7 mm were machined from 4340 steel. The notch was made as illustrated in Fig. 11, partly using an electrical discharge machining technique, and was then extended to a fatigue crack by repeated internal pressure.

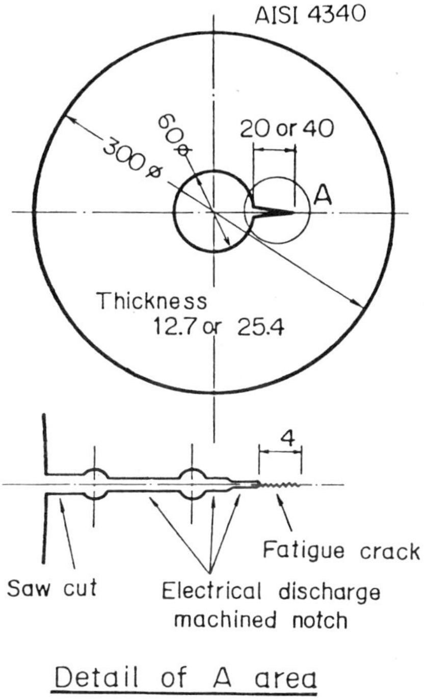

FIG. 11—*Configuration of rotating disk and crack.*

Test Procedure

Compact specimens were tested following the standard procedures [*9*]. The specimens were checked for stable crack growth by the electric potential technique in order to ascertain that crack extension did not take place during the test. The test procedure for rotating disks was described previously [*6*]. Spin tests were carried out in a vacuum vessel using a machine with an air-turbine-type drive unit. Test temperature was controlled within 20 ± 2°C. Disks were rotated up to the specified speed and then decelerated. The cracked part was cut out of the disk and was fractured at liquid nitrogen temperature after heat tinting. The amount of crack growth in the disk was determined by the same method as employed in the compact testing.

Results and Discussion

Compact Specimen

Compact Specimens with Varying Thicknesses—The experimental results for the load per unit thickness versus load-line displacement for varying-thickness compact specimens of 304 steel with machine notch are shown as the solid curves in Fig. 12, while the results for plane-strain and plane-stress FEM computations are depicted as the dotted curves. As reported by Andrews and Shih [*2*], the load per unit thickness versus load-line-displacement curves span a range between the plane-strain and plane-stress limit lines that were computed by 2DFEM. In order to check the effects of crack-tip severity on the load per unit thickness versus load-line displacement curves, experiments were conducted with precracked specimens. Both results are compared in Fig. 13. It may be concluded that the effects of crack-tip severity are not significant on the deformation properties.

The solid curves in Fig. 14 represent the J-Δ relations determined from the P/B-Δ curves shown in Fig. 12 using the Merkle-Corten formula. Figure 15 shows the J-P/B curves determined as in the foregoing. The dotted curves in these figures indicate the limit lines computed by 2DFEM. It is observed that for constant deflection, Δ, the thicker the specimen, the larger the J-value, while for constant load per unit thickness, P/B, the thicker the specimen, the smaller the J-value.

In order to determine the numerical value of the constant α in the empirical formula, the following equation is derived from Eqs 3, 5, 6, and 7

$$\frac{(P/B)/(P/B)_\sigma - 1}{m - 1} = \exp[-\alpha(b/B)^{1/2}] \tag{22}$$

The experimental results shown in Fig. 12 were replotted in Fig. 16 for given values of Δ, and $\alpha = 1.0$ for 304 steel was determined from the best fit line indicated in the figure.

Predictions by the empirical formula are shown as the dashed curves in

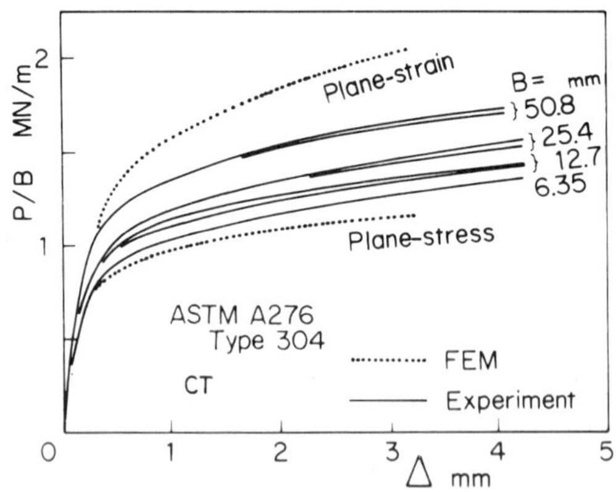

FIG. 12—*Load versus load-line displacement curves for compact specimens of 304 steel with various thicknesses.*

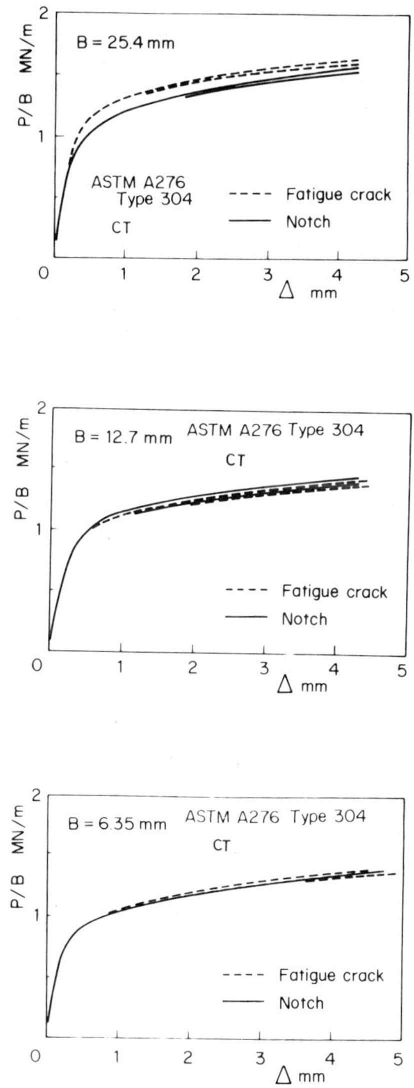

(a) $B = 25.4$ mm

(b) $B = 12.7$ mm

(c) $B = 6.35$ mm

FIG. 13—*Comparison of deformation properties of notched compact specimens with those of cracked specimens (304 steel).*

Figs. 14 and 15. A fairly good agreement is seen between the empirical and experimental results.

Compact Specimens with Varying Dimensions—Experimental results for deformation properties of the 1T and ½T compact specimens of 4340 steel are shown in Fig. 17 as the relation between the normalized load, P/B^2, ver-

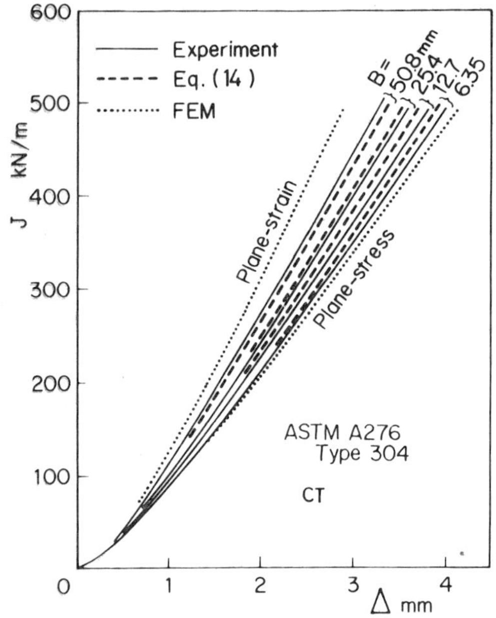

FIG. 14—J *versus load-line displacement curves for compact specimens with various thicknesses (304 steel) showing the comparison of* J-*values estimated by the empirical formula with the experimental results.*

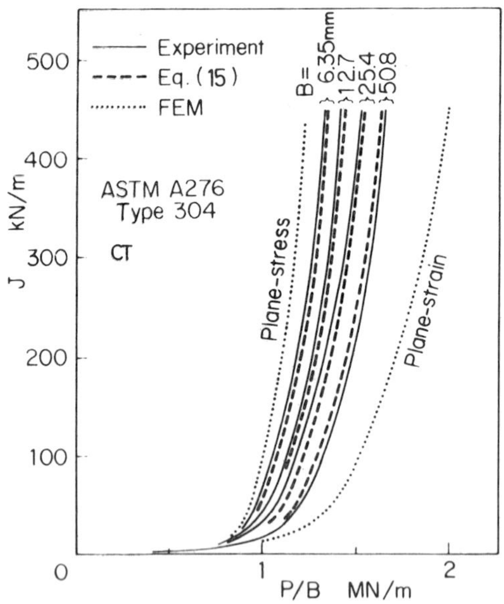

FIG. 15—J *versus load per unit thickness curves for compact specimens with various thicknesses (304 steel) showing the comparison of* J-*values estimated by the empirical formula with the experimental results.*

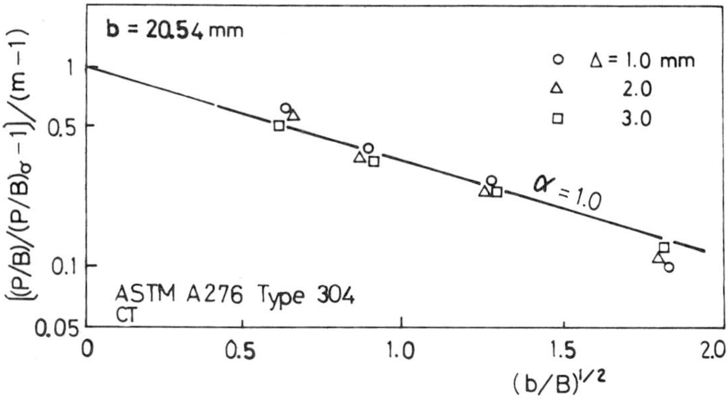

FIG. 16—*[(P/B)/(P/B)$_\sigma$ − 1]/(m − 1) versus (b/B)$^{1/2}$ relation to determine the constant* α.

sus normalized load-line displacement, Δ/B, together with the plane-strain and plane-stress limiting lines obtained by 2DFEM. The computed values by the empirical formula are shown as the dashed curves. The same numerical value of $\alpha(=1.0)$ in the empirical formula for 4340 steel was used. Here again, a fairly good agreement is seen between the empirical and experimental results.

Rotating Disks

The experimental results for the crack resistance curves of rotating disks in terms of J versus crack growth Δa are summarized in Fig. 18. The plane-

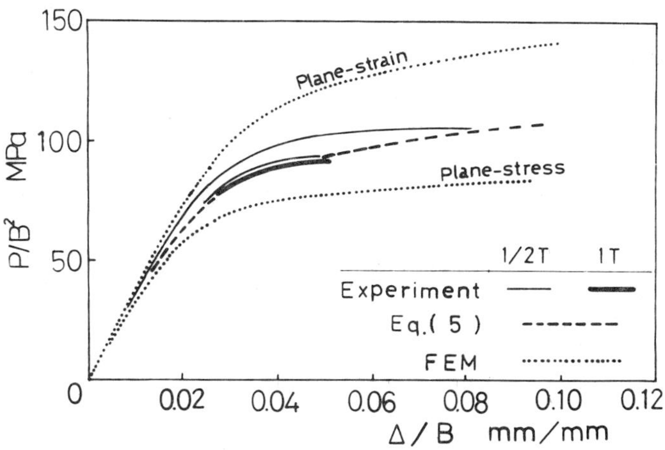

FIG. 17—*Normalized load versus normalized load-line displacement curves showing the size effect of compact specimens of ½T and 1T (4340 steel).*

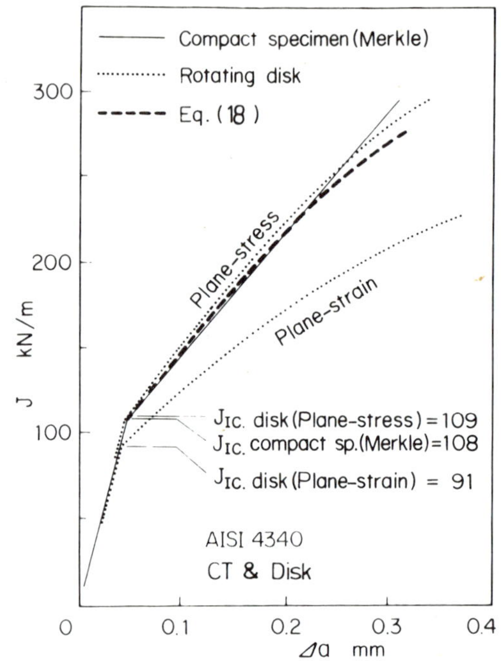

FIG. 18—*Comparison of the* J-R *curves of compact specimens with those of rotating disks. The result for the empirical formula is indicated by the dashed line.*

strain and plane-stress limiting lines were determined by 2DFEM and are shown as the dotted lines. The result for the empirical formula is depicted as the dashed line. The R-curve obtained by compact testing is drawn for the purpose of comparison. It is found from this figure that the onset of crack growth in a rotating disk with finite thickness can be estimated by using the empirical formula and the J_{Ic}-value obtained by the compact specimens.

Conclusions

1. The distribution of $\hat{J}$ was calculated along the crack front of compact specimens with varying thicknesses by using 3DFEM. It has been shown that $\hat{J}$ takes the maximum value at the midthickness and smaller values near the plate surfaces and that the $\hat{J}$-value at the midthickness coincides with the J-value determined from the computed load versus load-line displacement curve using the Merkle-Croten formula. This may explain the fact that the plane-strain fracture toughness value J_{Ic} can be experimentally determined by using specimens of moderate thickness.

2. An empirical formula based on 2DFEM computations was proposed to determine J-value which coincides with $\hat{J}$-value at the midthickness of struc-

ture with finite thickness. The validity of the empirical formula was demonstrated by comparison with the experimental results conducted on compact specimens of 304 and 4340 steels.

3. The experimental results of the rotating disk suggest that the onset of crack growth in a structure with finite thickness can be estimated by using the empirical formula and the J_{Ic}-value determined from the compact specimens.

References

[1] Begley, J. A. and Landes, J. D. in *Fracture Toughness, ASTM STP 514*, American Society for Testing and Materials, 1972, pp. 1-20.

[2] Andrews, W. R. and Shih, C. F. in *Elastic-Plastic Fracture, ASTM STP 668*, J. D. Landes, J. A. Begley, and G. A. Clarke, Eds., American Society for Testing and Materials, 1979, pp. 426–450.

[3] Shiratori, M. and Miyoshi, T. in *Proceedings*, Annual Meeting of Japan Society of Mechanical Engineers, No. 790-3, April 1979, pp. 160–162.

[4] Kishimoto, K., Aoki, S., and Sakata, M., *Engineering Fracture Mechanics*, Vol. 13, 1980, pp. 841–850.

[5] Merkle, J. G. and Corten, H. T., *Journal of Pressure Vessel Technology, Transactions*, American Society of Mechanical Engineers, Vol. 96, No. 4, Nov. 1974, pp. 286–291.

[6] Sakata, M., Aoki, S., Kanzawa, M., Ogure, N., and Tateishi, K., *Journal of Engineering Materials and Technology, Transactions*, American Society of Mechanical Engineers, Vol. 100, No. 2, April 1978, pp. 128–133.

[7] Sakata, M., Aoki, S., Kishimoto, K., and Takagi, R., *International Journal of Fracture*, to be published.

[8] Zienkiewcz, O. C., *The Finite Element Method in Engineering Science*, McGraw-Hill, New York, 1971, p. 148.

[9] Clarke, G. A., Andrews, W. R., Begley, J. A., Donald, J. K., Embley, G. T., Landes, J. D., McCabe, D. E., and Underwood, J. H., *Journal of Testing and Evaluation*, American Society for Testing and Materials, Vol. 7, No. 1, Jan. 1979, pp. 49–56.

[10] Rice, J. R. in *Fracture*, Vol. 2, H. Liebowitz, Ed., Academic Press, New York, 1968, pp. 191–311.

[11] Rice, J. R., Paris, P. C., and Merkle, J. G. in *Progress in Flaw Growth and Fracture Toughness Testing, ASTM STP 536*, American Society for Testing and Materials, 1973, pp. 231–245.

Richard B. King,[1] Yi-wen Cheng,[1] David T. Read,[1] and Harry I. McHenry[1]

J-Integral Analysis of Surface Cracks in Pipeline Steel Plates

REFERENCE: King, R. B., Cheng, Y. W., Read, D. T., and McHenry, H. I., **"J-Integral Analysis of Surface Cracks in Pipeline Steel Plates,"** *Elastic-Plastic Fracture: Second Symposium, Volume I—Inelastic Crack Analysis, ASTM STP 803*, C. F. Shih and J. P. Gudas, Eds., American Society for Testing and Materials, 1983, pp. I-444–I-457.

ABSTRACT: A capability for direct experimental evaluation of the J-integral in surface-cracked members under elastic-plastic deformation is useful for providing understanding of the driving force for fracture. In addition, such a capability makes it possible to evaluate analytical and numerical predictions of J. Experimental procedures have been developed for direct evaluation of J at the root of surface cracks from quantities measured at the specimen surfaces. These procedures rely on the assumption that the line J-integral around the root of a surface flaw is sufficiently path-independent for practical purposes. Arguments in favor of this assumption are made.

Experimentally measured J-integral results are presented for specimens containing surface cracks. Several specimens of pipeline steel 16 mm thick and 75 to 100 mm wide, containing surface cracks varying in length from 31 to 48 mm and in depth from 5.6 to 10.5 mm, were loaded in tension, and the J-integral was evaluated. Plots of measured J as a function of applied stress show a parabolic dependence followed by a nonlinear region and a vertical asymptote at the stress corresponding to net section yield. The experimental results are used to evaluate a simplified line-spring model for predicting J, and the model results compare favorably with the experimental data.

KEY WORDS: cracks, elastic-plastic, fracture mechanics, J-integral, surface cracks, surface flaws

Application of fracture mechanics to the determination of acceptable crack sizes in structures containing surface cracks requires a capability of evaluating a parameter that represents the driving force for fracture. The intent of this work is to extend the utility of the J-integral, which has proven to be useful as a fracture characterizing parameter for through cracks, to sur-

[1]Materials research engineers, physicist, and supervisory metallurgist, respectively, Fracture and Deformation Division, National Bureau of Standards, Boulder, Colo. 80303.

face cracks. Direct measurements of J in tensile panels containing surface cracks are used to provide an understanding of the driving force behavior for surface cracks and to evaluate analytical models for predicting J.

The procedures for direct measurement of J in through-cracked tensile panels have previously been developed by Read [*1*].[2] The theoretical background for direct evaluation of J in surface-cracked panels has been investigated. On the basis of this theory, the experimental procedures for evaluation of J in surface-cracked tensile panels have been developed and are described in this paper. Experimental results for J are presented for several crack sizes, and are compared with theoretical results using the simplified line-spring model presented in Ref *2*.

Theory and Analysis

The geometry of a surface crack is shown in Fig. 1. The contour J-integral is defined as

$$J = \int_c w dy - T_k \frac{\partial u_k}{\partial x_1} ds$$

where

c = contour surrounding crack tip at a point along the crack front,
s = measures arc length along c,
x_1 = direction normal to crack front,
w = strain energy density,
u_k = displacement vector, and
T_k = traction vector acting on outer side of c.

Sufficiently close to the tip of a surface crack, the deformation approaches a state of plane strain (except near the front surface, where it approaches plane stress); so if c is in the near-tip region at a point along the front of a surface crack, then J represents the driving force associated with extension of the crack tip at that point in the x_1-direction [*3*]. To evaluate the J-contour integral conveniently around a path further from the tip, it is necessary to rely upon its path-independence. In Ref *4* it is demonstrated that the J-integral is path-independent when evaluated around a surface crack if the usual line integral expression for J is modified by the addition of an area integral term, given by

$$J_A = -\int_A \frac{\partial}{\partial x_3} \left[\sigma_{3K} \frac{\partial u_k}{\partial x_1} \right] ds \tag{1}$$

[2]The italic numbers in brackets refer to the list of references appended to this paper.

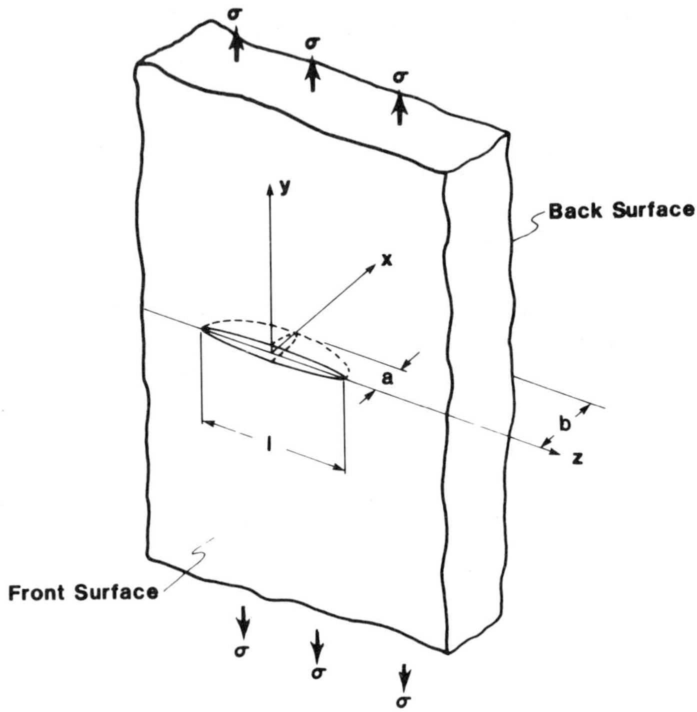

FIG. 1—*Geometry of surface flaw.*

where

x_1 = local normal to crack front,
x_2 = parallel to y-direction in Fig. 1, and
A = area enclosed by line integral contour.

If the contour is evaluated in the x-y plane of Fig. 1 (that is, the contour encloses the root of the crack), then x_1, x_2, x_3 in Eq 1 are parallel to x, y, z, respectively. The authors independently arrived at Eq 1, and attempted to prove that the value of J_A is zero symmetry in the x-y plane. A rigorous proof that J_A vanishes is not possible because, although the quantity in brackets does vanish in the symmetry plane, its out-of-plane derivative $\partial/\partial x_3$, is not necessarily zero. Nevertheless it can be reasoned that this derivative is small. The terms σ_{xz}, σ_{yz}, and $\partial u_z/\partial x$ all vanish on the symmetry plane. The constraint provided by the large width (z-dimension) of the specimens considered in this study should cause plane-strain conditions to be closely approximated near $z = 0$ so that all the foregoing terms should be negligible near $z = 0$, resulting in a negligible value of the out-of-plane derivative. In addition, the foregoing terms would be identically zero throughout the specimen were it not for the pertubation of the deformation fields caused by the presence of the crack, and therefore they should become small as the distance from the crack is increased. Near the crack tip, the deformation approaches plane strain so that the out-of-plane derivative is small. Finally, it is noted that the term J_A should also be present for edge-cracked specimens. It vanishes identically only in the limits of plane stress and plane strain. The usual line integral definition of J is thus path-independent for surface cracks to the same degree of approximation that it is for intermediate thickness edge-cracked specimens. A numerical estimation of the importance of the J_A term is possible from consideration of the finite-element method study of intermediate thickness through-cracked specimens presented in Ref [*5*]. It was found that for compact-tension specimens the value of $\hat{J}$, an integral which is equal to the sum of the usual line J-integral and J_A as long as no unloading occurs, reduces to the value of the line integral as given by the Merkle-Corten expression [*3*] (that is, the contribution of J_A is negligible) when evaluated in the centerplane of the specimen.

An analytical model for the J-integral based on line-spring analysis was described in Ref *2* and is briefly reviewed here. In line-spring analysis a surface crack is treated as being equivalent to a through crack with closure forces and moments acting on its faces due to the presence of the ligament [*6*]. These forces and moments are related to the crack opening displacement (COD) of the crack and the rotation of the crack faces by treating the ligament as a spring. The stiffness coefficients of this spring at any section through the surface crack are determined from the solution for an edge crack in plane strain of the same depth as the surface crack and in a plate of the same thickness, and the forces and moments in the spring are found by enforcing displacement and rotation compatibility between edge crack and ligament. Because the crack depth (and hence the stiffness coefficients) varies with position, this results in a pair of coupled integral equations along the length of the crack, which are solved numerically. Extension into the elastic-plastic regime is accomplished by rewriting the integral equations in incremental form and using tangent stiffnesses [*7,8*]. Once the force and moment in the spring are known, the values of parameters such as J are found by considering the plane-strain edge crack subject to these loads.

As discussed in Ref *2*, if one is concerned solely with predicting the value of J at the root (that is, deepest point) of a surface crack, then as a simplifying assumption the actual crack front may be replaced by a crack of constant depth, which permits the simplification of the model to a purely analytical one. Application of the simplified model to the prediction of J is summarized in the Appendix. The assumptions used in this model are similar to those used in deriving the modified critical-COD model presented by Cheng et al [*9*].

As presented, the model is directly applicable to flat plates in tension (it may easily be modified to include out-of-plane bending). It may be approximately applied to actual structures if the crack length is short in comparison with a characteristic dimension of the structure, and the crack may be treated as existing in a tension region of a flat plate. A more severe limitation is placed on crack length if the model is applied to pipes and other shell structures because curvature effects have been neglected and they become important as the crack becomes longer in comparison with the pipe diameter. Plasticity at the sides of the crack is accounted for only in the r_y correction discussed in the Appendix, and it is not reasonable to expect this approach to be accurate when the correction becomes large, say when r_y becomes comparable to the crack length. (Nevertheless, as seen in the following, good agreement with experimental data was obtained with this model right up to the next section yield.)

Direct Measurement of *J*-Contour Integral in Surface-Cracked Panels

Direct evaluation of the *J*-contour integral involves measurement of the terms in the integrand and determination of the integral by numerical quadrature. The experimental procedures used for evaluating J for through cracks have been described in detail in Ref *1*; those needed for surface cracks are quite similar. A schematic of a typical surface cracked specimen is shown in Fig. 2. The *J*-contour integral is evaluated around a contour *c*, which lies in the *xy*-plane of Fig. 3. Because of symmetry, only the top half of the contour needs to be evaluated. The contour proceeds vertically up the back sur-

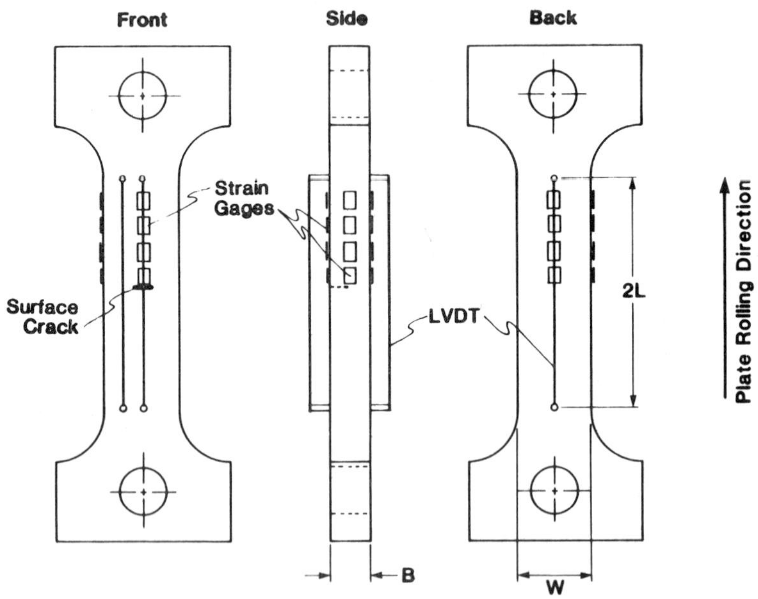

FIG. 2—*Typical specimen used for experimental evaluation of the J-integral.*

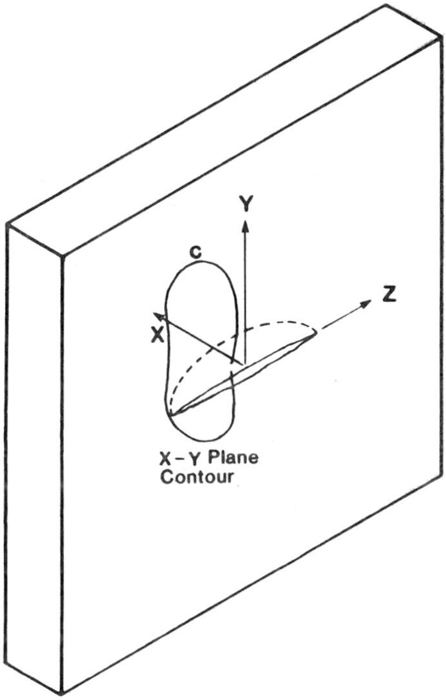

FIG. 3—*Plane in which the J-contour integral is approximately path-independent.*

face of the specimen, then horizontally across to the front of the specimen at a distance L above the crack, then vertically down the front of the specimen until it encounters the crack. Along the portions of c that follow the vertical edges of the specimen, T_k is zero and there are only two nonzero components of stress, σ_{yy} and σ_{zz}. The contribution of the latter component to the strain energy density is assumed to be negligible. This was confirmed experimentally using some transversely mounted strain gages.

A detailed calculation of the strain energy density term was made using the measured values of ϵ_{zz} *and* ϵ_{yy} and deformation plasticity theory. It was found that the error introduced by neglect of the ϵ_{zz} contribution is small. Thus w is calculated from

$$w = \int_0^{\epsilon_{yy}} \sigma_{yy} \, d\epsilon_{yy} \tag{2}$$

which is simply the area under the uniaxial stress-strain curve for the material, evaluated using the strain gages shown in Fig. 2. Along the horizontal portion of the contour, dy is zero. The shear stresses σ_{yz} and σ_{xy} are negligible in comparison with the applied tensile stress σ_{yy}, so the only contribution is from the traction term

$$T_2 \frac{\partial u_2}{\partial x} = \sigma_{yy} \frac{\partial u_y}{\partial x} \tag{3}$$

σ_{yy} is determined from the average strain readings from gages on the front and back surface near the horizontal path. The displacement gradient, $\partial u_y/\partial x$, is assumed to be constant along the horizontal path and is evaluated from values of displacement at the front and back surface measured using linear variable differential transformers (LVDT's). An alternative procedure is to evaluate the displacement by integrating the strain-gage readings. Thus

$$u_y \big|_{y=L} = \int_0^L \epsilon_{yy} \, dy + \delta \tag{4}$$

where δ is zero on the back (uncracked) surface of the specimen and is equal to half the crack mouth opening displacement (obtained using a clip-on gage) on the front (cracked) surface. This procedure for evaluating the displacement was found to be somewhat more reliable than the use of LVDT's.

An apparatus has been set up for automatically performing the direct measurement of J in tensile panels. A servohydraulic testing machine with a capacity of 1 MN is used, and the tests are run in displacement control. Additional instrumentation includes the load cell of the testing machine, stroke LVDT, and a clip-on gage for measuring crack mouth opening displacement. (Experimental results for crack mouth opening displacement are presented in Ref *9*.) All instruments are recorded with a commercial analog-to-digital conversion system with 16-bit resolution that is connected to a commercially obtained minicomputer. The minicomputer system consists of a processor, dual floppy disk data storage unit, a digital plotter, and a printer. In the tests, the testing machine is controlled manually. For each deformation increment, all instruments are automatically read and the information is stored. The J-integral is calculated as described in the preceding, and plots are made of J-integral around contour c as a function of average strain, defined as the average of the LVDT readings divided by the gage length. From the stored data, other plots, such as J as a function of applied stress, may be made if desired. An error analysis of the procedure for direct measurement of J in through-cracked panels was presented in Ref *1*. It was estimated that the measured value of J should be in error by no more than 8 percent in the elastic-plastic range. This same analysis applies to the evaluation of the contour integral around surface cracks by the procedure described in the foregoing.

Experimental Results and Discussion

Several experiments have been conducted in which the J-contour integral has been directly measured using the procedure described in the foregoing.

The specimens were of thickness 16 mm and width ranging from 75 to 100 mm, with surface cracks of depths ranging from 5.6 to 10.5 mm (depth to thickness ratio from 0.35 to 0.66) and length from 31 to 48 mm. The crack geometries are summarized in Table 1. The material was API 5LX-70 pipeline steel, with a modulus of elasticity of 2.07×10^5 MPa, yield strength 510 MPa, and ultimate tensile strength 570 MPa. The flow stress, σ_F was taken to be the average of the yield and ultimate strength, resulting in $\sigma_F = 540$ MPa. The J-integral around a contour along the front and back of the specimen (giving the driving force at the root of the crack) was evaluated, and the results are plotted in Fig. 4. The plots of J as a function of applied stress in Fig. 4 show a region in which J increases relatively slowly with increasing stress, then a sharp rise in J, and finally, a vertical asymptote. The material tested exhibited little strain hardening, and the asymptote occurs at the stress level that caused net section yielding in the specimens.

A typical result for J as a function of average strain, defined as the remote displacement divided by the gage length, $2L$, is shown in Fig. 5. After the strain corresponding to net section yield, the dependence of J on average strain is linear. The remote displacement contributes directly to crack opening once net section yield has occurred, and thus high values of J result.

For comparison, predictions of J using the analytical model are shown in Figs. 4 and 5. The predicted results show a region in which J has a parabolic dependence on stress (the linear elastic region), followed by a nonlinear region (the post-ligament-yield region; the nonlinearity is caused by the r_y correction), and finally, a vertical asymptote at the stress corresponding to net section yield. There is good qualitative agreement with the experimental data, and the net section yield asymptote is accurately modeled. The maximum discrepancy between the predicted and observed values of J is approximately 18 percent. The model predicts that after net section yield, the incremental remote displacement is equal to the incremental crack opening displacement, which results in the predicted slope

$$dJ/d\epsilon_{avg} = \sigma_F(2L) \tag{5}$$

which agrees well with the experimentally determined slope shown in Fig. 5.

The agreement of the model predictions with the experimental results pro-

TABLE 1—*Specimen geometries.*

Specimen No.	Crack Length, mm	Crack Depth, mm	Plate Thickness, mm	Plate Width, mm
1	31.2	5.6	15.9	101.6
2	48.3	6.9	15.9	101.6
3	31.2	10.5	15.9	76.2

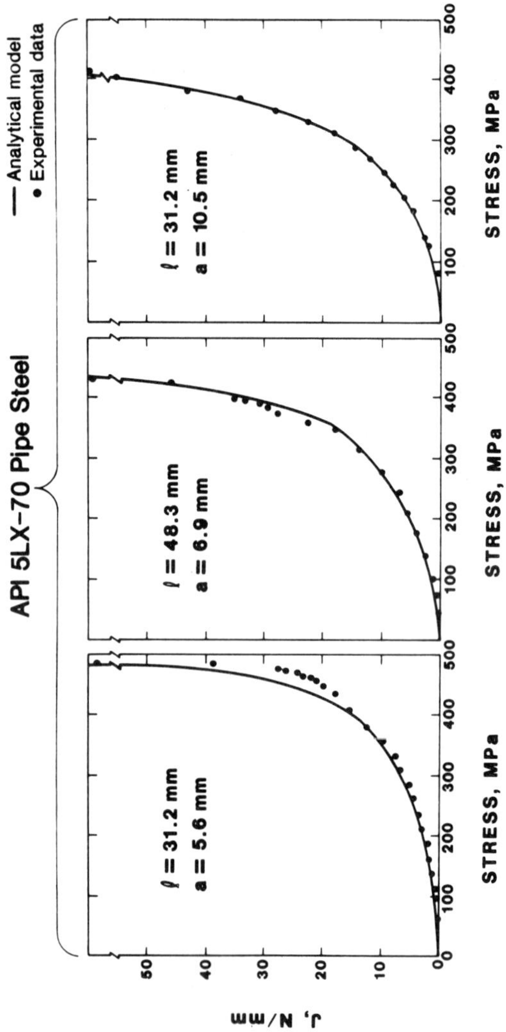

FIG. 4—Comparison of model predictions and experimental results for J versus applied stress.

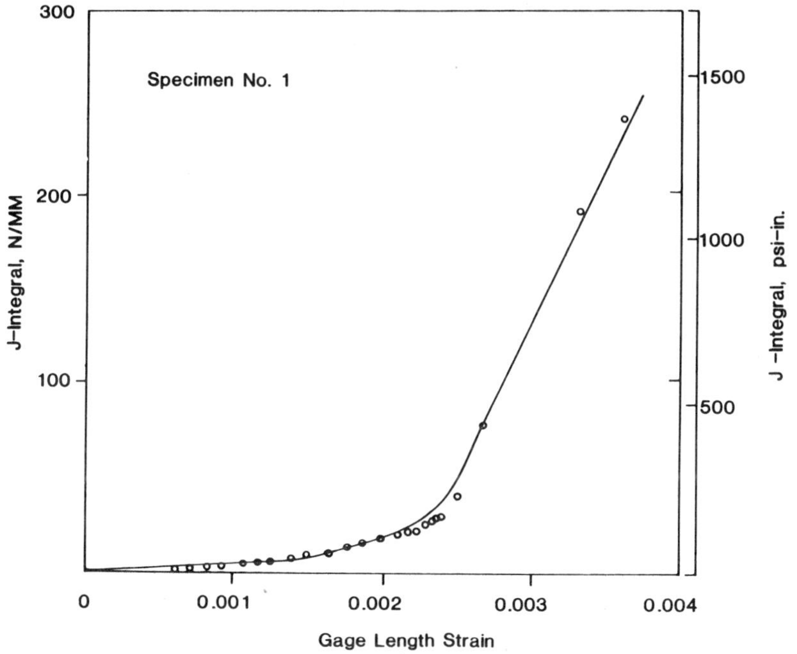

FIG. 5—*Comparison of model predictions and experimental results for* J *versus average strain.*

vides confidence that the various assumptions made in deriving the simplified line-spring model are reasonable, and that the physical meaning of the various regions of the J-integral plots is well understood.

Summary and Conclusions

Founded on the path-independence arguments for the J-integral discussed in the preceding, experimental procedures have been developed for the direct measurement of J in surface-cracked specimens.

With this experimental capability, analytical and numerical calculations of J for surface flaws can be evaluated, as has been demonstrated using the simplified line-spring analysis presented in Ref *2*. In addition, the ability to measure J directly can provide useful information about how the driving force for fracture, as characterized by J, is influenced by various factors, such as strain-hardening and size effects in bodies containing surface flaws and under elastic-plastic deformation.

Acknowledgment

The authors would like to express their appreciation to David McColskey for help with experimental work, and to James C. Newman for useful suggestions concerning the manuscript. This work was supported by the U.S. Department of Transportation, Office of Pipeline Safety Regulation.

APPENDIX

Description of Line Spring Model For the J-Integral

The following describes the application of the simplified line-spring model to predicting J. It is convenient to introduce the notation average closure stress, $\sigma_c = N/B$ and bending stress $m = 6(M/B^2)$, where N and M are the force per unit length and moment per unit length, respectively, and B is the plate thickness. The history of σ_c and m as functions of applied stress, σ, is first determined. In the elastic range, σ_c and m are related to the displacement, δ, and rotation, θ, in the ligament spring by

$$\begin{aligned} \delta &= \frac{2(1-\nu^2)B}{E}(a_{11}\sigma_c + a_{12}m) \\ \theta &= \frac{12(1-\nu^2)}{E}(a_{12}\sigma_c + a_{22}m) \end{aligned} \tag{6}$$

where the a_{ij} are compliance coefficients obtained from an elasticity solution for the plane-strain edge-cracked strip. Values of the a_{ij} as a function of crack depth, a (valid until $a/B = 0.8$), were presented by Delale and Edrogan [10], and are shown in Fig. 6. Displacement and rotation in the through crack are given by

$$\delta = \frac{2l}{E}(\sigma - \sigma_c) \tag{7}$$

and

$$\theta = \frac{-4(1+\nu)l}{(3+\nu)EB}\,m \tag{8}$$

where l is the crack length.

Equating δ and θ in Eqs 7 and 8 with δ and θ in Eqs 6 leads to two equations that are easily solved for σ_c and m, giving

$$\sigma_c = \alpha\sigma$$

$$m = \beta\sigma$$

where

$$\alpha = \frac{l}{(1-\nu^2)B}\left(a_{22} + \frac{l}{(3+\nu)(1-\nu)B}\right)\Big/D$$

$$\beta = \frac{l}{(1-\nu^2)B}\,a_{12}/D$$

$$D = \left(a_{11} + \frac{l}{(1-\nu^2)B}\right)\left(a_{22} + \frac{l}{3(1-\nu)(3+\nu)\beta}\right) - a_{12}{}^2$$

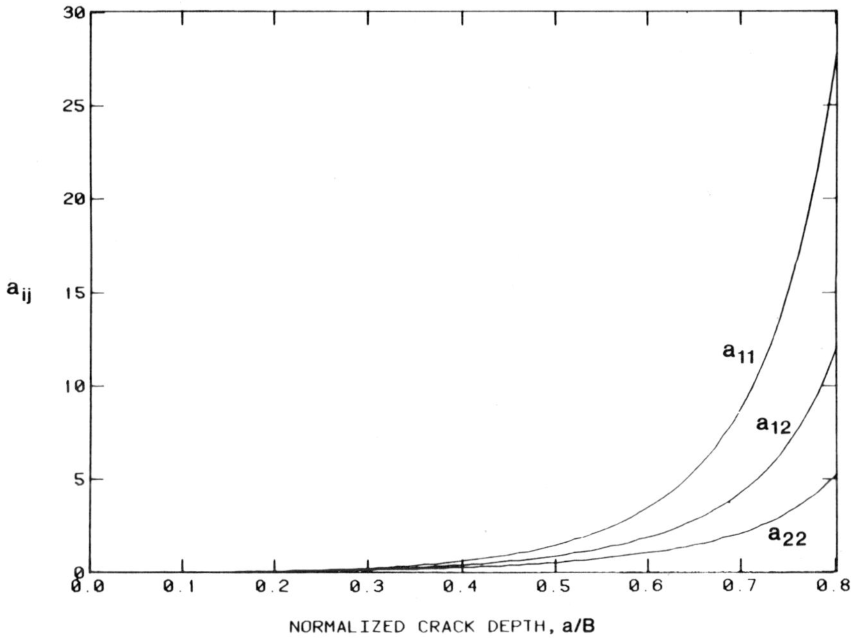

FIG. 6—*Compliance coefficients* a_{11}, a_{12}, *and* a_{22} *as a function of crack depth.*

Ligament yield is then assumed to occur when a yield condition is satisfied by σ_c and m. Here the yield condition used was the average stress condition

$$\sigma_c = \left(1 - \frac{a}{B}\right)\sigma_F \tag{9}$$

where σ_F is the flow stress of the material. From this condition, the applied stress at ligament yield, σ_{LY}, is calculated

$$\sigma_{\mathrm{LY}} = \frac{1}{\alpha}\left(1 - \frac{a}{B}\right)\sigma_F$$

Beyond this point σ_c and m are assumed to remain constant. The value of applied stress at which net section yield occurs is denoted by σ_{NSY} and is calculated from equilibrium:

$$\sigma_{\mathrm{NSY}} = \sigma_F\left(1 - \frac{l}{W}\frac{a}{B}\right) \tag{10}$$

where W is the plate width.

δ and θ can be calculated from σ_c and m using Eqs 7 and 8, in which to account for plasticity at the sides of the crack, the crack length, l, is replaced by the "effective length," $l + 2r_y$. The r_y correction is determined from the strip-yield solution for an infinite plate presented in Ref *2*.

To calculate the J-integral, attention is now returned to the plane-strain edge crack that represents the ligament. In the elastic range

$$J = \frac{1 - \nu^2}{E} K_1{}^2 \tag{11}$$

The stress-intensity factor, K_1, is computed from an elasticity solution for the edge crack and is of the form

$$K_1 = \sqrt{\beta}\,(\sigma_c g_t + m g_b)$$

where g_t and g_b are nondimensional functions of a/B.

Values of g_t and g_b as presented in Ref *9* are shown in Fig. 7. In the elastic-plastic range, J is calculated from the loads and load-point displacement and rotation [*3*] by

$$J = -\int_0^{\delta} \frac{\partial N}{\partial c}\, d\sigma - \int_0^{\theta} \frac{\partial M}{\partial a}\, d\theta$$

$$= J_{\mathrm{el}} - \int_{\delta}^{\delta} \frac{\partial N}{\partial a}\, d\delta \tag{12}$$

where J_{el} is the value of J in the elastic range (from Eq 11) evaluated at σ_{LY}, and δ_{LY} is the displacement corresponding to σ_{LY}. The moment term does not contribute after

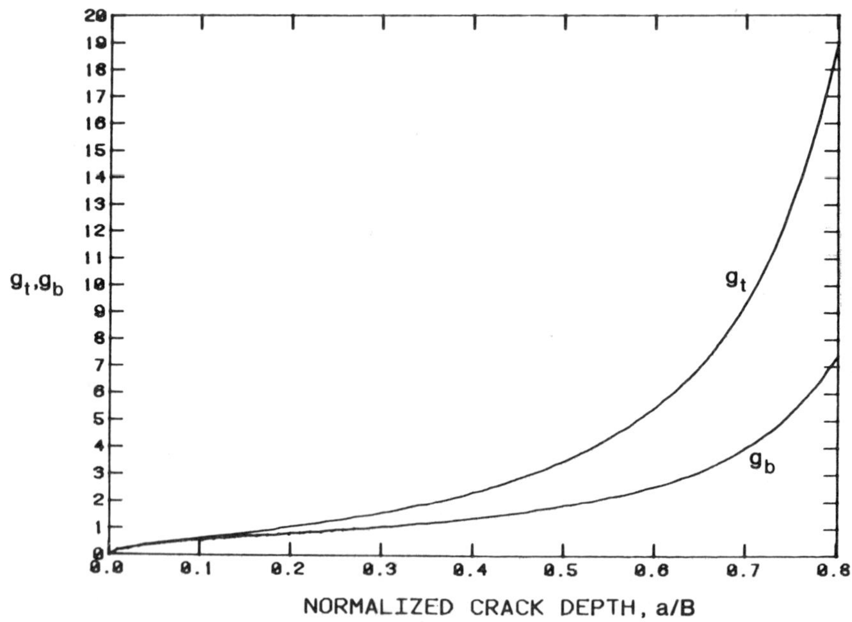

FIG. 7—*Functions* g_t *and* g_b *used to evaluate* K_I.

ligament yield because θ does not increase beyond its value corresponding to ligament yield, as is seen by considering Eq 8 and noting that m is constant after ligament yield. The value of $\partial N/\partial a$ is determined using the yield criterion, which for the simple condition (Eq 9) results in

$$\frac{\partial N}{\partial a} = -\sigma_F \tag{13}$$

The plastic part of J becomes

$$J_p = \sigma_F(\delta - \delta_{\text{LY}}) \tag{14}$$

The value of δ is evaluated from Eq 6, in which the effective crack length $l + 2r_y$ is used. The result for J is

$$J = J_{\text{el}} + \frac{2\sigma_F}{E}\left[(l + 2r_y)\,\sigma - l\sigma_{\text{LY}} - 2\left(1 - \frac{a}{B}\right) r_y\sigma_F\right]$$

where r_y is evaluated at σ. A similarity exists between Eq 14 and the often-stated relation

$$J = \mu\sigma_F\delta_t \tag{15}$$

where δ_t is the crack-tip opening displacement, and the value of μ here would be 1 because no constraint was included in the yield condition (Eq 9). A little caution must be taken in making this comparison because δ in Eq 15 is not the crack-tip opening displacement but is instead the opening displacement at the centerline of the plate, and its relation to δ_t is determined by geometry to be

$$\delta_t = \delta + \theta(B - 2a)$$

These differ only by a constant after ligament yield.

References

[*1*] Read, D. T. in *Fracture Mechanics: Fourteenth Symposium, ASTM STP 791*, American Society for Testing and Materials, 1983, pp. II-199–II-213.

[*2*] King, R. B., "Elastic Plastic Analysis of Surface Flaws Using Simplified Line Spring Model," *Engineering Fracture Mechanics*, in press.

[*3*] Rice, J. R., Paris, P. C., and Merkle, J. G. in *Progress in Flaw Growth and Fracture Toughness Testing, ASTM STP 536*, American Society for Testing and Materials, 1973, pp. 231–245.

[*4*] Amestoy, M., Bui, H. D., and Labbens, R., *Mechanics Research Communications*, Vol. 36, 1981, p. 231.

[*5*] Sakata, M. et al, this publication, pp. I-425–I-443.

[*6*] Rice, J. R. and Levy, N., *Journal of Applied Mechanics*, Vol. 39, 1972, pp. 185–194.

[*7*] Parks, D. M. in *Advances in Fracture Research, Proceedings*, Fifth International Conference on Fracture, Pergamon Press, Oxford, U.K., 1981, pp. 2589–2598.

[*8*] Rice, J. R. in *The Surface Crack: Physical Problems and Computational Solutions*, J. L. Swedlow, Ed., American Society of Mechanical Engineers, 1972.

[*9*] Cheng, Y. W., McHenry, H. I., and Read, D. T., in *Fracture Mechanics: Fourteenth Symposium, ASTM STP 791*, American Society for Testing and Materials, 1983, pp. II-214–II-231.

[*10*] Delale, F. and Erdogan, F., "Line-Spring Model for Surface Cracks in a Reissner Plate," Interim Report to U.S. Department of Transportation, Washington, D.C., Oct. 1980.

George Green[1] *and Laurence Miles*[1]

Use of J-Integral Estimation Techniques to Determine Critical Fracture Toughness in Ductile Steels

REFERENCE: Green, G. and Miles, L., **"Use of J-Integral Estimation Techniques to Determine Critical Fracture Toughness in Ductile Steels,"** *Elastic-Plastic Fracture: Second Symposium, Volume I—Inelastic Crack Analysis, ASTM STP 803*, C. F. Shih and J. P. Gudas, Eds., American Society for Testing and Materials, 1983, pp. I-458–I-479.

ABSTRACT: The use of J-integral estimation techniques is discussed with reference to the evaluation of fracture toughness values for ductile materials. Two applications are considered, tensile plates containing elliptical surface cracks and two specimens designed to produce Mode II sliding shear.

In all cases good agreement was obtained between critical fracture toughness values obtained using J-integral estimation techniques and those obtained in established specimen geometries or by using fractographic techniques. Caution is advised in ensuring that the specimen geometry utilized for J-integral analysis is sufficiently constrained to produce a crack-tip stress-strain field uniquely characterized by the J-integral.

KEY WORDS: ductile fracture, stable crack growth, Mode II sliding shear, J-integral, J-integral estimation techniques, elastic-plastic fracture

When gross yielding precedes fracture, elastic analyses and fracture toughness testing techniques are not applicable and the determination of critical fracture toughness values is difficult. For ductile materials instability is often preceded by substantial stable crack growth and techniques have been developed to determine critical values of fracture toughness under these conditions. These techniques are based on the measurement of the J-integral in fully plastic bend specimens [*1*][2] and utilize the interpretation of the

[1]Group leader and research officer, respectively, Central Electricity Generating Board SW Region, Scientific Services Department, Bristol, U.K.

[2]The italic numbers in brackets refer to the list of references appended to this paper.

J-integral as a potential energy release rate [*2*]. However, when determining laboratory data for use in the defect tolerance assessment of structures, it is desirable to model the defect geometry in the structure as closely as possible in the laboratory specimen. When the material in question deforms in an elastic plastic manner, considerable experimentation or analysis is required to determine the J-integral crack length relationships needed for data analysis.

The use of J-integral estimation techniques [*3*] allows fracture toughness data to be obtained from load displacement records obtained in fracture toughness tests utilizing specimens whose *J-a* relationships are not well characterized. Where possible, comparison with results obtained in well characterized specimens permits the accuracy of the estimated J-integral values to be established. It may not always be possible to make such comparisons, and some alternative method of checking the accuracy of estimation formula must be used. This paper illustrates the application of J-integral estimation techniques to determine critical J-integral values corresponding to the onset of cleavage crack growth in large surface-notched specimens when general yielding preceded fracture, and in fully plastic Mode II sliding shear specimens. In the former case, measurements of crack-tip blunting prior to fracture (the stretch zone) [*4*] and stretch zone/stress-intensity correlations [*5*] were used to establish the accuracy of derived fracture toughness values, while in the latter case detailed studies in two Mode II sliding shear specimen geometries and comparison with Mode I toughness data formed the basis of the validation of the estimation technique.

J-Integral Estimation Procedures

The estimation procedure [*3*] is based on the relationship between the J-integral and absorbed energy. For the purpose of analysis the load displacement record obtained during a test is idealized as being composed of a linear elastic portion with potential energy U_e and a rigid plastic portion with potential energy U_p. The J-integral is given by

$$J = J_e + J_p = \frac{\eta_e U_e}{B(W - a)} + \frac{\eta_p U_p}{B(W - a)} \tag{1}$$

where

a = crack length,
B = thickness, and
W = width.

η_e and η_p are geometry dependent and are calculated from the compliance function and limit load, respectively [*3*].

Experimental Details

Composition and Mechanical Properties of Carbon-Manganese Steels

Specimens were prepared from two plates of carbon-manganese steel designated Plates A and B. The chemical composition and mechanical properties of these plates are given in Table 1.

Details of Specimens

A series of wide-plate specimens 900 mm square by 25 mm thick was prepared from Plate A. A number of semi-elliptical defects were introduced into the centers of the wide plates using saw cuts and extending these by fatigue. Small 25-mm by 25-mm bend specimens were prepared with the notches in the same orientation as the notches in the wide plates. The notches were extended by fatigue to give a final crack length-to-width ratio of approximately 0.5.

Two types of Mode II sliding shear specimens were manufactured from Plate B. The specimen designs are shown in Fig. 1 and are referred to as the compact double shear specimen (CDSS) first introduced by Jones and Chisholm [*6*] and the novel double-punch specimen (DPS) developed as a more compact alternative. Note that all Mode II specimens were side-grooved, the final notching being achieved by electro-discharge machining. A short (10 min) stress relief at 630°C was given to anneal out the Mode I plastic zone associated with the fatigue precracking operation. In addition, 25-mm compact tension specimens were manufactured in an identical orientation.

TABLE 1—*Chemical composition and mechanical properties of carbon-manganese steel plates.*

	Chemical Composition, weight %								
	C	Si	Mn	P	S	Cr	Al	Sn	Cu
Plate A	0.23	0.24	0.78	0.03	0.019	0.02	0.067	0.005	0.013
Plate B	0.16	0.27	1.13	0.015	0.04	0.07	...	...	0.14

	Mechanical Properties			
	Temperature, °C	Lower Yield Stress, MNm^{-2}	Ultimate Tensile Strength, MNm^{-2}	Elongation, %
Plate A	20	321	510	28
	−80	400	600	23
Plate B	20	255	429	42

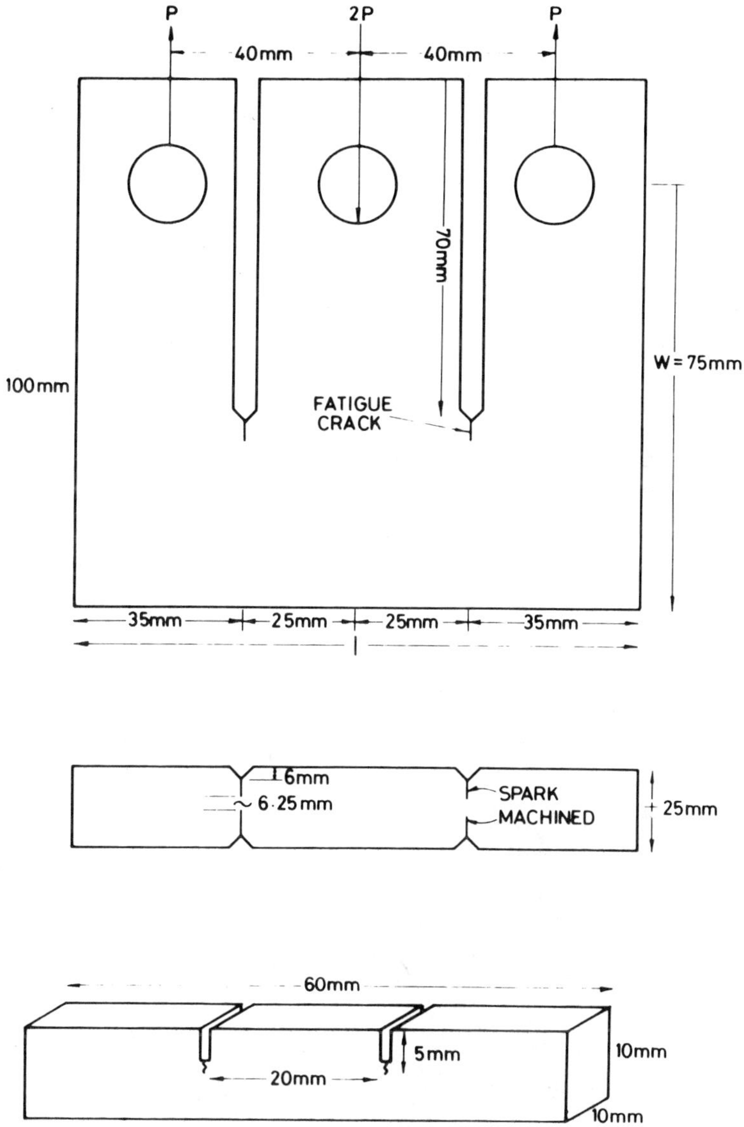

FIG. 1—*Mode II sliding shear specimens:* (a) *compact double shear specimen;* (b) *double-punch specimen.*

Mechanical Testing Details

The wide plates were tested in tension in the wide-plate test rig at the British Steel Corp. General Steels Division, Motherwell, U.K. Strain was monitored using linear variable differential transformers (LVDT's) at gage lengths of 500 and 900 mm and clip gages were attached across the notch centers and at the ends of the notch. The gage length for the clip gages was 30 mm. Temperature was monitored by thermocouples attached in the vicinity of the notch. Load and clip gage displacement and load and LVDT displacement were monitored during the tests. The 25-mm-square bend specimens were loaded in three-point bending and clip gage displacement and load-monitored during testing.

The loading arrangements for the experiment with CDSS and DPS specimens are given in Figs. 2 and 3. Tests were performed under displacement control and a loading rate of 1.0 mm/min was used. Load-point displacement (measured off the loading train) against load was monitored across both cracks. A series of CDS and DP specimens was loaded to different displacements and unloaded. Specimens were subsequently cooled in liquid nitrogen and broken open under Mode I loading, producing a brittle cleavage fracture and revealing the extent of stable crack growth. This was measured at seven points across the test specimen thickness and an average value recorded. Crack growth was taken as the average of the individual growth at each crack tip. Provided that alignment was good prior to testing, crack lengths were found to be reasonably balanced. All specimens experienced general yielding prior to crack growth.

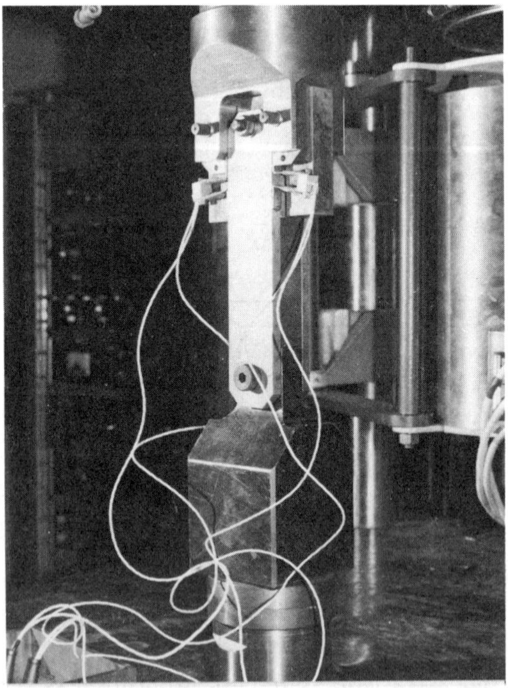

FIG. 2—*Loading arrangement for compact double shear specimen.*

FIG. 3—*Loading arrangement for double-punch specimen. Note one clip gage has been removed to show loading arrangement more clearly.*

Mode I fracture toughness tests were carried out on compact tension specimens, the load-point displacement being monitored against load for a series of specimens loaded to different displacements. The extent of stable crack growth was marked by heat-tinting and breaking open in liquid nitrogen. Crack growth was measured as a seven-point average across the test specimen thickness and recorded.

Fractographic Examination

After testing, the fracture surfaces of the wide plate and bend specimens were prepared and examined in a scanning electron microscope. Mode I and Mode II specimens of Plate B were also subjected to detailed fractographic investigation.

Results

Development of the Double-Punch Specimen

Although experience with the CDS specimen was encouraging, it was decided to develop a more compact version based on a Charpy-sized specimen.

Preliminary experiments with Perspex specimens indicated that extremely close tolerances on clearances in the fully encapsulated rig and on the punch, and close tolerances on the alignment of the cracks in the specimen, were required in order to promote crack growth in the direction of the applied forces. Indeed, it proved impossible to achieve this without side-grooving the specimens. A similar experience had been observed in the steel CDS specimens, particularly in ensuring that the fatigue cracks were in plane, although the more severe side-grooving in these specimens compared with the DP specimens was present to simulate a service defect. Experimentation revealed the optimum conditions to be an overall clearance of 0.05 mm and the start cracks and side grooves had to be coplanar to within 0.025 mm.

The much greater ratio of fatigue crack area to ligament area in the DP specimens compared with the larger CDS specimens implied that frictional forces due to the fatigue crack faces "keying" together could be significant. This effect was demonstrated by producing a number of specimens with identical overall crack lengths, but with different combinations of spark-machined notch plus fatigue crack depths. These specimens were loaded to the elastic limit and the increase in energy above that required to deform a specimen with no fatigue crack measured. The results are shown in Fig. 4 and clearly indicate an increase in the friction effect with increasing fatigue crack area.

The following approximate correction procedure was employed to estimate contributions from friction in generally yielded specimens. A series of specimens containing combinations of machined notches and fatigue cracks was loaded to produce stable crack growth. The value of J-integral was calculated and compared with that obtained in CDS specimens at equivalent amounts of stable crack growth. The deviation from the latter was found to increase with increasing fatigue crack area. The additional work due to friction for a given fatigue crack area was estimated from this calibration and subsequent load displacement records obtained from DP specimens were corrected in this way. In general the frictional corrections amounted to eight percentage points of the total energy absorbed in the specimens.

Calculation of the J-Integral Estimates for the Compact Double Shear and Double-Punch Specimens

Finite-element calculations of the compliance function (Y-function) of the CDS specimen [*7*] and the double-punch specimen [*8*] were utilized to calculate η_e-values for the two geometries from the relationship

$$\eta_e = \frac{Y^2(1 - a/W)}{\int_0^a Y^2 da}$$

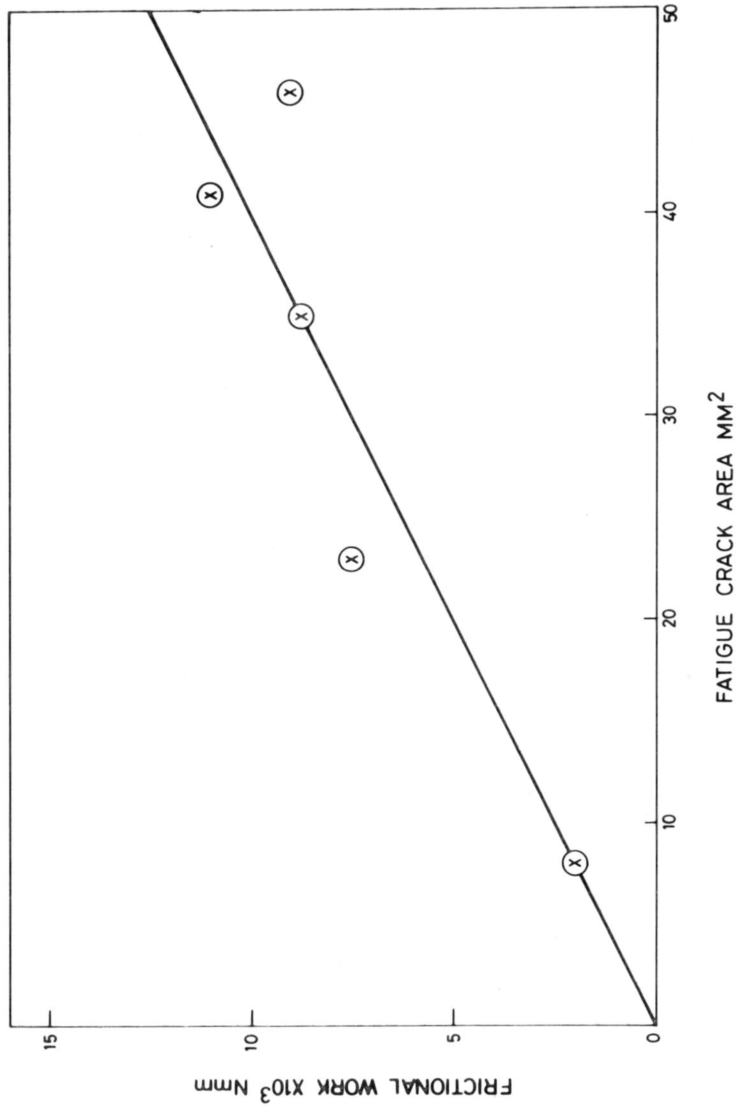

FIG. 4—*Relationship between frictional work and fatigue crack area.*

Values calculated as in the preceding [*9*] are shown in Fig. 5 for the range of crack length ratios of interest. Note that the original compliance function calculations for the Mode II specimens were performed for plane-sided specimens. The compliance function of the side-grooved specimens is assumed to vary as $(B/B_N)^{1/2}$ where B_N is the net remaining thickness between the side grooves. This was validated by measuring the fracture loads for specimens containing different crack lengths manufactured from Perspex and from a high-strength steel which fractured in an elastic manner. Comparison of the ratio of respective Y-functions and fracture loads for different crack lengths confirmed the finite-element predictions and the $(B/B_N)^{1/2}$ relationship.

The plastic η-factor, being derived from the relationship between limit

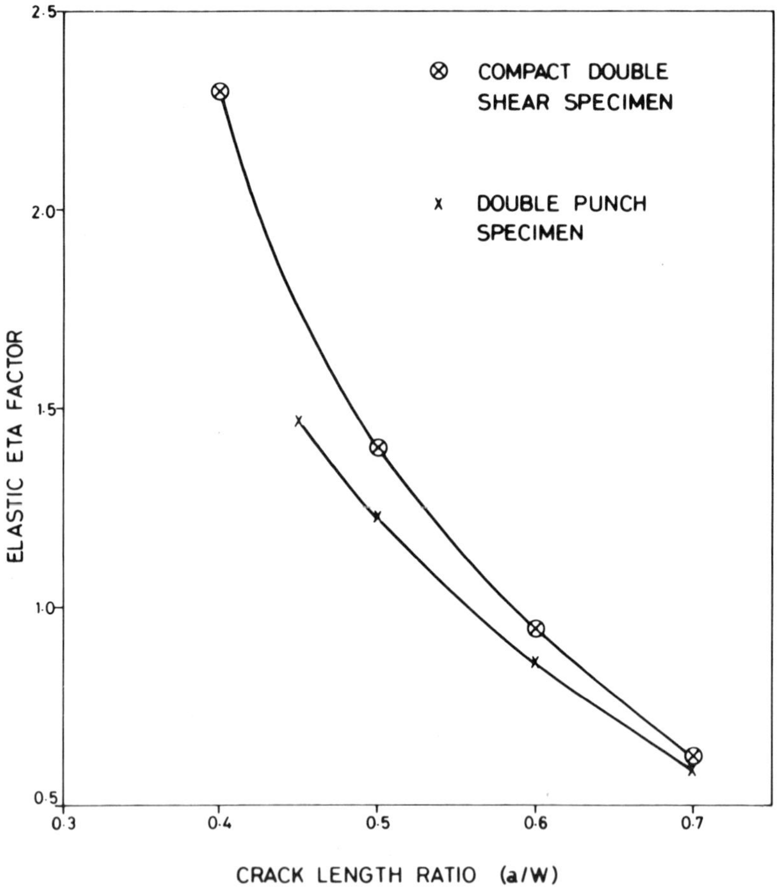

FIG. 5—*Elastic η-factors as a function of crack length ratio for the Mode II geometries.*

load and crack length, reduces to unity for the case of pure shear, in agreement with values for equivalent tension geometries

The load-displacement records must be corrected for extraneous elastic deflections and plastic indentations and, in the case of DP specimen, for frictional effects. The corrected area under the load-displacement record is then used to estimate the J-integral from Eq 1 noting that the initial crack length and the net thickness, B_N, is used to evaluate the J-integral. These values are not corrected for stable crack growth.

Crack Growth Resistance Curves in Mode II and Mode I

Crack growth resistance curves developed using the foregoing estimation formula for the CDSS and DPS tests are shown in Fig. 6. Also shown for comparison are data obtained in 25-mm Mode I compact tension specimens. The J-integrals in this case are calculated using the Merkle-Corten [*10*] relationship. The data are presented in Table 2 and the intercepts and slopes of the crack growth resistance curves are given in Table 3. Note that the intercepts defined at zero crack extension are very close. However, this is not the conventional definition of J_{Ic} for Mode I ductile tearing, this usually being defined as the point at which crack-tip blunting is superceded by ductile tearing. This will be the subject of later discussion.

Figure 7 shows the sliding displacements measured at the loading points (corrected for extraneous deflections) as a function of stable crack extension in both Mode II specimens. Note that the intercepts at the onset of stable crack extension are again close, although the rate of increase of sliding displacement with crack extension is different for the two geometries. To determine how the sliding displacement relates to the crack-tip displacement, a series of parallel lines was scribed onto the surface of double-punch specimens and the offset of the lines plotted against sliding displacement. In addition, metallographic studies of the offset of ferrite grains in the vicinity of the crack tip indicated the limits of accuracy of this procedure. Although there is a considerable scatter, a least-squares linear regression line through the data indicates that the sliding displacement D is related to the crack-tip sliding displacement S by $S = 0.6D$.

Although calibrations are limited to displacements prior to the onset of stable crack growth, the correction has been assumed to apply to sliding displacements throughout the range studied. Figure 8 shows the "true" sliding displacement as a function of crack growth. Note the similarity to the reported variation of Mode I crack-opening displacement with crack extensions [*11,12*]. Note also the one-to-one relationship between sliding displacement and crack growth indicated by this procedure. Clearly similar critical sliding displacements would be predicted for the CDS and DP specimens if

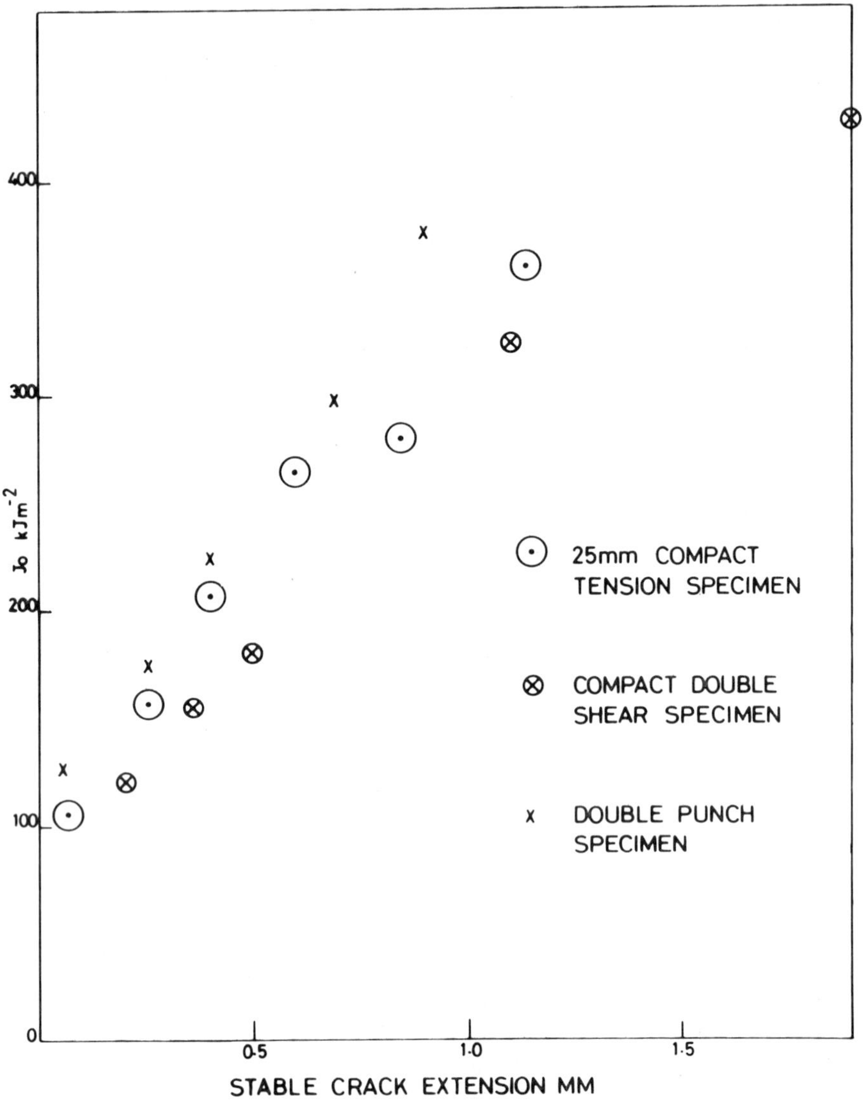

FIG. 6—*Crack growth resistance curves for Mode I and Mode II.*

the same correction procedure were employed, although the crack growth resistance would be different in the two geometries.

Fractographic and Metallographic Studies of Crack-Tip Deformation

Microhardness surveys were conducted on both the CDS and DP specimens along the direction of cracking. The results are shown in Fig. 9. Similar

TABLE 2—*Crack growth resistance data for carbon-manganese steel B.*

Specimen Geometry	J_0, kJm^{-2}	Stable Crack Growth, Δa, mm
Double-punch specimen	126	0.06
	175	0.25
	223	0.40
	296	0.68
	395	0.09
Compact double shear specimen	120	0.19
	154	0.36
	180	0.50
	326	1.09
	428	1.90
Compact tension specimen	107	0.07
	154	0.23
	210	0.41
	268	0.59
	282	0.86
	363	1.13

TABLE 3—*Intercepts and slopes of crack growth resistance curves.*

Specimen Geometry	Intercept at $\Delta a = 0$, kJm^{-2}	Slope of Resistance Curve, dJ/da, MJm^{-3}
Double-punch specimen	99.5	313
Compact double shear specimen	92.0	185
Compact tension specimen	105	230

strain levels were found in both specimen geometries with a focusing of the strain in a region approximately 200 μm from the crack tip. Also shown in Fig. 9 is a microhardness survey ahead of a crack tip in a Mode I specimen. Note a similar crack-tip strain to that found in Mode II and focusing of strain in a region approximately 200 μm ahead of the crack tip. Quantitative metallographic studies of this material indicated a major inclusion spacing of approximately 150 μm, which suggests that the intense strain region is microstructurally controlled. Note that in the case of Mode II crack growth, this occurred in the plane of the crack.

Fractographic examination of the crack-tip region proved unrewarding in observing any fractographic features associated with possible crack-tip blunting prior to the onset of stable tearing. No obvious change in fracture mode was evident between the original fatigue crack and the onset of deformation resulting from Mode II sliding. This is in direct contrast to the transitional stretch zone representing crack-tip blunting observed in Mode I.

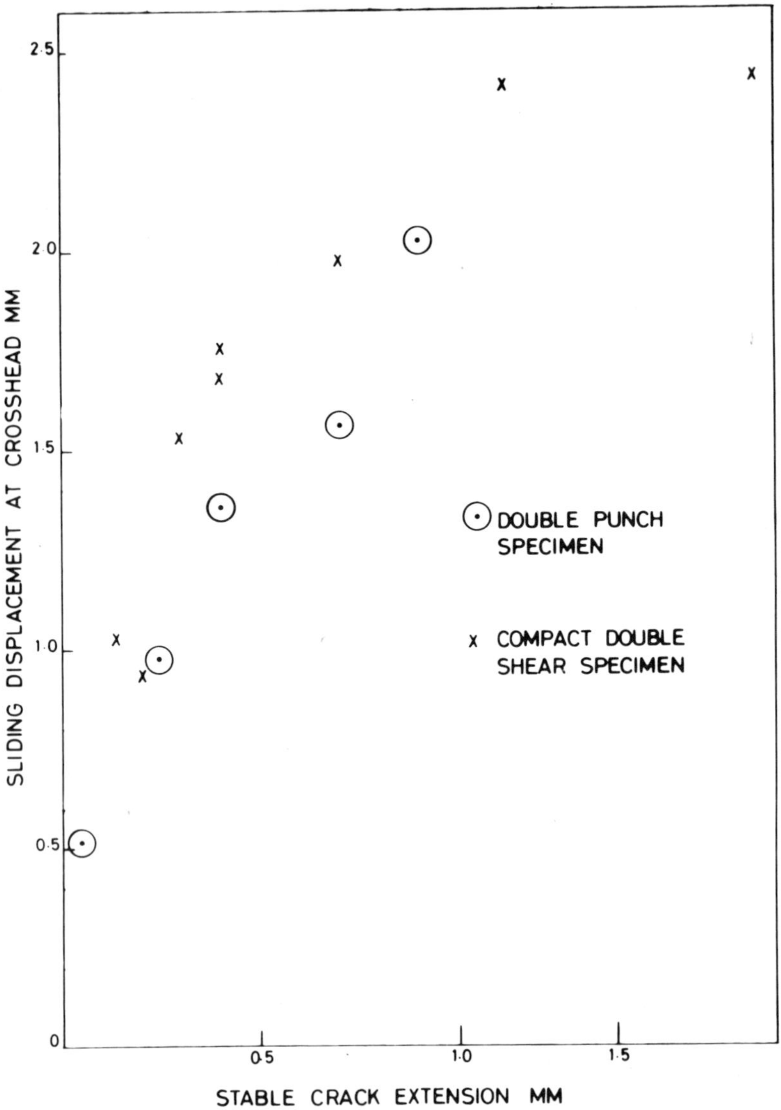

FIG. 7—*Sliding displacement (measured at testing machine crosshead) as a function of stable crack growth for Mode II specimens.*

Results of the Surface-Cracked Plate Tests

The results reported here are extracted from a more comprehensive study of fracture in surface-cracked plates [*13*]. Only those results relevant to using the J-integral to characterize fracture in such specimens are reported in this paper. The clip gage and LVDT readings at fracture are given in Table 4. In

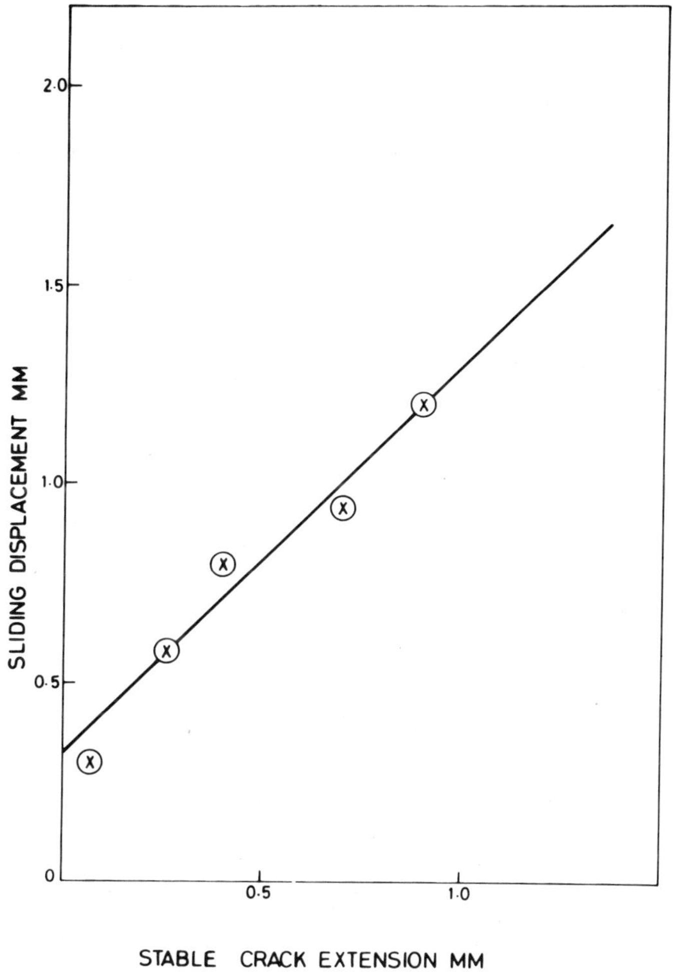

FIG. 8—*Corrected sliding displacement at specimen as a function of stable crack growth for the double-punch specimen.*

all cases fracture was by cleavage after general yielding of the specimen. The figures shown in brackets are the nominal strains corresponding to the measured displacements divided by the respective gage lengths. In general there is a similarity in strain irrespective of the position of measurement.

Calculation of the J-Integral Estimates for the Surface-Cracked Plates

Because of the three-dimensional nature of the geometry of a surface elliptical defect, the calculation of an elastic η-factor is not straightforward. Following the procedures outlined previously, the energy dissipated at the crack

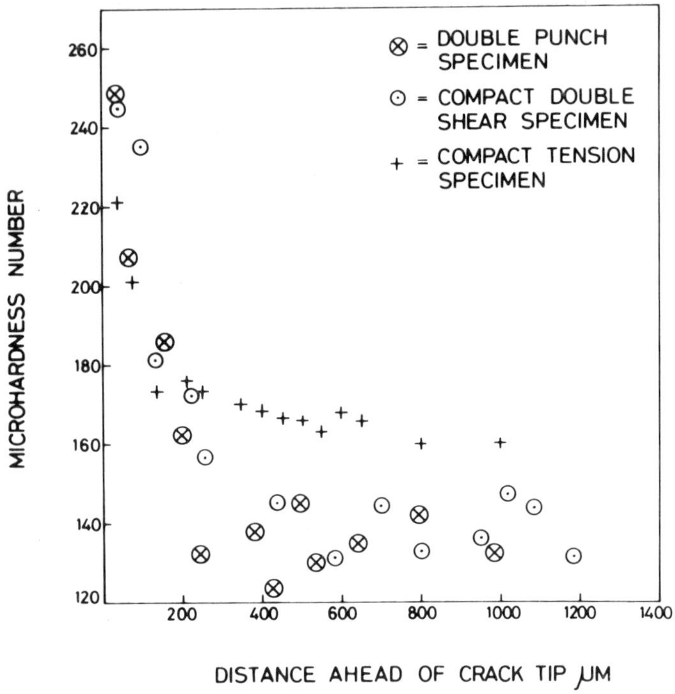

FIG. 9—*Microhardness profiles ahead of crack tips in Mode I and Mode II specimens.*

TABLE 4—*Clip gage and LVDT measurements at fracture in wide-plate tests.*

Specimen	Temperature, °C	*a*, mm	2*c*, mm	Clip Gage 1,[a] mm	Clip Gage 2,[b] mm	LVDT 1,[c] mm	LVDT 2,[d] mm
1014	−80	6	36	0.69 (2.3%)	...	4.3 (0.9%)	8.8 (1.0%)
1015	−80	4.5	60	0.29 (1.0%)	0.38 (1.3%)	6.7 (1.3%)	11.5 (1.3%)
1016	−80	5	100	0.24 (0.8%)	0.35 (1.2%)	6.3 (1.3%)	...
1017	−80	5	100	0.24 (0.8%)	0.31 (1.0%)	3.6 (0.7%)	6.4 (0.7%)

[a]Clip gage across ends of surface notch, gage length 30 mm.
[b]Clip gage across surface notch center, gage length 30 mm.
[c]LVDT gage length 500 mm.
[d]LVDT gage length 900 mm.

tip must be distributed across the crack front and this is not simply related to the crack depth as in the case of a through-thickness crack. The elastic energy is therefore estimated by calculating the stress intensity at the crack tip at its maximum depth using the Kobayashi and Moss analysis [*14*] and relating this to the elastic J-integral through (assuming plane strain)

$$J_e = \frac{K^2}{E}(1 - \nu^2)$$

The plastic part of the energy absorbed is calculated by considering the rate of change of limit load with respect to crack area for an elliptical defect. This procedure can be shown to reduce to an η_p value of unity, in agreement with other tension geometries. Note that this procedure differs from that due to Garwood [*15*], who used the differential of the limit load with respect to the crack length, appropriate to through-thickness defects. Such analyses are likely to be in error.

The J-integral estimates have been calculated from the plastic areas under the clip gage and LVDT records for the wide-plate tests and are given in Table 5. Note the decreasing J-integral values with decreasing gage lengths.

Fractographic Examination of Surface-Cracked Plates

In all the specimens examined final fracture was found to have occurred by cleavage, although appreciable blunting of the fatigue crack tip had occurred. The blunted crack tip is called the stretch zone and can be directly related to the crack-opening displacement (COD) and hence stress intensity by

$$\delta = \sqrt{2}w$$

where w is the stretch zone width and it is assumed that deformation occurs along slipbands inclined at 45 deg to the principal tensile stress.

Observations of the stretch zone width enable estimates of the COD at fracture to be made [*4*]. Measurements were made at least six points along the crack length in all specimens. The average values measured and estimated COD's are given in Table 6 compared with the clip gage readings across the edges of the defect. It is clear that the latter overestimate the crack-tip opening. This table also includes stretch zone measurements made on deeply cracked bend specimens tested at the same temperature. Agreement of estimated and measured COD values in the latter case is good and provides evidence to support the estimates of crack-tip COD at fracture made in the surface-cracked specimens.

TABLE 5—*J-Integral values estimated from displacement measurements.*

Specimen	$J_e = \frac{K^2}{E}$, KJm^{-2}	LVDT 900-mm G1, J_p, kJm^{-2}	LVDT 500-mm G1, J_p, kJm^{-2}	Clip gage 30-mm G1, J_p, kJm^{-2}
1014	24.3	2830	1330	...
1015	20.5	3430	2490	139
1016	22.6	1730	1121	95
1017	28.8	2350	...	111

TABLE 6—*Comparison of clip gage readings at fracture and COD's predicted from stretch zone measurements in wide-plate tests and bend tests.*

Specimen	Clip gage Reading, mm	Stretch Zone Measurement, mm	COD Estimated from Stretch Zone, mm
1014	0.69	0.13	0.18
1015	0.29	0.18	0.25
1016	0.24	0.12	0.17
1017	0.24	0.14	0.19
Bend specimen[a]	. . .	0.13	0.18

[a]Measured critical COD = 0.17 mm.

Estimates of Crack-Tip Stress Intensity in the Surface-Cracked Plates

Estimates of crack-tip stress intensity were made from stretch zone measurements using an empirical relationship [5]

$$w = 3.9 \times 10^{-4} (K/\sigma_y)^{1.6}$$

which has been shown to hold for a range of structural steels and aluminum alloys. Note that this is a best-fit line to a range of data. The values of crack-tip stress intensity (K_{SZemp}) obtained from this relationship are given in Table 7.

It is also possible to estimate the stress intensity using the theoretical relationship between COD and strain energy release rate G

$$\lambda \delta \sigma_y = G = \frac{K^2}{E}(1 - \nu^2)$$

where λ is a material dependent parameter assumed to be 2 for materials of this strength level. Using the relationship between COD and stretch zone width, estimates of the stress intensity ($K_{SZtheor}$) can be made. These are given in Table 7 and are calculated from

$$W = 0.35 \left(\frac{K}{\sigma_y}\right)^2 \frac{\sigma_y}{E}$$

Table 7 also includes estimates of stress intensity derived from J-integral values assuming

$$\frac{K^2}{E}(1 - \nu^2) = J$$

Note the close agreement between stress intensities estimated from stretch zones and those estimated from the clip gage readings on the surface-cracked specimens.

TABLE 7—*Comparison of stress intensities computed from load-displacement records with stress intensities estimated from stretch zone measurements in surface-cracked specimens.*

Test Specimen	LVDT 900-mm G1, K_J, MNm$^{-3/2}$	LVDT 500-mm G1, K_J, MNm$^{-3/2}$	Clip Gage, K_J, MNm$^{-3/2}$	$K_{SZtheor}$, MNm$^{-3/2}$	K_{SZemp}, MNm$^{-3/2}$
1014	756	522	...	169	196
1015	831	710	187	200	242
1016	594	498	160	177	208
1017	690	...	175	166	191

Discussion

Two examples have been chosen to illustrate the use of J-integral estimation techniques to establish critical fracture toughness values in ductile materials. In both cases the estimation techniques produced toughness values in good agreement with those established by other methods.

Estimation of a Critical Stress Intensity in Mode II Sliding Shear

The agreement between the J-integral values at the onset of stable crack growth obtained in two dissimilar specimen geometries indicates that the J-integral characterizes the crack-tip stress-strain field in spite of the specimens being generally yielded. Estimates of plane-stress plastic zone sizes in Modes I and II [*16*] indicate a similar size immediately ahead of the crack, although plane-strain plastic zone sizes are widely different [*17*]. If the requirement for plane strain is considered to be that specimen dimensions exceed the plane-stress plastic zone size, then the requirements for plane strain in Modes I and II should be identical. This is expressed for Mode I as

$$B, (W - a) > \frac{25J}{\sigma_{flow}}$$

Taking values from Tables 2 and 3, specimen dimensions should exceed 7.5 mm, which is satisfied in the Mode I specimens but not satisfied in all cases of the Mode II specimens. The presence of the side grooves by analogy with Mode I behavior [*18,19*] can be considered to promote plane strain at least in the center of the specimen.

The similarity in strain distribution and level revealed by microhardness measurements in the Mode II specimens assists in confirming identical crack-tip stress/strain fields and strongly suggests the concept of a process zone similar to that observed in Mode I [*20*].

Agreement between J_c-values in Mode II and the value obtained in Mode I when this value relates to the onset of stable crack growth is encouraging, but it should be noted that the conventional definition of Mode I J_c-values relates to the onset of tearing. Measured blunting in this case corresponds to a crack growth of 0.085 mm and to a J_{Ic}-value of 125 kJm^{-2}, which is some 25 percentage points greater than the J_c-value in Mode II. This result appears to be in conflict with published data [*6,21*] obtained in materials which fractured in an elastic manner. The bulk of experimenters indicate a high toughness in Mode II compared with that in Mode I. The increase in toughness observed in Mode II is attributed to the reduced hydrostatic tension at the crack tip compared with that in Mode I [*17*].

The discrepancy between the present results and those referred to in the foregoing may be due to differences in the stress/strain fields in fully plastic specimens deforming in Modes I and II or due to a difference in the definitions of critical J-values. Sliding displacement and J-integral measurements indicate that some finite deformation is required prior to the onset of crack growth, but detailed fractography failed to detect the equivalent of blunting in Mode I. This latter observation may not be surprising, however, when it is considered that blunting in Mode I takes place by a shear mechanism and tearing is visible as a different mode of fracture. Clearly, in Mode II the processes of blunting and stable crack growth take place by the same mechanism and may not be resolvable. The critical sliding displacement in the DP specimen is estimated as approximately 0.31 mm. This is in contrast with a crack-tip opening of approximately 0.16 mm (estimated from stretch zone measurements) in Mode I. Microhardness measurements indicate that the Mode II deformation is localized within a region of the order of the microstructural dimensions and therefore the crack tip strain for fracture in Mode II appears greater than that for Mode I, consistent with the reduction in hydrostatic stress ahead of the crack tip [*17*], and perhaps the absence of localized necking in Mode II.

A crude estimate of blunting in Mode II can be obtained as follows. The critical sliding displacement can be considered to be equal to the crack growth during blunting. Taking the J-resistance curve for the CD specimen as the true resistance curve in Mode II (in view of the greater constraint compared with the DP specimens), the J-integral value corresponding to blunting is 147 kJm^{-2}, some 15 percent greater than the Mode I value.

One point of interest to emerge from this study is that the slopes of the crack growth resistance curves in Modes I and II and the J_c-values at the onset of stable crack growth are similar. In addition, the region of intense strain at the crack tip appears similar in Mode I and II. These observations suggest that ductile fracture may be controlled by a critical value of strain energy density, which must be achieved over a microstructurally determined volume.

Estimation of J-Integral Values in Surface-Cracked Plates

The close agreement between measured COD values and those estimated from stretch zones in bend test specimens implies that reliable crack-tip

COD values, and hence stress-intensity values, can be obtained in the surface-cracked plates. The estimated COD-values are clearly lower than the measured openings at the surface of the plates and suggest that such methods of toughness measurement may not be reliable under these conditions.

An approximate analysis of the load-displacement records (Table 7) indicated that the closest agreement between stress intensities obtained from stretch zones and J-integral estimation techniques is obtained using the energy evaluated from the clip gage. This agreement is encouraging in the light of the approximate analysis employed.

It is not surprising that the displacements measured closest to the crack tip reflect the energy dissipated in the crack-tip region. However, the specimens under consideration all experienced general yielding prior to fracture. Consideration of the LVDT and clip gage measurements given in Table 3 indicates that the nominal strain in the wide-plate specimens is approximately uniform across the specimen length. This is consistent with the anticipated lack of constraint in tension geometries. McMeeking and Parks [*22*] have recently shown that the crack-tip stress-strain field is uniquely characterized by the J-integral for tension geometries provided that B, $(W - a)$ is greater than 200 J/σ_{flow}. Taking a value of J at the onset of fracture derived from stretch zone measurements of 162 kJm^{-2} and a σ_{flow}-value of 500 MNm^{-2}, the requirement becomes B, $(W - a)$ should exceed 65 mm. Clearly this is satisfied by the specimen thickness, but not by the ligament in these wide plates. Values of J-integrals measured would therefore be expected to be dominated by remote plasticity, and agreement between clip gage derived J-integral values and those estimated from stretch zone measurements may be fortuitous.

Conclusions

J-integral estimation techniques have been used to determine critical fracture toughness values in ductile materials in specimen geometries for which detailed elastic-plastic analyses are not available. These specimen geometries are plates containing surface elliptical defects and two specimen geometries designed to produce Mode II sliding shear. Agreement between the estimated J-integral value at the onset of stable crack growth in Mode II with that obtained in Mode I in a compact tension was good, as was agreement between crack growth resistance curves. Comparison of J_{Ic}- and J_{IIc}-values is complicated by the need to establish a blunting criterion for Mode II. An approximate blunting criterion indicated an increase in J_{IIc} compared with J_{Ic} in agreement with other data obtained in materials fracturing in an elastic manner.

Detailed fractographic investigations of crack-tip blunting in surface-cracked plates, and the use of empirical relationships between crack-tip blunting and stress intensity, established agreement between J-integral estimates and critical fracture toughness values. Although agreement was good when the work done local to the crack tip was evaluated, the general lack of constraint in the plate may mean that this agreement was fortuitous.

From these two examples it appears that J-integral estimation techniques can be a powerful method of obtaining fracture toughness values for ductile materials. However, caution must be exercised in ensuring that the energy measured in the test can be related to the crack-tip stress-strain field through the calculated η-factors. This can be done only by comparison with data obtained in well-characterized geometries or by comparison with other toughness estimation techniques.

Acknowledgments

The authors wish to acknowledge the experimental assistance of Mr. D. Gladwin and valuable discussion with Mr. D. C. Williams and Dr. R. Bradford of the Central Electricity Generating Board (CEGB) SW Region, Scientific Services Department. The contribution of Mr. D. Harris of the British Steel Corp. Sheffield Laboratories to the fractographic studies of the surface-cracked plates is gratefully acknowledged.

This paper is published by permission of the director general of CEGB, South Western Region.

References

[1] Clarke, G. A. et al, *Journal of Testing and Evaluation*, Vol. 7, No. 1, Jan. 1979, pp. 49–56.
[2] Rice, J. R., *Journal of Applied Mechanics, Transactions*, American Society of Mechanical Engineers, June 1968, pp. 379–386.
[3] Sumpter, J. D. G. and Turner, C. E. in *Cracks and Fracture, ASTM STP 601*, American Society for Testing and Materials, 1976, pp. 3–18.
[4] Harris, D. and Elliott, D. in *Proceedings*, Institute of Physics/Metals Society Conference on the Mechanics and Physics of Fracture, Cambridge, U.K., 1975.
[5] Bates, R. C. and Clark, W. G. in *Transactions*, American Society of Metals, Vol. 62, 1969, p. 380.
[6] Jones, D. L. and Chisholm, D. B., *Engineering Fracture Mechanics*, Vol. 7, 1975, pp. 261–270.
[7] Williams, D. C., Central Electricity Generating Board, South West Region, Bristol, Avon, U.K., Unpublished work, 1981.
[8] Williams, D. C., Report SSD/SW/81/N62, Central Electricity Generating Board, South West Region, Bristol, Avon, U.K., 1981.
[9] Bradford, R., Central Electricity Generating Board, South West Region, Bristol, Avon, U.K., Unpublished work, 1981.
[10] Merkle, J. G. and Corten, H. T., *Journal of Pressure Vessel Technology, Transactions*, American Society of Mechanical Engineers, Vol. 96, No. 4, Nov. 1974, pp. 286–292.
[11] Green, G. and Knott, J. F., *Journal of the Mechanics and Physics of Solids*, Vol. 23, 1975, p. 167.
[12] Shih, C. F., deLorenzi, H. G., and Andrews, W. R. in *Elastic-Plastic Fracture, ASTM STP 668*, J. D. Landes, J. A. Begley, and G. A. Clarke, Eds., American Society for Testing and Materials, 1979, pp. 65–120.
[13] Green, G. and Harris, D., CEGB Report SSD/SW/R360, Central Electricity Generating Board, South West Region, Bristol, Avon, U.K., 1981.
[14] Kobayashi, A. S. and Moss, W. L. in *Proceedings*, Second International Conference on Fracture, Chapman and Hall, London, 1969, p. 31.

[*15*] Garwood, S. J. in *Fracture Mechanics, ASTM STP 700*, American Society for Testing and Materials, 1980, pp. 271-295.
[*16*] McClintock, F. A. and Irwin, G. R. in *Fracture Toughness Testing and Its Application, ASTM STP 381*, American Society for Testing and Materials, 1965, pp. 84-113.
[*17*] Shih, C. F. in *Fracture Analysis, ASTM STP 560*, American Society for Testing and Materials, 1974, pp. 187-210.
[*18*] Green, G. and Knott, J. F., *Metals Technology*, Vol. 2, 1975, p. 422.
[*19*] Shih, C. F., deLorenzi, H. G., and Andrews, W. R., *International Journal of Fracture*, Vol. 13, 1977, pp. 544-548.
[*20*] Shoji, T., *Metal Science*, Vol. 10, 1976, p. 165.
[*21*] Wilson, W. K., "On Combined Mode Fracture Mechanics," Research Report 69-IE7-FMECH-R1, Westinghouse Research Laboratories, Pittsburgh, Pa., June 1969.
[*22*] McMeeking, R. M. and Parks, D. M. in *Elastic-Plastic Fracture, ASTM STP 668*, J. D. Landes, J. A. Begley, and G. A. Clarke, Eds., American Society for Testing and Materials, 1979, pp. 175-194.

Walter G. Reuter,[1] Donald T. Chung,[1] and Cheryl R. Eiholzer[2]

Evaluation of Plate Specimens Containing Surface Flaws Using J-Integral Methods

REFERENCE: Reuter, W. G., Chung, D. T., and Eiholzer, C. R., **"Evaluation of Plate Specimens Containing Surface Flaws Using J-Integral Methods,"** *Elastic-Plastic Fracture: Second Symposium, Volume I—Inelastic Crack Analysis, ASTM STP 803*, C. F. Shih and J. P. Gudas, Eds., American Society for Testing and Materials, 1983, pp. I-480–I-502.

ABSTRACT: Using annealed Type 304 stainless steel plate specimens containing ellipsoidal, rectangular, or triangular-shaped surface flaws, critical values of J are calculated for three separate situations: (1) the assumption that the specimen consists of a single-edge notch (SEN) and two unflawed specimens, (2) the prediction of an equivalent-length through-thickness defect for the surface flaw, and (3) the use of an equation developed by Paris. It was concluded that J_{init} (Charpy specimen) and J_{Ic} cannot be used to successfully predict structural integrity using any of the three approaches for the 6.4-mm-thick (0.25-in.) annealed Type 304 stainless steel.

Since it was not possible to develop a comparison between the absolute values of J_{init} (from the surface-flawed specimen) and J_{Ic}, an attempt was made to evaluate a possible relative agreement. Values of J_{init} and J_{pen} from only those specimens containing rectangular-shaped defects were used to calculate $\Delta J/\Delta a$. The average values agree reasonably well with those obtained in the open literature.

KEY WORDS: surface flaws, J-integral, Type 304 stainless steel, $\Delta J/\Delta a$ elastic-plastic fracture

Nomenclature

- a Depth of surface flaw
- a' Half-crack length
- Δa Change of flaw depth
- b' Half-width of plate

[1]Senior engineering specialist, and engineer, Materials Technology Division, respectively, EG&G Idaho, Inc., Idaho Falls, Idaho 83415.

[2]Graduate student, University of Illinois, formerly, engineer, Materials Technology Division, EG&G Idaho, Inc., Idaho Falls, Idaho 83415.

b Remaining ligament
$2c$ Length of surface flaw
c' Length of a through-thickness flaw
cF^2 Parameter quantifying the severity of a surface flaw
E Young's modulus
EDM Electrical-discharge machined
e_t True strain
e_{ys} Yield strain, σ_{ys}/E
F Combined net-section shape factor and magnification factor for front and back faces

$$F = M_c\sqrt{\frac{a}{cQ}}\left(1 - \frac{\text{crack area}}{\text{total area}}\right)$$

$\{F(\sigma/\sigma_{ys})\}$ Stress correction term where σ and σ_{ys} are based on true stress
J J-integral
J_{Ic} Value of J corresponding to initiation of crack growth satisfying plane-strain conditions
J_{init} Value of J corresponding to initiation of subcritical crack growth as detected by acoustic emission technique
J_{pen} Value of J corresponding to penetration of surface crack through the thickness
ΔJ Change of J-integral
K_{Ie} Elastic stress-intensity factor
M Correction factor for front and back faces
n Strain-hardening term in Eq 3
Q Defect geometry correction factor
r^2 Regression correlation coefficient
SEN Single-edge notch
t Thickness
U Potential energy
U_d Energy in region containing defect
U_t Total energy
W Specimen width
α' Slope of line in Eq 3
$\bar{\sigma}_{calc}$ Flow stress calculated $= (\sigma_{ys} + \sigma_{uts})/2$
$\bar{\sigma}_{test}$ Flow stress from test results
σ_t True stress
σ_{ys} Yield strength
σ_{uts} Ultimate tensile strength
$(\sigma_{ys})_{ten}$ Yield strength from uniaxial tension specimen
$(\sigma_{ys})_{eff}$ Effective yield strength at crack tip
ω Nondimensional parameter; see Eq 6
$\propto$ Proportioned to

Extensive research has been conducted in the analytical and experimental development and verification of the J-integral concept. The greatest amount of activity, by far, has involved those specimens containing through-thickness cracks. In contrast, little analytical and experimental research has been directed to the surface flaw. This paper describes an attempt to use the J-integral approach to predict the behavior of specimens containing surface flaws. A test program was conducted using surface-flawed plate specimens fabricated from annealed Type 304 stainless steel. This paper presents an analysis of the data based on using the J-integral. The general approach is to calculate values of J_{init} corresponding to initiation of subcritical crack growth and compare these values with J_{Ic} or other critical values of J obtained from other sources. An evaluation is then made as to the applicability of using J_{Ic} and simple methods of calculating the J-integral to predict initiation of crack growth for specimens containing surface flaws.

Test Program

Thirty-six test specimens were fabricated from 12.7-mm-thick (0.5-in.) plate; see Fig. 1. The chemistry and tensile properties for the annealed Type 304 stainless steel plate material used in the test program are given in Table 1.

Surface defects of rectangular, triangular, or ellipsoidal configuration were electrical-discharge machined (EDM) in the center of the gage length, with the crack length ($2c$) transverse to the gage section. Figure 2 shows the defect configurations. The depths of the EDM defects represented 28, 56, and 80 percent of the thickness (t) in the gage sections. The aspect ratios ($a/2c$) were nominally either 0.1 or 0.5. The specimens were tested at room temperature, under stroke control, in straight rising load to instability or to the 127-mm (5-in.) stroke limitation of the test machine.

Data

The collected numerical data consisted of

1. load versus loading pinhole displacement, to allow estimation of the energy supplied to the specimen;
2. strain (from five strain gages) versus load;
3. crack-mouth-opening displacement versus load; and
4. acoustic emission versus load, to determine the load corresponding to initiation of subcritical crack growth.

Motion pictures, still photographs, and visual observations were made to enhance the understanding of the crack initiation and growth process. In ad-

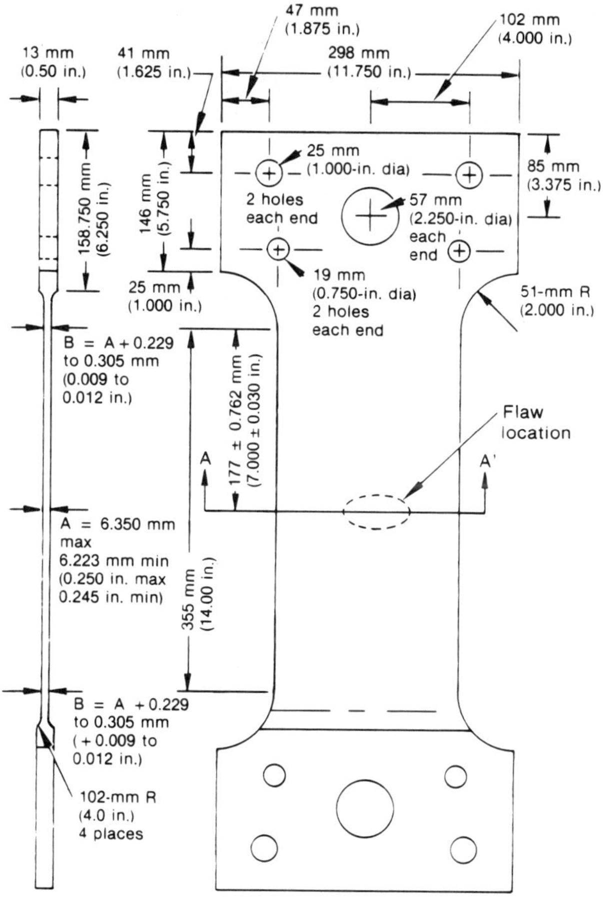

FIG. 1—*Configuration of flat-plate specimens.*

dition, reflective photoelasticity techniques were used to further monitor and record strain behavior.

Loads Corresponding to Initiation and Penetration

Acoustic emission data were used to establish the loads corresponding to initiation of subcritical crack growth. The load corresponding to initiation was established as the intersection of the line representing noise and the line representing noise plus acoustic emission; see Fig. 3.

Defect penetration of the wall thickness was detected either visually or by

TABLE 1—*Room-temperature mechanical properties and chemistry for the Type 304 stainless steel used in test program.*

Material Identification	σ_{ys}		σ_{uts}				% Chemical Composition								
	MPa	ksi	MPa	ksi	El[a]	RA[b]	Cr	Ni	Mn	C	Si	Mo	P	S	Fe
Flat plate—1	253	36.7	663	96.1	71.3	67.8	18.35	9.00	1.33	0.04	0.53	0.21	0.016	0.010	bal
Flat plate—2	260	37.6	603	87.5	70.6	76.4	18.75	9.60	1.70	0.06	0.49	0.48	0.020	0.019	bal
Flat plate—3	254	36.9	621	90.1	80.0	59.8	18.15	9.05	1.09	0.05	0.62	0.18	0.020	0.013	bal

[a] El = elongation.
[b] RA = reduction in area.

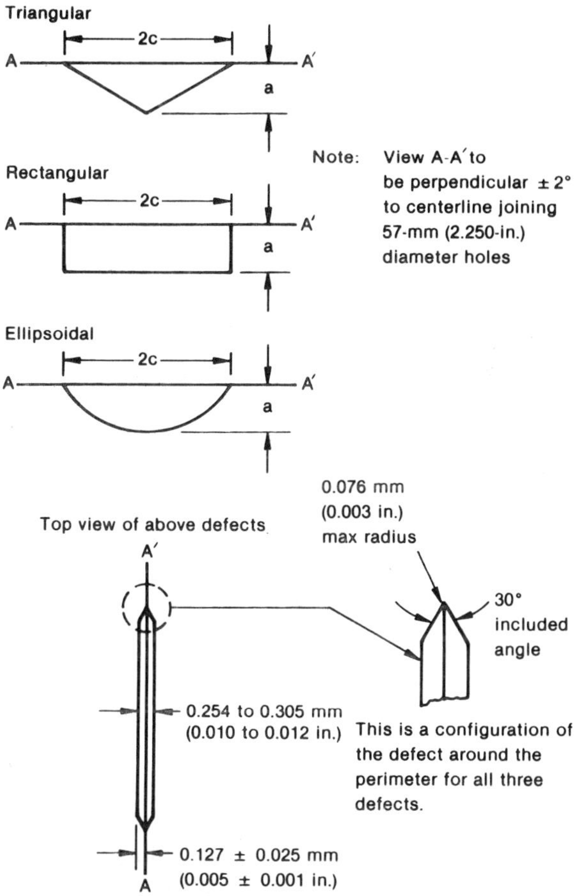

FIG. 2—*Configuration of surface defects.*

an air bulb device attached to the back surface of the specimens, noting the corresponding loads.

Analysis of Results

Two approaches are used to calculate *J* (the J-integral) for the specimens containing surface flaws. The first approach is based on the Begley-Landes technique [*1*][3] and is concerned only with the conditions associated with initiation of subcritical crack growth. The second approach is based on an equation developed by Paris [*2*] and is used to calculate *J* for initiation and penetration.

[3]The italic numbers in brackets refer to the list of references appended to this paper.

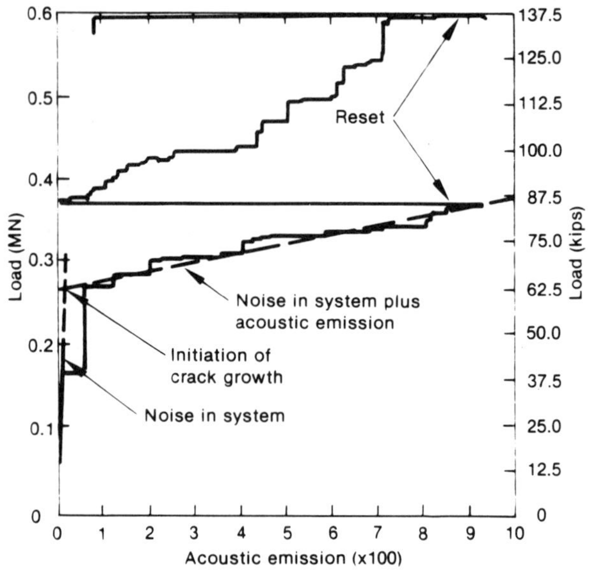

FIG. 3—*Plot of load versus acoustic emission.*

Begley-Landes Approach [1]

This approach is summarized as follows:

1. Test several specimens, each with a different length of crack, recording load versus load-point displacement.
2. At a given total displacement, measure the area under the load-displacement curve. Repeat this measurement for several different increments of displacement.
3. Plot the energy, at constant displacement, as a function of crack length. The slope of this curve is equal to $\partial U/\partial c'$, where U is potential energy. $J = \partial U/\partial c'$, normalized to thickness.
4. J_{Ic} is estimated on the basis of the average deflection at failure.

The foregoing approach can be very useful for specimens containing through-thickness cracks such as compact tension or bend specimens where it is relatively easy to identify crack length and appropriate thickness. But, for the surface flaw where there is a crack length ($2c$) and a crack depth (a), it is not straightforward as to what thickness or what value of crack length is required. Therefore, two approaches were evaluated based on (1) modifying geometry (modification of specimen geometry based on observations of tests), and (2) equivalent through-crack length (predicting an equivalent through-crack length based on the dimensions of the surface flaw).

Modifying Geometry—Visual observation, stop-action photographs, and motion pictures showed that the defects grew through the thickness prior to growing in the length ($2c$) direction. It was therefore assumed that the surface-flawed specimen consisted of a single-edge notch (SEN) specimen of thickness ($2c$) and width (t) and two rectangular unflawed specimens. It was then assumed that the energy is distributed uniformly across the 178 mm (7.0 in.) width, or U proportional to W.

A plot of the energy per unit thickness versus crack depth (a), for all data, was meaningless due to the substantial scatter. Linear regression analysis was then used to establish the best fit to the test data in the plot of energy per unit thickness versus crack depth, for each of the three defect configurations. Figure 4 shows some of the linear plots and regression correlation coefficients (r^2) for each configuration. There is a substantial increase in r^2 for the rectangular-shaped defects. The slopes of the plots in Fig. 4 are the changes in potential energy per unit thickness per unit crack length, which is equivalent to J, and are plotted versus deflection in Fig. 5.

An alternative approach is to assume that the energy is distributed as a function of cross-sectional area or U proportional to A. The energy (U_d) carried by the region containing the defect is

$$U_d = \frac{U_t(2ct - \text{crack area})}{tW - \text{crack area}}$$

Figure 6 is a plot of J versus deflection, which follows a straight line and shows no tendency to approach a maximum value as is the case in Fig. 5.

Equivalent Through-Crack Length—Two approaches [*3,4*] were used to estimate the equivalent through-thickness crack length for a given surface flaw. Specimens with surface flaws converted to equivalent through-thickness defects per the approach in Ref *3* were used in conjunction with the approach from Ref *1* to obtain the potential energy per unit thickness per unit crack length. Since the regression coefficients were so low (r^2 averaged 0.02) based on all data, a second attempt was made using only the specimens containing rectangular-shaped defects. The regression coefficients were somewhat higher for this attempt (r^2 averaged 0.35).

A similar result was seen when Ref *4* was used to convert the surface flaws to equivalent through-crack lengths. The regression coefficients based on the Ref *4* approach are in general agreement with those based on the Ref *3* approach; see Table 2. The r^2-values are too low to give additional consideration to the approach based on an equivalent through-thickness crack length.

Prediction of J *Based on Paris Equation*

The equation developed by Paris [*2*]

$$J = \frac{\sigma^2_{ys}}{E} a \left\{ F\left(\frac{\sigma}{\sigma_{ys}} \right) \right\} \left[\frac{M^2}{Q} \right] \tag{1}$$

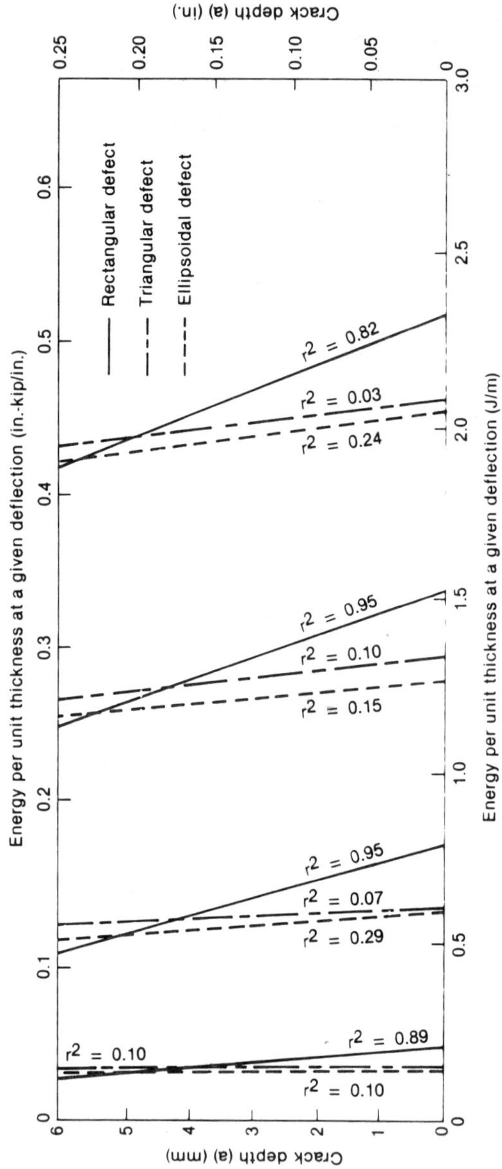

FIG. 4—*Plot of crack depth versus energy per unit thickness at a given deflection.*

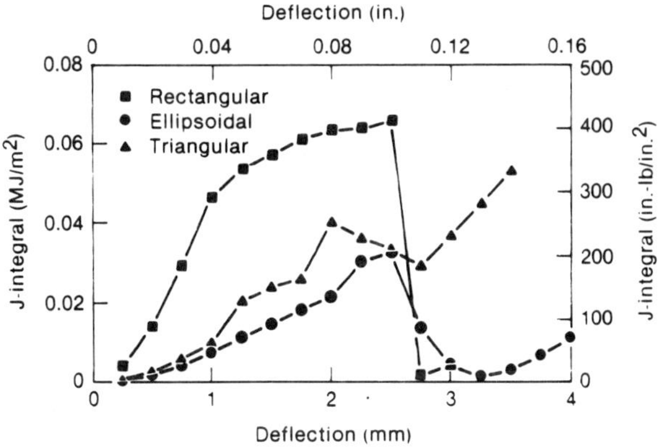

FIG. 5—*Plot of J-integral (energy proportional to width) versus deflection for an assumed SEN specimen configuration.*

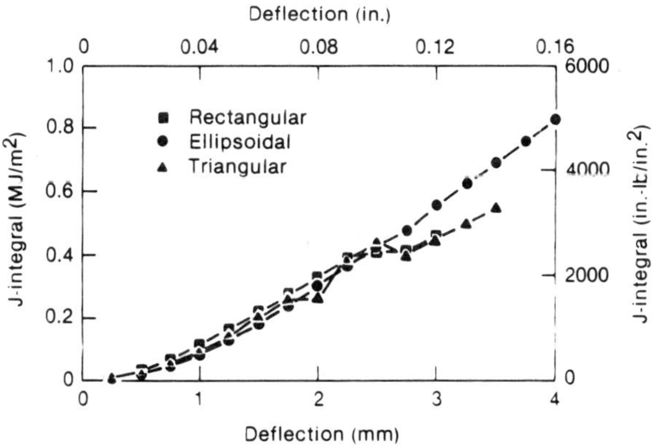

FIG. 6—*Plot of J-integral (energy proportional to area) versus deflection for an assumed SEN specimen configuration.*

Equation 1 was developed for application to reactor pressure vessels containing surface flaws, for example, thick components [203 to 305 mm (8 to 12 in.)] loaded to stress levels which exceed σ_{ys} only slightly. Equation 1 is being extended, in this evaluation, to study the effect of surface flaws in the much thinner [6.4 mm (0.25 in.)] more ductile, annealed Type 304 stainless steel. This is an attempt to empirically evaluate the effectiveness of Eq 1 when extended well outside the analytical limit.

Reference 5 evaluates the sensitivity of the stress term as a function of the stress-strain curves for ferritic steels. The curves evaluated were obtained from weld metal tested at room temperature and from base metal tested at

TABLE 2—*Values of J and some linear regression correlation coefficients (r^2) corresponding to specific analysis methods.*

Source of Data	J	Range of J MJ/m^2	Range of J (in.·lb/in.^{2})	r^2 avg
SEN where $U \propto W$				
rectangular defects	J_{init}	0.07 to 0.08	(400 to 440)	0.76
all defects	J_{init}	0.03 to 0.05	(180 to 280)	0.15
triangular	J_{init}	...	...	0.18
ellipsoidal	J_{init}	...	...	0.17
SEN where $U \propto A$				
rectangular	J_{init}	0.35 to >0.47	2000 to >2700	0.97
all defects	J_{init}	0.32 to 0.79	1750 to 4500	0.85
triangular	J_{init}	0.28 to >0.56	1600 to>3200	0.97
ellipsoidal	J_{init}	0.32 to 0.79	1800 to 4500	0.97
Equivalent through-crack				
rectangular	J_{init}			
Ref *3*		0.17 to ≥0.18	950 to ≥1050	0.35
Ref *4*		0.40 to ≥0.44	2300 to ≥2500	0.47
all defects	J_{init}			
Ref *3*		0.30 to 0.11	150 to 600	0.02
Ref *4*		0.20 to 0.77	1500 to 4400	0.13
triangular	J_{init}			
Ref *3*		0.12 to 0.77	700 to 4400	0.08
Ref *4*		0.04 to ≥0.68	200 to ≥3900	0.05
ellipsoidal	J_{init}			
Ref *3*		0.12 to 0.74	700 to 4200	0.40
Ref *4*		0.21 to 0.53	1200 to 3000	0.30
Paris equation				
	J_{init}	0 to 0.17	0 to 950	...
	J_{pen}	0.18 to 1.75	1000 to 10 000	...
Paris equation				
plus constraint	J_{init}	0.05 to 1.40	300 to 8000	...
	J_{pen}	0.35 to 1.75	2000 to 10 000	...
Precracked Charpy (slow bend), Fig. 9	J_{init}	0.03 to 0.24	150 to 1400	...
Ref *16*				
forged 304 SS				
2T compact	J_{Ic}	0.78	4450	...
3-point bend	J_{Ic}	≥0.70	≥4000	...

both room temperature and 561 K (550°F). The stress term is separated into an elastic and a plastic component [*6*] as shown in

$$\left\{F\left[\frac{\sigma}{\bar{\sigma}_{calc}}\right]\right\} = \left\{\frac{b'}{a'}\Psi\left[\frac{a'}{b'},\ n=1\right]\left[\frac{\sigma}{\bar{\sigma}_{calc}}\right]^2 + \alpha'\frac{b'}{a'}G_{In}\left[\frac{a'}{b'},n\right]\left[\frac{\sigma}{\bar{\sigma}_{calc}}\right]^{n+1}\right\} \qquad (2)$$

The approach of Zahoor [*7*] was used to solve for the elastic component, using data from Ref *8* and *9* to determine the constants for plane-stress and plane-strain conditions, respectively.

The plastic component is solved in two steps. The first solves for α' and n in the Ramberg-Osgood equation

$$e_t/e_{ys} = \sigma_t/\sigma_{ys} + \alpha' (\sigma_t/\sigma_{ys})^n \tag{3}$$

The second step uses the approach suggested in Ref *7* to solve for $b'/a'G_{In}$ $(a'/b', n)$. These analyses showed a correlation between the ratio of applied strain to yield strain and the stress function.

The same approach was used for annealed Type 304 stainless steel and the results are shown in Fig. 7. These results are based on true-stress/true-strain data. In order to solve for α' and n, it was necessary to remove the elastic term in Eq 3 because the solution process involved taking the natural log of both sides. In addition, the modified Eq 3 was solved using the yield strength in the denominator; therefore, the yield strength was used in Eq 2. In discussion with Paris [*10*], it was suggested that Eq 2 might need to be modified to reflect the modification in Eq 3. This was not done, however, since the elastic portion is responsible for an error of less than 1 and 10 percent for $\{F(\sigma/\sigma_{ys})\}$ of 4000 and 50, respectively.

The geometric correction factors M and Q were calculated using equations

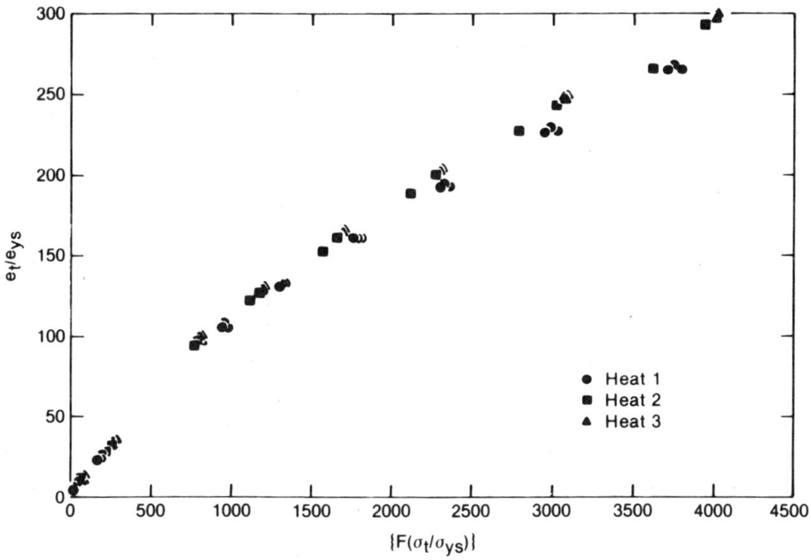

FIG. 7—*Plot of* e_t/e_{ys} *versus the stress correction* $\{F(\sigma_t/\sigma_{ys})\}$.

from Ref *11* and *12*, respectively. It has been suggested in Ref *13* that these factors might be useful for very ductile materials, based on a correlation observed between K_{Ie} and cF^2, both of which are developed from the two-parameter fracture criterion of Newman [*14,15*]. (The cF^2 term is an attempt to quantify the severity of the surface flaw in that c is the half-crack length and F is the combined net-section conversion, shape factor, and magnification factor for front and back face [*15*].) Figures 8*a* and 8*b* present plots of J versus cF^2 for initiation of crack growth, and for the crack to penetrate the 6.4 mm (0.25 in.) wall thickness, respectively. These values of J were calculated from initial defect sizes, the strains corresponding to initiation and to penetration, and to Fig. 7 data. Values of J were calculated for penetration even though unloading had occurred due to subcritical crack growth.

Discussion

The appropriate values of J_{init}, from Charpy specimens, and J_{Ic} for comparison with J-values calculated in this study will be discussed first. The Begley-Landes approach and the Paris equation used in this paper for calculating J are then discussed in order to provide the most likely values. Comparisons are then made as to the ability of predicting accurate values of stresses for initiation and penetration, if appropriate, based on the specific analytic technique and J_{Ic} or J_{init}.

Applicable Values of J-*Critical*

Values of $J_{Ic} > 0.70$ MJ/m^2 (4000 in.·lb/in.^{2}) are given in Table 2 for forged 304 stainless steel [*16*]. In addition, precracked Charpy specimens were tested, in slow bend, from the materials in Table 1. For the Charpy specimen, the precrack was oriented the same as the defect in the specimens containing surface flaws. The test data were analyzed using the approach in Ref *1*; the results are shown in Fig. 9. The deflection corresponding to initiation of subcritical crack growth, using acoustic emission techniques, varied from 0.5 to 2.3 mm (0.02 to 0.09 in.). Therefore, the value of J_{init} varied from 0.03 to 0.24 MJ/m^2 (150 to 1400 in.·lb/in.^{2}). These values are summarized in Table 2.

Differences between J_{Ic} and J_{init} are explained by estimating the maximum value of J that can be measured with the Charpy specimen and still satisfy the conditions for plane strain and a J-controlled field, for initiation, based on the requirements in Ref *17*

$$t, b \geq \frac{25\,J}{\bar{\sigma}_{calc}} \quad \text{for bending} \tag{4}$$

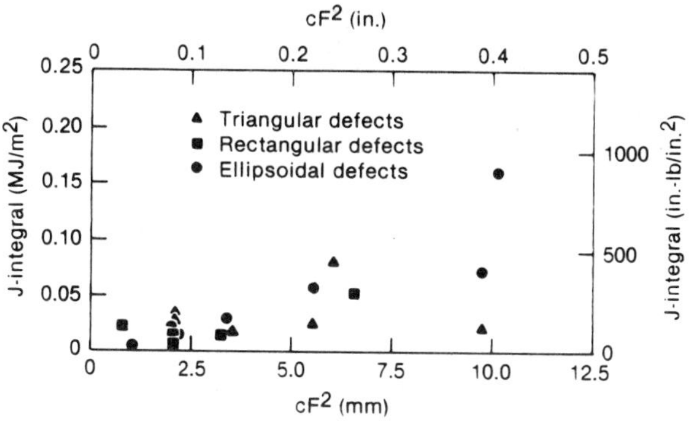

(a) Initiation

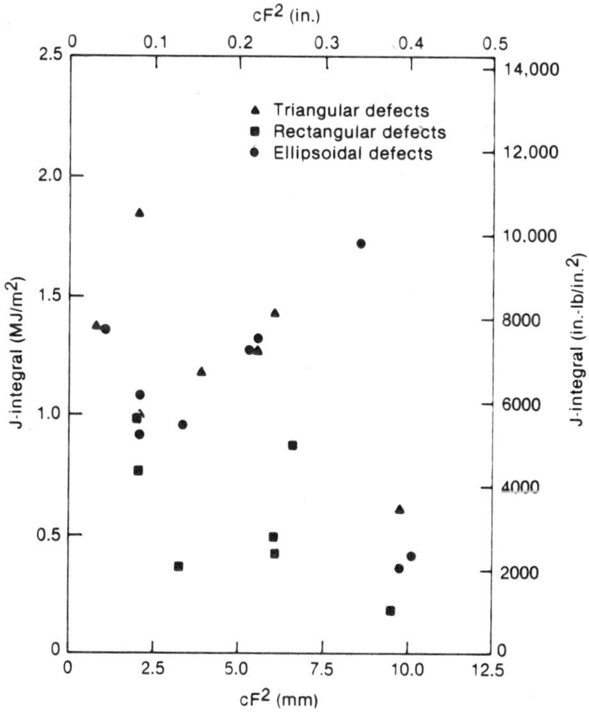

(b) Penetration

FIG. 8—*Plot of J-integral, based on Paris equation, versus flaw severity* (cF^2).

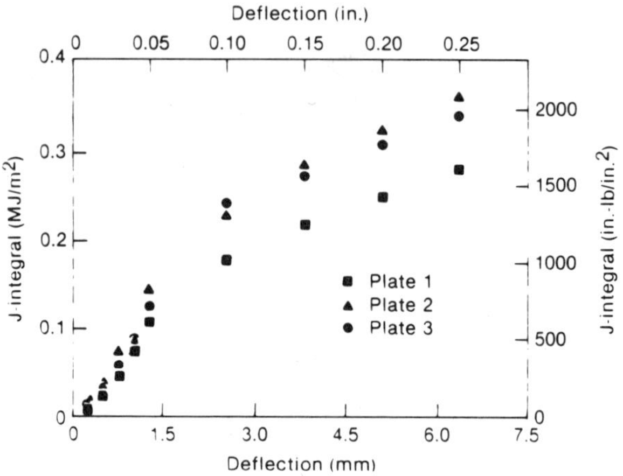

FIG. 9—*Plot of J-integral versus deflection based on precracked Charpy specimens.*

For $t = 9.9$ mm (0.39 in.) the upper limit for J will be 0.18 MJ/m^2 (1053 in.·lb/in.^{2}), and for b varying from 7.6 to 2.9 mm (0.30 to 0.11 in.), the upper limit will range from 0.14 to 0.05 MJ/m^2 (810 to 308 in.·lb/in.^{2}).

It is evident that it would not be possible to measure J_{Ic}-values of the magnitude reported in Ref *16* using Charpy-sized specimens tested in bending.

Therefore, both $J_{Ic} > 0.70$ MJ/m^2 (4000 in.·lb/in.^{2}) and J_{init} ranging from 0.03 to 0.24 MJ/m^2 (150 to 1400 in.·lb/in.^{2}) for Charpy specimens will be compared with data generated from surface-flawed specimens. The latter comparison is of particular interest since both the Charpy and surface-flawed specimens were taken from the same material. Since it has already been shown that the Charpy specimen cannot be used to measure J_{Ic}, in this particular application, it seems reasonable to evaluate the surface-flawed specimen's ability to satisfy conditions for plane strain and a J-controlled field. It is expected that the requirement for plane strain and a J-controlled field for initiation would fall somewhere between those given in Eqs 4 and 5 where

$$t, b \geq \frac{200\,J}{\bar{\sigma}_{calc}} \quad \text{tensile loading of center-cracked panel [17]} \qquad (5)$$

If the specimen width [nominally 177.8 mm (7.0 in.)] is assumed to be the thickness, since the crack first penetrated the 6.4-mm-thick (0.25-in.) wall before growing in the 2c direction, then Eq 5 is satisfied for $J_{init} \leq 0.38$ MJ/m^2 (2170 in.·lb/in.^{2}). To satisfy the requirements for b, the upper limit of J will range from 0.9 MJ/m^2 (516 in.·lb/in.^{2}) for $b = 5.1$ mm (0.2 in.) using Eq 4 to 0.02 MJ/m^2 (92 in.·lb/in.^{2}) for $b = 1.3$ mm (0.05 in.) for Eq 5. Some of the surface-flawed specimens calculated with U proportional to W conformed with the requirement in Eqs 4 and 5. But, none of the surface-flawed specimens calculated with U proportional to A conform with the requirements in Eqs 4 and 5.

The requirements based on Eqs 4 and 5 may be too severe, as indicated by observations of a reflective photoelastic coating on the back surface of a specimen containing a crack 80 percent of the way through the thickness. On the back surface, the coating showed two strain concentrations each parallel to the crack length with one above and one below where the surface crack would penetrate if extended through to the back surface. This strain concentration, on the back surface, is very similar to that for plane-strain slip depicted in Ref *18*. If the strain concentrations are associated with slip field lines for unconstrained deformation (low triaxiality due to dimpling), then it would be expected that the two parallel lines would tend to meet each other as the crack grew through the wall thickness. This did not occur; instead, concentric shapes, similar to topographic lines, formed about each of the first two parallel lines. This suggested that substantial constraint existed, possibly sufficient to satisfy conditions for plane strain for these surface-flawed specimens even though the requirements of Eqs 4 and 5 were not met. At this time it is not certain if a plane-strain condition exists or not. In addition, there is no experimental evidence to refute the probability that a J-controlled field did not exist. But, this does not pose a problem since the purpose of this investigation was to evaluate a 6.4-mm-thick (0.25-in.) specimen of annealed Type 304 stainless steel which did not meet the requirements of Eqs 4 and 5.

Begley-Landes Approach

Modifying Geometry—The test results and analysis, based on initiation, suggest that the defect configuration had a substantial effect on some of the results. The linear regression correlation coefficients given in Table 2 for the column based on U proportional to W show a substantial increase in r^2-values for specimens containing rectangular shaped defects compared with those containing the other shapes. With the exception of the rectangular-shaped defects the values of r^2, in Table 2, suggest no correlation.

Therefore, the rectangular-shaped defects in Fig. 5 show that J_{init} is nearly a constant value as indicated by its asymptotic approach to a limit. The results in Figs. 5 and 6 are limited to three or more specimens for each displacement for each defect configuration. The useful results for the rectangular-shaped defects are limited to a displacement ≤ 3.05 mm (0.12 in.) because only two specimens are available beyond this point. The two low J-values at displacements of 2.79 and 3.05 mm (0.11 and 0.12 in.) had very low values of $r^2 = 0.01$. No explanation exists for the low values. The deflection corresponding to the load associated with initiation of subcritical crack growth varies from 2.0 to over 4.1 mm (0.08 to 0.16 in.); the value of J, in Fig. 5, varies from 0.07 to 0.08 MJ/m^2 (400 to 440 in. · lb/in.^{2}).

The linear regression coefficients given in Table 2, for the column based on U proportional to A, show that the results are essentially independent of the defect configuration as evidenced by r^2-values ranging from 0.85 to 0.97. These are the highest values of r^2 obtained for the methods given in Table 2. The value of J_{init} in Table 2, for all defect configurations combined together, varies from 0.21 to 0.79 MJ/m^2 (1750 to 4500 in. · lb/in.^{2}).

J_{init}, based on U proportional to W, was about 10 percent of J_{Ic} and had no agreement with J_{init} from the Charpy specimens. This suggests that the latter two sources of critical values of J cannot be used successfully to predict structural integrity using the simplistic approach based on U proportional to W.

J_{init}, based on U proportional to A, is a variable but the range includes J_{Ic}. This suggests the possibility of a link, but it is not readily evident at this time. J_{init}, based on U proportional to A or based on Charpy specimens, shows a range of values but it is not readily evident how the two can be used to predict structural integrity.

Paris Equation

Values of J_{init} and J_{pen} calculated using Eq 1 are plotted versus cF^2 in Fig. 8*a* and 8*b*. The intent was to plot J versus some parameter that quantified the severity of the surface flaw which would provide a means for comparing defects of various aspect ratios ($a/2c$), crack depth (a), etc. A statistical analysis has been conducted of various parameters used to describe the severity of the surface flaw, and cF^2 was shown to provide a substantial improvement over using single parameters such as a/t, etc. [*19*].

The results in Figs. 8*a* and 8*b* show substantial scatter, especially for the latter. Figure 8*a* shows J_{init} to be less than 0.17 MJ/m^2 (950 in. · lb/in.^{2}) with the majority of the values less than 3.5×10^4 J/m^2 (200 in. · lb/in.^{2}). Figure 8*b* shows considerable scatter, but with an apparent trend of decreasing J with increasing cF^2. A similar observation has been made in Ref *20*, where the flow stress ratio $(\bar{\sigma}_{test}/\bar{\sigma}_{calc})_{eng}$ decreases with increasing cF^2, suggesting that the flow stress varies as a function of constraint. True stress constraint effects are taken into consideration by assuming that $\sigma_{ys(ten)}$ is equivalent to $\bar{\sigma}_{calc}$ and $\sigma_{ys(eff)}$ is equivalent to $\bar{\sigma}_{test}$. Therefore, in calculating the constraint effects, the ratio $(\sigma_{pen}/\sigma_{ys})_{true}$ is divided by $(\bar{\sigma}_{test}/\bar{\sigma}_{calc})_{true}$ and the effective yield strength is obtained by multiplying $(\sigma_{ys})_{eng}$ by $(\bar{\sigma}_{test}/\bar{\sigma}_{calc})_{eng}$.

Calculations of J are made using Eq 1, but including the constraint effects on $(\sigma_{ys})_{eng}$ and $(\sigma_{ys})_{true}$, and the results are shown in Figs. 10*a* and 10*b*. The shapes of these figures are consistent with those of Figs. 6 and 9 as well as K_{Ie} versus cF^2 in Ref *13*. The use of the constraint effect does not reduce the total scatter of J-values but does reduce the range of J-values from a factor of 5 to less than a factor of 2 at a constant value of cF^2.

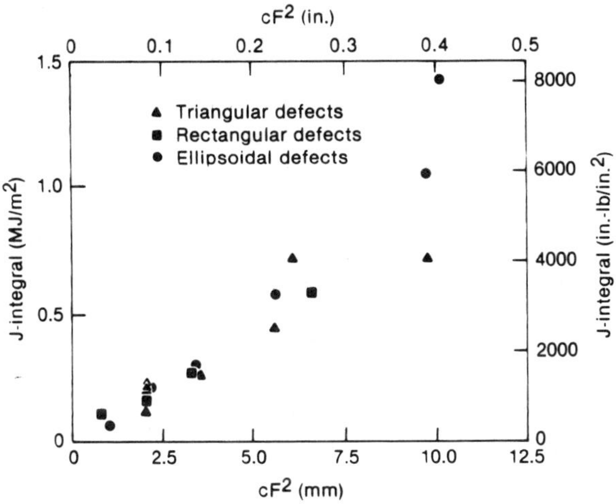

(a) Initiation

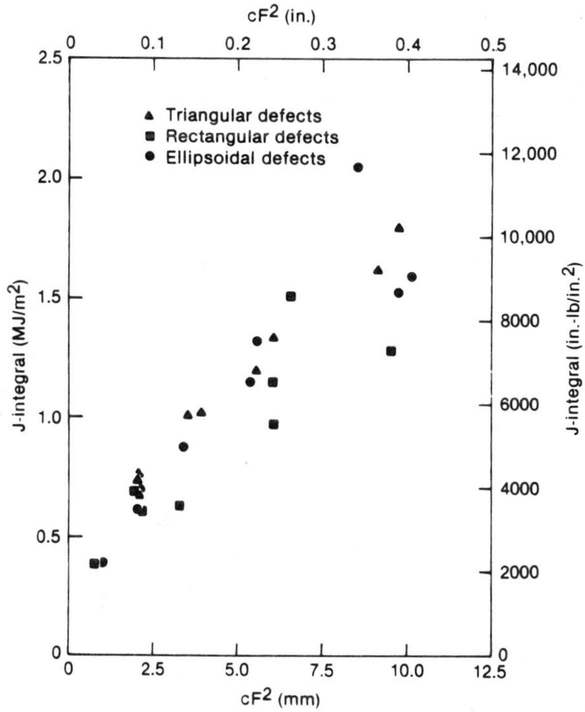

(b) Penetration

FIG. 10—*Plot of J-integral, based on Paris equation plus constraint, versus flaw severity* (cF^2).

For those calculations where the Paris equation is used without constraint effect, the values of J_{init} increase from almost zero to 0.17 MJ/m^2 (950 in.·lb/in.^{2}) with increasing cF^2. But, the majority of the data are nominally <0.04 MJ/m^2 (<200 in.·lb/in.^{2}). This follows the nominal trend for J_{init} based on Charpy specimens but no correlation has been made between deflection and cF^2. Therefore, there is no apparent correlation between J_{init} (Charpy specimens) and J_{init} (surface-flawed specimens).

For those calculations where Eq 1 was used in conjunction with constraint effects, the value of J_{init} ranges from 0.05 to 1.40 MJ/m^2 (300 to 8000 in.·lb/in.^{2}) with decreasing cF^2. But, the more realistic upper value appears to be nominally 1.05 MJ/m^2 (6000 in.·lb/in.^{2}). This encompasses J_{Ic}, which hints of a possible link though it is not readily evident at this time. There appear to be some similarities in the shape of the plots for J_{init} (Charpy specimens) and J_{init} (surface-flawed specimens). But, there is a substantial difference in the range of J_{init}. A possible correlation exists but at the present time neither J_{Ic} nor J_{init} (Charpy specimens) can be used with the Paris equation (with or without constraint effects) to predict structural integrity for the 6.4-mm-thick (0.25-in.) annealed Type 304 stainless steel.

Even though the use of Eq 1 to analyze defects in annealed Type 304 stainless steel was a substantial extrapolation of the intention of its originator, it was thought that J_{init} could be a constant value. It was then intended to compare this constant value with values of J_{Ic} obtained from compact specimens. This is not possible since J_{init} varies as a function of initial crack size, especially if apparent constraint effects are taken into consideration. This suggests that either J_{init} varies as a function of flaw size (cF^2) or that the values for M and Q need to be modified.

But, values of J_{init} calculated from the Paris equation with the constraint effect range from 0.05 to 1.05 MJ/m^2 (300 to 6000 in.·lb/in.^{2}), which agrees reasonably well with J_{init}, based on U proportional to area, which ranges from 0.32 to 0.79 MJ/m^2 (1750 to 4500 in.·lb/in.^{2}). This suggests a link between the two approaches, but at this time it is not known how J_{Ic} or J_{init} (Charpy specimen) can be used with either approach. But, the nominal agreement suggests that a constant value of J_{Ic} would not be applicable and the relative values of M and Q are not in error.

Even though it was not possible to develop a comparison between the absolute value of J_{init}, from the surface-flawed specimens, and J_{Ic}, it was desired to evaluate a possible relative fit. This was done by using Eq 1 to calculate J_{pen} (even though subcritical crack growth had occurred) and J_{init}.

Values of J_{init} and J_{pen} were used to calculate $\Delta J/\Delta a$ ratios where $\Delta a = b$ for specimens containing rectangular-shaped defects; see Table 3. The ratios varied from 138 to 325 MJ/m^3 (20.1 to 47.1 in·kips/in.^{3}) and averaged 223.4 MJ/m^3 (32.4 in.·kips/in.^{3}). For calculations of J where constraint effects were taken into consideration, the ratios varied from 55 to 239 MJ/m^3 (8.0 to 34.6 in.·kips/in.^{3}) and averaged 146.3 MJ/m^3 (21.2 in.·kips/in.^{3}). Both of

TABLE 3—*Calculation of ΔJ/Δa for rectangular-shaped defects.*

Specimen No.	J_{init} MJ/m^2	J_{init} in.·lb/in.^{2}	J_{pen} MJ/m^2	J_{pen} in.·lb/in.^{2}	Δa mm	Δa in.	$\Delta J/\Delta a$ MJ/m^3	$\Delta J/\Delta a$ in.·kip/in.^{3}
			EQUATION 1					
18	...	...	0.42	2400	2.77	0.109	...	...
27	0.052	299	0.87	4993	3.25	0.128	253	36.7
28	...	...	0.19	1064	1.60	0.063	...	...
29	0.0086	49	0.77	4394	4.60	0.181	165	24.0
30	0.014	81	0.37	2089	1.50	0.059	138	20.1
31	...	...	0.49	2813	2.72	0.107	...	...
32	0.012	71	0.99	5674	3.02	0.119	325	47.1
35	0.024	137	1.11	6338	4.60	0.181	236	34.3
			EQUATION 1 PLUS CONSTRAINT EFFECT					
18	...	...	...	...	...	...	...	...
27	0.93	5324	1.51	8596	3.25	0.128	176	25.6
28	...	...	1.28	7300	1.60	0.063	...	...
29	0.16	908	0.61	3494	4.60	0.181	99	14.3
30	0.27	1554	0.63	3597	1.50	0.059	239	34.6
31	...	...	1.14	6528	2.72	0.107	...	...
32	0.18	1031	0.67	3830	3.02	0.119	162	23.5
35	0.11	644	0.37	2100	4.60	0.181	55	8.0

these average values agree reasonably well with 197.4 MJ/m^3 (28.6 in.·kips/in.^{3}) for Type 304 stainless steel compact specimens in Ref *16*. The rectangular-shaped defects were evaluated, since the crack front was a constant distance from the back face. Ratios calculated from the surface-flawed specimen should be less than those in Ref *16*, because J_{pen}, in the analysis, was calculated from the original defect size, resulting in a low value of J_{pen} and subsequently a low value of $\Delta J/\Delta a$.

Thus, Eq 1 may not provide absolute values of J for the thin, ductile annealed Type 304 stainless steel, but it appears to provide reasonably accurate relative values.

For J-controlled crack growth, it is necessary to maintain a J-dominant field which requires [*21*]

$$\omega = \frac{dJ}{da}\,\frac{(t-a)}{J} \gg 1.0 \tag{6}$$

where

$$\frac{dJ}{da} > 197.04\ \frac{\text{MJ}}{\text{m}^3}\left(28\,600\ \frac{\text{in}\cdot\text{lb}}{\text{in.}^3}\right) \quad \text{Ref } 16$$

and $t - a$ varies from 1.27 to 5.08 mm (0.05 to 0.20 in.).

For J_{Ic} = 0.78 MJ/m^2 = (4450 in.·lb/in.^{2}), Ref *16*, ω will vary from 0.32 to

1.29. For J_{init} ranging from 0.05 to 1.05 MJ/m^2 (300 to 6000 in. · lb/in.^{2}), per Table 2, ω will vary from 0.95 to 4.77.

The ω-value of 4.77 is not sufficiently greater than 1.0; therefore, these tests of crack growth were probably not conducted in a J-dominant field.

Summary and Conclusions

In order to develop a comparison between J_{init} from the surface-flawed specimens, and J_{init} from Charpy specimens, or J_{Ic}, it was necessary to identify appropriate values for the latter two. Values of $J_{Ic} > 0.70$ MJ/m^2 (4000 in. · lb/in.^{2}) are given for forged 304 stainless steel [*16*]. In addition, precracked Charpy specimens were tested, in slow bend, from the materials in Table 1. The values of J_{init} varied from 0.03 to 0.24 MJ/m^2 (150 to 1400 in. · lb/in.^{2}) and are less than J_{Ic}, probably because of noncompliance with requirements for plane strain and J-controlled field.

Two analytical techniques based on either the Begley-Landes [*1*] approach or an equation developed by Paris [*2*] were used to evaluate the data generated from specimens containing surface flaws.

Two approaches were used to evaluate the Begley-Landes approach; one assumed that the surface-flawed specimen was a single-edge notch plus two unflawed strips and the other approach was based on converting the surface flaw to an equivalent through-crack. The regression correlation coefficients (r^2) were too low for the latter approach (see Table 2) and further investigation was discontinued. For the assumed single-edge notch plus two unflawed strips, the energy absorbed by each of the three sections was assumed to be a function of either the specimen width or the cross-sectional area. For the former, the regression correlation coefficients (r^2) increased substantially for specimens containing rectangular-shaped defects compared with those specimens containing the triangular, ellipsoidal, or all three shapes evaluated together. The values of J_{init}, for rectangular-shaped defects, was nearly constant at 0.07 to 0.08 MJ/m^2 (400 to 440 in. · lb/in.^{2}). For the approach based on cross-sectional area, the linear regression coefficients (r^2) show that the results are independent of defect configurations. The value of J_{init} for all defect configurations combined together varies from 0.21 to 0.79 MJ/m^2 (1750 to 4500 in. · lb/in.^{2}).

The results based on the Begley-Landes approach are then compared with J_{Ic} and J_{init} (Charpy specimens). It was concluded that J_{init} (Charpy specimen) and J_{Ic} cannot be used to successfully predict structural integrity using the simplistic approach based on either U proportional to W or U proportional to A.

Values of J_{init} were calculated using Eq 1, developed by Paris, and there is substantial scatter in the results as shown in Figs. 8*a* and 8*b*. Equation 1 was modified to take into consideration possible constraint effects and the results are shown in Fig. 10*a* where values of J_{init} nominally range from 0.05 to 1.05

MJ/m^2 (300 to 6000 in.·lb/in.^{2}). At the present time, neither J_{init} (Charpy specimens) nor J_{Ic} can be used in conjunction with Eq 1 to predict structural integrity for the 6.4-mm-thick (0.25-in.) annealed Type 304 stainless steel.

It was observed that the range of J_{init}-values [0.05 to 1.05 MJ/m^2 (300 to 6000 in.·lb/in.^{2})] based on Eq 1 plus constraint was comparable to that observed for calculations based on U proportional to A where J_{init} ranged from 0.32 to 0.79 MJ/m^2 (1750 to 4500 in.·lb/in.^{2}). This suggested possible experimental verification of Eq 1. Since it was not possible to develop a comparison between the absolute value of J_{init} (from the surface-flawed specimens) and J_{Ic}, it was desired to evaluate a possible relative fit. Equation 1 was used to calculate J_{pen}. Values of J_{init} and J_{pen} from only those specimens containing rectangular-shaped defects were used to calculate $\Delta J/\Delta a$. The average values agree reasonably well with those obtained in Ref *16*.

Acknowledgment

This work was performed in the Materials Application Section of the Materials Technology Division of EG&G Idaho, Inc., under the support of the U.S. Department of Energy, Office of Energy Research, Office of Basic Energy Sciences, under Department of Energy Contract No. DE-AC07-76ID01570.

References

[*1*] Begley, J. A. and Landes, J. D. in *Fracture Toughness (Proceedings of the 1971 National Symposium on Fracture Mechanics) Part II, ASTM STP 514*, American Society for Testing and Materials, 1972, pp. 1–20.

[*2*] Paris, P. C. in "Resolution of the Reactor Vessel Materials Toughness Safety Issue," NUREG-0744, Vol. II, U.S. Nuclear Regulatory Commission, Washington, D.C., Aug. 1981.

[*3*] Harrison, J. D., Dawes, M. G., Archer, G. L., and Kamath, M. S. in *Elastic-Plastic Fracture, ASTM STP 668*, J. D. Landes, J. A. Begley, and G. A. Clark, Eds., American Society for Testing and Materials, 1979, pp. 606–631.

[*4*] Kuhn, P., "Residual Tensile Strength in the Presence of Through Cracks or Surface Cracks," NASA Technical Note, NASA TDN-5432, National Aeronautics and Space Administration, March 1970.

[*5*] "Resolution of the Reactor Vessel Materials Toughness Safety Issue, Appendix I, Resolution of Some Problems Related to NRC's Task Action Plan A-11," NUREG-0744, Vol. 2., U.S. Nuclear Regulatory Commission, Washington, D.C., Aug. 1981.

[*6*] Tada, H. and Paris, P. C., *Tearing Instability Analysis Handbook*, NUREG/CR-1221, U.S. Nuclear Regulatory Commission, Washington, D.C., Jan. 1980.

[*7*] Zahoor, A., "Tearing Instability of Elastic-Plastic Crack Growth," Ph.D. Dissertation, George Washington University, St. Louis, Mo., Aug. 1978.

[*8*] Paris, P. C. et at, "A Treatment of the Subject of Tearing Instability," NUREG-0311, U.S. Nuclear Regulatory Commission, Washington, D.C., Aug. 1977.

[*9*] Shih, C. F. and Hutchinson, J. W., *Transactions*, American Society of Mechanical Engineers, *Journal of Engineering Materials and Technology*, Oct. 1976, pp. 289–295.

[*10*] Paris, P., George Washington University, St. Louis, Mo., personal communication, 8 July, 1981.

[*11*] Newman, J. C., Jr. in *Part-Through Crack Fatigue Life Predictions, ASTM STP 687*, J. B. Chang, Ed., American Society for Testing and Materials, 1979, pp. 16–42.

[*12*] Merkle, J. G., "A Review of Some of the Existing Stress Intensity Factor Solutions for Part-Through Surface Cracks," ORNL-TM-3983, Oak Ridge National Laboratory, Oak Ridge, Tenn., Jan. 1973.

[*13*] Reuter, W. G. and Place, T. A., *International Journal of Pressure Vessels and Piping*, Vol. 10, No. 1, Jan. 1982, pp. 55–67.

[*14*] Newman, J. C., Jr., *Engineering Fracture Mechanics*, 1973, pp. 667–689.

[*15*] Newman, J. C., Jr. in *Properties Related to Fracture Toughness, ASTM STP 605*, American Society for Testing and Materials, 1976, pp. 104–123.

[*16*] Bamford, W. H. and Bush, A. J., *Elastic-Plastic Fracture, ASTM STP 668*, J. D. Landes, J. A. Begley, and G. A. Clark, Eds., American Society for Testing and Materials, 1979, pp. 551–577.

[*17*] Shih, C. F., "Methodology for Plastic Fracture," EPRI NP-1735, Electric Power Research Institute, Palo Alto, Calif., March 1981, pp. 5–11.

[*18*] Hahn, G. T. and Rosenfield, A. R., *Acta Metallurgica*, Vol. 13, 1965, p. 293.

[*19*] Reuter, W. G., Matthews, S., and Smith, F. W., "Critical Parameters of Surface Flaws in Annealed Type 304 Stainless Steel," *Engineering Fracture Mechanics*, to be published.

[*20*] Reuter, W. G. and Place, T. A., *International Journal of Pressure Vessels and Piping*, Vol. 9, No. 5, Sept. 1981, pp. 385–395.

[*21*] Paris, P. C. and Toda, H., "Further Results on the Subject of Tearing Justability-1," NUREG/CR-1220 Vol. 1, U.S. Nuclear Regulatory Commission, Washington, D.C., Jan. 1980.

Viscoplastic Analysis and Correlation

Hermann Riedel[1]

Crack-Tip Stress Fields and Crack Growth Under Creep-Fatigue Conditions

REFERENCE: Riedel, H., **"Crack-Tip Stress Fields and Crack Growth Under Creep-Fatigue Conditions,"** *Elastic-Plastic Fracture: Second Symposium, Volume I—Inelastic Crack Analysis, ASTM STP 803*, C. F. Shih and J. P. Gudas, Eds., American Society for Testing and Materials, 1983, pp. I-505–I-520.

ABSTRACT: Fatigue crack growth at elevated temperatures is a complicated problem since widespread creep deformation, in addition to elastic deformation, can accompany crack extension, and since the micromechanisms of crack growth comprise grain boundary cavitation, the striation mechanism, and corrosive effects. In the present paper, the time-dependent crack-tip stress fields in an elastic-nonlinear viscous body are analyzed for variable load. From the results, it is concluded that for slow cycling the C*-integral is the load parameter that determines the crack-tip fields and therefore the crack growth rate. If rapid load variations occur within otherwise slow cycles, the stress intensity factor, ΔK_{I}, also affects the crack-tip fields; ΔK_{I} determines the severity of the stress peaks that follow rapid load changes. For rapid cycling, the time-dependent stressing and straining near the crack tip is governed by ΔK_{I}. If rapid cycling is unbalanced, a time-dependent component is superimposed on the rapidly varying ΔK_{I}-controlled field. This time-independent component can be described by the C*-integral only for R-ratios near unity (that is, for high mean stress and small amplitude). The general case of unbalanced rapid cycling remains unresolved. Characteristic frequencies ω_1 are developed to distinguish slow and rapid cycling, and a characteristic time t_1 is given to decide whether a given load increase within a cycle is rapid or slow.

The stress fields are combined with simple models for the striation and cavitation mechanisms of crack growth. The resulting crack growth rates, da/dN, are compared with a small number of experimental results from the literature.

KEY WORDS: creep crack growth, creep-fatigue interaction, fracture mechanics at elevated temperature, elastic-plastic fracture

Fatigue crack growth in metallic materials is a much more complicated process at elevated temperature than it is at room temperature. At room temperature, fatigue crack extension is merely cycle-dependent, as opposed to time-dependent. Crack growth rates can very often be correlated between dif-

[1]Max-Planck-Institut für Eisenforschung GmbH, Max-Planck-Str. 1, D-4000 Düsseldorf, West Germany.

ferent types of test specimens using the stress-intensity factor as the correlating load parameter. This holds as long as the plastic zones near the crack tip are small enough. The micromechanism by which the crack advances is usually the striation mechanism, which results from the irreversible cyclic opening and closing of the crack tip. At elevated temperature, fatigue crack growth depends on the number of cycles as well as on the loading frequency and waveform [*1–8*].[2] Since considerable creep deformation, in addition to elastic deformation, often accompanies crack extension, the stress-intensity factor will not always suffice to correlate crack growth rates [*9*]. Further, a greater variety of micromechanisms for crack extension is observed at high temperature. The most common ones are probably the striation mechanism, which predominates at high loading frequencies, and intergranular creep crack growth at low frequencies [*10–15*]. The latter usually occurs by the initiation, growth, and coalescence of grain boundary cavities ahead of the main crack, just as it does under sustained load. Finally, the effects of corrosion on crack growth are almost ubiquitous at elevated temperature [*16–22*]. More references to the experimental literature are given in the review-type papers [*16,23–25*].

The present paper is primarily concerned with the question of which are the right, macroscopically measurable, load parameters to describe crack growth for various loading histories, including cyclic loading. The answer is based on a continuum-mechanical analysis of the deformation fields in a cracked specimen. The stress analysis is done ignoring the micromechanisms of crack growth. This is justified if the fracture process zone, where the micromechanisms are localized, is sufficiently small. Then the crack-tip fields that are calculated neglecting the process zone determine the development of the process zone and therefore crack growth behavior.

Description of the Problem

A cracked body is subjected to a time-dependent load, $P(t)$, in plane strain tension. The cyclic loading frequency is denoted by ω. The analysis is based on the equilibrium and compatibility equations for infinitesimally small strain and displacement. In order to delimit the complexity of the problem, the material behavior is modeled as elastic-nonlinear viscous, according to

$$\dot{\epsilon}_{ij} = \frac{1+\nu}{E}\,\dot{\sigma}_{ij} - \frac{\nu}{E}\,\dot{\sigma}_{kk}\,\delta_{ij} + \frac{3}{2}\,B\sigma_e^{n-1}\,\sigma'_{ij} \tag{1}$$

where σ_{ij} and ϵ_{ij} are the stress and strain tensors, respectively. The equivalent stress is defined as $\sigma_e = (3\,\sigma'_{kl}\,\sigma'_{kl}/2)^{1/2}$. The prime denotes the deviatoric stress, the summation convention is applied, δ_{ij} is the unit tensor, and the

[2]The italic numbers in brackets refer to the list of references appended to this paper.

superposed dot denotes the time-derivative. E, ν, B, and n are Young's modulus, Poisson's ratio, the creep coefficient, and the stress exponent, respectively. The nonlinear viscous term in Eq 1 represents power-law, secondary creep.

The stress field in such a material is known to be dominated asymptotically near the crack tip by the creep term of Eq 1 [*26*]. It must be an HRR-type field (named after Hutchinson [*27*] and Rice and Rosengren [*28*]), namely

$$\sigma_{ij} = \pm \left[\frac{C(t)}{I_n \, Br} \right]^{1/(n+1)} \tilde{\sigma}_{ij}(\theta) \tag{2}$$

Here, r and θ are polar coordinates centered at the crack tip. The plus, or minus, sign applies if the near-tip field is tensile, or compressive, respectively. The dimensionless factor I_n and the dimensionless angular function $\tilde{\sigma}_{ij}(\theta)$ have been given graphically by Shih [*29*]. The amplitude of the near-tip stress field, $C(t)$, depends on the load, on the specimen geometry, and on the time. For sustained load, for example, it has been shown [*26,30*] that for short times after load application, the crack-tip field is determined by the stress-intensity factor K_{I} according to

$$C(t) = \frac{K_{\text{I}}^2(1 - \nu^2)}{(n + 1)\, Et} \tag{3}$$

whereas for long times, $C(t) = C^*$. The C*-integral is the creep analog of the J-integral [*31–34*]. It is related to the applied load, or to the load-line displacement rate, $\dot{\Delta}$, through

$$C^* = aB\, \sigma_{\text{net}}^{n+1} g_1\left(\frac{a}{W}, n\right) = \sigma_{\text{net}}\, \dot{\Delta}\, g_2\left(\frac{a}{W}, n\right) \tag{4}$$

where

a = crack length,
σ_{net} = load divided by area of uncracked ligament (net section stress),
W = specimen width, and
g_1 and g_2 = dimensionless functions of the specimen geometry and of the stress exponent n.

Recalling the analogy between the C* and J-integral, numerical values of g_1 and g_2 can be taken from Refs *31* and *35*. Under constant load conditions, C^* has been successfully applied to the more ductile structural alloys [*36–38*], and these results have been discussed in Ref *39*. The characteristic time

$$t_1 = \frac{K_{\text{I}}^2(1 - \nu^2)}{(n + 1)EC^*} \tag{5}$$

characterizes the transition from the short-time, K_I-controlled, to the long-time, C^*-controlled limit.

Correspondingly, the applicability of ΔK_I and C^* in the case of fatigue loading at elevated temperature is examined in the present paper. As one might suspect, C^* dominates for very slow cycling, whereas at high frequencies the elastic response of the specimen, and therefore ΔK_I, dominates. If ω denotes the cyclic loading frequency ($\omega = 2\pi/\text{cycle time}$), then the characteristic frequency to distinguish between slow and fast cycling will be of the order $\omega_1 \approx 1/t_1$. It is clear that ω_1 characterizes the transition from creep-dominated to elastically dominated overall specimen behavior. It has nothing to do with the transition from the cavitation to the striation micromechanism, which will, in general, occur at another characteristic frequency which is not the object of the present paper.

Slow Cycling Without Rapid Load Variations

If the cyclic loading frequency ω is small compared with ω_1 and if no rapid load variations occur within the cycles, then the elastic strain rate, $\dot{\sigma}/E \approx \omega\,\sigma/E$, is everywhere small compared with the creep rate, $B\sigma^n$. If the elastic strain rate is neglected, Eq 1 represents a purely nonlinear viscous material law. Under these circumstances, it is known that the stress field varies in time in proportion to the applied load. The near-tip field amplitude $C(t)$ is then given by $C^*(t)$. Since C^* is the load parameter that determines the near-tip field, it is concluded that C^* governs the crack growth behavior for slow cycling.

Although the micromechanisms of crack growth are not of primary interest in the present paper, a small excursion is made here. Creep crack growth by grain boundary cavitation in the C^*-controlled limit has been studied theoretically in Refs *39* and *40* for constant or variable load. The principal result for the crack growth rate is that $da/dt \propto C^{*n/(n+1)}$ ([*40*], Eq 13). In this limiting case, crack growth is purely time-controlled, rather than cycle-controlled. If the load varies cyclically in time such that $C^* = C^*_{max} f(\omega t)$, integration of da/dt over one cycle leads to the crack growth increment per cycle

$$\frac{da}{dN} \propto \frac{{C^*_{max}}^{n/(n+1)}}{\omega} \int_0^{2\pi} [f(\tau)]^{n/(n+1)} d\tau \tag{6}$$

The use of Eq 6 should be confined to tension-tension tests, since the behavior of the crack in compression has not been modeled in Ref *40*. Moderate compressive loads, however, within a mainly tensile cycle will have little effect on the crack growth rate.

Next, crack growth by the striation mechanism is considered. Here, the crack growth increment per cycle is of the order of the cyclic crack-opening displacement in some cases [*14,41*]. The opening of the crack-tip at a fixed distance behind the tip is computed from the known HRR-type stress field by time-integration of the corresponding strain rate. This leads to the same dependence of da/dN on C^* and ω as the cavitation mechanism, Eq 6, except that now the time integral in Eq 6 is to be extended over the compressive part of the cycle only. For tension-tension-type loading, no crack growth by the striation mechanism would be predicted since then no cyclic opening and closing of the crack tip occurs in the limit of slow cycling.

Rapid Load Variations

In this section, rapid load variations within the load cycles are examined. The prototype of a rapid load variation is a sudden load step as it occurs periodically in square-wave loading.

The instantaneous response of an elastic-nonlinear viscous material to a sudden load change is purely elastic since creep deformation requires time. This means that the stress field immediately after the load step is given by

$$\sigma_{ij}(t = 0^+) = \sigma_{ij}(t = 0^-) + \Delta\sigma_{ij} \tag{7}$$

Here $t = 0^-$ and $t = 0^+$ indicate the times immediately before and after the load step, and $\Delta\sigma$ is the linear elastic field in the cracked body subject to the load step ΔP. In particular, the near-tip field is the well-known elastic singular field. Since the elastic singularity is stronger than the HRR-type singularity, the near-tip field is dominated by $\Delta\sigma_{ij}$ alone. For $r \to 0$

$$\sigma_{ij}(t = 0^+) = \Delta\sigma_{ij} = \frac{\Delta K_{\mathrm{I}}}{\sqrt{2\pi r}} f_{ij}(\theta) \tag{8}$$

where ΔK_{I} is the stress-intensity factor associated with ΔP. Such an elastic field also represents the initial condition for the development of stress and strain after a load step from zero load [*26,30*]. Therefore the leading term of $C(t)$ for short times after the load step can be taken from Refs *26* and *30*; $C(t)$ is given by Eq 3 with K_{I} replaced by ΔK_{I}. If the load is kept constant long enough after the load step, the specimen arrives at a steady state of extensive creep. Elastic strain can then again be neglected and $C(t)$ becomes equal to C^* on the new load level.

This behavior of the stress near the crack tip is shown schematically in Fig. 1 for square wave loading. The stress peaks that are observed after each load step may have significant effects on the micromechanisms of crack growth, and therefore on the crack growth rate. For example, Raj and Ashby [*42*] and Raj [*43*] have shown that the nucleation of microcavities is extremely sensitive to the stress. At low stress levels, the time for cavity nucleation can be comparable or greater than the time for cavity growth and coalescence. In this case, the enhancement of void nucleation ahead of the crack by the stress

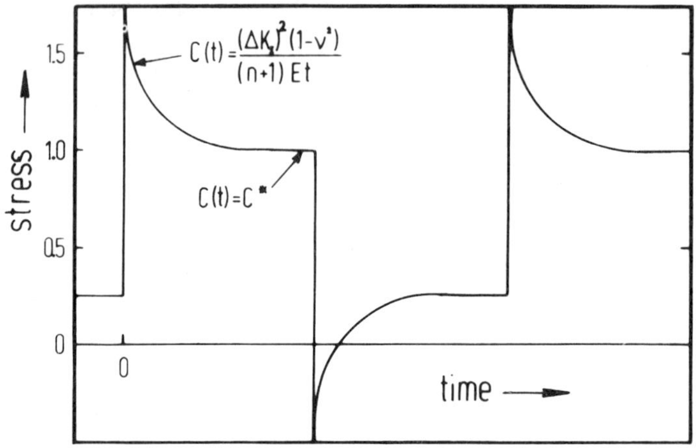

FIG. 1—*Near-tip stress field versus time for square-wave loading with hold-times greater than the characteristic time* t_l. *Schematic drawing.*

peaks will accelerate crack growth markedly compared with the case of sustained load where the stress peaks are absent. Another possible effect of the stress peaks is due to the rapid opening of the crack tip during the tensile stress peaks. It is known that protective oxide films on metals rupture if the metal is deformed too rapidly [*44*]. So it appears possible that a stable oxide film covers the crack tip at low opening rates, but breaks at the higher rates occurring during the stress peaks. This would allow the chemical atmosphere to react more easily with the metal. This can lead, for instance, to internal oxidation or carburization along grain boundaries [*16,45*].

Whatever the micromechanisms of crack growth might be, the conclusion of the stress analysis is that two macroscopic load parameters are needed to describe the near-tip stress fields for square-wave loading with long hold times. The stress-intensity factor ΔK_I determines the stress peaks, whereas the C*-integral governs the behavior for the longer times after the load steps. Which one has the greater influence on the crack growth rate depends on the micromechanisms of crack growth and on the test conditions.

In practice, load variations occur with finite velocity and the question arises which load variations can be considered as being effectively a load step and which ones are slow. To examine this question we consider cases where the load increases in proportion to a power β of time. According to Refs *30* and *46*, loading according to a power-law in time can be treated in the same way as step loading, which represents the special case $\beta = 0$. For short times, the stress-intensity factor is the relevant load parameter and it increases according to

$$K_I(t) = k_I t^{\beta} \tag{9}$$

where k_I is constant in time. Employing the same method as in Ref *26*, the amplitude of the HRR-field is found to vary for short times as

$$C(t) = \alpha(n, \nu, \beta)\,\frac{1 + 2\,\beta n}{n + 1}\;\frac{(1 - \nu^2)\,K_I^2(t)}{Et} \tag{10}$$

The dimensionless factor α is estimated to be $\alpha = 1$ if path-independence of the J-integral is assumed as in Ref *26*. Finite-element calculations have been done for $\beta = 0$ (step loading) [47], and for $\beta = 1$. The numerical method reproduces the stress field calculated with $\alpha = 1$ within 5 percent. A creep zone has been defined in Ref *46*. It grows around the crack tip according to

$$r_{cr} = \frac{K_I^2}{2\pi}\left(\frac{EBt}{1 + 2\,\beta n}\right)^{2/(n-1)} F_{cr}(\theta) \tag{11}$$

as long as it is small enough compared with the crack length and ligament width. The dimensionless shape functions $F_{cr}(\theta)$ are shown in Fig. 2. They have been calculated using the finite-element program described in Refs *47* and *48*.

For large times, the whole specimen creeps extensively. Then $C(t)$ becomes equal to $C^*(t)$. The characteristic time t_1 for the transition from short to long times is given by

$$t_1 = \frac{1 + 2\beta n}{n + 1}\;\frac{K_I^2(t_1)\,(1 - \nu^2)}{C^*(t_1)E} \tag{12}$$

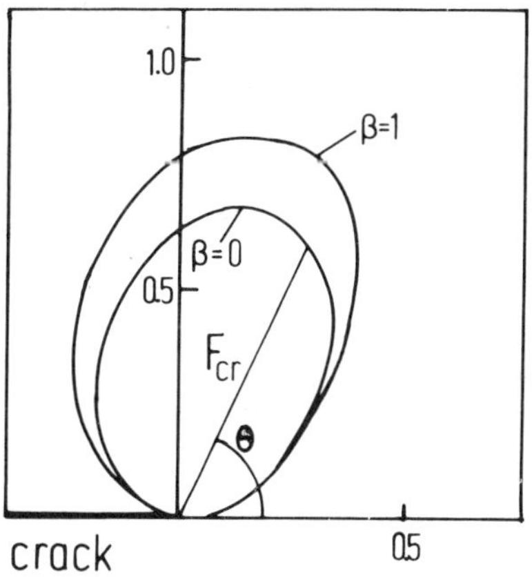

FIG. 2—*Shape of the creep zone,* $F_{cr}(\theta)$, *for step loading* ($\beta = 0$) *and linear load increase* ($\beta = 1$). *Finite-element result for* $n = 5$, $\nu = 0.3$.

This is the generalization of Eq 4 for arbitrary β. The development of the HRR-field amplitude in the whole range from short to long times can approximately be described by the interpolation formula

$$C(t) \approx \frac{1 + 2\beta n}{n + 1} \frac{K_I^2(t)(1 - \nu^2)}{Et} + C^*(t) \tag{13}$$

This simple additive superposition of the short-time and long-time results reproduces the stresses within 5 percent compared with finite-element calculations for different types of cracked specimens [*47,48*].

Figure 3 shows the development of the near-tip stress for a linear load increase ($\beta = 1$) followed by a hold time. The stress is normalized by the value which is assumed after a long time. The behavior during the rise time, t_r, is described by Eq 10. If the rise time is long compared with t_1, then no stress peak is observed (Fig. 3*a*). For short rise times $t_r \ll t_1$, however, the stress at the end of the rise time is enhanced compared with its long-time value (Fig. 3*b*). Thus the ratio of the rise time, t_r, and the characteristic time t_1 determines whether or not the near-tip stress field exhibits a peak at the end of the rise time. The transient behavior of the HRR-amplitude during the hold time following a short linear rise can be approximated by the formula

$$C(t) \approx \frac{K_I^2(1 - \nu^2)/E}{(n + 1)[t - 2nt_r/(1 + 2n)]} + C^*, \qquad \text{for } t > t_r \tag{14}$$

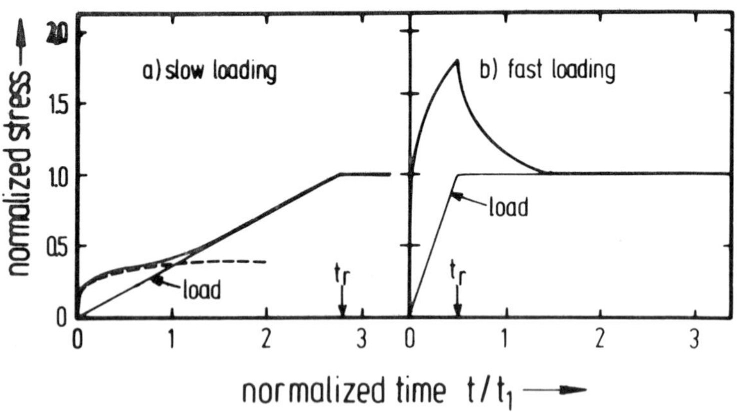

FIG. 3—*Crack-tip stress in response to a linear load increase followed by a hold, for short and long rise times. The stress is normalized by its long-time value. The dashed line represents the short-time solution,* n = *4.*

Rapid Cycling

Rapid cycling is characterized by a predominantly elastic response of the cracked body within a cycle. For unbalanced cycles, however, also creep strain is accumulated over many cycles, and it is not a simple matter to solve the long-term creep problem in the presence of superimposed elastic cyclic stressing. Therefore, the special case of balanced cycling is considered first.

Balanced Cycling

A cyclic load $P(t)$ is called balanced here if the time-average of $|P|^{n-1}P$ over a cycle vanishes. Then no creep strain is accumulated over long times in a proportional stress field. Throughout the present paper, the possibility of crack closure is neglected. This assumption may appear critical for balanced cycling, but it is likely that, because of crack closure, the crack in a creeping body will open up within the first few cycles so that crack closure becomes less and less important.

The problem of rapid balanced cycling can be treated just like the short-time problem for step loading using the boundary-layer approach [*26*]. The remote boundary condition for the near-tip stress field is the requirement of asymptotic approach to the elastic singular stress field, which is characterized by the stress-intensity factor range ΔK_{I}, by the cyclic frequency ω and by the waveform. Dimensional analysis of the boundary-value problem shows that the near-tip HRR-field must have the form

$$\sigma_{ij} = S(\omega t)\left[\frac{(\Delta K_{\mathrm{I}})^2\,(1-\nu^2)\omega}{I_n EBr}\right]^{1/(n+1)}\tilde{\sigma}_{ij}(\theta) \tag{15}$$

The dimensionless function $S(\omega t)$ depends on n, ν, and on the wave shape. Figure 4 shows the results of a finite-element calculation for sinusoidal, triangular, and trapezoidal waveforms. The numerical program proceeds stepwise in time, as described in Ref *47*, with typically 120 time-steps within a loading cycle. An interesting feature of the solutions shown in Fig. 4 is that $S(\omega t)$ reaches a steady cyclic state very quickly. The curves for the second and all following cycles are indistinguishable, and even the first cycle deviates only in its initial parts, since it starts from a state that is free from any internal stresses. Figure 4 also shows the expected stress peaks for the trapezoidal wave shape with short rise time.

The results given in Eq 15 and in Fig. 4 are valid as long as the creep zones that grow and shrink cyclically around the crack tip are small enough compared with the specimen dimensions. The smallness of the creep zone is ensured by large enough loading frequencies according to

$$\omega \gtrsim \omega_1 = \pi/t_1 = \frac{\pi(n+1)\,EC^*}{(1+2\beta n)\,K_{\mathrm{I}}^2(1-\nu^2)} \tag{16}$$

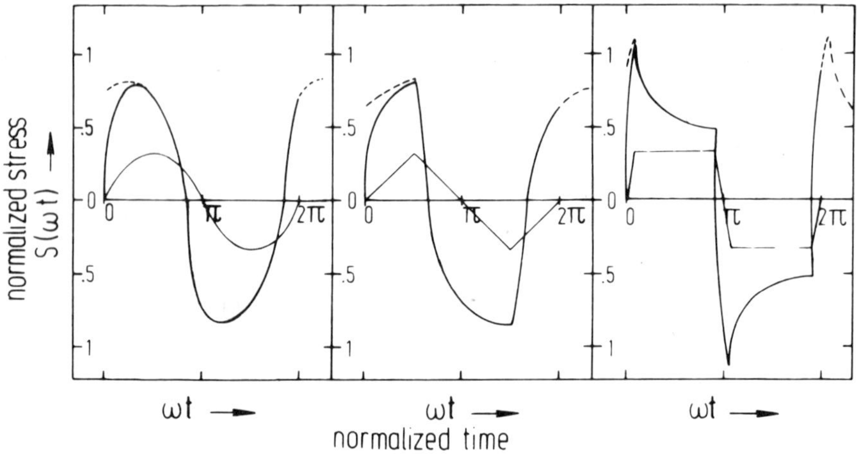

FIG. 4—*Time dependence of the near-tip stress field,* S(ωt), *for rapid balanced cycling for sinusoidal, triangular, and trapezoidal loading waveforms. Solid lines: first loading cycle; dashed lines: second and all following cycles; light lines: load versus time. Finite-element calculations for* n = 5, ν = *0.3.*

Here, K_I and C^* are to be computed for the maximum value of the applied load and β is equal to zero for square-wave loading and equal to unity for triangular waveform; for sinusoidal or trapezoidal wave shapes β-values between 0 and 1 should be chosen.

Unbalanced Rapid Cycling

Following Ponter [*49,50*] and Ponter and Brown [*51*], the stress field is decomposed according to

$$\sigma_{ij} = \rho_{ij} + \sigma_{ij}^{\text{el}} \tag{17}$$

Here, σ_{ij}^{el} is the linear-elastic stress field associated with the applied load, and ρ_{ij} is the internal stress field which varies slowly in time over a cycle. The increment in a cycle, $\Delta\rho_{ij}$, can be calculated by solving a linear elastic problem with a prescribed inelastic strain distribution. This is an efficient and relatively simple method to obtain rapid cycling solutions. It works as long as the elastic strain rate, $\dot{\sigma}/E \approx \omega\sigma^{\text{el}}/E$, is large enough compared with the creep rate, $B\sigma^n$. This prerequisite breaks down in a region near the crack tip since the creep rate diverges more strongly as $r \to 0$ than does the elastic strain rate. However, if this region is small enough, Ponter's solution is still valid in the bulk of the specimen and it sets the remote boundary conditions for the near-tip zone. Since ρ_{ij} is practically time-independent, the cyclic stressing and straining history of the crack-tip region is then uniquely determined by the elastic part, σ^{el}, that is, by the stress-intensity factor range ΔK_I.

The analytic form of ρ_{ij} in Ponter's solution is unknown so far. The present author surmises that in the steady cyclic state, ρ_{ij} cancels the time-average of the elastic singular field and further contains an HRR-type term. The amplitude of the latter, $C(t)$, must be calculated numerically using Ponter's method excluding the crack tip region. Since this has not yet been done it is further assumed that $C(t)$ is of the order of the C*-integral associated with the mean load. This assumption can be proven true if the cyclic load range ΔP is small enough compared with the mean load. With this estimate for ρ_{ij} the limits can be derived within which Ponter's solution can serve as a remote boundary condition for the near-tip region. The elastic strain rate has the order of magnitude $\omega\Delta K_I/(E\sqrt{r})$ and the creep rate is approximately $B[C^*/(Br)]^{n/(n+1)}$. Equating the two rates leads to the size of the zone within which the creep rate is larger than the elastic rate, Ponter's method is not applicable. This zone must be small enough compared with the crack length if the boundary-layer approach is to work properly. This leads to the condition

$$\omega \gg \omega_1 = \frac{BE}{a^{(n-1)/2(n+1)}\Delta K_I}(C^*/B)^{n/(n+1)} \approx EB\sigma^n/\Delta\sigma \qquad (18)$$

where in the second form, ΔK_I and C^* have been replaced by the crack length and some reference stress setting the geometry functions equal to 1.

In conclusion, the time-dependent near-tip stress fields are determined by the stress-intensity factor if the cyclic frequency is large enough to statisfy Eq 16 for balanced cycling, or Eq 18 for unbalanced cycling with high mean load. This means that crack growth rates should correlate with ΔK_I for rapid balanced cycling. The same is expected for unbalanced cycling provided that the micromechanism of crack growth responds primarily to cyclic straining (like, probably, the striation mechanism).

For unbalanced cycling, there is a time-independent stress superimposed on the cyclic stressing. Its magnitude is characterized by C^* if the mean load is large compared with the load range. Otherwise no analytic expression for the time-independent component has been found.

Making an excursion to the micromechanisms we assume as before that the crack growth increment per cycle is proportional to the cyclic crack-opening displacement at some fixed distance behind the crack tip. Calculating the opening displacement from the stress field given in Eq 15 leads to

$$\frac{da}{dN} \propto \frac{(\Delta K_I)^{2n/(n+1)}}{\omega^{1/(n+1)}}\int_0^{\pi} |S(\omega t)|^n d(\omega t) \qquad (19)$$

The predicted dependence on ω is now weak and ΔK_I enters with a power that is slightly less than 2. The numerical value of the time-integral in Eq 19 depends on the wave shape. For the trapezoidal wave shape shown in Fig. 4, the integral and therefore da/dN is some 40 percent greater than for the sinusoidal waveform and 50 percent greater than for the triangular waveform.

Discussion

Numerical values for the characteristic frequencies given in Eqs 16 and 18 vary greatly with the material parameters and with the load level. For 304 stainless steel at 650°C, the relevant material parameters have been taken from the tabulations of Frost and Ashby [*52*] as $E = 150$ GPa; $B = 2 \times 10^{-23}\text{MPa}^{-n}/\text{s}$; and $n = 7.5$.

If a balanced cyclic load with an amplitude of 20kN is applied to a standard (compact tension) 1CT-specimen with $a/W = 0.5$, C^* and K_I at maximum load are calculated as $C^* = 0.44$ N/(ms) and $K_I = 34.5$ MNm$^{-3/2}$ using Ref [*35*]. Then Eq 16 yields the characteristic frequency for triangular waveform as $\omega_1 = 10^{-4}$/s, which corresponds to a cycle time of 17 h. For unbalanced cycling with a mean load of 18 kN and a load amplitude of 2 kN, Eq 18 yields $\omega_1 = 7.5 \times 10^{-4}$/s, which corresponds to a cycle time of 2 h. The characteristic frequencies calculated in the same way for the more creep-resistant superalloys are usually much slower, corresponding typically to one cycle in several years. These examples illustrate that the loading frequencies used in the laboratory are usually higher than the characteristic frequencies ω_1. Under low-cycle fatigue conditions in service, however, cycle times of a day are not untypical.

The aim of this paper was to identify the macroscopically measurable load parameters that determine the crack-tip fields and therefore the crack growth rates. The conditions have been defined under which the stress-intensity factor and the C*-integral are applicable. There are, however, some obvious deficiencies. Firstly, the practically important case of rapid unbalanced cycling could not be resolved completely. If the mean load is of the same order as is the load amplitude, the cyclic component of the near-tip fields is determined by ΔK_I but the time-independent component could not be related to the average C^* or any other load parameter. Here, more work is needed to understand the effect of the R-ratio on the near-tip fields and on crack growth behavior.

Further, instantaneous plasticity has been neglected in the present analysis and the creep behavior has been described simply by a nonlinear viscous term in Eq 1. For step loading, the effect of instantaneous plasticity on the crack-tip fields has been studied in Refs *53* and *54*. One of the effects of instantaneous plasticity will be a purely cycle-dependent fatigue crack growth rate at very high frequencies. Strain-hardening creep has been considered in Refs *30* and *53*, and creep-recovery models have been examined in Refs *55–57*. Although stresses and strain rates can be considerably affected in detail, it is hoped that, in practice, many situations can be dealt with by using the stress-intensity factor and C^* rather than other parameters. In Ref *53* it has been attempted to define the limits of this approach in the presence of strain-hardening primary creep.

The crack growth rates calculated in the preceding sections are not intended to predict experimental values accurately since this would require a much more detailed knowledge of the micromechanisms. Nevertheless, in some cases a good coincidence of theoretical and observed crack growth rates seems to indicate that the simple models for the striation and cavitation mechanism used here are not completely unrealistic. Saxena and Shih [*7,8*], in a certain regime, find a dependence of da/dN on ΔK_I and ω which is well represented by the result for rapid cycling, Eq 19. In the slow-cycling limit, Eq 6 predicts time-controlled crack extension which is usually found in experiments. For sustained load ($\omega = 0$), a detailed comparison of observed crack growth rates with calculated rates (based on cavity growth models) [*39,40*] showed good agreement for a number of materials.

Conclusions

Based on an elastic-nonlinear viscous material behavior, the conditions have been indentified under which the stress-intensity factor and the C*-integral determine the crack-tip fields and should therefore be used to correlate crack growth rates.

A characteristic frequency, ω_1, has been derived to distinguish between slow and rapid cycling (Eqs 16 and 18).

For slow cycling without rapid load variations, the C*-integral determines the near-tip fields.

For slow cycling with rapid load variations within the cycles both, ΔK_I and C^* affect the near-tip fields. A load variation is to be considered rapid if the rise time is short compared with the characteristic time t_1 given in Eq 12.

For rapid balanced cycling, the stress-intensity factor is the only load parameter that enters into the crack-tip fields (Eq 15).

The important case of rapid unbalanced cycling could not be treated completely. The rapidly varying component of the near-tip field is determined by the stress-intensity factor range ΔK_I, but the time-independent component is not given by C^* except in the limiting case of R-ratios close to unity, that is, for high mean load and small load amplitude.

Little effort has been expended in the present paper to model the complicated micromechanisms of crack growth. A very simple model for the striation mechanisms based on the cyclic opening and closing of the crack tip leads to crack growth rates da/dN which agree with a few experimental data. A more elaborate model of crack growth by grain boundary cavitation has been applied to the slow-cycling limit where few experimental data are available except for sustained load ($\omega = 0$). For this case good agreement between predicted and observed crack growth rates has been obtained.

Acknowledgments

Part of this work has been supported by the Deutsche Forschungsgemeinschaft under Contract No. Ri 329/7. The author is indebted to Dr. R. Ehlers, who developed the finite-element program used in this study. Valuable discussions with Professor J. R. Rice are gratefully acknowledged.

References

[*1*] James, L. A. in *Stress Analysis and Growth of Cracks, ASTM STP 513*, American Society for Testing and Materials, 1972, pp. 218–229.

[*2*] Solomon, H. D. and Coffin, L. F., Jr., in *Fatigue at Elevated Temperatures, ASTM STP 520*, American Society for Testing and Materials, 1973, pp. 112–122.

[*3*] Koterazawa, R. and Iwata, Y., *Transactions*, American Society of Mechanical Engineers, *Journal of Engineering Materials and Technology*, Vol. 98, Oct. 1976, pp. 296–304.

[*4*] Clavel, M. and Pineau, A., *Metallurgical Transactions*, Vol. 9A, April 1978, pp. 471–480.

[*5*] James, L. A., *Transactions*, American Society of Mechanical Engineers, *Journal of Pressure Vessel Technology*, Vol. 101, May 1979, pp. 171–176.

[*6*] Sadananda, K. and Shahinian, P., *Transactions*, American Society of Mechanical Engineers, *Journal of Engineering Materials and Technology*, Vol. 101, July 1979, pp. 224–230.

[*7*] Saxena, A., *Fatigue of Engineering Materials and Structures*, Vol. 3, No. 3, 1980, pp. 247–256.

[*8*] Saxena, A. and Shih, T. T. in *Advances in Fracture Research, Proceedings*, Fifth International Conference on Fracture, Cannes, France, 1981, Vol. 5, D. Francois, Ed., Pergamon Press, Oxford, U.K., 1980, pp. 2403–2412.

[*9*] Taira, A., Ohtani, R., and Komatsu, T., *Transactions*, American Society of Mechanical Engineers, *Journal of Engineering Materials and Technology*, Vol. 101, April 1979, pp. 162–167.

[*10*] Tomkins, B. and Wareing, J., *Metal Science*, Vol. 11, Aug./Sept. 1977, pp. 414–424.

[*11*] Runkle, J. C. and Pelloux, R. M. in *Fatigue Mechanisms, ASTM STP 675*, J. T. Fong, Ed., American Society for Testing and Materials, 1979, pp. 501–527.

[*12*] Yamaguchi, K. and Kanazawa, K., *Metallurgical Transactions*, Vol. 10A, Oct. 1979, pp. 1445–1451.

[*13*] Plumtree, A., and Lemaitre, J. in *Advances in Fracture Research, Proceedings*, Fifth International Conference on Fracture, Cannes, France, 1981, Vol. 5, D. Francois, Ed., Pergamon Press, Oxford, U.K., 1980, pp. 2379–2384.

[*14*] Pedron, J. P. and Pineau, A. in *Advances in Fracture Research, Proceedings*, Fifth International Conference on Fracture, Cannes, France, 1981, Vol. 5, D. Francois, Ed., Pergamon Press, Oxford, U.K., 1980, pp. 2385–2392.

[*15*] Rezgui, B., Petrequin, P. and Mottot, M. in *Advances in Fracture Research, Proceedings*, Fifth International Conference on Fracture, Cannes, France, 1981, Vol. 5, D. Francois, Ed., Pergamon Press, Oxford, U.K., 1980, pp. 2393–2402.

[*16*] Ericsson, T., *Canadian Metallurgical Quarterly*, Vol. 18, 1979, pp. 177–195.

[*17*] Speidel, M. O. in *Advances in Fracture Research, Proceedings*, Fifth International Conference on Fracture, Cannes, France, 1981, Vol. 6, Pergamon Press, Oxford, U.K., 1981, pp. 2685–2704.

[*18*] Skelton, R. P., *Materials Science and Engineering*, Vol. 32, 1978, pp. 211–219, and Vol. 35, 1978, pp. 287–298.

[*19*] Teranishi, H. and McEvily, A. J., *Metallurgical Transactions*, Vol. 10A, Nov. 1979, pp. 1806–1808.

[*20*] Floreen, S. and Kane, R. H., *Fatigue of Engineering Materials and Structures*, Vol. 2, pp. 401–412.

[*21*] Sadananda, K. and Shahinian, P., *Metallurgical Transactions*, Vol. 11A, Feb. 1980, pp. 267–276.

[*22*] Shahinian, P. and Sadananda, K. in *Engineering Aspects of Creep, Proceedings*, Interna-

tional Conference, Sheffield, U.K., 1980, Vol. 2, The Institution of Mechanical Engineers, London, 1980, pp. 1-8.

[23] Ellison, E. G. and Harper, M. P., *Journal of Strain Analysis*, Vol. 13, No. 1, 1978, pp. 35-51.

[24] Chen, W.-C., "A Literature Review on Fatigue and Creep Interaction," Report to National Aeronautics and Space Administration, NASA NGR 33-022-157, 1979.

[25] Tomkins, B. in *Creep and Fatigue in High Temperature Alloys*, J. Bressers, Ed., Applied Science Publishers, London, 1981, pp. 73-110 and pp. 111-143.

[26] Riedel, H. and Rice, J. R. in *Fracture Mechanics: Twelfth Conference, ASTM STP 700*, American Society for Testing and Materials 1980, pp. 112-130.

[27] Hutchinson, J. W., *Journal of the Mechanics and Physics of Solids*, Vol. 16, No. 1, 1968, pp. 13-31.

[28] Rice, J. R. and Rosengren, G. F., *Journal of the Mechanics and Physics of Solids*, Vol. 16, No. 1, 1968, pp. 1-12.

[29] Shih, C. F. in *Fracture Analysis, ASTM STP 560*, American Society for Testing and Materials, 1974, pp. 187-210.

[30] Ohji, K., Ogura, K., and Kubo, S., *Journal of the Society for Materials Science*, Japan, Vol. 29, No. 320, 1980, pp. 465-471.

[31] Goldman, N. L. and Hutchinson, J. W., *International Journal of Solids and Structures*, Vol. 11, 1975, pp. 575-591.

[32] Landes, J. D. and Begley, J. A. in *Mechanics of Crack Growth, ASTM STP 590*, American Society for Testing and Materials, 1976, pp. 128-148.

[33] Ohji, K., Ogura, K., and Kubo, S., *Transactions*, Japanese Society of Mechanical Engineers, Vol. 42, 1976, pp. 350-358.

[34] Nikbin, K. M., Webster, G. A., and Turner, C. E. in *Cracks and Fracture, ASTM STP 601*, American Society for Testing and Materials, 1976, pp. 47-62.

[35] Kumar, V., German, M. D., and Shih, C. F., "An Engineering Approach for Elastic-Plastic Fracture Analysis," Report to Electric Power Research Institute, Project No. 1237-1, Palo Alto, Calif., July 1981.

[36] Taira, S., Ohtani, R., and Kitamura, T., *Transactions*, American Society of Mechanical Engineers, *Journal of Engineering Materials and Technology*, Vol. 101, April 1979, pp. 154-161.

[37] Saxena, A. in *Fracture Mechanics: Twelfth Conference, ASTM STP 700*, American Society for Testing and Materials, 1980, pp. 131-151.

[38] Ohji, K., Ogura, K., and Kubo, S. in *Engineering Aspects of Creep, Proceedings*, International Conference, Sheffield, U.K., 1980, Vol. 2, The Institution of Mechanical Engineers, London, 1980, pp. 9-16.

[39] Riedel, H. in *Proceedings*, Advanced Seminar on Fracture Mechanics, Ispra, Italy, Oct. 1981, to be published by Applied Science Publishers, London, 1983.

[40] Riedel, H. in *Creep in Structures, Proceedings*, 3rd IUTAM-Symposium, Leicester, U.K., Sept. 1980, A. R. S. Ponter and D. R. H. Hayhurst, Eds., Springer-Verlag, Berlin-Heidelberg-New York, 1981, pp. 504-519.

[41] Wareing, J. and Vaughan, H. G., *Metal Science*, Vol. 11, Oct. 1977, pp. 439-446.

[42] Raj, R. and Ashby, M. F., *Acta Metallurgica*, Vol. 23, June 1975, pp. 653-666.

[43] Raj, R., *Acta Metallurgica*, Vol. 26, 1978, pp. 995-1006.

[44] Berchtold, L., Sockel, H. G., and Ilschner, B. in *Behaviour of High-Temperature Alloys in Aggressive Environments, Proceedings*, International Conference, Petten, The Netherlands, 1979, I. Kirman et al, Eds., The Metals Society, London, 1980, pp. 927-937.

[45] Schnaas, A. and Grabke, H. J., *Oxidation of Metals*, Vol. 12, No. 5, 1978, pp. 387-404.

[46] Riedel, H., *Zeitschrift für Metallkunde*, Vol. 69, No. 12, 1978, pp. 755-760.

[47] Ehlers, R. and Riedel, H. in *Advances in Fracture Research, Proceedings*, Fifth International Conference on Fracture, Cannes, France, D. Francois, Ed., Vol. 2, Pergamon Press, Oxford, U.K., 1981, pp. 691-698.

[48] Ehlers, R., "Stress Analysis in the Vicinity of a Stationary Crack under Creep Conditions Using the Finite Element Method," Ph.D. Dissertation, Fakultät für Bergbau and Hüttenwesen, Technische Hochschule Aachen, Aachen, Germany, 1981.

[49] Ponter, A. R. S., *Transactions*, American Society of Mechanical Engineers, *Journal of Applied Mechanics*, Vol. 40, June 1973, pp. 589-594.

[*50*] Ponter, A. R. S., *International Journal of Solids and Structures*, Vol. 12, 1976, pp. 809–825.
[*51*] Ponter, A. R. S. and Brown, P. R., *International Journal for Numerical Methods in Engineering*, Vol. 12, 1978, pp. 1001–1024.
[*52*] Frost, H. J. and Ashby, M. F. in *Fundamental Aspects of Structural Alloy Design*, R. I. Jaffee and B. A. Wilcox, Eds., Plenum Press, New York and London, 1977, pp. 27–58.
[*53*] Riedel, H., *Journal of the Mechanics and Physics of Solids*, Vol. 29, No. 1, 1981, pp. 35–49.
[*54*] Leckie, F. A. and McMeeking, R. M., *International Journal of Fracture*, Vol. 17, No. 5, 1981, pp. 467–476.
[*55*] Riedel, H. in *Engineering Aspects of Creep, Proceedings*, International Conference, Sheffield, U.K., 1980, Vol. 1, The Institution of Mechanical Engineers, London, 1980, pp. 215–218.
[*56*] Kubo, S., Brown University Report MRL E-109, Providence, R. I., Dec. 1980.
[*57*] Kubo, S., Brown University Report MRL E-133, Providence, R. I., March 1981.

Edward W. Hart[1]

Stable Crack Extension Rates in Ductile Materials: Characterization by a Local Stress-Intensity Factor

REFERENCE: Hart, E. W., **"Stable Crack Extension Rates in Ductile Materials: Characterization by a Local Stress-Intensity Factor,"** *Elastic-Plastic Fracture: Second Symposium, Volume I—Inelastic Crack Analysis, ASTM STP 803*, C. F. Shih and J. P. Gudas, Eds., American Society for Testing and Materials, 1983, pp. I-521–I-531.

ABSTRACT: A theory is presented for the steady-state propagation of a Mode III crack in a ductile material. The inelastic deformation of the matrix material and the cracking mechanism are both described by real-time rate constitutive laws. The theory leads to realistic nonsingular stresses and relates the crack-tip velocity v to the *apparent stress-intensity factor* K_A that characterizes the remote loading. It is shown that the crack-tip behavior is characterized by a *local stress-intensity factor* K and that the inelastic deformation is described by a *plastic stress-intensity factor* K_P such that $K = K_A + K_P$. These stress-intensity factors are all well defined even though the stress is nonsingular.

KEY WORDS: crack propagation, ductile fracture, creep fracture, stress-intensity factors in ductile fracture, elastic-plastic fracture

In a recent publication [*1*][2] it was shown that the stress field of a sharp crack that was propagating at a finite velocity in a creeping material could be characterized by a *local stress-intensity factor* K. It was further shown that the value of K was not equal to the *apparent stress-intensity factor* K_A that would be given by the remote loading and an assumption of linear elastic behavior. In that paper the value of K was computed as a function of K_A, the crack-tip velocity v, and the material flow law. For the purposes of computation, a calculated dependence of v on K due to Lawn [*2*] was adopted. It was then possible to eliminate K between the two relationships and to determine a dependence of v on K_A as it would be presented by experiment. This resulting relationship had many of the usual properties of measured v versus K_A curves.

[1]Professor, Department of Theoretical and Applied Mechanics, Cornell University, Ithaca, N.Y. 14853.

[2]The italic numbers in brackets refer to the list of references appended to this paper.

The calculation just described was for a *sharp* Mode III crack and it admitted the presence of a stress singularity at the crack tip. Because of that, the calculation depended somewhat strongly on the stress dependence of the creep rate at very high stress levels. In particular, if the creep rate stress dependence is too strong at high stress, the calculation diverges in the form given. A similar conclusion was reached by Hui and Riedel [*3*], who considered the same problem by a method different from that used in Ref *1*. However, if a more realistic model is employed for the cracking zone, there can be no stress singularity, and the calculation has no threat of divergence. In this paper we shall show that the local stress-intensity factor K is then still defined and characterizes the crack even though no stress singularity is present.

The formulation of the theory that follows is done for the semi-infinite Mode III crack. We shall omit the customary subscribt III on the variables. In the present treatment we restrict ourselves to the consideration of real-time steady-state crack propagation in materials with real-time inelastic flow behavior.

Basis for the Theory

We distinguish strongly in the theory between the inelastic deformation of the material and the cracking process. As was shown in Ref *1*, the inelastic deformation of the material serves principally to screen the crack tip from the full effects of the remote loading. The cracking process itself, which results in discontinuous separation of the material, can depend on a variety of mechanisms that range from the singular atomistic cracking process described by Lawn [*2*] to the various void growth models that can involve diffusive processes.

In order to include both material inelastic deformation *and* crack growth processes in a consistent mechanical description of a cracking body it is necessary to represent those processes in terms of constitutive laws among the kinematic and mechanical variables of a continuum model of the system. It is not our goal in this paper to detail those constitutive laws but rather to show how they enter into the mechanical calculation. We shall, therefore, discuss the constitutive laws only in general terms, and will be more concerned about the nature of the variables that are related through the constitutive relations rather than the relations themselves.

For the general formulation of the theory we shall allow the constitutive relations to be, in principle, as complex as may be desired. Of course, they may then be restricted or specialized in any way that may be necessary for computational practicality.

Inelastic Flow Law

We shall consider the inelastic deformation to be that of a real time inelastic strain rate in response to the deviator of the stress tensor. We employ the same notation and type of constitutive law as used in Ref *1*.

We assume specifically that the inelastic strain rate $\dot{\boldsymbol{\epsilon}}$ and the stress deviator $\boldsymbol{\sigma}$ are related through a simple isotropic relation

$$\dot{\boldsymbol{\epsilon}} = (\dot{\epsilon}/\sigma)\boldsymbol{\sigma} \tag{1}$$

where the light-faced symbols $\dot{\epsilon}$ and σ stand for the simple second invariant of their respective tensors of the form

$$\dot{\epsilon} \equiv +\sqrt{(\dot{\epsilon}_{ij}\dot{\epsilon}_{ij})} \tag{2}$$

and the summation convention is implied for repeated indices.

The remainder of the constitutive law consists of scalar relations. Thus, for a reasonably effective law involving a single internal scalar state variable σ^*, we have the flow law

$$\dot{\epsilon} = \dot{\epsilon}(\sigma, \sigma^*) \tag{3}$$

and the evolution law (hardening law)

$$\dot{\sigma}^* = \dot{\epsilon}\Gamma(\sigma, \sigma^*) \tag{4}$$

and the functions $\dot{\epsilon}(\sigma, \sigma^*)$ and $\Gamma(\sigma, \sigma^*)$ must be specified.

For the special case that σ^* is constant the flow model becomes effectively that of a non-Newtonian viscous body, and it is for that case that the explicit calculations of Hart [*1*] and Hui and Riedel [*3*] were carried out.

Because of the complexity of the crack propagation problem we do not at present consider more complicated inelastic constitutive equations than those given by the preceding equations.

Cracking Process

We shall be considering principally the steady-state propagation of a crack through a body from left to right at a velocity v as illustrated in Fig. 1. The origin of the coordinates is at the "crack tip." The negative x-axis coincides with the separated crack faces, and those faces are traction-free. The portion of the x-axis to the right of the origin contains the *cracking zone*.

We assume that any real micromechanical cracking process can be reasonably well modeled on a continuum level by an atomically thin plane extending from the crack tip into the material along which there is a displacement discontinuity at each point x between the upper and lower halves of the body, and on which the stress is continuous. The regular matrix material elastic and inelastic properties are assumed to extend right up to the plane of discontinuity.

Since we are describing the antiplane strain Mode III case, the displacement w is normal to the plane of the figure. The displacement discontinuity q is given by

$$q \equiv w(x, +0) - w(x, -0) \tag{5}$$

We define also the *lineal dislocation density* κ associated with q as

$$\kappa \equiv -dq/dx \tag{6}$$

It is κ that generates the stress produced in the matrix due to the crack zone.

Now, depending on the nature of the micromechanical cracking process, the constitutive relation between q or κ and the stress σ can be of various kinds. If, for example, the relation is a static one of the form

$$q = q(\sigma_{yz}) \tag{7}$$

we have the type of relation presumed by Barenblatt [*4*]. We shall consider also relations of the kinetic type. One such relation might be

$$\dot{q} = \dot{q}(\sigma_{yz}; T) \tag{8}$$

where T is the absolute temperature. Of course, the best way to find an appropriate constitutive relation is to reduce a more detailed micromechanical model to a "thin zone" continuum model. If extensive diffusive behavior is required it may even be necessary to employ constitutive relations that are nonlocal.

We do not investigate the details of any model in this paper. We present

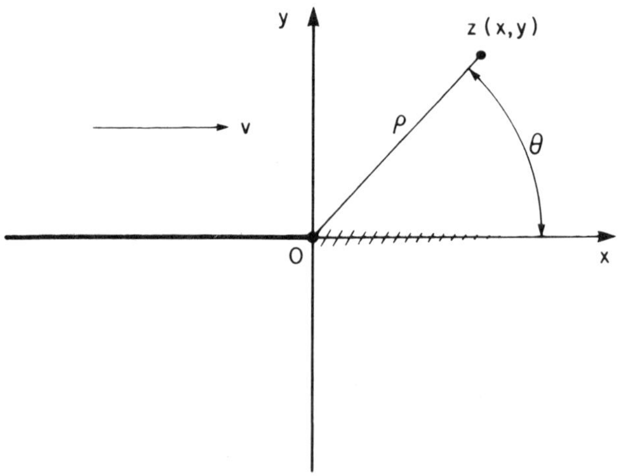

FIG. 1—*Crack geometry for the semi-infinite Mode III crack. The "crack" occupies the negative* x*-axis, and the "cracking zone" lies on the positive* x*-axis. The "crack tip," located at the origin, is moving to the right with velocity* v.

only an idea of the types of relations that can enter the crack theory that we develop in the following.

Mechanical Description

The mechanical description employed here is substantially the same as in Ref *1*. We briefly summarize the relations of importance.

As is well-known, the antiplane problem requires only two stress components σ_{xz} and σ_{yz} which we treat as two-dimensional vector components σ_x and σ_y (or σ_1 and σ_2 when numerical indices are used). We also use the vector representation for the inelastic strain rate $\dot{\epsilon}$.

The mechanical effects of self-stress due to incompatibility of the inelastic flow are dependent on the dislocation density α that is computed from the relation

$$\dot{\alpha} = -\nabla \times \dot{\epsilon} \tag{9}$$

In this equation the curl is two-dimensional and $\dot{\alpha}$ is a z-component.

The problem is most conveniently formulated in terms of the familiar complex stress

$$\tilde{\sigma} \equiv \sigma_x - i\sigma_y \tag{10}$$

As was discussed in Ref *1*, the stress $\tilde{\sigma}^{\alpha}$ due to a distributed dislocation density $\alpha(z)$, where $z \equiv x + iy$, is given by

$$\tilde{\sigma}^{\alpha} = \frac{G}{2\pi i}\int ds'\alpha(z')\frac{1}{z - z'} \tag{11}$$

for an infinite uncracked body. The stress due to α in a body with a semi-infinite crack along the negative x-axis, and for the case that $\alpha(\bar{z}) = \alpha(z)$, is given by

$$\tilde{\sigma}^{\alpha} = \frac{G}{2\pi i}\int ds'\alpha(z')\sqrt{\frac{z'}{z}}\,\frac{1}{z - z'} \tag{12}$$

where the integral is extended over the entire body and $ds \equiv dxdy$ is the differential area element.

We distinguish between the dislocation density due to the inelastic deformation of the matrix, which we shall continue to denote simply by α, and the dislocation density due to the thin cracking zone, which is given by $\kappa(x)\delta(y)$, where $\delta(y)$ is the one-dimensional Dirac delta function. Then the self-stress $\tilde{\sigma}^{\kappa}$ due to the cracking zone is

$$\tilde{\sigma}^{\kappa} = \frac{G}{2\pi i} \int ds' \cdot \delta(y') \kappa(x') \sqrt{\frac{z'}{z}} \frac{1}{z - z'} \tag{13}$$

$$= \frac{G}{2\pi i} \int_0^{\infty} dx' \kappa(x') \sqrt{\frac{x'}{z}} \frac{1}{z - x'} \tag{14}$$

The singular integral of Eq 14 is to be defined in the sense of the Cauchy principal value.

It is convenient for what is to follow to transform Eqs 12 and 14 to the forms:

$$\tilde{\sigma}^{\alpha} = -\frac{G}{2\pi i} \frac{1}{\sqrt{z}} \int ds' \alpha(z') \frac{1}{\sqrt{z'}} + \frac{G}{2\pi i} \sqrt{z} \int ds' \alpha(z') \frac{1}{\sqrt{z'}} \frac{1}{z - z'} \tag{15}$$

$$\tilde{\sigma}^{\kappa} = -\frac{G}{2\pi i} \frac{1}{\sqrt{z}} \int_0^{\infty} ds' \kappa(x') \frac{1}{\sqrt{x'}} + \frac{G}{2\pi i} \sqrt{z} \int_0^{\infty} dx' \kappa(x') \frac{1}{\sqrt{x'}} \frac{1}{z - x'} \tag{16}$$

The stress due to the remote loading of the cracked body $\tilde{\sigma}^A$, which is the "applied" stress, will be taken to be

$$\tilde{\sigma}^A = \frac{1}{i} \frac{K_A}{\sqrt{2\pi}} \frac{1}{\sqrt{z}} \tag{17}$$

where K_A is the "apparent" stress-intensity factor. Then the total stress $\tilde{\sigma}$ is

$$\tilde{\sigma} = \tilde{\sigma}^A + \tilde{\sigma}^{\alpha} + \tilde{\sigma}^{\kappa} \tag{18}$$

Formulation of the Theory

We shall now formulate a theory for a semi-infinite crack propagating in the positive x-direction with a constant velocity v as shown in Fig. 1. We seek the steady-state configuration at this velocity and the value of K_A that is appropriate to v. This is best done by formulating the problem in the coordinate frame translating with the crack tip.

In the moving frame α and q are stationary, and so, if $\dot{\alpha}$ and $\dot{q}$ are the time rates of change of α and q in the fixed frame, then the following relations are valid:

$$0 = \dot{\alpha} + v\, \partial_x \alpha \tag{19}$$

$$0 = \dot{q} + v\, \partial_x q \tag{20}$$

where the derivatives are evaluated in the material frame. Of course the spatial derivatives are the same in the moving frame and we may use the same notation for the derivatives in the moving frame. Since q is defined only on the positive x-axis, however, the spatial derivative of q in the moving frame is a total derivative, and so we may write for Eq 20

$$0 = \dot{q} + v dq/dx \tag{21}$$

and referring to Eq 6 we obtain

$$\kappa = -dq/dx = \dot{q}/v \tag{22}$$

The Stress Equation

The stress equation for this problem is Eq 18. Noting, first, that the right-hand side (RHS) of Eqs 15 and 16 includes a term in each that is analogous to Eq 17, we define the quantities K_P and K_Q as

$$K_P \equiv -\frac{G}{\sqrt{2\pi}} \int ds \frac{\alpha(z)}{\sqrt{z}} \tag{23}$$

$$K_Q \equiv -\frac{G}{\sqrt{2\pi}} \int_0^\infty dx \frac{\kappa(x)}{\sqrt{x}} \tag{24}$$

Now we may write Eq 18 explicitly as

$$\begin{aligned}\tilde{\sigma}(z) = {} & \frac{1}{i\sqrt{2\pi z}} (K_A + K_P + K_Q) \\ & + \frac{G}{2\pi i} \sqrt{z} \int ds' \frac{\alpha(z')}{\sqrt{z'}} \frac{1}{z - z'} + \frac{G}{2\pi i} \sqrt{z} \int_0^\infty dx' \frac{\kappa(x')}{\sqrt{x'}} \frac{1}{z - x'}\end{aligned} \tag{25}$$

A necessary condition for the stress to be nonsingular is that K_P and K_Q be finite and that the condition

$$K_A + K_P + K_Q = 0 \tag{26}$$

be satisfied. Furthermore, so long as $\dot{\epsilon}$ and $\dot{q}$ are not singular for finite σ, the condition is also a sufficient one for the problem of steady-state crack propagation.

The stress equation at steady state is then equivalent to the two equations

$$\tilde{\sigma}(z) = \frac{G}{2\pi i}\sqrt{z}\int ds'\,\frac{\alpha(z')}{\sqrt{z'}}\,\frac{1}{z-z'} + \frac{G}{2\pi i}\sqrt{z}\int_0^\infty dx'\,\frac{\kappa(x')}{\sqrt{x'}}\,\frac{1}{z-x'} \quad (27)$$

and Eq 26.

In order to introduce the constitutive equations for flow and cracking into these equations we must now transform them into integrals that involve $\dot{\alpha}$ and $\dot{q}$.

Transformation of the Equations

The transformation we seek employs the steady-state conditions and partial integration where necessary.

The most direct and simplest transformation is to replace κ by $\dot{q}/v$ in Eq 24 and the second integral on the RHS of Eq 27.

The transformation of Eq 23 was already carried out in Ref *1*, and we quote the result from that paper, namely

$$K_P = -\frac{2}{v}\,\frac{G}{\sqrt{2\pi}}\int ds\,\sqrt{z}\,\dot{\alpha}(z) \quad (28)$$

Consider next the remaining integral containing α. Define the auxiliary function $L(z, z')$ as

$$L(z, z') \equiv \frac{1}{2}\,\ell n\left[\frac{\sqrt{z'} + \sqrt{z}}{\sqrt{z'} - \sqrt{z}}\right]^2 \quad (29)$$

Now

$$\begin{aligned}\partial_{z'}L &= \sqrt{\frac{z}{z'}}\,\frac{1}{z-z'}\\ &= \partial_{x'}L\end{aligned} \quad (30)$$

Then

$$\begin{aligned}\sqrt{z}\int ds'\,\frac{\alpha(z')}{\sqrt{z'}}\,\frac{1}{z-z'} &= \int ds'\alpha(z')\,\partial_{x'}L\\ &= -\int ds'L\,\partial_{x'}\alpha(z') + \int_C dl\hat{\mathbf{x}}\cdot\hat{\mathbf{n}}\,L(z, z')\alpha(z')\end{aligned}$$

where the contour integral is extended over a contour C that is expanded to infinite radius, and $\hat{n}$ is the unit outward normal to the line element dl. It is readily shown that the contour integral goes to zero as C is expanded, and so, making use of the relation $\dot{\alpha} = -v\partial_x\alpha$

$$\sqrt{z}\int ds'\,\frac{\alpha(z')}{\sqrt{z'}}\,\frac{1}{z-z'} = \frac{1}{v}\int ds'\dot{\alpha}(z')L(z, z') \quad (31)$$

Now the resulting equations of the theory are as follows

$$K_P = -\frac{2}{v}\frac{G}{\sqrt{2\pi}}\int ds\,\sqrt{z}\,\dot{\alpha} \tag{28}$$

$$K_Q = -\frac{1}{v}\frac{G}{\sqrt{2\pi}}\int_0^\infty dx\,\frac{\dot{q}(x)}{\sqrt{x}} \tag{32}$$

$$K_A + K_P + K_Q = 0 \tag{26}$$

$$\begin{aligned}\tilde{\sigma}(z) = {} & \frac{1}{v}\frac{G}{2\pi i}\int ds'\dot{\alpha}(z')\cdot\frac{1}{2}\,\ell\text{n}\left[\frac{\sqrt{z'}+\sqrt{z}}{\sqrt{z'}-\sqrt{z}}\right]^2 \\ & + \frac{1}{v}\frac{G}{2\pi i}\sqrt{z}\int_0^\infty dx'\,\frac{\dot{q}(x')}{\sqrt{x'}}\,\frac{1}{z-x'}\end{aligned} \tag{33}$$

$$\dot{\alpha} = -\nabla\times\dot{\epsilon} \tag{9}$$

and the constitutive equations for $\dot{\epsilon}$ and $\dot{q}$.

Techniques for Solution

The general strategy for solution of these rather complex equations is as follows:

1. Substitute for $\dot{\alpha}$ and $\dot{q}$ in Eq 33 their dependence on σ from the constitutive equations.
2. Solve Eq 33 for the stress field $\boldsymbol{\sigma}$ at each value of v. The solution must be one for which $\boldsymbol{\sigma}$ is nonsingular and which behaves at distances remote from the crack tip like $\boldsymbol{\sigma}^A$. There may be intervals of v for which no such "proper" solutions exist.
3. For each solution determine the values for K_A, K_P, and K_Q.

This formal solution program is not very transparent because of the apparent complexity of Eq 33. Fortunately, for most cases of interest, there is available a simplification by which that equation can be handled as two separate equations.

If the size of the cracking zone is small compared with the size of the region over which $\dot{\alpha}$ is significant, then in the cracking zone Q the contributions to $\boldsymbol{\sigma}$ from the first term of the RHS of Eq 33 can be neglected in the evaluation of the second term. This comes about as follows.

For $z \ll z'$, the kernel function $L(z, z')$ is given by

$$L = \frac{1}{2}\,\ell\text{n}\,\frac{1+2\sqrt{z/z'}+(z/z')}{1-2\sqrt{z/z'}+(z/z')} \cong 2\sqrt{z/z'} \tag{34}$$

and so the first integral has only small contribution to σ in the region Q. Then, in the region Q, the equation becomes simply

$$\tilde{\sigma}_Q = \frac{1}{\nu} \frac{G}{2\pi i} \sqrt{z} \int_0^{\infty} dx' \frac{\dot{q}(x')}{\sqrt{x'}} \frac{1}{z - x'} \tag{35}$$

and the solution $\tilde{\sigma}_Q$ has the form for large z of

$$\begin{aligned} \tilde{\sigma}_Q &\cong \frac{1}{\nu} \frac{G}{2\pi i} \frac{1}{\sqrt{z}} \int_0^{\infty} dx' \frac{\dot{q}(x')}{\sqrt{x'}} \\ &= -\frac{1}{i\sqrt{2\pi z}} K_Q \end{aligned} \tag{36}$$

To this approximation, the equation for the cracking zone, Eq 35, appears as the equation for the crack behavior in an apparent applied field characterized by $K \equiv -K_Q$. The solution of that problem then determines an effective *kinetic law* for the crack velocity v of the form

$$v = v(K) \tag{37}$$

where K is the local apparent stress-intensity factor that drives the cracking. This is the type of kinetic law employed by Hart [*1*] in the earlier treatment of the sharp crack, which is of course the limit in which Q becomes negligibly small.

Now in the region outside of Q, the form of Eq 33 is

$$\tilde{\sigma}(z) = \frac{1}{\nu} \frac{G}{2\pi i} \int ds' \dot{\alpha}(z') L(z, z') + \frac{K}{i\sqrt{2\pi z}} \tag{38}$$

This is just the form of the equations for the problem of Ref *1*. In that earlier treatment Hart evaluated the effect of inelastic deformation by approximating σ by the dominant term that depends on K.

The conclusion of this section, then, is that, for sufficiently small cracking zone Q, the crack behavior can be deduced under the action of a local stress-intensity factor K without further attention to the inelastic flow of the matrix. The effect of the inelastic deformation is then determined separately, as though the cracking zone were confined to the origin, with a stress field σ that is apparently singular. The actual non-singularity of σ can then be approximated by a cutoff in the integral of Eq 28 at the vicinity of the crack tip.

Discussion and Conclusions

The main point of the theory presented here is that crack propagation is well characterized by the "stress intensity factors" K_A, K_P, and K_Q even though there is no actual stress singularity. In fact, it is just the nonsingular nature of σ that insures the existence of the integrals that define the K's. We can see neither need nor applicability in this problem for the new singularity deduced by Hui and Riedel [3] since for realistic nonsingular stress fields there remains no vestige of the new singularity.

Experimental curves of v versus K_A typically exhibit a nonanalytic behavior with "threshold" and "critical" values of K_A for which the slope of the v versus K_A curve is vertical and between which values the slope is very shallow. Hart [1] showed that such behavior would arise from K_P even for crack kinetic laws that were perfectly regular and monotonic. The nonanalytic behavior appeared at values of K_A that separated stable and unstable branches of the steady-state solutions. Because of the approximate separability of Eq 33 this behavior is preserved in the present more-detailed theory. It is necessary then, in experimentally testing kinetic laws deduced from micromechanical models, to account as well for the effects of the matrix inelastic flow. Those effects can be profound even at low loading levels.

Acknowledgments

The research reported here was supported in part by the National Science Foundation through the Materials Science Center, Cornell University. The preparation of the paper was carried out in part during the author's stay at the U.S. National Bureau of Standards as a visiting scientist in the Center for Materials Science.

References

[1] Hart, E. W., *International Journal of Solids and Structures*, Vol. 16, 1980, pp. 807-823.
[2] Lawn, B. R., *Journal of Materials Science*, Vol. 10, 1975, pp. 469–480.
[3] Hui, C. Y. and Riedel, H., *International Journal of Fracture*, Vol. 17, No. 1, 1981.
[4] Barenblatt, G. I., *Advances in Applied Mechanics*, Vol. 7, 1962, pp. 55–129.

John L. Bassani[1]

Cracks in Materials with Hyperbolic-Sine-Law Creep Behavior

REFERENCE: Bassani, J. L., "**Cracks in Materials with Hyperbolic-Sine-Law Creep Behavior,**" *Elastic-Plastic Fracture: Second Symposium, Volume I—Inelastic Crack Analysis, ASTM STP 803*, C. F. Shih and J. P. Gudas, Eds., American Society for Testing and Materials, 1983, pp. I-532–I-550.

ABSTRACT: At high stress levels the creep strain rate for many materials varies as the exponential of stress while at low stress levels it varies as stress to some power. An analysis is presented for a stationary crack under antiplane shear or Mode III loading in a material that deforms both elastically and by hyperbolic-sine-law creep, $\dot{\epsilon}^c \propto [\sinh(\sigma/\sigma_0)]^n$. The asymptotic crack-tip stress intensification is weaker and the strain-rate intensification is stronger than for a power-law creeping material. The elastic stress-intensity factor K_{III} (for low applied loads) and the path-independent integral C^* are shown to be the relevant short-time (small-scale yielding) and long-time (extensive creep) loading parameters as in the power-law case.

KEY WORDS: elevated temperature fracture, creep fracture, hyperbolic-sine creep law, small-scale yielding, stress analysis, elastic-plastic fracture

Recent analyses of the mechanics of creep cracks [*1–7*][2] have been based upon a power-law stress dependence of the creep strain rate, $\dot{\epsilon}^c \propto (\sigma/\sigma_0)^n$, even when the effects of hardening and recovery are included. There is experimental evidence and a theoretical basis that suggest a stronger stress dependence (for example, exponential) under conditions of high stress (for example, around a crack) and temperature. Some investigators have interpreted this effect as a stress dependence on a creep exponent n, or more simply and generally as "power-law breakdown" [*8,9*]. A hyperbolic-sine stress dependence, $\dot{\epsilon}^c \propto [\sinh(\sigma/\sigma_0)]^n$, has been proposed [*8,9*] which reduces at low stress to a power law and at high stress to an exponential law.

In this paper a Mode III analysis of a stationary crack tip is presented for a

[1]Department of Mechanical Engineering and Applied Mechanics, University of Pennsylvania, Philadelphia, Pa. 19104.

[2]The italic numbers in brackets refer to the list of references appended to this paper.

material that deforms both elastically and by hyperbolic-sine secondary creep. The asymptotic crack-tip fields are analyzed using a hodograph-like transformation (that can be adapted for any secondary creep flow relation). The crack-tip stress singularity is found to be weaker than the power-law case, while the strain-rate singularity is stronger. With r denoting the radial coordinate measured from the crack tip, the power-law solutions predict $\sigma \propto r^{-1/(n+1)}$ and $\dot{\epsilon} \propto r^{-n/(n+1)}$ as $r \rightarrow 0$, whereas, from the exponential-law solutions that follow, $\sigma \propto \ell n\ (1/r)$ and $\dot{\epsilon} \propto [r\ \ell n\ (1/r)]^{-1}$. For either case situations exist for which the proportionality (or amplitude) factor for σ and $\dot{\epsilon}$ near the crack tip is given in terms of the elastic stress-intensity factor K_{III} or the path-independent integral C^*.

The influence of elasticity is accounted for in the spirit of the Riedel and Rice [*1-3,5*] small-scale-yielding analysis for the situation where the load is suddenly applied to a cracked and heated specimen. For short times after load application, the stresses relax from the elastic distribution due to creep deformation, and the stress-intensity factor K_{III} is the relevant loading parameter that also depends on specimen geometry. After long times under constant load a steady-state stress distribution is reached and then the path-independent integral C^* is the relevant parameter. The effect of specimen geometry in these circumstances enters into the solution only in terms of K_{III} or C^*. The physical phenomenon of a growing creep zone around the crack tip is observed, as in the power-law analysis [*1,2,5*], but the time-dependent extent of this zone is different.

Under relatively low applied loads, the spatial and temporal variations of the crack-tip fields are predicted to enter into the small-scale-yielding solution in the similarity variable K_{III}^2/rt, as in the power-law case, where t denotes time. The predictions of the small-scale-yielding analysis for low applied loads are found to agree with preliminary finite element calculations.

The influence of K_{III} and C^* on the time to initiate crack growth is discussed based on a critical strain fracture criterion that under certain conditions can be used to model the growth and coalescence of microcavities on the grain boundaries ahead of the crack. The initiation time has a stronger dependence on the loading parameters K_{III} and C^* than in the power-law case.

Constitutive Relations

In the present analysis, only elastic and secondary (or steady-state) creep deformations are considered; the effects of hardening and recovery are neglected. At constant temperature the secondary creep rate can be expressed as a function of the uniaxial stress σ in terms of constant normalizing quantities $\dot{\epsilon}_0$ and σ_0, and possibly other (constant) material parameters α_m as

$$\dot{\epsilon}^c = \dot{\epsilon}_0 f(\sigma/\sigma_0; \alpha_m) \tag{1}$$

The isotropic, multiaxial generalization of the uniaxial law (Eq 1) is conventionally obtained in terms of the Mises effective stress

$$\bar{\sigma} = \left(\frac{3}{2}\,\sigma'_{ij}\,\sigma'_{ij}\right)^{1/2} \tag{2}$$

where the prime denotes the stress deviator and summation over repeated indixes is implied. The creep strain-rates $\dot{\epsilon}_{ij}{}^{c}$ are taken to be parallel to σ'_{ij}. With isotropic elasticity represented in terms of the shear modulus G, the total deviatoric strain rates that are the sum of the elastic and creep strain rates are

$$\dot{\epsilon}'_{ij} = \dot{\sigma}'_{ij}/2G + \frac{3}{2}\,\dot{\epsilon}_0 f(\bar{\sigma}/\sigma_0;\,\alpha_m)\sigma'_{ij}/\bar{\sigma} \tag{3}$$

Dilatation is purely elastic

$$\epsilon_{ii} = \left(\frac{1-2\nu}{E}\right)\sigma_{ii} \tag{4}$$

where $E = 2(1 + \nu)G$ is Young's modulus and ν is Poisson's ratio. Equation 3 implies that the effective creep strain rate $\dot{\bar{\epsilon}}^{c}$ and the effective stress $\bar{\sigma}$ are related by Eq 1, where the effective strain rate is

$$\dot{\bar{\epsilon}} = \left(\frac{2}{3}\,\dot{\epsilon}_{ij}\,\dot{\epsilon}_{ij}\right)^{1/2} \tag{5}$$

Based on creep tests that, for most alloys, seldom span more than a factor of 2 or 3 in stress, the power-law form $f = (\sigma/\sigma_0)^n$ is most often assumed [*10*]. Many of the pure metals and alloys that have been tested over a wider range of stress exhibit, at high stress, strain rates that vary exponentially with stress [*8,9,11–14*]. The hyperbolic-sine form for f, which can correlate these creep tests over a wide range of stress, has been suggested by many investigators

$$f = [\sinh(\sigma/\sigma_0)]^n \tag{6a}$$

$$\simeq \begin{cases} (\sigma/\sigma_0)^n, & \sigma < \sigma_0 \quad (6b) \\ 2^{-n}\exp(n\sigma/\sigma_0), & \sigma > \sigma_0 \quad (6c) \end{cases}$$

The low stress limit of the hyperbolic-sine form is the power law of Eq 6*b*; the exponential form given in Eq 6*c* is the high stress limit. These limiting forms of the hyperbolic-sine creep function suggest that the reference stress σ_0 can be interpreted as the power-law breakdown stress. The exponential-law behavior at high stress is consistent with theoretical models of creep flow that are based on stress-assisted, thermally activated motion of dislocations (see, for example, Ref *15*).

In closing this section, a case study is presented for 304 stainless steel (SS), which is one of the few creep-resistant structural alloys that has been extensively tested. The Nuclear Systems Material Handbook [*16*] design code for 304 SS is based on a hyperbolic-sine law that is fit to the data of Blackburn [*11*]. At 650°C the material constants in Eqs 3 and 6*a* are given as [*16*]

$$\sigma_0 = 125 \text{ MN/m}^2, \qquad \dot{\epsilon}_0 = 5.2 \times 10^{-9}\,\text{s}^{-1}, \qquad n = 6$$
$$E = 145\,000 \text{ MN/m}^2, \qquad \nu = 0.31$$

A comparison of these material parameters with the Mode I creep fracture experiments of Koterazawa and Iwata [*17*] and Koterazawa and Mori [*18*] indicates that their experiments were performed at relatively high applied loads. The average or net section stress over the uncracked ligament, σ_{net}, in their tests was in the range $1 \leq \sigma_{\text{net}}/\sigma_0 \leq 4$. Therefore, local crack-tip stresses, which generally are much larger than σ_{net}, were considerably greater than σ_0 in their tests. This raises the question as to whether accelerated laboratory tests that are conducted at high stress levels provide reliable information about the actual cracks under relatively lower stress service conditions. The importance of exponential stress dependence for predicting crack growth behavior will depend on the magnitude of the applied load (which in turn determines the size of the region around the crack tip where exponential-law creep dominates) and the size of the fracture process zone. The analysis of the spatial and temporal variations of the crack-tip fields that follows points out differences in the low and high stress behavior.

Mode III Crack-Tip Fields

Consider a two-dimensional crack lying in the $x_2 = 0$ plane with r and θ denoting the crack-tip polar coordinates in the $x_1 - x_2$ plane and $\theta = 0$ ahead of the crack along $x_2 = 0$ (see Fig. 1). The antiplane shear or Mode III problem corresponds to zero in-plane displacements, $u_1 = u_2 = 0$, and nonzero u_3 (x_1, x_2, t), where t denotes time. The normalized and nondimensional nonzero strain rates are $\gamma_1 = 2\dot{\epsilon}_{31}/\dot{\gamma}_0 = \dot{u}_{3,1}/\dot{\gamma}_0$ and $\gamma_2 = 2\dot{\epsilon}_{32}/\dot{\gamma}_0 = \dot{u}_{3,2}/\dot{\gamma}_0$ (where a comma denotes a partial derivative) and the corresponding stresses are $\tau_1 = \sigma_{31}/\tau_0$ and $\tau_2 = \sigma_{32}/\tau_0$. Here $\dot{\gamma}_0$ and τ_0 replace the material parameters $\dot{\epsilon}_0$ and σ_0 in Eq 3. The rate form of the compatibility condition is

$$\gamma_{1,2} - \gamma_{2,1} = 0 \tag{7a}$$

or in vector notation in terms of the vector cross product

$$\nabla_x \times \gamma = \mathbf{0} \tag{7b}$$

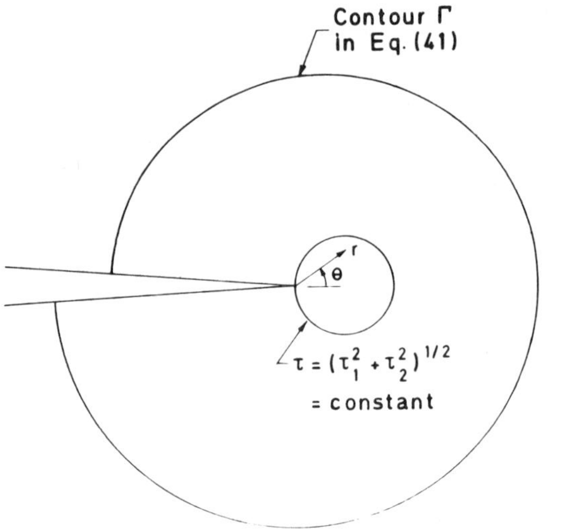

FIG. 1—*Crack-tip coordinates, locus of constant effective stress* τ, *and an integration contour for Eq 41.*

where γ is a two-dimensional vector with Cartesian components γ_1 and γ_2, and ∇_x is the gradient operator with respect to the physical coordinates x_1 and x_2. The equilibrium condition is

$$\tau_{1,1} + \tau_{2,2} = 0 \tag{8a}$$

or in terms of the scalar product

$$\nabla_x \cdot \boldsymbol{\tau} = 0 \tag{8b}$$

where $\boldsymbol{\tau}$ has Cartesian components τ_1 and τ_2. The boundary conditions of stress-free crack faces are $\tau_2 = \tau_\theta = 0$ on $\theta = \pm\pi$.

For the small-strain analysis presented here, if we anticipate unbounded stresses as $r \rightarrow 0$, then with Eq 6 the nonlinear term on the right-hand side of Eq 3 will dominate the linear one; that is, as $r \rightarrow 0$ the elastic strain-rates can be neglected at the tip of a stationary crack (see, for example, Ref *2*). Furthermore, for $t > 0$ the form of the spatial variation of the crack-tip fields as $r \rightarrow 0$ is independent of time. A form of Eq 3 that is useful for the present analysis is

$$\gamma_\alpha = f(\tau)\,\frac{\tau_\alpha}{\tau} \tag{9}$$

where the superscript α takes on the values 1 and 2. Here the nondimensional effective stress and strain rates are, respectively

$$\tau = (\tau_1{}^2 + \tau_2{}^2)^{1/2} \tag{10}$$

$$\gamma = (\gamma_1{}^2 + \gamma_2{}^2)^{1/2} \tag{11}$$

so that $\gamma = f(\tau)$.

Equations 7–9 form a quasi-linear system. They are simplified using a hodograph-like transformation [*19–21*] whereby the coordinates x_α are regarded as functions of the stresses τ_α or strain rates $\dot{\gamma}_\alpha$. The presentation given in the following closely follows the one given by Rice [*21*].

Let $\mathbf{x}$ be considered a function of $\boldsymbol{\gamma}$. Then the compatibility Eq 7 can be rewritten as [*21,22*]

$$\nabla_\gamma \times \mathbf{x} = \mathbf{0} \tag{12}$$

when the Jacobian $\partial(x_1, x_2)/\partial(\gamma_1, \gamma_2) \neq 0$. Here, $\boldsymbol{\Delta}_\gamma$ is the two-dimensional gradient vector with respect to the strain-rate vector $\boldsymbol{\gamma}$. Similarly, let $\mathbf{x}$ be considered a function of $\boldsymbol{\tau}$. Then the equilibrium Eq 8 can be rewritten as

$$\nabla_\tau \cdot \mathbf{x} = 0 \tag{13}$$

where the Jacobian $\partial(x_1, x_2)/\partial(\tau_1, \tau_2) \neq 0$. Here, ∇_τ is the gradient vector in the stress variables. With the constitutive relation Eq 9 it is easy to verify that (summation over repeated subscript $\alpha = 1,2$ is implied and $f' = df/d\tau$)

$$\begin{aligned} \frac{\partial}{\partial \gamma_\beta} &= \frac{\partial \tau_\alpha}{\partial \gamma_\beta} \frac{\partial}{\partial \tau_\alpha} \\ &= \left[\frac{\tau}{f} \delta_{\beta\alpha} + \left(\frac{1}{f'} - \frac{\tau}{f} \right) \frac{1}{\tau^2} \tau_\beta \tau_\alpha \right] \frac{\partial}{\partial \tau_\alpha} \end{aligned} \tag{14}$$

and therefore

$$\nabla_\gamma = \left[\frac{\tau}{f} \mathbf{I} + \left(\frac{1}{f'} - \frac{\tau}{f} \right) \frac{1}{\tau^2} \boldsymbol{\tau\tau} \right] \cdot \nabla_\tau \tag{15}$$

where $\mathbf{I}$ is the two-dimensional second-order identity tensor with components given in terms of the Kronecker delta $\delta_{\beta\alpha}$.

The compatibility condition Eq 12 implies the existence of a scalar potential ψ such that

$$\mathbf{x} = \nabla_\gamma \psi \tag{16}$$

identically satisfies Eq 12. With Eq 15 the equilibrium condition Eq 13 takes the form

$$\nabla_\tau \cdot \left[\frac{\tau}{f}\mathbf{I} + \left(\frac{1}{f'} - \frac{\tau}{f}\right)\frac{1}{\tau^2}\boldsymbol{\tau}\boldsymbol{\tau}\right] \cdot \nabla_\tau \psi = 0 \tag{17}$$

This is the governing linear second-order partial differential equation for ψ that, as Neuber [*20*] and Rice [*21*] observe, is valid for any constitutive relation of the form of Eq 9. Rice [*21*] first used this approach to study the crack-tip fields in a power-law material of the form of Eq 6*b*.

Equation 17 takes a particularly simple form if polar coordinates (τ, ϕ) are used in the stress plane, with

$$\boldsymbol{\tau} = \tau \mathbf{e}_\tau \tag{18a}$$

$$\phi = \tan^{-1}(-\tau_1/\tau_2) = \tan^{-1}(-\gamma_1/\gamma_2) \tag{18b}$$

$$\nabla_\tau = \mathbf{e}_\tau \frac{\partial}{\partial \tau} + \mathbf{e}_\phi \frac{1}{\tau}\frac{\partial}{\partial \phi} \tag{18c}$$

where $\mathbf{e}_\tau$ and $\mathbf{e}_\phi$ are the orthonormal polar base vectors. Noting that $\mathbf{e}_{\tau,\tau} = 0$, $\mathbf{e}_{\phi,\tau} = 0$, $\mathbf{e}_{\tau,\phi} = \mathbf{e}_\phi$, and $\mathbf{e}_{\phi,\phi} = -\mathbf{e}_\tau$, we can show that with Eqs 18 the equilibrium condition Eq 17 is reduced to

$$\frac{1}{f'}\psi_{,\tau\tau} + \left[\frac{1}{\tau f'} - \frac{f''}{(f')^2}\right]\psi_{,\tau} + \frac{1}{\tau f}\psi_{,\phi\phi} = 0 \tag{19}$$

The physical coordinates derived from ψ using Eq 16 and Eqs 9–11 and 18 are [*21*]

$$x_1 = \frac{\partial \psi}{\partial \gamma_1} = -(\sin\phi/f')\,\psi_{,\tau} - (\cos\phi/f)\,\psi_{,\phi} \tag{20a}$$

$$x_2 = \frac{\partial \psi}{\partial \gamma_2} = (\cos\phi/f')\,\psi_{,\tau} - (\sin\phi/f)\,\psi_{,\phi} \tag{20b}$$

The boundary condition of traction-free crack surfaces is $\tau_2 = 0$ on $\theta = \pm\pi$ or $x_2 = 0$ with $x_1 < 0$. From Eq 18*b*, $\tau_2 = 0$ corresponds to $\phi = \pm\pi/2$, so that with Eq 20*b*, $x_2 = 0$ implies

$$\psi_{,\phi} = 0 \quad \text{on } \phi = \pm\pi/2 \tag{21}$$

This completes the formulation for the crack-tip problem using the hodograph transformation.

Exponential Stress Dependence

With unbounded stresses at the crack tip, the form Eq 6c of the hyperbolic-sine creep function will dominate there. In this case $f(\tau)$ in Eq 9 is given as

$$f(\tau) = e^{\tau} \tag{22}$$

The form of the exponential creep-law given by Eqs 9 and 22 is equivalent to Eqs 3 and 6c with the substitutions $\dot{\gamma}_0 = \sqrt{3}\, 2^{-n}\, \dot{\epsilon}_0$ and $\tau_0 = \sigma_0/(n\sqrt{3})$.

For the exponential-law behavior, Eq 19 reduces to

$$\tau \psi_{,\tau\tau} + (1 - \tau)\, \psi_{,\tau} + \psi_{,\phi\phi} = 0 \tag{23}$$

The separable solution to Eq 23 that satisfies the boundary conditions of Eq 21 has a trigonometric ϕ dependence and is of the form

$$\psi(\tau, \phi, t) = - \sum_{k=0}^{\infty} A_k(t) \sin[(2k + 1)\phi]\, S_k(\tau) \tag{24}$$

Determination of the time-dependent amplitude factors $A_k(t)$ from the initial conditions of a particular problem is discussed in the next section. Under a constant applied load, after long times, steady-state conditions ($\dot{\sigma}_{ij} = 0$) are approached everywhere in a cracked specimen and the A_k's approach a constant value, independent of time.

From Eqs 23 and 24 the governing differential equation for $S_k(\tau)$ is

$$\tau S_k'' + (1 - \tau)\, S_k' - m^2 S_k = 0 \tag{25}$$

where $m = 2k + 1$. The linearly independent solutions to this equation can be expressed in terms of the confluent hypergeometric or Kummer functions [23], $M(m^2, 1, \tau)$ and $U(m^2, 1, \tau)$.

As $\tau \rightarrow \infty$ [23]

$$M(m^2, b, \tau) = \frac{\Gamma(b)}{\Gamma(m^2)}\, e^{\tau} \tau^{m^2 - b} \tag{26a}$$

$$U(m^2, b, \tau) = \tau^{-m^2} \tag{26b}$$

where Γ denotes the gamma function and $\Gamma(m^2) = (m^2 - 1)!$ Formulas for derivatives are

$$\frac{dM(m^2, 1, \tau)}{d\tau} = m^2\, M(m^2 + 1, 2, \tau) \tag{27a}$$

$$\frac{dU(m^2, 1, \tau)}{d\tau} = -m^2 U(m^2 + 1, 2, \tau) \tag{27b}$$

Now, for the crack-tip problem if $\tau \to \infty$ only at $r = 0$, then from Eqs 20, 24, 26, and 27, $M(m^2, 1, \tau)$ is not an admissible solution, and therefore

$$\psi(\tau, \phi, t) = -\sum_{k=0}^{\infty} A_k(t) U[(2k+1)^2, 1, \tau] \sin[(2k+1)\phi] \tag{28}$$

The radial coordinate measured from crack tip, $r = (x_1{}^2 + x_2{}^2)^{1/2}$, is given in terms of this series solution from Eqs 20. The leading terms for r as $\tau \to \infty$ correspond to $k = 0$ in Eq 28 with the $(1/f)\psi_{,\phi}$ contributions to Eqs 20; with Eqs 26b and 27b, as $\tau \to \infty$

$$r = \frac{A_0 \cos\phi}{\tau e^\tau} \tag{29}$$

for $-\pi/2 \leq \phi \leq \pi/2$. The denominator of Eq 29 can also be expressed as $\tau\gamma = (\ell n\, \gamma)\gamma = \tau e^\tau$.

The r^{-1} radial variation of $\tau\gamma$ as given in Eq 29 is consistent with an observation due to Rice and Rosengren [*24*]. Based upon path independence of the J-integral they argued that all two-dimensional, small-strain crack-tip solutions for nonlinear elastic materials have the property that the product of stress and strain has a singularity of r^{-1} at $r = 0$. For the nonlinearly viscous materials considered here, strain rate replaces strain and the C*-integral [*5*] replaces J in their argument.

The leading terms for x_1 and x_2 are found in a similar way from Eq 20

$$x_1 = \frac{A_0 \cos^2\phi}{\tau e^\tau} = \frac{A_0 (\tau_2/\tau)^2}{\tau e^\tau} \tag{30a}$$

$$x_2 = \frac{A_0 \sin\phi \cos\phi}{\tau e^\tau} = \frac{-A_0 \tau_1 \tau_2/\tau^2}{\tau e^\tau} \tag{30b}$$

where, from Eq 18b, $\cos\phi = \tau_2/\tau$ and $\sin\phi = -\tau_1/\tau$ has been used. Furthermore, the polar angle θ in the physical plane satisfies $x_2/x_1 = \tan\theta = \tan\phi$, so that Eqs 30 are valid for $-\pi/2 \leq \theta \leq \pi/2$. Replacing $\cos\phi$ in Eq 29 with $\cos\theta$ we obtain

$$\tau e^\tau = \tau\gamma = A_0 \cos\theta / r \tag{31}$$

With Eqs 10 and 18b

$$\tau_1 = -\tau \sin\theta \quad \text{and} \quad \tau_2 = \tau \cos\theta \tag{32}$$

Loci of constant effective stress or strain rate in the asymptotic solution are circles centered straight ahead of the crack tip at a distance equal to their radius as shown in Fig. 1. These loci resemble the ones that Riedel [*1*] and Rice [*21*] found in the $n \rightarrow \infty$ limit of the power-law analysis, and Hult and McClintock [*19*] found in their elastic-perfectly plastic analysis.

Equation 31 cannot be solved explicitly for $\tau(r, \theta)$, although an approximation to any desired accuracy for small r can be found. Letting $R = A_0 \cos \theta / r$ and taking logarithms of both sides of Eq 31, we obtain

$$\tau = \ell n\, R - \ell n\, \tau \tag{33}$$

Now successive substitution of $\ell n\, R - \ell n\, \tau$ for τ on the right side of Eqs 33 gives[3]

$$\tau = \ell n\, \{R/\ell n[R/\ell n\, (\ldots)]\} \tag{34}$$

The leading term in this equation gives the effective stress as $r \rightarrow 0$ ($R \rightarrow \infty$) as

$$\tau = \ell n\, (A_0 \cos \theta / r) \tag{35}$$

The leading term for the effective creep strain rate is found in a similar manner

$$\gamma = \frac{A_0 \cos \theta}{r\, \ell n (A_0 \cos \theta / r)} \tag{36}$$

The reader is cautioned that together Eqs 35 and 36 are consistent with Eq 31 although Eq 36 is not obtained directly from Eq 35 using the constitutive relation of $\gamma = e^{\tau}$.

Small-Scale Yielding

The time-dependent coefficients $A_k(t)$ in Eq 24 and, in particular, A_0 in Eq 31 depend on the initial conditions or more generally on the history of loading for a given problem. Here we will focus on the short-time response after a load is suddenly applied to a cracked body and thereafter held constant. The instantaneous response is purely elastic and the creep deformation that accumulates with time fastest around the crack tip *relaxes* the elastic stress concentration. The phenomenon of growing creep zones (that is, regions around the crack tip where creep strains exceed elastic strains) is observed. This problem has been widely studied for power-law creep behav-

[3]This successive approximation was suggested by Professor J. W. Hutchinson of Harvard University.

ior (see, for example, Refs *1, 2,* and *5*). The hyperbolic-sine form of the creep equation is used throughout this section.

The Riedel and Rice boundary-layer formulation of the small-scale-yielding (SSY) problem [*1,2*] is based on a semi-infinite crack in an infinite body with the boundary condition that the stresses decay as $r^{-1/2}$ as $r \to \infty$. This formulation is justified for the power-law creep behavior at any magnitude of the applied load because of the scaling laws discussed by Bassani and McClintock [*5*]. The existence of the scaling laws is based on a homogeneity property of the creep function f in Eq 1, where, at two stress levels σ_a and σ_b, $f(\sigma_a/\sigma_0)/f(\sigma_b/\sigma_0) = f(\sigma_a/\sigma_b)$. The power-law function of Eq 6*b* has this property: neither the hyperbolic-sine function nor the exponential-law function of Eqs 6*a* and 6*c* are homogeneous.

The proper SSY problem for the hyperbolic-sine creep law depends on the magnitude of the effective stress at the boundary of the region where, at $t = 0$, the singular elastic crack-tip field dominates. This magnitude, in turn, depends upon the magnitude of the applied load. Let the radial extent of this region from the crack tip be approximately d_E, which typically is a small fraction of the crack length; then from the elastic solution [*25*]

$$\overline{\sigma}_E = \sqrt{3}\, K_{\mathrm{III}}/\sqrt{2\pi d_E} = \sqrt{3}\, \overline{\tau}_E \tag{37}$$

where K_{III} is the elastic stress-intensity factor under Mode III or antiplane shear loading conditions. In general $\overline{\sigma}_E$ is on the order of, but greater than, the net section stress in the uncracked ligament as determined from global equilibrium considerations. If $\overline{\sigma}_E \ll \sigma_0$, then the region around the crack tip where $\overline{\sigma} > \sigma_0$ (the power-law breakdown stress) is small compared with d_E and an SSY problem of the Riedel-Rice type, with stresses tending to zero as $r \to \infty$, is justified. On the other hand, if $\overline{\sigma}_E \gg \sigma_0$, then the semi-infinite crack problem is inappropriate. Recall that in the experiments of Koterazawa et al [*17,18*], $\overline{\sigma}_E \geq \sigma_0$.

In the rest of this section the SSY problem for low applied loads is discussed, and numerical results are presented that verify the predicted asymptotic behavior. The high-load cases are currently being studied numerically and the results will be reported elsewhere.

Low Applied Loads

The problem of a semi-infinite crack in an infinite body subject to the boundary condition

$$\begin{Bmatrix} \tau_1 \\ \tau_2 \end{Bmatrix} = \frac{K_{\mathrm{III}}/\tau_0}{(2\pi r)^{1/2}} \begin{Bmatrix} -\sin(\theta/2) \\ \cos(\theta/2) \end{Bmatrix} \tag{38}$$

as $r \to \infty$ and $t \geq 0$ is considered. At $t = 0$ the response is purely elastic and the stresses are given everywhere by Eq 38. The problem is to determine $A_0(t)$ in Eqs 31 or 35.

With all stress-like quantities normalized by τ_0, the variables and material parameters that enter into this Mode III SSY problem are

$$r, \theta, t, K_{\text{III}}/\tau_0, G/\tau_0, \dot{\gamma}_0, n$$

where the hyperbolic-sine form of the creep function, Eq 6*a*, is used in Eq 3, and, once again, τ_0 and $\dot{\gamma}_0$ replace σ_0 and $\dot{\epsilon}_0$ as discussed earlier. Arguments based on dimensional consistency will now be used to determine the functional form of $A_0(t)$ (see, for example, Refs *1* and *2*).

If Eq 3, which gives the total strain rates as the sum of the elastic and creep rates, is substituted into the compatibility Eq 7, thereby eliminating the strain-rate quantities from the problem, we see that G/τ_0, $\dot{\gamma}_0$, and t can enter into the solution only in the nondimensional group $G\dot{\gamma}_0 t/\tau_0$. Since A_0 must have the dimension of length but be independent of r and θ, we conclude that $A_0 \propto K_{\text{III}}^2/\tau_0^2$, and

$$A_0 = \frac{K_{\text{III}}^2}{\tau_0^2}\, g\left(\frac{\tau_0}{G\dot{\gamma}_0 t}, n\right) \tag{39}$$

where g is an unknown function of time, that, in general, must be determined numerically. In the SSY problem for a power-law creeping material [*1,2,5*], for $r \ll$ and $t \gg$, the product of stress times strain-rate ($\sigma\dot{\epsilon}$) relaxes like $1/t$. Our preliminary finite-element results, for $n = 1$, suggest the same relaxation for the hyperbolic-sine-law creeping material so that, comparing Eqs 31 and 39 we tentatively suggest that

$$A_0 = \beta_n \frac{K_{\text{III}}^2}{G\tau_0\dot{\gamma}_0 t} \tag{40}$$

where β_n is a numerical factor. Calculations for other values of n (as well as the high-load case) currently are underway.

The validity of Eq 40 can be motivated by consideration of the region of path-independence of the C-integral, which is defined for Mode III conditions as (refer to Ref *26* with strain-rates in place of strains)

$$C = \int_\Gamma [W_{\dot{\epsilon}} dx_2 - \dot{u}_{3,1} (\sigma_{31} dx_2 - \sigma_{32} dx_1)] \tag{41}$$

where Γ is a path surrounding the crack tip as shown in Fig. 1. The strain-rate potential in Eq 41 is

$$W_{\dot{\epsilon}} = \int_0^{\dot{\bar{\epsilon}}} \bar{\sigma}\, d\dot{\bar{\epsilon}} \tag{42}$$

for the creep part of the constitutive relation of Eq 3 with Eqs 2 and 5. In a region around the crack tip where the creep strain rates are much larger than the elastic ones, the C-integral is path-independent (or nearly so) [5,26]. If part of this region includes low enough stresses so that the creep rates are given (approximately) by a power law, that is, Eq 6*b*, then C, which involves terms of order $\sigma\dot{\epsilon}$, must decay like $1/t$. Now, shrinking the path of integration for C to include the crack-tip region where Eq 31 holds, we conclude that Eq 40 is valid.

Preliminary Numerical Results

A finite element calculation has been carried out for the SSY problem that incorporates a semi-circular mesh ($x_2 \geq 0$) of 162 four-node isoparametric elements and the constitutive relation of Eq 3 with the hyperbolic-sine creep function of Eq 6*a* for $n = 1$. Nine sectors, each with circumferential extent of $\pi/9$, were radially focused at the crack tip. The crack-tip elements were chosen to give a $1/r$ strain-rate singularity at $r = 0$, which closely approximates Eq 36. The outer radius of the mesh, r_{max}, was approximately 400 times the radial extent of the crack-tip elements. This mesh reproduced known elastic and power-law solutions to within 20 percent accuracy. Explicit time integration with automatic time step control was incorporated as in Ref 5; for each step $(\Delta t)\dot{\epsilon}_0 \simeq (\sigma_0/E)/\cosh(\overline{\sigma}_{max}/\sigma_0)$, where the maximum effective stress is found at one of the integration points nearest to the crack tip. All results presented in the following are for $\sigma_0/E = 5 \times 10^{-4}$ and $\nu = 0.3$, or $\tau_0/G = 7.5 \times 10^{-4}$, and time is normalized with respect to $\dot{\gamma}_0 = \sqrt{3}\dot{\epsilon}_0/2$.

Figure 2 is a plot of the radial variation of $\tau = \overline{\tau}/\tau_0$ near the crack tip along $\theta = 4$ deg, for $\overline{\sigma}_E/\sigma_0 = \sqrt{3}\,K_{III}/(\sqrt{2\pi r_{max}}\,\sigma_0) = 0.069$. The results plotted are for a relatively long time after load application when the creep strains were roughly one half the elastic strains at the boundary of the mesh ($r = r_{max}$). As seen in the figure, the radial variation of τ for $r \ll$ is in good agreement with Eq 35.

The short-time relaxation of the crack-tip fields are plotted in Figs. 3 and 4 for $\overline{\sigma}_E/\sigma_0 = 0.14$. In Fig. 3, $\tau = \overline{\tau}/\tau_0$ at $r/r_{max} = 5 \times 10^{-4}$ and $\theta = 4$ deg is shown to relax for short times according to Eqs 35 and 40 from the initial value $\tau = 7.44$ at $t = 0$. The effective stress at this integration point determined the allowable stable time step as discussed in Ref 5.

For the hyperbolic-sine function of Eq 6*a*, with $n = 1$, the strain-rate energy potential in Eqs 41 and 42 is

$$W_{\dot{\epsilon}} = \overline{\sigma}\,\dot{\overline{\epsilon}} - \sigma_0\,(\dot{\overline{\epsilon}}^2 + \dot{\epsilon}_0^2)^{1/2} \qquad (43)$$

where $\overline{\sigma}$ and $\dot{\overline{\epsilon}}$ are given in Eqs 2 and 5. Figure 4 is a plot of $C(t)$ of Eq 40 versus $1/t$ for short times. The integral was calculated numerically on circular

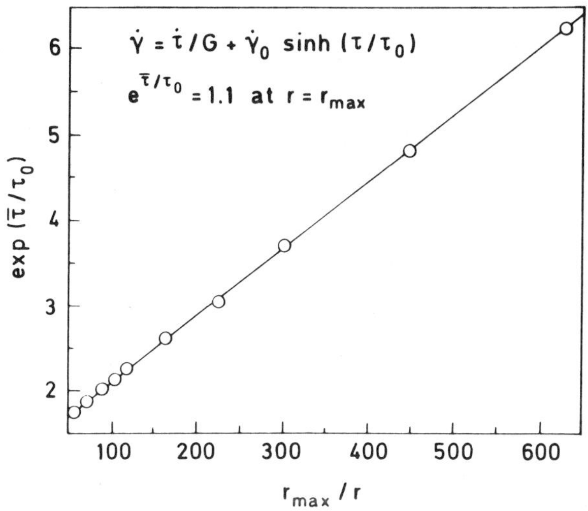

FIG. 2—*Radial variation of the effective stress along $\theta = 4$ deg, for* t ≫, *from the numerical small-scale-yielding calculation. Note agreement with Eq 35.*

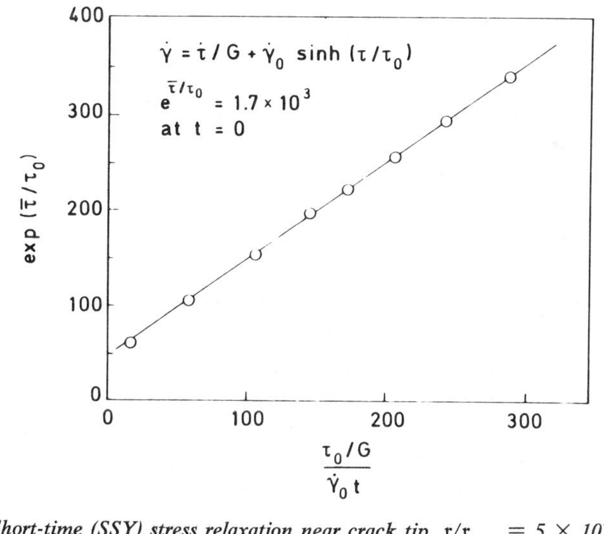

FIG. 3—*Short-time (SSY) stress relaxation near crack tip,* $r/r_{max} = 5 \times 10^{-4}$, $\theta = 4$ *deg, from the numerical small-scale-yielding calculation. Note agreement with Eqs 35 and 40.*

paths around the crack tip that connected element integration points. The values plotted for C are averages from several paths around the crack tip where C was nearly path-independent [5]. For very short times, path-dependent values of C were extrapolated to $r = 0$; this extrapolation connected smoothly with the path-independent values. The numerical results of

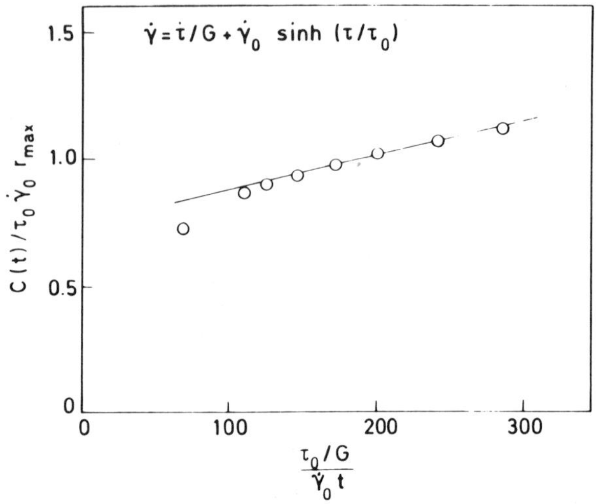

FIG. 4—*Short-time (SSY) relaxation of the integral in Eq 41 calculated from the numerical solution in regions where* $\dot{\gamma}^c \gg \dot{\gamma}^e$. *Same applied* K_{III} *as in Fig. 3.*

Fig. 4 are consistent with the prediction that $C \propto 1/t$ for low values of K_{III} or the applied load.

Again, it is emphasized that these numerical results are preliminary. Currently, calculations for a wider range of applied K_{III}-values are being performed. Other calculations with a finer mesh will be used to study convergence. These results will be reported in a subsequent paper.

Long-Time Steady-State Amplitude

Under a constant applied load after long times ($t \rightarrow \infty$), steady-state conditions corresponding to $\dot{\sigma}_{ij} = 0$ will be approached everywhere in the cracked specimen. Furthermore, $C \rightarrow C^* =$ constant, and the integral of Eq 41 is path-independent on all paths surrounding the crack tip. If the integral is evaluated on a circular path around the crack tip in the region where the asymptotic solution of Eqs 32, 34, and 35 holds, then it is straightforward to verify from Eqs 41 and 43 that, as $t \rightarrow \infty$ and $r \rightarrow 0$

$$A_0 = \frac{2}{\pi} \frac{C^*}{\tau_0 \dot{\gamma}_0} \tag{44}$$

The integration is performed over $-\pi/2 \leq \theta \leq \pi/2$, since that is the angular extent of the region of dominance of the most singular terms in the crack-tip solution.

In general, values of C^* will have to be found numerically for different geometries using the constitutive relations of Eqs 3 and 6*a*. If the region of power-law breakdown is small compared with all other characteristic dimensions of a problem, then C^* will be given approximately by solutions for power-law behavior of Eq 6*b*. Some relevant solutions for this case are given by Amazigo [*27*] and Shih [*28*] for nonlinear elastic materials; these solutions are valid for creeping materials if strains are replaced by strain rates.

Discussion

Although the present analysis is limited to stationary cracks under Mode III loading conditions, the important conclusions concerning the radial dependence of the stresses and strain rates as well as the small-scale-yielding time dependence probably hold as well as Modes I (tensile loading) and II (in-plane shear) for isotropic materials characterized by a hyperbolic-sine creep law. This is certainly the case for materials with the power-law stress dependence of Eqs 3 and 6*b*. The r^{-1} variation of $\sigma\dot{\epsilon}$ near the crack tip can be argued for all two-dimensional cracks in a general material governed by Eqs 1 and 3 based on path-independence of the C-integral (see Eqs 41 and Ref 5) in regions where the creep strain-rates are much larger than the elastic strain-rates. The influence of other secondary-creep stress dependencies (see Ref *29*, for an example) on Mode III fracture problems could also be investigated using the hodograph transformation. Recall that the tensorial Eq 17 and its polar form, Eq 19, are valid for any creep law of the form given in Eqs 1 and 3.

At the tip of a stationary crack (in a material with a stronger than linear stress dependence on the creep strain rate) the creep strain rates dominate the elastic ones. The significance of exponential-law stress dependence at high stress, when contrasted with power-law stress dependence, is the weaker stress intensification and stronger strain intensification near the crack tip. Qualitatively, the present results lie between the high n limit of the power law and the nonhardening limit of time-independent plasticity. This latter case is considered in the creep crack context by Leckie and McMeeking [*6*].

The hyperbolic-sine (or exponential) creep function is not homogeneous in stress. This fact probably leads to different SSY relaxation behavior under low and high applied loads, that is, different functional forms for g in Eq 39. Under a constant but low applied load the prediction that $\sigma\dot{\epsilon} \propto K_{III}^2/t$, as in the power-law case, has been verified numerically for $n = 1$ in Eq 6*a*. For the low-stress SSY problem the scaling of the amplitude of $\sigma\dot{\epsilon}$ with the elastic stress-intensity factor K_{III} is based on dimensional considerations alone. Under high applied loads another length scale, namely, the size of the region dominated by the elastic singular crack-tip field at $t = 0$, can enter the solution and possibly invalidate the dependence on K_{III} given in Eq 39. The case of high applied loads is being studied numerically, and a more definitive account of the time dependency in SSY will be presented in a separate paper.

For long times after load application, in either case, steady-state conditions are approached and $\sigma\dot{\epsilon} \propto C^*$, which is a constant given by the integral in Eq 41 that is path independent everywhere.

The implications of the hyperbolic-sine stress dependence on fracture predictions depend on the micromechanisms causing fracture. At temperatures roughly greater than one-third the melting temperature of a metal, creep fracture generally is found to be intergranular; that is, it proceeds by material separation along the boundary between crystals (grains). Often this separation is associated with the formation and growth of cavities on the grain boundary [*30–32*] that may be accelerated by environmental attack [*33*]. Another observed mechanism is the embrittling of the grain boundary by an oxide layer [*34,35*] or by some other environmentally caused contamination.

Grain boundary cavities can grow by a diffusion of atoms or by creep deformation or both [*30–32*]. The grain-boundary diffusion-controlled process is linearly dependent on the local crack-tip stress. In this case, if the initiation of crack growth is associated with some cavity near the crack tip attaining a critical size, then from Eqs 35, 40, and 44, the initiation time would vary inversely with $\log(K_{III}^2)$ during SSY and vary inversely with $\log(C^*)$ after extensive creep of an entire specimen. If creep deformation dominates the growth of cavities, then the initiation of crack growth is likely to be associated with the criterion of a critical strain at some distance ahead of the crack. For extensive creep of a specimen before crack growth, from Eqs 36 and 44, the initiation time would vary inversely with $\log(C^*)/C^*$. This is a stronger dependence on C^* than for the power-law materials [*2*]. Under SSY conditions this criterion is difficult to evaluate from Eqs 36 and 40, but a stronger dependence on K_{III} than for the power-law materials can be demonstrated.

Other mechanisms causing fracture can be considered (that is, surface diffusion-controlled cavity growth [*31*] or grain-boundary oxidation [*35*]) that would lead to other variations of the initiation time with loading parameters K_{III} and C^*. The general conclusion based on mechanisms of crack growth that are most sensitive to the local crack-tip stress is a weaker dependence on the loading parameters for hyperbolic-sine-law materials as compared with power-law materials. Stronger dependencies would be predicted from mechanisms that are most sensitive to the local crack-tip strains.

Acknowledgments

The author is grateful to Mr. Giai Van Do for his help with the numerical calculations and to Professor Pierre Burgers for helpful discussions. This work was supported by the Department of Mechanical Engineering and Applied Mechanics, University of Pennsylvania.

References

[1] Riedel, H., *Zeitschrift für Metallkunde*, Vol. 69, 1978, pp. 755–760.
[2] Riedel, H. and Rice, J. R. in *Fracture Mechanics: Twelfth Conference, ASTM STP 700*, American Society for Testing and Materials, 1980, pp. 112–130.
[3] Riedel, H., *Journal of the Mechanics and Physics of Solids*, Vol. 29, No. 1, Feb. 1981, pp. 35–49.
[4] Hui, H., and Riedel, H., *International Journal of Fracture*, Vol. 17, No. 4, Aug. 1981, pp. 409-425.
[5] Bassani, J. L. and McClintock, F. A., *International Journal of Solids and Structures*, Vol. 17, No. 5, 1981, pp. 479–492.
[6] Leckie, F. A. and McMeeking, R. M., *International Journal of Fracture*, Vol. 17, No. 4, Oct. 1981, pp. 467-476.
[7] Kubo, S., "Effects of Creep Recovery and Hardening on the Stress and Strain-Rate Fields near a Crack Tip in Creeping Materials," MRL E-109, Materials Research Laboratory, Brown University, Providence, R.I., Dec. 1980.
[8] Finnie, I. and Heller, W. R., *Creep of Engineering Materials*, McGraw-Hill, New York, 1959.
[9] Garofalo, F., *Fundamentals of Creep and Creep-Rupture in Metals*, Macmillan, New York, 1965.
[10] Takeuchi, S. and Argon, A. S., *Journal of Materials Science*, Vol. 2, 1976, pp. 1542–1566.
[11] Blackburn, L. D. in *Generation of Isochronous Stress-Strain Curves*, A. O. Schaefer, Ed., American Society of Mechanical Engineering, 1972, pp. 15–48.
[12] Gilbert, E. R. and Blackburn, L. D., *Journal of Engineering Materials Technology*, Vol. 99, 1977, pp. 168–180.
[13] Laks, H., Wiseman, C. D., Sherby, O. D., and Dorn, J. E., *Journal of Applied Mechanics*, Vol. 24, 1957, pp. 207-213.
[14] Mukherjee, A. K., Bird, J. E., and Dorn, J. E., *Transactions*, American Society for Metals, Vol. 62, 1969, pp. 155–179.
[15] Barrett, C. R. and Nix, W. D., *Acta Metallurgica*, Vol. 13, 1965, pp. 1247-1258.
[16] *Nuclear Systems Materials Handbook*, J. C. Spanner, Chairman, Part 1, Group 1, Section 2 on 203 Stainless Steel in Vol. 1 on Design Data, 1980.
[17] Koterazawa, R. and Iwata, Y., *Journal of Engineering Materials Technology*, Vol. 98, Oct. 1976, pp. 296-304.
[18] Koterazawa, R. and Mori, T., *Journal of Engineering Materials Technology*, Vol. 99, Oct. 1977, pp. 298-305.
[19] Hult, J. A. H. and McClintock, F. A., *Proceedings*, Ninth International Congress of Applied Mechanics, Vol. 8, Brussels, 1956.
[20] Neuber, H., *Journal of Applied Mechanics*, Vol. 28, 1961, pp. 544–550.
[21] Rice, J. R., *Journal of Applied Mechanics*, Vol. 34, 1967, pp. 287-298.
[22] Hildebrand, F. B., *Advanced Calculus for Applications*, Prentice-Hall, Englewood Cliffs, N.J., 1976, Chapter 7.
[23] *Handbook of Mathematical Functions*, M. Abramowitz and I. A. Stegun, Eds., Dover, New York, 1970, Chapter 13.
[24] Rice, J. R. and Rosengren, G. F., *Journal of the Mechanics and Physics of Solids*, Vol. 16, 1968, pp. 1-12.
[25] Rice, J. R. in *Fracture*, H. Liebowitz, Ed., Vol. 2, 1968, Academic Press, New York. pp. 131-311.
[26] Rice, J. R., *Journal of Applied Mechanics*, Vol. 35, 1968, pp. 379-386.
[27] Amazigo, J., *International Journal of Solids and Structures*, Vol. 11, 1975, pp. 1291–1299.
[28] Shih, F. in *Mechanics of Crack Growth, ASTM STP 590*, American Society for Testing and Materials, 1976, pp. 3–26.
[29] Valanis, K. C. and Lee, C. F., *International Journal of Solids and Structures*, Vol. 17, 1981, pp. 589-604.
[30] Needleman, A. and Rice, J. R., *Acta Metallurgica*, Vol. 28, 1980, pp. 1315–1332.
[31] Takasugi, T. and Vitek, V., *Metallurgical Transactions*, Vol. 12A, April 1981, pp. 659-667.

[*32*] Bassani, J. L. in *Creep and Fracture of Engineering Materials and Structures*, B. Wilshire and D. R. J. Owens, Eds., Pineridge Press, Swansea, U.K., 1981, pp. 329–344.
[*33*] Florren, S. in *Creep-Fatigue Environment Interactions*, R. M. N. Pelloux and N. Stoloff, Eds., American Institute for Mining, Metallurgical and Petroleum Engineers, AIME, 1980.
[*34*] McMahon, C. J., Jr., *Materials Science and Engineering*, Vol. 13, 1974, pp. 295–297.
[*35*] McClintock, F. A. and Bassani, J. L. in *Three-Dimensional Constitutive Relations and Ductile Fracture*, J. Zarka and S. Nemat-Nasser, Eds., North Holland Publishing Co., Dourdan, France, 1981, pp. 119–141.

Chun W. Lau,[1] Ali S. Argon,[1] and Frank A. McClintock[1]

Stress Concentrations Due to Sliding Grain Boundaries in Creeping Alloys

REFERENCE: Lau, C. W., Argon, A. S., and McClintock, F. A., "**Stress Concentrations Due to Sliding Grain Boundaries in Creeping Alloys,**" *Elastic-Plastic Fracture: Second Symposium, Volume I—Inelastic Crack Analysis, ASTM STP 803*, C. F. Shih and J. P. Gudas, Eds., American Society for Testing and Materials, 1983, pp. I-551–I-572.

ABSTRACT: The most dominant singular stress distributions caused by grain-boundary sliding accommodated by power-law creep have been determined at hard grain-boundary particles and at triple grain junctions, where cavitation and eventual creep cracking are frequently observed. Finite element computations have been used to connect these local stresses to the far-field stresses. These finite element analyses incorporate a specially formulated power-type singularity element, a new method of simulating periodicity and symmetry boundary conditions, a scheme for automatic variable step selection for stable time integration, and a method to bypass the influence of incompressibility. The deviatoric stresses obtained from the finite element results show very good agreement with the dominant singularity solutions, and provide accurate estimates of the generalized stress intensity factors.

KEY WORDS: intergranular fracture, creep cracking, grain-boundary sliding, grain-boundary stress concentration, generalized stress intensity factor, elastic-plastic fracture

Fracture in high-temperature alloys under service loading above 0.4 T_m usually involves the nucleation of intergranular cavities, their growth and linking to form macrocracks, and the eventual propagation of these macrocracks to final fracture. Experiments have shown that creep cavities form on grain-boundary particles and grow preferentially near triple grain junctions [*1–3*].[2] It is now clear that most cavities are nucleated by locally accentuated stresses on interfaces of grain-boundary particles. These stresses are caused by grain-boundary sliding, accommodated principally by power-law creep within the grains and around the particles, but also by diffusional flow over much shorter ranges [*4,5*]. Although many creep fracture mechanisms have

[1]Post-doctoral associate and professors, respectively, Department of Mechanical Engineering, Massachusetts Institute of Technology, Cambridge, Mass. 02139.

[2]The italic numbers in brackets refer to the list of references appended to this paper.

been proposed [*2,3,6–8*], there have been few quantitative descriptions of such mechanisms incorporating the appropriate concentrated stress and strain fields which are essential for rigorous mechanistic analysis of intergranular creep crack nucleation.

Here we present a summary of some new methods used for determining the concentrated stresses near apexes of hard, grain-boundary particles and at triple grain junctions, in creeping alloys where grain-boundaries have little resistance to sliding—conditions that are widely encountered above 0.4 T_m. The results of this study have made it possible to calculate the rate of cavity nucleation in a creeping alloy with better accuracy [*4,5*].

This stress analysis is in two parts. First, the dominant singular nature of the stresses is determined near apexes of grain-boundary particles and triple grain junctions with freely sliding boundaries in a power-law creeping alloy by solving the appropriate local boundary-value problems using methods from isotropic, incompressible, plane-strain, power-law nonlinear elasticity. Since a preliminary version of this analysis has already been published [*9*] and a complete description of it will be published elsewhere [*10*], we will summarize here only the more important results. This analysis is a generalization of the approaches of Williams [*11*] and Hutchinson [*12,13*]. It can give the distributions of stress and strain to within a scaling factor K, the generalized stress intensity factor. In the two boundary-value problems discussed in this paper, K cannot be determined by the conventional approach of a contour integral because the energy and traction integrals along parts of the contour cannot be eliminated, as is possible in problems of sharp cracks.

Therefore, in the second part of the analysis, a special finite element program has been developed to provide the connection between the far-field stresses and the local singular stresses so as to determine the generalized stress intensity factor, K, from the complete stress solutions at the vicinity of the singularity. This finite element computation incorporates specially formulated power-type singularity elements, a novel method of kinematically simulating boundary conditions having periodicity and symmetry, a scheme for automatic variable time-step selection for stable explicit time integration, and a method of bypassing the influence of incompressibility to determine the creep stress intensity factor accurately.

Dominant Singularity

Figure 1 shows schematically two apparent sources of intergranular stress concentration: nondeformable grain-boundary particles, and triple grain junctions, both of which have been amply demonstrated to be of importance in creep fracture [*1,3*]. We have discussed earlier [*4*] that these two problems are complementary in nature and in timing. While the first problem of stress concentration around grain-boundary particles is the dominant one, it exists with full intensity only during short, transient periods of rapid sliding. The

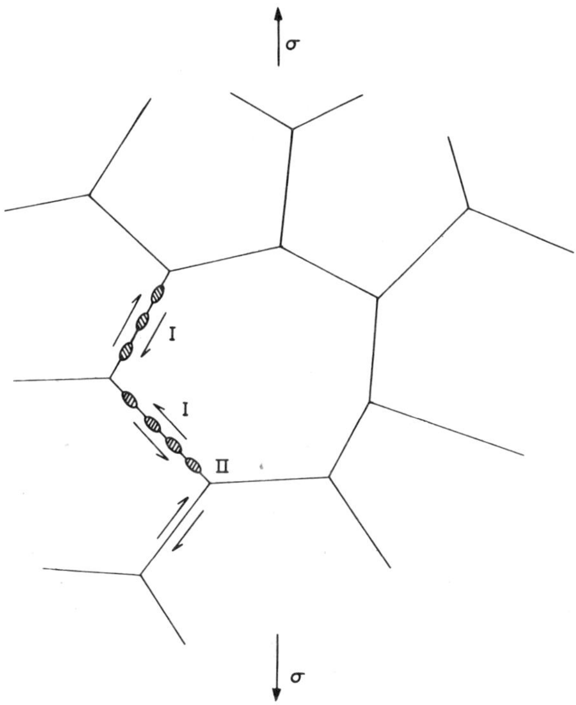

FIG. 1—*Two boundary-value problems important in intergranular cavitation:* I. *Stress concentrations around grain boundary particles;* II. *Stress concentrations at triple grain junctions.*

second problem of stress concentration at triple grain junctions, on the other hand, builds up gradually as the shear tractions on slanted boundaries are reduced to zero. Both analysis [*4*] and experiments [*1,4*] indicate that the first problem is important in cavity nucleation and the second in rapid coalescence of already-formed cavities to form wedge cracks.

Figures 2*a* and 2*b* show the boundary conditions at the apex of the grain-boundary particle and the triple grain junction under far fields of shear and tension, respectively [*9*]. In both cases, we have chosen conditions to obtain the most severe stress singularity that correspond to freely sliding boundaries supporting no shear traction. This results in a discontinuity in tangential velocity across the sliding boundaries but does not affect continuity of normal tractions and normal velocities across the boundaries. Only plane strain is considered. The accommodating flow inside the grains is modeled by a steady-state power-law connection between equivalent strain rate and equivalent stress that holds for structural alloys above 0.4 T_m, at intermediate stresses of $0.5 \times 10^{-4}\ \mu$ to $10^{-3}\ \mu$ (where μ is the shear modulus), and is understood to result from the glide and climb motion of lattice dislocations. This steady-state creep rate in tensile deformation is conveniently taken to be of the form

$$\dot{\epsilon}_s = A\sigma^m \tag{1a}$$

$$A = C\frac{D_v \mu b}{kT}\frac{1}{\mu^m} \tag{1b}$$

where

C, m = material constants,
D_v = volume self-diffusion constant,
μ = shear modulus,
b = interatomic distance,
k = Boltzmann's constant, and
T = absolute temperature.

Typical values for m are 3 for Class I alloys and 5 for Class II alloys and pure metals[3] [*14*]. Although here we will present only results for these two representative powers of 3 and 5, we have obtained detailed results for m up to 10 [*10*], thus covering not only steady-state creep in single-phase alloys, but also nonsteady state behavior of many engineering alloys which show effective creep exponents $m > 5$ at their respective minimum creep rates. These results will be presented elsewhere. Solutions for even higher values of m that might be useful to model nonsteady flow behavior of a nearly plastic type can be generated by the same methods we have developed.

For multiaxial deformation, Eq 1 is generalized to

$$\dot{\epsilon}_{ij} = \frac{3}{2}A{\sigma_e}^{m-1}s_{ij} \tag{2a}$$

where

$$s_{ij} = \sigma_{ij} - \delta_{ij}\,\sigma_{kk}/3 \tag{2b}$$

and

$$\sigma_e = \left[\frac{3}{2}s_{ij}s_{ij}\right]^{1/2} \tag{2c}$$

For monotonic increments of creep strain under steady applied tractions, where Hoff's [*15*] analogy is valid, the approach of Hutchinson's [*12,13*] deformation theory analysis for the stress singularity at a crack tip in power-law hardening plasticity can be directly applied to the problems of grain-boundary sliding in power-law creeping alloys.

The method of solution of this problem, through the use of stress functions, has been discussed earlier [*9*], and additional applications will be discussed in

[3]Both Class I and Class II alloys are solid-solution alloys. The former is distinguished from the latter by an absence of subgrains at steady state and a low apparent internal stress [*14*].

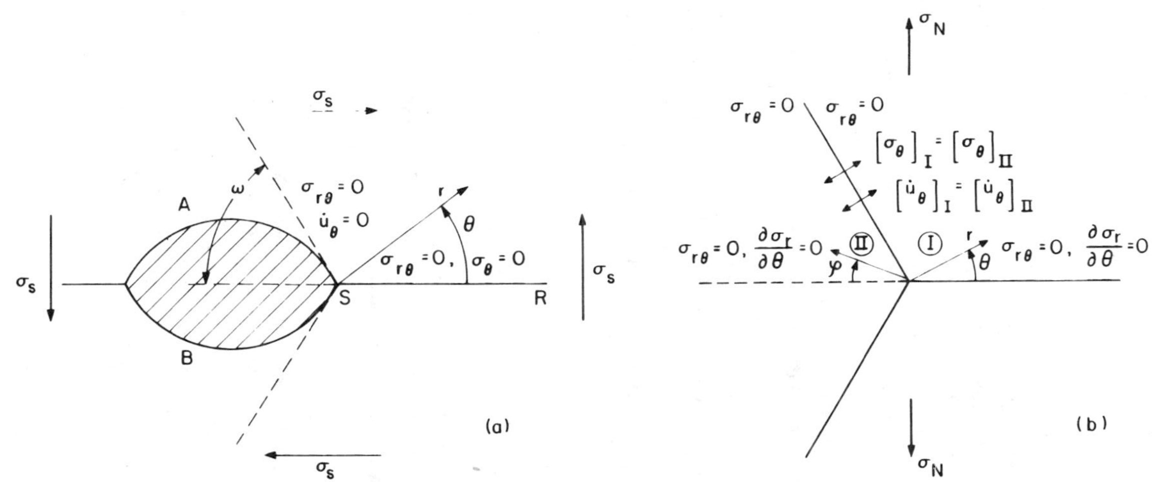

FIG. 2—*Boundary conditions at:* (a) *apex of hard particle of half-angle* ω *embedded in a sliding grain boundary subject to far field shearing; and* (b) *triple grain junction subject to far-field tensile loading.*

greater detail elsewhere [*10*]. It is sufficient to state here that the solutions of these boundary-value problems for stress and strain are of the separable type

$$\sigma_{ij} = K r^{\lambda} \hat{\sigma}_{ij}(\theta) \tag{3}$$

$$\dot{\epsilon}_{ij} = A K^{m} r^{m\lambda} \hat{\epsilon}_{ij}(\theta) \tag{4}$$

where the radius-dependent term incorporates a singularity with a power $\lambda(m)$ and the separate angle-dependent term shows regular behavior. The exponents λ, obtained as the eigenvalues in a prescribed range of a characteristic equation, are functions of the creep exponent and depend also on the geometry and the nature of loading of the boundary value problem. For rigid particles with 90-deg apex angle, and for normal hexagonal grains with 120-deg angles, the exponents λ, for the two boundary-value problems, for $m = 3$ and 5, are given in Table 1. For these specific cases of $m = 3$ and 5, Figs. 3*a* and 3*b* give the angular distributions of the total stresses, $\hat{\sigma}_{ij}(\theta)$, for the square-shaped grain-boundary particles, while Figs. 4*a* and 4*b* give the angular distributions of deviatoric stresses, $\hat{s}_{ij}(\theta)$, for the same problem. The corresponding distributions of $\hat{\sigma}_{ij}(\theta)$ and $\hat{s}_{ij}(\theta)$ for the triple grain junction for $m = 3$ and 5 are given in Figs. 5*a* and 5*b*, and Figs. 6*a* and 6*b*, respectively. In all cases, the angular distributions have been normalized such that the maximum in the angular distribution of the equivalent stress, $\hat{\sigma}_e$, has the value of unity.

Examination of the results given in Figs. 3 and 4 shows the development of an extremum in normal tractions $\sigma_{\theta\theta}$ across the interface of the particle where eventual cavitation can be expected and a maximum in the shear component $\sigma_{r\theta}$ or $s_{r\theta}$ halfway in between the sliding horizontal boundary and the slanting particle interface. For the case of the triple grain junction, Figs. 5 and 6 show that the normal tractions $\sigma_{\theta\theta}$ reach extremum values across the symmetrical

TABLE 1—*Values of* λ(m) *and* mλ(m) *for singularities at the apexes of square particles and at triple grain junctions.*

	I. Square Particle ($\omega = 45$ deg)	
	λ	$m\lambda$
m	$(\sigma \propto r^{\lambda})$	$(\dot{\epsilon} \propto r^{m\lambda})$
3	−0.225	−0.675
5	−0.152	−0.760
	II. Triple Grain Junction	
3	−0.231	−0.693
5	−0.150	−0.750

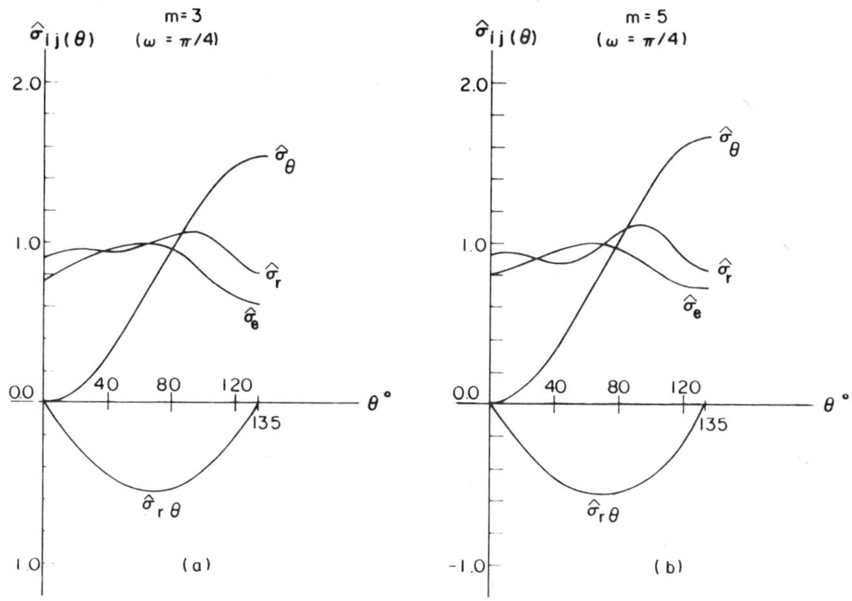

FIG. 3—*Angular distribution of stresses,* $\hat{\sigma}_{ij}(\theta)$, *at the apex of a hard particle of* $\omega = \pi/4$: (a) m = *3;* (b) m = *5.*

horizontal boundary and across the slanting boundary. We note further that in the angular range $(2\pi/3) \leq \theta \leq \pi$, $\sigma_{r\theta}$ is practically zero and $\sigma_{\theta\theta}$ is close to σ_{rr}. Thus, this region is acting as a rigid wedge, subjected to a pressure gradient, making the symmetrical triple-junction problem in tension equivalent to the lateral insertion of a rigid, lubricated wedge.

As remarked already, the solutions discussed in the preceding have been determined only to within a scale factor K, the generalized creep stress intensity factor. This factor needs to be determined by nesting the local singularity solutions into far-field solutions which incorporate the detailed geometry of the microstructure as well as the distant boundary conditions. This is discussed in the next section.

Finite Element Analysis

Statement of the Problem and Special Features of the Solution

Statement of the Problems—To determine the generalized creep stress intensity factor K for the two problems discussed in the foregoing we used an incremental initial strain finite element analysis for an elastic and power-law creeping continuum [*16,17*].

Figures 7 and 8 state the geometry and boundary conditions, and the finite

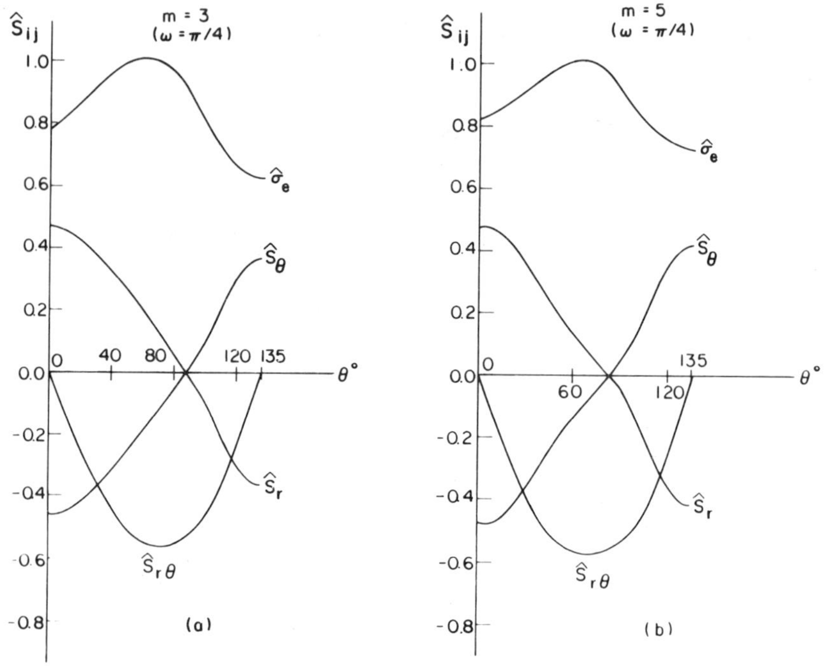

FIG. 4—*Angular distribution of deviatoric stresses, $\hat{s}_{ij}(\theta)$, at the apex of a hard particle of $\omega = \pi/4$:* (a) m = *3;* (b) m = *5.*

element mesh used in the problem of the rigid, square particles on the sliding grain boundaries for the specific case of the ratio $p/\ell = 1/3$, of the particle size to particle spacing. The geometry, boundary conditions, and finite element mesh for the corresponding problem of the triple grain junctions with sliding boundaries are stated in Figs. 9 and 10. In both cases, solutions have been obtained for $m = 3$ and $m = 5$.

The principal boundary conditions for the two problems are somewhat different but complementary. For the rigid particle at time $t = 0$, a far-field shear stress $\sigma_s = 10^{-3}\,E$ is applied and kept constant. For each exponent m, a convenient set of values is assigned to E, v, and A, which are, respectively, the Young's modulus, Poisson's ratio, and the scale factor of the creep law in Eq 1. The actual values for these parameters do not enter the final steady-state stresses and may be regarded as mere vehicles to be used to reach the final stress solution. For the triple grain boundary junction problem, the principal boundary condition is the imposed vertical velocity $\dot{\bar{V}}$ in Fig. 9, which was assigned a constant value at $t = 0$. The velocity $\dot{\bar{V}}$ was chosen to generate a nominal stress, $\sigma_N = 10^{-3}\,E$ in steady-state creep without the sliding grain bound-

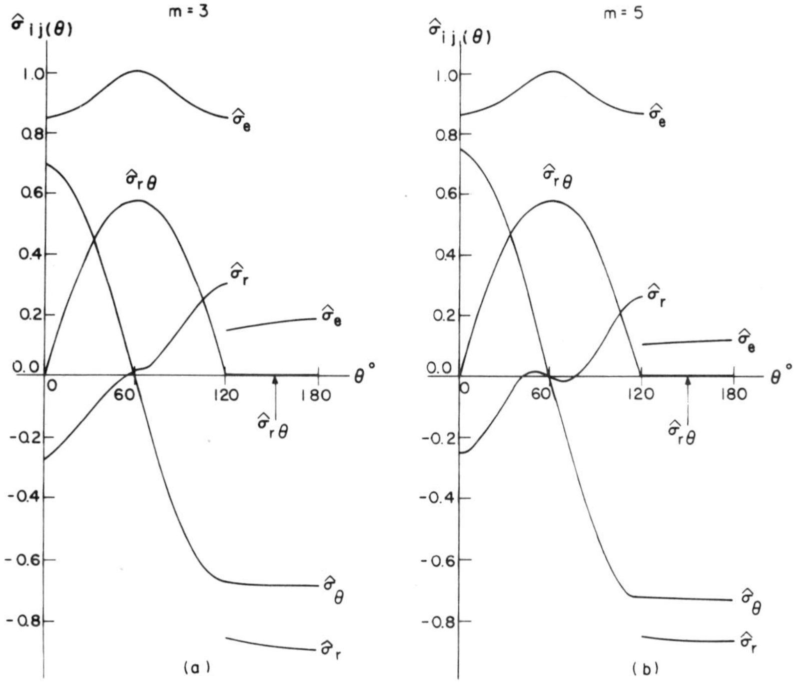

FIG. 5—*Angular distribution of stresses,* $\hat{\sigma}_{ij}(\theta)$, *at the triple grain junction:* (a) m = *3;* (b) m = *5.*

aries. Time quantities are scaled by t_N, which is the time required by the creep strain to equal the elastic strain in the nominal field, that is

$$t_N = \epsilon^e(\sigma_N)/\dot{\epsilon}^c(\sigma_N) = 1/(AE{\sigma_N}^{m-1}) \quad (5)$$

The first time step was assigned as $\Delta t = t_N$. This produces an elastic response of the boundary moving at $\bar{V}$ during t_N. The very high stresses then activate massive creep deformation in the second and subsequent steps to quickly reach steady-state creep. Useful information for the determination of K is furnished by those stress sampling points in which the accumulated creep strain dominates over the elastic strain.

Before we present specific results for the two problems, we shall point out several novel features of our finite element computations.

Novel Features of the Finite Element Method (FEM) Solutions—1. The Power-Type Singularity Element: The power-type singularity element of Tracey and Cook [*18*] was further developed to add both an originally lacking radial linear displacement term and a quadratic circumferential displacement

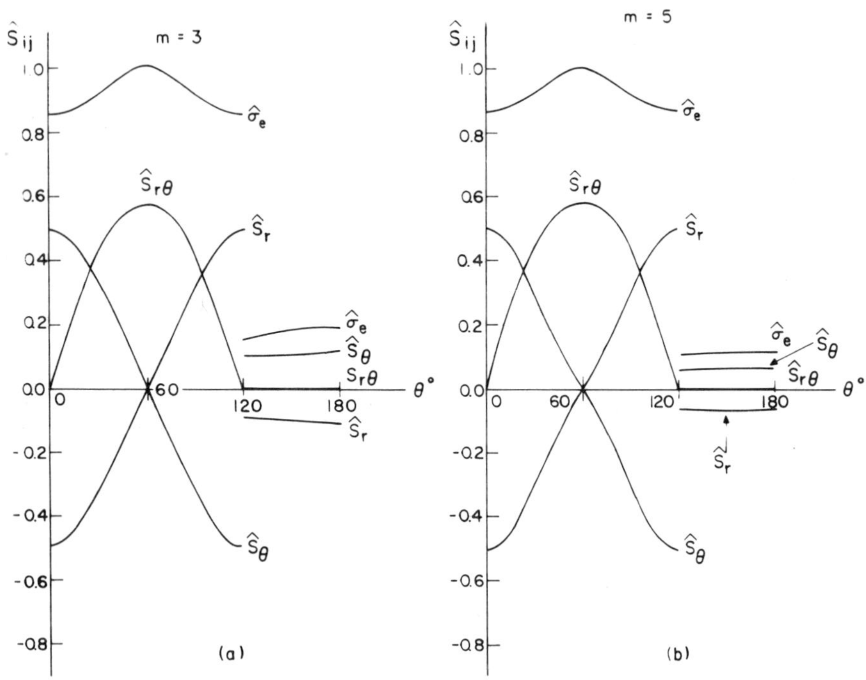

FIG. 6—*Angular distribution of deviatoric stresses,* $\hat{s}_{ij}(\theta)$, *at the triple grain junction:* (a) m = 3; (b) m = 5.

term. The displacement interpolation functions of this new element (Fig. 11*a*) are different from its coordinate mapping functions, while the latter are identical to those of the serendipity family of isoparametric elements. For an element mapping into a unit square of $0 \leq \xi \leq 1$ and $0 \leq \eta \leq 1$ in the natural space (ξ, η) (Fig. 11*b*), these special displacement interpolation functions that give rise to the power-type distribution in strain ($\epsilon \sim r^{q-1}$), are

$$N_1 = \frac{\xi + \alpha\xi^q}{1 + \alpha}(1 - 2\eta)(1 - \eta) \tag{6a}$$

$$N_2 = \frac{\xi + \alpha\xi^q}{1 + \alpha}(-1 + 2\eta)\,\eta \tag{6b}$$

$$N_3 = 1 - \frac{\xi + \alpha\xi^q}{1 + \alpha} \tag{6c}$$

$$N_4 = 4\,\frac{\xi + \alpha\xi^q}{1 + \alpha}(\eta - \eta^2) \tag{6d}$$

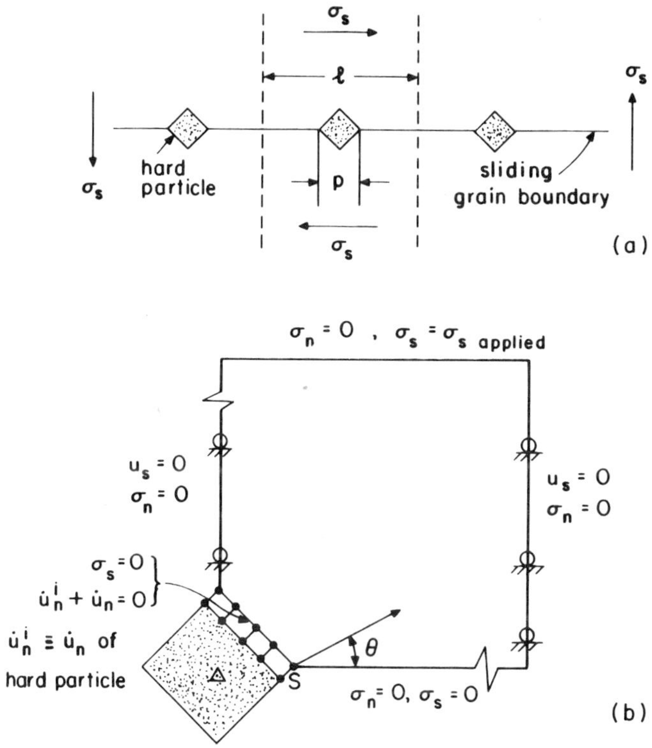

FIG. 7—*Geometry and boundary conditions in the finite-element analysis of the grain-boundary particle.*

When the parameter q takes the value of 1, or when the parameter α takes the value of 0, this special element is reduced to a standard five-node isoparametric element with the two corner nodes of the side opposite to the three-node side collapsed into one physical node.

In the radial direction in physical space, this special element is capable of representing constant displacement, constant strain, and power-type singular strain. Thus, it covers not just the most dominant local stress singularity but also the constant stress term in the field solution. In the circumferential direction, it can represent constant displacement, constant strain and linear strain, and is compatible with adjoining eight-node isoparametric elements. Further details of the special element are described elsewhere [*16,17*].

When applied to solve for the creep stress intensity factors, a fan-shaped array of these isosceles-triangular special elements, with their singularity nodes at the origin of the strain singularity, was connected to a surrounding mesh of eight-node isoparametric elements as shown in Fig. 11*c*. Special quadratures have been developed to integrate the stiffness matrix, equivalent force vector,

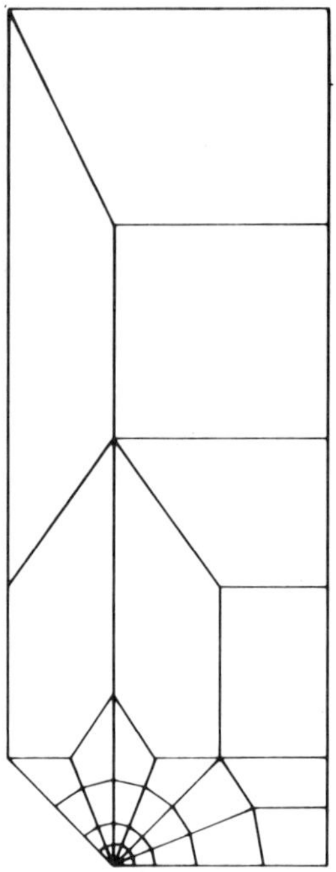

FIG. 8—*Finite-element mesh for the grain-boundary particle problem.*

and equivalent creep force vector of the special elements [*16*]. A two-by-two Gauss-Legendre quadrature was used for the eight-node isoparametric elements.

2. Kinematic Modeling of Boundary Conditions: Both the rigid grain boundary particle as well as the hexagonal grains are characteristic microstructure problems having special periodicity and symmetry features. These give rise to boundary conditions that are outside the capabilities of common general-purpose finite element programs. In small-strain finite element analysis, however, it is possible to simulate the desired symmetry and periodicity in the boundary conditions purely by kinematic modeling, without special formulation or programming.

Figure 12*a* shows an arrangement consisting of three vertical bars connected to a freely rotating horizontal beam. Equilibrium at the mid-point, E, of the beam, dictates that the normal force at A, $(F_n)_A$, must be equal to the normal force at B, $(F_n)_B$. Also, if the material of the bars AF, CE, BG is rigid compared with the material of RACBT, then the normal displacements at A and B, $(U_n)_A$ and $(U_n)_B$, respectively, are antisymmetric about C. Since the analysis is for small strain and the bars are perpendicular to ACB, the shear force, F_s, is zero at Nodes A, C, and B.

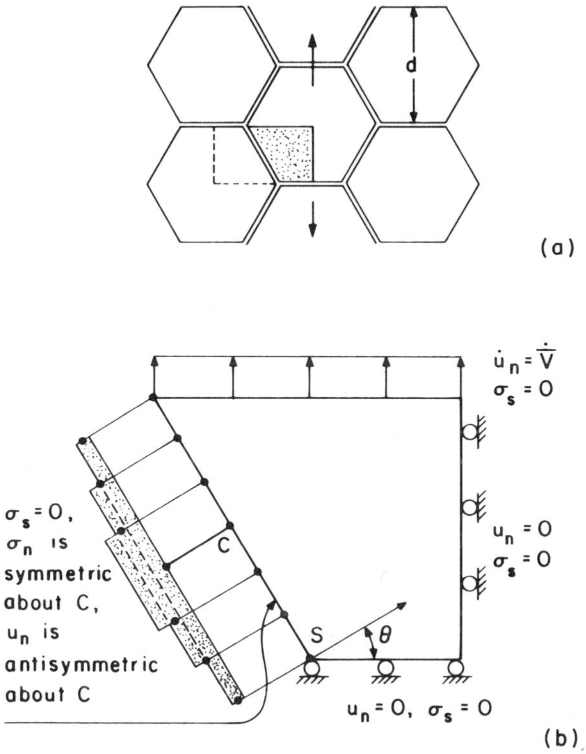

FIG. 9—*Geometry and boundary conditions in the finite-element analysis of the triple grain junction.*

Such a beam can be modeled by a three-node beam element, or equally well by a conventional two-dimensional element with high Young's modulus. We have used a five-node isoparametric element to model it. Other combinations of any available two-dimensional elements giving rise to the required three nodes on the line FEG can also be used.

When several units of these arrangements are used, one on top of the other, with Node E in common, as is shown schematically in Fig. 12*b*, the conditions of $\sigma_s = 0$, σ_n symmetric about C, and U_n antisymmetric about C on boundary ACB can be simulated. This concept is applied to model that part of the boundary conditions arising from symmetry and geometric repetition in the triple grain junction problem as shown in Fig. 9. In the solution of the rigid grain boundary particle, no special kinematic linkages are required anywhere except on the interface of the particle with the matrix. Here the condition of no shear traction on the interface and the freedom of the particle to rotate about its center is modeled by means of a rigid plate pivoting at its center and

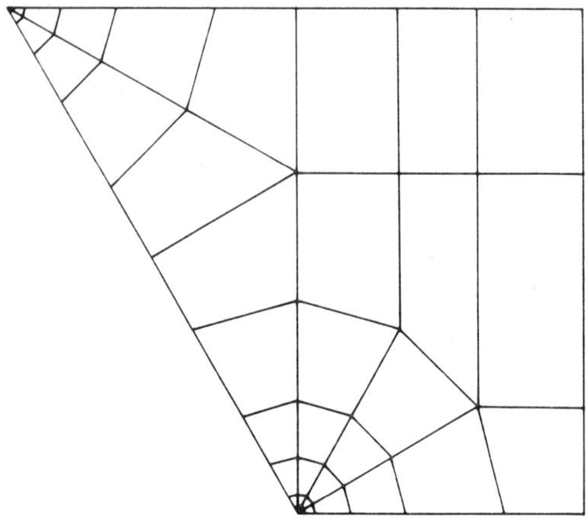

FIG. 10—*Finite-element mesh for the triple grain junction problem*

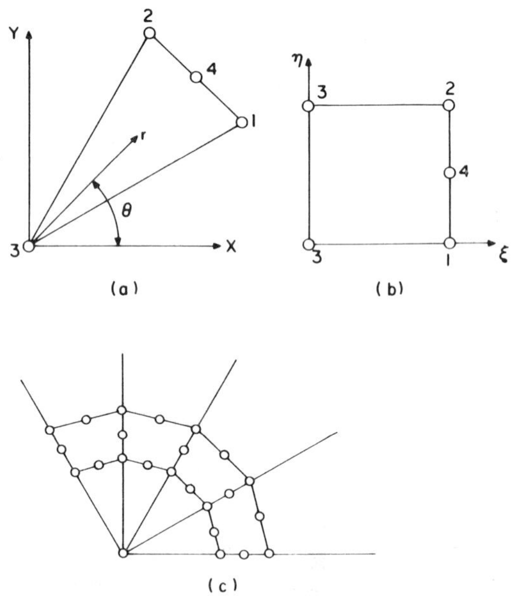

FIG. 11—*Power-type singularity element in:* (a) *physical space;* (b) *natural space; and* (c) *application.*

being attached to the interface of the deforming matrix by slender bars of high axial stiffness but negligible bending stiffness as shown in Fig. 7*b*.

3. Bypassing the Effect of Incompressibility Constraints: In our displacement-based, plane-strain, elasto-creep, finite element computation, the effect

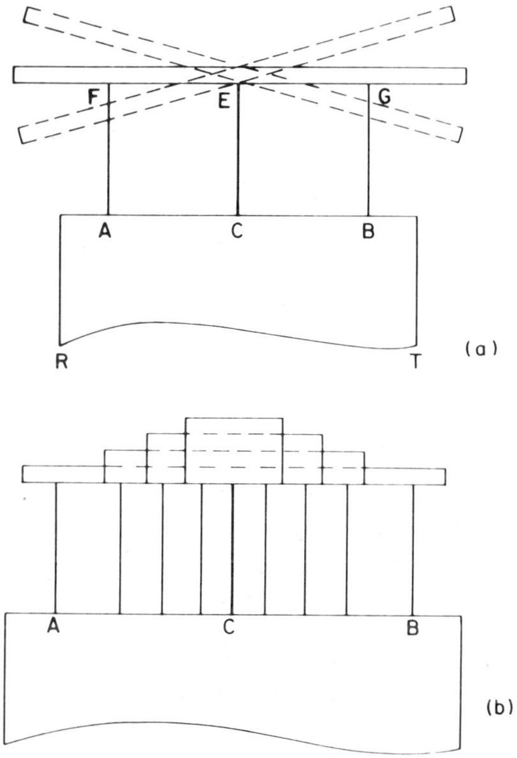

FIG. 12—*Kinematic simulation of boundary conditions by:* (a) *a simple kinematic unit;* (b) *multiple kinematic units.*

of the incompressibility constraint is felt when the accumulated creep strains dominate over the elastic strains. The finite element solution was found to produce stresses that deviate from the field solution of the preceding section by a hydrostatic component. This condition is not alleviated by conventional remedies such as reduced-integration. However, we observed that the angular distributions of the effective stress and the deviatoric stresses were not subject to this artifact and agreed very well with the field solution of the preceding section. It has been appreciated that over-constraint can cause the finite element mesh to "lock" to produce erroneous displacement solutions [*19*]. However, we observed from our incremental, initial-strain finite element results that the incompressibility constraint does not cause a general locking of the mesh to produce a randomly erroneous displacement (and hence strain) field. Instead, it causes a special form of locking of the mesh to produce a displacement field that deviates from the correct solution such that only the volumetric strain derived from this displacement is in error, while the deviatoric strains are given correctly. Therefore, we have found that the total stresses are subject to error but the deviatoric stresses are not. This behavior has also been encountered by Naylor [*20*], who has reported that oscillatory error exists only in the hydrostatic pressure, but not in the deviatoric stresses in his iterative, tangent stiffness, finite element analysis of an elastic-plastic sphere under internal pressure.

This characteristic of our solution enables the creep stress intensity factors to be determined from the distributions of the equivalent stress σ_e or any other deviatoric stress quantities, free from artifacts due to incompressibility

$$\sigma_e = K r^{\lambda} \hat{\sigma}_e \tag{7a}$$

and

$$s_{ij} = K r^{\lambda} \hat{s}_{ij} \tag{7b}$$

4. Self-Adjusting Time Step Incrementation: In incremental, initial-strain, elasto-creep finite element computation, the stiffness matrix needs only to be assembled and decomposed once and stored in high-speed storage to be used in all subsequent steps, if one uses an explicit time integration scheme such as Euler's method, which is a conditionally stable scheme. To stably march to steady state in the shortest time, the time steps Δt are automatically self-adjusted in our computations according to the latest available largest equivalent stress $(\sigma_e)_{\max}$ of all the spatial integration points by

$$\Delta t = f \cdot \frac{2(1+\upsilon)}{3} \, \frac{1}{m} \, \frac{1}{\mathrm{EA}\,(\sigma_e)_{\max}{}^{m-1}} \tag{8}$$

where f is an adjustable parameter that can be prescribed to have different values for each time step. Physically, Eq 8 requires that $\Delta\epsilon^c/\epsilon^e$ be $\sim f/m$.

If f is always set equal to 2, Eq 8 becomes the stability criterion developed by Cormeau [*21*], which has been found to be conservative for some problems [*22*]. The time-step increments in other than the first step are automatically chosen by Eq 8. For $m = 3$, f in Eq 8 was set equal to 1.0. For $m = 5$, the f's were assigned values which consisted mostly of repeating sequences of 5, 5, 2, 2, 2.

The Solutions

The Rigid Grain-Boundary Particle—Figures 13*a* and 13*b* show the comparison of the finite element results for the θ-variation of the deviatoric stresses, $\hat{s}_{ij}$, with the dominant singular solution shown in Figs. 4*a* and 4*b*, for $m = 3$ and $m = 5$, respectively, for the geometry of the grain boundary particle problem with $p/l = 1/3$. In Figs. 13*a* and 13*b*, the discrete points are finite element results of $r^{-\lambda} s_{ij}$ $(=K\,\hat{s}_{ij})$ divided by a scaling constant h. This scaling constant h is determined by requiring that one particular point of the FEM result, that

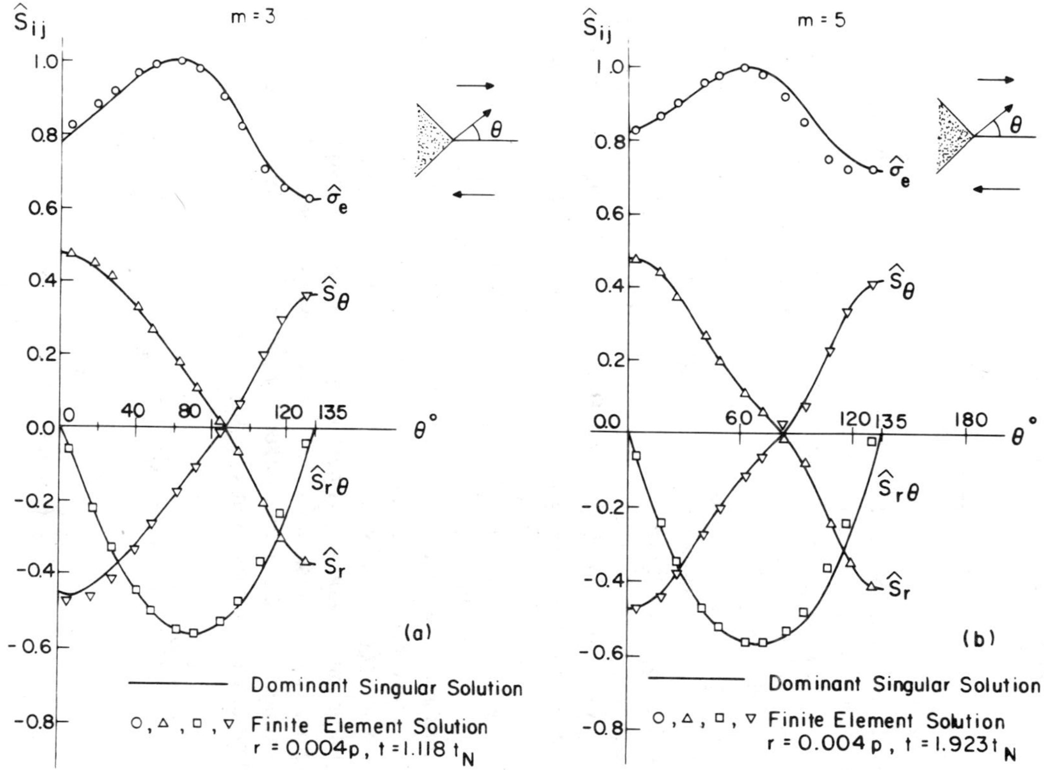

FIG. 13—*Comparison of the finite-element solution of* $\hat{s}_{ij}(\theta)$ *with the dominant singular solution for the hard particle:* (a) m = *3;* (b) m = *5.*

of $r^{-\lambda}\sigma_e/h$ $(=\hat{\sigma}_e K/h)$ at $\theta = 62.8$ deg, to have the same value as $\hat{\sigma}_e$ (62.8 deg) of the dominant singular solution. Thus by just "attaching" this one point of the FEM solution to the singular solution, it is observed that all other points of the FEM solution fall on the solid lines representing the dominant singular solution. This indicates that the sampling points of these finite element results are within the dominant singularity zone, and these finite element results in the limit of $r \to 0$ do indeed obey Eq 7. Then K is given by

$$K = \frac{r^{-\lambda}\sigma_e\,(62.8\text{ deg})}{\hat{\sigma}_e\,(62.8\text{ deg})} = h \tag{9}$$

Since all components of $\hat{s}_{ij}$ of the FEM results agree with the singularity solution very well for all θ, K can, in principle, be determined from any $\hat{s}_{ij}$ distribution at any θ. For highest accuracy we choose to determine K from $\hat{\sigma}_e$ at a θ-value where $\hat{\sigma}_e$ is close to maximum. Thus, for $m = 3$, as determined from Fig. 13*a*, ($r/p = 0.004$, and $t/t_N = 1.118$), $Kp^{\lambda}/\sigma_s = 0.965$. Similarly, for $m = 5$, using Fig. 13*b*, ($r/p = 0.004\,p$, and $t/t_N = 1.923$), the value of Kp^{λ}/σ_s is determined to be 1.109. Examination of the results at the same time but at different radii, r/p, has shown that $\hat{s}_{ij}$ starts to deviate appreciably from the singular solution for $r/p \geq 0.177$. For $r/p < 0.177$, both $\hat{\sigma}_e$ and $\hat{s}_{r\theta}$ attain their maximum values around $\theta \approx \pi/3$. Hence values of K determined from these stresses are subject to the least error and are therefore most trustworthy. Furthermore, examination of the solutions at $r/p = 0.004$ as a function of time indicates that even as early as $t/t_N = 0.304$ the deviatoric stresses are already dominated by the steady-state singular solution, and that after that there are only minimal fluctuations in Kp^{λ}/σ_s over time. This indicates that the self-adjusting time-step selection scheme discussed earlier has worked well.

With the values of Kp^{λ}/σ_s obtained for $m = 3$ and $m = 5$, and resorting to overall equilibrium considerations pertaining to the characteristic repeat unit of the geometry, it is possible to give the generalized creep stress intensity factor as a function of the main geometrical parameters, including the particle apex half-angle $\omega(\omega = \pi/4$ in Fig. 13)

$$\frac{Kp^{\lambda}}{\sigma_s}\,\frac{p}{l}\tan\omega = C(m)\frac{(1+\lambda)\,(2\cos\omega)^{\lambda}}{\hat{\sigma}_{\theta\theta}\,(\pi-\omega)} \tag{10}$$

where

$$C(m) = 0.689 + 0.039\,(m-3) \tag{11}$$

is determined on the basis of the results for the finite element computations for $m = 3$ and $m = 5$. Clearly $C(m)$ is valid only for m-values near the range of 3 to 5. In Eq 10, $\hat{\sigma}_{\theta\theta}(\pi - \omega)$ is the θ-dependent component of the normal stress evaluated on the slanting interface of the particle.

The Triple Grain Junction—Figures 14*a* and 14*b* show, for $r/d = 0.001$ at

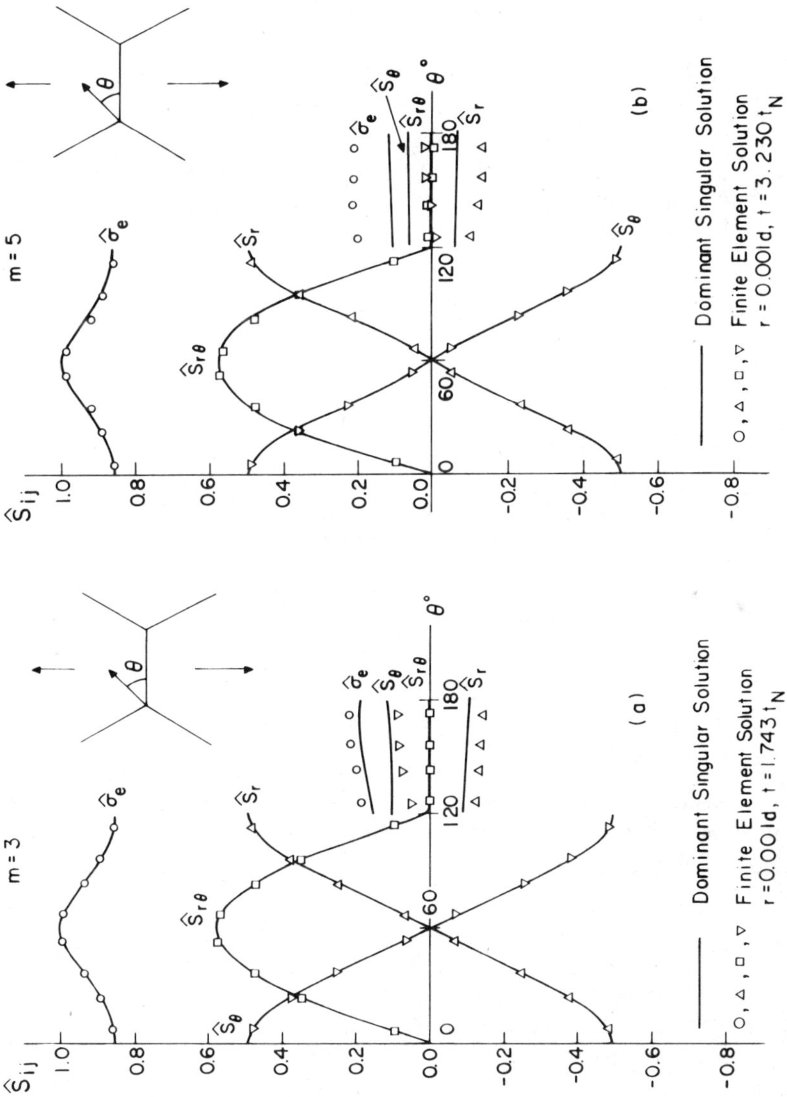

FIG. 14—*Comparison of the finite-element solution of* $\hat{s}_{ij}(\theta)$ *with the dominant singular solution for the triple grain junction:* (a) m = 3; (b) m = 5.

$t/t_N = 1.743$ ($m = 3$) and $t/t_N = 3.230$ ($m = 5$), the comparison of the finite element results for the θ-variation of the deviatoric stresses $\hat{s}_{ij}$ with the dominant singular solution given earlier in Figs. 6*a* and 6*b* for $m = 3$ and $m = 5$, respectively. In this case again, one particular point from the finite-element solution for $r^{-\lambda}\sigma_e$ (that is, $= K\,\hat{\sigma}_e$) at $\theta = 66.22$ deg was used to match the previously determined singular field solution to obtain the generalized stress intensity factor K. Examination of Figs. 14*a* and 14*b* show that the agreement is very good for $0 < \theta < 2\pi/3$, but not as good over the range $2\pi/3 < \theta < \pi$. The poor agreement in the latter range is attributed to the fact that the latter data are taken from a corner of the mesh where displacement-rate boundary conditions are enforced. Evaluation of the generalized stress intensity factor from σ_e at $\theta = 66.22$ deg gave

$$\frac{Kd^{\lambda}}{\sigma_N} = 0.753 \qquad (\text{for } m = 3) \tag{12a}$$

and

$$\frac{Kd^{\lambda}}{\sigma_N} = 0.848 \qquad (\text{for } m = 5) \tag{12b}$$

respectively. Once again, on the basis of these specific results and overall equilibrium, a more general expression of the following form is suggested

$$\frac{2}{3}\,\frac{Kd^{\lambda}}{\sigma_N} = C(m)\,\frac{(1+\lambda)\,(2\sqrt{3})^{\lambda}}{\hat{\sigma}_{\theta\theta}\,(0)} \tag{13}$$

where

$$C(m) = 0.613 - 0.010\,(m - 3) \tag{14}$$

which is considered valid for m-values not substantially differing from the range investigated. In Eq 13, $\hat{\sigma}_{\theta\theta}\,(0)$ is the θ-dependent component of the normal stress evaluated on the symmetrical horizontal grain boundary.

Discussion

The two problems treated by us are generalizations of the classical linear wedge problem discussed first by Williams [*11*]. Here we have given these solutions for an incompressible, nonlinearly creeping continuum in a microstructural context for the purpose of understanding the cavitation phenomenon on sliding grain boundaries in structural alloys, where we have introduced a new method of simulating boundary conditions arising from periodicity and symmetry. The kinematical devices which accomplish these constraints are quite effective with existing programs. The special singularity element that we have introduced enables us to make the match between the local singular field and the distant background with good precision. These new techniques have much more general applicability to problems of this type involving space-filling microstructural forms where local interface tractions are of great interest in the modeling of damage.

In the special case of modeling intergranular cavitation that we are interested in, it is necessary to apply some modifications to these stress fields by considering the stress-smoothing effect of diffusional flow between portions of a boundary under high tensile stress and adjacent portions under high compressive stress. The methods for such corrections and the eventual utility of these modified solutions to explain quantitatively the rate of cavity production were discussed by us earlier [*4,5*].

Acknowledgment

This research was supported by the U.S. Department of Energy under Contract No. DE-AC02-77ERO4461. We are grateful to Professors D. M. Parks and J. L. Bassani for many useful discussions on computational techniques.

References

[*1*] Chen, I. W. and Argon, A. S., *Acta Mettallurgica*, Vol. 29, 1981, pp. 1321–1333.
[*2*] Perry, A. J., *Journal of Materials Science*, Vol. 9, 1974, pp. 1016–1039.
[*3*] Grant, N. J. in *Fracture*, Vol. 3, H. Liebowitz, Ed., Academic Press, New York, 1971, pp. 483–533.
[*4*] Argon, A. S., Chen, I. W., and Lau, C. W. in *Creep-Fatigue-Environment Interactions*, R. M. Pelloux and N. S. Stoloff, Eds., American Institute for Mining, Metallurgical, and Petroleum Engineers, Warrendale, Pa., 1980, pp. 46–85.
[*5*] Argon, A. S., Chen, I. W., and Lau, C. W. in *Three-dimensional Constitutive Relations and Ductile Fracture*, S. Nemat-Nasser, Ed., North-Holland Publishing Co., Amsterdam, 1981, pp. 23–49.
[*6*] Morris, D. G., *Metal Science*, Vol. 12, 1978, pp. 19–29.
[*7*] Raj, R., *Acta Metallurgica*, Vol. 26, 1978, pp. 995–1006.
[*8*] Heald, P. T. and Williams, J. A., *Philosophical Magazine*, Vol. 22, 1970, pp. 1095–1100.
[*9*] Lau, C. W. and Argon, A. S. in *Fracture 1977*, Vol. 2, D. M. R. Taplin, Ed., University of Waterloo Press, Waterloo, Ont., Canada, 1977, pp. 595–601.
[*10*] Lau, C. W. and Argon, A. S., to be published.
[*11*] Williams, M. L., *Transactions*, American Society of Mechanical Engineers, *Journal of Applied Mechanics*, Vol. 74, 1952, pp. 526–528.
[*12*] Hutchinson, J. W., *Journal of the Mechanics and Physics of Solids*, Vol. 16, 1968, pp. 13–31.
[*13*] Hutchinson, J. W., *Journal of the Mechanics and Physics of Solids*, Vol. 16, 1968, pp. 337–347.
[*14*] Takeuchi, S. and Argon, A. S., *Journal of Materials Science*, Vol. 11, 1976, pp. 1542–1566.
[*15*] Hoff, N. J., *Quarterly of Applied Mathematics*, Vol. 12, 1954, pp. 49–55.
[*16*] Lau, C. W., "Dominant Singularity and Finite Element Analyses of Plane-Strain Stress Fields in Creeping Alloys with Sliding Grain Boundaries," Ph.D. thesis, Department of Mechanical Engineering, Massachusetts Institute of Technology, Cambridge, Mass., 1982.
[*17*] Lau, C. W., Argon, A. S., and McClintock, F. A., *Computers & Structures*, Vol. 17, No. 5-6, 1983, pp. 923–931.
[*18*] Tracey, D. M. and Cook, T., *International Journal for Numerical Methods in Engineering*, Vol. 11, 1977, pp. 1225–1233.

[*19*] Nagtegaal, J. C., Parks, D. M., and Rice, J. R., *Computer Methods in Applied Mechanics and Engineering*, Vol. 4, 1974, pp. 153–177.
[*20*] Naylor, D. J., *International Journal for Numerical Methods in Engineering*, Vol. 8, 1974, pp. 443–460.
[*21*] Cormeau, I., *International Journal for Numerical Methods in Engineering*, Vol. 9, 1975, pp. 109–127.
[*22*] Bassani, J. L. and McClintock, F. A., *International Journal of Solids and Structures*, Vol. 17, 1981, pp. 479–492.

Chung Yuen Hui[1]

Steady-State Crack Growth in Elastic Power-Law Creeping Materials

REFERENCE: Hui, C. Y., **"Steady-State Crack Growth in Elastic Power-Law Creeping Materials,"** *Elastic-Plastic Fracture: Second Symposium, Volume I—Inelastic Crack Analysis, ASTM STP 803,* C. F. Shih and J. P. Gudas, Eds., American Society for Testing and Materials, 1983, pp. I-573–I-593.

ABSTRACT: The problem of a quasi-statically growing crack in an elastic power-law creeping material is considered for the case of plane-strain Mode I under small-scale yielding conditions. A finite-element method is used to compute the complete stress and strain field. A crack growth criterion is used to derive a relation between the crack growth rate $\dot{a}$ and the stress-intensity factor K.

KEY WORDS: steady-state crack growth, power-law creep, stress-intensity factor, small-scale yielding, void growth, elastic-plastic fracture

Cracking under creep conditions is observed to occur in two stages: an initiation period and a growth period. During the initiation period, creep deformation develops ahead of the crack until a critical amount of damage has accumulated in the creep zone and the crack starts to propagate. This stage has been analyzed by both analytical and numerical methods [*1–3*].[2] One important result is that the time-dependent stress and strain field of a stationary crack in an elastic power-law creeping material is governed by the Hutchinson-Rice-Rosengren (HRR) singularity, that is, the C^* singular field [*4*]. Often it is possible to divide the growth period into two stages: Stage 1 in which the crack advances due to continuous time-dependent creep damage accumulated ahead of the crack tip, and Stage 2 in which the crack has attained steady-state growth. In this paper we will consider in detail the problem of steady-state quasi-static crack growth under small-scale yielding con-

[1]Assistant professor, Theoretical and Applied Mechanics, Cornell University, Ithaca, N.Y. 14850.

[2]The italic numbers in brackets refer to the list of references appended to this paper.

ditions in an elastic power-law creeping material. Under uniaxial tension σ, the strain rate $\dot{\epsilon}$ of such a material is given by

$$\dot{\epsilon} = \frac{\dot{\sigma}}{E} + B\sigma^n \tag{1}$$

where

E = Young's modulus,
B = creep coefficient, and
n = creep exponent.

Under multiaxial stress conditions Eq 1 can be generalized to

$$\dot{\epsilon}_{ij} = \frac{(1+\nu)}{E}\dot{S}_{ij} + \frac{(1-2\nu)}{3E}\dot{\sigma}_{pp}\delta_{ij} + \frac{3}{2}B\sigma_e^{\,n-1}S_{ij} \tag{2}$$

where ν is Poisson's ratio, S_{ij} is the stress deviator defined by

$$S_{ij} = \sigma_{ij} = \frac{1}{3}\sigma_{kk}\delta_{ij} \tag{3}$$

and σ_e is the equivalent stress defined by

$$\sigma_e^{\,2} = \frac{3}{2}S_{ij}S_{ij} \tag{4}$$

In a previous paper [5], the asymptotic near-tip field of a quasi-statically growing crack in an elastic power-law creeping material was derived under conditions of antiplane shear, plane strain, and plane stress. It was shown that the functional dependence of the crack-tip singularity changes at $n = 3$. For $n < 3$, the asymptotic stress field has an inverse square-root singularity. For $n > 3$, a new type of singular field is found which requires the balance of elastic and creep strain near the crack tip. In this case, the strength of the near-tip fields is completely specified by the current growth rate $\dot{a}$ of the crack (besides material parameters). With respect to a polar coordinate system (r, θ) attached to and moving with the crack tip, the near-tip stress and strain fields of a plane-stress or plane-strain crack are given by

$$\begin{bmatrix} \sigma_{ij} \\ E\epsilon_{ij} \end{bmatrix} = \alpha_n \left(\frac{\dot{a}}{EBr}\right)^{1/(n-1)} \begin{bmatrix} \hat{\sigma}_{ij}(\theta) \\ \hat{\epsilon}_{ij}(\theta) \end{bmatrix} \tag{5}$$

where α_n is a numerical constant depending on n and is given in Ref 5; $\hat{\sigma}_{ij}$ and $\hat{\epsilon}_{ij}$ are nondimensional angular variations of the near-tip stress and strain fields. Notice that the singularity in r is more singular than the HRR singularity but less singular than the elastic inverse square-root singularity. In general, the near-tip fields are dependent on Poisson's ratio, ν. In the case of plane stress, the near-tip stresses are independent of ν. In the case of plane strain, numerical results are obtained only for the case of elastic incompressibility.

The in-plane near-tip displacement fields u_i can be determined by integration of the near-tip strain field [*6*]. This leads to

$$u_i = \frac{\alpha_n}{E}\left(\frac{\dot{a}}{EB}\right)^{1/(n-1)} r^{(n-2)/(n-1)}\,\hat{u}_i(\theta) \tag{6}$$

The crack opening displacement $\delta \equiv 2u_y(r, \theta = \pi)$ is seen to be

$$\delta = \frac{2\alpha_n}{E}\left(\frac{\dot{a}}{EB}\right)^{1/(n-1)} r^{(n-2)/(n-1)}\,\hat{u}_y(\theta = \pi) \tag{7}$$

The total strain ϵ_{ij} can be written as a sum of its elastic component $\epsilon_{ij}{}^e$ and creep component $\epsilon_{ij}{}^c$, that is

$$\epsilon_{ij} = \epsilon_{ij}{}^e + \epsilon_{ij}{}^c$$

Close to the crack tip, both $\epsilon_{ij}{}^e$ and $\epsilon_{ij}{}^c$ have the same asymptotic behavior as given by Eq 5 if $\hat{\epsilon}_{ij}$ is replaced by $\hat{\epsilon}_{ij}{}^e$ and $\hat{\epsilon}_{ij}{}^c$, respectively. The nondimensional functions $\hat{\epsilon}_{ij}{}^e$ and $\hat{\epsilon}_{ij}{}^c$ are given in Ref *5*.

Formulation of the Steady-State Crack Growth Problem

We will consider the problem of steady-state crack growth under small-scale yielding conditions. As mentioned in Ref *5*, steady-state crack growth in a material governed by Eq 2 necessarily requires small-scale yielding conditions. A finite-element method is used to compute the complete stress and strain field for the case of plane-strain Mode I and antiplane shear. In this paper, we will present results only for the case of plane strain, although computations for the case of antiplane shear have been carried out completely in Ref *6*.

In the small-scale yielding limit the elastic singular field is imposed as the far field on a semi-infinite crack as shown in Fig. 1. Thus, as $r \to \infty$

$$\sigma_{ij} = \frac{K}{\sqrt{2\pi r}}\hat{f}_{ij}(\theta) \tag{8}$$

where K is the elastic stress intensity factor. The angular variation $\hat{f}_{ij}$ depends on the symmetry of the field with respect to the crack plane $y = 0$ and these variations are given in most texts on fracture mechanics.

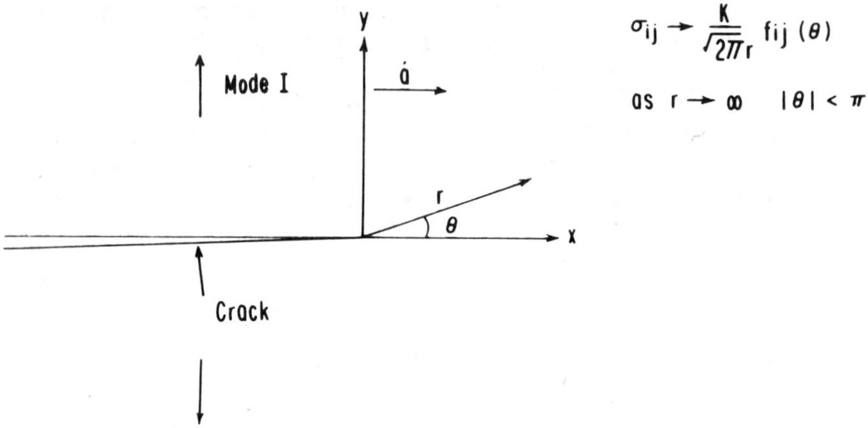

FIG. 1—*Steady-state crack growth under small-scale yielding.*

The small-scale yielding problem is well defined only if the creep strain rate caused by the imposed elastic singular K field is small compared with the elastic strain rate at infinity. This requires the elastic terms in Eq 2, calculated using Eq 8, to be asymptotically dominant as $r \to \infty$ in the far field, that is

$$\lim_{r\to\infty} \frac{BK^{n-1}r^{-(n-3)/2}}{(\sqrt{2\pi})^{n-1}\dot{a}} = 0 \tag{9}$$

Equation 9 requires that the creep exponent $n > 3$ in order for small-scale yielding to exist. Recall that $n > 3$ is the condition for the existence of the new singular field.

Physically, it is possible to explain why steady-state crack growth exists only for $n > 3$. The result of Riedel and Rice [*1*] indicates that, for a stationary crack in an elastic power-law creeping material under step loading, the creep zone boundary $r(t, \theta)$ is proportional to $t^{2/(n-1)}$, where t is the time elapsed after the start of loading. This implies that the creep zone expands at a rate proportional to

$$[2/t^{(n-1)}]^{-1} = t^{-(n-3)/(n-1)} \tag{10}$$

Equation 10 indicates that the rate at which the creep zone expands increases with time for $n < 3$, while the reverse is true for $n > 3$. Thus, for $n < 3$, the creep zone cannot be confined and hence one would anticipate that steady state may not be reached.

The foregoing discussion leads to the observation that, when the time taken for crack initiation is large compared with the stress relaxation time derived by Riedel and Rice [*1*] or when crack advance becomes so slow that creep can no longer be confined to a zone small compared with relevant specimen dimension, the effect of the C^* field will be dominant and steady-state crack growth will not be attained. Under such conditions, the crack growth rate can no longer be related to the stress-intensity factor and a better correlating factor is C^*. This possibility was recently examined by Leckie and McMeeking [*16*].

Steady-State Crack Growth in Mode I

We will consider steady state crack growth in plane strain under small-scale yielding conditions. The discussion will be limited to the case of elastic incompressibility in order to make contact with the incompressible asymptotic results of Ref *5*. The region of asymptotic validity of the near-tip stress fields is estimated using the finite-element results. It will be seen that the region of asymptotic validity of the near-tip stress and strain fields is dependent on both the crack growth rate $\dot{a}$ and the applied stress-intensity factor K, although the asymptotic near-tip fields themselves depend only on $\dot{a}$. A definition of an effective creep zone is given in the following. Plots of the crack opening displacement are provided and an interpolation scheme for the stresses discussed.

Nondimensional Equations and Finite-Element Method

As in the case of antiplane shear [*5,6*], the steady-state stress field under small-scale yielding conditions can be expressed in terms of the nondimensional variation Σ_{ij} by

$$\sigma_{ij} = \left(\frac{\dot{a}}{EB_1K^2}\right)^{1/(n-3)} \Sigma_{ij}(\tilde{R}, \theta) \tag{11}$$

where the nondimensional distance from the tip $\tilde{R}$ is given by

$$\tilde{R} = r(EB_1K^{n-1}/\dot{a})^{-2/(n-3)}, \qquad B_1 \equiv \frac{3}{2}B \tag{12}$$

The asymptotic behavior of Σ_{ij} for large and small $\tilde{R}$ can be obtained from Eqs 8 and 5, that is

$$\Sigma_{ij}(\tilde{R} \to \infty, \theta) = \frac{1}{\sqrt{2\pi}} \tilde{R}^{-1/2} \hat{f}_{ij}(\theta) \tag{13a}$$

$$\Sigma_{ij}(\tilde{R} \to 0, \theta) = \alpha_n \left(\frac{2}{3}\tilde{R}\right)^{-1/(n-1)} \hat{\sigma}_{ij}(\theta) \tag{13b}$$

From the constitutive law, Eq 2, it is clear that the strain rate and strain tensor everywhere can be expressed in terms of Σ_{ij}.

A displacement-based finite-element method is used to solve the steady-state crack growth problem. Details of the mesh used and method of implementation are discussed in the Appendix. Nondimensional variables are introduced into the field equations to expedite the analysis and are given in the Appendix. The parameters appearing in the plane-strain dimensional equations are the velocity of the crack $\dot{a}$, the stress-intensity factor K, the Young's modulus E, Poisson's ratio ν, and the creep coefficient B and exponent n. Only ν and n appear in the nondimensional plane-strain equations. The nondimensional equivalent of the far-field condition, Eq 8, is given in the Appendix and is found to be independent of K and $\dot{a}$. Thus, both the nondimensional equations and the boundary conditions are independent of K and $\dot{a}$, which implies that to relate $\dot{a}$ and K an extra condition such as a growth criterion must be imposed. The results obtained by the finite-element computation are presented next, along with a formula that approximates the stresses obtained from the finite-element calculation. Two length scales are introduced to facilitate the presentation of the finite-element results.

Creep Zone

The distance to the creep zone boundary $r(\theta)$ will be estimated by setting the equivalent creep strain rate

$$\sqrt{\frac{2}{3}\dot{\epsilon}_{ij}{}^{c}\,\dot{\epsilon}_{ij}{}^{c}}$$

calculated using just the stresses of the elastic singular field equal to the equivalent elastic strain rate calculated from the same field. This results in

$$r(\theta) = h_n\left(\frac{EB_1(K\sin\theta)^{n-1}}{\dot{a}}\right)^{2/(n-3)} \tag{14}$$

where

$$h_n \equiv \frac{1}{4}\left(\frac{1}{2\pi}\right)^{(n-1)/(n-3)}$$

Let L denote the estimate of the maximum creep zone extent which is attained at $\theta = \pi/2$, that is

$$L = h_n\left(\frac{EB_1K^{n-1}}{\dot{a}}\right)^{2/(n-3)} \tag{15}$$

We introduce a second length scale which will be used in the approximation formulas for the stresses in the following. Let D_y be the crossover value of the x-coordinate where the σ_{yy}-component of the far elastic field equals that of the asymptotic field (Eq 5) ahead of the crack. D is found to be

$$D_y = 4\left[\frac{3(\alpha_n \hat{\sigma}_{yy})^{n-1}}{2}\right]^{-2/(n-3)} L \tag{16}$$

where $\hat{\sigma}_{yy} \equiv \hat{\sigma}_{yy}\,(\theta = 0)$. For $n = 4$, $\hat{\sigma}_{yy} = 1.695$ and $\alpha_4 = 1.042$. For $n = 6$, $\hat{\sigma}_{yy} = 2.113$ and $\alpha_6 = 1.237$. For $n = 4$, $D_y = 5.86 \times 10^{-2}L$, and for $n = 6$, $D_y = 1.24 \times 10^{-1}L$. Another length scale, D_x, can be defined in exactly the same way if the σ_{xx} component is used instead.

Stresses and Region of Asymptotic Validity of the Near-Tip Stresses

Riedel and Wagner [7] suggested an approximation formula for the stresses ahead of the crack tip which interpolates beteen the near-tip asymptotic field and the far elastic stress field. For the component σ_{yy}, this is

$$\Sigma_{yy}(\tilde{R}, \theta = 0) = \left[\frac{2}{3}\,(\alpha_n \hat{\sigma}_{yy})^{1-n}\,\tilde{R} + (2\pi\tilde{R})^{(n-1)/2}\right]^{-1/(n-1)} \tag{17}$$

The first term of Eq 17 in the square brackets is responsible for the asymptotic near-tip behavior while the second term gives the asymptotic behavior for large $\tilde{R}$. Using the length scale D_y discussed previously, this interpolation formula can be re-expressed in either of the following two forms

$$\Sigma_{yy}(\tilde{R}, \theta = 0) = \frac{3}{2}\,\alpha_n \hat{\sigma}_{yy}\,\tilde{R}^{-1/(n-1)}\left[1 + \left(\frac{r}{D_y}\right)^{(n-3)/2}\right]^{-1/(n-1)} \tag{18}$$

$$= \frac{\tilde{R}^{-1/2}}{\sqrt{2\pi}}\left[1 + \left(\frac{D_y}{r}\right)^{(n-3)/2}\right]^{-1/(n-1)} \tag{19}$$

Similarly, Σ_{xx} can be expressed exactly as Eqs 18 and 19 with $\hat{\sigma}_{yy}$ and D_y replaced by $\hat{\sigma}_{xx}$ and D_x.

It should be noted that the first terms on the outside of the square brackets in Eqs 18 and 19 give the asymptotic behavior of the interpolation formula (Eq 15) for $\tilde{R} \rightarrow 0$ and $\tilde{R} \rightarrow \infty$, respectively. Furthermore, the asymptotic behavior of the interpolation formula for small and large $\tilde{R}$ is in exact agreement with the asymptotic results of Eqs 13*a* and 13*b*. Equations 18 and 19 are found to approximate the σ_{yy} component of the finite-element computation to within 10 percent everywhere for both cases of $n = 4$ and $n = 6$. For the case of $n = 4$, the interpolation formula approximates the σ_{xx}-component of the finite-element result to within 10 percent for all $\tilde{R} > \tilde{D}_x$, whereas for $n = 6$ the interpolation formula approximates the same component of stresses to within 10 percent for all $\tilde{R} > 2\tilde{D}_x$. In both cases of $n = 4$ and $n = 6$, the finite-element result for σ_{xx} falls below the results predicted by the interpolation formula outside of the indicated range described in the preceding.

Nondimensional plots of $\sigma_{yy}/\sigma_{yy}^{asy}$ and $\sigma_{yy}/\sigma_{yy}^{e}$ ahead of the crack tip as determined by the finite-element calculation are given in Fig. 2 for both cases of $n = 4$ and $n = 6$. The component in the y-direction of the near-tip asymptotic stress field and the elastic stress field are denoted by σ_{yy}^{asy} and σ_{yy}^{e}, respectively. Their nondimensional representations are given in the Appendix. Figure 2 shows that, close to the crack tip, the finite-element result falls within 16 and 7 percent of the asymptotic result of Eq 5. It should be recognized that this is the region of high triaxiality in which σ_{yy} is about four times that of σ_e.

The region of asymptotic validity will be defined as the region where the finite-element results deviate by no more than 10 percent from the asymptotic results calculated using Eq 5. Ahead of the crack tip, the region of asymptotic validity is estimated to be $0.02\,L = 0.35\,D_y$ and $0.08\,L = 0.64\,D_y$ based on the component of stress σ_{yy} for $n = 4$ and $n = 6$, respectively. Directly above the crack tip, at $\theta = \pi/2$, the region of asymptotic validity is larger and is estimated to be $0.12\,L = 2.0\,D_{yy}$ and $0.17\,L = 2.85\,D_y$ for the components of stress σ_e and σ_{yy}, respectively, when $n = 4$. The region of asymptotic validity is estimated to be $0.16\,L$ and $1.29\,D_y$ and $0.43\,L = 3.84\,D_y$ for the same components of stress for $n = 6$. It should be noted that the region around $\theta = \pi/2$

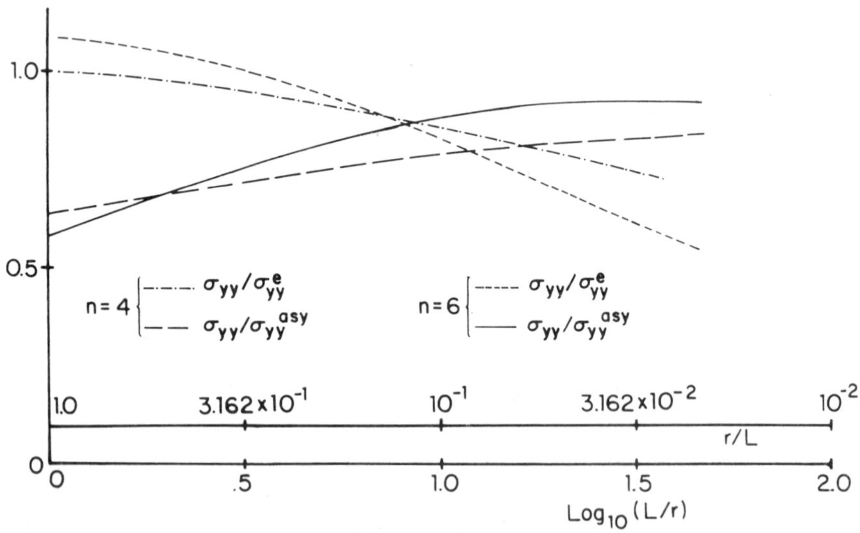

FIG. 2—*Plots of* $\sigma_{yy}/\sigma_{yy}^{asy}$ *and* $\sigma_{yy}/\sigma_{yy}^{e}$ *ahead of a steadily growing plane-strain crack as a function of* r/L *for* n = *4.6.*

is where the strain rates are the highest. An interpolation formula, identical in form to Eq 17, can be written as a function of θ [7] and we have found that they work as well for $\theta = \pi/2$ as for $\theta = 0$. The region of asymptotic validity of σ_{xx} and σ_e ahead of the crack tip cannot be resolved as successfully as σ_{yy}. Close to the crack tip, the finite-element result for σ_{xx} deviates about 20 percent from the asymptotic result given by Eq 5.

We will next examine the variation of the equivalent stress field σ_e in a region close to the crack tip. Contours of constant nondimensional equivalent stress $\hat{\sigma}_e$ are given in Fig. 3 for the cases of $n = 4$ and $n = 6$. The constant chosen is representative of typical values of equivalent stresses in the creep zone. The solid curves in Fig. 3 are obtained from the finite-element computation of $\hat{\sigma}_e$, while the dotted curve in the same figure is obtained using the nondimensional asymptotic equivalent stress, shown in the Appendix, for $\hat{\sigma}_e$. Finally, one notices that there is considerable similarity between the curves of $\hat{\sigma}_e$ = constant in Fig. 3 and the same curves obtained from the analysis of a stationary crack in an elastic-plastic material.

Crack Opening Displacement

Plots of the y-component of the nondimensional displacement on the crack face, $\delta = 2u_\theta\,(r, \theta = \pi)$, are given in Fig. 4 for the case of $n = 4$ and $n = 6$, respectively. The nondimensional elastic crack opening displacement calculated using the same applied stress-intensity factor is included in both plots to serve as a comparison. Figure 4 shows that the crack opening displacement during crack growth is considerably less than that of a stationary crack in an elastic material. Riedel and Rice [*1*] show that the crack opening displacement of a stationary crack in an elastic power-law creeping material varies as $r^{+1/(n+1)}$ close to the crack tip after the initial elastic response. The asymptotic behavior of a steadily growing crack, given by Eq 7, shows that the crack opening displacement varies as $r^{+(n-2)/(n-1)}$ as $r \rightarrow 0$. The foregoing consideration shows that the crack opening displacement during crack growth is always less singular than when the crack is stationary. Similar behavior is also known in elastic-plastic crack growth. A log-log plot of the crack opening displacement versus r from Eq 6 should have a slope of $(n - 2)/(n - 1)$ close to the crack tip. Using results from the four elements closest to the crack tip on the crack face, this slope is found to be 0.66 and 0.76 for the case of $n = 4$ and $n = 6$, respectively. This compares well with the theoretical values of ⅔ and 0.8.

Imposing a Crack Growth Criterion on a Steadily Growing Crack Under Plane-Strain Conditions

As discussed in the previous section, a fracture criterion is needed to relate the steady-state velocity to the applied stress-intensity factor. In this section,

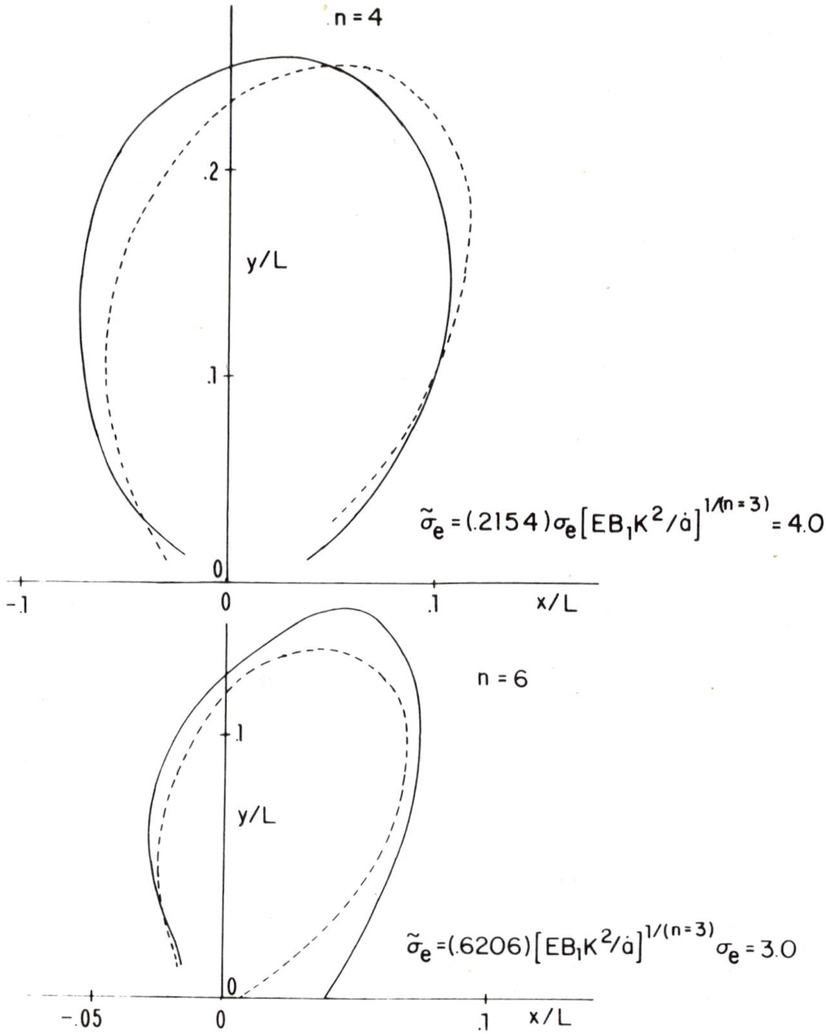

FIG. 3—*Plots of contour of* σ_e = *constant for a steadily growing plane strain crack for* n = *4,6.*

an analysis is carried out for a particular fracture criterion discussed in the following. The analysis reveals that, at a high growth rate, $\dot{a} \propto K^n$; furthermore, there exists a threshold stress intensity factor, K_{min}, below which no stable steady crack growth is possible. The dependence of the steady creep zone size on K is also discussed.

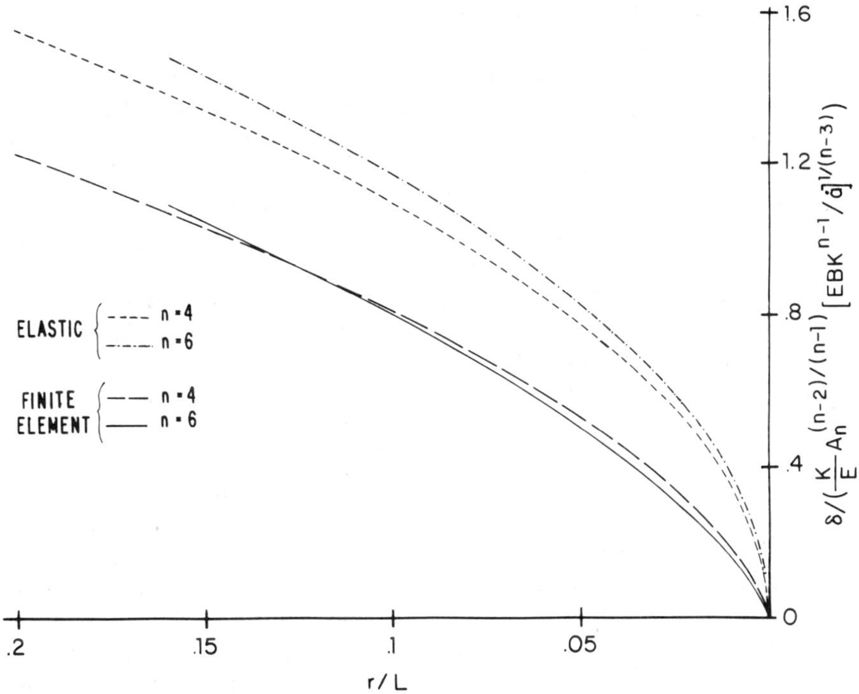

FIG. 4—*Plots comparing normalized elastic and asymptotic crack opening displacement for a steadily growing plane strain crack for* n = *4, 6.*

Crack Growth Criterion

Under conditions of high temperature, one common mechanism of crack growth in structural materials is growth and coalescence of grain boundary voids. The growth of boundary voids has recently been a subject of intensive investigation. A number of theories have been proposed [*8–10*] and pursued to various degrees, yet comprehensive understanding is still in a fragmentary stage. Above about half the melting point of most structural metals, voids grow by diffusion, by power-law creep, or by a combination of both mechanisms. In general, void growth is determined by the coupling of the mechanisms of diffusion and power-law creep. However, under conditions of sufficiently low stresses, power-law creep becomes less important as far as the kinetics of void growth is concerned; that is, locally, the void grows by the mechanism of diffusion. The local dominance of the diffusion mechanism, however, does not necessarily imply that the actual void growth rate can be calculated on the basis of diffusion alone. As Dyson first pointed out [*11*], growth of a void is necessarily dependent on the surrounding grains which have to deform continuously to accommodate the matter diffusing away from the growing void. Rice [*12*] quantified Dyson's concept and showed that, in a material where cavitated grain boundaries are relatively isolated from each other, the overall growth rate is determined in the low stress limit by the overall creep rate and not by the kinetics of the diffusive cavitation process. At the other extreme, when stresses are sufficiently high, the mechanism of void growth is controlled by power-law creep. Under such conditions, the overall void growth rate is dependent on geometrical factors such as void shape and the interaction between neighboring voids as well as the state of stress near the void [*13*]. So far, most investigations of void growth have assumed relatively simple geometrical and loading conditions; therefore, very little is known about void growth near the tip of a growing crack.

In light of this, results of void growth derived under relatively simple geometrical conditions are used in the ensuing analysis. We further assume that the conditions near the steadily growing crack are such that the void growth mechanism is controlled by power-law creep. Later, brief comments are made about the possibility of including different mechanisms of void growth in the analysis of creep crack growth.

In the following analysis, which is similar to that recently worked out by Riedel and Wagner [*7*], we use a result obtained by Budiansky, Hutchinson, and Slutsky [*13*]. In their recent paper, they found that the volume growth rate $\dot{V}$ of a single isolated spherical void of radius b in a power-law creeping material under conditions of high triaxiality and axial symmetric loading is very well approximated by

$$\frac{\dot{V}}{V} = B_1 \left[\frac{3}{2n} \sigma_m + g_n \sigma_e \right]^n \tag{20}$$

where

$$g_n = \frac{(n-1)\,[n + 0.4319]}{n^2} \tag{21}$$

where

σ_m = remote applied mean stress,
σ_e = remote equivalent stress, and
n = creep exponent.

It should be noted that Eq 20 ignores the interactive effects between voids which can be significant even at relatively low void volume fraction for nonlinear materials. Equation 20 does not take into account the effects of shape change during void growth which may accelerate coalescence. Hence, the following analysis, which is based on Eq 20, should be regarded as a first attempt to deduce some qualitative features of creep crack growth.

Following Nix, Matlock, and Dimelfi [*14*], the steady-state velocity $\dot{a}$ is chosen such that a spherical cavity of radius b_0 far from the crack will have grown to half the cavity spacing λ by the time the cavity is within a distance of λ from the crack tip (that is, $b = \lambda$). Using Eq 20, the crack growth rate of a spherical void of radius b ahead of the crack tip is

$$\frac{\dot{b}}{b} = \frac{B_1}{3}\left[\frac{3}{2n}\sigma_m + g_n\sigma_e\right]^n \tag{22}$$

Assuming that σ_m and σ_e are given by the steady-state stress distribution (Eq 11) and using the incompressibility condition, we have

$$\begin{aligned}\sigma_m(r,\theta) &= \left(\frac{\dot{a}}{EB_1K^2}\right)^{1/(n-3)}[\Sigma_{xx}(\tilde{R},\theta) + \Sigma_{yy}(\tilde{R},\theta)]/2 \\ \sigma_e(r,\theta) &= \left(\frac{\dot{a}}{EB_1K^2}\right)^{1/(n-3)}\frac{\sqrt{3}}{2}[\Sigma_{yy}(\tilde{R},\theta) - \Sigma_{xx}(\tilde{R},\theta)]\end{aligned} \tag{23}$$

Using Eq 23 for $\theta = 0$ and the expression $dx = -\dot{a}\,dt$, Eq 22 can be integrated subject to the condition $b = \lambda$ with the result

$$\frac{B_1}{3\dot{a}}\ln\left(\frac{\lambda}{b_0}\right) = \int_\lambda^\infty\left(\frac{3}{2n}\sigma_m + g_n\sigma_e\right)^n dx \tag{24}$$

Equation 24 can be rearranged to

$$\frac{E\sqrt{\lambda}\gamma_c}{K} = F(\tilde{\lambda}) \tag{25}$$

where

$$F(\tilde{\lambda}) = \frac{\tilde{\lambda}^{1/2}}{3}\int_{\tilde{\lambda}}^\infty\left(\frac{3}{2n}\Sigma_m + g_n\Sigma_e\right)^n d\tilde{r} \tag{26}$$

with

$$\lambda_c \equiv \ln\left(\frac{\lambda}{b_0}\right) \tag{27}$$

and where $\tilde{\lambda}$ is the nondimensional cavity spacing given by

$$\tilde{\lambda} = \left(\frac{EB_1K^{n-1}}{\dot{a}}\right)^{-2/(n-3)}\lambda$$

Equation 25 is a mathematical statement of our fracture criterion and it gives $\dot{a}$ implicitly as a function of K. The behavior of $F(\tilde{\lambda})$ for small $\tilde{\lambda}$ can be obtained by the near-tip asymptotic behavior of σ_e and σ_m given by Eq 13*b*. Similarly, the behavior of $F(\tilde{\lambda})$ for large $\tilde{\lambda}$ can be obtained from the far elastic behavior of σ_e and σ_m given by Eq 13*a*. This is found to be

$$F(\tilde{\lambda}) = \frac{(n-1)\alpha_n{}^n}{3}\left(\frac{3}{2}\right)^{n/(n-1)}\left[\frac{3}{2n}\,\sigma_m + g_n\hat{\sigma}_e\right]^n \tilde{\lambda}^{(n-3)/2(n-1)} \qquad \tilde{\lambda} \to 0 \tag{28}$$

$$= \left(\frac{2}{3}\right)^{n-1}\left(\frac{1}{n\sqrt{2\pi}}\right)^n \frac{\tilde{\lambda}^{-(n-3)/2}}{(n-2)} \qquad \tilde{\lambda} \to \infty \tag{29}$$

with $\hat{\sigma}_m = (\hat{\sigma}_{xx}(\theta = 0) + \hat{\sigma}_{yy}(\theta = 0))/2$.

Thus we see that the behavior of $F(\tilde{\lambda})$ for small $\tilde{\lambda}$ is controlled by the near-tip field and that of large $\tilde{\lambda}$ is controlled by the far elastic field. Furthermore, Eqs 28 and 29 show that $F(\tilde{\lambda}) \to 0$ for large and small $\tilde{\lambda}$; hence $F(\tilde{\lambda})$ attains a maximum F^* at some $\tilde{\lambda}$ which will be denoted by $\tilde{\lambda}^*$. A plot of $F(\tilde{\lambda})$ versus $\tilde{\lambda}$ is given in Fig. 5 for the case of $n = 6$. The value of $F(\tilde{\lambda})$ was obtained by

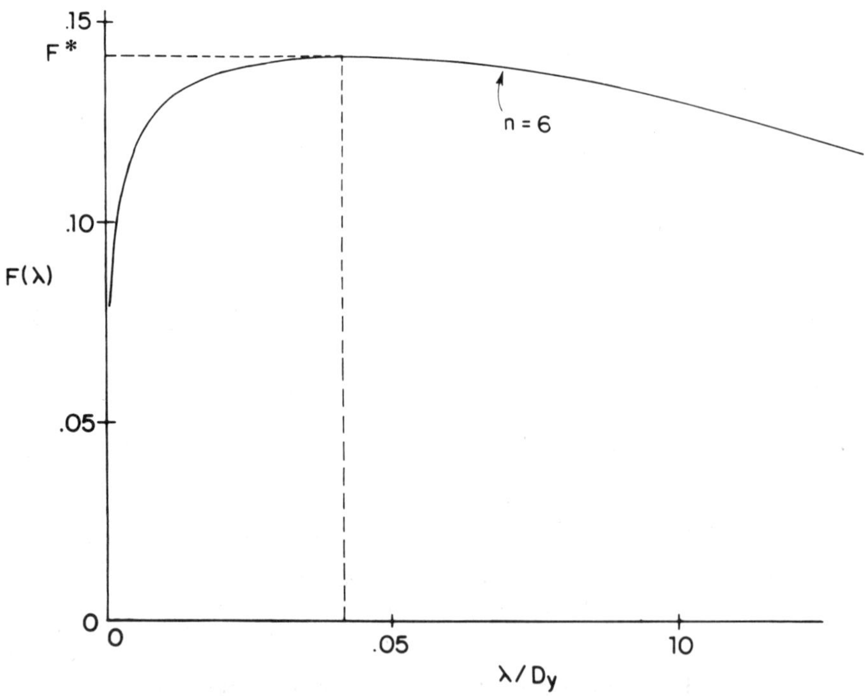

FIG. 5—*Plot of* F($\tilde{\lambda}$) *defined by Eq 26 as a function of* $\tilde{\lambda}$ *for* n = 6.

numerically integrating Eq 26 using the interpolation formula. F^* and $\tilde{\lambda}^*$ are found to be 0.133 and $4.34 \times 10^{-3} h_n$, respectively, for the case $n = 6$. For the case $n = 4$, F^* and $\tilde{\lambda}^*$ are found to be 0.038 and $3.98 \times 10^{-4} h_n$. According to Eq 25, a maximum of $F(\tilde{\lambda})$ corresponds to a minimum value of K, that is

$$K_{\min} = \frac{E\gamma_c\lambda^{1/2}}{F^*} \tag{30}$$

below which no solution exists for Eq 25. Steady-state crack growth is not possible according to this model for $K < K_{\min}$. The minimum steady-state velocity $\dot{a}_{\min}$ corresponding to $K_{\min}$ can be obtained from the definition of $\tilde{\lambda}$

$$\dot{a}_{\min} = \left(\frac{\tilde{\lambda}^*}{\lambda}\right)^{(n-3)/2} EB_1K_{\min}^{n-1} = (\tilde{\lambda}^*)^{(n-3)/2}(\gamma_c/F^*)^{n-1}E^n B_1\lambda \tag{31}$$

Figure 5 shows that, for a given $K > K_{\min}$, two values of $\tilde{\lambda}$, and therefore two velocities $\dot{a}$, result from Eq 25. These give rise to the two branches of the $\dot{a}(K)$ versus K curve plotted in Fig. 6. The lower branch of the $\dot{a}(K)$ curve is obtained from solving Eq 25 in the region $\tilde{\lambda} < \tilde{\lambda}^*$ whereas the upper branch is

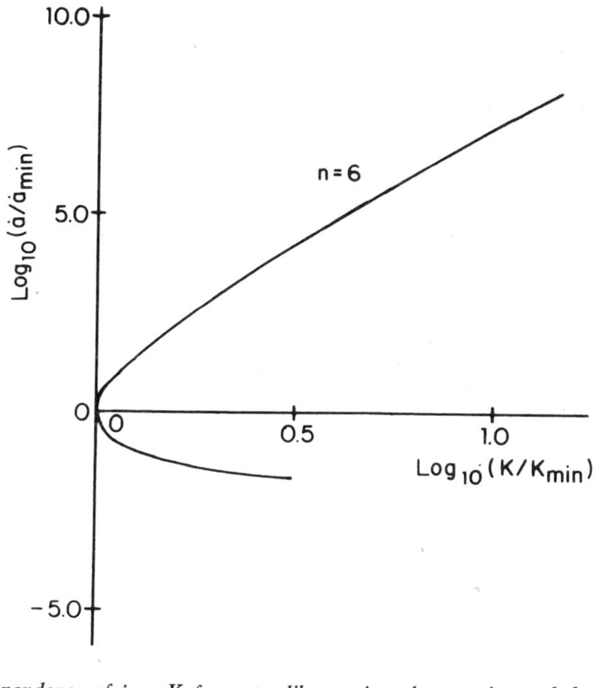

FIG. 6—*Dependence of* ȧ *on* K *for a steadily growing plane-strain crack based on a crack growth criterion for* n = 6.

obtained from solving the same equation in the region $\tilde{\lambda} > \tilde{\lambda}^*$. As in the case of Mode III [5], the lower branch is unstable in the sense that, for a fixed $K > K_{min}$, a slight increase in the velocity of the crack will cause a void to grow beyond its critical size, thus causing further acceleration of the crack until the higher velocity is reached on the upper stable branch. Using the definition of $\tilde{\lambda}$, the condition $\tilde{\lambda} > \tilde{\lambda}^*$ is equivalent to the following condition on cavity spacing

$$\lambda > \frac{\tilde{\lambda}^*}{h_n} L \tag{32}$$

where L is the maximum extent of the creep zone size defined by Eq 15. With the aid of the definition of D_y and L, Eq 32 can be rewritten as

$$\lambda > 6.8 \times 10^{-3} D_y = 3.98 \times 10^{-4} L \qquad n = 4 \tag{33}$$

$$\lambda > 3.5 \times 10^{-2} D_y = 4.34 \times 10^{-3} L \qquad n = 6 \tag{34}$$

Thus, Eqs 33 and 34 are necessary conditions for steady-state crack growth. Physically, Eq 33 or Eq 34 implies that λ cannot be too small compared with the region of asymptotic dominance for steady-state crack propagation.

The relationship between the steady-state creep zone size and the applied K field will now be considered. In general, Eq 15 indicates that the creep zone size depends on both $\dot{a}$ and K. The branch $\tilde{\lambda} > \tilde{\lambda}^*$ of Fig. 5 shows that, for $K > K_{min}$, the ratio $\dot{a}/K^{n-1}$ has to increase as K increases in order to satisfy our particular fracture criterion (Eq 25). This means that $\dot{a}$ varies faster than K^{n-1} on the upper branch of Fig. 6 and thus the creep zone size defined by Eq 15 must decrease as K increases in this model. Indeed, at $K = K_{min}$, the creep zone reaches its maximum extent. In the opposite limit when $K \gg K_{min}$, creep effects are confined to a zone small compared with λ; therefore, the far elastic field exclusively determines the strain at λ. Then, from Eq 29, a power-law relation between $\dot{a}$ and K follows

$$\dot{a} = \left(\frac{2}{3}\right)^{n-1} \left(\frac{1}{n\sqrt{2\pi}}\right)^{n} \frac{B_1 K^n}{(n - 2)\gamma_c} \lambda^{-(n-2)/2} \tag{35}$$

Figure 6 shows that the limit $\dot{a} \propto K^n$ is attained rather rapidly. At $K = 3.2\ K_{min}$, the slope of the $\log_{10} \dot{a}$-versus-$\log_{10} K$ curve differs no more than 4 percent from the asymptotic value of 6 as given by Eq 35.

In general, the size of the steady-state creep zone can be estimated by the dependence of $\dot{a}$ on K. Since many experiments report the result $\dot{a} = CK^m$, where m and C are taken to be material parameters, we will use it to estimate the dependence of creep zone size on the applied stress-intensity factor. With $\dot{a} = CK^m$, L, the maximum creep zone extent, becomes

$$L = h_n \left(\frac{EB_1 K^{n-m-1}}{C} \right)^{2/(n-3)} \tag{36}$$

This again shows that, if the crack growth rate $\dot{a}$ varies faster than K^{n-1}, the size of the creep zone decreases with increasing K while the converse is true if $\dot{a}$ varies slower than K^{n-1}.

Finally, we will consider briefly the effect of void interaction on the crack growth rate's dependence on K. The interactive effect between neighboring voids may be investigated by including a factor that depends on the current void volume fraction ρ in the volume growth rate equation (Eq 20). Guided by the result of Ref *13*, Eq 20 can be modified as

$$\frac{\dot{V}}{V} = B_1 \left(\frac{3}{2n} \sigma_m + g_n \sigma_e \right)^n f(\rho) \tag{37}$$

where the modification $f(\rho) = (1 - \rho^{1/2})^{-n}$ was obtained by Ref *13* for the special case of a spherical void geometry under radial loading conditions. Letting $\rho = (b/2\lambda)^3$, and replacing Eq 20 by Eq 37 in the derivation of Eq 25, it is found that all the results from Eq 25 to Eq 34 remain unchanged provided γ_c is redefined by

$$\gamma_c = \ell n \frac{\lambda}{b_0} - \frac{1}{3} \sum_{k=1}^{n} \frac{n}{(n-k+1)} \left(1 - \left(\frac{b_0}{2\lambda} \right)^{3/n} \right)^{2(n-k)+2} \left(1 - \left(\frac{1 - (\frac{1}{2})^{3/n}}{1 - (b_0/2\lambda)^{3/n}} \right)^{2(n-k)+2} \right) \tag{38}$$

It can be verified that γ_c defined by Eq 38 is positive and has a value smaller than the old $\gamma_c = \ell n\,(\lambda/b_0)$ defined previously by Eq 27. With $(\lambda/b_0) = 5$ and $n = 6$, it is seen that the ratio of the new γ_c defined by Eq 38 over γ_c defined by Eq 27 is approximately ⅓—a reduction of the old γ_c by 66 percent. The interaction effect of voids thus tends to decrease γ_c and thus decreases the value of $K_{\min}$ and $\dot{a}_{\min}$, as can be seen from Eqs 30 and 31. The value of γ_c may be further reduced if the change of void shape is taken into account, as Ref *13* points out that, even with the same volume growth rate, the process of coalescence would be accelerated if the void shape were an oblate spheroid.

In conclusion, the growth criterion in the foregoing lends some useful insight in the analysis of creep crack growth. It points out that, since a void in general spends a non-negligible amount of time outside the region of asymptotic dominance of the near-tip fields, the complete stress field as well as the near-tip field have to be included in the steady-state analysis. In the limiting case when the void spacing is so small such that Eq 32 is not satisfied, steady-state growth is not possible.

Finally, a few comments on the effect of diffusion on creep crack growth. As mentioned earlier in this section, neglect of diffusion would in general underestimate the void growth rate. In the extreme case where the void growth rate is completely determined by diffusion, the void growth rate is proportional to the applied stresses, as suggested by Hull and Rimmer [*8*]. In the previous discussion we have shown that the contribution to the void growth outside the region of dominance of the near-tip fields cannot be ignored in general. By assuming that the void growth rate is everywhere given by that of Hull and Rimmer and thus ignoring the "constraint effect" suggested by Dyson which was discussed earlier in this section, it is found that the void growth rate due to the far elastic stress field is not integrable. That is, it is not possible to integrate in from infinity the void growth rate due to the $1/\sqrt{r}$ stress field. Thus, steady-state crack growth is not possible due to the high void growth rate caused by the far field.

We therefore concluded that the "constraint effect" suggested by Dyson cannot be ignored. Rice [*12*] recently quantified Dyson's concept of constraint cavity growth by assuming that the material considered is in a state of uniaxial tension. However, no calculation has yet been done using a more complex state of stress. Further analysis of void growth in this direction will be extremely useful for gaining further insight into the problem of creep crack growth.

Acknowledgments

I am indebted to my advisor, John W. Hutchinson, who suggested this problem and provided excellent supervision. I would like to thank R. Dean, who kindly permitted me to use his finite-element program in this investigation. Finally, I would like to thank F. McClintock and John Bassani for their helpful comments during the initial stage of this work. This work was supported in part by the National Science Foundation under Grants ENG-78-10756 and DMR-79-23597, and by the Division of Applied Science of Harvard Univeristy.

APPENDIX

Nondimensional Variables

Nondimensional variables for plane strain are

$$x_i = A_n \left(\frac{EB_1K^{n-1}}{\dot{a}} \right)^{2/(n-3)} \tilde{x}_i \tag{39}$$

$$u_i = A_n{}^{(n-2)/(n-1)} \frac{K}{E} \left(\frac{EB_1K^{n-1}}{\dot{a}} \right)^{1/(n-3)} \tilde{u}_i \tag{40}$$

$$\epsilon_{ij} = \frac{A_n^{-1/(n-1)}}{E}\left(\frac{\dot{a}}{EB_1K^2}\right)^{1/(n-3)} \tilde{\epsilon}_{ij} \tag{41}$$

$$\sigma_{ij} = A_n^{-1/(n-1)}\left(\frac{\dot{a}}{EB_1K^2}\right)^{1/(n-3)} \tilde{\sigma}_{ij} \tag{42}$$

Numerical calculations were facilitated by the preceding scale factor A_n which allows the maximum nondimensional creep zone extent in the finite-element computation to be of order 0.1. (The nondimensional crack length is of order 1.)

Using Eqs 39 through 41, the far-field boundary condition (Eq 8) is nondimensionalized as

$$\tilde{\sigma}_{ij}(r \to \infty, \theta) = \frac{A_n^{-(n-3)/2(n-1)}}{\sqrt{2\pi\tilde{r}}} \hat{f}_{ij}(\theta) \tag{43}$$

where

$$\tilde{r} \equiv (\tilde{x}_i\tilde{x}_i)^{1/2}$$

The value of $A_n = 0.010$ is used for the case $n = 4$ whereas $A_n = 0.092$ is used for the case of $n = 6$.

The nondimensional near-tip asymptotic stresses are independent of A_n, and they are

$$\tilde{\sigma}_{ij}(\tilde{r}, \theta) = \left(\frac{3}{2}\right)^{1/(n-1)} \alpha_n \tilde{r}^{-1/(n-1)} \hat{\sigma}_{ij}(\theta)$$

$$\tilde{\sigma}_{ij}(\tilde{r}, \theta) = \left(\frac{3}{2}\right)^{1/(n-1)} \alpha_n \tilde{r}^{-1/(n-1)} \hat{\sigma}_{e}(\theta) \tag{44}$$

The finite-element method used is discussed briefly next; most details remain exactly the same as those of Dean [15]. A displacement-based finite-element method is employed in the analysis. In standard vector notation, the variational equations of equilibrium are

$$\int \vec{\sigma}^T \, \partial\vec{\epsilon} dA = \int \vec{F}^T \, \partial\vec{u} ds \tag{45}$$

where dA is an area element and the traction from Eq 43 is applied in a circuit with a line element ds far from the crack tip. The stresses may be replaced in Eq 45 by

$$\vec{\sigma} = \mathbf{D}^e(\vec{\epsilon} - \vec{\epsilon}^{\,cp}) \tag{46}$$

where $\mathbf{D}^e$ is the elastic constitutive matrix, and $\vec{\epsilon}^{\,cp}$ is the creep strain vector. The strains and displacements in Eq 45 are given in terms of nodal displacements $\vec{u}$ by

$$\vec{\epsilon} = \mathbf{B}\vec{u} \quad \text{and} \quad \vec{u} = \mathbf{R}\vec{u} \tag{47}$$

where small-strain theory is assumed. With

$$\mathbf{K}^e = \int \mathbf{B}^T\mathbf{D}^e\mathbf{B}\, dA$$

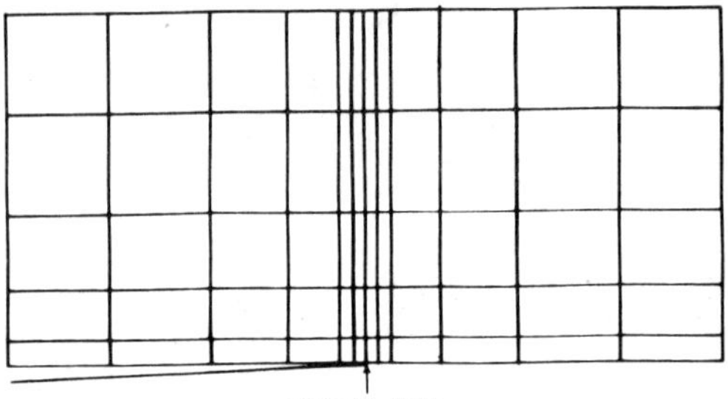

FIG. 7—*Coarse representation of finite-element mesh.*

as the elastic stiffness matrix, the nodal displacement must satisfy

$$\mathbf{K}^e u = \int \mathbf{B}^T \vec{F}\, ds + \int \mathbf{B}^T \mathbf{D}^e \vec{\epsilon}^{cp}\, dA \tag{48}$$

The procedure for solving Eq 45 is as follows:

1. Given the estimate of $\vec{\epsilon}^{cp}$ from the $(i - 1)$th iteration, use it in Eq 48 to compute the ith estimate of $\vec{u}$, $\vec{u}^{(i)}$.
2. Compute $\vec{\epsilon}^{(i)}$ everywhere from $u^{(i)}$.
3. Using the steady-state relation $(\partial/\partial t) = -\dot{a}(\partial/\partial x)$ and the constitutive law (Eq 2), integrate in the negative x-direction through the elastic creep elements to obtain $\vec{\sigma}^{(i)}$ where $\delta\epsilon_{mn}/\delta x$ is obtained by taking approximate differences in $\vec{\epsilon}^{(i)}$ between elements.
4. Compute the ith estimate of the creep strain using

$$\vec{\epsilon}^{cp} = \vec{\epsilon} - (\mathbf{D}^e)^{-1}\vec{\sigma}$$

5. Repeat steps 1 through 4 until convergence is achieved.

The basic element used in the finite-element analysis is the constant-strain triangle. The finite-element mesh, shown schematically in Fig. 7, is composed entirely of rectangles formed from triangles with the center node condensed out. The mesh is symmetrical about the y-axis. The first element on the right-hand side of each row is taken to be purely elastic. In Mode I, the smallest element increases in both horizontal and vertical directions by a ratio of about 1.4 after the second element. The length of the smallest element is chosen to be 0.01 h_n for the case of $n = 4$, where h_n is the nondimensionalized maximum creep extent defined by Eq 15. For $n = 6$, the length of the smallest element is chosen to be 0.008 h_n.

References

[1] Riedel, H. and Rice, J. R. in *Fracture Mechanics: Twelfth Conference, ASTM STP 700*, American Society for Testing and Materials, 1980, pp. 112–130.

[2] Riedel, H., *Journal of the Mechanics and Physics of Solids*, Vol. 29, 1981, pp. 35–49.

[3] Bassani, J. L. and McClintock, F. A., *Reports of Research in Mechanics and Materials*, Massachusetts Institute of Technology, Cambridge, Mass., 1980.

[4] Hutchinson, J. W., *Journal of the Mechanics and Physics of Solids*, Vol. 16, 1968, pp. 13–21.

[5] Hui, C. Y. and Riedel, H., *International Journal of Fracture*, Vol. 17, 1981, pp. 409–425.
[6] Hui, C. Y., Ph.D. Thesis, Harvard University, Cambridge, Mass., 1981.
[7] Riedel, H. and Wagner, W. in *Proceedings*, Fifth International Conference of Fracture, Cannes, France, D. Francois, Ed., Pergamon Press, New York, 1981, pp. 683–690.
[8] Hull, D. and Rimmer, D. E., *Philosophical Magazine*, Vol. 4, 1959, pp. 673–684.
[9] Chuang, T. J., Kagawa, K. I., Rice, J. R., and Sills, L. B., *Acta Metallurgica*, Vol. 27, 1979, pp. 265–284.
[10] Edward, G. H. and Ashby, M. F., *Acta Metallurgica*, Vol. 27, 1979, pp. 1505–1518.
[11] Dyson, B. F., *Metal Science*, Vol. 10, 1976, pp. 349–353.
[12] Rice, J. R., *Acta Metallurgica*, Vol. 29, 1981, pp. 675–681.
[13] Budiansky, B., Hutchinson, J. W., and Slutsky, S., Harvard University Report, Division of Applied Sciences, MECH-15, Feb. 1980. To be published in *Mechanics of Solids, The Rodney Hill 60th Anniversary Volume*, H. G. Hopkins and M. J. Sewell, Eds., Pergamon Press, New York 1981.
[14] Nix, W. D., Matlock, D. K. and Dimelfi, R. J., *Acta Metallurgica*, Vol. 25, 1977, pp. 495–503.
[15] Dean, R. H. and Hutchinson, J. W. in *Fracture Mechanics: Twelfth Conference, ASTM STP 700*, American Society for Testing and Materials, 1980, pp. 383–405.
[16] McMeeking, R. M. and Leckie, F. A in *Proceedings*, Fifth International Conference on Fracture, Cannes, France, D. Francois, Ed., Pergamon Press, New York, 1981, pp. 699–704.

Shiro Kubo[1]

Effects of Creep Recovery and Hardening on the Stress and Strain-Rate Fields Near a Crack Tip in Creeping Materials

REFERENCE: Kubo, S., **"Effects of Creep Recovery and Hardening on the Stress and Strain-Rate Fields Near a Crack Tip in Creeping Materials,"** *Elastic-Plastic Fracture: Second Symposium, Volume I—Inelastic Crack Analysis, ASTM STP 803*, C. F. Shih and J. P. Gudas, Eds., American Society for Testing and Materials, 1983, pp. I-594–I-614.

ABSTRACT: It is thought that creep recovery may be one of the main causes of acceleration of crack growth rates under variable loads. To explore this possibility, the effects of creep recovery and hardening on the near-tip fields are examined in this paper. Robinson's constitutive relation, which is based on the Bailey-Orowan model, is used to take these effects into account. In this model, the strain rate is proportional to the nth power of an effective stress, which is the difference between applied stress and internal stress. The rate of the internal stress is given by the difference between a hardening term and a recovery term, which are proportional to the strain rate divided by the βth power of the internal stress and the $(n - \beta)$th power of the internal stress, respectively.

It is found that the Hutchinson-Rice-Rosengren-type singular fields prevail near a creep crack tip irrespective of the combination of the material parameters n and β. When $\beta < n - 1$, the "steady-state creep conditions" prevail near the crack tip; when $\beta > n - 1$, the "initial-state creep conditions" prevail, under which the internal stress is small compared with the applied stress.

The intensities of the singular stress and strain-rate fields are obtained rigorously for the case of $\beta = n - 1$. For the case of $\beta < n - 1$ and $\beta > n - 1$, the intensities are estimated by using a hypothesized path independence of the J-integral or the C*-integral. For any combination of n and β, the intensity of strain rate is very high immediately after load increase. Then the strain rate decreases and approaches its steady-state value. On the other hand, for load decrease the strain rate is low immediately after load decrease, and increases until it attains the steady-state value.

From a survey of the values of n and β, the case of $\beta < n - 1$ seems to be prevailing for metallic creep. In this case C^* can be a characterizing parameter of the near-tip fields. The observed acceleration and deceleration of crack growth rates under variable loads are found to be explained by the present analysis.

[1]Visiting research associate (1980–81), Division of Engineering, Brown University, Providence, R.I. 02912; also a permanent affiliation with Department of Mechanical Engineering, Osaka University, Yamada-oka, Suita, Osaka 565 JAPAN.

KEY WORDS: elastic-plastic fracture, nonlinear fracture mechanics, creep, recovery and hardening, stress and strain analysis, Hutchinson-Rice-Rosengren (HRR) singular fields, fracture (materials), variable loads, crack growth

During the past ten years increasing attention has been paid to the crack growth problems at elevated temperatures [*1–13*].[2] The crack stress analysis and crack growth characterization under the extensive and steady-state creep conditions are the easiest to handle in these problems and are treated successfully by using the nonlinear fracture mechanics approach.

Under the extensive and steady-state creep conditions the strain rate $\dot{\boldsymbol{\epsilon}}$ is given by a function of the current stress $\boldsymbol{\sigma}$[3]

$$\dot{\boldsymbol{\epsilon}} = \frac{3}{2} B \sigma_e^{n-1} \boldsymbol{\sigma}' \tag{1}$$

Here the equivalent tensile stress σ_e is given by

$$\sigma_e = \left(\frac{3}{2} \boldsymbol{\sigma}' : \boldsymbol{\sigma}' \right)^{1/2} \tag{2}$$

where B and n are material constants. The stress and strain-rate fields near a crack tip are given by the Hutchinson-Rice-Rosengren (HRR)-type singular fields [*14–16*] and the intensities are expressed in terms of the path-independent integral C^* [*5–7*][4]

$$\boldsymbol{\sigma} = K_\sigma \tilde{\boldsymbol{\sigma}}(\theta, n)\, r^{-1/(n+1)} \tag{3}$$

$$\dot{\boldsymbol{\epsilon}} = K_{\dot{\epsilon}} \tilde{\boldsymbol{\epsilon}}(\theta, n)\, r^{-n/(n+1)} \tag{4}$$

$$\begin{aligned} K_\sigma &= [C^*/(I_n B)]^{1/(n+1)} \\ K_{\dot{\epsilon}} &= B[C^*/(I_n B)]^{n/(n+1)} \end{aligned} \tag{5}$$

Here, K_σ and $K_{\dot{\epsilon}}$ are the plastic stress-intensity factor and strain-rate intensity factor, respectively. Polar coordinates (r, θ) are taken with the origin at the crack tip. The angular functions $\tilde{\boldsymbol{\sigma}}(\theta, n)$ and $\tilde{\boldsymbol{\epsilon}}(\theta, n)$ are given schematically for the Mode I crack, plane-stress and plane-strain cases [*14–16*]. An exact solution for the Mode III crack cases is given by Rice [*17*]. I_n is a constant which depends on n only.

[2]The italic numbers in brackets refer to the list of references appended to this paper.

[3]In this paper, bold face quantities are tensors. A dot between two tensors indicates summation over one index; a double dot indicates summation over two indices. A prime denotes the deviatoric part of a three-dimensional tensor.

[4]The term "modified J-integral" [*6, 7*] is used in Japan instead of the C*-integral.

It is becoming recognized that the creep crack growth rates in ductile materials under static loads correlate well with the experimental value of C^* [4,5,8–10]. In many experiments the C^* is calculated from the load point displacement rate $\dot{\delta}$ (or notch-opening displacement rate) and the net section stress σ_{net} by using formulas proposed by Ohji et al [18]

$$C^* = e(n)\,\dot{\delta}\,\sigma_{net} \tag{6}$$

Here, $e(n)$ is a function of the creep exponent n. For example, $e(n)$ is given by $e(n) = (n - 1)/(n + 1)$ for plate specimens with deep double-edge cracks. The C^*-values obtained experimentally using formulas of this kind coincide reasonably well with the analytically determined values [19–22] based on the assumption of steady-state creep [23].

Crack growth under variable loads is also one of the most important problems in the field of strength of materials at high temperatures. This problem is important especially in practical applications. However, the causes of acceleration of crack growth rates under variable loads have long been in dispute. Recently some researchers examined the applicability of C^* to time-dependent crack growth under variable loads stimulated by the success in the application of C^* to the creep crack growth, in spite of the fact that C^* proposed so far is based on the assumption of steady-state creep and is not formulated for variable loads.

Taira et al [11], Saito et al [12], and Koterazawa and Mori [13] conducted crack growth experiments under cyclic, two-step, and intermittent loads, and reported that time-dependent crack growth rates are accelerated 10 to 1000 times by the variation of loads. However, the correlation between the crack growth rates and the experimentally determined C^* under variable loads is good and agrees reasonably well with this same correlation under static creep conditions. Since the experimental value of C^* is obtained by using formulas such as Eq 6, the good C^*-correlation means that the acceleration of crack growth rates is accompanied by a comparable degree of acceleration of deformation rate $\dot{\delta}$. Furthermore, this experimental result suggests that knowledge about the deformation behavior of cracked specimens under variable loads is essential to the understanding of the crack growth under variable loads, and that creep recovery, which causes accelerated creep deformation, can be one of the main causes of the acceleration of crack growth rates.

To explore this possibility, the effects of creep recovery and hardening on the stress and strain-rate fields near the crack tip are discussed.

Constitutive Equations

We consider a two-dimensional crack embedded in a material which undergoes infinitesimal and incompressible creep deformations.

In many creep analyses, a Norton-type creep law, Eq 1, is used. This creep law prescribes the creep behavior under steady-state creep conditions; Eq 1 predicts that the strain rate is expressed in terms of the current stress $\boldsymbol{\sigma}$ only, which is not true for variable loads. One possible way to describe the creep behavior under variable loads is to employ a creep recovery-hardening model.

In the model the "internal stress" variable α is introduced, which is interpreted as the back stress induced by the dislocation structures. The strain rate is assumed to be a function of the effective stress Σ, which is defined as the difference between the applied stress σ and the internal stress α. Bailey [*24*] and Orowan [*25*] explained the creep as the result of two competing mechanisms, that is, the strain hardening process caused by distortion and recovery caused by thermal action. The rate of change of α is given by the following Bailey-Orowan equation

$$\dot{\alpha} = h(\alpha)\dot{\epsilon} - s(\alpha) \tag{7}$$

Here h and s are functions of α only. This model accounts for recovery and hardening which are observed in creep tests. The creep rate in the primary stage as well as that in the secondary stage is described by this model.

Mathematical representation of the model under multiaxial stress states was given by Pugh and Robinson [*26,27*]. Based on the experimental results of Mitra and McLean [*28*], they assumed that the functions $h(\alpha)$ and $s(\alpha)$ are given by power functions of α. Their representation, with the assumption that the yield stress is negligibly small, reduces to

$$\dot{\boldsymbol{\epsilon}} = \frac{3}{2} A\Sigma_e^{n-1}\boldsymbol{\Sigma} \tag{8}$$

$$\dot{\boldsymbol{\alpha}}' = \frac{2}{3} C\dot{\boldsymbol{\epsilon}}\alpha_e^{-\beta} - D\alpha_e^{n-\beta-1}\boldsymbol{\alpha}' \tag{9}$$

where the second-order tensor $\boldsymbol{\alpha}$ is the "internal stress" variable. The effective stress tensor $\boldsymbol{\Sigma}$ is defined by

$$\boldsymbol{\Sigma} = \boldsymbol{\sigma}' - \boldsymbol{\alpha}' \tag{10}$$

Scalar quantities Σ_e and α_e are

$$\Sigma_e = \left(\frac{3}{2}\,\Sigma:\Sigma\right)^{1/2}, \qquad \alpha_e = \left(\frac{3}{2}\,\alpha':\alpha'\right)^{1/2} \tag{11}$$

A, C, D, n, and β are positive material constants. Elastic strains are ignored for simplicity. The contribution of the elastic strains is discussed in a separate paper [*29*].

As was pointed out by Pugh and Robinson [26,27], the flow law, Eq 8, and the growth law, Eq 9, with $n = 4$ and $\beta = 1$ coincide with the form of physical model formulated on the basis of dislocation theory by Lagneborg [30].

It is useful to consider some special cases of this model. One special case corresponds to the "initial state." In this case the internal stress variable $\boldsymbol{\alpha}$ is equal to zero or very small compared with the applied stress $\boldsymbol{\sigma}'$; accordingly, Eq 8 reduces to

$$\dot{\epsilon} = \frac{3}{2} A \sigma_e^{n-1} \sigma' \tag{12}$$

Another special case is associated with the "steady state." The steady-state creep prevails when $\dot{\alpha} = 0$ under constant applied stress. Then, Eqs 8 and 9 lead to

$$\dot{\epsilon} = \frac{3}{2} (A/H_1)\, \sigma_e^{n-1} \sigma' \tag{13}$$

$$\alpha' = H_2 \sigma' \tag{14}$$

Here

$$H_1 = [1 + (AC/D)^{1/n}]^n, \qquad H_2 = (AC/D)^{1/n}/[1 + (AC/D)^{1/n}] \tag{15}$$

If the factor A/H_1 is replaced by B, Eq 13 is coincident with the Norton-type steady-state creep law, Eq 1.

Stress and Strain Analysis in the Case of $\beta = n - 1$

The solution of stress and strain varies depending on whether β is greater than, equal to, or less than $(n - 1)$. In what follows, therefore, analyses are made for these three separate cases.

In this section stress and strain distribution in the case of $\beta = n - 1$ is discussed.

Consider a cracked specimen subjected to load $P(t)$ varying with time t. Denote steady-state stress and strain-rate solution for unit load as $[\sigma(\boldsymbol{x})]_{SS1}$ and $[\dot{\epsilon}(\boldsymbol{x})]_{SS1}$. The plastic stress-intensity factor and the strain-rate intensity factor under steady-state and unit load conditions are denoted by $(K_\sigma)_{SS1}$ and $(K_{\dot{\epsilon}})_{SS1}$, respectively. When load P is applied, the steady-state solution is given by

$$\sigma = P(\sigma)_{SS1}, \qquad \dot{\epsilon} = P^n(\dot{\epsilon})_{SS1}$$
$$K_\sigma = P(K_\sigma)_{SS1}, \qquad K_{\dot{\epsilon}} = P^n (K_{\dot{\epsilon}})_{SS1} \tag{16}$$

Now we assume that the distribution of $\boldsymbol{\alpha}'(\boldsymbol{x})$ at $t = t_0$ is proportional to $(\boldsymbol{\sigma}')_{SS1}$ at $t = t_0$

$$\boldsymbol{\alpha}'(\boldsymbol{x}) = R_\alpha(\boldsymbol{\sigma}')_{SS1} \tag{17}$$

Here R_α is a constant which does not depend on coordinates. The following stress distribution is examined as a candidate solution at $t = t_0$

$$\boldsymbol{\sigma} = P(t_0)(\boldsymbol{\sigma})_{SS1} \tag{18}$$

This solution satisfies equilibrium equations in the body since $(\boldsymbol{\sigma})_{SS1}$ does. Traction boundary conditions are also satisfied. Substitution of Eqs 17, 18, and 13 into Eqs 8 and 9 gives

$$\dot{\epsilon}(t_0) = \frac{3}{2}A|P - R_\alpha|^{n-1}(P - R_\alpha)\,(\sigma_e)_{SS1}{}^{n-1}(\boldsymbol{\sigma}')_{SS1}$$

$$= H_1|P - R_\alpha|^{n-1}(P - R_\alpha)(\dot{\epsilon})_{SS1} \tag{19}$$

$$\dot{\boldsymbol{\alpha}}'(t_0) = AC|P - R_\alpha|^{n-1}(P - R_\alpha)(\boldsymbol{\sigma}')_{SS1}/|R_\alpha|^{n-1} - DR_\alpha(\boldsymbol{\sigma}')_{SS1} \tag{20}$$

From Eq 19, $\dot{\epsilon}(t_0)$ satisfies compatibility equations of strain, since $(\dot{\epsilon})_{SS1}$ does. Therefore, stress distribution Eq 18 and associated strain-rate distribution Eq 19 are true solutions. Equation 20 shows that the rate of internal variable $\dot{\boldsymbol{\alpha}}'$ is equal to $(\boldsymbol{\sigma}')_{SS1}$ multiplied by a factor which does not depend on coordinates. Then, once the condition Eq 17 is met, this equation holds from that time on even if the scalar quantity R_α varies with time t. Equation 17 is satisfied if we start with the initial state; that is, $R_\alpha = 0$. Under steady-state creep conditions, Eq 17 holds and R_α is given by $R_\alpha = H_2P$. Therefore, important cases—for example, the transition from the initial state or the steady state—can be treated under the condition of Eq 17.

The differential equation for R_α is obtained from Eq 20 as

$$\dot{R}_\alpha = AC|P - R_\alpha|^{n-1}(P - R_\alpha)/|R_\alpha|^{n-1} - DR_\alpha \tag{21}$$

The internal variable parameter R_α, which is defined by Eq 17, can be obtained by solving the differential Eq 21.

From Eq 19 the strain-rate tensor $\dot{\epsilon}$ is parallel to $(\dot{\epsilon})_{SS1}$. Therefore, it is possible to introduce the strain-rate parameter $R_{\dot{\epsilon}}$ and strain parameter R_ϵ, which are defined by

$$\dot{\epsilon} = R_{\dot{\epsilon}}(\dot{\epsilon})_{SS1}, \qquad \epsilon = R_\epsilon(\dot{\epsilon})_{SS1}/(AC) \tag{22}$$

Combining Eq 22 with Eq 19 we obtain

$$R_{\dot{\epsilon}} = H_1|P - R_\alpha|^{n-1}(P - R_a) \tag{23}$$

$$R_\epsilon = H_1 \int |P - R_\alpha|^{n-1}(P - R_\alpha)AC\,dt \tag{24}$$

Near-Tip Fields

In the vicinity of the crack tip the steady-state solutions $(\sigma)_{SS1}$ and $(\dot{\epsilon})_{SS1}$ are expressed by the HRR-type singular fields (Eqs 3 and 4). Therefore, from Eqs 18 and 22 the HRR fields prevail near the crack tip when $\beta = n - 1$. The plastic stress-intensity factor K_σ and strain-rate intensity factor $K_{\dot{\epsilon}}$ under variable loads are expressed by

$$\begin{aligned} K_\sigma &= P(K_\sigma)_{SS1} \\ K_{\dot{\epsilon}} &= H_1|P - R_\alpha|^{n-1}(P - R_\alpha)(K_{\dot{\epsilon}})_{SS1} = R_{\dot{\epsilon}}(K_{\dot{\epsilon}})_{SS1} \end{aligned} \tag{25}$$

Variation of Internal Variable, Strain Rate, and Strain for Step Loading

From the standard dimension analysis, one can conclude that R_α/P_0, $R_{\dot{\epsilon}}/P_0{}^n$, and $R_\epsilon/P_0{}^n$ are functions of ACt, AC/D, and n only, where P_0 is a level of applied load. The variation of these parameters is calculated numerically using the Runge-Kutta method for some representative cases.

The first example is step loading

$$P = 0 \quad \text{for } t < 0, \qquad P = P_0 \quad \text{for } t \geq 0 \tag{26}$$

An example of the solutions of Eqs 21, 23, and 24 with the initial condition $R_\alpha = 0$ at $t = 0$ is shown in Fig. 1. For small t, R_α is small compared with P; then the solution is approximated as

$$R_\alpha/P_0 = (nACt)^{1/n} \tag{27}$$

$$R_{\dot{\epsilon}}/P_0{}^n = H_1[1 - (nACt)^{1/n}]^n \tag{28}$$

$$R_\epsilon/P_0{}^n = H_1 \int_0^t [1 - (nACt)^{1/n}]^n AC\,dt \tag{29}$$

Equations 27–29 are shown by chained lines in Fig. 1. For large t, R_α/P_0 and $R_{\dot{\epsilon}}/P_0{}^n$ approach their steady-state values H_2 and 1, respectively. A characteristic time t_1 of the transition from the initial state to the steady state is defined by equating Eq 27 with the steady-state value of R_α/P_0

$$t_1 = 1/(nDH_1) \tag{30}$$

It can be seen from the figure that the value of $R_{\dot{\epsilon}}$ is very high immediately after the load application. Then the value decreases rapidly and approaches its steady-state value.

Variation of Internal Variable, Strain Rate, and Strain for Step Unloading

The second example concerns step unloading

$$P = P_0 \quad \text{for } t < 0, \qquad P = 0 \quad \text{for } t \geq 0 \tag{31}$$

The exact solution of this problem is

$$R_\alpha/P_0 = H_2 e^{-(AC+D)t} \tag{32}$$

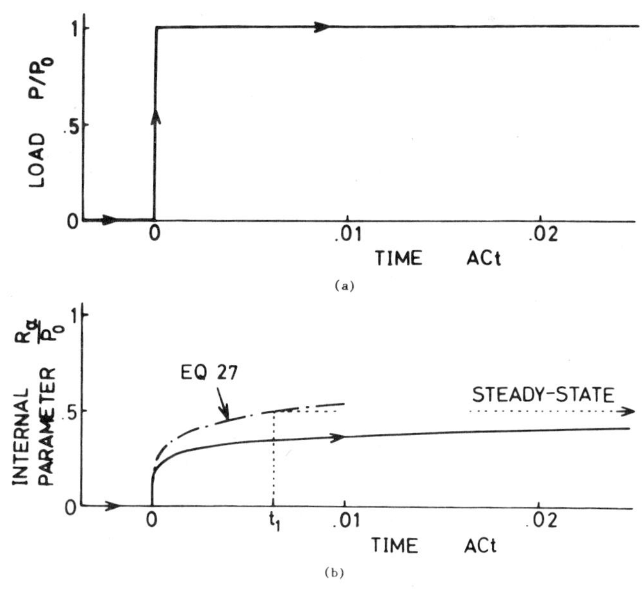

FIG. 1—*Variation of* R_α/P_0, $R_{\dot{\epsilon}}/P_0^n$ *and* R_ϵ/P_0^n *for step loading* ($n = 5$, $\beta = 4$, AC/D = 1).

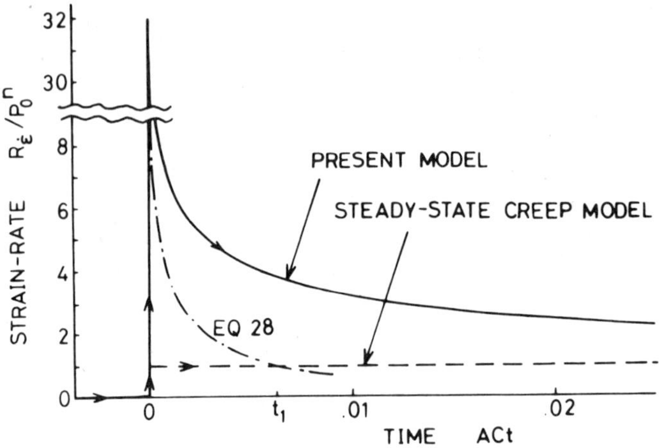

(c)

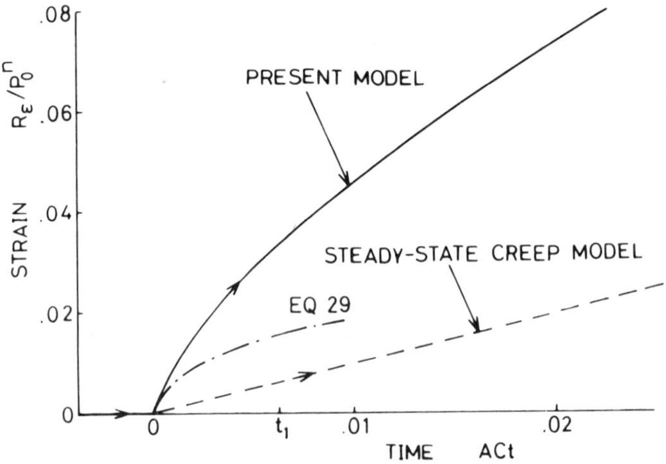

(d)

FIG. 1—*Continued.*

$$R_{\dot{\epsilon}}/P_0{}^n = -(AC/D)e^{-n(AC+D)t} \tag{33}$$

$$R_\epsilon/P_0{}^n = -(AC/D)^2[1 - e^{-n(AC+D)t}]/[(1 + AC/D)n] \tag{34}$$

An example of the variation of these parameters is shown in Fig. 2. As can be seen from the figure, step unloading is followed by reversed deformation. A characteristic time t_2 can be defined as the relaxation time of R_α

$$t_2 = 1/(AC + D) \tag{35}$$

Stress and Strain Analysis in the Case of $\beta < n - 1$

Singular Behavior at the End of a Crack Tip

In this subsection singular asymptotic stress and strain behavior are examined. Now we assume that

$$\dot{\epsilon} \sim r^{-a}, \qquad \Sigma \sim r^{-b}, \qquad \dot{\alpha} \sim r^{-c} \tag{36}$$

near the crack tip ($r \to 0$), where a, b, and c are unknown constants and $a > 0$. Then

$$\begin{gathered} A\,\Sigma_e^{n-1}\Sigma \sim r^{-nb}, \qquad \alpha \sim r^{-c}, \qquad C\dot{\epsilon}/\alpha_e{}^{\beta} \sim r^{-a+\beta c}, \\ D\alpha_e^{n-\beta-1}\alpha \sim r^{-(n-\beta)c} \end{gathered} \tag{37}$$

Comparing the exponents involved in each term of Eq 8, one obtains $a = nb$. For Eq 9 to hold near the crack tip, the smallest two exponents involved in the terms of Eq 9 are equal. Then, $a = nc > 0$. In this case $\dot{\alpha}$ in Eq 9 is small in comparison with the terms on the right-hand side of Eq 9. Then the relation between the strain rate and the stress very near the crack tip is given by the

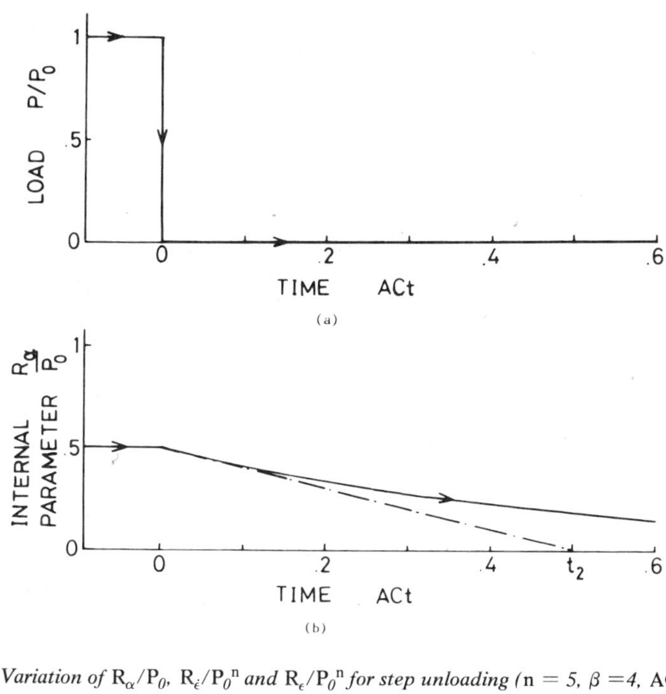

FIG. 2—*Variation of* R_α/P_0, $R_{\dot{\epsilon}}/P_0{}^n$ *and* $R_\epsilon/P_0{}^n$ *for step unloading* ($n = 5$, $\beta = 4$, $AC/D = 1$).

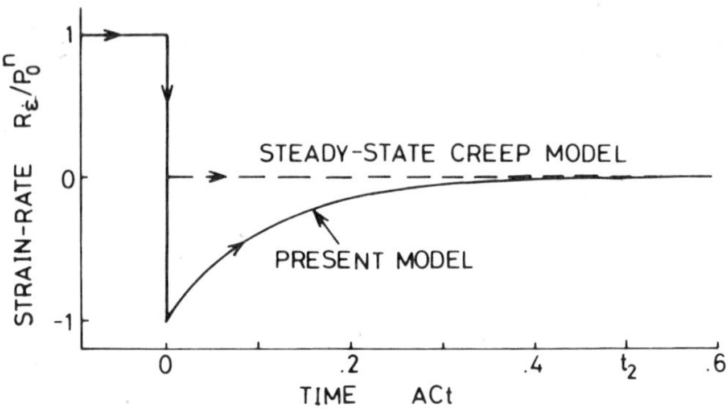

(c)

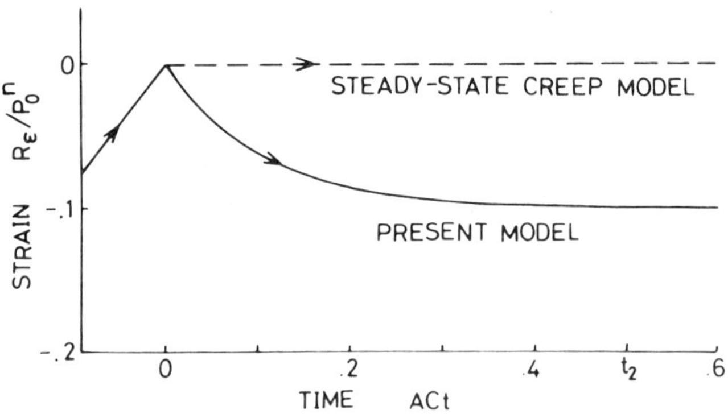

(d)

FIG. 2—*Continued.*

power-law relation Eq 13, which is deduced for the steady-state creep conditions. Accordingly, we can conclude that the HRR singular fields (Eqs 3 and 4), which are valid under steady-state creep conditions, prevail for the case of $\beta < n - 1$ also. Then, $a = n/(n + 1)$ and $b = c = 1/(n + 1)$ for $\beta < n - 1$.

Self-Similar Solutions

Consider a crack embedded in a specimen subjected to a step load (Eq 26). It is assumed that the internal stress variable is equal to zero at $t = 0$. Under this condition, the strain rate is given by Eq 12. Accordingly, the distributions of stress and strain-rate near the crack tip at $t = 0$ are given by the HRR fields

$$\sigma = [C^*_{\text{far}}/(I_n Ar)]^{1/(n+1)}\tilde{\sigma}(\theta, n) \tag{38}$$

$$\dot{\epsilon} = A[C^*_{\text{far}}/(I_n Ar)]^{n(n+1)}\tilde{\epsilon}(\theta, n) \tag{39}$$

Here C^*_{far} is the value of C^* far from the crack tip.[5] The C^*-integral for a power-law creep is defined by the following line integral along a path Γ which surrounds the crack tip

$$C^* = \int_{\Gamma} [\{n/(n+1)\}\boldsymbol{\sigma}:\dot{\boldsymbol{\epsilon}}\nu_1 - \boldsymbol{T}\cdot\partial\dot{\boldsymbol{u}}/\partial x_1]d\Gamma \tag{40}$$

The Cartesian coordinates x_1, x_2 are taken so that the x_1-axis is parallel to the crack surfaces; $\dot{\boldsymbol{u}}$ is the displacement rate and $\boldsymbol{T}$ is the traction vector defined using the unit normal vector $\boldsymbol{\nu}$ to Γ; $\boldsymbol{T} = \boldsymbol{\sigma}\cdot\boldsymbol{\nu}$.

For small t, the "steady-state creep region" is confined to a small region which surrounds the crack tip, although the size of the steady-state creep region increases with increasing time (Fig. 3). This situation can be termed "small-scale steady-state creep." In this case one can expect that the HRR singular field Eq 38 governs stresses at distances from the crack tip which are large compared with the size of the steady-state creep region. Then a boundary-layer

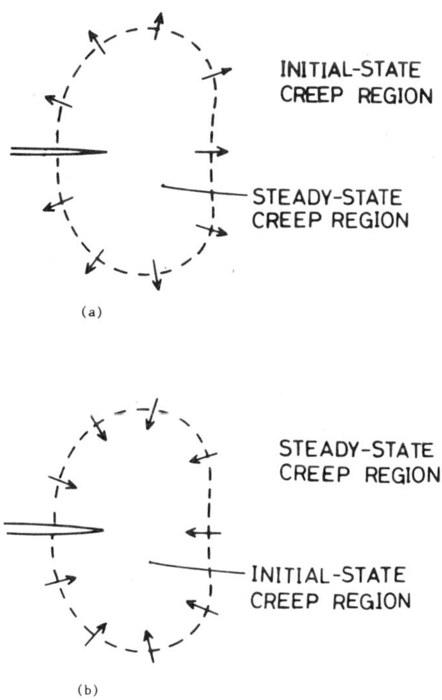

FIG. 3—*Development of the "steady-state creep region": (a) $\beta < n - 1$, (b) $\beta > n - 1$.*

[5] C^* can be calculated by using the numerical results in Refs *19–22* with the stress/strain-rate relation, Eq 12.

approach similar to that for the "small-scale yielding" [*31*] can be employed with the asymptotic boundary condition; Eq 38 holds as $r \to \infty$.

The basic equations of the stress analysis are constitutive Eqs 8 and 9, the boundary Eq 38 at $r \to \infty$, the equilibrium equation

$$\nabla \cdot \sigma = 0 \tag{41}$$

and the following compatibility equation

$$\dot{\epsilon}_{ij,kl} + \dot{\epsilon}_{kl,ij} - \dot{\epsilon}_{ik,jl} - \dot{\epsilon}_{jl,ik} = 0 \tag{42}$$

If we assume that C^*_{far} is constant after load application, the stress field at any point (r, θ) is given by a function of the following set of variables and material parameters

$$r, \theta, t, C^*_{\text{far}}, A, C, D, n, \beta$$

Substitution of Eq 8 into Eq 42 gives an equation which does not contain A, C, D, and time t. The equilibrium Eq 41 also does not contain these parameters. Substituting Eq 8 into Eq 9, we obtain

$$\dot{\alpha}' = AC\ \Sigma_e^{n-1}\ \Sigma/\alpha_e^{\beta} - D\alpha_e^{n-\beta-1}\alpha' \tag{43}$$

From Eqs 38 and 43 we observe that A, C, D, t, and C^*_{far} can enter as the product ACt, Dt, and C^*_{far}/A only. The $1/(\beta + 1 - n)$th power of ACt or Dt has the same dimension as the stress. Then the standard dimension analysis requires that the stress and the internal stress be given by

$$\sigma = (ACt)^{1/(\beta+1-n)}F[(Dt)^{(n+1)/(n-\beta-1)}\{C^*_{\text{far}}/(Ar)\}, \theta, AC/D, n, \beta] \tag{44}$$

$$\alpha = (ACt)^{1/(\beta+1-n)}G[(Dt)^{(n+1)/(n-\beta-1)}\{C^*_{\text{far}}/(Ar)\}, \theta, AC/D, n, \beta] \tag{45}$$

where F and G are dimensionless functions of their arguments. It is evident that the distributions of stress and internal stress have self-similar forms. Full detailed stress analysis can be completed by combining Eqs 44 and 45 with the basic equations (that is, Eqs 8, 9, 38, 41 and 42) and solving them for the Mode I plane stress, Mode I plane strain, and the Mode III cases. However, no solutions are available yet. An approximate description of the near-tip fields is given in the following.

Stress and Strain-Rate Intensities Under Small-Scale Steady-State Creep Conditions

From the discussion in the foregoing, the relation between the strain rate and the stress is given by Eq 13 very near the crack tip, and the HRR singular fields prevail for $t > 0$

$$
\begin{aligned}
\sigma &= [M(t)H_1/(I_n Ar)]^{1/(n+1)}\tilde{\sigma}(\theta, n) \\
\dot{\epsilon} &= (A/H_1)[M(t)H_1/(I_n Ar)]^{n/(n+1)}\tilde{\epsilon}(\theta, n) \\
K_\sigma &= [M(t)H_1/(I_n A)]^{1/(n+1)} \\
K_{\dot{\epsilon}} &= (A/H_1)[M(t)H_1/(I_n A)]^{n/(n+1)}
\end{aligned}
\tag{46}
$$

where $M(t)$ is an undetermined function.

From Eq 46, σ is proportional to $r^{-1/(n+1)}$ as $r \to 0$. Accordingly, Eq 44 reduces to

$$
\sigma = (AC/D)^{1/(\beta+1-n)}[C^*_{\text{far}}/(Ar)]^{1/(n+1)} f(\theta, AC/D, n, \beta) \quad \text{as } r \to 0 \tag{47}
$$

Here f is a function of θ, AC/D, n, and β only. Equation 47 should be compatible with Eq 46. Then

$$
M(t) = C^*_{\text{far}} m(AC/D, n, \beta) \tag{48}
$$

where m is a function of AC/D, n, and β only. Equation 48 says that the stress-intensity factor and strain-rate intensity factor are constant under the small-scale steady-state creep conditions, although the size of the steady-state creep region increases with time.

The function m involved in Eq 48 can be estimated by using an approximation. Riedel and Rice [*32*] and Ohji et al [*33*] employed a hypothesis of path-independence of J-integral [*31*] to estimate the intensities of stress and strain-rate for an elastic nonlinear viscous material under the small-scale creep conditions. The J-integral is given by

$$
J = \int_\Gamma (W\nu_1 - \boldsymbol{T}\cdot\partial\boldsymbol{u}/\partial x_1)\,d\Gamma \tag{49}
$$

Here $\boldsymbol{u}$ denotes displacement and the quantity

$$
W = \int \sigma : d\epsilon \tag{50}
$$

is defined as an integral over the deformation history at each point. In general, the J-integral is path-dependent for this material. However, the plausibility of the path-independence of the J-integral for the elastic nonlinear viscous material under the small-scale creep conditions was confirmed numerically by Ohji et al [*33*], and Bassani and McClintock [*34*]. The path-independence of J as an approximation is used in this paper also.

Since the far fields are given by Eqs 38 and 39, and the near-tip fields are given by Eq 46, the path-independence of J reduces to

$$M(t) = C^*_{\text{far}} \tag{51}$$

The C*-integral, which is defined by Eq 40, is also path-dependent in general. The hypothesized path-independence of C^* leads to Eq 51 again.

Next a rough estimation of the size of the steady-state creep region is made. Since Eq 38 holds at $t = 0$, and $\tilde{\sigma}_e(\theta, n)$ is the order of 1, the internal stress variable is roughly given by

$$\boldsymbol{\alpha}' = [(\beta + 1)ACt]^{1/(\beta+1)}[C^*_{\text{far}}/(I_n At)]^{n/\{(n+1)(\beta+1)\}}\boldsymbol{\sigma}'(\theta, n) \tag{52}$$

from Eq 9 for small t if $\boldsymbol{\alpha}'$ is assumed to be small compared with $\boldsymbol{\sigma}$. On the other hand, the steady-state value of $\boldsymbol{\alpha}'$ is obtained by combining Eq 14 with Eqs 46 and 51

$$\boldsymbol{\alpha}' = H_2[C^*_{\text{far}}H_1/(I_n Ar)]^{1/(n+1)}\tilde{\boldsymbol{\sigma}}'(\theta, n) \tag{53}$$

The size of the steady-state creep region r_{SI} can be defined by equating Eqs 52 and 53

$$r_{\text{SI}} = \frac{C^*_{\text{far}}}{I_n A} \frac{[(\beta + 1)ACt]^{(n+1)/(n-\beta-1)}}{H_2^{(n+1)(\beta+1)/(n-\beta-1)}H_1^{(\beta+1)/(n-\beta-1)}} \tag{54}$$

Transition from Small-Scale Steady-State Creep to Large-Scale Steady-State Creep

For small load duration t, the size r_{SI} of the steady-state creep region is small, which is termed the "small-scale steady-state creep." The size r_{SI}, however, increases with time as can be seen from Eq 54 (Fig. 3). Therefore, the steady-state creep region can develop extensively in the specimen in a finite time, which can be termed "large-scale steady-state creep." In this subsection, the transition from the small-scale steady-state creep to the large-scale steady-state creep is discussed.

We consider a cracked specimen subjected to a step stress far from the crack

$$\sigma_\infty = 0 \quad \text{for } t < 0, \qquad \sigma_\infty = \sigma_0 \quad \text{for } t \geq 0 \tag{55}$$

For the small-scale steady-state creep, the hypothesis of path-independence of J-integral was used. The hypothesis of path-independence of C^*-integral was found to give the same result for the problem. In this subsection, the hypothesized path-independence of C^* is used to estimate the intensities of stress and strain-rate fields near the crack tip in the transition from small-scale steady-state creep to large-scale steady-state creep.

From Eq 9 the ratio of the internal stress to the applied stress far from the crack tip for small t is roughly given by

$$[(\beta + 1)AC\sigma_\infty^n t]^{1/(\beta+1)}/\sigma_\infty$$

when the internal stress is small compared with the applied stress. Then the strain rate, which is proportional to the nth power of the difference between the applied stress and the internal stress is

$$[1 - \{(\beta + 1)ACt\}^{1/(\beta+1)}\sigma_\infty^{n/(\beta+1)-1}]^n$$

multiplied by that at $t = 0$. Therefore, the C*-integral far from the crack tip is expressed in terms of its initial value C_0^* as follows

$$C_{\text{far}}^* = [1 - \{(\beta + 1)ACt\}^{1/(\beta+1)}\sigma_\infty^{n/(\beta+1)-1}]^n C_0^* \tag{56}$$

The stress and strain-rate fields near the crack tip are given by Eq 46. Only the amplitude $M(t)$ of the singular fields is unknown. The hypothesis of path-independence of the C*-integral predicts that

$$M(t) = [1 - \{(\beta + 1)ACt\}^{1/(\beta+1)}\sigma_\infty^{n/(\beta+1)-1}]^n C_0^* \tag{57}$$

for small t.

When the steady-state creep region develops extensively, the C*-integral far from the crack tip approaches its steady-state value

$$C_{\text{far}}^* = C_0^*/H_1 \tag{58}$$

By equating Eq 57 with Eq 58, a characteristic time t_3 can be obtained

$$t_3 = H_2^{(\beta+1)}\sigma_\infty^{\beta+1-n}/[(\beta + 1)AC] \tag{59}$$

This quantity is a measure of the time of the transition from the small-scale steady-state creep to the large-scale steady-state creep.

Stress and Strain Analysis in the Case of $\beta > n - 1$

Singular Behavior at the End of a Crack Tip

For the discussion of singular behavior, Eqs 36 and 37 are used again. From the comparison of exponents involved in the terms of Eqs 8 and 9, $a = nb = (\beta + 1)c > nc > 0$. In this case, the internal stress α is very small compared with the effective stress Σ in the vicinity of the crack tip ($r \to 0$). Therefore, the power-law relation Eq 12 between the strain rate and the stress for the initial-state creep can be applied in the vicinity of the crack tip. It has been reported [*5–7*] that the HRR singular fields prevail if the strain rate is given by a power function of current stress only. Then Eqs 3 and 4 hold in the case of $\beta > n - 1$ as well. Then

$$a = n/(n+1), \quad b = 1/(n+1), \quad c = n/[(n+1)(\beta+1)] \quad \text{for } \beta > n-1.$$

Self-Similar Solutions and Intensities of Stress and Strain Rate Under Small-Scale Initial-State Creep Conditions

From the discussion in the foregoing, the "initial-state," under which the relation between the strain rate and the stress is given by Eq 12, prevails very near the crack tip. Accordingly, the near-tip fields are given by [*14–16*]

$$\begin{aligned} \sigma &= [M(t)/(I_n A r)]^{1/(n+1)} \tilde{\sigma}(\theta, n) \\ \dot{\epsilon} &= A[M(t)/(I_n A r)]^{n/(n+1)} \tilde{\epsilon}(\theta, n) \\ K_\sigma &= [M(t)/(I_n A)]^{1/(n+1)} \\ K_{\dot{\epsilon}} &= A[M(t)/(I_n A)]^{n/(n+1)} \end{aligned} \tag{60}$$

As time goes on, however, the steady-state creep region develops in the far region and comes to confine the initial-state creep region (Fig. 3). This situation can be termed "small-scale initial-state creep," in contrast to the "small-scale steady-state creep" introduced for $\beta < n - 1$. Then we anticipate that the singular fields for the steady-state creep condition, which are given by the following equations, prevail at distances from the crack tip which are large compared with the initial-state creep region

$$\sigma = [C^*_{\text{far}} H_1/(I_n A r)]^{1/(n+1)} \tilde{\sigma}(\theta, n) \tag{61}$$

$$\dot{\epsilon} = (A/H_1)[C^*_{\text{far}} H_1/(I_n A r)]^{n/(n+1)} \tilde{\epsilon}(\theta, n) \tag{62}$$

Then a boundary-layer approach, which uses Eq 61 as a boundary condition, can be employed.

A discussion similar to that given in the preceding section leads to the solution that the stress and the internal stress are expressed by the same form as those of Eqs 44 and 45. The compatibility of Eq 44 with Eq 60 near the crack tip requires Eq 48 again. The assumption of the path-independence of the J or C^* leads to Eq 51 as for the case of $\beta < n - 1$.

The size of the initial-state creep region r_{SI} can be estimated as was done in the foregoing section. It is shown that Eq 54 is valid for the case of $\beta > n - 1$ as well. From the equation, the size decreases with increasing time (Fig. 3).

Transition from Large-Scale Initial-State Creep to Small-Scale Initial-State Creep

Suppose that a cracked specimen is subjected to the step stress (Eq 55) far from the crack. The discussion given for $\beta < n - 1$ in the preceding section can be used for $\beta > n - 1$. Equation 56 stands for small load duration t. The hypothesized path-independence of the C*-integral leads to Eq 57. The characteristic time t_3, which is given by Eq 59, is introduced again. In this case, the value of t_3 represents the time at which the steady-state creep region prevails far from the crack tip. For $t \ll t_3$, the large-scale initial-state creep condition prevails; for $t \gg t_3$, the small-scale initial-state creep condition prevails.

Discussion

Mitra and McLean [*28*] and McLean [*35*] determined the values of n and β from experimental data for nickel, aluminum, and zinc. The results are given in Table 1. From the table, n is nearly equal to 5, and β is 1 to 2 for these materials. Lagneborg [*30*] considered a recovery-creep dislocation model. As pointed out in the foregoing, this theory coincides with Eqs 8–10 in the case of $n = 4$, $\beta = 1$, using the condition that the internal stress is proportional to the square root of the average dislocation density. Theoretically predicted values ($n = 4$, $\beta = 1$) are close to the experimental values.

From Table 1, $\beta - (n - 1)$ is negative in all cases. Accordingly, the case of $\beta < n - 1$ seems to be prevailing for metallic creep between the three cases: $\beta = n - 1$, $\beta < n - 1$, and $\beta > n - 1$.

When $\beta < n - 1$, the steady-state creep prevails and C^* can be defined very

TABLE 1—*Material parameters* n *and* β.

	Author	Material	n	β	$\beta - (n - 1)$
Experiment	Mitra and McLean [*28*]	Ni	4.7	1.7	−2
		Al	4.6	1.1	−2.5
	McLean [*35*]	Zn[a]	5	1.5	−2.5
Theory	Lagneborg [*30*]	...	4	1	−2

[a]Only rough values of n and β can be determined from the data available.

near the crack tip. In other words the stress and strain-rate fields near the crack tip are characterized by C^*. These can be recognized by comparing Eqs 3–5 with Eq 46 and by setting $A/H_1 = B$.

From Eq 58 the ratio of C^* far from the crack tip at load application to that under the steady-state creep conditions is expressed by H_1. The ratio of strain-rate under the initial-state creep conditions to that under the steady-state creep conditions, which is equal to H_1 from Eqs 12 and 13, is reported to be 30 to 1000 for some representative materials [35]. Therefore, C^* near the crack tip at load application is higher than its steady-state value by several orders, if the assumption of the path-independence of C^* is employed.

Now we consider a cracked specimen subjected to a zero-tension cyclic load. We consider a case in which tension-hold time is much shorter than t_3. In this case the internal stress is not high compared with the applied stress, just the same as under the initial-state creep conditions. Then, the value of C^* near the crack tip during tension hold is expected to be several orders of magnitude higher than its steady-state value. This acceleration of deformation can enhance crack growth rates and susceptibility of crack initiation. Under tension-compression cyclic load the acceleration is more pronounced.

Taira et al [*11*] conducted creep crack growth experiments and reported that sudden load increase and decrease are followed by accelerated crack growth and decelerated crack growth, respectively, for Type 316 stainless steel. From creep crack growth experiments conducted by Koterazawa and Mori [*13*], creep crack growth rates under intermittent loads are 20 to 100 times larger than those under constant loads. Furthermore, time-dependent crack growth under cyclic tension-compression loads is accelerated [*11,12*]. This acceleration and deceleration of crack growth rates is accompanied by a comparable degree of acceleration and deceleration of deformation of the cracked specimen. Accordingly, the tendencies appearing in the experimental results can be explained qualitatively by the foregoing discussion. Quantitative prediction of the acceleration and deceleration is not available because values of material parameters are not fully known.

If the strain rate far from the crack is accelerated or decelerated due to creep recovery and hardening by a factor of p compared with that under the steady-state creep condition, it is plausible that the load point displacement rate $\dot{\delta}$ is also accelerated or decelerated by the same factor p. Therefore, C^* evaluated experimentally from Eq 6 may give a good approximate value of C^*_{far}. If we use the assumption of the path-independence of C^* as an approximation, experimentally determined C^* gives a good measure of the near-tip stress and strain-rate fields. This is thought to support the experimental evidence [*11–13*] that time-dependent crack growth rates under variable loads correlate well with the experimental C^*-value and that the correlation agrees reasonably with the same correlation under static creep conditions.

In this paper the hypothesized path-independence of the J-integral or the

C^*-integral was used to estimate the intensities of the singular stress and strain-rate fields for $\beta \neq n - 1$. The validity of the hypothesis should be checked by numerical calculations.

Conclusions

The effects of creep recovery and hardening on the stress and strain-rate fields near the creep crack tip are discussed. Robinson's constitutive relation, which is based on the Bailey-Orowan model, is used to take these effects into account. It is found that the Hutchinson-Rice-Rosengren-type singular fields prevail near the crack tip for any combination of material parameters. The intensities of the singular fields are obtained by using the hypothesized path-independence of the J-integral or the C*-integral. The intensity of strain rate after load increase is very high compared with its steady-state value; after load decrease, it is low. The observed acceleration and deceleration of crack growth rates are found to be explained by the present analysis.

Acknowledgment

This work was supported by the U.S. Department of Energy and, in addition, the author was supported by a visiting appointment in the National Science Foundation Materials Research Laboratory program and by the Murata Oversea Scholarship Foundation. The author is very grateful to Professor James R. Rice of Brown University for his valuable advice and discussions.

References

[*1*] Ohji, K., *Theoretical and Applied Mechanics, Proceedings*, 27th Japan National Congress for Applied Mechanics, 1977, Vol. 27, 1979, pp. 3–20.

[*2*] Van Leeuwen, H. P., *Engineering Fracture Mechanics*, Vol. 9, No. 4, 1977, pp. 951–974.

[*3*] Siverns, M. J. and Price, A. T., *International Journal of Fracture*, Vol. 9, No. 2, 1973, pp. 199–207.

[*4*] Nikbin, K. M., Webster, G. A., and Turner, C. E. in *Cracks and Fracture, ASTM STP 601*, American Society for Testing and Materials, 1976, pp. 47–62.

[*5*] Landes, J. D. and Begley, J. A. in *Mechanics of Crack Growth, ASTM STP 590*, American Society for Testing and Materials, 1976, pp. 128–148.

[*6*] Ohji, K., Ogura, K., and Kubo, S., Japan Society of Mechanical Engineers Preprint No. 740-11, 1974, pp. 207–210.

[*7*] Ohji, K., Ogura, K., and Kubo, S., *Transactions*, Japan Society of Mechanical Engineers, Vol. 42, 1976, pp. 350–358.

[*8*] Koterazawa, R. and Mori, T., *Journal of Engineering Materials and Technology, Transactions*, American Society of Mechanical Engineers, Vol. 99, No. 4, 1977, pp. 298–305.

[*9*] Ohji, K., Ogura, K., Kubo, S., and Katada, Y., *Engineering Aspects of Creep*, Vol. 2, Institution of Mechanical Engineers, 1980, pp. 9–16.

[*10*] Taira, S., Ohtani, R., and Kitamura, T., *Journal of Engineering Materials and Technology, Transactions*, American Society of Mechanical Engineers, Vol. 101, No. 2, 1979, pp. 154–161.

[*11*] Taira, S., Ohtani, R., and Komatsu, T., *Journal of Engineering Materials and Technology, Transactions*, American Society of Mechanical Engineers, Vol. 101, No. 2, 1979, pp. 162–167.

[*12*] Saito, H., Ohji, K., Ogura, K., Kubo, S., and Fukumoto, M., Japan Society of Mechanical Engineers Preprint No. 804-2, 1980, pp. 64–67.

[*13*] Koterazawa, R. and Mori, T., *Engineering Aspects of Creep*, Vol. 1, Institution of Mechanical Engineers, 1980, pp. 219–224.

[*14*] Hutchinson, J. W., *Journal of the Mechanics and Physics of Solids*, Vol. 16, No. 1, 1968, pp. 13–31.

[*15*] Hutchinson, J. W., *Journal of the Mechanics and Physics of Solids*, Vol. 16, No. 5, 1968, pp. 337–347.

[*16*] Rice, J. R. and Rosengren, G. F., *Journal of the Mechanics and Physics of Solids*, Vol. 16, No. 1, 1968, pp. 1–12.

[*17*] Rice, J. R., *Journal of Applied Mechanics, Transactions*, American Society of Mechanical Engineers, Vol. 34, No. 2, 1967, pp. 287–298.

[*18*] Ohji, K., Ogura, K., and Kubo, S., *Transactions*, Japan Society of Mechanical Engineers, Vol. 44, No. 382, 1978, pp. 1831–1838.

[*19*] Goldman, N. L. and Hutchinson, J. W., *International Journal of Solids and Structures*, Vol. 11, No. 5, 1975, pp. 575–591.

[*20*] Shih, C. F. and Hutchinson, J. W., *Journal of Engineering Materials and Technology, Transactions*, American Society of Mechanical Engineers, Vol. 98, No. 4, 1976, pp. 289–295.

[*21*] Hutchinson, J. W., Needleman, A., and Shih, C. F., *Fracture Mechanics*, N. Perrone, H. Liebowitz, D. Mulville, and W. Pilkey, Eds., University Press of Virginia, Charlottesville, Va., 1978, pp. 515–528.

[*22*] Ohji, K., Ogura, K., and Kubo, S., *Transactions*, Japan Society of Mechanical Engineers, Series A, Vol. 47, No. 416, 1981, pp. 400–405.

[*23*] Ohji, K., Ogura, K., Kubo, S., and Katada, Y., Japan Society of Mechanical Engineers Preprint No. 780-9, 1978, pp. 123–128.

[*24*] Bailey, R. W., *Journal of the Institute of Metals,* Vol. 35, 1926, pp. 27–40.

[*25*] Orowan, E., *Journal of the West of Scotland Iron and Steel Institute*, Vol. 54, 1946, pp. 45–53.

[*26*] Pugh, C. E. and Robinson, D. N., *Nuclear Engineering and Design*, Vol. 48, No. 1, 1978, pp. 269–276.

[*27*] Robinson, D. N., Oak Ridge National Laboratory Report No. TM-5969, Oak Ridge, Tenn., 1978.

[*28*] Mitra, S. K. and McLean, D. in *Proceedings*, Royal Society of London, Vol. 295A, 1966, pp. 288–299.

[*29*] Kubo, S., Brown University Report No. MRL E-133, Providence, R.I., 1981.

[*30*] Lagneborg, R., *Metal Science Journal*, Vol. 6, 1972, pp. 127–133.

[*31*] Rice, J. R., *Journal of Applied Mechanics, Transactions*, American Society of Mechanical Engineers, Vol. 35, No. 2, 1968, pp. 379–386.

[*32*] Riedel, H. and Rice, J. R. in *Fracture Mechanics, ASTM STP 700*, American Society for Testing and Materials, 1980, pp. 112–130.

[*33*] Ohji, K., Ogura, K., and Kubo, S., *Journal of the Society of Materials Science Japan*, Vol. 29, No. 320, 1980, pp. 465–471.

[*34*] Bassani, J. L. and McClintock, F. A., MIT Report, Massachusetts Institute of Technology, Cambridge, Mass., 1980.

[*35*] McLean, D., *Reports on Progress in Physics*, Vol. 29, Part 1, 1966, pp. 1–33.

Mark M. Little,[1] *Erhard Krempl,*[1] *and C. Fong Shih*[2]

On the Time and Loading Rate Dependence of Crack-Tip Fields at Room Temperature—A Viscoplastic Analysis of Tensile Small-Scale Yielding

REFERENCE: Little, Mark M., Krempl, E., and Shih, C. F., **"On the Time and Loading Rate Dependence of Crack-Tip Fields at Room Temperature—A Viscoplastic Analysis of Tensile Small-Scale Yielding,"** *Elastic-Plastic Fracture: Second Symposium, Volume I—Inelastic Crack Analysis, ASTM STP 803*, C. F. Shih and J. P. Gudas, Eds., American Society for Testing and Materials, 1983, pp. I-615–I-636.

ABSTRACT: A finite-element analysis of stationary cracks in Mode I small-scale yielding is presented. It is asserted that the deformation of structural metals at room temperature and quasi-static loading rates can be significantly time dependent and so the theory of viscoplasticity based on total strain and over-stress is employed. Loading levels are controlled by the elastic stress-intensity factor K while loading rates are governed by the time rate of change of K. For loading rates varying over six orders of magnitude in the quasi-static range followed by subsequent periods of sustained loading, significant time dependence of the crack-tip fields is predicted. Interpreting the results using two different fracture criteria, it is predicted that the value of K at which crack growth is initiated is loading-rate dependent. Moreover, it also follows that strain-induced crack growth could be initiated during periods of sustained load. The results are strongly suggestive that time-dependent deformation should be considered in design against fracture even at room temperature. Appropriate design approaches, based on the special features of the viscoplasticity model employed, are proposed.

KEY WORDS: fracture, cracks, viscoplasticity, creep, relaxation, time dependence, finite-element analysis, Type 304 stainless steel, room temperature, elastic-plastic fracture

[1]Research assistant and professor of mechanics, respectively, Department of Mechanical Engineering, Aeronautical Engineering and Mechanics, Rensselaer Polytechnic Institute, Troy, N. Y. 12181.

[2]Visiting associate professor of engineering, Division of Engineering, Brown University, Providence, R.I. 02912.

In recent years considerable progress has been made in the analysis of cracked bodies. Within the framework of power-law characterizations of inelastic deformation, the structures of the stress and strain fields asymptotically near crack tips were revealed by Hutchinson [*1*][3] and Rice and Rosengren [*2*] for time-independent materials and by Riedel and Rice [*3*], Riedel [*4*], and Hui and Riedel [*5*] for time-dependent materials. Continuing in the power-law framework, the more general stress and strain fields of cracked bodies were displayed by numerous workers; for instance, by Shih [*6*] for time independence and by Bassani and McClintock [*7*] for time dependence. In departing from power-law characterizations, state variable models showing time-dependence were used by Hart [*8*], Morjaria and Mukherjee [*9*], and Kubo [*10*] to study crack-tip fields while loading rate-sensitive constitutive laws were employed by Aboudi and Achenbach [*11*] and by Freund and Douglas [*12*] in their dynamic analyses.

It is significant that the results of the time-independent analyses have been generally accepted for use in theories of fracture of metals at low homologous temperatures and at loading rates below the dynamic range. This is evidenced by the widespread use of the time-independent formulations in the evaluation of experiments [*13–16*] as well as by the recent development of procedures, based on time-independent plasticity, for estimating fracture controlling parameters [*17*].

A substantial body of evidence [*18–21*] is now available showing that the inelastic deformation of metals—even at room temperature and quasi-static strain rates (below 10^{-3}/s)—may be time dependent. For example, in numerous tests conducted by Krempl et al [*22–25*], significant loading rate sensitivity, creep, and relaxation were revealed in Type 304 stainless steel, Ti-7Al-2Cb-1Ta, and A533B steel at room temperature. Typical experimental results for Type 304 stainless steel are shown in Fig. 1. These and other time-dependent phenomena cannot be adequately modeled by the classical power-law formulations and so the theory of viscoplasticity based on total strain and over-stress was developed by Cernocky and Krempl [*26*]. It is this constitutive theory which was used herein to explore the influences of time and loading rate on cracked structures in Mode I small-scale yielding.

Theory of Viscoplasticity Based on Total Strain and Overstress

Detailed descriptions of the theory of viscoplasticity based on total strain and overstress and supporting experimental evidence have been presented [*22–29*]. Herein only a brief overview of the fundamental concepts of the theory is offered to motivate interpretation of the analytical predictions.

It is instructive to first consider the application of this theory to a uniaxial state of stress. The following decomposition is employed

[3]The italic numbers in brackets refer to the list of references appended to this paper.

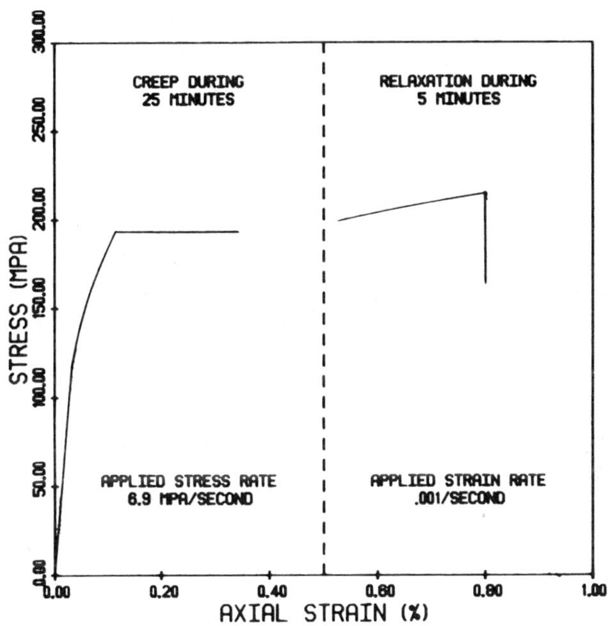

FIG. 1—*Typical results of uniaxial tension tests on annealed Type 304 stainless steel at room temperature.*

$$\dot{\epsilon} = \dot{\epsilon}^{el} + \dot{\epsilon}^{in} \tag{1}$$

where ϵ, ϵ^{el}, and ϵ^{in} are the total, elastic, and inelastic infinitesimal axial strains, respectively, and differentiation with respect to time is indicated by a superposed dot. The viewpoint is taken that the inelastic strain rate is driven by the overstress[4]; thus

$$\dot{\epsilon} = \frac{\dot{\sigma}}{E} + \frac{\sigma - g[\epsilon]}{Ek[\sigma - g[\epsilon]]} \tag{2}$$

Arguments of functions are enclosed in brackets. The applied engineering stress is σ; $g[\epsilon]$ is the equilibrium stress associated with the current strain state[5] (that is, it is the stress level which would be obtained at the given strain state either when all time-dependent processes have terminated or when loading is done at a vanishingly small rate); $\sigma - g[\epsilon]$ is the overstress; k is the viscosity function which is positive, has the dimension of time, and is typically a strongly decreasing function of increasing argument; E is Young's modulus.

[4]This terminology was introduced by Malvern [*30*] in a similar context.

[5]In fact, g is a function of a strain invariant, but it is convenient and reasonable to use axial strain as the argument here.

In a uniaxial creep test, with σ_c as the fixed stress level, Eq 2 can be rewritten as

$$\dot{\epsilon} = \frac{\sigma_c - g[\epsilon]}{Ek[\sigma_c - g[\epsilon]]} \tag{3}$$

If $\sigma_c - g[\epsilon] > 0$, then $\dot{\epsilon} > 0$. If g is an increasing function of ϵ, then as strain is accumulated the strain rate is diminished as $g[\epsilon]$ approaches σ_c asymptotically in time. This is typical of primary creep. If $g[\epsilon]$ is constant, then secondary creep is exhibited.

In a uniaxial relaxation test with ϵ_r as the fixed strain level, Eq 2 can be rewritten as

$$\dot{\sigma} = -\frac{\sigma - g[\epsilon_r]}{k[\sigma - g[\epsilon_r]]} \tag{4}$$

Clearly, if $\sigma - g[\epsilon_r] > 0$, then $\dot{\sigma} < 0$. As the stress is relaxed, so is the overstress and the stress rate approaches zero as the equilibrium stress is approached asymptotically in time.

It may be seen from Eq 2 that the limiting cases of infinite and infinitesimal loading rates are explicitly contained in the theory. Specifically, for an infinite loading rate, the response is elastic, while for an infinitesimal loading rate the equilibrium response—similar to that of a deformation theory plastic metal—is developed. The availability of these limiting cases may be used to advantage as will be demonstrated in the Discussion section. These ideas are also discussed in Refs *26* and *29*.

The generalization[6] of these concepts to multidimensional states using conventional tensor notation with summation over repeated indices is

$$\dot{\epsilon}_{ij} = C^{\text{el}}_{ijmn}\,\dot{\sigma}_{mn} + C^{\text{in}}_{ijmn}\left\{\frac{\sigma_{mn} - G_{mn}[\epsilon_{rs}]}{k[\Gamma]}\right\} \tag{5}$$

where

ϵ_{ij} = infinitesimal strain tensor,
σ_{mn} = engineering stress tensor,
G_{mn} = equilibrium stress tensor,
Γ = overstress invariant given by

$$\Gamma = \{(\sigma_{ij} - G_{ij})(\sigma_{ij} - G_{ij})\}^{1/2} \tag{6}$$

[6]Equation 5 cannot be reduced to Eq 2 exactly; however, the reduction is extremely close with slight differences resulting only in the transition from elastic to elastic-inelastic behavior. For all practical purposes the predictions of Eq 2 and the reduction of Eq 5 are the same.

and

$$C_{ijmn} = \frac{(1+\kappa)}{E}\left\{\frac{1}{2}(\delta_{im}\,\delta_{jn} + \delta_{\text{in}}\,\delta_{jm}) - \frac{\kappa}{1+\kappa}\,\delta_{ij}\,\delta_{mn}\right\} \tag{7}$$

such that

$$\kappa = \nu \text{ in } C^{\text{el}}_{ijmn}$$

$$\kappa = \alpha \text{ in } C^{\text{in}}_{ijmn}$$

and α is a compressibility factor such that $\nu \leq \alpha \leq 1/2$. *For inelastic incompressibility* $\alpha = 1/2$. While there is some evidence [*31,32*] to support the assertion that inelastic deformation may not be completely incompressible, α *is taken to equal 1/2 in the present study.*

For $\alpha = 1/2$, Eq 5 can be rewritten as

$$\dot{\epsilon}_{ij} = C^{\text{el}}_{ijmn}\,\dot{\sigma}_{mn} + \frac{3}{2}\,\frac{(\sigma'_{ij} - G'_{ij})}{E\,k[\Gamma]} \tag{8}$$

where the prime denotes a deviatoric part. This form must be augmented if the over-stress changes sign; however, no sign changes occurred in the analysis presented herein.

The equilibrium stress tensor G_{mn} may be represented by any suitable function. In Ref *29* some suitable representations were given. To facilitate comparison with existing solutions [*1,2,6*] a power-law representation is employed here. Thus

$$\epsilon_{ij} = \frac{(1+\nu)}{E}G'_{ij} + \frac{(1-2\nu)}{3E}G_{mm}\,\delta_{ij} + \beta\left\{\frac{G_e}{g_0}\right\}^{n-1} \times \left\{\frac{(1+\alpha)}{E}G'_{ij} + \frac{(1-2\alpha)}{3E}G_{mm}\,\delta_{ij}\right\} \tag{9}$$

where

$$G_e = \left\{\frac{3}{2}G'_{ij}G'_{ij}\right\}^{1/2} \tag{10}$$

and β, g_0, and n are constants in a uniaxial Ramberg-Osgood relationship of

$$\epsilon = \frac{g}{E} + \beta\frac{g}{E}\left(g/g_0\right)^{n-1} \tag{11}$$

Of course, in application Eq 9 must be inverted to obtain equilibrium stress components in terms of strain components. The form Eq 9 is presented here to facilitate recognition of the power-law plastic strain terms.

The following constants were used in order to represent annealed American Iron and Steel Institute (AISI) Type 304 stainless steel at room temperature

$$\begin{aligned} \nu &= 0.3 & g_0 &= 70 \text{ MPa} \\ \alpha &= 0.5 & n &= 5 \\ E &= 194\,536 \text{ MPa} & \beta &= 1 \end{aligned} \tag{12}$$

and

$$k[\Gamma] = C1\left(1 + \frac{\Gamma}{C2}\right)^{C3} \tag{13}$$

where

$$C1 = 3.14 \times 10^5/\text{s} \qquad C2 = 7.14 \times 10^1/\text{MPa} \qquad C3 = -2.20 \times 10^1$$

Numerical Solution Strategy

The constitutive Eqs 5 are in rate form and so total quantities can be determined only by integration over time. Due to the complex composition of these equations, closed-form integration is generally not possible and recourse must be made to numerical methods. Moreover, Eqs 5 are "stiff" differential equations; large changes (or errors) in some variables can be induced by relatively small changes (or errors) in others. An appropriate solution strategy is illustrated in the following by considering the integration of Eq 2 for a fixed strain rate, $\dot{\epsilon}_f$. Then, from Eqs 1 and 2

$$\dot{\sigma} = E\{\dot{\epsilon}_f - \dot{\epsilon}^{\text{in}}\} \tag{14}$$

For a small time increment, Δt

$$\Delta\sigma \cong \Delta t\, E\{\dot{\epsilon}_f - \dot{\epsilon}^{\text{in}}[t]\} \cong E\{\Delta\epsilon - \Delta\epsilon^{\text{in}}\} \tag{15}$$

where a finite change is indicated by Δ and

$$\Delta\epsilon^{\text{in}} \cong \Delta t\, \dot{\epsilon}^{\text{in}}[t] \tag{16}$$

If Δt is sufficiently small, Eqs 15 and 16 are good approximations and only small errors and a stable solution are generated. If Δt is not sufficiently small, larger errors and an unstable solution may be developed.

In practice, it is found that very small time increments are required to make

Eqs 15 and 16 sufficiently good approximations. An alternative approach, permitting significantly larger time steps, is the so-called forward gradient method [*33*]. In that method, an improved estimate of the inelastic strain increment is made by replacing Eq 16 by

$$\Delta\epsilon^{in} = \Delta t\{(1-\theta)\,\dot{\epsilon}^{in}[t] + \theta\dot{\epsilon}^{in}[t+\Delta t]\} \tag{17}$$

where θ is a weighting factor such that $0 \leq \theta \leq 1$ and the inelastic strain rate at the end of the time increment is estimated by a truncated Taylor series expansion. Thus

$$\dot{\epsilon}^{in}[t+\Delta t] \cong \dot{\epsilon}^{in}[t] + \left.\frac{\partial\dot{\epsilon}^{in}}{\partial\sigma}\right|_t \Delta\sigma + \left.\frac{\partial\dot{\epsilon}^{in}}{\partial\epsilon}\right|_t \Delta\epsilon \tag{18}$$

Substituting Eqs 18 and 17 into Eq 15, it is found that

$$\Delta\sigma \cong \frac{E\left\{\Delta\epsilon\left(1-\theta\Delta t\left.\dfrac{\partial\dot{\epsilon}^{in}}{\partial\epsilon}\right|_t\right) - \Delta t\,\dot{\epsilon}^{in}[t]\right]}{\left\{1+\theta\Delta t E\left.\dfrac{\partial\dot{\epsilon}^{in}}{\partial\sigma}\right|_t\right\}} \tag{19}$$

With this method, time steps much larger than those possible with Eq 16 can be employed. The penalty is that derivatives must be evaluated at each time step. Still, our experiences is that the forward gradient method is considerably advantageous.

The forward gradient method may be embodied in a finite-element technique for the solution of complex boundary and initial value problems using the theory of viscoplasticity based on total strain and overstress, as demonstrated by Zirin and Krempl [*34*]. That approach was followed here with a weighting factor of $\theta = 1$. Our experience is that the method greatly enhances solution stability. The only disadvantage is that a nonsymmetric stiffness matrix must be constructed and decomposed at each time step.

In the work presented here, a finite-element model with 606 degrees of freedom was analyzed. It was found by numerical experimentation that approximately 70 time steps were required to achieve stability up to maximum load while about 170 time steps were needed for the sustained loading sequence. Our experience is that accuracy and stability are essentially coincident in problems of this type. A Digital Equipment Corp. VAX 11/780 computer was employed for the computations. About 12 h of central processing unit (CPU) time were used for each loading-plus-sustained-loading sequence.

Construction and Loading of the Small-Scale Yielding Model

In the present viscoplastic analysis[7] Rice's [*36*] small-scale yielding (SSY) hypothesis was assumed. Accordingly, the load level was prescribed by the value of K, the elastic stress-intensity factor. Stationary cracks under conditions of plane strain and Mode I loading were analyzed using the finite-element model shown in Fig. 2. Eight-noded isoparametric elements with straight sides [*37*] were employed. The elements were arrayed in eleven circumferential layers from the crack tip to the outer boundary with eight elements per layer. Element size was increased with distance from the crack tip.

Strain singularities, with strength $1/r$, were produced in the crack-tip elements by collapsing the three crack side nodes onto one point and by then permitting them to displace independently [*38,39*]. Although, due to the complexity of the constitutive equations, no asymptotic solution for the crack-tip fields has been obtained and hence the nature of any strain singularity is unknown, the degenerate elements were used with confidence. They have been used successfully [*40,41*] in modeling singularities of the Hutchinson-Rice-Rosengren (HRR) type where for various hardening materials the strength of the singularity is between $1/\sqrt{r}$ and $1/r$. Based on the nature of the fast and slow loading limit cases in our viscoplasticity model as discussed, it is expected that any singularities would lie in the HRR range.

In keeping with the SSY hypothesis, the plastic zone size was kept small compared with the model dimensions. Specifically, with the length of the crack-tip elements as the unit length, the distance from the crack tip to the

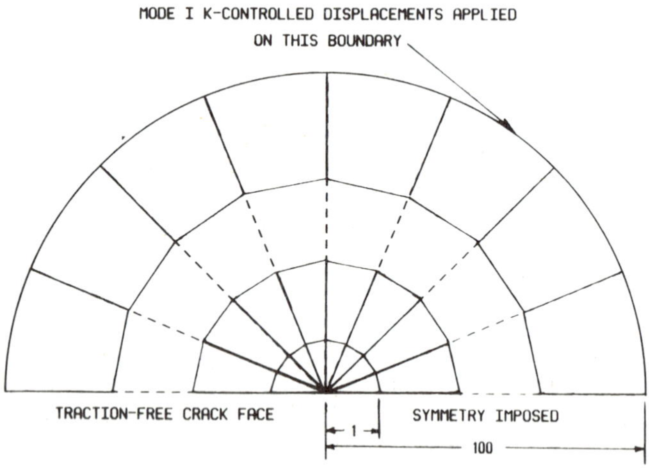

FIG. 2—*The small-scale yielding model.*

[7]In a subsequent work, a compact-tension specimen will be analyzed to explore the limitations of the small-scale yielding model as suggested by Larsson and Carlsson [*35*].

remote boundary was 100 units and the maximum extent of the plastic zone was about 26 units. Plastic zone boundaries were taken to be at locations where the effective strain $(2/3\ \epsilon_{ij}'\ \epsilon_{ij}')^{1/2}$ equalled g_0/E.

Displacements corresponding to the K-controlled elastic singular field were imposed at the remote boundary. The crack face was traction-free. From symmetry, only half of the K-controlled zone was modeled, and, along the line ahead of the crack, vertical displacements were prohibited.

The numerical integration required by the finite-element method was performed with two-by-two Gaussian integration. Two line integrals, J [*42*] and the strain-rate version of J, C^* [*43–45*], were evaluated along the 22 approximately circular paths surrounding the crack tip and through the Gauss points. Since these integrals have been employed extensively in fracture mechanics studies, an attempt was made to evaluate them in the present analysis.

The SSY model was loaded with constant loading rates (constant dK/dt) until the maximum load level at K_{max} was achieved. That load level was then sustained until the equilibrium (nonchanging) state was approximately attained.

A loading time of 33 s up to K_{max} was taken to correspond to the American Society for Testing and Materials (ASTM) Test for Plane-Strain Fracture Toughness of Metallic Materials (E 399-78a) loading rate. Other quasi-static loading rates varying by several orders of magnitude were also applied as shown in Table 1. The symbols shown in the table are used in the figures only to indicate the loading rate; they do not necessarily represent data points.

Results

Loading rate and time dependence of the stress and deformation fields near crack tips was evidenced in the results. Pertinent results indicative of general trends are summarized in the following.

Stresses

The influence of loading rate on the crack opening stresses, which are ahead of the normal to the plane of the crack, is shown in Fig. 3. The opening stress development is shown at three locations during the constant dK/dt ramps. The locations are identified here and in every figure by the non-dimensional

TABLE 1—*Loading rate description.*

Time to Maximum Load, s	$\frac{dK/dt}{dK/dt_{E\ 399}}$	Symbol in Figures
0.33	10^2	*
33	1	○
3300	10^{-2}	×
330 000	10^{-4}	□

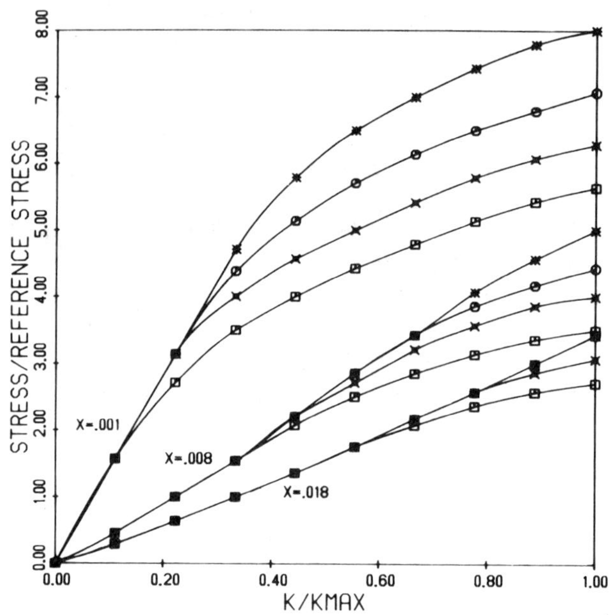

FIG. 3—*Influence of loading rate on the development of crack opening stresses at three locations. The reference stress is* g_0.

distance $X = xg_0^2/K_{max}^2$ where x is the distance from the crack tip along the crack plane. It is apparent that the stresses nearest the crack tip are most affected by loading rate. Furthermore, it is clear that, as the loading rate is increased, increased crack opening stress levels near the crack tip are associated with a given K level.

For comparison purposes, the crack opening stress distributions at the end of selected constant dK/dt ramps ($K = K_{max}$) and at the equilibrium state are shown in Fig. 4. Again it is clear that higher near tip stress levels are achieved with higher loading rates.

Finally, the relaxation of the opening stresses during sustained loading is demonstrated in Fig. 5. The general trend is illustrated by showing the relaxation of the opening stress at a particular location. Two observations may be made. First, the influence of the initial loading rate is not felt after about one day of sustained load time. Second, the equilibrium state is nearly attained in about one year.

Line Integrals

The typical effective path-independence of the calculated J-integral is illustrated in Fig. 6. This pattern was found at all times during loading and sustained loading. The slight variation of J over the first two paths is typical of variations associated with the use of singular finite elements.

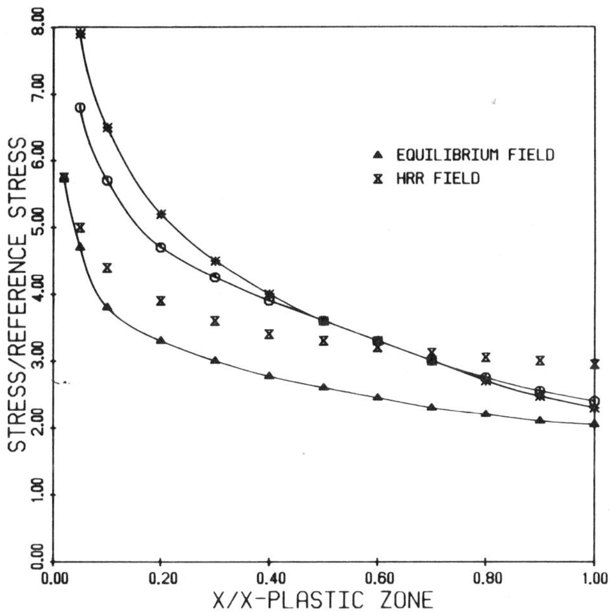

FIG. 4—*Crack opening stress distributions at the ends of two loading ramps compared with the equilibrium and HRR fields. The reference is* g_0.

The value of the J-integral taken on a path through the purely elastic part of a plane-strain K-controlled singular field is given by

$$J = \frac{K^2(1 - \nu^2)}{E} \tag{20}$$

In our SSY model this was the value of J associated with the applied K level. In fact, the line integral value of J was found to deviate from the applied J-value typically by less than 6 percent at the end of the loading ramps and by less than 10 percent at the end of sustained loading. The corresponding deviations in K as calculated from J compared with the applied K were 3 percent and 5 percent, respectively.

No comparable path independence in the C^*-integral was found. In fact, the variation of the calculated C^*-integral was so great that no meaningful interpretation of C^* could be made.

Deformation

A convenient and appropriate measure of crack-tip deformation—the crack opening displacement—is chosen to elucidate the effects of loading rate and time. Following Rice (in Tracey [*46*]), the crack opening displacement (COD)

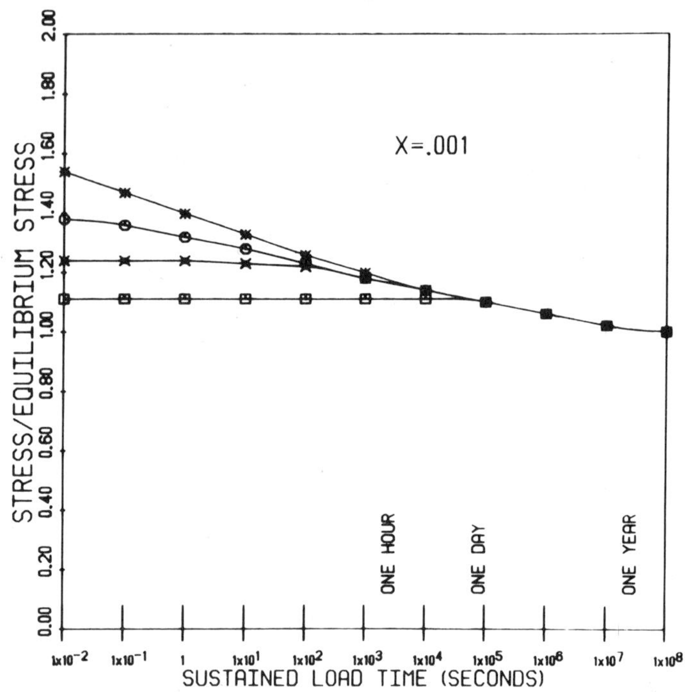

FIG. 5—*Relaxation of crack opening stresses during sustained loading.*

is operationally defined as the vertical distance between the crack faces at the points where the faces are intersected by lines drawn back from the deformed crack tip at 45 deg to the original crack line. This is illustrated in Fig. 7.

Using the applied J-value from Eq 20 as a loading parameter, the effect of loading rate on the development of the COD is shown in Fig. 7. The relationships were, to a very good approximation, linear and are linearized in the figure. The equilibrium COD is also given and an "equilibrium" loading line has been inferred. The slight offset of the lines near the origin is associated with the "inelastic singularity" element's response to the elastic effect at very low levels of loading.

The influence of sustained loading is shown in Fig. 8. It is clear that the COD undergoes continued growth during sustained loading. As in the relaxation of the opening stresses, the influence of prior loading rate is not significant after about 1000 s and again the equilibrium state is approximately attained in about one year. Finally it is noted that at the ASTM Method E 399 loading rate, only about 75 percent of the equilibrium COD was developed at the initial achievement of maximum load.

The plastic zone size is also affected by sustained loading. While the length of the plastic zone directly ahead of the crack tip appears to remain constant

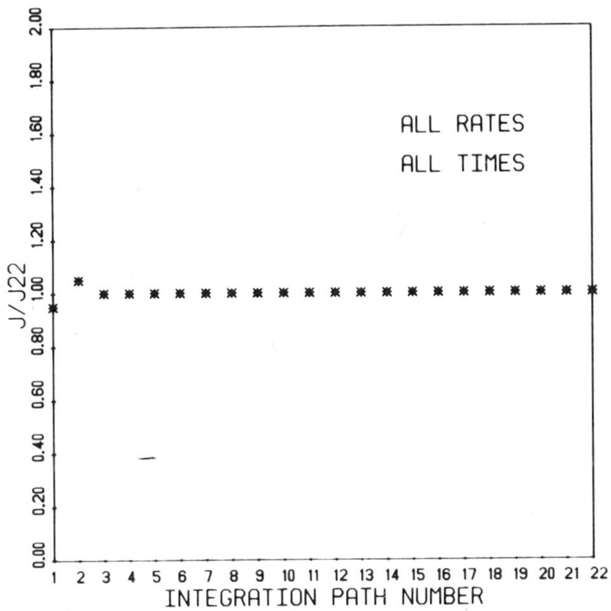

FIG. 6—*A typical lack of variation of the line integral value of* J *over the integration paths. Paths are numbered in ascending order from the crack tip out.*

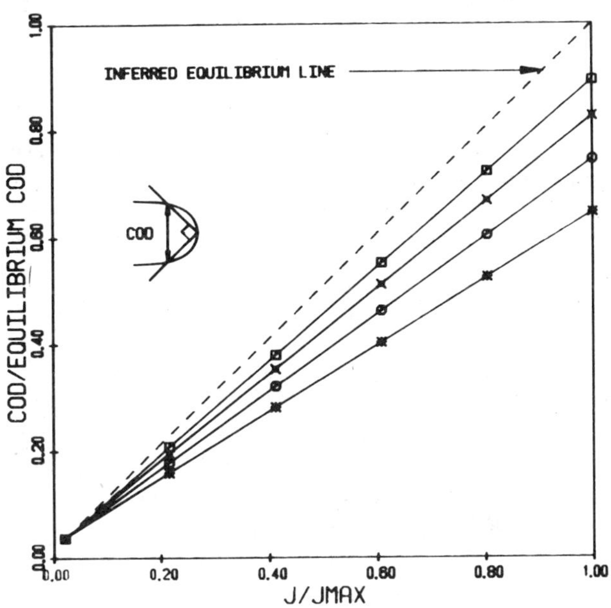

FIG. 7—*Loading-rate-dependent relationships between the applied* J *and the crack opening displacement.*

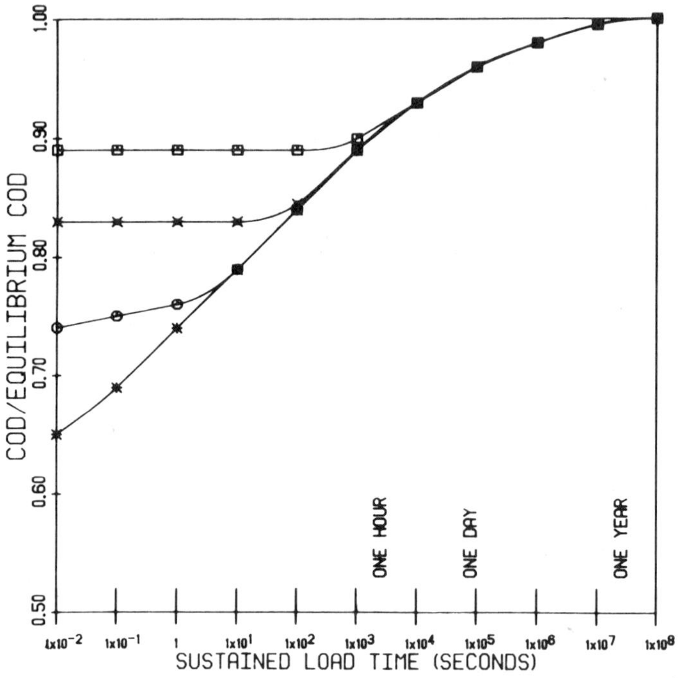

FIG. 8—*Growth of crack opening displacement during sustained loading.*

during sustained loading, the maximum extent of the plastic zone is increased. For example, at the end of the 33 s loading ramp, the maximum extent of the plastic zone was 16 units and it was expanded to 26 units during sustained loading up to the equilibrium state.

Discussion

Stresses

In Fig. 4, the equilibrium crack opening stress field is presented along with the HRR stress field based on the equilibrium material properties. A direct comparison of these two fields may be made because the equilibrium stress-strain relations are of the type used in the HRR solutions. It can be seen that the equilibrium field is in general agreement with the trend of the HRR field and is quite close to the HRR field in a region adjacent to the crack tip. This result is in excellent agreement with the results obtained by Shih and German [*41*] with similar material properties. Furthermore, it is consistent with the concept of a *J*-controlled zone at the tip of a crack in a power-law material.

The J-Integral

The effective path-independence of the J-integral is displayed in Fig. 6; however, a rigorous proof of the path independence of J for our constitutive law has proved to be elusive. Still, the effective path-independence J might be anticipated by considering the fast and slow loading limit states. In the limit of very fast loading, the material response is linear elastic and path-independence of J is assured. In the limit of very slow loading, the response is like that of a nonlinear elastic material and again path-independence is required. By imagining that any given nonequilibrium state is "bracketed" by associated fast and slow loading limit states in which path-independence is a property of J, it might seem that near path-independence of J could be exhibited in the intermediate states as well. This is indeed demonstrated by the computations.

Deformation

In Fig. 7, loading-rate-dependent linear relationships between the applied J-integral and the COD are displayed. The J-COD relationship may thus be cast in the form

$$\text{COD} = d\frac{J}{g_0} \tag{21}$$

where d is a function of the loading rate, dK/dt. This relationship is valid only during the constant dK/dt loading ramps. However, the inferred equilibrium relationship—the limiting case as $dK/dt \rightarrow 0$—may be compared to results based on the HRR fields for the reasons mentioned in the foregoing. As discussed by Shih [*47*], the relationship required by the HRR field is

$$\text{COD} = d_n\frac{J}{\sigma_0} \tag{22}$$

where d_n is strongly dependent on the hardening index n and mildly dependent on the ratio σ_0/E, and σ_0 is a reference stress. Appropriately equating the reference stress in Eq 22 to g_0, it is found that the inferred equilibrium result of d from Eq 21 is about 25 percent higher than the value of d_n from the HRR result, Eq 22. This discrepancy is perhaps associated with the influence of g_0/E on COD. From the results of time-independent plasticity studies, it is clear that the equilibrium plastic zone size r_p is proportional to $(K/g_0)^2$ in SSY. From Eq 20, then

$$r_p \propto \frac{J}{g_0}\left(\frac{E}{g_0}\right) \tag{23}$$

Substituting Eq 23 into Eq 22, it follows that

$$\frac{\text{COD}}{r_p} \propto d_n \left(\frac{g_0}{E}\right) \tag{24}$$

Using the expression for d_n from Ref *47*, it may be seen that

$$\frac{\text{COD}}{r_p} \propto \left(\frac{g_0}{E}\right)^{n+1/n} \tag{25}$$

The value of g_0/E in our material model is very small (0.0004)[8]—an order of magnitude smaller than those studied in Ref *47*—and so the COD in our model is very small with respect to the plastic zone size. Recall that the SSY model has been scaled by the plastic zone size. Thus even though the element sizing in our model was comparable to that used in Ref *47*, the size of the crack-tip elements may not be sufficient to precisely model effects on the order of our small COD. Nevertheless, the trend is unmistakable—fast loading inhibits the development of COD. Moreover, the ratio of d for a given loading rate to the inferred equilibrium d, as illustrated in Fig. 7, is in fairly good agreement with the results of our as yet unpublished analysis of a compact tension specimen. In that work, a finer mesh was employed in the crack-tip region and excellent agreement with the HRR field results was obtained. Thus, it appears that the results presented here may simply be scaled by using the HRR field value of d_n for the inferred equilibrium d-value.

Implications for Fracture Analysis

When these results are interpreted by some assumed, yet reasonable, fracture criteria, some conclusions about the influences of time and loading rate on fracture may be reached. Two separate criteria—a stress-based criterion and a strain-based criterion—are employed in the following to interpret the results.

Fracture Stress Criterion—The viability of using a stress-based criterion to predict slip-induced cleavage fracture has been demonstrated by Ritchie, Knott, and Rice [*48*]. In their work it is argued that a critical crack opening stress level, which is independent of temperature, must be achieved over a microstructurally significant distance ahead of the crack in order for slip-induced cleavage fracture to occur. Adopting this viewpoint and assuming that the critical stress level is independent of loading rate, the results in Fig. 4 may be used to predict the influence of loading rate on fracture toughness.

It is noted that the present analysis was based on the room temperature properties of annealed 304 stainless steel. This material would not be ex-

[8]This is an equilibrium property of the low-strength stainless steel.

pected to fail by slip-induced cleavage. Nevertheless, the interpretation of the results with the stress-based criterion is carried out as an exercise to suggest the trends that could occur in metals that would fail in slip-induced cleavage.

In order to apply the stress-based criterion, a critical stress level and a critical distance must be selected.[9] The critical stress level is taken to be $5g_0$ and the critical distance is taken to be at about 1/35th of the equilibrium plastic zone length ahead of the crack. While it may appear that the selected critical distance is small, it is actually quite large compared with the equilibrium COD. With these assumed values, the variation in the critical value of K is developed from Fig. 3 and is shown in Fig. 9. With other reasonable choices of the critical stress and critical distance, a similar pattern would result. Since the fast[10] and slow loading limits are well defined, they are also included in Fig. 9. The trend is obvious. As loading rate is increased, the apparent tough-

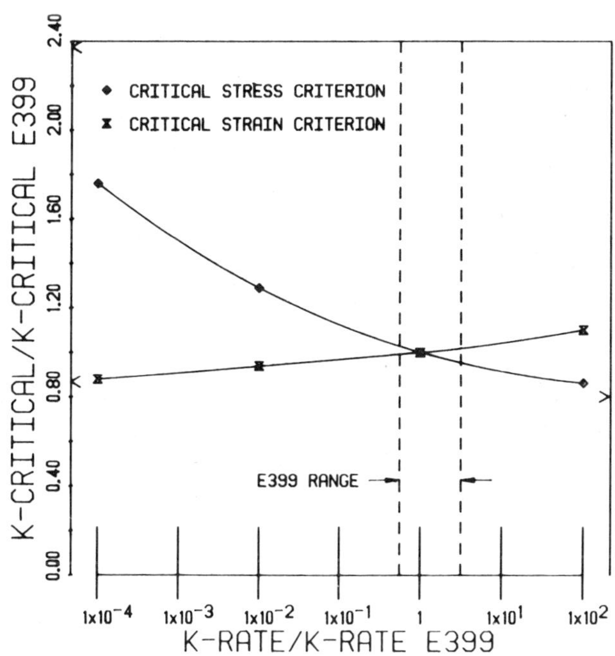

FIG. 9—*Variation of* K-*critical with loading rate for two hypothetical fracture criteria. Infinitesimal and infinite loading rate limits are indicated by arrowheads. The strain-criterion infinite loading rate limit is well off the graph.*

[9]Alternatively, a critical value of K, at say an ASTM rate, could be assumed. This could then be used to evaluate the critical stress if the critical distance is known or guessed. Subsequently, all results could be normalized with respect to the assumed ASTM critical K-value.

[10]The fast loading limit is taken to be the extrapolated linear response at the distance being considered.

ness is decreased. It is noted that over the ASTM Method E 399 loading rate range, only a minor variation in toughness is predicted.

Fracture Strain Criterion—The crack opening displacement has been considered in the foregoing as an appropriate measure of the strength of the crack-tip deformation field. It will be used next in a strain-based fracture criterion. It is a strain measure in that it is the integral of the opening strain at the crack tip. Moreover, its viability for use in ductile fracture has been well demonstrated [*49,50*]. Since at room temperature 304 stainless steel would be expected to fracture in an at least locally ductile mode, the interpretation of the analytical results by means of a COD criterion is perhaps most meaningful.

The lengths in the SSY model are arbitrary so that any value of COD may be selected as the critical value. Accordingly, a convenient one is chosen and it is assumed to be independent of loading rate, that is to say, that a critical deformation (strain) must be reached at the crack tip, regardless of loading rate. The relationship between applied load and this critical deformation state is clearly not loading rate independent. With the assumed critical value taken to be one half of the equilibrium COD, it is clear from Fig. 7 that for different loading rates, different values of J(or K through Eq 20) are associated with the critical COD. The resulting variation in K-critical along with the fast and slow loading limits is shown in Fig. 9. This value is an initiation value only; significant stable crack growth could be evidenced prior to final fracture.

The trend for a strain-based criterion is clearly opposite to that for a stress-based criterion. An apparent increase in toughness is caused by an increase in loading rate. Again only a minor variation in toughness is predicted over the ASTM range.

The implications of the effects of loading rate on apparent toughness are obvious. In stress-controlled fracture, failure could be induced at applied K-levels below the E 399 level by loading faster than the E 399 rate. On the other hand, in strain-controlled fracture, crack growth could be initiated at K-levels below the E 399 level by loading more slowly than the E 399 level. Moreover, from the predicted growth of COD during sustained loading, as shown in Fig. 8, it is possible that crack growth could be initiated at some time subsequent to the initial attainment of maximum load. In fact, such behavior was revealed in an experiment by Ingham and Morland [*51*] on A533B steel at room temperature. Other examples of crack growth during sustained load have been given in Refs *49, 52,* and *53*.

These results are strongly suggestive that safe design approaches can be developed only with consideration of the possible time-dependent response of the material. However, a rather simple design approach, accounting for material time dependence, can be rationalized by appealing to the cases of the fast and slow loading limits available in the theory of viscoplasticity based on total strain and overstress. The approach is as follows.

In slip-induced cleavage fracture, a minimum toughness is associated with the fast loading limit. Thus, if the critical opening stress and the critical distance in the Ritchie-Knott-Rice criterion were known, predictions of minimum toughness could be made using known K-controlled elastic crack-tip stress field solutions. Inertia effects are not included in such an approach, but the result should still be a useful lower bound for quasi-static loading rates.

In strain-induced ductile fracture, a minimum *initiation* toughness is associated with the slow loading limit, that is, with the equilibrium state. Due to the correspondence of the equilibrium stress-strain relations and commonly used power-law plasticity formulations, it would be possible to use available solutions advantageously to predict lower bounds on toughness. The critical COD could be measured in a conventional loading rate test if it is indeed rate-independent. However, the minimum K-value that could be applied to achieve the critical COD would be the one associated with the equilibrium state. The *equilibrium* properties could then be used to determine the minimum applied K-level by using Eqs 20 and 22 to find

$$K_{\mathrm{crit}_{\min}} = \left\{ \frac{\mathrm{COD}_{\mathrm{crit}}\, g_0 E}{d_n (1 - \nu^2)} \right\}^{1/2} \tag{26}$$

where $\mathrm{COD}_{\mathrm{crit}}$ is the critical crack opening displacement.

It is premature to draw conclusions about the influence of material time dependence on final ductile fracture. Analyses of crack growth would be required. Nevertheless, it is hoped that the equilibrium solutions could still be used in a design approach. Such an approach would be extremely attractive since then schemes like those of Kumar, German, and Shih [*17*] could be used to make conservative estimates of fracture-controlling parameters using the *equilibrium* material properties g_0 and d_n.

In closing, some speculation with regard to further implications of our results is offered. First, the prediction of time dependence of plastic deformation at crack tips may be of interest in studies of stress corrosion cracking. If some average strain measure, such as the crack opening displacement, can be related to the rupturing of protective layers [*54*] at crack tips, then the time dependence of that measure could be important. There may be some impact of the continued development of this strain measure during sustained loading as well as the loading rate dependence of this measure in problems involving corrosion fatigue.

Second, it is predicted that crack opening stress is promoted by fast loading and that crack-tip deformation is inhibited by fast loading. If, in ferritic steels, there is a competition between stress-controlled (brittle) fracture and strain-controlled (ductile) fracture, then it would seem possible that a shift in the brittle-to-ductile transition temperature could be induced by an increase in loading rate.

Conclusions

The following conclusions may be drawn from the analyses of the influence of time and loading rate on the crack-tip fields of Type 304 stainless steel at room temperature:

1. In stress-controlled fracture, apparent toughness should be decreased by loading rate increases.
2. In strain-controlled fracture, apparent toughness should be decreased by loading rate decreases.
3. In strain-controlled fracture, crack growth could be initiated at time subsequent to the attainment at maximum load if the load is sustained.
4. The time-dependent response of the material should be taken into account in design against fracture. It appears that conservative design approaches could be developed by using the fast and slow loading limiting states available in the theory of viscoplasticity based on total strain and overstress.

Acknowledgments

The financial support of the first two authors by the National Science Foundation and the Electric Power Research Institute is gratefully acknowledged. The first author would also like to thank Mr. R. Zirin of the General Electric Co. (Schenectady, N.Y.) for supplying his finite-element code, Dr. G. Green of the Central Electricity Generating Board (U.K.) for stimulating discussions, and Dr. T. Ingham of the U.K. Atomic Energy Authority for an advance copy of his manuscript on time-dependent fracture.

References

[1] Hutchinson, J. W., *Journal of the Mechanics and Physics of Solids*, Vol. 16, 1968, pp. 13–31.
[2] Rice, J. R. and Rosengren, G. F., *Journal of the Mechanics and Physics of Solids*, Vol. 16, 1968, pp. 1–12.
[3] Riedel, H. and Rice, J. R. in *Fracture Mechanics: Twelfth Conference, ASTM STP 700*, American Society for Testing and Materials, 1980, pp. 112–130.
[4] Riedel, H., *Journal of the Mechanics and Physics of Solids*, Vol. 29, No. 1, 1981, pp. 35–49.
[5] Hui, H. and Riedel, H., *International Journal of Fracture*, Vol. 17, 1981, pp. 409–425.
[6] Shih, C. F. in *Fracture Analysis, ASTM STP 560*, American Society for Testing and Materials, 1974, pp. 187–210.
[7] Bassani, J. L. and McClintock, F. A., *International Journal of Solids and Structures*, Vol. 17, 1981, pp. 479–492.
[8] Hart, E. W., *International Journal of Solids and Structures*, Vol. 16, 1980, pp. 807–823.
[9] Morjaria, M. and Mukherjee, S., *International Journal of Solids and Structures*, Vol. 17, 1981, pp. 127–143.
[10] Kubo, S., this publication, pp. I-594–I-614.
[11] Aboudi, J. and Achenbach, J. D., "Transition from Brittle to Ductile Fracture for a Rapidly Propagating Crack," American Society of Mechanical Engineers, Paper 81-PVP-16, 1981.
[12] Freund, L. B. and Douglas, A. S., this publication, pp. I-5–I-20.
[13] Begley, J. A. and Landes, J. D. in *Fracture Toughness Proceedings of the 1971 National*

Symposium on Fracture Mechanics, Part II, ASTM STP 514, American Society for Testing and Materials, 1972, pp. 1-20.

[*14*] Broberg, K. B., *Journal of the Mechanics and Physics of Solids*, Vol. 19, 1971, pp. 407–418.

[*15*] Shih, C. F., Andrews, W. R., deLorenzi, H. G., German, M. D., Van Stone, R. H., and Mowbray, D. F., "Methodology for Plastic Fracture," EPRI NP-1735, Final Report, Electric Power Research Institute, Palo Alto, Calif., March 1981.

[*16*] Kanninen, M. F., Hahn, G. T., Breck, D., Stonesifer, R. B., Marschall, C. W., Abou-Sayed, I. S., and Zahoor, A., "Development of a Plastic Fracture Methodology," EPRI NP-1734, Final Report, Electric Power Research Institute, Palo Alto, Calif., March 1981.

[*17*] Kumar, V., German, M., and Shih, C. F., "An Engineering Approach for Elastic-Plastic Fracture Analysis," EPRI NP-1931, Electric Power Research Institute, Palo Alto, Calif., July 1981.

[*18*] Ellis, J. R., Robinson, D. N., and Pugh, C. E., *Nuclear Engineering Design*, Vol. 47, 1978, pp. 115–123.

[*19*] Albertini, C. and Montagnani, M., *Nuclear Engineering Design*, Vol. 37, 1976, pp. 115–124.

[*20*] Maiden, C. J. and Green, S. J., *Transactions*, American Society of Mechanical Engineers, *Journal of Applied Mechanics*, Vol. 33, 1966, pp. 496–504.

[*21*] Karnes, C. H. and Ripperger, E. A., *Journal of the Mechanics and Physics of Solids*, Vol. 14, 1966, pp. 75–88.

[*22*] Krempl, E., *Journal of the Mechanics and Physics of Solids*, Vol. 27, 1979, pp. 363–375.

[*23*] Kujawski, D., Kallianpur, V., and Krempl, E., *Journal of the Mechanics and Physics of Solids*, Vol. 28, 1980, pp. 129–148.

[*24*] Kujawski, D. and Krempl, E., *Transactions*, American Society of Mechanical Engineers, *Journal of Applied Mechanics*, Vol. 48, 1981, pp. 55–63.

[*25*] Krempl, E. and Kallianpur, V., "On the Room Temperature Rate (Time) Dependent Behavior of A533B Class I Steel," Rensselaer Polytechnic Institute Report CS 81-1, Troy, N.Y., Feb. 1981.

[*26*] Cernocky, E. P. and Krempl, E., *International Journal of Nonlinear Mechanics*, Vol. 14, 1979, pp. 183–203.

[*27*] Liu, M. C. M. and Krempl, E., *Journal of the Mechanics and Physics of Solids*, Vol. 27, 1979, pp. 377–391.

[*28*] Krempl, E. *Mechanical Testing for Deformation Model Development ASTM STP 765*, American Society for Testing and Materials, 1982, pp. 5–28.

[*29*] Cernocky, E. P. and Krempl, E., *Acta Mechanica*, Vol. 36, 1980, pp. 263–289.

[*30*] Malvern, L. E., *Journal of Applied Mechanics*, Vol. 18, 1951, pp. 203–208.

[*31*] Krempl, E. and Hewelt, P., *Mechanics Research Communications*, Vol. 7, 1980, pp. 282–288.

[*32*] Asada, Y., Kargarnovin, M. H., and Krempl, E., Rensselaer Polytechnic Institute Report RPI CS 81-3, Troy, N.Y., 1981.

[*33*] Argyris, J. H., Vaz, L. E., and William, K. J., *Computer Methods in Applied Mechanics and Engineering*, Vol. 16, 1978, pp. 231–277.

[*34*] Zirin, R. M. and Krempl, E., "A Finite Element Time Integration Method for the Theory of Viscoplasticity Based on Infinitesimal Total Strain," American Society of Mechanical Engineers, Paper 80-C2/PVP-33, 1980.

[*35*] Larsson, S. G. and Carlsson, A. J., *Journal of the Mechanics and Physics of Solids*, Vol. 21, 1973, pp. 263–277.

[*36*] Rice, J. R. in *Fatigue Crack Propagation, ASTM STP 415*, American Society for Testing and Materials, 1967, pp. 247–309.

[*37*] deLorenzi, H. G. and Shih, C. F., *International Journal of Fracture*, Vol. 13, 1977, pp. 507–511.

[*38*] Henshell, R. D. and Shaw, K. G., *International Journal for Numerical Methods in Engineering*, Vol. 9, 1975, pp. 495–507.

[*39*] Barsoum, R. S., *International Journal for Numerical Methods in Engineering*, Vol. 11, 1977, pp. 85–98.

[*40*] Shih, C. F., *Journal of the Mechanics and Physics of Solids*, Vol. 29, No. 4, 1981, pp. 305–326.

[*41*] Shih, C. F. and German, M. D., *International Journal of Fracture*, Vol. 17, No. 1, Feb. 1981, pp. 27–43.

[*42*] Rice, J. R., *Transactions*, American Society of Mechanical Engineers, *Journal of Applied Mechanics*, Vol. 35, 1968, pp. 379–386.
[*43*] Goldman, N. L. and Hutchinson, J. W., *International Journal of Solids and Structures*, Vol. 11, 1975, pp. 575–591.
[*44*] Landes, J. D. and Begley, J. A. in *Mechanics of Crack Growth, ASTM STP 590*, American Society for Testing and Materials, 1976, pp. 128–148.
[*45*] Nikbin, K. M., Webster, G. A., and Turner, C. E. in *Cracks and Fracture, ASTM STP 601*, American Society for Testing and Materials, 1976, pp. 47–62.
[*46*] Tracey, D. M., *Transactions*, American Society of Mechanical Engineers, *Journal of Engineering Materials and Technology*, Vol. 98, 1976, pp. 146–151.
[*47*] Shih, C. F., *Journal of the Mechanics and Physics of Solids*, Vol. 29, 1981, pp. 305–326.
[*48*] Ritchie, R. O., Knott, J. F., and Rice, J. R., *Journal of the Mechanics and Physics of Solids*, Vol. 21, 1973, pp. 395–410.
[*49*] Green, G. and Knott, J. F., *Journal of the Mechanics and Physics of Solids*, Vol. 23, 1975, pp. 167–183.
[*50*] Andrews, W. R. and Shih, C. F. in *Elastic-Plastic Fracture, ASTM STP 668*, J. D. Landes, J. A. Begley, and G. A. Clarke, Eds., American Society for Testing and Materials, 1979, pp. 426–450.
[*51*] Ingham, T. and Morland, E., this publication, pp. I-721–I-746.
[*52*] Tsuru, S. and Garwood, S. J. in *Proceedings*, International Conference on Fracture Mechanics, Cambridge, U.K., 1979, pp. 519–528.
[*53*] Schulze, H. D. and Fuhlrott, H., *International Journal of Pressure Vessels and Piping*, Vol. 8, 1980, pp. 131–142.
[*54*] Ford, F. P., "A Mechanism of Environmentally Controlled Crack-Growth of Structural Steels in High Temperature Water," General Electric Co. Report 81CRD125, Schenectady, N.Y., Aug. 1981.

Vinod Banthia[1] and Subrata Mukherjee[1]

Boundary-Element Analysis of Stresses in a Creeping Plate with a Crack

REFERENCE: Banthia, V. and Mukherjee, S., **"Boundary-Element Analysis of Stresses in a Creeping Plate with a Crack,"** *Elastic-Plastic Fracture: Second Symposium, Volume I—Inelastic Crack Analysis, ASTM STP 803*, C. F. Shih and J. P. Gudas, Eds., American Society for Testing and Materials, 1983, pp. I-637–I-653.

ABSTRACT: The boundary-element method is used to determine the solution of the problem of a creeping plate with a crack. Particular attention is focussed on the numerical evaluation of a certain line integral—called the C-integral—over various paths around a crack tip, as functions of time for short times. The growth of the creep zone around a crack tip is also studied numerically.

KEY WORDS: fracture mechanics, stress analysis, elevated temperature, creep, C-integral, elastic-plastic fracture

The study of stresses near crack tips is of fundamental importance in fracture mechanics. These stresses are relatively easy to determine if the behavior of the cracked solid is linear elastic, but becomes much more complicated if inelastic (plastic or creep) deformation is included in the analysis. Plastic deformation generally must be included in the study of cracked metallic solids and creep deformation becomes important in metals subjected to mechanical and thermal loads at elevated temperatures.

There has been a considerable amount of recent effort in the determination of stresses in creeping solids that contain cracks. The stress distribution near a crack tip in such a solid has been obtained by asymptotic analysis by Riedel and Rice [*1–3*],[2] by finite-element analysis by Ehlers and Riedel [*4*] and Bassani and McClintock [*5*], and by boundary-element methods by Mukherjee

[1]Graduate student and associate professor, respectively, Department of Theoretical and Applied Mechanics, Thurston Hall, Cornell University, Ithaca, N.Y. 14853.

[2]The italic numbers in brackets refer to the list of references appended to this paper.

and Morjaria [6–9]. The asymptotic approach gives elegant closed-form expressions for the stress field near the crack tip, but neglects elastic strain rates in the vicinity of a crack tip relative to the nonelastic ones. While this assumption is asymptotically correct as one approaches the crack tip, its validity at even a small distance away from the crack tip needs to be carefully examined. The finite-element approach, on the other hand, requires the use of special elements to reflect the proper singular behavior of stresses at the tip of a crack. The form of this stress singularity is the well known $r^{-1/2}$ for the elastic case (where r is the distance from a crack tip) but changes to $r^{-1/n+1}$ (in terms of the power-law creep exponent n) in a creeping plate. Thus, in a creep problem one should be able to change, in a consistent time-dependent fashion, from the elastic singularity at zero time to the proper nonelastic singularity. This appears difficult to achieve by the finite-element approach [5].

It is desirable, therefore, to devise a numerical method that is capable of solving the complete transient problem in a consistent manner. The boundary-element method, as used by Mukherjee and Morjaria [6–9], offers such a possibility. In this formulation, the usual kernels of the integral equations are augmented in such a way that the new kernels satisfy the traction-free boundary conditions on the crack surface. Thus, the proper boundary conditions on the crack surface are satisfied exactly in an implicit manner. Because of this, the crack surface need not even be discretized in a numerical solution procedure in many cases. In other cases, only a known function needs to be evaluated on the crack surface. Thus, in all cases, the unknowns of the problem lie only on the outside boundary of the body.

The primary advantages of this boundary-element method (BEM) over the finite-element method (FEM) lie in the facts that numerical modeling of the crack surface is easier in this case and that no *a priori* assumptions need to be made about the stress field near a crack tip. As mentioned before, such assumptions are usually made in the FEM in order to design special crack-tip elements. The BEM, on the other hand, is able to calculate the changing stress field near the crack tip in a completely consistent fashion. Also, the elastic and non-elastic strain rates are allowed to coexist and the full governing equations are solved in the entire region at all times.

Path-independent integrals have played an invaluable role in fracture mechanics in recent years. The most well known of these is the J-integral due to Rice [10]. This line integral is path-independent for a purely elastic solid. It retains its path-independent character in an elastic-plastic solid if plastic deformation is modeled by the deformation theory of plasticity. An analogous energy rate line integral C has been defined by Landes and Begley [11] for the case of a creeping solid. This integral characterizes the near-tip stress field in a cracked plate in steady-state creep when the stress field becomes independent of time [2] and is path-independent in this limit. It appears to be a viable parameter for characterizing creep crack growth under steady-state conditions [12]. However, its role in the transient small-scale creep regime (analogous to small-scale yielding in elastic-plastic fracture) is not fully understood. One immediate problem is that this integral is not path-independent when elastic strain rates are of significant magnitude, as might be the case, for example, under time-varying loads. The primary purpose of this paper is to investigate this question for a Mode III crack in a creeping plate.

The paper begins with a review of the boundary-element method as applied to study stresses in a creeping plate with a crack. A discussion of the C-integral follows. One must be careful as to whether the nonelastic or the total strain rate is used in the definition of the rate potential. Finally, numerical results are presented for the values of the C-integral on several paths around a crack tip, as functions of time. These results are obtained for Mode III deformation with the plate subjected to a remote stress that starts at zero and increases at a steady rate.

Determination of Stresses by the Boundary-Element Method

The essence of the method is an integral formulation of the problem and the use of modified kernels that satisfy traction-free boundary conditions on the surface of a crack. The method is described here for Mode III deformation of a creeping plate with a crack in it.

It is assumed that the total strain rate $\dot{\epsilon}_{ij}$ can be decomposed into an elastic and nonelastic rate

$$\dot{\epsilon}_{ij} = \dot{\epsilon}_{ij}^{(e)} + \dot{\epsilon}_{ij}^{(n)} \tag{1}$$

The elastic strain components are related to the stress components by Hooke's law and the nonelastic strain rates are assumed to follow non-linear power-law viscous behavior of the type

$$\dot{\epsilon}_{ij}^{(n)} = \frac{3}{2}\frac{\dot{\epsilon}^{n}}{\sigma} s_{ij} \tag{2}$$

$$\dot{\epsilon}^{(n)} = \dot{\epsilon}_c \left(\frac{\sigma}{\sigma_c}\right)^{n} \tag{3}$$

where the stress and strain rate invariants are defined as

$$\sigma = \sqrt{(3/2)s_{ij}s_{ij}}\ , \qquad \dot{\epsilon}^{(n)} = \sqrt{(2/3)\dot{\epsilon}_{ij}^{(n)}\,\dot{\epsilon}_{ij}^{(n)}}$$

where

s_{ij} = deviatoric part of the stress tensor σ_{ij},
$\dot{\epsilon}_c$ and σ_c = reference strain rate and stress, respectively, and
n = index of the power law.

The governing differential equation for the rate of the stress function, $\dot{\Phi}$, for inelastic deformation of a planar body in antiplane shear (Mode III), can be obtained as [*8*]

$$\nabla^2\dot{\Phi} = F^{(n)}, \quad \text{where } F^{(n)} = -2G\left[\frac{\partial\dot{\epsilon}_{31}{}^{(n)}}{\partial x_2} - \frac{\partial\dot{\epsilon}_{32}{}^{(n)}}{\partial x_1}\right] \tag{4}$$

where

G = shear modulus,
$\dot{\epsilon}_{31}{}^{(n)}$ $(= \dot{\epsilon}_{13}{}^{(n)})$ and $\dot{\epsilon}_{32}{}^{(n)}$ $(= \dot{\epsilon}_{23}{}^{(n)})$ = only nonzero nonelastic strain rates,
∇^2 = Laplacian operator in two-dimensions, and
x_i = set of orthogonal cartesian axes with x_1 and x_2 in the plane of the body and x_3 normal to it.

The traction $\tau_3 = \tau$ at any point on the boundary of the body equals the tangential derivative of the stress function, that is, $\tau = d\Phi/dc$, where c is the distance measured along a boundary in the anticlockwise sense. The stress components are given by spatial derivatives of Φ; that is, $\sigma_{31} = \Phi_{,2}$, $\sigma_{32} = -\Phi_{,1}$.

A simply connected region is considered first. In this case, Eq 4 can be transformed into an integral equation

$$2\pi\dot{\Phi}(p) = \oint_{\partial B} (\ell n\, r)_{pQ} F(Q) dc_Q + \int_B (\ell n\, r)_{pq} F^{(n)}(q)\, dA_q \tag{5}$$

where F is the boundary density function to be determined from the boundary conditions and r is the distance between the source point p (or P) and the field point q (or Q) where the lower-case letters denote points inside the body B and capital letters denote points on its boundary ∂B. This equation can be alternatively written as

$$2\pi\dot{\Phi}(p) = \oint_{\partial B} K(p, Q)F(Q)dc_Q + \int_B K(p, q)F^{(n)}(q)dA_q \tag{6}$$

where

$$K = \text{Re}\,[\hat{\phi}(z,\ z_0)] = \text{Re}[\ell\text{n}\,(z - z_0)] \tag{7}$$

with Re the real part of the complex argument and z and z_0 the source and field points, respectively (see Fig. 1 in Ref *8*).

Augmented kernels for a body with a cutout are obtained in the following manner. First it is assumed that the body B, with outer boundary ∂B_2, has a *cir-*

cular cutout with boundary ∂B_1. If the stress function Φ is constant on ∂B_1, the traction on it is zero. This constant can be taken to be zero without loss of generality.

A function ϕ^* is defined as

$$\phi^*(\bar{z}, z_0) = -\ell n \left(\frac{1}{\bar{z}} - z_0 \right) \tag{8}$$

This function satisfies Laplace's equation and is regular outside the unit circle $|z| = 1$. Further, $\phi = \hat{\phi} + \phi^*$ vanishes on the unit circle $|z| = 1$. Thus, if a kernel defined as the real part of ϕ is used in a formulation analogous to Eq 6 for a body with a circular cutout, the cutout will be traction-free.

The function ϕ^* for an elliptic cutout is obtained by using the mapping function

$$z = \omega(\xi) = \frac{1}{\xi} + m\xi \tag{9}$$

where $m = (a - b)/(a + b)$ (with $a + b = 2$) in terms of the semimajor and semiminor axes a and b, respectively, of the ellipse. Now

$$\phi = \ell n (1 - r_i/\xi) - \ell n (1 - r_i \bar{\xi}) \tag{10}$$

where

$$r_i = (z_0 \pm \sqrt{(z_0^2 - 4m)})/2m, \qquad |r_i| \leq 1$$

and

$$\xi = (z \pm \sqrt{(z^2 - 4m)})/2m, \qquad |\xi| < 1.$$

This function vanishes on the surface of the elliptical cutout.

The stress rates for a body with an elliptical cutout become (with $j, k = 1, 2$)

$$2\pi\dot{\sigma}_{3j}(p) = \oint_{\partial B_2} H_{3j}(p, Q) F(Q) dc_Q + \int_B H_{3j}(p, q) F^{(n)}(q) dA_q - \oint_{\partial B_1} H_{3j}(p, Q) D_k^{(n)}(Q) n_k(Q) dc_Q \tag{11}$$

where

$$H_{31} = \text{Re}\left[\frac{\partial \phi}{\partial x_2}\right] \quad \text{and} \quad H_{32} = -\text{Re}\left[\frac{\partial \phi}{\partial x_1}\right]$$

ϕ is given by Eq 10 and

$$D_1^{(n)} = 2G\dot{\epsilon}_{32}^n \quad \text{and} \quad D_2^{(n)} = -2G\dot{\epsilon}_{31}^n$$

The last term in Eq 11 is necessary to obtain single valued displacements on the cutout boundary [*8*].

The traction rate $\dot{\tau}$ at a point P on the boundary ∂B_2, where it is locally smooth, is obtained by taking the limit of Eq 11 as an inside point p approaches a boundary point P. This gives

$$2\pi\dot{\tau}(P) = \oint_{\partial B_2} H_{3j}(P, Q)n_j(P)F(Q)dc_Q + \int_B H_{3j}(P, q)n_j(P)F^{(n)}(q)dA_q - \oint_{\partial B_1} H_{3j}(P, Q)n_j(P)D_k^{(n)}(Q)n_k(Q)dc_Q \quad (12)$$

The boundary integrals in the preceding equation must be interpreted in the sense of Cauchy principal values.

Path-Independence of Line Integrals

The best known path-independent line integral in fracture mechanics is the J-integral due to Rice [*10*]

$$J = \int_\Gamma \left(Wn_1 - \tau_j \frac{\partial u_j}{\partial x_1} \right) dc \quad (13)$$

where

$$W = \int_0^\epsilon \sigma_{ij} d\epsilon_{ij},$$

u_j and τ_j = displacement and traction components at a point on Γ, and
$n_1 = \mathbf{n} \cdot \mathbf{i}$ with $\mathbf{n}$ the unit outward normal to Γ at a point on it and $\mathbf{i}$ a unit vector along the x_1 axis.

In linear elasticity, this integral is path-independent and W is the strain energy density. In fact, this is one of a class of path-independent integrals that can be derived systematically for linear elastic materials (see, for example, Ref *13*). The J-integral is also path-independent in a deformation-type elastic-plastic material where $d\epsilon_{ij}$, in the definition of W, is the total (sum of elastic and plastic) strain increment.

An analogous integral suitable for the study of cracks in creeping plates was first defined by Landes and Begley [*11*]. This is the C-integral defined as

$$C(t) = \int_\Gamma \left(W^* n_1 - \tau_j \frac{\partial \dot{u}_j}{\partial x_1} \right) dc \quad (14)$$

where $W^* = \int_0^{\dot{\epsilon}} \sigma_{ij} d\dot{\epsilon}_{ij}$ can be interpreted as a rate potential. There is some question as to whether the increment of the total or the nonelastic strain rate should be used in the expression for W^*. In this paper, $d\dot{\epsilon}_{ij}$ is taken to be the increment in the total strain rate, in a manner analogous to $d\epsilon_{ij}$ in the J-integral. A similar integral can be defined, instead, with $d\dot{\epsilon}_{ij}{}^{(n)}$ in the expression for W^*, as has been done by Bassani and McClintock [5]. Such an integral is here called $\bar{C}(t)$. The two integrals C and $\bar{C}$ become identical in the limit of steady-state creep where zero stress rates lead to zero elastic strain rates, and this limiting value, called C*, can be shown to be path-independent. This integral C^* has been shown to characterize the stress field near a crack tip in a plate undergoing steady-state creep deformation [2]. In the presence of elastic strain rates, however, as is the case, for example, during transient creep or with time-dependent loading, the integrals C and $\bar{C}$ have different values and both are path-dependent. Their values are obtained numerically for different paths, as functions of time, later in this paper.

The special case of Mode III loading is of particular interest in this paper. In this case, the two parts of the integrand in Eq 14 assume simple forms. These are

$$\tau_j \frac{\partial \dot{u}_j}{\partial x_1} = 2\left[\sigma_{31}n_1 + \sigma_{32}n_2\right]\left[\frac{\dot{\sigma}_{31}}{2G} + \dot{\epsilon}_{31}{}^{(n)}\right] \tag{15}$$

$$W^* = W_1{}^* + W_2{}^*$$

$$W_1{}^* = \int_0^{\dot{\epsilon}} \sigma_{ij} d\dot{\epsilon}_{ij}{}^{(n)} = \frac{n}{n+1}\,\frac{\dot{\epsilon}_c}{(\sigma_c)^n}\,\sigma^{n+1} \tag{16}$$

$$W_2{}^* = \int_0^{\dot{\epsilon}} \sigma_{ij} d\dot{\epsilon}_{ij}{}^{(e)}$$

$$= \frac{1}{G}\left[(\sigma_{31}\dot{\sigma}_{31} + \sigma_{32}\dot{\sigma}_{32})\Big|_0^t - \int_0^t (\dot{\sigma}_{31}{}^2 + \dot{\sigma}_{32}{}^2)dt'\right] \tag{17}$$

The components of the outward unit normal to Γ are denoted by n_1 and n_2 in Eq 15.

Numerical Implementation

The deformation of a center-cracked plate in Mode III is considered in this paper. The crack is modeled as a thin ellipse of axis ratio 1999. A line crack is avoided since the nonelastic strain rates would become singular at a crack tip in such a case, and this would require an analytical description of $\dot{\epsilon}_{31}{}^{(n)}$ and $\dot{\epsilon}_{32}{}^{(n)}$ very near the crack tip in order to evaluate the area integrals in Eqs 11 and 12. The goal in this paper is numerical evaluation of the various quantities without any *a priori* analytical prescriptions regarding stress or strain fields.

The last integral (over the crack boundary ∂B_1) in Eqs 11 and 12 can be shown to vanish for the examples considered in this paper. The outer boundary ∂B_2 is divided into N_2 straight boundary segments using N_b ($N_b = N_2$) boundary nodes. A small part of the interior of the plate near a crack tip (see Figs. 1 and 2) is divided into n_i triangular internal cells. This is a particular advantage of the boundary-element method in that the integral over the body B in Eqs 11 and 12 involves known integrands at any time step and only that part of B with significant values of nonelastic strain rate gradients (that is, $F^{(n)}$) needs to be included. The choice of a mesh can be justified by checking these strain rates *a posteriori* at the final time of integration. Thus, in the numerical results discussed in the next section, for example, the nonelastic strain rates (with $n = 7$) at the outskirts of the mesh of Fig. 1, at the final time, are about nine orders of magnitude lower than the same quantities near the crack tip. Using such a discretization, a discretized version of Eq 12 is obtained as

$$2\pi\dot{\tau}(P_M) = \sum_{N_2} \int_{\Delta c_i} H_{3j}(P_M, Q) n_j(P_M) F(Q) dc_Q + \sum_{n_i} \int_{\Delta A_i} H_{3j}(P_M, q) n_j(P_M) F^{(n)}(q) dA_q \tag{18}$$

where $\tau(P_M)$ are the traction components at a point P on the boundary which coincides with the node M at the center of a segment on ∂B_2, and Δc_i and ΔA_i are boundary elements and internal cells, respectively.

A very simple numerical scheme is used in which the concentrations F are assumed to be piecewise uniform on each boundary segment with their values assigned at the nodes which lie at the centers of each segment. The nonelastic strain rates $\dot{\epsilon}_{3j}^{(n)}$ are interpolated linearly over each triangular internal element so that $F^{(n)}$ is uniform within each element. The integrals of H_{3j} on boundary elements are evaluated analytically for the singular and by Gaussian quadrature (six Gauss points) for the regular portions. Integrals of H_{3j} on triangular internal cells are obtained by Gaussian quadrature (seven Gauss points). This leads to an algebraic system of the type

$$\{\tau\} = [A]\{F\} + \{d\} \tag{19}$$

The coefficients of the matrix $[A]$ contain boundary integrals of the kernels. The traction rates are prescribed, the vector $\{d\}$ contains the contributions from the area integral, and the vector $\{F\}$ contains the unknown source strengths at the boundary nodes. The dimension of the unknown vector C depends only on the number of boundary elements on ∂B_2.

Equation 11 for the stress rates at an internal point p are discretized in a similar fashion.

The solution strategy is as follows. The initial stress field is obtained by solving the corresponding elastic problem. The initial rates of the nonelastic strains are obtained from the constitutive model (2). The vector **d** in Eq 19 is calculated next, and this, together with the prescribed rates of boundary tractions, is used to calculate the initial distribution of concentration F on ∂B_2. These concentrations are now used in the discretized version of Eq 11 to calculate the initial stress rates throughout the body. These rates are used to determine the values of the variables after a small time interval Δt and so on, and in this way the time histories of the relevant variables are obtained. Time integration is carried out by a Euler-type stepwise procedure with automatic time-step control. This method is described in Refs *7* and *14*.

The line integrals C and $\bar{C}$ are evaluated at each time through Eqs 15–17. Each of the three paths—small (S), medium (M), and large (L) (see Fig. 1)—are decomposed into a sufficient number of straight segments and the integral over each segment is obtained by Gaussian quadrature (six Gauss points).

Numerical Results and Discussion

Material Parameters

The material parameters for stainless steel at 400°C, used in these numerical calculations, are [*15*]

$$G = 0.646 \times 10^5 \text{ MPa}$$
$$\dot{\epsilon}_c = 0.277 \times 10^{-3}\,\text{s}^{-1}, \qquad \sigma_c = 0.125 \times 10^4 \text{ MPa}$$
$$n = 2 \quad \text{and} \quad 7$$

Computer Program Validation

The nonelastic stress histories obtained from the boundary-element computer calculations have already been compared against the asymptotic solutions of Riedel and Rice in a previous publications [*9*] and the correlation has been shown to be satisfactory. In order to check the numerical line integration, the J-integral has been computed on the three paths (Fig. 1) for the elastic solution due to an applied remote stress $\sigma_{32}{}^{\infty} = 6.895$ MPa. The values obtained are $J_s = 60.7$ N/m, $J_M = 60.62$ N/m, and $J_L = 60.57$ N/m, which show a numerical scatter within about 0.1 percent. The corresponding theoretical value of J for a path around a line crack is $J = K_{\text{III}}{}^2/2G = 58.507$ N/m [*10*], the difference being due to the different geometries of the cracks in the two cases.

C-Integral for Short Times

The computed values of the C-integral, for the three paths, at different times, for two values of n, are given in Table 1. The remote loading rate is constant, $\dot{\sigma}_{32}{}^{\infty} = 6.895$ MPa/s, with $\sigma_{32}{}^{\infty}\,(0) = 0$. The numbers are considerably

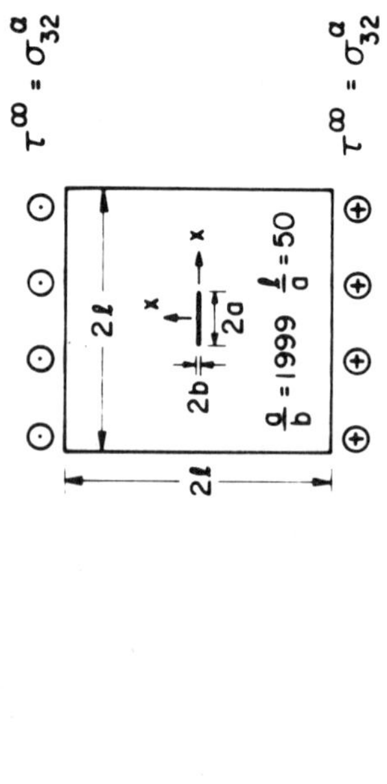
τ^∞ = σ^a_32
τ^∞ = σ^a_32
2ℓ
2ℓ
2b
2a
x
x
a/b = 1999
ℓ/a = 50

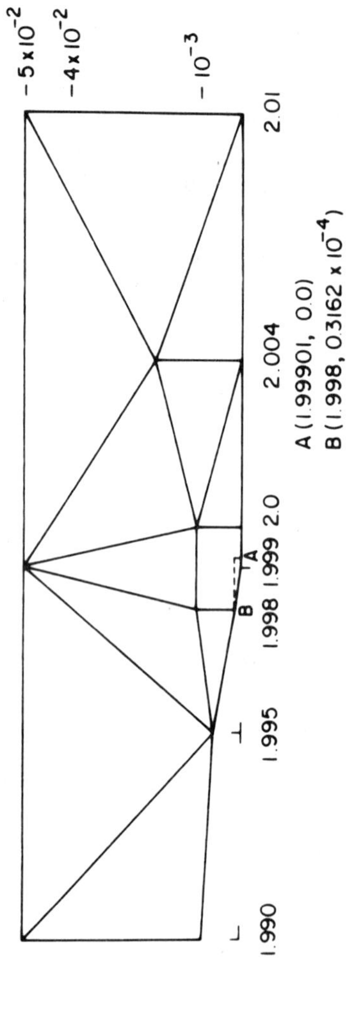
−5 x 10^-2
−4 x 10^-2
−10^-3
2.01
2.004
2.0
1.999
1.998
1.995
1.990
A
B
A (1.99901, 0.0)
B (1.998, 0.3162 x 10^-4)

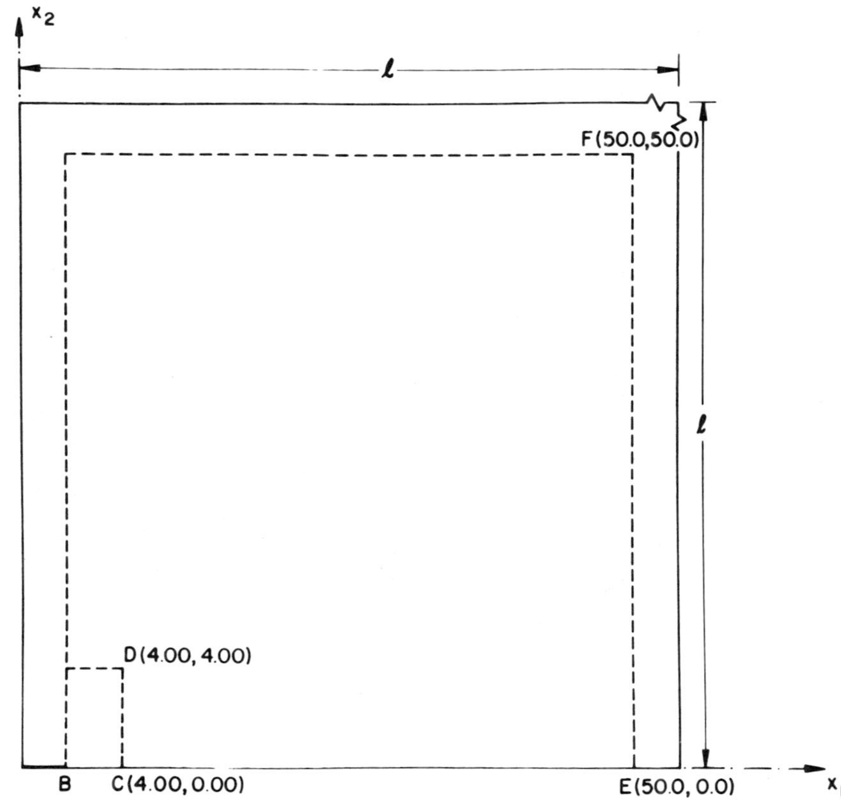

FIG. 1—*Internal mesh with integration paths: small (AB), medium (CDB), and large (EFB).*

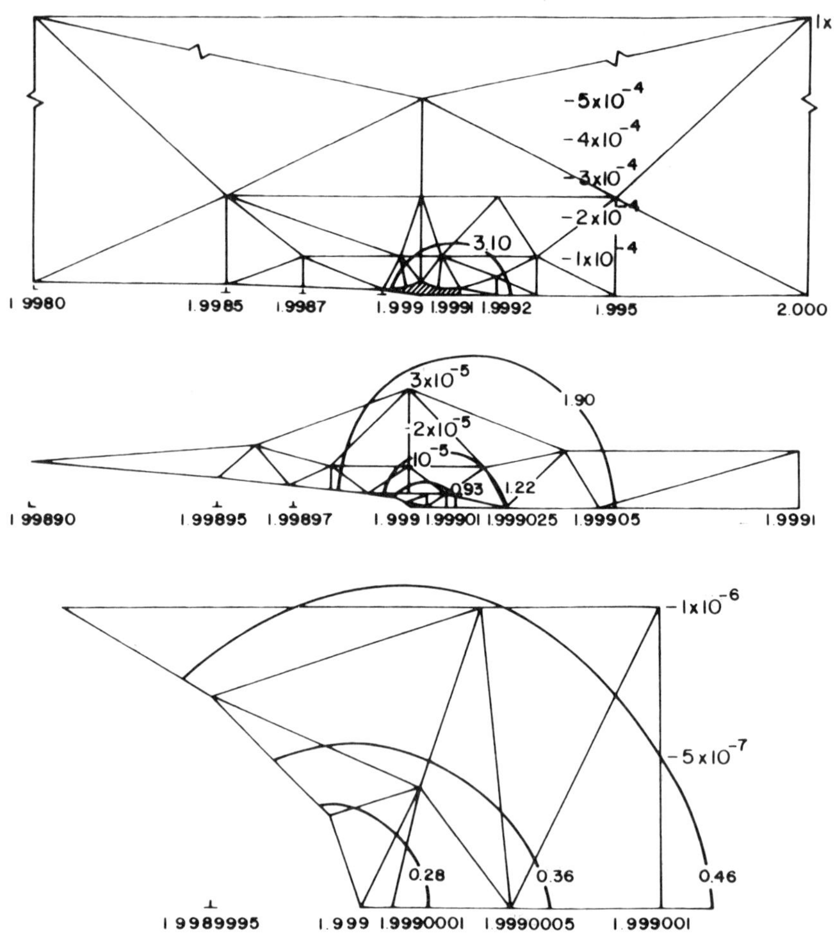

FIG. 2—*Creep zone as function of time.*

different on the three paths, showing that the C-integral is path-dependent in this case. The path-dependence of the C-integral is due to the presence of significant elastic strain rates (relative to nonelastic ones) in these calculations. While it is clear that the C-integral is path-independent during steady-state creep, it otherwise might differ considerably from one path to another. A calculation with a suddenly applied load of $\sigma_{32}^{\infty} = 6.895$ MPa, for example, yields strain rates of $\dot{\epsilon}_{32}^{(e)} = 2905\ \text{s}^{-1}$ and $\dot{\epsilon}_{32}^{(n)} = 1843\ \text{s}^{-1}$ at the crack tip at time 0^+. The comparable magnitude of the two strain rates is expected to make the C-integral strongly path-dependent at short times, even for small paths near a crack tip.

The computed values of the $\overline{C}$-integral (with $d\dot{\epsilon}_{ij}^{(n)}$ in the expression for W^*)

TABLE 1—*Values of* C(t) *for different paths for two values of* n.

	$n = 2$				$n = 7$		
Time	C_s(N·m/s)	C_M(N·m/s)	C_L(N·m/s)	Time	C_s(N·m/s)	C_M(N·m/s)	C_L(N·m/s)
1×10^{-5}	2.47×10^{-4}	8.09×10^{-4}	2.95×10^{-4}	1×10^{-5}	2.47×10^{-4}	8.09×10^{-4}	2.95×10^{-4}
0.57	14.80	46.38	16.94	0.60	25.21	48.87	17.84
1.19	34.50	96.12	35.14	1.23	92.47	99.43	36.32
1.80	61.15	145.95	53.45	1.82	132.00	147.55	53.95
2.62	110.39	212.60	78.11	2.51	321.35	203.64	74.54
2.91	132.84	235.94	86.81	3.02	791.01	244.54	89.62

for two values of n are given in Tables 2 and 3, respectively. The $\bar{C}$-integral is also path dependent since, although $d\dot{\epsilon}_{ij}^{(n)}$ is used in W^*, the values of stresses, traction, and displacement rates are those obtained from the complete solution, and are the same quantities that are used before in C. This, in fact, leads to very nearly equal values of C and $\bar{C}$ over the medium and large paths where the contribution of W^* in C is very small. The integrals over the smallest path start diverging fairly quickly as soon as the contribution of W^* in C starts becoming significant. This difference is much more pronounced for $n = 7$ than for $n = 2$.

The overall variation of the dimensionless Ĉ-integral on the different paths, with time, is shown in Figs. 3 and 4 for two values of n. The Ĉ-integral is defined as

$$\hat{C}(t) = \frac{C(t)}{\sigma_\infty \dot{\epsilon}_\infty{}^{(n)} a} \tag{20}$$

where σ_∞ and $\dot{\epsilon}_\infty{}^{(n)}$ are the remote stress and nonelastic strain rate invariants, respectively, and a is half the crack length. This integral decreases in time owing to the strong increase of the denominator in Eq 20 as a function of time. The overall qualitative behavior of $\hat{C}$ is seen to be similar in all cases and shows the expected relaxation with time.

As stated earlier, the C and $\bar{C}$-integrals must both become path-independent and converge to C^* in the limiting case of steady-state creep where the stress rates, and therefore the elastic strain rates, vanish. The numerical results presented in this paper are obtained for time-dependent remote loading for which exact steady-state through-out the plate cannot ever be achieved.

TABLE 2—*Comparison of* C(t) *and* $\bar{C}$(t) *for different paths,* n = 2.

Time	C_s(N·m/s)	$\bar{C}_s$(N·m/s)	C_M(N·m/s)	$\bar{C}_M$(N·m/s)	C_L(N·m/s)	$\bar{C}_L$(N·m/s)
1×10^{-5}	2.47×10^{-4}	2.47×10^{-4}	8.09×10^{-4}	8.09×10^{-4}	2.95×10^{-4}	2.95×10^{-4}
0.57	14.80	14.84	46.38	46.38	16.94	16.94
1.19	34.50	35.22	96.12	96.11	35.14	35.14
2.91	132.84	154.83	235.94	235.74	86.81	86.70

TABLE 3—*Comparison of* C(t) *and* $\bar{C}$(t) *for different paths,* n = 7.

Time	C_s(N·m/s)	$\bar{C}_s$(N·m/s)	C_M(N·m/s)	$\bar{C}_M$(N·m/s)	C_L(N·m/s)	$\bar{C}_L$(N·m/s)
1×10^{-5}	2.47×10^{-4}	2.47×10^{-4}	8.09×10^{-4}	8.09×10^{-4}	2.95×10^{-4}	2.95×10^{-4}
0.60	25.21	33.77	48.87	48.86	17.84	17.84
1.23	92.47	150.63	99.43	99.37	36.32	36.29
1.82	132.00	199.07	147.55	147.36	53.95	53.84
2.24	206.59	257.09	181.60	181.26	66.44	66.24

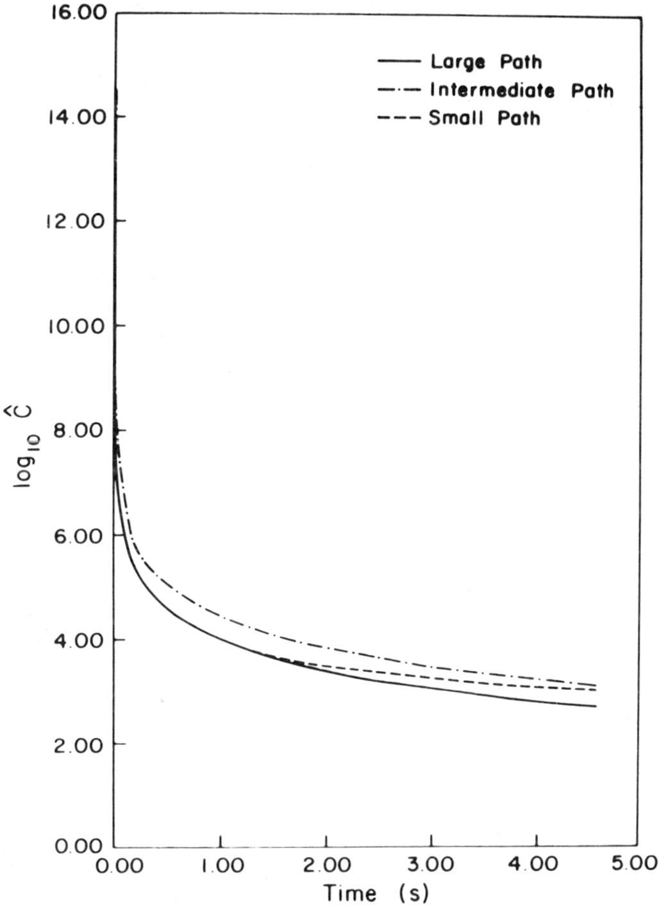

FIG. 3—*Dimensionless C-integral as function of time over three paths:* $\dot{\sigma}_{32}^{\infty} = 6.895$ *MPa/s,* n = *2.0.*

Similar results for time-independent remote loading are presented in a recent paper by the same authors [*16*]. There it is shown that the numerically calculated values of $C(\tau)$ and $\bar{C}(\tau)$, over the three paths, tend to converge within numerical accuracy. The value of the time τ in Ref *16* is a thousand seconds.

Stress Redistribution

Numerical results for the redistribution of stresses near a crack tip, with time, have been presented previously for several examples [*8*, *9*]. They are not repeated in this paper.

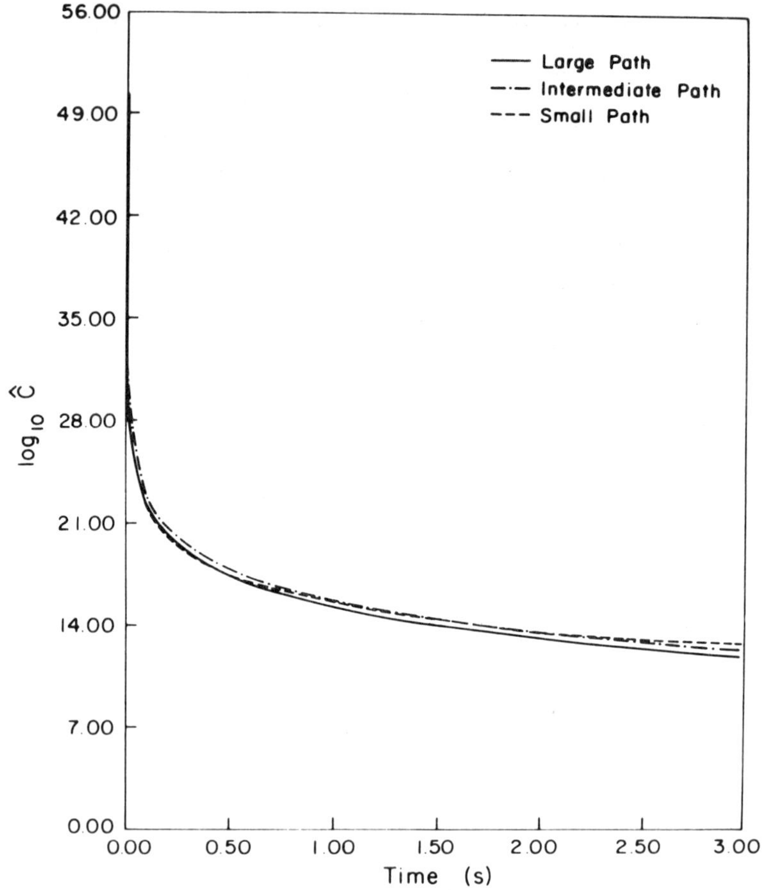

FIG. 4—*Dimensionless C-integral as function of time over three paths:* $\dot{\sigma}_{32}{}^{\infty} = 6.895$ *MPa/s,* n = *7.0.*

Growth of Creep Zone

The boundary of the creep zone is defined in the usual way as the locus of points where the elastic and nonelastic strain invariants are equal; that is $\epsilon^{(e)} = \epsilon^{(n)}$. The growth of the creep zone for the problem discussed in the preceding section is shown in Fig. 2 for the case $n = 7$. The creep zones are seen to be nearly circular as has been suggested by Riedel [*1*].

Conclusions

The important conclusions derived from this paper are as follows:

1. The boundary-element method as used here is very well suited for the determination of quantities of interest near crack tips in a creeping plate, with-

out any *a priori* assumptions whatsoever about the nature of the stress field or relative magnitudes of the elastic and nonelastic strain rates.

2. The C-integral is numerically determined to be path-dependent around a crack tip in a creeping solid at short times. Thus, except for the case of steady-state creep, the usefulness of the C-integral as a controlling parameter for crack growth rate in such a solid remains an open question.

Acknowledgments

This research was supported by Grant No. CME-7918658 of the National Science Foundation to Cornell University. Sincere thanks are expressed to Dr. Mahesh Morjaria for many valuable discussions during the course of this research.

References

[*1*] Riedel, H., *Zeitschrift fur Metallkunde*, Vol. 69, No. 12, 1978, pp. 755–760.

[*2*] Riedel, H. and Rice, J. R. in *Fracture Mechanics: Twelfth Conference, ASTM STP 700*, American Society for Testing and Materials, 1980, pp. 112–130.

[*3*] Riedel, H., "Creep Deformation at Crack Tips in Elastic-Viscoplastic Solids," Technical Report MRL E-114, Materials Research Laboratory, Brown University, Providence, R.I., June 1979.

[*4*] Ehlers, R. and Riedel, H., "A Finite Element Analysis of Creep Deformation in a Specimen Containing a Macroscopic Crack," Presented at ICF5, 5th International Conference on Fracture, Cannes, France, 1981.

[*5*] Bassani, J. L. and McClintock, F. A., *International Journal of Solids and Structures*, Vol. 17, No. 5, 1981, pp. 479–492.

[*6*] Mukherjee, S. and Morjaria, M., *International Journal of Solids and Structures*, Vol. 17, No. 1, 1981, pp. 115–126.

[*7*] Morjaria, M. and Mukherjee, S., *International Journal of Solids and Structures*, Vol. 17, No. 1, 1981, pp. 127–143.

[*8*] Mukherjee, S. and Morjaria, M., *International Journal of Solids and Structures*, Vol. 17, No. 8, 1981, pp. 753–763.

[*9*] Morjaria, M. and Mukherjee, S., "Numerical Solutions for Stresses Near Crack Tips in Time-Dependent Inelastic Fracture Mechanics," *International Journal of Fracture* (in press).

[*10*] Rice, J. R., *Journal of Applied Mechanics*, Vol. 35, No. 2, June 1968, pp. 379–386.

[*11*] Landes, J. D. and Begley, J. A. in *Mechanics of Crack Growth, ASTM STP 590*, American Society for Testing and Materials, 1976, pp. 128–148.

[*12*] Kumar, V. and Shih, C. F. in *Proceedings*, of Second International Conference on Engineering Aspects of Creep, Institute of Mechanical Engineers, London, 1980, pp. 211–214.

[*13*] Delph, T. J., "Conservation Laws in Linear Elasticity Based Upon Divergence Transformations," *Journal of Elasticity* (in press).

[*14*] Kumar, V., Morjaria, M., and Mukherjee, S., *Transactions*, American Society of Mechanical Engineers, *Journal of Engineering Materials and Technology*, Vol. 102, No. 1, Jan. 1980, pp. 92–96.

[*15*] Odquist, F. K. G., *Mathematical Theory of Creep and Creep Rupture*, Clarendon Press, Oxford, U.K., 1966.

[*16*] Banthia, V. and Mukherjee, S., *Res Mechanica*, Vol. 5, 1982, pp. 151–158.

David J. Smith[1] *and George A. Webster*[1]

Estimates of the C^* Parameter for Crack Growth in Creeping Materials

REFERENCE: Smith, D. J. and Webster, G. A., **"Estimates of the C^* Parameter for Crack Growth in Creeping Materials,"** *Elastic-Plastic Fracture: Second Symposium, Volume I—Inelastic Crack Analysis, ASTM STP 803,* C. F. Shih and J. P. Gudas, Eds., American Society for Testing and Materials, 1983, pp. I-654–I-674.

ABSTRACT: This paper discusses methods of determining the creep equivalent of the J-contour integral termed C^* for a cracked body subjected to creep deformation. It is shown that C^* can be expressed in terms of a creep displacement rate, which can be measured either at the loading points, or locally across the crack plane and an appropriate nondimensional factor. It is found that this factor can be estimated from numerical calculations for a material undergoing power-law hardening plastic deformation or from approximate limit analysis techniques. With either approach the nondimensional factor is found to be dependent on specimen geometry and crack length but to be relatively insensitive to the creep stress index n.

The form of expression developed for C^* is particularly suitable when experimental measurements of displacement rate can be made since it enables C^* to be determined with little knowledge of the creep properties of the material. Comparisons with experimental creep crack growth data have indicated that the approximate limit analysis estimates of C^* are satisfactory for the most part.

KEY WORDS: creep crack growth, fracture mechanics, C^* parameter, limit analysis, numerical methods, elastic-plastic fracture

The need to assess the integrity at elevated temperatures of components containing crack-like defects has caused attention to be focused on methods of characterizing creep crack growth. The relevance of linear elastic fracture mechanics (LEFM) concepts, nonlinear fracture mechanics, and reference stress σ_{ref} techniques have all been considered with varying degrees of success [*1–14*].[2] Generally, comparisons of creep cracking with stress-intensity factor K or reference stress (or net section stress) have been satisfactory only in restricted

[1]Research assistant and reader, respectively, Mechanical Engineering Department, Imperial College of Science & Technology, London SW7, U.K.

[2]The italic numbers in brackets refer to the list of references appended to this paper.

circumstances [*1–6*]. Because of the highly nonlinear nature of the stress dependence of creep deformation in metals, there has been an increasing trend towards the adoption of a creep equivalent of the J-contour integral, termed C^* [*9–14*] or $\dot{J}$ [*7,8*], as a characterizing parameter for creep crack growth. Although sometimes there is appreciable scatter in the data, a considerable body of results has been amassed over a range of materials and testing conditions [*7–14*] to indicate that crack growth rate $\dot{a}$ can often be expressed as

$$\dot{a} \propto C^{*\phi} \tag{1}$$

where ϕ is a number slightly less than one.

A difficulty with making comparisons with the C^* approach has been uncertainty over the accuracy of estimates of this parameter for a sufficiently wide range of test specimen geometries. An analytical expression for C^* for double-cantilever beam (DCB) test specimens was developed by Webster [*7*] and a data reduction method of experimental results was employed by Landes and Begley [*9*], but for the most part approximate estimates of C^* have been used. The approximate estimates have been obtained from dimensional analysis arguments [*11*] or limit analysis methods [*10,13,14*].

Recently [*15–17*] numerical estimates of J have been produced for a range of shapes of test specimens made of materials which undergo power-law hardening deformation. In this paper it is shown how these plasticity solutions can be used to estimate C^* provided the power-law hardening stress index is the same as that for creep. Comparisons are made between the numerical calculations and approximate estimates of C^* based on limit analysis techniques. The influence of calculating C^* by both methods on correlations of creep crack growth data is examined.

Analysis

It is possible to define a contour integral around a crack tip in a creeping material in the same way as for a time-independent material. In general, a material may exhibit primary, secondary, and tertiary creep. When secondary creep dominates, as is assumed in many practical circumstances, creep strain rate $\dot{\epsilon}$ can be expressed in terms of stress σ by

$$\dot{\epsilon} = A\sigma^n \tag{2}$$

where A is a proportionality factor which will depend on temperature and n is the stress sensitivity of creep. Expressed in multiaxial stress form Eq 2 becomes

$$\dot{\epsilon}_{ij} = 3/2\, A \cdot \sigma_e^{(n-1)}\, \sigma'_{ij} \tag{3}$$

where

$\dot{\epsilon}_{ij}$ = the creep strain rate tensor,
σ'_{ij} = deviatoric stress, and
σ_e = equivalent stress which may be defined in terms of the Tresca or von Mises equivalent stress criteria.

Consider a contour Γ surrounding a crack in a two-dimensional body of unit thickness. Then an integral C^* can be defined such that

$$C^* = \int_{\Gamma} W_s^* \, dy - T_i \left(\frac{\partial \dot{u}_i}{\partial x} \right) ds \tag{4}$$

where W_s^* is strain energy rate density and is given by

$$W_s^* = \int \sigma_{ij} \, d\dot{\epsilon}_{ij} \tag{5}$$

T_i and $\dot{u}_i$ are components of the traction and displacement rate vectors, respectively, and s is the arc length along Γ. Clearly, when secondary creep dominates, for a given crack length and loading conditions, C^* will be independent of time. The integral is equivalent to the Rice [*18*] J-contour integral but with strain replaced by strain rate in Eq 4. C^* will be subject to the same restrictions as J with regard to path independence. When path independence is valid, C^* can be used to characterize the stress and strain rate fields around a crack tip in a material in which secondary creep dominates such that at a distance r ahead of the crack tip [*9–13*],

$$\sigma_{ij} \, \alpha \left(\frac{C^*}{rA} \right)^{1/(n+1)} \tag{6}$$

and

$$\dot{\epsilon}_{ij} \, \alpha \, A \left(\frac{C^*}{rA} \right)^{n/(n+1)} \tag{7}$$

These equations are of the same form [*19,20*] as those obtained for a power-law hardening material deforming according to the law

$$\epsilon = A' \sigma^N \tag{8}$$

where A' and N are material coefficients. Strict comparison is provided when A' is replaced by A, N by n, and ϵ_{ij} by $\dot{\epsilon}_{ij}$. Therefore, around a crack in a creeping material a stress state will be found that is identical to that in a power-law hardening material when $n = N$.

Following Rice [*21*], C^* can be obtained from the load P/load-point displacement rate $\dot{\Delta}$ diagram in the same way that J can be determined from a load-displacement Δ diagram. Consequently, from Fig. 1, for a body of thickness B

$$C^* = -\frac{1}{B}\frac{dU^*}{da} \tag{9}$$

where dU^* is the change in energy rate for a crack extension da. When the secondary creep law (Eq 2) applies, the deformation and stress histories at all points in the body will be proportional, and it can be shown from Ilyushin's theorem [22] that

$$\dot{\Delta} = C_n P^n \tag{10}$$

where C_n is a measure of the creep compliance of the body. This will in general depend upon the test specimen dimensions, crack length, A, and n. Substituting Eq 10 into Eq 9 gives

$$C^* = \frac{P\dot{\Delta}}{BW}\left[\frac{W}{(n+1)} \cdot \frac{1}{C_n} \cdot \frac{dC_n}{da}\right] \tag{11}$$

which can be written in the form

$$C^* = \frac{P\dot{\Delta}}{BW} \cdot F \tag{12}$$

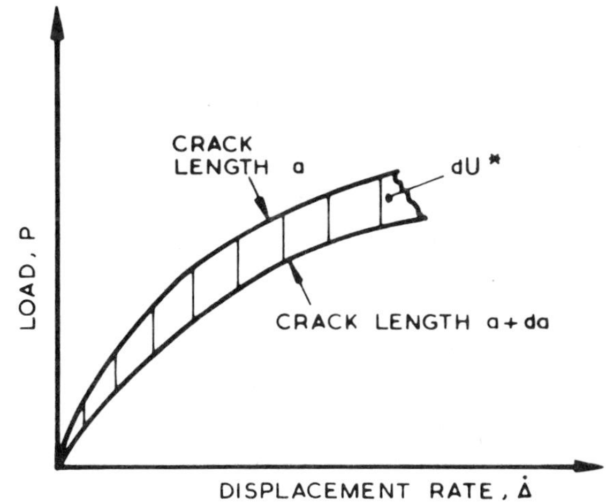

FIG. 1—*Load-displacement rate curve for a cracked body exhibiting steady-state creep deformation.*

where F is a nondimensional factor given by the square brackets in Eq 11. For crack extension at constant load, Eq 11 reduces to

$$C^* = \frac{P\dot{\Delta}}{BW}\left[\frac{W}{(n+1)}\,\frac{1}{\dot{\Delta}}\,\frac{d\dot{\Delta}}{da}\right] \tag{13}$$

or if crack growth takes places at constant displacement rate.

$$C^* = \frac{P\dot{\Delta}}{BW}\left[\frac{n}{(n+1)}\cdot\frac{W}{P}\cdot\frac{dP}{da}\right] \tag{14}$$

The square brackets in any of Eqs 11, 13, or 14 can be used to calculate F. Analytical solutions to these equations are possible only in very restricted circumstances [*7*]. In most cases approximate procedures or numerical computations have to be employed.

The approximate procedures are based on the experimental observation that $n \gg 1$ in Eq 2 for most metals. In this circumstance limit analysis techniques can be applied to express P as a fraction of the plastic collapse load of the cracked body P_{LC} [*23*]. Provided the proportionality factor is only weakly dependent on crack length, it is possible to write [*13*]

$$\frac{1}{P}\,\frac{dP}{da} \approx \frac{1}{P_{LC}}\,\frac{dP_{LC}}{da} \tag{15}$$

so that from Eq 14

$$F \approx \left[\left(\frac{-n}{n+1}\right)\cdot\frac{W}{P_{LC}}\,\frac{dP_{LC}}{da}\right] \tag{16}$$

Substitution of Eq 16 into Eq 12 allows approximate estimates of C^* to be obtained for a range of values of n and test specimen configurations, provided appropriate limit load solutions are available.

For any particular specimen the deflection rate $\dot{\Delta}$ can be separated into a component due to the uncracked body $\dot{\Delta}_0$ and a component due to the crack alone, $\dot{\Delta}_c$, so that

$$\dot{\Delta} = \dot{\Delta}_0 + \dot{\Delta}_c \tag{17}$$

In this analysis it is assumed that creep terms dominate so that elastic displacements can be ignored. As a consequence, C^* can be expressed in terms of either of the preceding two displacement rates using nondimensional factors F_0 and F_c in Eq 12 so that

$$C^* = \frac{P\dot{\Delta}_0}{BW}F_0 = \frac{P\dot{\Delta}_c}{BW}F_c \tag{18}$$

where $\dot{\Delta}F = \dot{\Delta}_0 F_0 = \dot{\Delta}_c F_c$. This condition is satisfied with Eq 17 when

$$\frac{1}{F} = \frac{1}{F_0} + \frac{1}{F_c} \tag{19}$$

Consequently, either Eq 12 or Eq 18 can be used to evaluate C^*.

Equation 12 is written in a convenient form for obtaining C^* from experimental data when the precise creep behavior of the material is not known. Provided the load-line displacement rate $\dot{\Delta}$ can be measured as a function of crack length and the dependence of F on crack length is also available, C^* can be determined directly from Eq 12. In the next section, approximate estimates of F are compared with numerical solutions for a selection of test specimen configurations.

Estimates of C* for Selected Test Specimens

The approximate estimates have all been made by determining F from Eq 11 using limit analysis techniques. The numerical computations have been arrived at by making use of the similarity in the forms of Eqs 2 and 8.

Finite-element calculations of J and the corresponding displacement Δ_c and Δ_0 have been made for power-law hardening materials by a number of authors [*15–17*]. Using the analogy therefore between creep and power-law hardening behavior referred to previously, it is possible by substituting $N = n$ to obtain

$$F_c = \frac{JBW}{P\Delta_c} = \frac{C^*BW}{P\dot{\Delta}_c} \tag{20}$$

and

$$F_0 = \frac{JBW}{P\Delta_0} = \frac{C^*BW}{P\dot{\Delta}_0} \tag{21}$$

The term F is then determined from Eq 19.

In the comparisons which follow, attention has been confined to plane-strain conditions. Also, for consistency, the von Mises criterion has been adopted throughout even though F is independent of the yield criterion chosen.

Center-Cracked Plate Specimen (CCP)

The appropriate dimensions of this test specimen are shown in Fig. 2. The term F can be estimated from the collapse load for the cracked plate P_{LC} from Haigh and Richards [*23*] using

$$P_{LC} = \frac{4\sigma_y}{\sqrt{3}} B(W - a) \tag{22}$$

where σ_y is the uniaxial yield stress. This gives from Eq 16[3]

$$F = \frac{n}{n + 1} \cdot \frac{1}{2[1 - a/W]} \tag{23}$$

The numerical results for F for the CCP geometry can be obtained from the power-law hardening solutions of Kumar et al [17] and Hutchinson et al [16] by rewriting their expression for J and plastic displacements in terms of C^* and creep displacement rates as previously indicated. The creep equivalents of their equations then become

$$C^* = Aa(1 - a/W)g_1\left[\frac{\sqrt{3}}{4} \cdot \frac{P}{B} \cdot \frac{1}{(W - a)}\right]^{(n+1)} \tag{24}$$

and

$$\dot{\Delta}_c = Aag_3\left[\frac{\sqrt{3}}{4} \cdot \frac{P}{B(W - a)}\right]^n \tag{25}$$

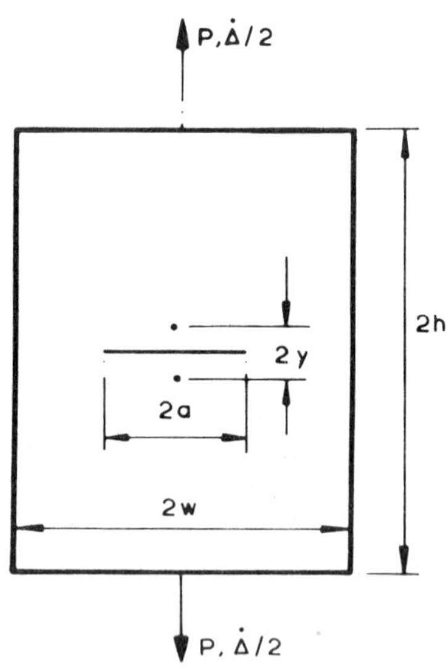

FIG. 2—*Center-cracked plate (CCP) geometry.*

[3]Where the factor of 2 arises because a symmetrical crack of length $2a$ was assumed.

where g_1 and g_3 are functions of n and geometry alone. Substituting in Eq 20 gives

$$F_c = \frac{\sqrt{3}}{4}\frac{g_1}{g_3} \tag{26}$$

For the plane-strain displacement rate of the un-cracked plate (Fig. 2)

$$\dot{\Delta}_0 = \sqrt{3}\,hA\left[\frac{\sqrt{3}}{4}\frac{P}{W}\right]^n \tag{27}$$

giving from Eq 21

$$F_0 = \frac{g_1}{8}\frac{a}{W}\left[\frac{1}{1 - a/W}\right]^n \tag{28}$$

Values of g_1 and g_3 have been calculated by Kumar et al [17] for a range of values of n. The corresponding values of F_0, F_c, and F for a typical creep exponent of $n = 7$ are listed in Table 1. In Fig. 3, a comparison is made between the numerical estimates of F and the approximate estimates of F from Eq 23 for $n = 3$, 7, and 20. The significance of the differences is discussed later.

Single-Edge-Notch Bend Specimen (SENB)

Both pure bend and three-point bend conditions have been considered. The appropriate specimen dimensions are shown in Fig. 4. For pure bend the collapse moment is [23]

$$M_{LC} = 0.364\,\sigma_y(W - a)^2 \tag{29}$$

TABLE 1—*Numerical solutions of* F_c, F_0 *and* F *for CCP, SENB (pure bend and three-point bend loading), and compact tension specimens for* $n = 7$.

	CCP			SENB (pure bend)			SENB (3-point bend)			
a/W	F_c	F_0	F	F_c	F_0	F	F_c	F_0	F	CT, F
0.25	0.747	0.745	0.373	2.186	3.901	1.401	2.077	6.511	1.575	1.717
0.375	0.708	2.870	0.568	...	...	...	2.701	110.13	2.636	2.736
0.5	0.787	14.07	0.745	3.297	1433.1	3.289	3.273	$>10^3$	3.268	4.115
0.625	0.970	87.58	0.959	...	...	...	4.475	$>10^3$	4.475	5.286
0.75	1.526	$>10^3$	1.524	7.084	$>10^3$	7.082	6.860	$>10^3$	6.86	6.913

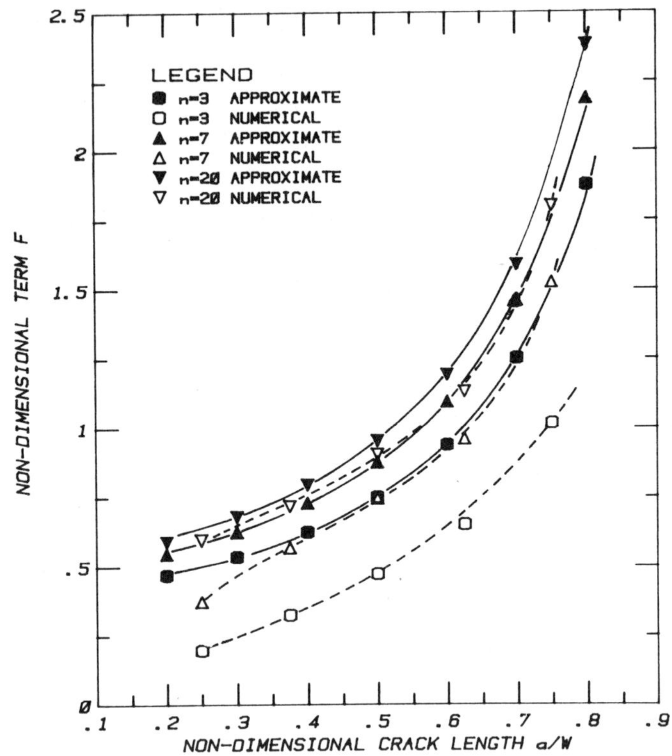

FIG. 3—*Numerical and approximate solutions for* F *for CCP geometry*, n = *3, 7, and 20.*

and for the three-point bending the collapse load is [*23*]

$$P_{LC} = \frac{0.364}{L} \sigma_y (W - a)^2 \tag{30}$$

For both cases the term F becomes

$$F = \frac{2n}{n+1} \cdot \frac{1}{1 - a/W} \tag{31}$$

For pure bending when the angle of rotation $\dot{\theta}$ is measured, C^* can be expressed in equivalent form to Eq 12 as

$$C^* = \frac{M\dot{\theta}}{BW} \cdot F \tag{32}$$

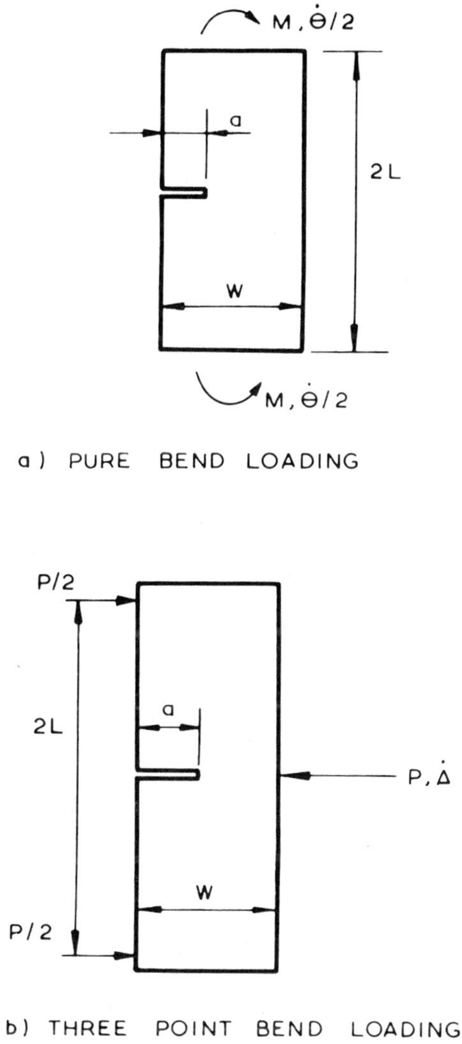

FIG. 4—*Single-edge-notch bend (SENB) geometry:* (a) *pure bend loading;* (b) *three-point bend loading.*

The corresponding numerical solutions can be obtained from Hutchinson et al [*16*] for pure bend and Kumar et al [*17*] for the three-point bend. For pure bending

$$C^* = A(W-a)\,h_1\left[\frac{M}{B\,0.364\,(W-a)^2}\right]^{(n+1)} \tag{33}$$

and

$$\dot{\theta}_c = A\, h_3 \left[\frac{M}{B\, 0.364\, (W - a)^2} \right]^n \tag{34}$$

where h_1 and h_3 are functions of n and geometry alone. The nondimensional term F_c then becomes

$$F_c = \frac{1}{1 - a/W} \cdot \frac{1}{0.364} \cdot \frac{h_1}{h_3} \tag{35}$$

The rotation rate $\dot{\theta}_0$ for the uncracked geometry can be shown from creep analysis to be

$$\dot{\theta}_0 = \frac{2\sqrt{3}\, Ah}{W - a} \left[\frac{2n + 1}{n} \frac{\sqrt{3}\, M}{(W - a)^2} \right]^n \tag{36}$$

For $h = 2W$, F_0 therefore is given by

$$F_0 = \left[\frac{n}{2n + 1} \right]^n \cdot \left[\frac{1}{0.364\sqrt{3}} \right]^{(n+1)} \cdot \left[\frac{1}{1 - a/W} \right]^{(2n+1)} \cdot \frac{h_1}{4} \tag{37}$$

Numerical values of F_0, F_c, and F have been estimated from the results of Hutchinson et al [*16*] and Kumar et al [*17*] for a stress exponent of $n = 7$ and are listed in Table 1. A comparison with the limit analysis estimates of F from Eq 31 for $n = 3$, 7, and 10 is shown in Fig. 5.

The respective values for three-point bend loading obtained from Kumar et al [*17*] are

$$C^* = A(W - a) j_1 \left[\frac{P \cdot L}{0.728\, B\, (W - a)^2} \right]^{(n+1)} \tag{38}$$

$$\dot{\Delta}_c = A a j_3 \left[\frac{P \cdot L}{0.728 B (W - a)^2} \right]^n \tag{39}$$

where j_1 and j_3 are again functions of n and geometry alone. For a length-to-width ratio of 2, F_c is given by

$$F_c = \frac{W}{0.364 a} \cdot \frac{j_1}{j_3} \left[\frac{1}{1 - a/W} \right] \tag{40}$$

The displacement rate of the uncracked geometry neglecting any shear component is

$$\dot{\Delta}_0 = A\left[\frac{2n+1}{n} \cdot \frac{P \cdot L}{BW^2}\right]^n \frac{2}{W} \cdot \frac{L^2}{n+2} \tag{41}$$

For a length-to-width ratio of 2

$$F_0 = \frac{h_1}{2.912} \cdot \frac{n+2}{1-a/W} \cdot \left[\frac{2n}{(2n+1)\,1.456\,(1-a/W)^2}\right]^n \tag{42}$$

Values of F_0, F_c, and F determined from the numerical results of Kumar et al [*17*], again for $n = 7$, are given in Table 1. A comparison of the numerical and approximate estimates of F for $n = 3$, 7, and 20 is shown in Fig. 6.

Compact Tension Specimen (CT)

The nondimensional term F may be determined from limit loads for this geometry from several sources. In particular, F can be derived from the col-

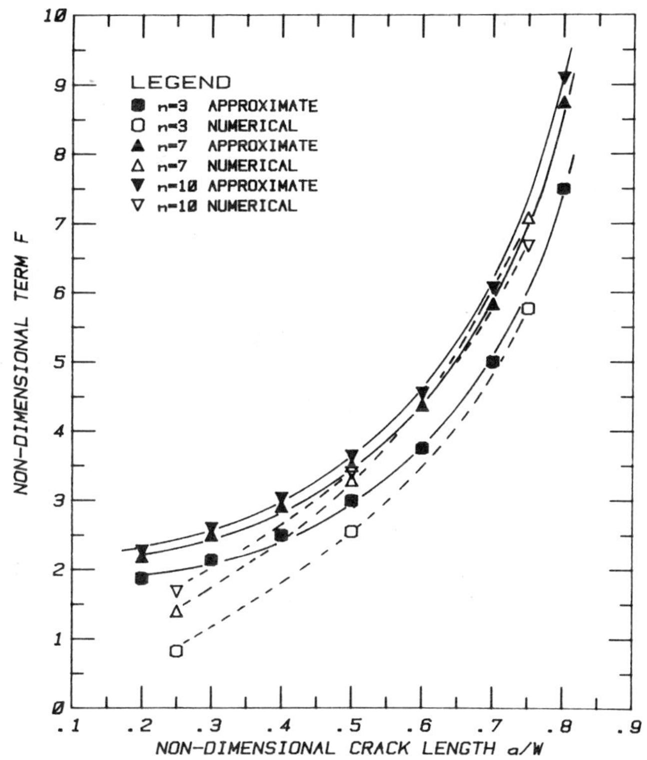

FIG. 5—*Numerical and approximate solutions for* F *for SENB geometry, pure bend loading,* n = *3, 7, and 10.*

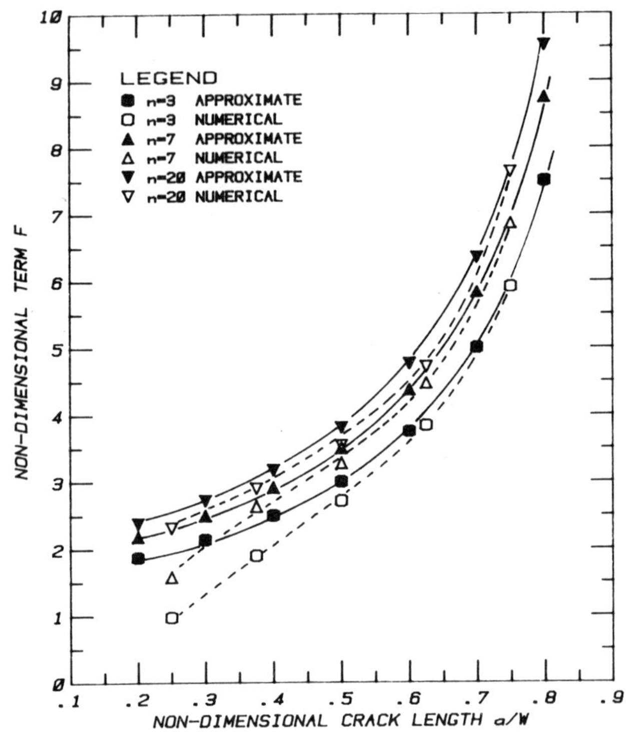

FIG. 6—*Numerical and approximate solutions for* F *for SENB geometry, three-point bend loading,* n = *3, 7, and 20.*

lapse solutions of Haigh and Richards [*23*] and Merkle and Corten [*24*], as shown in the Appendix. A comparison between the two solutions for F, Eqs 49 and 54, is shown in Fig. 7 for the standard CT geometry. Close agreement is obtained with the Haigh and Richards solution giving the slightly higher values.

The numerical results of Kumar and Shih [*25*] are used for comparison with the approximate estimates. Since it is unclear how uncracked displacement rates may be defined for this geometry, only total displacement rates have been calculated. The equations of Kumar and Shih [*25*] couched in terms of C^* and $\dot{\Delta}$ then become

$$C^* = A(W - a)\, k_1 \left[\frac{P}{1.455\, \eta\, B(W - a)} \right]^{(n+1)} \tag{43}$$

and

$$\dot{\Delta} = A\, a\, k_3 \left[\frac{P}{1.445\, \eta\, B(W - a)} \right]^{n} \tag{44}$$

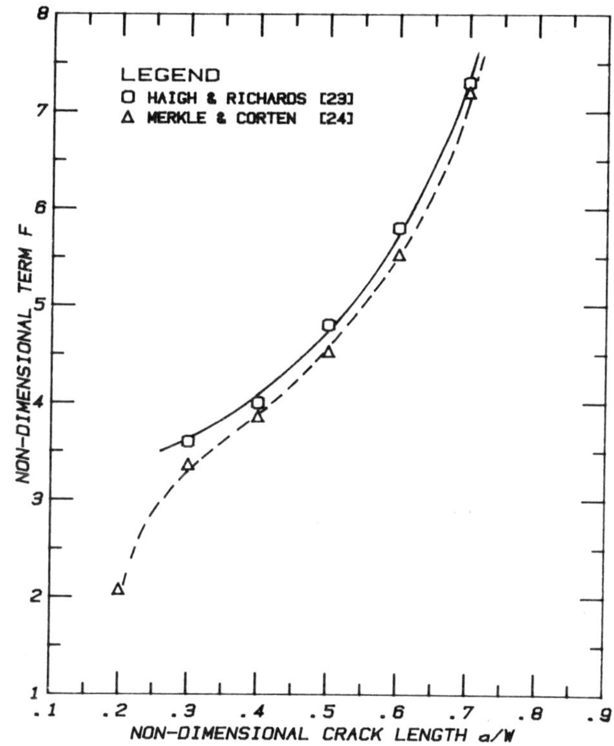

FIG. 7—*Collapse load solution (*n = *20) for* F *for CT geometry.*

where k_1 and k_3 are functions of n and geometry and η is as defined in the Appendix. The nondimensional term F then becomes

$$F = \frac{W}{a} \cdot \frac{k_1}{k_3} \frac{1}{1.455\eta} \tag{45}$$

Values of F for a stress exponent of $n = 7$ are tabulated in Table 1 and compared with the limit analysis estimates of Merkle and Corten [*24*] in Fig. 8 for $n = 3$, 7, and 20 since this analysis gives closer agreement with the numerical calculations at short crack lengths.

Discussion

An approximate procedure has been derived for estimating the parameter C^* associated with the creep deformation of a body containing a sharp crack. The procedure is based on limit analysis techniques and identifies a dimensionless factor which can be determined from nonlinear elastic or deforma-

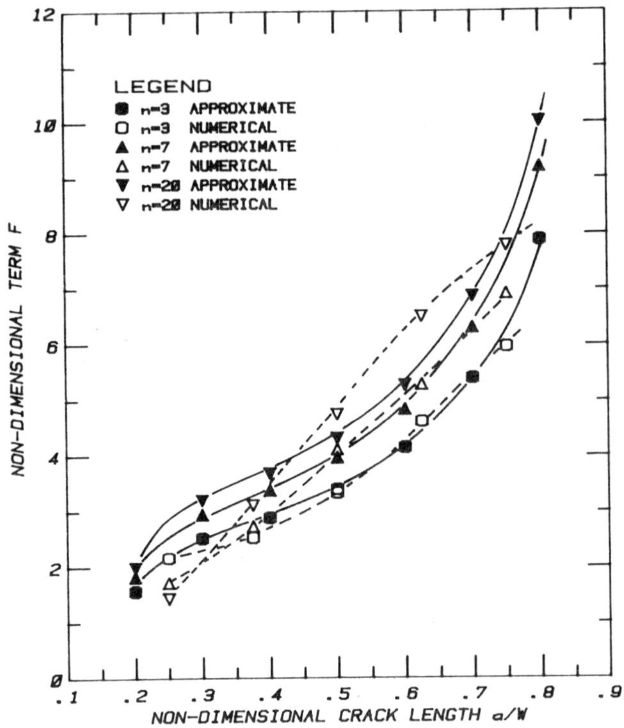

FIG. 8—*Numerical and approximate solutions for* F *for CT geometry*, n = *3, 7, and 20.*

tion theory plastic behavior using the appropriate stress index. The factor can also be calculated by numerical methods. Whichever method is used, it has been shown that the factor is a function only of the stress index n, crack length, and test specimen geometry.

Comparisons between the approximate and numerical estimate of the non-dimensional factor F can be obtained by making reference to Figs. 3, 5, 6, and 8. In these figures both estimates are plotted against a/W for a selection of values of n. Figures 3, 5, and 6 show appropriate data for CCP and SENB specimens. In all cases the approximate estimates are greater than the numerical calculations with the discrepancy decreasing with increase in crack length. Closer agreement is achieved with the SENB than the CCP geometry. Generally, for $a/W > 0.375$ and $n > 7$, a difference of less than 25 percent is obtained for the CCP test specimen and less than 10 percent for SENB specimens. For the CT geometry (Fig. 8) the approximate procedure underestimates F for $0.45 < a/W < 0.7$ and the error increases with increase in n, contrary to experience with the other shapes. A maximum difference in this region of about 12 percent is noted for $n = 20$. At shorter and longer crack lengths the approximate procedure overestimates F.

All the approximate estimates of F (Eqs 23, 31, and 49) contain the term $n/(n + 1)$ and a geometrical function which depends upon crack length and specimen shape only. Consequently, for a given crack length and geometry, F is proportional to $n/(n + 1)$. For $n > 7$ this term varies only from 0.875 to 1. If an accurate value of n for the specimen material is not known, estimates of F to within ± 6 percent can be determined by making $n/(n + 1) = 0.94$. Values of $n > 7$ are often experienced so that this approximation is of practical relevance. It is also particularly useful for those materials where n does not remain constant over the range of stress of interest.

The technique proposed in this paper for estimating F by limit analysis methods is in direct agreement with the procedure of Harper and Ellison [*10*] and the dimensional analysis arguments of Nikbin et al [*11*]. A further approximate solution for F can be obtained by considering the approach of Rice, Paris, and Merkle (RPM) [*26*]. Applied to the CCP geometry, and expressing their equations in a form relevant to a material creeping according to Eq 2, leads to

$$\dot{\Delta} \propto (W - a)\left(\frac{P}{P_{LC}}\right)^n \tag{46}$$

Substitution in Eq 11 yields

$$F_{RPM} = \frac{(n - 1)}{(n + 1)} \frac{1}{2(1 - a/W)} \tag{47}$$

which should be compared with Eq 23. These two equations differ only by the ratio $(n - 1)/n$. A comparison between the two approximate estimates of F and the numerical calculations of Kumar et al [*17*] and Hutchinson et al [*16*] is given in Table 2 for a range of values of n and a/W. The greatest discrepancies between the approximate and numerical estimates occur at low n and short crack lengths. At large n the estimates due to Rice, Paris, and Merkle [*26*] are gen-

TABLE 2—*Comparison of approximate solutions with numerical solutions for CCP specimen.*

n	1		3		7		20	
a/W	F_L/F	F_{RPM}/F	F_L/F	F_{RPM}/F	F_L/F	F_{RPM}/F	F_L/F	F_{RPM}/F
0.25	3.34	0	2.54	1.69	1.56	1.34	1.06	1.01
0.375	2.55	0	1.85	1.23	1.23	1.06	1.06	1.01
0.5	2.23	0	1.59	1.06	1.17	1.01	1.05	1.00
0.625	2.12	0	1.54	1.02	1.22	1.04	1.12	1.07
0.75	2.19	0	1.48	0.98	1.15	0.98	1.06	1.01

NOTE—F_L refers to limit analysis estimate of F.

erally in closer agreement with the numerical calculation. Application of the RPM method to the SENB specimen leads to an identical estimate for F to that presented in Eq 31.

The forms of Eqs 12 and 18 are particularly suitable for experimental determinations of C^*. C^* can be obtained from displacement rates measured at the loading points, or corresponding values measured across the crack, provided the appropriate value of F or F_c is used. This is of importance with regard to the position at which displacements (or rotations) are measured in practice. In addition, very little information need be known about the creep properties of the material since it has already been shown that F is insensitive to the value of n or equivalent stress criterion assumed. This is not true of the numerical expressions for C^* (Eqs 24, 33, 38, and 43) which involve calculated determinations of displacement rate. It can be seen from the forms of Eqs 25, 34, 39, and 44 that these are particularly sensitive to the values assumed for A and n and to the equivalent stress definition adopted. Furthermore, side grooves are often introduced in practice to promote flat fractures and control the direction of crack growth. Accurate estimates of C^* by numerical means will need to be able to allow for the influence of the side grooves on creep displacement rates. Additional complications may arise with those materials which do not exhibit a constant stress index n since the numerical expressions for C^* are especially sensitive to this term.

The influence of the method of calculating F or F_c on correlations of creep crack growth data with the parameter C^* can be gaged from Fig. 9. This figure shows experimental results obtained by Smith and Webster [*13*] and Christian et al [*14*] for a 1Cr-Mo-V steel tested at 565°C. The estimates of C^* have been made from experimental measurements of displacement rate. These were obtained by subtracting elastic displacements from the total displacements recorded to arrive at the creep displacement rate $\dot{\Delta}$ required in the expressions for C^*. The hollow and solid points on the figure compare predictions of C^* made for CT and SENB specimens using the limit analysis and numerical estimates of F. The error bands show the spread in a wider range of data. It is apparent that the differences between the assessments are well within the scatter exhibited by the results. This indicates that either method of determining F is satisfactory. Furthermore, no difference could be seen between an assessment made using local displacement measurements across the crack and F_c, implying that either local or remote displacement rates can be employed to determine C^* provided they are used in conjunction with the appropriate nondimensional factor. Use of the factor F, rather than F_c, avoids the need for making the corrections, often employed in time-independent J testing, for uncracked body displacements or rotations.

From uniaxial creep data for this material, Ellison and Paterson [*27*] have reported that the stress-sensitivity index for creep varies from about 10 to 30 over the range of interest. This leads to uncertainty in the choice of appropriate n for substitution into Eqs 38 and 44 for calculating C^* entirely by numerical

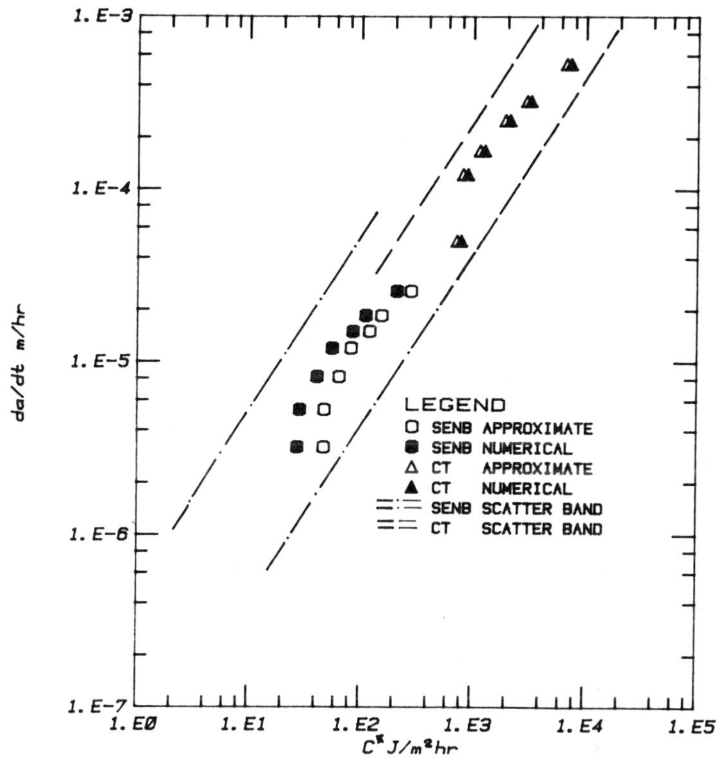

FIG. 9—*Dependence of creep crack growth on C* for 1Cr-Mo-V steel.*

means. For this reason, comparisons with this method of determining C^* are not shown in Fig. 9. Much greater variations in C^* are possible with such a wide range of n than when experimental determinations of $\dot{\Delta}$ are employed in conjunction with numerical or approximate estimates of F.

Conclusions

Approximate and numerical methods of determining the creep parameter C^* for center-crack plate (CCP), single-edge-notch bend (SENB), and compact tension (CT) test specimens subjected to plane-strain conditions have been compared. In each case C^* has been expressed in terms of creep displacement rate and a nondimensional factor. It has been shown that this nondimensional factor can be obtained from power-law hardening deformation theory or approximate limit analysis techniques. The nondimensional factor has been found to be dependent on specimen geometry and crack length but to be relatively insensitive to creep stress index n.

The greatest discrepancies between the numerical and approximate estimates of C^* were found to occur at low n and short crack lengths. For the CCP and SENB specimens the limit analysis approximation always overestimated C^*. For the CT test specimen it underestimated C^* at intermediate crack lengths, where most experiments are carried out.

When displacement rates are measured experimentally it has been shown that C^* can be determined without an accurate knowledge of the creep properties of the material of interest. This is not true when numerical calculations of displacement rate are employed. Determinations of C^* from experimental displacement rates are particularly suitable for those situations where n is not strictly constant. The analysis is applicable whether displacement rates are measured at the loading points or locally across the crack plane provided the appropriate nondimensional factor is used.

Comparisons of the procedures for estimating C^* have been made with experimental creep crack growth data on SENB and CT specimens of a 1Cr-Mo-V steel. These have indicated that the approximate limit analysis estimates of C^* are satisfactory for the most part when the scatter of experimental results is taken into account.

APPENDIX

Approximate Estimate of *F* for the CT Specimen

The nondimensional term F is determined for the CT specimen from collapse solutions provided by Haigh and Richards [*23*] and Merkle and Corten [*24*]. The former give the plane strain yielding equations as

$$(1 - a/W) = 1.26(2m + m^2)^{1/2} - m \tag{48}$$

where m is the ratio of the collapse load of a cracked test specimen to that of the uncracked body. Differentiating and rearranging into the appropriate form gives

$$F = \frac{n}{n+1}\left[\frac{5.404\,a}{Wr} - 1.702\right]\bigg/\left[0.851\,r - \left(1 + \frac{1.702\,a}{W}\right)\right] \tag{49}$$

where

$$r = \left[3.732 + 6.350\left(\frac{a}{W}\right)^2\right]^{1/2} \tag{50}$$

The collapsed load P_{LC} derived by Merkle and Corten [*24*] can be expressed as

$$P_{LC} = 1.455\,\sigma_y\,\eta[1 - a/W] \tag{51}$$

where η is a nondimensional function of a/W such that

$$\eta = (S^2 + 2S + 2)^{1/2} - S - 1.0 \tag{52}$$

and

$$S = \frac{2a}{W}/(1 - a/W) \tag{53}$$

Differentiating and rearranging gives

$$F = \frac{n}{n+1}\left[\left(-\frac{1}{1-a/w}\right) + \frac{1}{\eta}\cdot\frac{d\eta}{d(a/W)}\right] \tag{54}$$

where

$$\frac{d\eta}{d(a/W)} = \frac{1}{2}(S^2 + 2S + 2)^{-1/2}\cdot\left[2S\,\frac{dS}{d(a/W)} + 2\,\frac{dS}{d(a/W)}\right] - \frac{dS}{d(a/W)} \tag{55}$$

and

$$\frac{dS}{d(a/W)} = \frac{2a/W}{(1-a/W)^2} + \frac{2}{1-a/W} \tag{56}$$

References

[*1*] James, L. A. *International Journal of Fracture Mechanics*, Vol. 8, 1972, pp. 347–349.

[*2*] Siverns, M. J. and Price, A. T., *International Journal of Fracture*, Vol 9, No. 2, June 1973, pp. 199–207.

[*3*] Neate, G. J. and Siverns, M. J., "The Application of Fracture Mechanics to Creep Crack Growth," International Conference on Creep and Fatigue in Elevated Temperature Applications, Philadelphia, 1973; Sheffield, U.K., 1974; Institution of Mechanical Engineers, London, 1975.

[*4*] Freeman, B. L., "The Estimation of Creep Crack Growth Rates by Reference Stress Methods," Report No. SSD/MID/R54, Central Electricity Generating Board, Leatherhead, Surrey, U.K., 1977.

[*5*] Freeman, B. L. and Neate, G. J., *Materials Science and Engineering*, Vol. 36, 1978, pp. 241–251.

[*6*] Nicholson, R. D. and Formby, C. L., *International Journal of Fracture*, Vol. 11, 1975, pp. 595–604.

[*7*] Webster, G. A., "The Application of Fracture Mechanics to Creep Cracking," Conference on Mechanics and Physics of Fracture, Institute of Physics, Cambridge, U. K., Jan. 1975.

[*8*] Koterazawa, R. and Mori, T., *Journal of Engineering Materials and Technology, Transactions*, American Society of Mechanical Engineers, Vol. 99, Series H, No. 4, Oct. 1977, pp. 298–305.

[*9*] Landes, J. D. and Begley, J. A., "A Fracture Mechanics Approach to Creep Crack Growth," Report 74-1E7-FESGT-P1, Westinghouse Research Laboratory, Pittsburgh, Pa., Dec. 1974.

[*10*] Harper, M. P. and Ellison, E. G., *Journal of Strain Analysis*, Vol. 2, 1977, pp. 167–179.

[*11*] Nikbin, K. M., Webster, G. A. and Turner, C. E. in *Proceedings*, Fourth International Conference on Fracture, Waterloo, Ont., Canada, Vol. 2, 19–24 June 1977.

[*12*] Ellison, E. G. and Harper, H. P. *Journal of Strain Analysis*, Vol. 13, 1978, pp. 35–51.

[*13*] Smith, D. J. and Webster, G. A., *Journal of Strain Analysis*, Vol. 16, No. 2, 1981, pp. 137–143.

[*14*] Christian, E. M., Smith, D. J., Webster, G. A., and Ellison, E. G. in *Proceedings*, Fifth International Conference on Fracture, Cannes, France, 1980.

[*15*] Goldman, N. L. and Hutchinson, J. W., *International Journal of Solids and Structures*, Vol. 11, 1975, pp. 575–591.

[*16*] Hutchinson, J. W., Needleman, A., and Shih, C. F. in *Proceedings*, Office of Naval Research International Symposium on Fracture Mechanics, Washington D.C., 11–13 Sept. 1978.

[*17*] Kumar, V., German, M. D., and Shih, C. F., "Estimation Technique for the Prediction of Elastic-Plastic Fracture of Structural Components of Nuclear Systems," Report No. SRD-80-094, Electric Power Research Institute, Palo Alto, Calif., June 1981.

[*18*] Rice, J. R., *Journal Applied Mechanics, Transactions*, American Society of Mechanical Engineers, American Society of Mechanical Engineers, Vol. 35, 1968, pp. 379–386.

[*19*] Hutchinson, J. W., *Journal of the Mechanics and Physics of Solids*, Vol. 16, 1968, pp. 13–31.

[*20*] Rice, J. R. and Rosengren, G. F., *Journal of the Mechanics and Physics of Solids*, Vol. 16, 1968, pp. 1–12.

[*21*] Rice, J. R. in *Mathematical Analysis in the Mechanics of Fracture*, H. Liebowitz, Ed., Vol. 2, Academic Press, New York, 1968.

[*22*] Ilyushin, A. A., *Prikadnaia Matematika i Mekhanka*, Vol. 10, 1946, p. 347.

[*23*] Haigh, J. R. and Richards, C. E., "Yield Point Loads and Compliance Functions of Fracture Mechanics," CERL Memo No. RD/L/M/461, Central Electricity Generating Board, Leatherhead, Surrey, U.K., 1974.

[*24*] Merkle, J. G. and Corten, H. T., *Journal of Pressure Vessel Technology, Transactions*, American Society of Mechanical Engineers, Vol. 96, 1974, pp. 286–292.

[*25*] Kumar, V. and Shih, C. F. in *Proceedings*, International Conference on the Engineering Aspects of Creep, Sheffield, U.K., 1980, Paper C228/80.

[*26*] Rice, J. R., Paris, P. C. and Merkle, J. G. in *Progress in Flaw Growth and Fracture Toughness Testing, ASTM STP 536*, American Society for Testing and Materials 1973, pp. 231–245.

[*27*] Ellison, E. G. and Paterson, A. J. F. in *Proceedings*, Institution of Mechanical Engineers, London, Vol. 190, No. 12, 1976, pp. 321–332.

Vellore M. Radhakrishnan[1] *and Arthur J. McEvily*[2]

Crack Growth in Creeping Solids

REFERENCE: Radhakrishnan, V. M. and McEvily, A. J., **"Crack Growth in Creeping Solids,"** *Elastic-Plastic Fracture: Second Symposium, Volume I—Inelastic Crack Analysis, ASTM STP 803*, C. F. Shih and J. P. Gudas, Eds., American Society for Testing and Materials, 1983, pp. I-675–I-689.

ABSTRACT: The creep crack growth behavior in a creep-ductile material, 6061 aluminum, and a creep brittle material, Ti-6242, has been investigated. In the creep ductile material an observed load dependency in the relation between the load-point displacement rate, $\dot{\Delta}$, and the crack growth rate, $\dot{a}$, precluded the straightforward application of the C^* approach. An alternative approach has been developed in which $\dot{\Delta}$, $\dot{a}$, and P are interrelated through an equation of form

$$\dot{\Delta} \propto P^{\alpha}\dot{a}^{\theta}$$

The exponent α determines the contribution of creep deformation to $\dot{\Delta}$ and therefore relates to the creep ductility or creep brittleness of the material. For both materials tested the foregoing relationship was able to describe the experimental results without any systematic scatter.

KEY WORDS: creep crack growth, creep ductility, creep brittleness, parameter approach, creep crack opening displacement, elastic-plastic fracture

The subject of creep crack growth is attracting increased attention in terms of experimental work and analytical treatment. Experimental results thus far have shown that we are dealing with a complex process of deformation and localized rupture which can differ significantly from one alloy to another. The terms "creep brittle" and "creep ductile" have been introduced to indicate the degree of plastic deformation at the crack tip during crack extension [*1*].[3] With respect to analytical approaches the energy line rate integral C^* appears to have the greatest appeal [*2*]. Other analytical approaches based upon the stress-intensity factor, net section stress, or reference stress

[1]Assistant director, Fatigue and Fracture Mechanics Group, Defense Metallurgical Research Laboratory, Hyderabad, India.

[2]Professor, Metallurgy Department, University of Connecticut, Storrs, Conn.

[3]The italic numbers in brackets refer to the list of references appended to this paper.

may find application to either very creep-brittle materials (stress-intensity factor approach) or very creep-ductile materials (net-section approach), but as Webster [3] has pointed out, the C^* method can be developed to encompass these situations as special cases.

However, while it is clear that the C^* approach is worthy of serious consideration, certain published results raise questions concerning limitations of the C^* approach in providing a unique correlation with the creep crack growth rate. For example, Fig. 1*a* shows the nature of the correlation between C^* and the creep crack growth rate as obtained by Sadananda and Shahinian [4] for Udimet 700, a nickel-base superalloy. In this case the value of C^* was determined using the graphical method proposed by Landes and Begley [2] based on the relation

$$C^* = -\frac{1}{B}\frac{d\dot{U}}{da} \tag{1}$$

where $\dot{U}$ is the power or energy rate defined for a load P and displacement rate $\dot{\Delta}$ by

$$\dot{U} = \int_0^{\dot{\Delta}} P d\dot{\Delta} \tag{2}$$

where B is the specimen thickness and a the crack length.

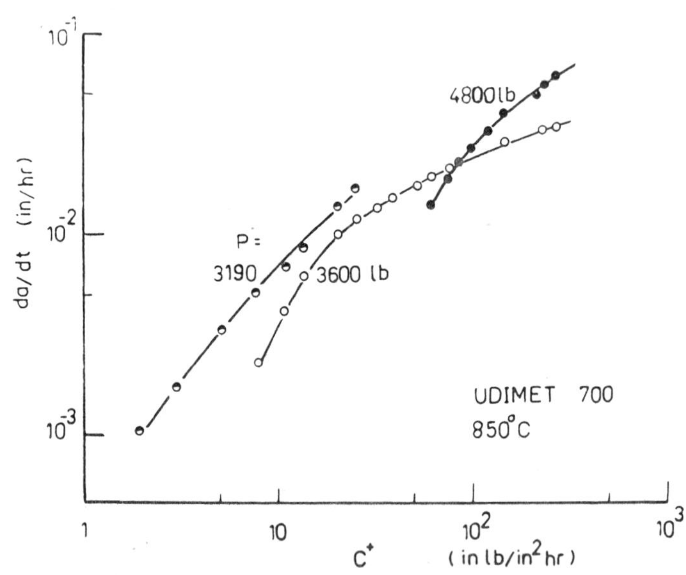

FIG. 1*a*—*Variation of* $\dot{a}$ *with* C* *[4] (1 in. = 25.4 mm; 1 lb = 0.45 kg; 1 in.-lb/in.*^{2} *= 0.17512 kJ/m*^{2}*).*

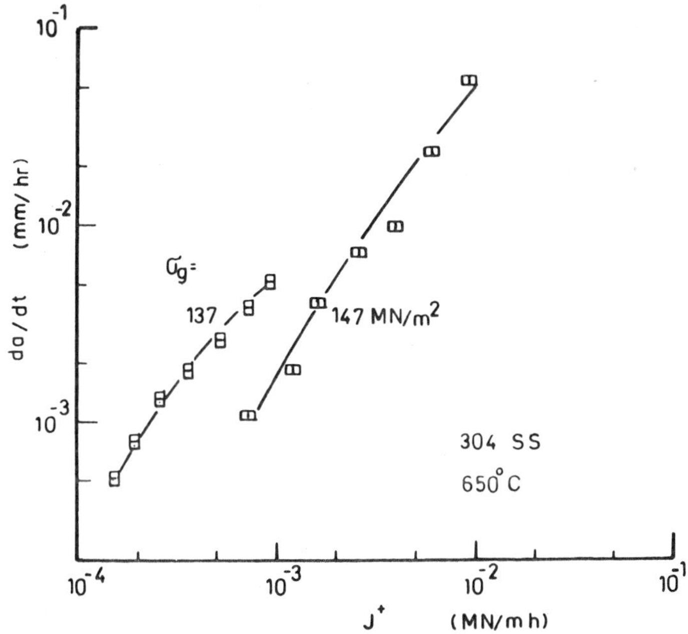

FIG. 1*b*—*Variation of* $\dot{a}$ *with* $\dot{J}$ *[5] (1 mm = 0.04 in.; 1 MN/m*^{2} *= 0.145 ksi).*

Figure 1*b* shows the nature of the correlation obtained by Taira et al [5] which is based upon the relation

$$C^* = A_1 \sigma_{net} \dot{\Delta} \tag{3}$$

where

A_1 = constant,
σ_{net} = net section stress, and
$\dot{\Delta}$ = load point deflection rate.

In both of these cases, although a general correlation between C^* and the crack growth rate is obtained, the relationship is not single-valued and a clear dependency on the load level, P, is evident. The reason for this load dependency is not apparent, but there are several possibilities. For example, the C^* approach is valid only during steady-state creep conditions with a relationship of the form

$$\dot{\epsilon} = D\sigma^n \tag{4}$$

being applicable. Here $\dot{\epsilon}$ is the strain rate, D and n are constants, and σ is the stress. Further, it is assumed that all transient effects as well as short-time plasticity effects can be ignored [*2*]. The approach is strictly valid only during monotonic loading conditions at a crack tip, and therefore the effect of unloading behind the advancing crack tip needs to be clarified. Further, C^* has been defined for the two-dimensional case (plane strain or plain stress) whereas in fact an increasing amount of out-of-plane lateral contraction occurs as the crack advances, particularly in a creep-ductile material.

Despite these many restrictions one can find examples of reasonably good correlation between C^* and $\dot{a}$, the creep crack growth rate. For example, Fig. 2 shows the nature of the correlation obtained for compact (CT) specimens of 6061 aluminum alloy at 200°C [*6*]. In this case the value of C^* was obtained using the graphical method of Landes and Begley [*2*], which is schematically illustrated in Fig. 3. In this method a straight line is fitted to two data points after each value of $\dot{\Delta}$ in the $\dot{U}$-a plot. An interesting aspect of this approach, which is evident in the plot of $\dot{U}$ versus a, is that if each of the lines at constant $\dot{\Delta}$ is indeed straight then the value of C^* is simply a function of $\dot{\Delta}$. It follows that if C^* is a measure of $\dot{a}$, the relation between $\dot{\Delta}$ and $\dot{a}$ must also be unique and not depend on the load level. However, as shown in Fig. 4, our results do reflect an influence of load level on the $\dot{\Delta}$-$\dot{a}$ relationship. Because of such observations of load dependency we have developed a different approach to the interrelationships between $\dot{a}$, $\dot{\Delta}$, and P for both creep ductile and creep brittle materials. The objective of the remainder of this paper will be to outline this approach and to demonstrate its applicability.

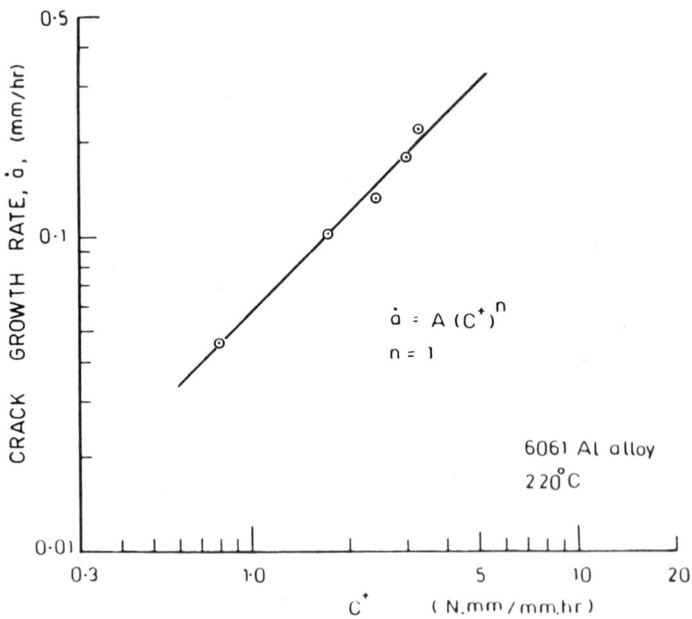

FIG. 2—*Relation between crack growth rate,* ȧ, *and* C* *[6] (1 mm = 0.04 in.).*

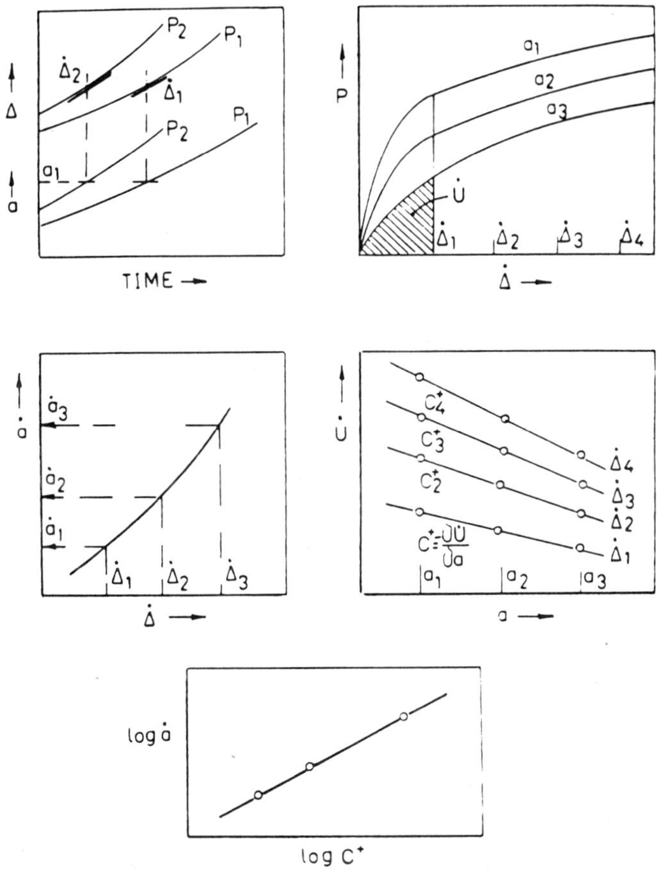

FIG. 3—*Schematic representation of procedures involved in determining* C* *[2].*

Experiments

One of the two alloys chosen for investigation was the aluminum alloy 6061 (0.25Zn-0.8Mg-0.7Fe-0.6Si), a creep-ductile material. This alloy was tested after overaging at 350°C for 24 h to stabilize the microstructure. The other alloy was a near-α titanium alloy designated Ti-6242 (6Al-2Sn-4Zn-2Mo-0.1Si), a creep-brittle material possessing a Widmanstetten-type microstructure. The aluminum alloy was tested in two specimen configurations, CT and center-cracked (CC). The titanium alloy was tested only in the CT configuration. These specimen types are shown in Fig. 5. It is noted that the elastic compliance of the CT specimens was approximately ten times that of the CC, which relates to the much lower rates of crack growth observed in the CC-type specimen.

A number of lines of known spacing were marked on the surface of each

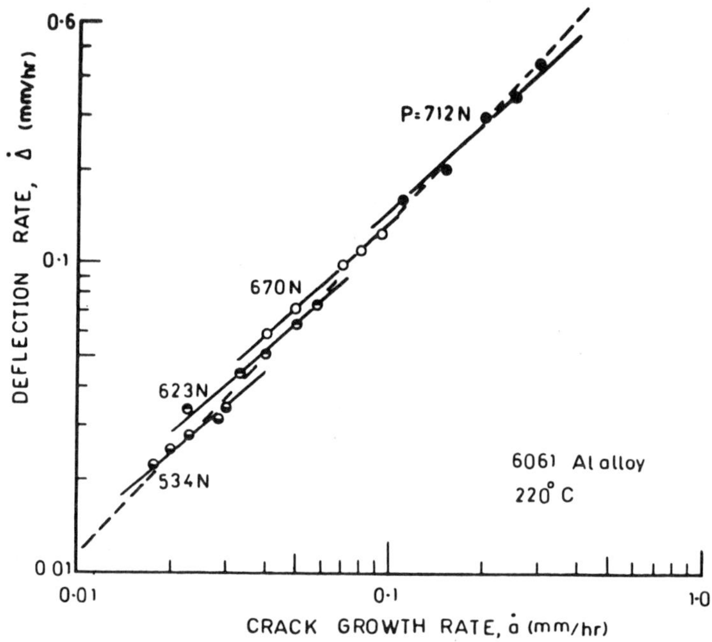

FIG. 4—*Relation between deflection rate,* $\dot{\Delta}$, *and crack growth rate,* $\dot{a}$ *(1 mm = 0.04 in.; 1 N = 0.225 lbf).*

specimen perpendicular to the direction of crack propagation so that the crack growth could be monitored by an optical method. The accuracy of this measurement was 0.25 mm. The experiments were carried out in a SATEC creep testing machine. The temperature was measured with a dummy specimen kept by the side of the test specimen and was controlled within ±2 deg C. The load-point deflection was measured by a dial gage with an accuracy of 0.025 mm. The load was applied on the specimen after a soaking period of one hour. The 6061 Al alloy was tested at a specimen temperatuare of 275°C (0.61 Tm) and the titanium specimens at 535°C (0.42 Tm). After allowing for the soaking time, the load was applied and the load-point deflection Δ and the crack length a from the tip of the machined notch were measured at regular intervals of time.

Results and Discussion

Figures 6*a*–6*c* show typical variations of crack length and load-point deflection with time as a function of load level. The figures by each curve give the deflection rate or the crack growth rate, as the case may be, at the respective time. It is observed that in general the nature of either the deflection

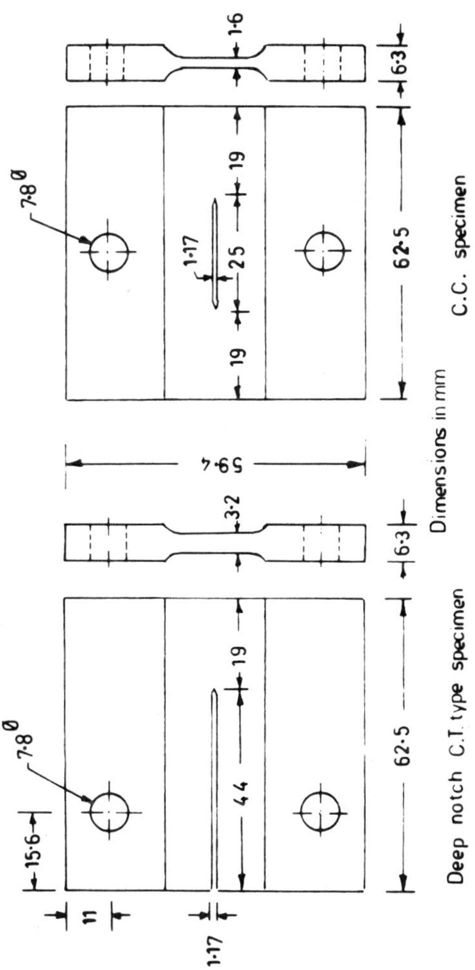

FIG. 5—*Specimen configurations (1 mm = 0.04 in.).*

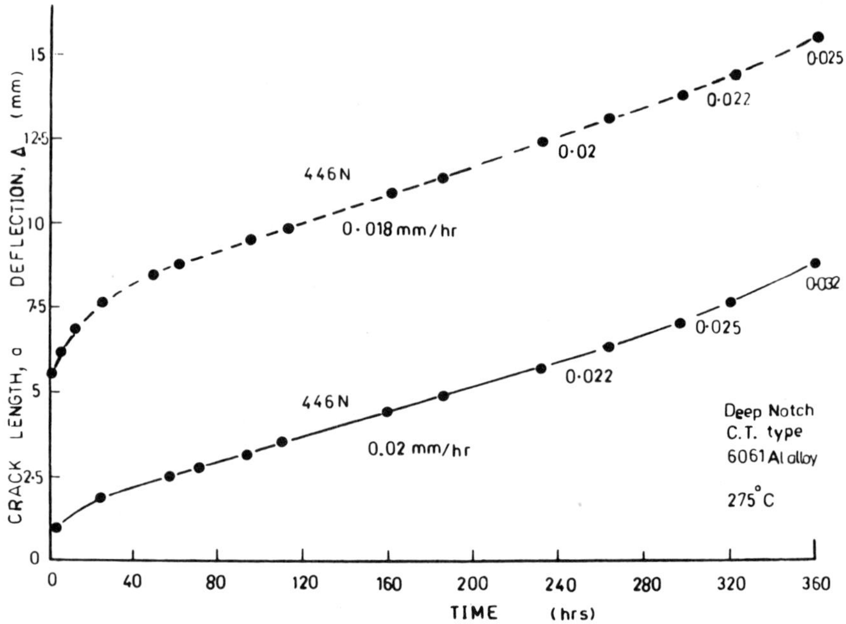

FIG. 6a—*Variation of crack length and load-point deflection with time (1 mm = 0.04 in.; 1 N = 0.225 lbf).*

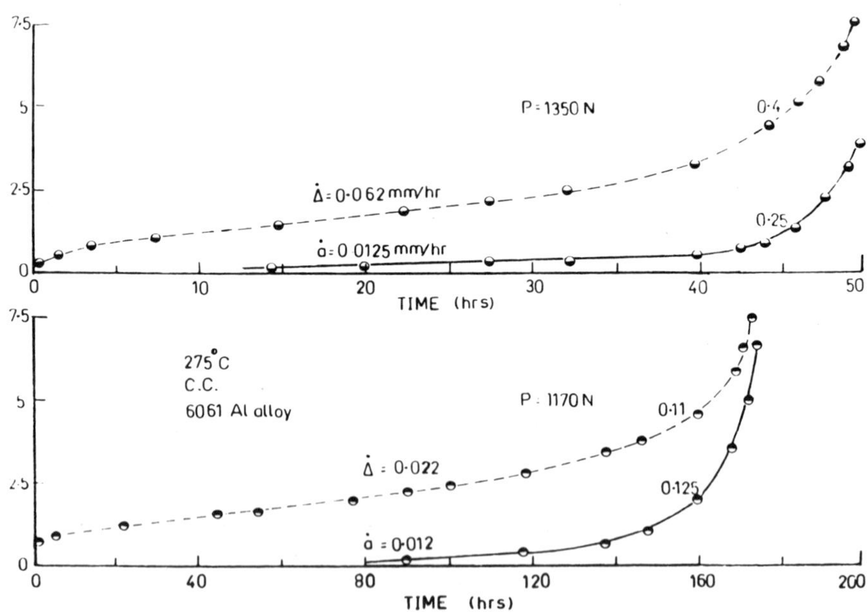

FIG. 6b—*Variation of crack length and load-point deflection with time (1 mm = 0.04 in.; 1 N = 0.225 lbf).*

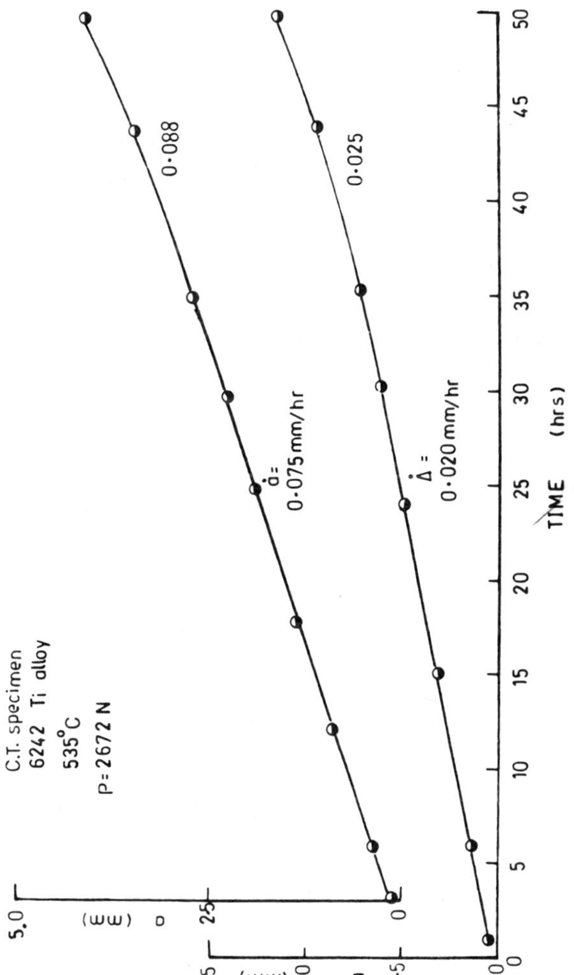

FIG. 6c—*Variation of crack length and load-point deflection with time (1 mm = 0.04 in.; 1 N = 0.225 lbf).*

rate or the crack growth rate with time is similar to the usual creep curve with its three different stages, that is, transient, steady state, and tertiary. The rate of crack growth is fairly constant over much of the crack growth period, as also observed by Jones and Pilkington [7], for example, in the case of a 0.5Cr-0.5Mo-0.25V steel. A considerable amount of plastic extension and lateral contraction was observed at the tip of the crack in the case of the aluminum alloy, with the total overall load-point deflection being of the order of 10 mm for a creep crack of 10 mm in length. On the other hand, the load-point deflection was only 2 mm from the titanium alloy at a creep-crack length of 5 mm, an indication of the more creep-brittle nature of the titanium alloy.

Figure 7 shows the relationship between a and $\dot{\Delta}$ for the two materials and different specimen geometries. In the case of the aluminum alloy it is seen that the $\dot{a}$ versus $\dot{\Delta}$ relation depends upon load level. On the other hand, for the titanium this relationship is load independent, and as discussed earlier this implies that C^* and $\dot{\Delta}$ are uniquely related. Therefore, with reference to Fig. 3, in the case of the titanium alloy in the plot of $\dot{U}$ versus a, the lines for a given $\dot{\Delta}$-value must be straight if C^* is to be a unique function of $\dot{\Delta}$.

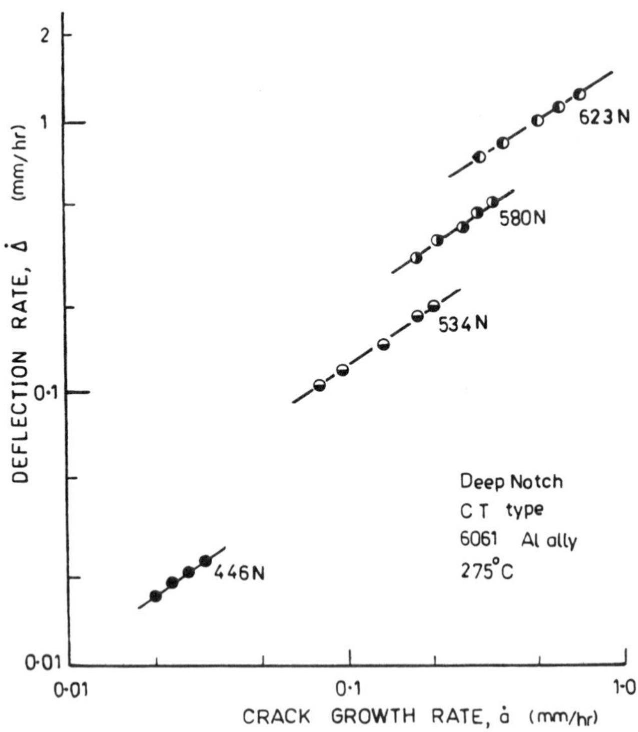

FIG. 7a—*Relation between* $\dot{a}$ *and* $\dot{\Delta}$*; CT, 6061 Al alloy (1 mm = 0.04 in.; 1 N = 0.225 lbf).*

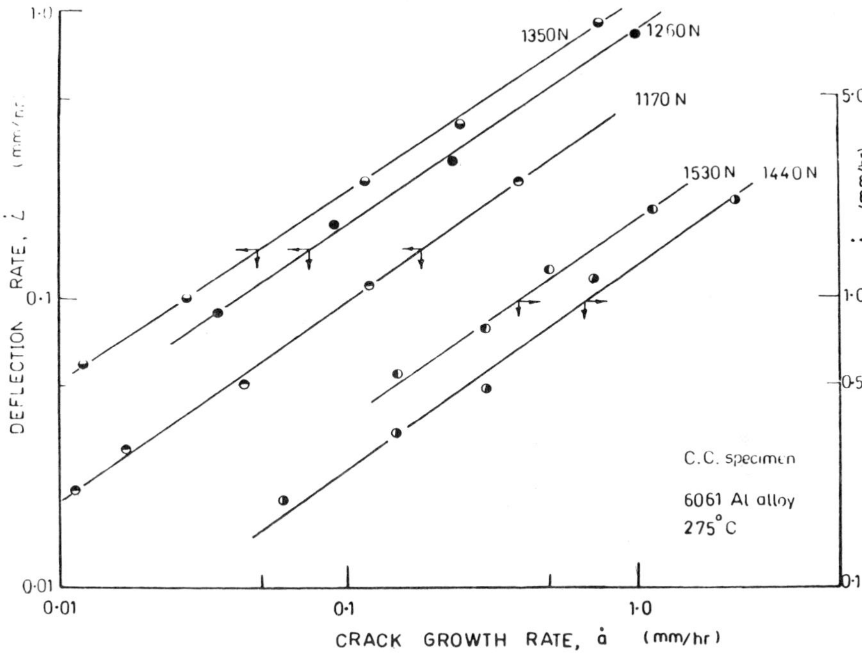

FIG. 7b—*Relation between* ȧ *and* $\dot{\Delta}$; *CC, 6061 Al alloy (1 mm = 0.04 in.; 1 N = 0.225 lbf).*

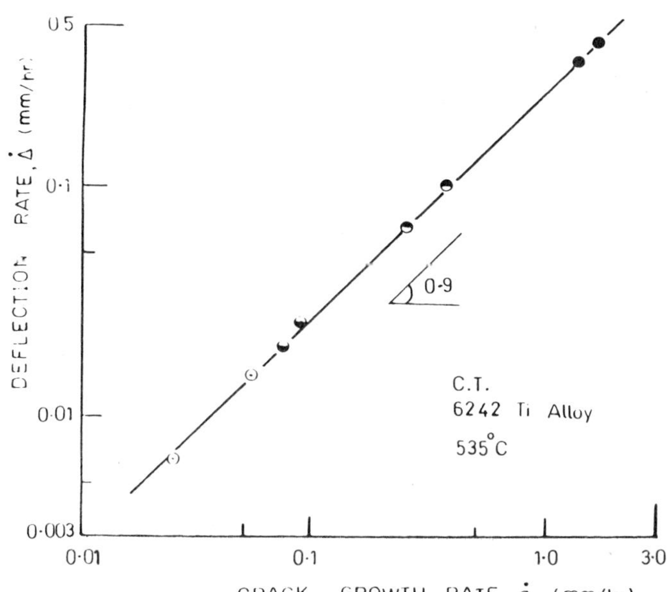

FIG. 7c—*Relation between* ȧ *and* $\dot{\Delta}$; *CT, Ti-6242 alloy (1 mm = 0.04 in.; 1 N = 0.225 lbf).*

However, in the case of the creep-ductile aluminum alloy the crack growth rate depends not only on $\dot{\Delta}$, but is also a function of load. This circumstance complicates the graphical C^* analysis for it indicates that the value of C^* is also dependent on both $\dot{\Delta}$ and P. Since at a given $\dot{\Delta}$ the crack growth rate increases with decrease in load, if the crack growth rate is a function of C^* this would imply that in moving from left to right in the $\dot{U}$ versus a plot of Fig. 3 along a line of constant $\dot{\Delta}$ the slope should increase; that is, the lines must be concave upward. Because of the experimental difficulties involved in establishing curved lines on the $\dot{U}$ versus a plot, we have developed an alternative approach.

Since the deflection rate at a given crack growth rate was found to be simply related to the load level for each specimen configuration as shown in Fig. 8, we can write

$$\dot{\Delta} \propto P^{\alpha} \tag{5}$$

where α has a value of 4. From Fig. 7 it can be seen that $\dot{\Delta} \propto (\dot{a})^{\theta}$, and since $\dot{\Delta}$ therefore depends both on the crack growth rate as well as the crack-tip creep extension, which is load dependent, we can write

$$\dot{\Delta} \propto (P/P_0)^{\alpha}(\dot{a})^{\theta} \tag{6}$$

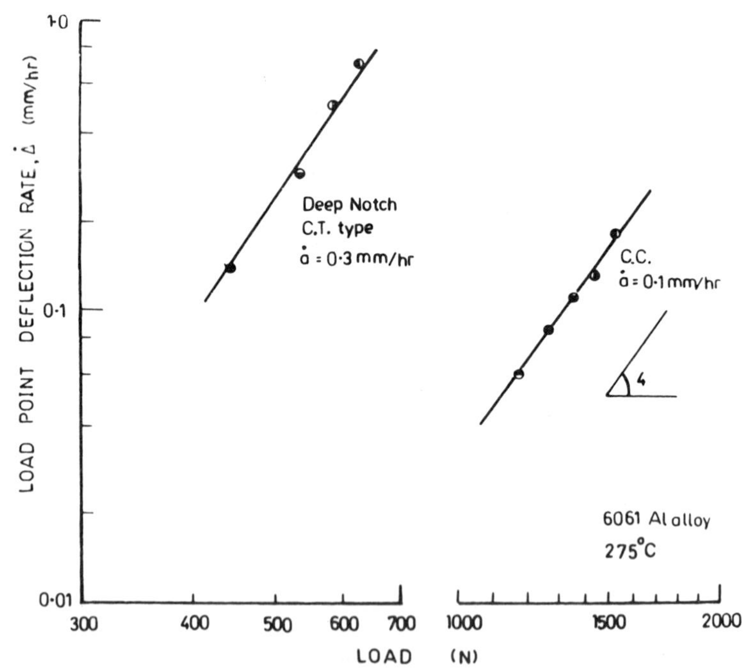

FIG. 8—*Dependence of $\dot{\Delta}$ on load* P *for 6061 aluminum alloy (1 mm = 0.04 in.; 1 N = 0.225 lbf).*

where P_0 is a unit load. The exponent α determines the contribution of the creep deformation to the load-point deflection rate. In brittle materials there is negligible creep deformation at the crack tip and the exponent α tends to zero as in the case of the titanium alloy where $\dot{a}$ is proportional to $\dot{\Delta}^{1.1}$. In such a case the load-point deflection rate will be independent of load and dependent only on crack growth rate and C^* will be a function of $\dot{\Delta}$ only. Haigh has also observed a relation of the type

$$\dot{\Delta} \propto (\dot{a})^{\theta} \tag{7}$$

for chromium-molybdenum-vanadium steel at 0.5 Tm with $\theta = 1.2$. Jones and Pilkington [7] have also found such a relationship with $\theta = 0.9$ for a 0.5Cr-0.5Mo-0.25V steel. The absence of load dependency in these tests suggest the materials were creep brittle under the test conditions used. On the other hand, in creep ductile materials where crack tip creep deformation contributes to $\dot{\Delta}$, C^* will be both load and $\dot{\Delta}$ dependent. For the general case Eq 6 can be rewritten as

$$a \propto \frac{\Delta^{1/\theta}}{(P/P_0)^{\alpha}} \tag{8}$$

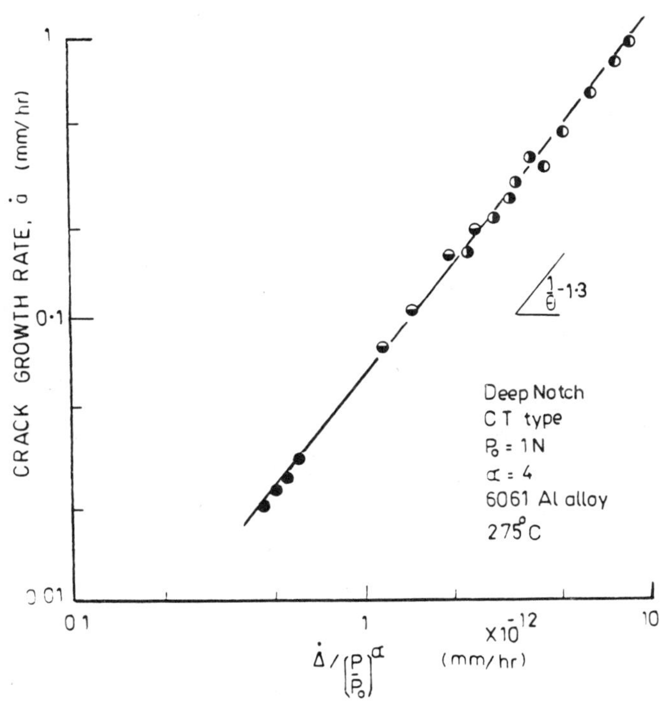

FIG. 9a—Relation between $\dot{a}$ and parameter for 6061 aluminum alloy, CT, specimen (1 mm = 0.04 in.; 1 N = 0.225 lbf).

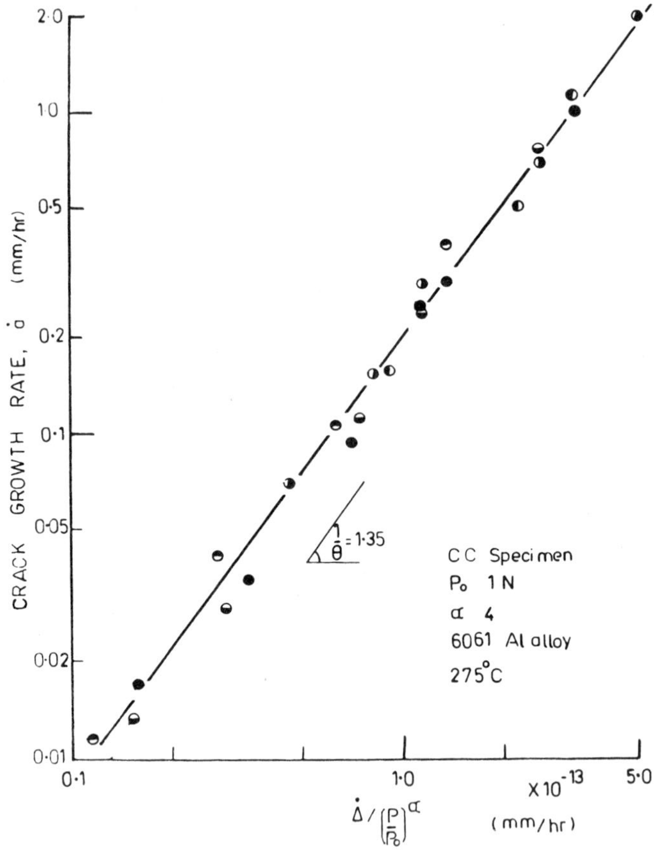

FIG. 9*b*—*Relation between* ȧ *and parameter for 6061 aluminum alloy, CC specimen (1 mm = 0.04 in.; 1 N = 0.225 lf).*

The relation between the creep crack growth rate and the parameter $\Delta(P/P_0)^\alpha$ is shown in Fig. 9 from the two geometries of the aluminum alloys. Identical values of θ and α apply to both cases. However, the specific effects of specimen geometry have not as yet been incorporated into this approach and obviously more work is needed.

We note also that the exponents α and θ have been found to be temperature dependent [6]. For 6061 below 0.5 Tm, the exponent α tends to zero, indicating that below this temperature level the contribution of plastic deformation to $\dot{\Delta}$ is negligible. Materials will therefore tend to be more creep brittle as the temperature is reduced below 0.5 Tm, as borne out by the results presented herein.

Conclusions

1. An equation of the form

$$\dot{\Delta} \propto (\dot{a})^{\theta}$$

has been found to describe the relationship between load-point deformation rate, applied load, and creep crack growth rate from both a creep ductile alloy (6061 aluminum alloy) and a creep brittle alloy (Ti-6242).

2. Further work is needed to be able to incorporate the effect of specimen geometry on this relationship and to establish the relationship to a generalized C^* approach.

3. The C^* approach as developed by Landes and Begley [*2*] is most readily applied to creep-brittle materials.

References

[*1*] Williams, J. A. and Price, A. T., *Transactions*, American Society of Mechanical Engineers, *Journal of Engineering Materials and Technology*, Vol. 97, 1975, pp. 214–222.

[*2*] Landes, J. D. and Begley, J. A. in *Mechanics of Crack Growth, ASTM STP 590*, American Society for Testing and Materials, 1976, pp. 128–148.

[*3*] Webster, G. A. and Nikbin, K. M. in *Creep in Structures*, Pouter and Hayburst, Eds., Springer-Verlag, Berlin, 1981, pp. 576–591.

[*4*] Sadananda, K. and Shahinian, P., *Metallurgical Transactions*, Vol. 9A, 1978, pp. 79–84.

[*5*] Taira, S., Ohtani, R., and Kitamura, T., *Transactions*, American Society of Mechanical Engineers, *Journal of Engineering Materials and Technology*, Vol. 101, 1979, pp. 154–161.

[*6*] Radhakrishnan, V. M. and McEvily, A. J. *Transactions*, American Society of Mechanical Engineers, *Journal of Engineering Materials and Technology*, Vol. 102, 1980, pp. 200–206.

[*7*] Jones, C. L. and Pilkington, R. in *Proceedings*, Fourth International Conference on Fracture, Waterloo, Ont., Canada, Vol. 2, 1977, pp. 641–647.

[*8*] Haigh, J. R., *Materials Science Engineering*, Vol. 20, 1975, pp. 213–235.

K. Sadananda[1] *and Paul Shahinian*[1]

Parametric Analysis of Creep Crack Growth in Austenitic Stainless Steel

REFERENCE: Sadananda, K. and Shahinian, P., "**Parametric Analysis of Creep Crack Growth in Austenitic Stainless Steel,**" *Elastic-Plastic Fracture: Second Symposium, Volume I—Inelastic Crack Analysis, ASTM STP 803*, C. F. Shih and J. P. Gudas, Eds., American Society for Testing and Materials, 1983, pp. I-690–I-707.

ABSTRACT: Creep crack growth behavior in Type 316 stainless steel at 593°C has been examined under displacement control and the results are compared with those under load control. The growth rates were analyzed in terms of linear elastic stress-intensity factor, K, reference stress, σ_r, and elastic-plastic energy rate parameter, J^*, to determine which one of the three is best suited for correlating creep crack growth rates in the stainless steel. The results indicate that the correlations in terms of the three parameters do not depend on the method of testing, that is, displacement of load-controlled tests. Furthermore, the data correlate best with J*-integral, with a spread in the data by a factor of 10 in comparison with a factor of 1000 in terms of the other two parameters. Since the dimensions and the range of values for K and J^* are significantly different, to facilitate a more meaningful comparison between elastic and elastic-plastic parameters, a linear elastic energy rate parameter K^* is defined which is equivalent to J^* in the linear elastic approximation. The analysis indicates that the crack growth data for the two types of tests correlate better in terms of K^* than in terms of J^* at low K^*-values. At high values of K^*, plasticity effects become pronounced, thus invalidating its applicability. Thus, in spite of the lack of correlation on a K basis, linear elastic fracture mechanics concepts may still be valid since the data can be correlated on a K^* basis. But the present work clearly shows that the J^*-integral is better than the stress-intensity factor K for creep crack growth in stainless steel.

KEY WORDS: creep crack growth, fracture mechanics, stress-intensity parameter, J*-integral for creep, elastic-plastic fracture mechanics, energy-rate parameter stainless steel

Recognition that growth of a crack can occur readily at low stress intensities and high temperatures under static load, and that the growth rates in some materials are comparable to those under cyclic load, has generated significant interest in creep crack growth [*1*].[2] From design considerations, time-

[1]Metallurgists, Material Science and Technology Division, Naval Research Laboratory, Washington, D.C. 20375.

[2]The italic numbers in brackets refer to the list of references appended to this paper.

dependent crack growth in some high-temperature structural materials is as important as cycle-dependent crack growth. Furthermore, with increasing temperature and decreasing frequency, there is an increasing time-dependent contribution to crack growth. Knowledge and understanding of time-dependent crack growth behavior is therefore important for safe design. Understanding and quantitative prediction of the growth rates, however, involve detailed examination of the growth behavior in several different specimen geometries and materials.

Examination of the time-dependent crack growth in many materials has shown that the growth occurs either by environmental effects, creep effects, or by a complex combination of the two. For example, crack growth rates in Alloy 718 and Inconel X-750 [*2*] are nearly two orders of magnitude greater in air than in vacuum, indicating that the environmental effects are more predominant than creep for these materials. On the other hand, the growth rates in austenitic stainless steels [*3*] are nearly equal in both air and in vacuum, indicating that the growth occurs essentially due to creep effects. The mechanism of crack growth, that is, whether it involves predominantly a creep or environmental effect, dictates the type of parameters that needs to be used to correlate crack growth rates [*1*].

There have been several parameters that have been suggested for correlation of crack growth rates under high-temperature and static load. These include linear elastic stress-intensity factor, K, reference stress, σ_r, elastic-plastic energy rate parameter, J^*, crack opening displacement, crack opening displacement rate, and combinations of these (see Ref *1* for details). The accumulated evidence [*4*] so far indicates that, for materials which are highly sensitive to environment in the temperature range of interest, the creep effects are minimal and crack growth can occur essentially under elastic conditions. Stress-intensity factor K can adequately describe the growth rates in such materials. On the other hand, if the material is very ductile and less sensitive to environment at the temperature of interest, creep occurs readily, contributing to significant crack-tip blunting instead of crack growth. The specimen then behaves similarly to a creep rupture specimen and failure life can be predicted better using reference stress, σ_r, [*5*]. Reference stress [*6*] is the equivalent uniaxial stress when applied to a uniformly stressed specimen that gives rise to the same displacement rate as that of a cracked body. Finally, for materials which fall in between the two extremes just discussed and where failure is due to the growth of a main crack, applicability of the linear elastic stress-intensity factor becomes questionable because of large plastic strain at the crack tip. For such cases, an energy-rate integral called the J*-integral appears to correlate the growth rates better than any other parameter [*4*]. The J*-integral is an extension of J-integral to creep where the integral provides a measure of the gradient of the strain energy [*7*] ahead of a slowly growing crack. Other parameters, such as displacement or displacement rate, have also been used to correlate creep crack growth. Their applicability, however, was limited to a specific material for selected specimen geometries. Hence, in the subsequent parametric analysis of creep crack growth, we will restrict it to the three parameters, K, σ_r, and J*-integral, to determine which one of them best correlates the growth rates and for an austenitic stainless steel.

Parametric analysis of creep crack growth in austenitic stainless steels has been done by several workers [*8–11*] using several different geometries. The results for Type 304 stainless steel indicate that J* is better than K or σ_r in correlating crack growth rates. But most of this work has been done at 650°C, a high temperature for stainless steel where steady-state conditions ahead of the growing crack are established relatively rapidly. It was not evident that such correlations are valid at lower temperatures, such as 593°C, which falls within the temperature range of application of stainless steel.

Saxena [*11*] has made a limited study of crack growth in Type 304 stainless steel at 593°C using compact tension and center-notched specimens and showed that J^* correlates the data better than K. All of the aforementioned tests were done under displacement rate-controlled conditions. More recently [*12*], extensive analysis of the crack growth in Type 316 stainless steel using compact tension and center-notched specimens was made under load-controlled conditions. There was a considerable spread in the crack growth rate data in terms of J^* parameter, but J^* still correlates the data better than K. Furthermore, the data in terms of J^* did not fall within the data band for Type 304 stainless steel at 650°C reported in the literature, and this was unexpected since most of the data of other materials fall within that band. It is not known whether this observed discrepancy is due to the difference in the method of testing since we have used load-controlled tests instead of displacement-controlled tests used by all others or due to the fact that the test temperature is lower and steady-state conditions ahead of a growing crack are not established in a reasonable time. The main objective of this study was to examine how the method of testing, that is, load controlled versus displacement controlled, affects the creep crack growth rates in Type 316 stainless steel in terms of the three parameters.

Experimental Details

The chemical composition of the alloy is listed in Table 1. Compact tension specimens of 1T size with nominal dimensions [*13*] but with 12.7-mm thickness and center-notched specimens with 5-mm thickness and 38.1-mm width were made with the crack plane parallel to the final rolling direction (T-L orientation). Compact tension specimens were side-grooved to a depth of 5 percent of the thickness. Notches in the center-notched specimens were made by spark-cutting using 0.25-mm wire. Both specimens were fatigue precracked at room temperature at stress intensities much lower than those used in the creep crack growth tests.

TABLE 1—*Chemical composition of Type 316 stainless steel.*

Element	Percent by Weight	Element	Percent by Weight
C	0.060	Mo	2.33
Mn	1.72	Cu	0.065
P	0.012	Co	0.030
S	0.007	Al	0.012
Si	0.40	Ti	0.003
Cr	17.30	B	0.0005
Ni	13.30	Fe	balance

The specimens were loaded in a servohydraulic machine under the displacement control mode. For center-notched specimens displacement along the loading line was measured using a strain-gage extensometer affixed to the specimen through quartz rods. For compact tension specimens, displacement was measured at the side surface and displacement along the loading line was computed by linear interpolation. The procedure is similar to that used earlier [*12*] for load-controlled tests. The specimens were heated by induction. Crack lengths in both type of specimens were measured using a low-magnification microscope. Corresponding load and displacement were also recorded. Crack growth rates and displacement rates were determined from the plots of crack length versus time and displacement versus time using a slope measuring device.

For compact tension specimens, stress-intensity factor K was determined for a/W ratio up to 0.7 using the ASTM K-calibration relation [*13*]:

$$K = \frac{P}{BW^{0.5}} [29.6(a/W)^{0.5} - 185.5(a/W)^{1.5} + 655.7(a/W)^{2.5} - 1017.0(a/W)^{3.5} + 638.9(a/W)^{4.5}] \quad (1)$$

and for a/W 0.7 from the relation [*14*]

$$K = \frac{P}{2B}\left(\frac{W+a}{(W-a)^{1.5}}\right)\left(4.0 + \frac{(W-a)}{(W+a)}\right) \quad (2)$$

and for center-notched specimens K is given by [*13*]

$$K = \frac{P}{BW}\left(1.77 + 0.277\left(\frac{2a}{W}\right) - 0.51\left(\frac{2a}{W}\right)^2 + 2.7\left(\frac{2a}{W}\right)^3\right) \quad (3)$$

where

P = applied load,
a = crack length from loading line,
W = specimen width, and
B = thickness.

Reference stress was calculated using the following equation [*6*]. For the compact tension specimen, it is given by

$$\sigma_r = \frac{2.02P}{BW}(1 + a/W)/(1 - a/W)^2 \tag{4}$$

and for center-notched specimen

$$\sigma_r = P/Bb \tag{5}$$

where b is $(W - 2a)$.

The value of J*-integral can be determined from crack length and displacement data using either an extensive data reduction procedure described earlier for both load-controlled [*15*] and displacement-controlled tests [*16*] or using an analytical estimation procedure which assumes that the material undergoes power-law creep of the form

$$\dot{\epsilon} = A(\sigma)^n \tag{6}$$

where $\dot{\epsilon}$ and σ are strain rate and stress and constants A and n depend on material and and temperature. It was shown [*12*] by detailed analysis using both center-notched and compact tension data that the J^*-values computed from the two procedures are nearly the same. Henceforth we will use the simple analytical estimation procedure to compute the J^*-value. Those for compact tension and center-notched specimens, respectively, are given by [*9*]

$$J^* = \frac{2(n-1)}{(n+1)} \frac{P\dot{\delta}}{Bb} \approx \frac{2P\dot{\delta}}{Bb} \tag{7}$$

and

$$J^* = \frac{(n-1)}{(n+1)} \frac{P\dot{\delta}}{Bb} \approx \frac{P\dot{\delta}}{Bb} \tag{8}$$

where n is the exponent in Eq 6 and $\dot{\delta}$ is the displacement rate along the loading line. For large values of n such as for annealed Type 316 stainless steel at 593°C [*8,10,17*], $(n - 1)/(n + 1)$ is assumed to be equal to 1.

Results and Discussion

For displacement-controlled tests, crack length increases with an increase in displacement while the load reaches a maximum and then decreases. Typical variation of crack length, displacement, and load as a function of time is shown in Figs. 1 and 2 for a compact tension specimen. Attempts to maintain displacement rate constant were not always successful, as can be seen in Fig. 1, where the rate increases with crack length and more so near the failure end. Using these plots, crack length and corresponding crack growth rate, load, displacement, and displacement rate were determined for the computation of K, σ_r, and J^*-values.

Stress-Intensity Factor

Crack growth rate as a function of linear elastic stress-intensity factor K is represented for displacement-controlled tests in Fig. 3 and the results are compared with those for load-controlled tests obtained previously and shown in Fig. 4. Under load control the data for the two geometries formed two separate bands with very little overlapping between the bands. The data for the displacement-controlled tests fall to the left of the data band for the corresponding specimen geometries, Fig. 3, that is, higher da/dt. This has the

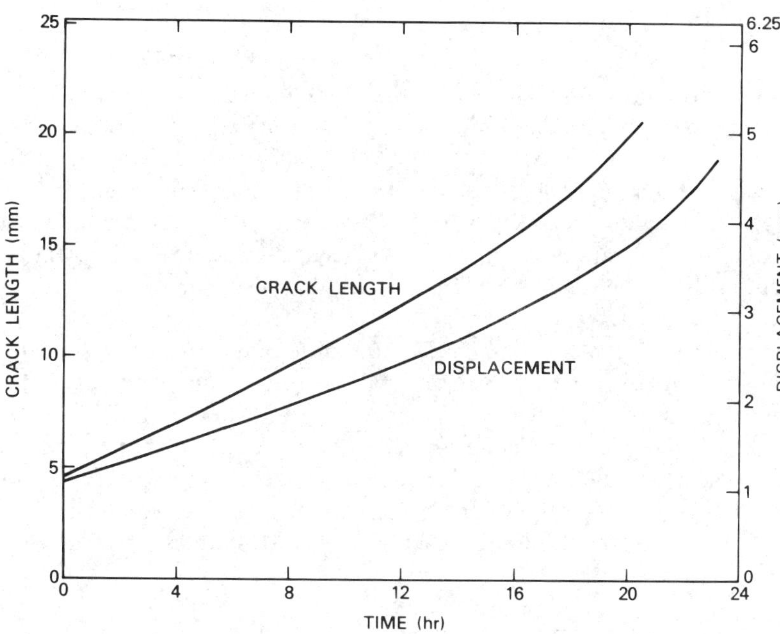

FIG. 1—*Typical plot showing the increase of crack length and displacement with time for a compact tension specimen.*

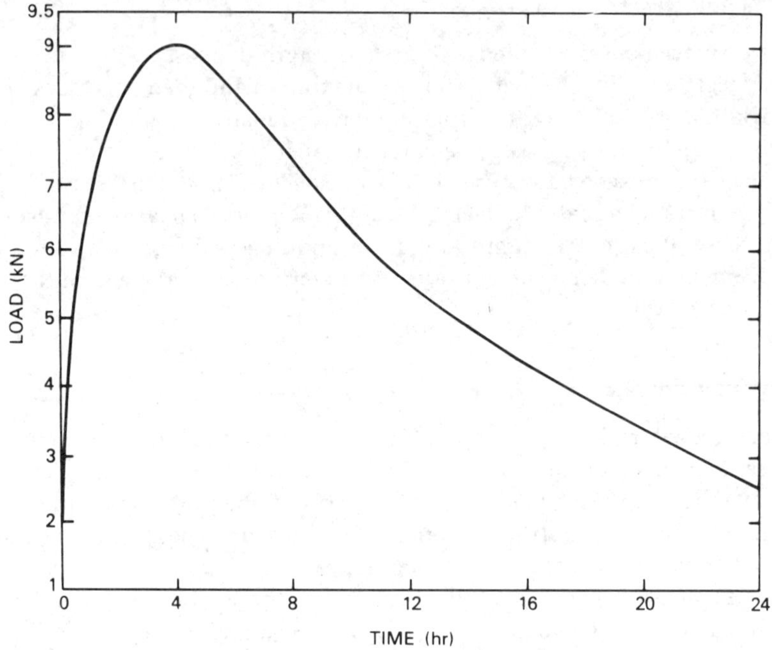

FIG. 2—*Typical variation of load with time for displacement-controlled test.*

effect of enlarging the total spread of the crack growth data in terms of K. Because of the large spread, K appears to be of little use in predicting growth rates in Type 316 stainless steel at 593°C. Also of importance to note is that under displacement control, crack growth rates for some specimens increase although the K-value decreases, which is physically inconceivable unless the computed linear elastic K-value is not an accurate measure of the existing stress intensity at the crack tip. This implies that K is not a controlling parameter for crack growth.

The increase of growth rate with decrease of K may also be accounted for by Hart's model [*18*] of creep crack growth where for each value of K there may be two possible branches in $\dot{a}$ versus K plots and changing the load during the displacement-controlled test could cause the crack to change from one branch to the other. This explanation presumes that K is the controlling factor for crack growth. But the foregoing data indicate that K cannot be safely used to predict the growth rates at least for Type 316 stainless steel at 593°C.

Reference Stress

Crack growth rate as a function of reference stress is shown for both specimen geometries for load-controlled tests, Fig. 5, and displacement-controlled

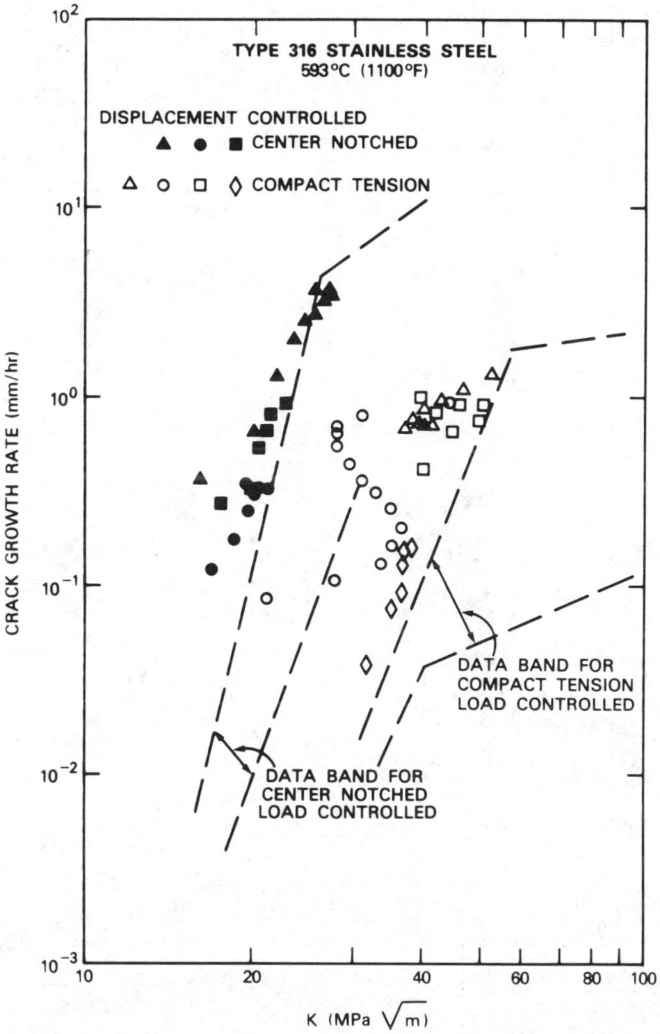

FIG. 3—*Crack growth rate in terms of stress-intensity factor for displacement-controlled tests.*

tests, Fig. 6. Just like the stress-intensity factor, σ_r correlates reasonably well if only the center-cracked specimens are considered. But the data spread is nearly three orders of magnitude in crack growth rates if both specimen geometries are considered. The data in fact diverge with increase in σ_r-value. For displacement-controlled tests, all of the data fall close to the left bound of the data band, Fig. 6. For some specimens crack growth rate increases with decrease in σ_r-value, which illustrates that σ_r may not be the controlling factor

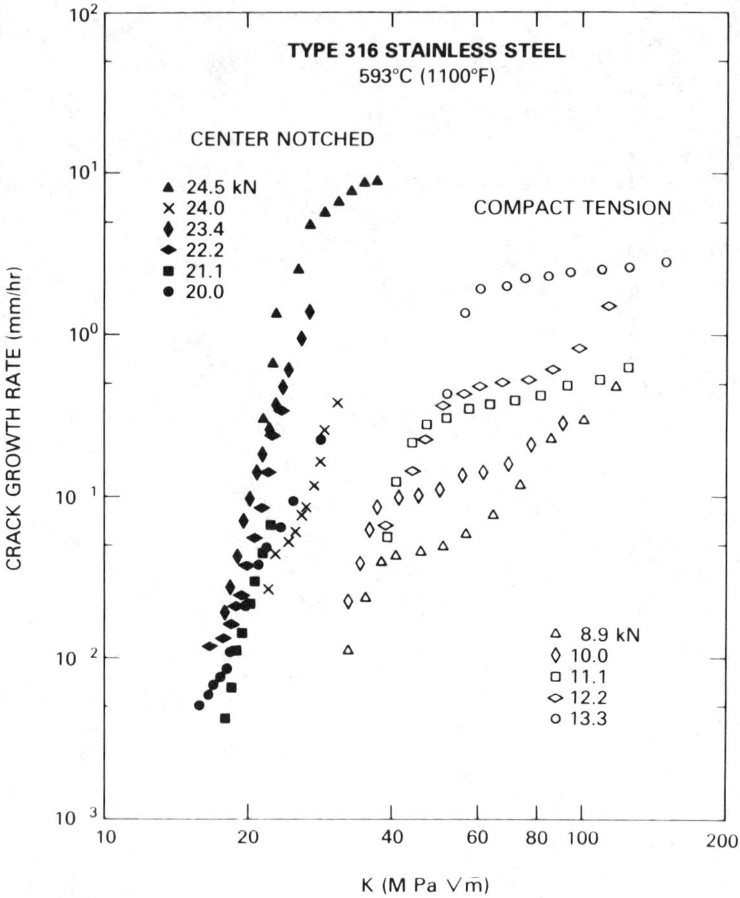

FIG. 4—*Crack growth rate in terms of stress-intensity factor for load-controlled tests.*

for crack growth. The significant data spread in terms of σ_r makes this parameter also unsuitable for predicting crack growth rates in a structure.

J-Integral*

Crack growth rate data for both load-controlled and displacement-controlled tests are next represented in terms of J*-integral in Figs. 7 and 8. The data for the load-controlled tests have been reported previously [*12*]. Examination of the two figures shows that the data for both specimen geometries overlap in the whole range of crack growth rate/J^*-values, indicating that J^* provides a better parameter than K or σ_r for correlating crack growth rates.

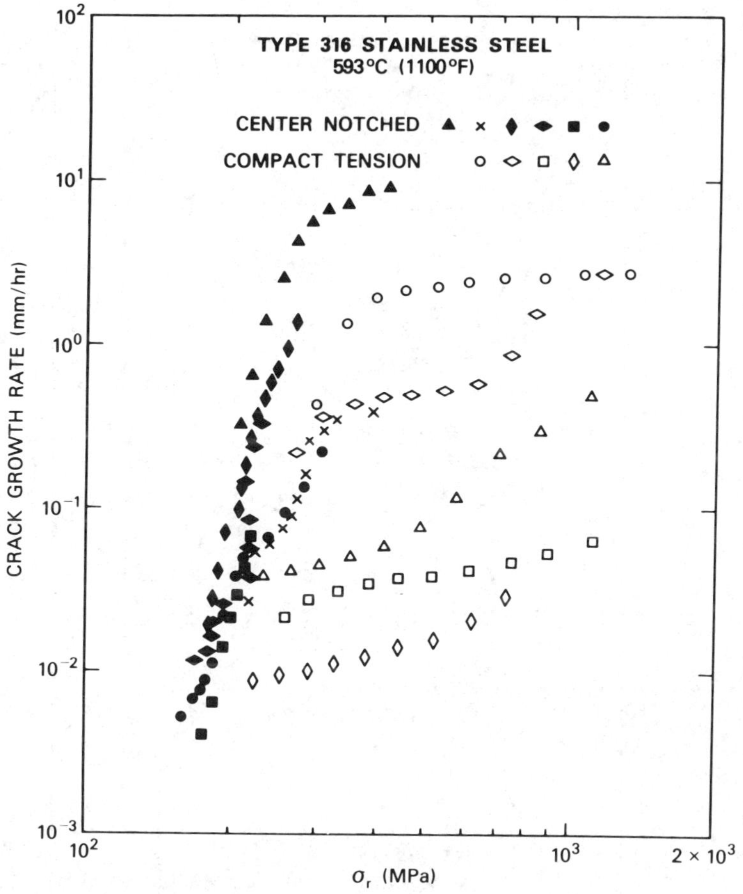

FIG. 5—*Crack growth rate in terms of reference stress for load-controlled tests.*

In response to the question whether the displacement-controlled tests are different from the load-controlled tests in terms of crack growth rate/J^* correlation, Fig. 8 shows that the data for displacement-controlled tests fall essentially within the data band for the load-controlled tests. Therefore, we conclude that both types of tests are equivalent in terms of $\dot{a}$ versus J^* relation, but from practical consideration load-controlled tests are much easier to perform and to control. Moreover, a larger range of crack growth rate data can be obtained from a single specimen under load-controlled conditions as can be seen from comparing Figs. 7 and 8.

Note that in Figs. 7 and 8 the spread in the data decreases with increase in crack growth rate or J^*-value, contrary to that observed either in terms of K

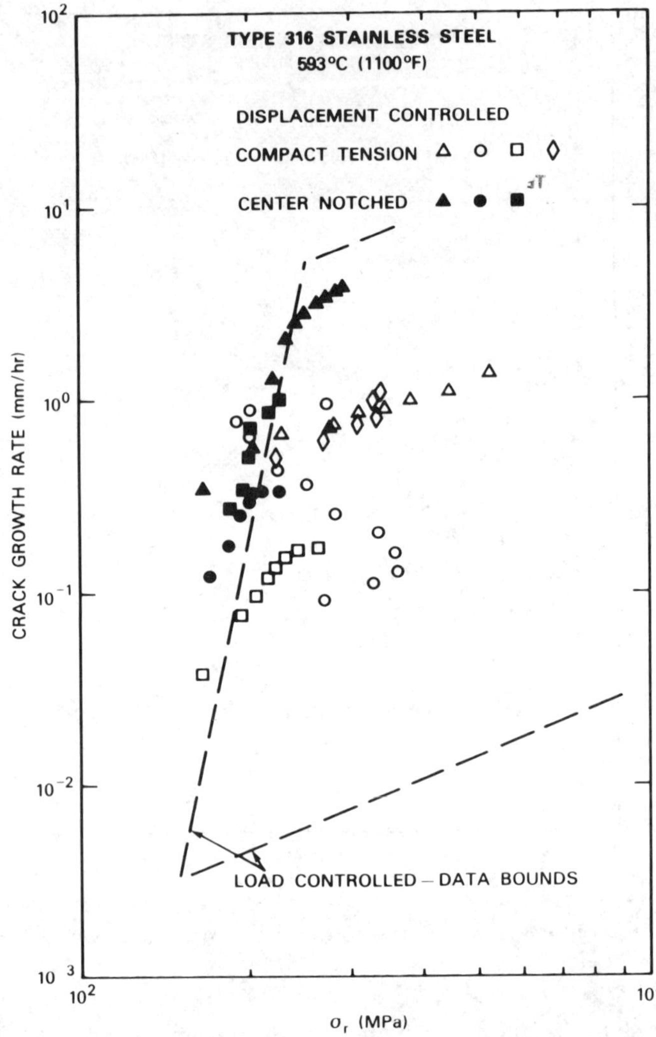

FIG. 6—*Crack growth rate in terms of reference stress for displacement-controlled tests.*

or σ_r. The spread is limited to a factor of 10 in contrast to a factor of 1000 in terms of K or σ_r. This demonstrates that J*-integral provides a better correlating parameter than K or σ_r. The reason for the decreasing spread in the data with increase in J^*-value is not known but may be the fact that steady-state crack growth conditions, which are a prerequisite for the validity of J^*, are not fully established during the early growth of a crack.

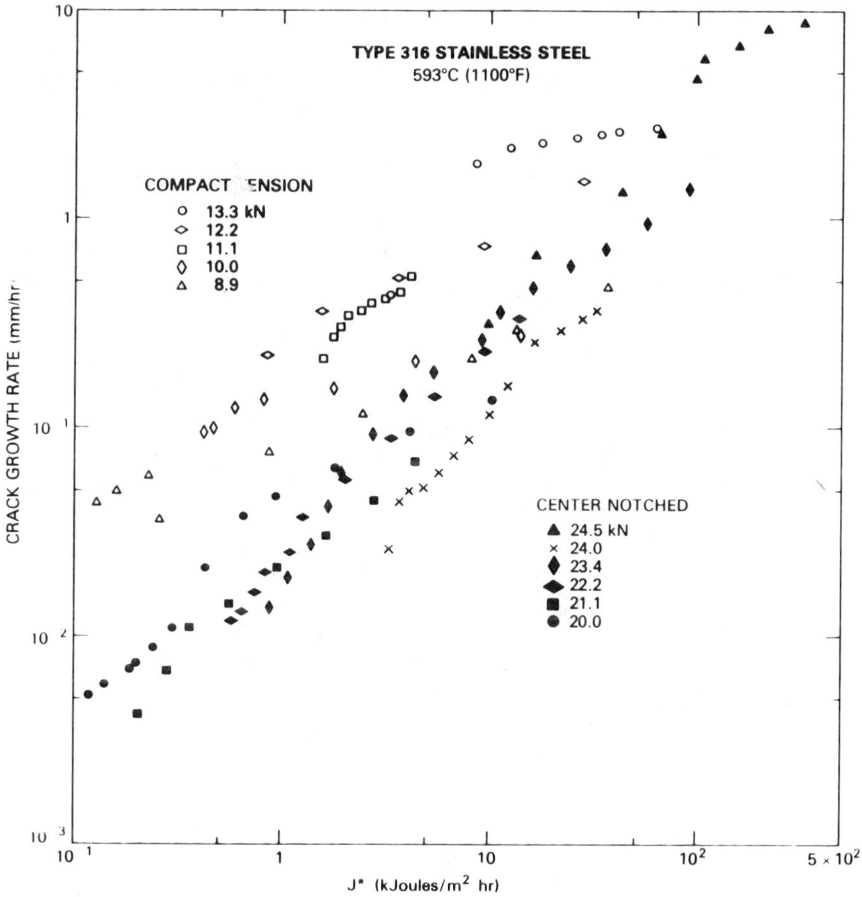

FIG. 7—*Crack growth rate in terms of J*-integral for load-controlled tests.*

Linear Elastic Energy Rate Parameter

Although the conclusion that the J^* parameter is better than K appears to be valid, a one-to-one comparison between the elastic-plastic and linear elastic parameter is difficult since K varies only by a factor of 10 whereas J^* varies by nearly a factor of 1000. Most importantly, the dimensions of the two parameters are also different, making it more difficult to compare the two.

To circumvent the foregoing problem, we defined earlier [*12*] a linear elastic energy rate parameter K^* which is dimensionally equivalent to J^* and is a time derivative of J in the limit of linear elastic behavior. Since the J-value in the elastic limit is equivalent to

$$J = K^2/E \tag{9}$$

its derivative with time is conceptually similar to J^* in the elastic limit. Thus

$$K^* = \frac{dJ}{dt} = \frac{2K}{E} \cdot \frac{dK}{dt} = \frac{2K}{E} \cdot \frac{dK}{da} \cdot \frac{da}{d\delta} \cdot \frac{d\delta}{dt} \tag{10}$$

where E is the Young's modulus. Just as for J^*-value, which requires the knowledge of displacement rate, $d\delta/dt$, K^* can be determined using the dis-

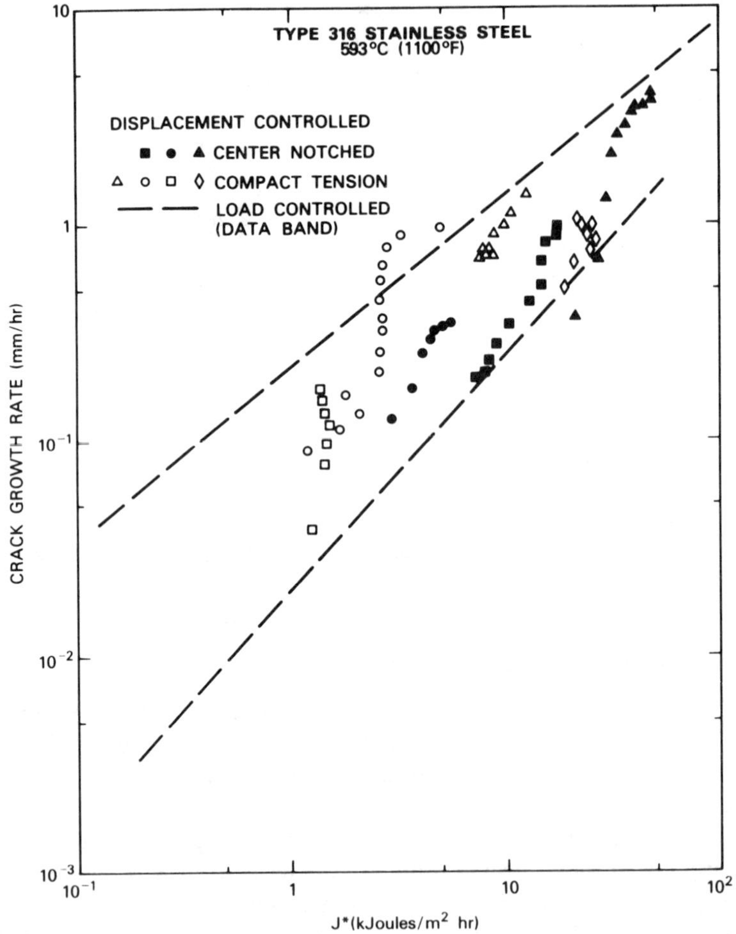

FIG. 8—*Crack growth rate in terms of J^*-integral for displacement-controlled tests.*

placement rate as in Eq 10. In the elastic range, crack length a and displacement δ are related by [19]

$$\delta = \sqrt{\frac{2K}{E}} \cdot \sqrt{\frac{a}{\pi}} = \frac{2\sigma\sqrt{\pi a}}{E} \cdot \sqrt{\frac{a}{\pi}} \tag{11}$$

Hence $d\delta/da$ can be obtained as $2K/E\sqrt{\pi a}$.

From this and from Eq 10, K^* is given by

$$K^* = \sqrt{a\pi}\,\frac{dK}{da} \cdot \frac{d\delta}{dt} \tag{12}$$

Using the compliance relations (Eqs 1–3), dK/da can be determined and K^* can be computed for each crack length increment. Important to note is that the use of Eq 11 implies that the measured displacement is essentially elastic and in reality the presence of plastic strain will overestimate the K^*-value.

Both J^* and K^* are energy rate integrals, but J^* has a nonzero value under constant load and crack length in a creeping material while K^* is zero. It may be stated, however, that if the material creeps at constant load and crack length, the J^*-da/dt relation will not be unique and thus the utility of J^* as a parameter for crack growth becomes questionable. Both parameters, however, describe the stress-strain rate behavior of the material ahead of the crack tip where one assumes a linear relation between the stress and strain while the other does not.

Crack growth rate as a function of K^* is represented in Fig. 9 for both specimen geometries and for both load- and displacement-controlled tests. Similar to J^*, K^* varies by a factor 1000 and the units for both are the same. A one-to-one comparison between the linear elastic and elastic-plastic parameters can be made without any ambiguity. In comparison with the data in terms of stress-intensity factor K, Figs. 3 and 4, crack growth rate data for the two geometries in terms of K^* overlap over a range of K^*-values.

Furthermore, in comparison with Fig. 8, Fig. 9 shows that the spread in the data in terms of K^* increases with increasing K^*-value, which is expected since at high K^*-value the plasticity effects become increasingly pronounced and the deviations in the computed K^*-value become increasingly larger. It appears from Figs. 8 and 9 that at low growth rates, K^* is better than J^*, but at high growth rates, J^* is better than K^*. The elastic-plastic parameter becomes more advantageous than the linear elastic parameter for crack growth rates higher than 10^{-1} mm/h.

The present results show that the lack of correlation of creep crack growth data on a stress-intensity-factor basis does not necessarily mean that linear elastic fracture mechanics is unsuitable for correlation of the data. It may imply that for time-dependent growth, K may not be the parameter, but a

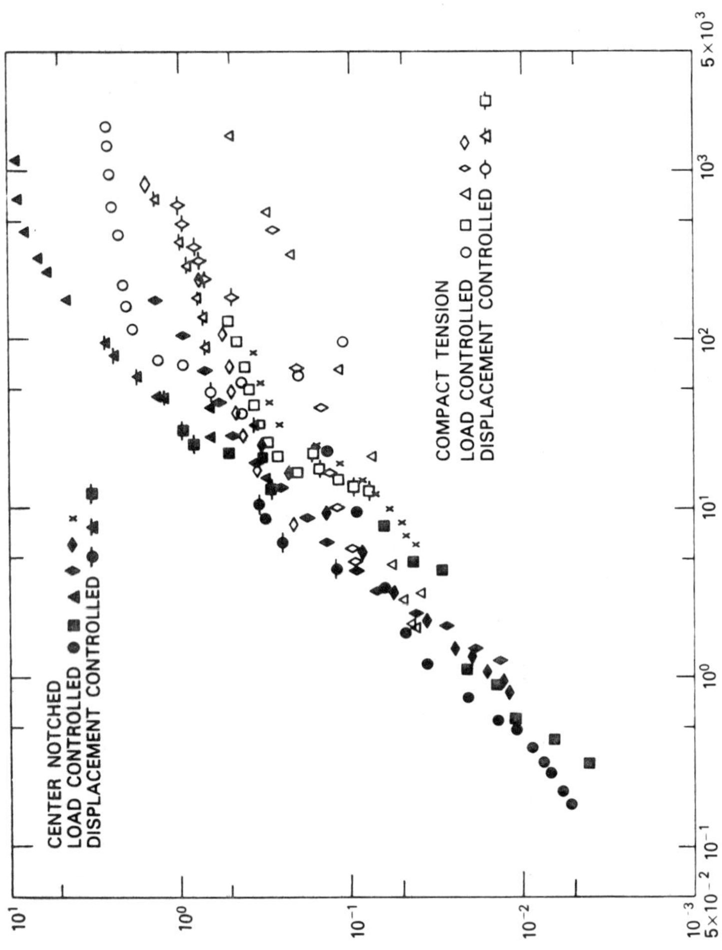

FIG. 9—*Crack growth rate in terms of linear elastic energy rate parameter* K* *for both load- and displacement-controlled tests.*

rate-dependent parameter is required, such as the linear elastic energy rate parameter, K^*. Under conditions of extensive plasticity, K^* is not valid and the elastic-plastic energy rate parameter J^* needs to be used to correlate the growth rate.

Comparison with Other Data

The present data on Type 316 stainless steel at 593°C are next compared with the data on Type 304 stainless steel at 650°C in Fig. 10, which also includes some limited data at 600 and 650°C on Type 316 stainless steel published earlier [*10*]. The data for Type 304 stainless steel are quite extensive in terms of different specimen geometries and loading constraints and were ob-

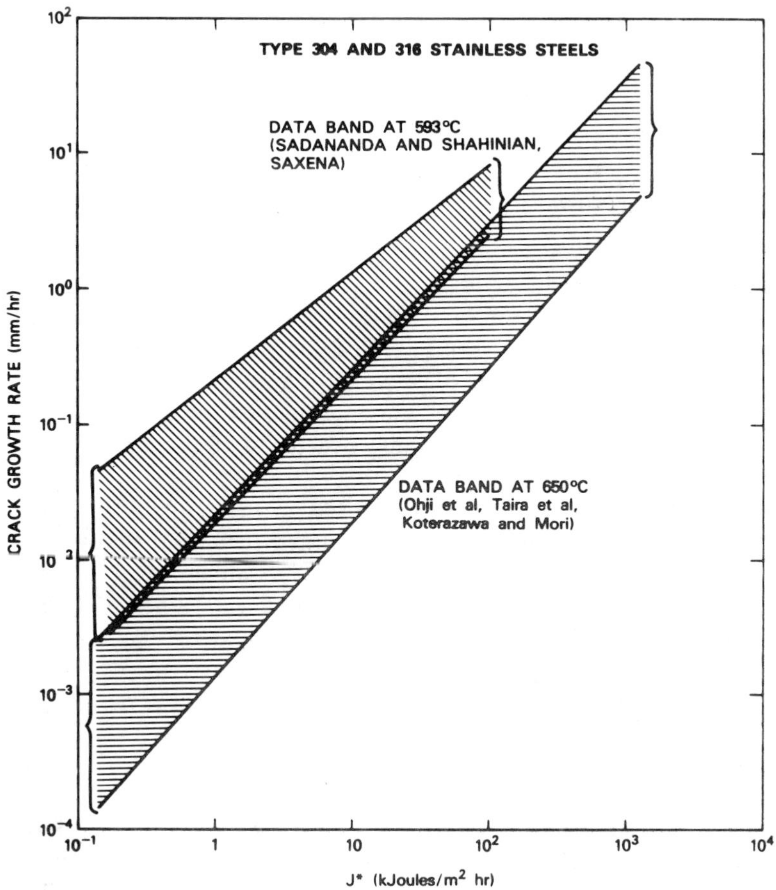

FIG. 10—*Comparison of crack growth data for Types 316 and 304 stainless steel at 593°C with those of Type 304 stainless steel at 650°C.*

tained independently by several investigators. Note that the growth rates for Type 316 stainless steel at 593°C are higher than those for Type 304 stainless steel. It appears therefore that a unique correlation of $\dot{a}$ versus J^* independent of temperature and material may not be always possible. This may be due to the fact that steady-state crack growth conditions may not always be achieved at the test temperature.

Finally, while ascertaining that the method of testing, that is, load-controlled versus displacement controlled, will not affect the crack growth rate/J^* correlation, the present analysis clearly points out that J^* parameter is much superior to linear elastic stress intensity factor K or reference stress σ_r. This better correlation with J^* is not necessarily related to the fact that J^* is an elastic-plastic parameter, but related to the fact that K is not a rate-dependent variable while J^* is. It is more advantageous to use linear elastic energy rate parameter K^* for small-scale creep and elastic-plastic energy rate parameter J^* for large-scale creep. Lack of correlation of creep crack growth on a stress-intensity-factor basis will have an important consequence even in terms of fatigue crack growth at high temperatures where creep effects become pronounced.

Summary and Conclusions

Parametric analysis of creep crack growth in Type 316 stainless steel at 593°C has been made under load- and displacement-controlled conditions using center-notched and compact tension specimens. In particular, crack growth rates were analyzed in terms of linear elastic stress intensity factor, K, reference stress, σ_r, and elastic-plastic energy rate parameter, J^*. Analysis shows that J^* parameter correlates the data better than either K or σ_r. The spread in the data in terms of J^* is only a factor of 10 in comparison with a factor of 1000 in terms of K or σ_r.

The spread in the data in terms of J^* is in fact slightly larger at low J^*-values and decreases with increase in J^*. Furthermore, the growth rates in terms of J^* are higher than those for Type 304 stainless steel at 650°C reported in the literature. This larger spread at lower values of J^* and the higher growth rates than Type 304 stainless steel may be due to the lack of fully steady-state conditions ahead of a moving crack.

Since K and J^* differ significantly in units as well as in the range of values, for a meaningful comparison of elastic and elastic-plastic parameters, a new parameter, the elastic energy rate parameter, K^*, is defined. Crack growth data in terms of K^* in fact show much better correlation than in terms of J^* particularly at low growth rates. At high growth rates the data spread in terms of K^* increases understandably due to the increasing contribution from plasticity effects. Thus, lack of correlation on a K basis does not necessarily imply that linear elastic fracture mechanics is invalid for creep crack growth, but implies only that the stress-intensity factor is not suitable for time-dependent growth and the rate-dependent parameter, K^*, needs to be used.

Finally, just on the basis of the comparison of crack growth data in terms of K, σ_r, and J^*, the present analysis clearly points out that J^* is better than the other two parameters in correlating the data. This will have a bearing not only on creep crack growth but on high-temperature fatigue crack growth as well, particularly when creep effects become important.

Acknowledgment

The experimental work was assisted by Messrs. H. Dan and T. Watson. This investigation was supported by the Office of Naval Research.

References

[*1*] Sadananda, K. and Shahinian, P. in *Cavities and Cracks in Creep and Fatigue*, J. Gittus, Ed., Applied Science, Barking, Essex, U.K., 1981, pp. 109-195.

[*2*] Sadananda, K. and Shahinian, P., *Materials Science and Engineering*, Vol. 43, 1980, pp. 159-168.

[*3*] Sadananda, K. and Shahinian, P., *Metallurgical Transactions*, Vol. 11A, 1980, pp. 267-276.

[*4*] Sadananda, K. and Shahinian, P., *Engineering Fracture Mechanics*, Vol. 15, 1981, pp. 327-342.

[*5*] Neate, G. J., *Materials Science and Engineering*, Vol. 34, 1978, pp. 165-173.

[*6*] Williams, J. A. and Price, A. T., *Transactions*, American Society of Mechanical Engineers, *Journal of Engineering for Materials and Technology*, Vol. 97, 1975, pp. 214-221.

[*7*] Hutchinson, J. W., *Journal of Mechanics and Physics of Solids*, Vol. 16, 1968, pp. 337-347.

[*8*] Ohji, K., *Theoretical and Applied Mechanics*, Vol. 27, 1977, pp. 3-20.

[*9*] Koterazawa, R. and Mori, T., *Transactions*, American Society of Mechanical Engineers, *Journal of Engineering Materials and Technology*, Vol. 99, 1977, p. 298.

[*10*] Taira, S., Ohtani, R., and Kitamura, T., *Transactions*, American Society of Mechanical Engineers, *Journal of Engineering Materials and Technology*, Vol. 101, 1979, pp. 154.

[*11*] Saxena, A. in *Fracture Mechanics: Twelfth Conference, ASTM STP 700*, American Society for Testing and Materials, 1980, pp. 131-151.

[*12*] Sadananda, K. and Shahinian, P., "Evaluation of J^* Parameter for Creep Crack Growth in Type 316 Stainless Steel," presented at the 14th National Fracture Mechanics Symposium, Los Angeles, July 1981.

[*13*] Annual Book of ASTM Standards, American Society for Testing and Materials, Part 10, 1974, pp. 432-451.

[*14*] Wilson, W. K., *Engineering Fracture Mechanics*, Vol. 2, 1970, pp., 169-171.

[*15*] Sadananda, K. and Shahinian, P., *Metallurgical Transactions*, Vol. 8A, 1977, pp. 439-449.

[*16*] Landes, J. D. and Begley, J. A. in *Mechanics of Crack Growth, ASTM STP 590*, American Society for Testing and Materials, 1976, pp. 128-148.

[*17*] Nicholson, R. D., *Materials Science and Engineering*, Vol. 22, 1976, pp. 1-6.

[*18*] Hart, E. W., Report No. 4111, Materials Science Center, Cornell University, Ithaca, N.Y., July 1979.

[*19*] Irwin, G. R., Kies, J. A., and Smith, H. L. in *Proceedings*, American Society for Testing and Materials, Vol. 58, 1958, p. 640.

Stephen Floreen[1]

Microstructural and Environmental Effects During Creep Crack Growth in a Superalloy

REFERENCE: Floreen, S., "**Microstructural and Environmental Effects During Creep Crack Growth in a Superalloy,**" *Elastic-Plastic Fracture: Second Symposium, Volume I—Inelastic Crack Analysis, ASTM STP 803*, C. F. Shih and J. P. Gudas, Eds., American Society for Testing and Analysis, 1983, pp. I-708–I-720

ABSTRACT: Two heats of a nickel-base superalloy were heat-treated to form three different grain boundary carbide microstructures, a blocky $M_{23}C_6$ carbide, a cellular $M_{23}C_6$ carbide, and a carbide-free structure. Creep crack growth rate tests were run at 650°C in four environments: high-purity helium, air, He-4CH_4 and He-3SO_2. Crack growth in all cases was intergranular and involved the nucleation and growth of micrometre-sized cavities on the grain boundaries. The air and He-3SO_2 environments were very aggressive, increasing the crack growth rates over those in helium by more than 100-fold. The grain boundary microstructures significantly affected the growth rates, and microstructures that were best in some environments and were worse in others. The results suggest that a single, generalized cavity nucleation and growth mechanism may be responsible for crack growth in superalloys, but environment-microstructure interactions will have to be accounted for to provide a satisfactory description of the cracking behavior. The results also suggest that some of the general ideas about the role of carbides in superalloys may not be true under crack growth conditions, and that alloys optimized by tests in air may not be the most suitable for use in other environments.

KEY WORDS: nickel alloys, heat-resistant alloys, chromium carbides, grain boundaries, creep strength, cavities, crack propagation, oxidizing atmospheres, sulfidation, elastic-plastic fracture

One of the newer applications of fracture mechanics is the study of crack growth during creep loading. In the past, creep usually was considered as a problem in plasticity. Starting about ten years ago, however, creep has come to be viewed as a fracture problem. This change in perspective seems to have arisen for two reasons. On the one hand, considerable evidence was developed

[1]Research fellow, Inco, Alloy Products Co. Research Center, Sterling Forest, Suffern, N.Y. 10901.

over the years that creep deformation is accompanied by cavitation, microcrack formation, and other small-scale fracture events. More recently, elevated temperature fatigue problems have generated considerable interest in creep fracture to help understand time-dependent fatigue behavior and creep/fatigue interactions.

A considerable amount of effort has been devoted to examining the properties of high-strength nickel-base superalloys. Most of this work has been summarized in several recent review papers [*1–3*].[2] At intermediate temperatures, on the order of 500 to 750°C, intergranular creep crack growth in some of these materials can take place at stress intensities as low as 20 $MPa\sqrt{m}$. It generally appears that linear elastic fracture mechanics provide a valid description of the loading conditions causing creep crack growth in these tests.

Creep crack growth in superalloys frequently appears to take place by a process involving cavity nucleation, growth, and microcrack formation on grain boundaries slightly ahead of the main crack. Grain boundaries oriented normal to the loading direction generally show the greatest damage. Cavities on the fractured grain boundaries are typically several micrometres in size and more or less equiaxed in shape. Some elongated cavities are usually present. Numerous attempts have been made to model cavity growth by diffusion-controlled processes, power-law creep controlled processes, or by models involving coupled stress and diffusion-controlled mechanisms. While some of these models appear correct in detail, none of them are adequate to predict the crack growth behavior of superalloys to any useful degree of accuracy.

One of the problems in trying to develop a general model of creep crack growth in these materials is their sensitivity to microstructural variations. It is now clear that order-of-magnitude changes in the creep crack growth rate can be achieved by changing the microstructure. As expected, grain size and grain shape are important. Coarser grain size, serrated grain boundaries, and elongated grains oriented perpendicular to the crack direction each significantly retard the crack growth rate. In addition, coarsening the gamma-prime (γ') precipitate so as to produce a more homogeneous slip distribution within the grains appears beneficial. Grain boundary chemistry is also important. Much of the work in developing superalloys involved removing deleterious impurities, and adding beneficial elements such as boron and zirconium. How grain boundary chemistry affects the creep crack growth behavior is only beginning to be explored.

A second major difficulty in developing a generalized creep crack growth model is the very potent role the environment may play. Air, or more specifically oxygen, can greatly accelerate the crack growth rate in creep or fatigue in many materials. High-strength superalloys appear more sensitive to oxygen, but not all superalloys are equally affected. There is also evidence

[2]The italic numbers in brackets refer to the list of references appended to this paper.

in fatigue that carburizing or particularly sulfidizing environments can act aggressively [*4–6*].

Almost all of the studies of the effects of microstructural variables have been done in air. Thus one question that remains unexplored is whether a microstructural change is beneficial because of some intrinsic reason independent of the environment, or whether the microstructural change is helpful because of extrinsic reasons; that is, because it minimizes a deleterious effect of the environment. An allied question is whether a microstructure optimized for service in air would provide the best properties when some other environment is present.

The purpose of the present study was to explore this question of microstructure versus environment. Two heats of a nickel-base superalloy, Inconel[3] alloy X-750, were processed so as to produce three different carbide morphologies in the grain boundaries. These three microstructures were then tested at 650°C in four different representative environments so that a total of 12 different microstructure-environment combinations was evaluated.

Experimental Procedure

Two heats of Inconel alloy X-750 were used for this study. The first, Heat A in Table 1, was commercially melted material that had been processed to 12.6-mm-thick plate stock. The second, Heat B, was a vacuum-melted laboratory heat made to duplicate Heat A, except that the carbon level was reduced to 0.001 weight percent. This heat was also processed to plate stock.

Specimen blanks were heat-treated based on the results of Raymond [*7*]. For Heat A the following two heat treatments were employed:

1. 1093°C/2 h/WQ + 885°C/24 h/AC + 704°C/20 h/AC
2. 1093°C/2 h/WQ + 704°C/20 h/AC

where WQ is water-quenched and AC is air-cooled.

Heat treatment No. 1 is a more typical one for this alloy in terms of producing better stress rupture properties. As shown in Fig. 1, this heat treatment produced discrete, blocky $M_{23}C_6$ grain boundary particles approximately

TABLE 1—*Chemical composition of Inconel alloy X-750 heats.*

Heat	Cr	Fe	Al	Ti	Cb	Mn	Si	C	Ni
A	14.8	6.6	0.75	2.46	0.97	0.52	0.35	0.08	balance
B	14.8	6.9	0.69	2.50	0.99	0.33	0.32	0.001	balance

[3]Trademark of the Inco family of companies.

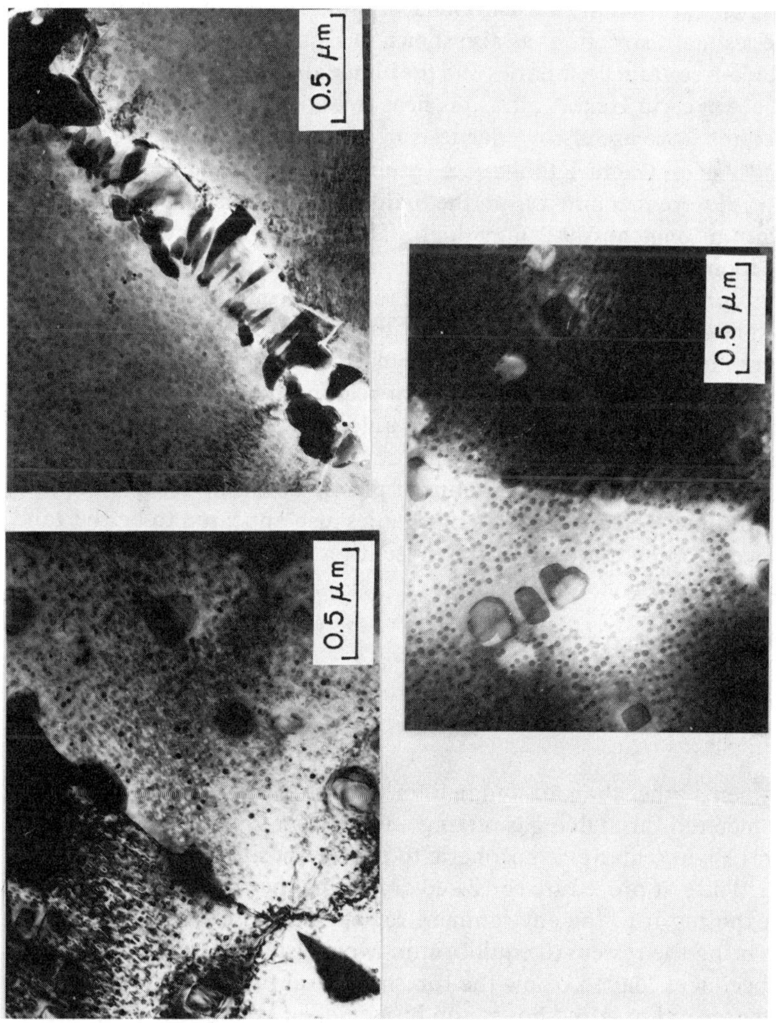

FIG. 1—*Transmission electron micrographs of grain boundary structures.* Upper left: *Heat A, heat treatment No. 1, blocky carbides.* Upper right: *Heat A, heat treatment No. 2, cellular carbides.* Bottom: *Heat B, heat treatment No. 1, carbide-free microstructure.*

0.3 μm in size and a bimodal γ' morphology consisting of 180 and 14.5 nm cuboidal particles.

Heat treatment No. 2 produced the cellular carbide microstructure shown in Fig. 1. The structure in this case consisted of $M_{23}C_6$ grain boundary carbide platelets about 0.4 μm long oriented perpendicular to the grain boundary and 14.5 nm cuboidal γ' particles within the grains.

Specimen blanks from the low-carbon Heat B were given only heat treatment No. 1. The resultant structure, as also shown in Fig. 1, consisted of essentially clean, carbide-free grain boundaries and the bimodal γ'-size distribution within the grains observed in Heat A after this heat treatment. Note that in all three microstructures there are also γ' particles in the grain boundaries.

Compact tension fracture toughness specimens of the type used in earlier studies [*3–6, 8*] were machined from the blanks. All specimens were 29.3 mm high, 30.5 mm long, and 9.2 mm thick. The specimens were fatigue precracked at room temperature at low load ranges to a/w ratios of 0.52 to 0.54. Testing was done in mullite environmental retorts to contain the test environment. A split-tube furnace surrounding the retort provided heat. Crack lengths were measured by the compliance technique suggested by Jones and West [*9*] using crack-opening displacement readings taken off the load line. Calibration tests on specimens which were unloaded prior to fracture showed that observed crack lengths agreed within ± 10 percent of calculated values. Because of the extremely minor amount of plastic flow that occurred during crack growth, linear elastic fracture mechanics also appeared to be the valid parameter controlling cracking in the present tests.

Four test environments were used:

1. 99.995 percent pure helium,
2. air,
3. He-4CH_4, and
4. He-3SO_2.

Environments other than air and helium were provided by high-purity bottled gases metered through a gas mixing and delivery system. Tests were run at pressures slightly above atmospheric to provide positive gas flow through the system. The test procedure consisted of heating the specimen to temperature and exposing it to the environment for approximately one hour before loading to bring the system to equilibrium. Most tests were run to failure. In cases of specimens loaded below the threshold level the specimens were unloaded after several hundred hours and broken open at ambient temperature to see if any creep crack growth had taken place. Some tests were interrupted short of failure and the specimens cross-sectioned to examine the cracks. Also, as discussed in the following, some specimens were exposed unloaded in air and then tested in helium. After failure the fracture surfaces of a number of specimens were examined by scanning electron microscopy.

Results

The results of the creep crack growth rate measurements are shown in Fig. 2. In the high-purity helium environment all three materials displayed a threshold stress-intensity value of approximately 23 MPa$\sqrt{m}$ below which no crack growth took place. At higher stress intensities the materials with blocky grain boundary carbides and carbide-free grain boundaries displayed essentially identical crack growth rates. With cellular grain boundary carbides the crack growth rate was about twice that of the other microstructures.

It should be noted that the high-purity helium environment is not an inert environment because even very small amounts of oxygen could provide a significant level of oxygen activity. Previous fatigue tests in this system showed that identical crack growth rates were obtained in high-purity helium, hydrogen, and nitrogen [*4*], suggesting that the present tests in helium could be considered as being done in a consistent, low-oxygen-activity environment. In this connection it also should be noted that a hard vacuum may not provide an inert environment because of volatilization of impurity or alloying elements from the specimen. Thus achieving a truly inert high-temperature environment may be a difficult task.

As shown in Fig. 2, testing in air significantly altered the crack growth behavior compared with helium. The threshold values for the cellular and carbide-free microstructures were decreased and the crack growth rates at higher stress intensities increased significantly. Note also that the best microstructures in helium were the worst ones in the air environment.

The He-4CH_4 carburizing environment had a slight effect on the material with cellular carbides, lowering the threshold slightly and producing about a threefold increase in crack growth rate over that in pure helium. The other two microstructures displayed results essentially identical to those of helium. This environment produced unit carbon activity and carbon sooting on the specimen surfaces. In fatigue this environment increased the fatigue crack growth rate in another superalloy by about three times that in pure helium [*5*]. In general both the present and the earlier tests suggest that carburizing environments do not seriously degrade the crack growth behavior of superalloys at intermediate temperatures.

In contrast to carburizing environments, sulfidizing environments are very aggressive. The results in Fig. 2 show that the He-3SO_2 environment increased the crack growth rate in all three microstructures by over 100-fold. Tests in this environment showed apparently high threshold values, but closer inspection revealed that crack growth slowed down and stopped due to blocking and blunting of the crack tip by sulfide particles. Figure 3 shows an example of this type of sulfide particle formation in a specimen that was unloaded and cross-sectioned. Sulfide particle formation near the crack tip has also been observed in fatigue tests, and severely perturbs the cracking behavior at low stress-intensity ranges because of the effects of the particles on crack closure [*4,6*]. At intermediate temperatures it seems to be generally characteristic of sulfidizing environments that cracking is a race between crack acceleration due to embrittlement and crack deceleration due to the effects of sulfide particle formation. At lower stress intensities crack growth can be stopped. Thus in theory it might be possible to design to fairly large loads in these environments, but whether the sulfide particles provide a stable means of preventing crack growth remains to be established.

To facilitate comparison of the results in Fig. 2, the crack growth rates at a stress intensity of 35 MPa$\sqrt{m}$ are summarized in Table 2. This relatively high stress-intensity value was used for the comparison because of the high apparent threshold values in the sulfidizing environment. As can be seen in Table 2,

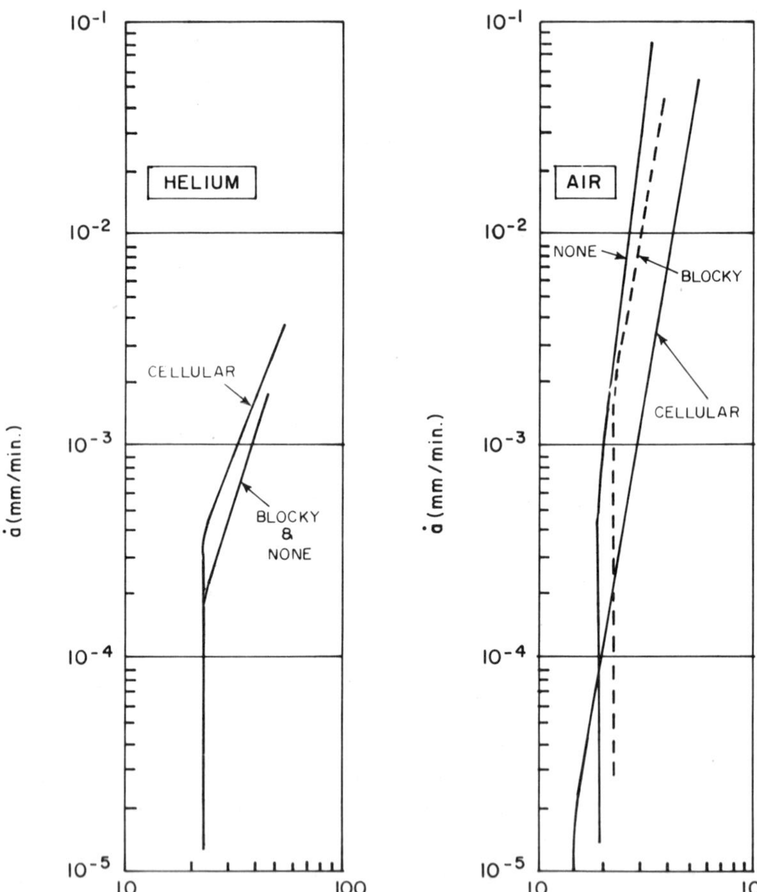

FIG. 2—*Creep crack growth rates at 650°C with different grain boundary carbide microstructures.*

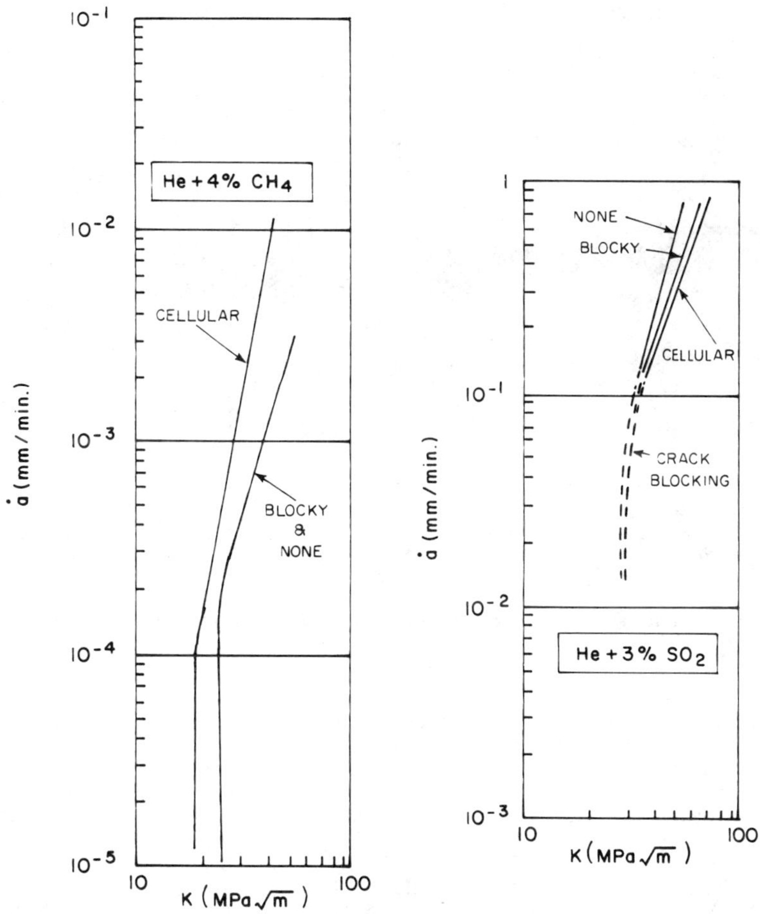

FIG. 2—*(Continued).*

crack growth rates vary by more than two orders of magnitude. Also, as noted previously, the best microstructure varies with environment. For example, the cellular grain boundary carbide structure provided the highest crack growth rates in high-purity helium and He-4CH_4, and the lowest crack growth rates in air and in He-3SO_2.

Earlier crack growth studies have indicated that environments that behaved very aggressively under mechanical loading produced no significant corrosion damage on unstressed materials [*4–6*]. Examination of the present specimens confirmed this pattern; the unstressed surfaces of all the specimens showed only slight discoloration and no signs of attack. To further explore this question a specimen with each microstructure was fatigue precracked and then exposed unloaded in air for 10 days at 650°C. These

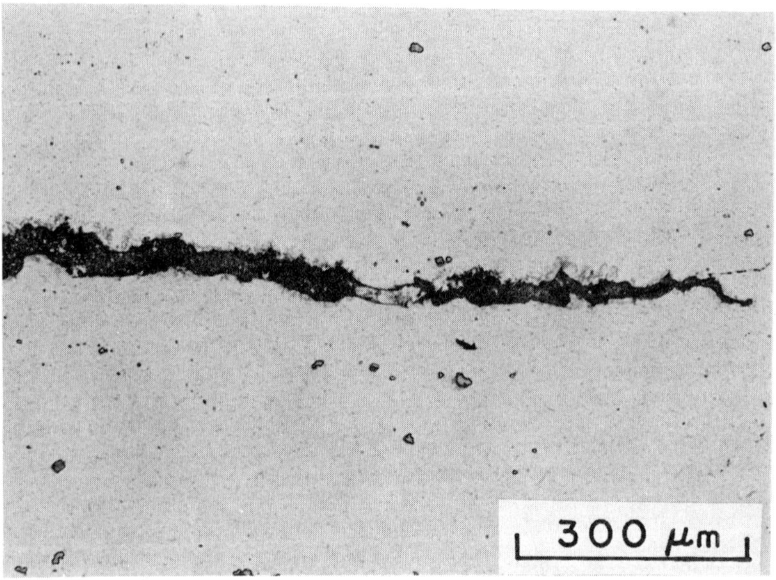

FIG. 3—*Example of crack tip filled with sulfide particles from test in He-3SO$_2$. (unetched).*

TABLE 2—*Comparison of creep crack growth rates at a stress intensity of 35 MPa$\sqrt{m}$.*

Grain Boundary Carbide Microstructure	*da/dt* (mm/min)			
	He	Air	He-4CH$_4$	He-3SO$_2$
Blocky	7.0×10^{-4}	1.95×10^{-2}	7.0×10^{-4}	1.5×10^{-1}
Cellular	1.15×10^{-3}	2.6×10^{-3}	3.1×10^{-1}	1.3×10^{-1}
None	7.0×10^{-4}	6.3×10^{-2}	7.0×10^{-4}	1.6×10^{-1}

specimens were then tested in high-purity helium. The creep crack growth data for these specimens were identical to those of fresh specimens. In these cases the prior exposure time in air was approximately ten times longer than the test duration in helium. Thus these results further indicate that environmental effects are dynamic in character; that is, stress is required to produce an environmental effect at these temperatures. It also follows that elevated-temperature corrosion tests may be nonconservative for predicting service performance.

Figure 4 shows some typical scanning electron micrographs (SEM's) of the fracture surfaces. Crack growth appeared to be intergranular in all cases. Cavities were commonly observed on the grain boundary surfaces. Both relatively equiaxed and also elongated cavities could be seen. Cavities nucleated

FIG. 4—*SEM's showing typical appearance of fracture surfaces. Specimens are from Heat B, tested in air.*

at carbides could occasionally be seen. However, there did not appear to be any significant effect of the carbide morphology or the appearance of the cavities. Also, allowing for some obscuring of the surface details in the tests run in the more aggressive environments, there did not appear to be any systematic effect of environment on the dimensions of the cavities. In all of the tests it seemed that crack growth involved the nucleation and growth of cavities to several micrometres in size. Earlier tests on Inconel alloy 718 in high-purity helium and in air also revealed virtually identically appearing cavities despite a 100-fold difference in crack growth rates in the two environments [*10*].

Discussion

The present results with Inconel X-750, as well as those of many of the earlier studies, suggest that intermediate-temperature creep crack growth in superalloys involves the nucleation and growth of micrometre-sized cavities at grain boundaries. Thus it is tempting to think that a single mechanism is responsible for crack growth in all of these cases. However, if this is so it is clear that microstructural and environmental variations can produce very large changes in the rates of cavity nucleation or growth or both.

Thus any generalized crack growth theory must deal with microstructural parameters such as those shown in Fig. 1. Of greater importance, and of much greater difficulty, environmental effects must be treated. To date there is no theory that even begins to predict environmental factors, yet it is becoming increasingly clear that the environment often may be the most important parameter controlling the crack growth rate in either creep or fatigue. Recent work makes it clear that the presence of oxygen on the grain boundaries can drastically reduce the ductility and promote intergranular crack growth at elevated temperatures [*11–13*]. Diffusion of oxygen, sulfur, or carbon into the grain boundaries and reaction with alloying elements or phases seem likely in superalloys, but the specific reactions are not known. The present results make it clear that environmental effects depend upon microstructure, and that one should consider the microstructure-environment system as a whole in discussing creep crack growth.

From the practical point of view, the role of the grain boundary carbides in crack growth is of interest. Most superalloys are designed to an optimum carbide concentration and morphology. Carbides are believed to play an important role in stabilizing the microstructure by minimizing grain boundary sliding or migration during creep, and also perhaps to help getter deleterious tramp elements in the grain boundaries. An excessive level of carbides, however, or a continuous carbide morphology is usually considered to be detrimental because cracking of the carbide particles could nucleate microcracks in the grain boundaries. In the present study the blocky carbide morphology shown in Fig. 1 would most closely approach the desired grain boundary structure. The carbide-free microstructure would be considered undesirable because grain boundary migration and sliding would be unimpeded, while the cellular morphology would be expected to promote microcracking.

In the present tests the results for the blocky carbide morphology and the carbide-free microstructure were quite similar to each other in all four environments. Thus under creep crack growth testing the blocky carbide particles were relatively ineffective, and thus impeding grain sliding or migration, or gettering tramp elements, evidently were not significant contributions of these grain boundary carbides. Also, it is apparent that the blocky carbides did not markedly alter the sensitivity to the air environment. Earlier studies [*14,15*] have shown that carbide particles can be preferentially oxidized, which would provide one explanation for the deleterious effects of oxygen in the environment. The present results show that the carbide-free and blocky carbide microstructures were comparably affected by air, implying that attack of the blocky carbide particles was not a significant factor.

The results with the cellular carbides show a different pattern. This microstructure was poorer in helium and in He-4CH_4, indicating that the more continuous carbide structure was deleterious when the environment either was nonaggressive or promoted carburization. In these cases the more continuous carbide structure may have promoted cracking by fracturing of the carbide particles. In air, however, the cellular microstructure provided the best properties. The implication is strong in this case that the cellular structure provided better resistance to the aggressive environment, that is, that the main role of carbides was an extrinsic one involving the environment.

Conclusions

The present results suggest, as has been observed in the past, that creep crack growth in superalloys takes place by a process of cavity nucleation and growth on grain boundaries. In principle it seems feasible to model this mode of crack growth by a single, generalized mechanism, but microstructural parameters and environmental factors must be accounted for before a useful description of cracking can be achieved. Furthermore, as illustrated in the present tests, microstructure and environmental effects are not independent of each other. One of the major needs at the present time is a clearer understanding of the ways in which an aggressive environment may react with a material undergoing creep deformation.

In the practical sense the results suggest that some of the general beliefs about the role of grain boundary carbides may not be true under creep crack growth conditions. The results also indicate that compositions and microstructures optimized by tests in air may not be best in other environments, and that materials with better resistance to these environments could be developed through a better understanding of the microstructure-environment interactions.

References

[*1*] Van Leuwen, W. P., *Engineering Fracture Mechanics*, Vol. 9, 1977, p. 951.
[*2*] Fu, L. S., *Engineering Fracture Mechanics*, Vol. 13, 1980, p. 307.
[*3*] Floreen, S., *Creep-Fatigue-Environment Interactions*, R. M. Pelloux and N. S. Stoloff, Eds., Metallurgical Society of the American Institute of Mining, Metallurgical, and Petroleum Engineers, Warrendale, Pa., publishers, 1980, p. 112.
[*4*] Floreen, S. and Kane, R. H., *Metallurgical Transactions*, Vol. 10A, 1979, p. 1745.
[*5*] Floreen, S. and White, C. J., *Metallurgical Transactions*, Vol. 12A, 1981, p. 1973.
[*6*] Floreen, S. and Kane, R. H., *Metallurgical Transactions*, Vol. 13A, 1982, p. 145.
[*7*] Raymond, E. L., *Transactions*, Metallurgical Society of the American Institute of Mining, Metallurgical and Petroleum Engineers, Vol. 239, 1967, p. 1415.
[*8*] Floreen, S., *Metallurgical Transactions*, Vol. 6A, 1975, p. 1741.
[*9*] Jones, A. T. and West, A. J., *Engineering Fracture Mechanics*, Vol. 9, 1977, p. 5.
[*10*] Floreen, S. and Kane, R. H., *Fatigue of Engineering Materials and Structures*, Vol. 2, 1980, p. 401.
[*11*] Woodford, D. A., *Metallurgical Transactions*, Vol. 12A, 1981, p. 299.
[*12*] Bricknell, R. H. and Woodford, D. A., *Metallurgical Transactions*, Vol. 12A, 1981, p. 425.
[*13*] Nieh, T. G. and Nix, W. D., *Metallurgical Transactions*, Vol. 12A, 1981, p. 893.
[*14*] Reuchet, J. and Remy, L., *Fatigue of Engineering Materials and Structures*, Vol. 2, 1979, p. 51.
[*15*] McMahon, C. J. and Coffin, L. F., *Metallurgical Transactions*, Vol. 1, 1970, p. 3443.

Terence Ingham[1] and Edward Morland[1]

Influence of Time-Dependent Plasticity on Elastic-Plastic Fracture Toughness

REFERENCE: Ingham, T. and Morland, E., "**Influence of Time-Dependent Plasticity on Elastic-Plastic Fracture Toughness**," *Elastic-Plastic Fracture: Second Symposium, Volume I—Inelastic Crack Analysis, ASTM STP 803*, C. F. Shih and J. P. Gudas, Eds., American Society for Testing and Materials, 1983, pp. I-721–I-746.

ABSTRACT: The paper describes the results of preliminary experiments to examine the effect of time-dependent plasticity on ductile crack growth, resistance to ductile tearing, and instability in laboratory fracture toughness tests. Experiments on a carbon-manganese (C-Mn) steel and two specimens of A533B-1 steel have shown that unstable ductile fractures can be induced at ambient temperature when 20- and 40-mm-thick compact specimens are held at constant loads between the load approximating that for crack initiation and the maximum load attained in a conventional displacement-controlled fracture toughness test; the time to failure after holding at constant load decreases with increasing constant load level and increasing strain rate prior to load-holding. The data suggest that final failure occurs under conditions paralleling "tertiary creep" behavior. Specimens were loaded at typical rates used in "static" fracture toughness testing and at a slower rate where the maximum load condition was reached in approximately 24 h. Crack extension under constant load has been related to *J*-resistance curves derived from multispecimen displacement-controlled tests.

The *J*-resistance curve for C-Mn steel specimens is dependent upon strain rate at rates of loading which are conventionally taken to be representative of "static" conditions. This effect was not observed in tests on A533B-1 steel.

KEY WORDS: constant load, creep, displacement control, ductile, initiation, *J*-controlled growth, load-control, plastic collapse, R-curves, strain rate, time dependence, upper shelf, carbon-manganese, A533B-1 steels, elastic-plastic fracture

Fracture toughness measurements at temperatures corresponding to upper-shelf conditions are obtained by determining the characteristic resistance to ductile crack growth (R) curve for the material. The R-curve, which can be determined using either a multiple-specimen interrupted test proce-

[1]Senior scientific officer and scientific officer, respectively, Risley Nuclear Power Development Laboratories, U.K. Atomic Energy Authority, Risley, Warrington, Cheshire, U.K.

dure or specially instrumented single-specimen methods [*1*],[2] is used to define both initiation toughness and resistance to crack growth (expressed in terms of the R-curve slope). Initiation toughness values will be conservative when assessing the conditions for structural failure under upper-shelf conditions, but use of post-initiation data is appropriate only if the R-curve defined in laboratory tests is either equivalent to or conservative with respect to the R-curve for the structure. Provided this condition is met, a firm foundation can be established for identifying the instability conditions at which stable ductile crack growth reverts to unstable ductile fracture. Apart from relatively simple structures—for example, pipework—the determination of characteristic R-curves for defects in structures is impracticable, and the use of post-initiation data must be justified on theoretical grounds. The theory of J-controlled growth [*2*] has provided this justification and has led to practical methods for defining plastic instability conditions in structures of, for example, Paris [*3*], Shih [*4*], and Chell and Milne [*5*]. System compliance plays a key role in these instability analyses, but the effect of time-dependent plasticity and crack growth, which is markedly dependent upon compliance, has not yet received detailed consideration.

The occurrence of time-dependent ductile crack growth under constant-load conditions has been recognized for many years, and workers in the United Kingdom [*6–8*] and Germany [*9*] have examined this phenomenon using small-scale laboratory fracture toughness specimens. Time-dependent plasticity is taken into account by introducing a load-hold period (typically 24 h) at the proof load during preservice proof-testing of pipework [*10*]. In most of the laboratory studies, time-dependent fractures were induced in relatively short times by holding specimens under constant load at combinations of load and specimen displacement which were close to the maximum load condition as defined from displacement-controlled tests. The present authors have extended the previous work by examining in more detail the stability of specimens held under constant load in relation to R-curve behavior. These preliminary tests have identified a high dependence of rate of loading on subsequent time-dependent crack growth, which also prompted the authors to examine R-curve behavior at testing rates lower than those used in conventional "static" fracture toughness testing.

Materials and Specimen Types

Time-dependent effects have been studied using the following three pressure vessel steels:

1. a 25-mm-thick, silicon-killed and aluminum-treated, normalized carbon-manganese (C-Mn) steel plate (BS 1501-224),
2. a 150-mm-thick quenched-and-tempered A533 Grade B Class 1 plate,

[2]The italic numbers in brackets refer to the list of references appended to this paper.

stress-relieved for 25 h at 615 ±10°C to simulate the post-weld heat treatment which might be applied during the production of a light-water reactor (LWR) pressure vessel, and

3. specimens from a 200-mm-thick quenched-and-tempered A533 Grade B Class 1 steel plate tested in the ASTM cooperative program on crack arrest toughness testing [*11*].

The chemical compositions and mechanical properties of the plates are given in Tables 1 and 2, respectively.

The R-curves and all subsequent time-dependent results were obtained using plane-sided 20- or 40-mm-thick compact specimens modified for measuring load-line displacement. Size(s), orientation, and location in the respective plates/specimens are given in Table 3.

Experimental Details

Determination of R-Curve Data

Values of initiation toughness and resistance to crack growth were derived using the multiple-specimen method. Nominally identical specimens were loaded to different levels of displacement at ambient temperature, unloaded, heat-tinted and then fractured at low temperature to reveal the extent of ductile crack growth. All tests were performed on a servohydraulic machine op-

TABLE 1—*Chemical compositions (weight %).*

	Steel		
	Co-op A533B-1	A533B-1 150-mm Plate	C-Mn
C	0.22	0.19	0.15
Si	0.26	0.20	0.22
S	0.011	0.006	0.024
P	0.011	0.010	0.014
Mn	1.56	1.40	1.10
Ni	0.6	0.66	0.05
Cr	0.1	0.13	<0.05
Mo	0.54	0.49	0.02
Cu	0.11	0.10	...
V	0.009	...	...
Sn	0.011	...	...

TABLE 2—*Tensile data.*

Steel	Temperature, °C	σ_y, Mpa	σ_u, Mpa
Co-op A533B-1	20	473	628
A533B-1 150-mm plate	20	442	570
C-Mn	20	284	466

TABLE 3—*Specimen details.*

Steel	Specimen Thickness, mm	Specimen Orientation	Location of Specimens in Plate or Sample
C-Mn	20	RW	center thickness of 25-mm plate
Co-op A533B-1	20	WR[a]	centered at T/4 and 3T/5 of broken 50-mm-thick weld-embrittled crack arrest specimens
150-mm A533B-1	40	RW	between ~T/4 and ~3T/4 of 150-mm plate

[a]Identical orientation to crack arrest toughness specimens tested in cooperative program.
NOTE—direction of mother plate. R = rolling direction of mother plate; W = transverse (width)

erating in displacement control. R-curves for all three steels were derived from specimens tested at ram-displacement rates conventionally used in "static" fracture toughness testing. In addition, a number of specimens from the C-Mn plate and the 150-mm-thick A533B-1 plate were tested at slower displacement rates chosen so that the maximum load instability condition was reached after approximately 24 h.

Constant-Load Tests

Specimens were held under constant-load conditions on either the servohydraulic machine or on a 50-kN-capacity Denison creep frame; the load capacity of the creep frame limited its use to tests on 20-mm-thick specimens. Long-term (>100 h hold times) experiments were performed only on 20-mm-thick specimens.

Specimens were held at constant load after attaining toughness levels below the maximum load values for each material. The load-hold conditions were related to the material R-curves but actual test conditions were specified in terms of specimen displacement rather than load (or toughness). This was necessary because of the small difference between initiation load and maximum load in the specimens tested and the variability in load arising from small differences in initial crack lengths of the specimens.

The tests were conducted at rates which were equivalent (in terms of specimen displacements) to the "conventional" or "slow" rates used in R-curve tests. Initially, specimens tested in the servohydraulic machine were loaded to the desired displacement in displacement control, at which point the "bumpless" transfer function was used to transfer to load control. To effect a transfer in this manner requires a finite time to "balance" the different control modes of the servohydraulic machine. Slight stress relaxation can occur during the time (a few seconds) that the specimen is held under displacement control but the original stress level is restored on transfer to load control. Because of this effect, tests where the bumpless transfer was used experience a marginal decrease in initial crack growth rate under constant load. Later, specimens were loaded to the holding condition in load control at rates which were selected to attempt to ensure equivalence between load and displacement-controlled loadings.

Only three specimens (two C-Mn steel, 1 Co-op A533B-1) were tested on the creep frame because of experimental difficulties in maintaining a constant rate of loading up to the load-hold condition. The two C-Mn specimens were manually loaded and loading rates were approximately "conventional" and 10 times "conventional." The Co-op A533B-1 specimen was loaded automatically at the "conventional" loading rate.

All load-hold experiments were conducted at ambient temperature. Load-line displacement (load) and time at constant load were monitored during each test.

The displacement rates used in the R-curve tests, the method of control, rates, and type of loading for all load-hold tests together with additional practical details concerned with tests on the creep frame are given in Appendix I.

Results

Time-Dependent Crack Growth Under Constant Load

Tests on C-Mn Steel—The results of load-hold tests on C-Mn steel specimens which were loaded to the load-hold condition at normal testing rates are presented in Fig. 1. The inset to this figure shows a sketch of a typical load versus load-line displacement (P, Δ) test record which illustrates the relationship between the conditions (A $\rightarrow$ G) at which specimens were held under constant load and (P, Δ)-values corresponding to the initiation of ductile crack growth [$(P, \Delta)_{init'n}$] and the maximum load condition [$(P, \Delta)_{max}$]. The toughness and thence (P, Δ) at ductile crack initiation was derived from the J_R-curve defined from multiple-specimen tests. The positions at which specimens were held at constant load with respect to the (displacement controlled) J_R-curve are shown in Fig. 12*a* of Appendix II, which also outlines the method of J_R analysis used to define initiation toughness.

The inset to Fig. 1 shows that, at crack initiation, the load and displacement are ~92 percent and ~40 percent, respectively, of the values attained at maximum load condition. All of the specimens which were held at constant load after loading to displacements greater than approximately 45 percent Δ_{max} (that is, Tests C $\rightarrow$ G) failed by unstable ductile fracture. Times-to-failure ranged from ~30 h (Test D; Δ hold $\simeq$ 50 percent Δ_{max}) to as low as 0.5 h (Test G; Δ hold $\simeq$ 65 percent Δ_{max}). Tests on the servohydraulic machine for Conditions A and B which corresponded to Δ_{hold} less than ~45 percent Δ_{max} showed no increase in time-dependent growth after holding for ~80 h and these tests were terminated. The apparent stability "threshold" indicated by Conditions A and B was examined by repeating Condition B on the creep frame. The first specimen (B_1) tested on this machine (incremental

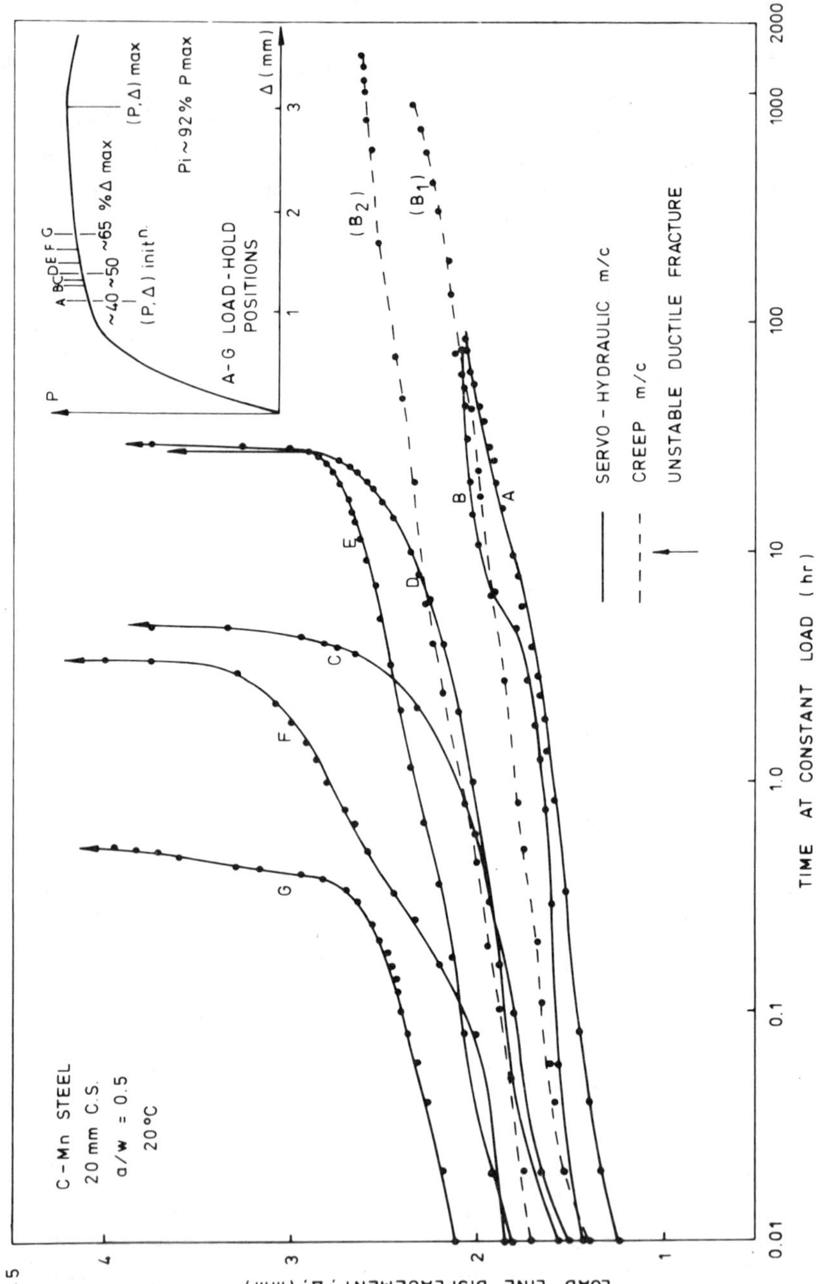

FIG. 1—*Constant-load data for carbon-manganese steel compact specimens.*

loading) was unloaded after holding at constant load for 850 h. The continuous increase in specimen displacement with time suggested that, eventually, this specimen could have failed. The specimen continuously loaded in the creep frame to the same load-hold condition (B_2) sustained a constant load for 1450 h without fracturing and the test was terminated.

Loading rates to the load-hold condition in the two creep frame tests differed by a factor of ~ 10 (see Appendix I), and higher rates of specimen displacement during the first few minutes of the load-hold period were observed for the specimen tested at the faster loading rate. The effect of prior loading rate on time-to-failure was examined by loading additional C-Mn specimens to Condition G (Δ_{hold} = 2 mm). The results of these experiments are shown in Fig. 2. There was no significant difference between the two conventional rate tests where the load-hold condition was reached in either load control or displacement control. Times-to-failure for the two slowly loaded specimens were approximately 20 times greater than those for specimens loaded at "conventional" rates. Several small load increases were observed during the load-hold period of both slowly loaded specimens which may have reduced the time to failure in these tests. The tests confirmed the marked difference in initial rates of crack opening during constant-load holding in specimens loaded at different rates. All four specimens were held under constant load after attaining a load-line displacement (Δ_{hold}) of 2 ± 0.01 mm. The data in Fig. 2 show that the load-line displacement in "conventional" rate tests is significantly higher than in slowly loaded specimens after holding under constant load for only 0.01 h. The difference in load-line displacement values at 0.01 h for the two conventional rate tests is probably related to the effect, described earlier, of using the bumpless transfer function during one of these tests. The effect of prior loading rate is examined further in Fig. 3, which compares initial rates of crack opening (measured as the average crack-opening rate over the first 10 min of load-holding) as a function of specimen displacement at the start of load-holding for all tests on the C-Mn steel. The scatterbands shown in the figure encompass all data from specimens loaded at conventional rates.

Figure 4 shows the influence of time-dependent crack growth under constant load on the stability of the J_R-curve for fully yielded 20-mm-thick C-Mn steel specimens. The ratios J_S/Ji were determined using an initiation value (Ji) of 0.17 MNm^{-1} and J-values at the start of load holding (J_S) which were either calculated directly after infiltrating the specimen with ink to identify the amount of ductile crack growth which had occurred prior to load-holding, or were estimated(e) after inferring a value of ductile crack growth appropriate to the displacement value at the point of load holding. The inferred pre-load hold crack growth values were average values derived from a statistical analysis of previous R-curve data.

Figure 4 indicates that 20-mm-thick C-Mn steel specimens loaded at a typical rate for static fracture toughness testing would fail by time-dependent

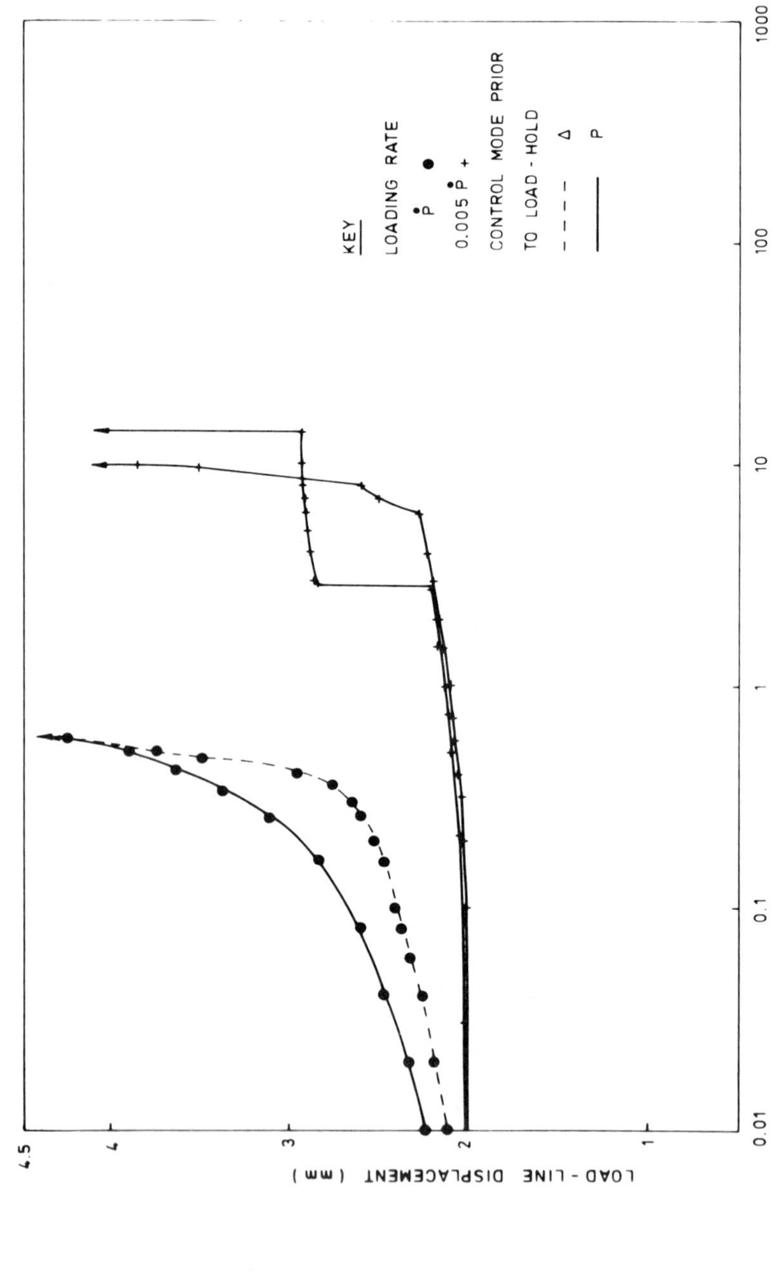

FIG. 2—*Effect of loading rate on time-to-failure at constant load for 20-mm-thick carbon manganese steel compact specimens.*

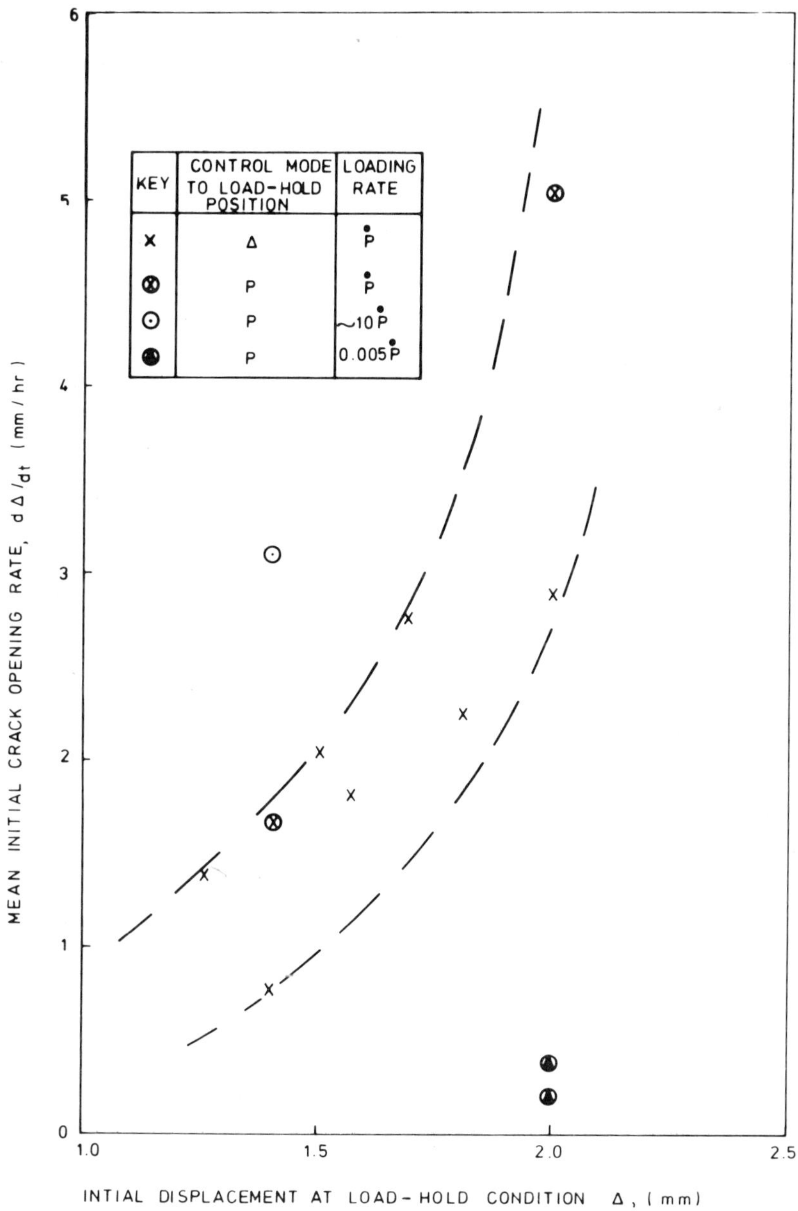

FIG. 3—*Effect of prior loading rate on the initial rate of crack opening after holding 20-mm-thick carbon-manganese compact specimens at constant load.*

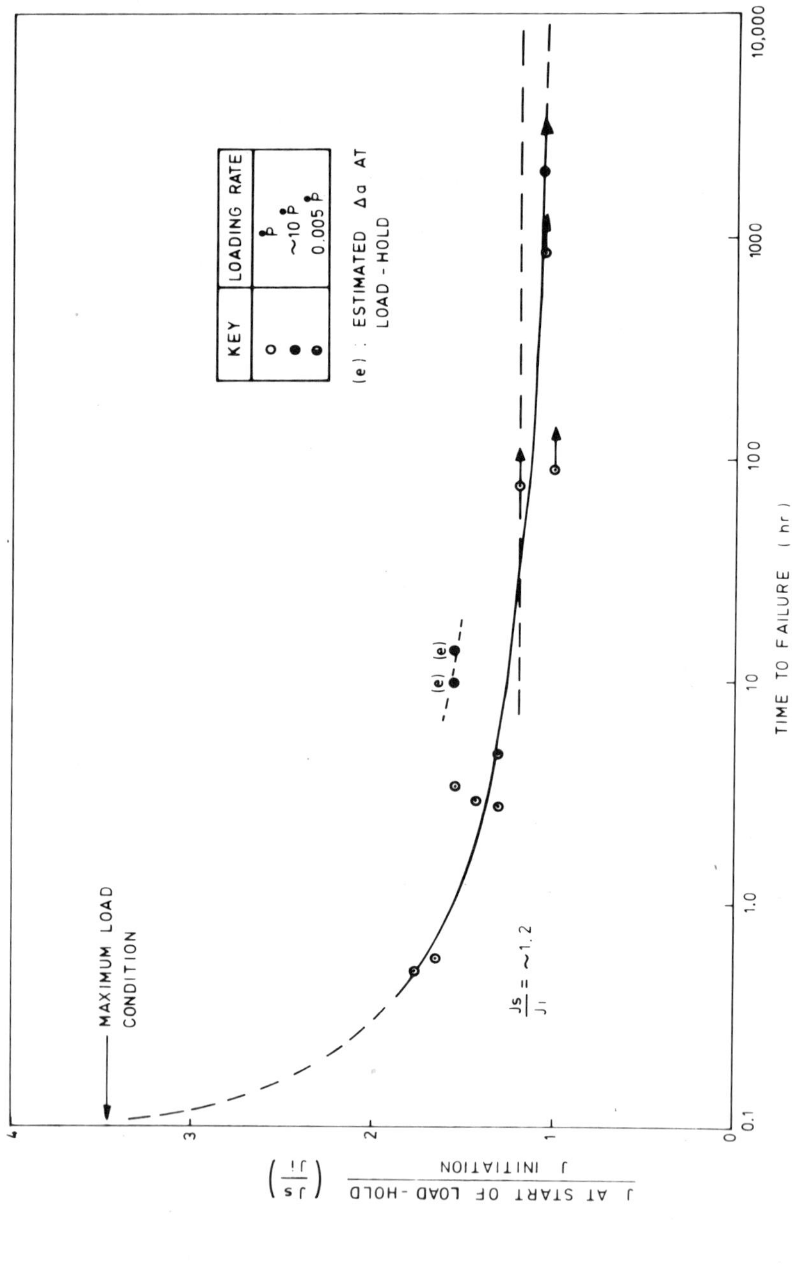

FIG. 4—*J at the onset of load-hold as a function of* Ji *versus time-to-failure under constant load for carbon-manganese specimens tested at different initial loading rates.*

crack growth when held at applied J-values 20 percent higher than the initiation toughness value, Ji. Any threshold value will be dependent upon the strain rate prior to load-holding, and the implicit assumption that for any particular loading rate, R-curve data up to the maximum load condition will be independent of control mode.

Tests on A533B-1 Steels—The results of tests on the 150-mm-thick A533B-1 steel followed the pattern seen in tests on the C-Mn steel. Figure 5 shows load-line displacement, as a function of time at constant load, for 40-mm-thick compact specimens of this steel loaded to the load-hold condition under load control at both conventional and slow loading rates. Three of the four specimens loaded at the conventional rate failed by unstable ductile fracture after less than 6 h at constant load. Both slowly loaded tests were terminated after the specimens had been held under constant load for sufficient time to confirm the prior strain rate effect. The load-hold conditions are related to the J_R-curve in Appendix II (Fig. 12*b*); using this information in conjunction with Fig. 5 suggests that threshold values, J_S/Ji, below which time-dependent failures would not occur in 40-mm-thick specimens loaded at conventional and slow loading rates would be 1.16 and 1.44, respectively.

The Co-op A533B-1 steel was tested only at conventional rates. Ductile initiation occurred much closer to maximum load in this steel ($Pi \approx 98$ percent P_{max}), and specimens were held under constant load at displacements corresponding to or less than the estimated displacement for crack initiation. The results of these tests are shown in Fig. 6. Figure 12*c* of Appendix II shows that, in fact, Specimens A and C were held at J-values just above initiation and Specimen B was held below initiation. Specimens A and C failed by cleavage fracture after sustaining ductile crack extensions of 3 and 5.1 mm, respectively. These values are well in excess of the amount of crack growth ($\sim$0.3 mm) associated with the maximum-load instability condition identified from the displacement-controlled R-curve measurements. Specimen A was tested in the servohydraulic machine and failure occurred under load control. Specimen C was tested in the automated creep frame and actually failed under displacement control because the control response was unable to maintain the load as the ductile crack extended prior to cleavage. The specimen loaded to a pre-initiation condition (B) was terminated after 165 h of load-holding. The instantaneous increase in load-line displacement after holding this specimen for 70 h at constant load could not be attributed to any visible cleavage; on examining the heat-tinted, post-test fracture it was observed that 0.60 mm of ductile crack growth occurred during the load-hold test. J at the termination of the test was 0.319 MNm^{-1}. However, the three results indicate that time-dependent crack growth would promote failures when J_S/Ji exceeds unity in 20-mm-thick specimens of this material tested at normal loading rates.

Typical fracture surfaces of time-dependent failures in the three steels tested are shown in Fig. 7.

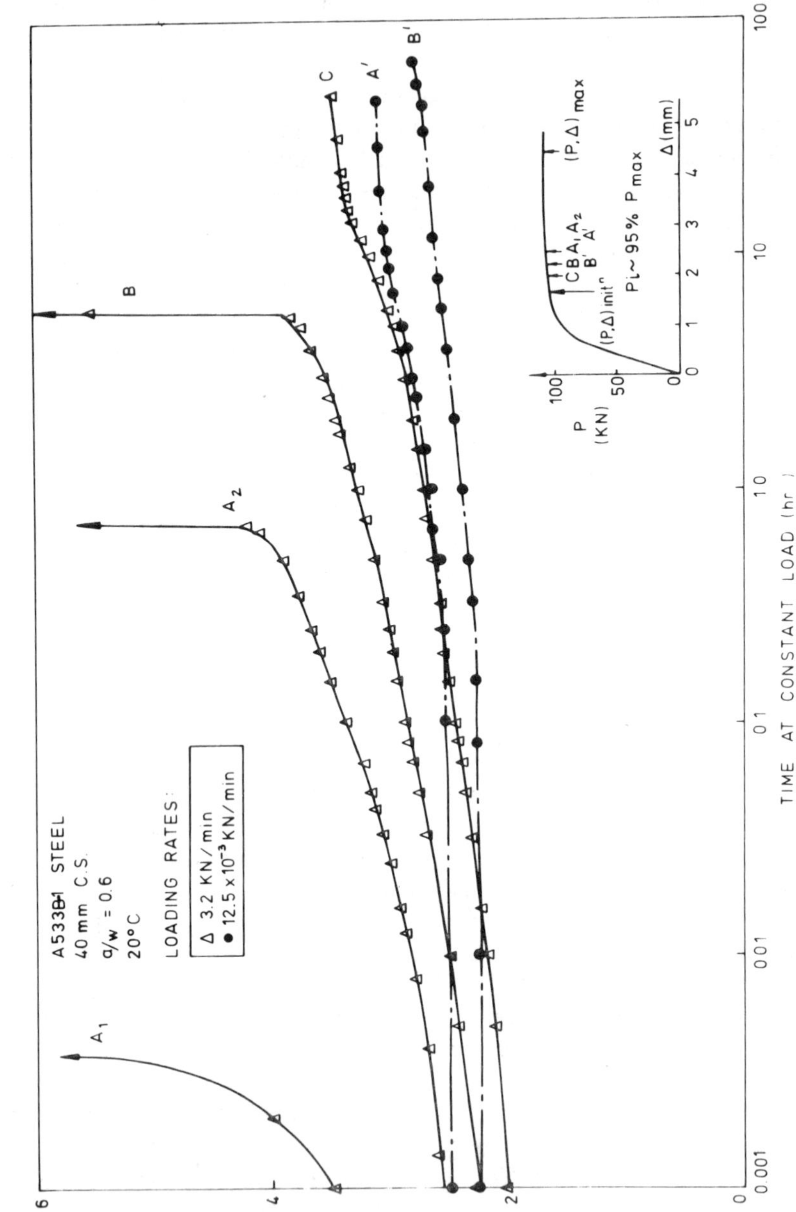

FIG. 5—*Constant-load data for 40-mm compact specimens from the 150-mm-thick A533B-1 plate.*

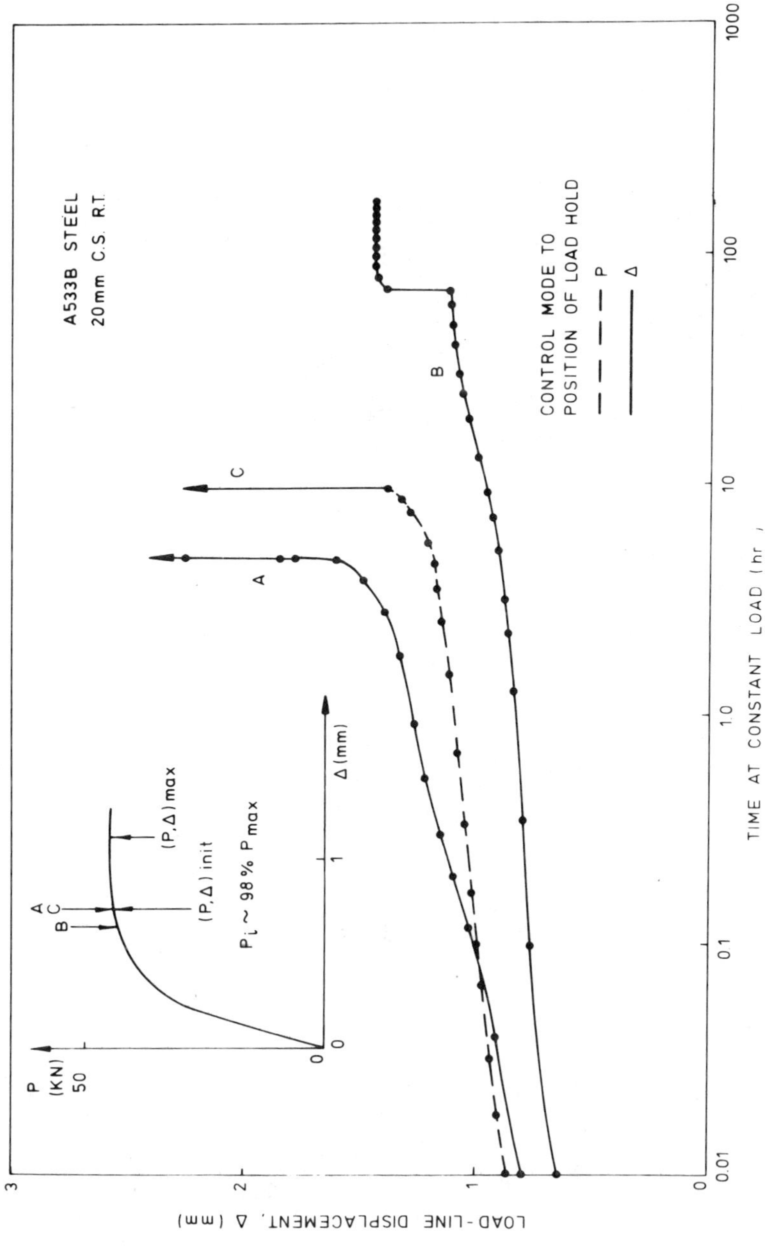

FIG. 6—*Constant-load data for 20-mm compact specimens, Co-op A533B-1.*

FIG. 7—*Fracture surfaces for specimens which failed under constant load.*

Strain-Rate Effects on R-Curves

The strain-rate sensitivity of the C-Mn steel can be seen from Fig. 8, which shows P-Δ test records, up to the maximum load conditions, for 20-mm compact specimens of nominally equal crack lengths tested at "conventional" static (0.3 mm/min) and "slow" static (2.4×10^{-3} mm/min) displacement rates. There is a clear distinction between P-Δ curves for the two rates, specimens tested at the faster rate attaining higher loads than the slow rate tests. J-Δa data for the slowly loaded specimens are compared in Fig. 9 with data

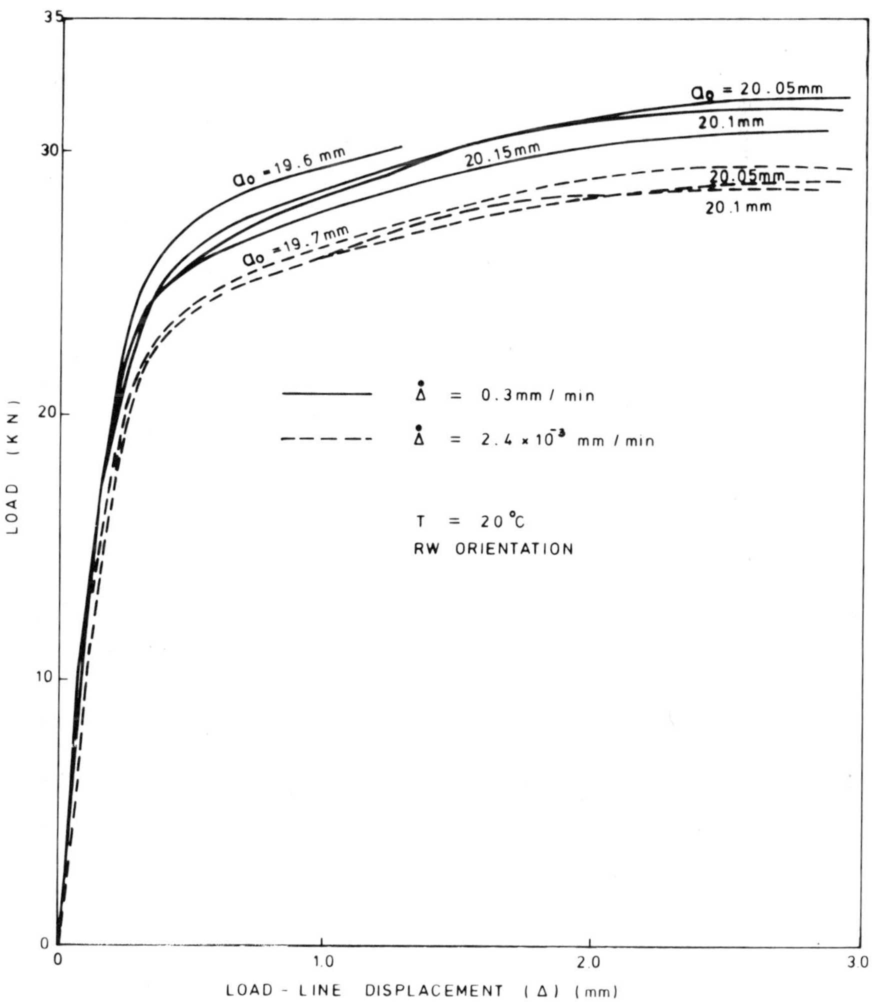

FIG. 8—*Force-displacement curves for 20-mm-thick carbon-manganese steel compact specimens tested at two "static" displacement rates.*

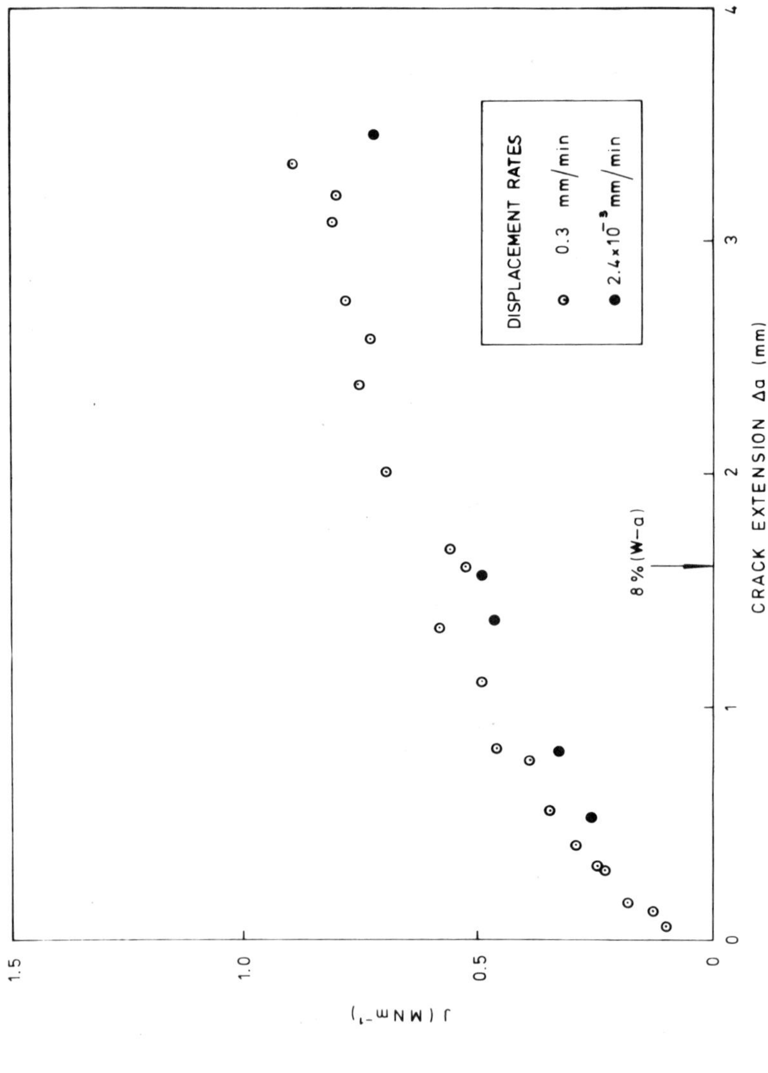

FIG. 9—*Comparison of* J-Δa *data for tests at two displacement rates, 20-mm compact specimens, carbon-manganese steel.*

obtained from multiple-specimen tests performed at the conventional rate. There are insufficient data to develop an R-curve within the range 0.15 mm $\leq \Delta a \leq 0.06b$ for the slow rate tests. To permit comparison of R-curve data for the two strain rates, J-Δa data for both rates were analyzed within the range 0.15 mm $\leq \Delta a \leq 0.08b$. Linear regression analyses over this Δa range provided values of initiation toughness (*Ji*) and resistance to crack growth (dJ/da) for the more slowly loaded specimens which were lower than the corresponding values for the conventional testing rate by 11 percent and 22 percent, respectively. In contrast, 40-mm compact specimens from the 150-mm-thick A533B-1 plate showed no effect of strain rate when tested at rates equivalent to those used in the experiments on C-Mn steel. Figure 10 shows there are no differences between P-Δ data from specimens of nominally identical initial crack lengths tested at both conventional and slow displacement rates, while Fig. 11 shows that testing at the slower rate has no effect on J-Δa data for this material.

Discussion

The experimental program has identified two separate effects of strain rate when fracture toughness specimens are tested at ambient temperature at strain rates where the authors would expect plastic deformation to be rate-independent. Firstly, tests on both C-Mn steel and A533B-1 steels have shown that, when specimens are held at loads approaching the maximum load condition, time-dependent crack growth is highly dependent on the rate of straining prior to load holding. Secondly, resistance to crack growth can decrease when specimens are tested under displacement control at slower rates than those normally used in static fracture toughness testing.

Load-Hold Tests

The results of constant-load tests on a C-Mn steel and two specimens of A533B-1 steel are similar to those obtained from creep experiments. Figures 1, 2, 5, and 6 are similar to creep curves for tests at increasing stress levels. Growth under constant load at ambient temperature and attainment of an instability condition (within the times allowed for the tests) is dependent upon the level of load at which the specimen is held. In all cases where the load hold conditions resulted in time-dependent fracture, specimen failure was preceded by a rapid increase in crack-opening rate resembling the tertiary stage in a creep test.

The results of the load-hold studies on the C-Mn steel are in broad agreement with Krempl's work on creep of 304 stainless steel at ambient temperature [*12,13*], which demonstrated that creep strains were significantly influenced by both stress level and stress rate preceding the creep test. Krempl's results are at variance with classical creep theory and support the theory of viscoplasticity based on total strain and overstress.

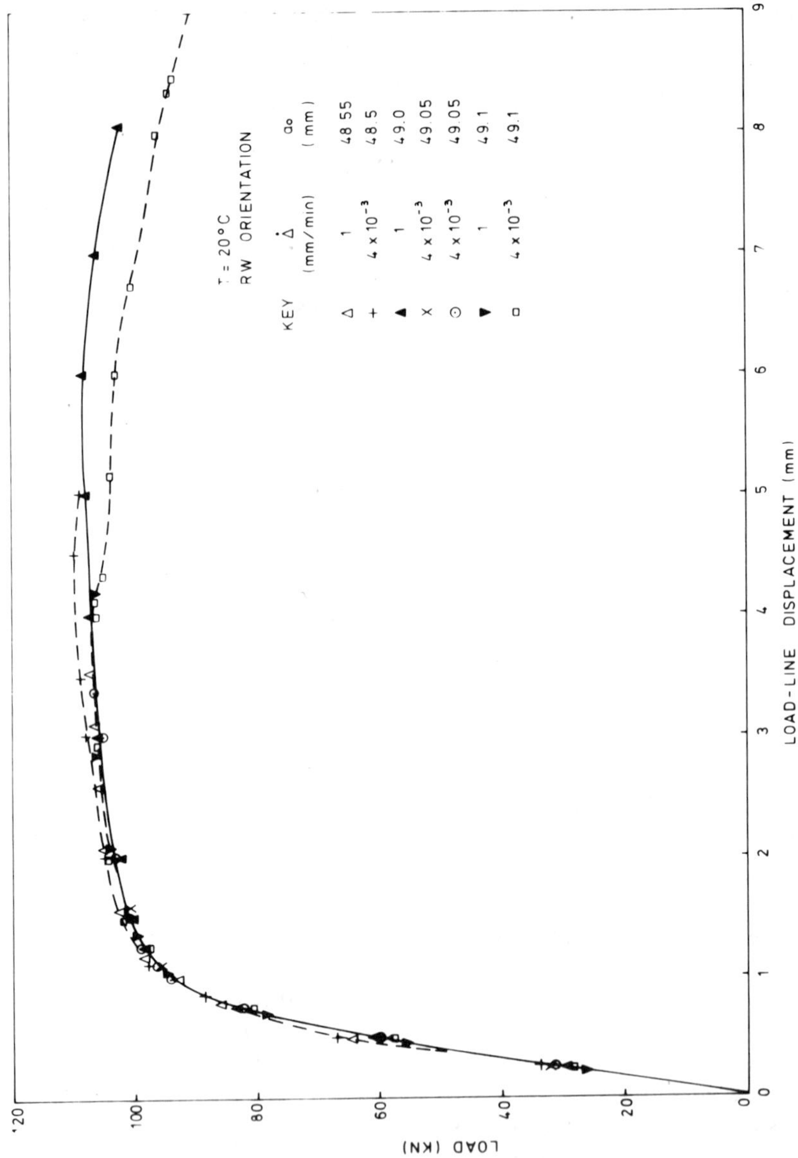

FIG. 10—*Force-displacement data for 40-mm compact specimens from 150-mm A533B-1 plate tested at two "static" displacement rates.*

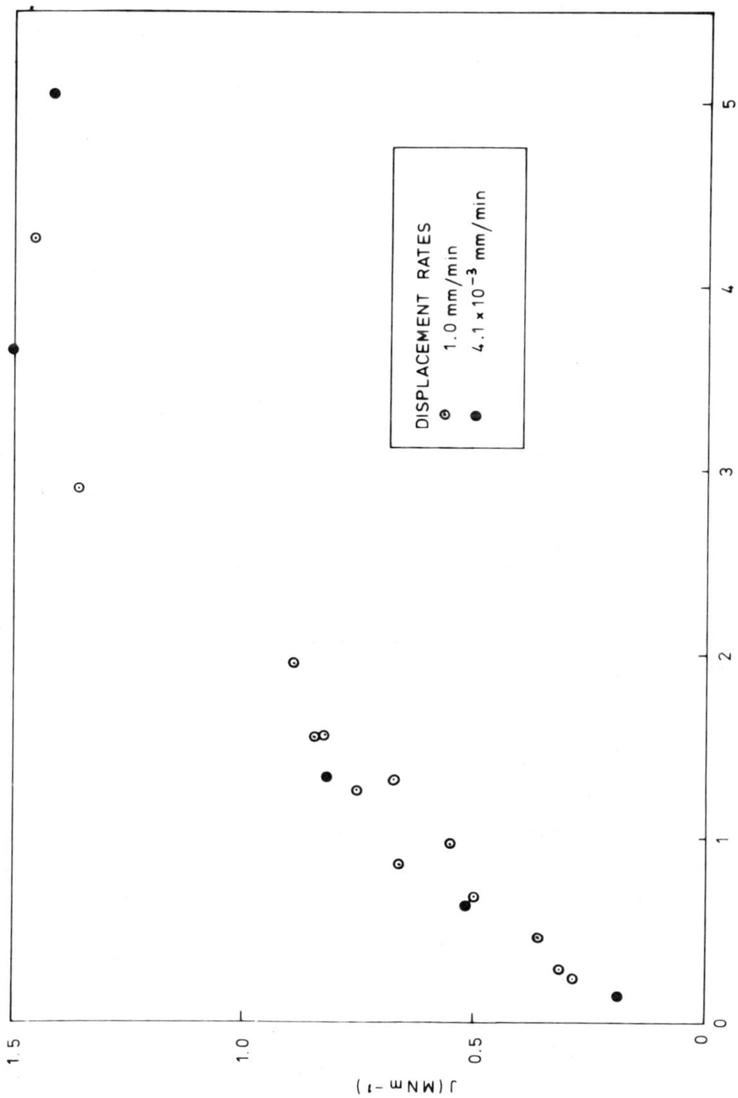

FIG. 11—*Comparison of J-Δa data for tests at two displacement rates, 40-mm compact specimens from 150-mm A533B-1 plate.*

Viscoplasticity theories admit the existence of an equilibrium stress-strain curve which corresponds to truly static (rate-independent) conditions, and predict rising stress levels (for a given strain level) with increasing loading rate. This concept was used by Tsuru and Garwood [*8*] to explain time-dependent crack growth under constant load in a line-pipe steel. If a specimen is loaded at "conventional" displacement rates up to a specified displacement and the load is held constant, then time-dependent crack growth can occur provided the specified displacement exceeds some threshold value. Crack growth will continue until the force-displacement (P-Δ) curve for the growing crack reaches the equilibrium P-Δ curve for the particular specimen, at which point time-dependent crack growth will cease. If the specimen is held at a constant load which is higher than the maximum load attainable on the equilibrium P-Δ curve, then failure will eventually occur. This model describes time-dependent crack growth purely in terms of the overstress associated with an increase in strain rate and is supported by the results from tests on the C-Mn steel.

Tests on the 150-mm-thick A533B-1 steel showed that P-Δ curves for 40-mm-thick specimens were independent of displacement (strain) rate when the rates differed by two orders of magnitude. The two rates, therefore, produced identical overstress levels with respect to the equilibrium P-Δ curve and the overstress model would predict identical crack growth rates under constant load for specimens tested at both strain rates. However, the results in Fig. 5 show that crack growth under constant load for a given load-hold condition is still dependent upon the prehold strain rate. The implication from these results is that for materials which are usually considered to be strain rate-insensitive (that is, do not exhibit overstress characteristics) small-scale specimens may still be susceptible to time-dependent plasticity and crack growth.

All of the load-hold conditions resulted in uncontained yielding ahead of the crack tip, and a dominant factor influencing failure due to time-dependent crack growth, independent of strain-rate effects, is the proximity of the load-hold condition to plastic collapse. Calculation of collapse loads for specimens which did not fail during load hold tests using the expression [*4*].

$$Po = 1.455\ \sigma_{\text{flow}}\ \eta\ \text{Bb}$$

where

$$\eta = [(2a/b)^2 + 2(2a/b) + 2]^{1/2} - [2a/b + 1]$$

showed that, for all specimens tested at the conventional rate, actual holding loads were ~12 percent lower than the collapse loads. Since collapse loads decrease with increasing crack length, specimen failure during load holding would have occurred very close to, or at, the collapse condition.

Comparison of (P, Δ) conditions for the three steels at ductile crack initiation with those at the maximum-load instability shows that the threshold value below which failures did not occur, J_S/Ji, decreases as ductile crack initiation approaches the maximum load condition. While the initiation toughness values and R-curve data for 20-mm-thick C-Mn and Co-op A533B-1 specimens are very similar (Figs. 12*a* and 12*c* of Appendix II), the time-dependent response of the Co-op A533B-1 is more severe (compare Figs. 1 and 6). This suggests that, for a given prehold strain rate, the ratio P_{hold}/P_{max} is a more important parameter than J_S/Ji in determining stability of time-dependent growth in fully yielded specimens. However, it is important to recognize that instabilities due to growth under constant load have not been observed in specimens held below the initiation toughness values defined using the method given in Appendix II.

Although only 20- and 40-mm-thick specimens have been tested, similar ambient temperature time-dependent failures would be expected in tests on larger deeply notched fracture toughness specimens (for example, 100-mm-thick compact specimens) where ductile initiation occurs after general yield and the load at initiation is within ~10 percent of the collapse load. These effects are dominated by the approach to plastic collapse and hence would not be expected in specimens (or cracked structures) where ductile crack initiation occurs under contained yielding and the ratio Pi/P_{max} is significantly lower than 0.9. Experimental validation of this viewpoint would require tests on either very large mild steel or A533B-1 specimens or, alternatively, tests on other materials with equivalent toughness/yield ratios to provide ductile initiation prior to general yield in a more practical specimen size.

Effect of Strain Rate on R-Curve Behavior

The investigation of the effect of strain rate on displacement-controlled R-curve behavior of both the C-Mn and A533B-1 steels indicates that care must be taken when defining "static" fracture toughness data. Decreasing the displacement rate by an order of magnitude in the tests on 20-mm-thick C-Mn steel specimens resulted in a decrease in both initiation toughness (Ji) and R-curve slope (dJ/da). The 11 percent reduction in Ji may not be significant, given the smaller number of data for the slower rate tests and error confidence limits on the regression analysis, but the reduction in resistance to crack growth is larger (22 percent) and better defined. Decreasing the displacement rate by two orders of magnitude in tests on the 40-mm-thick A533B-1 steel specimens had no effect on either initiation toughness or the J_R-curve. While an effect of strain rate on the ambient temperature R-curve for A533B-1 steel might be expected if much slower rates were examined, such rates are unlikely to be of any practical relevance.

Conclusions and Practical Implications

Tests on a C-Mn steel and two specimens of A533B-1 steel have shown that both prior strain rate and the level of overstress (relative to an equilibrium strain rate) influence the rate of time dependent ductile crack growth. These effects are seen in fully yielded fracture toughness specimens tested at ambient temperature at strain rates where plastic deformation is usually taken to be rate independent. Fully yielded fracture toughness specimens (and cracked structures) held at constant load above some threshold value will eventually fail by time-dependent crack growth after sustaining amounts of stable crack growth substantially smaller than those associated with the maximum-load instability condition. Threshold values, J_S/Ji (or $P_{\text{hold}}/P_{\text{max}}$), below which time dependent failures have not been observed are strain-rate dependent and, in any practical applications, such values would have to be determined at the appropriate loading rate.

Time-dependent failures have been induced only when specimens are held under constant loads within a few percent of the collapse load. For this reason, failure due to time-dependent crack growth would not be expected in specimens (or structures) where crack-tip yielding is contained by an elastic stress field, that is, all those situations where the applied load on the ligament will be below the general yield load.

The practical implications of these results are:

1. When the level of constant load is close to the collapse load, noninclusion of time-dependent material properties into fully plastic instability analyses of cracked structures will provide nonconservative predictions of the amount of ductile crack extension. For these conditions, however, inclusion of time-dependent properties will have only a small effect on predicted failure loads. Fully plastic conditions can arise in thin structures (for example, pipework) where a postulated large defect could cause failure by plastic collapse across the remaining ligament.
2. Time-dependent effects need not be considered when assessing defect significance in large structures operating in load control under a condition of contained yielding. This condition should be achieved if post-initiation fracture toughness data are restricted to the small amount of J-controlled crack growth associated with the initial steeply rising portion of the material J_R-curve.
3. Time-dependent crack growth is of relevance only to steady-state loadings.
4. Independent of time-dependent effects, displacement-controlled J_R-curves for the C-Mn steel specimens were strain rate dependent when tested at a rate conventionally used in "static" fracture toughness testing and at a slower rate more representative of service loading. A similar effect was not observed for A533B-1 steel.

APPENDIX I

Displacement and Loading Rates Used in the Test Program and Creep Frame Details

Displacement Rates for R-Curve Tests

These rates are given in Table 4*a*.

Loading Rates Prior to Load-Holding in the Servohydraulic Machine

These rates are given in Table 4*b*.

Tests on the Creep Frame

The creep-frame operated on the lever-beam principle. For tests of the C-Mn steel, specimens were loaded manually up to the desired load-hold condition. Constant-load conditions were then maintained manually by periodically trimming the balance of the load-applying lever beam as plastic deformation/crack growth occurred. Manual control was too coarse to maintain a uniform rate of loading, particularly at loads approaching the load-hold condition where the specimens were fully yielded. The first specimen was loaded incrementally so that the *average* loading rate matched the "conventional" rate of tests on the servohydraulic machine. The second specimen was loaded continuously to the load-hold condition and the resultant loading rate was approximately 10 times "conventional."

The Co-op A533B-1 specimen was tested after modifying the machine by adding a load cell and a motorized drive control unit which provided both a constant "conventional" loading rate and automatic load control during the load-hold period.

TABLE 4*a*—*Displacement rates for R-curve tests.*

		$\dot{\Delta}$ (mm/min)	
Steel	Specimen Size, mm	"Conventional"	"Slow"
C-Mn	20	0.3	2.4×10^{-3}
150-mm A533B-1	40	1.0	4.1×10^{-3}
Co-op A533B-1	20	0.3	. . .

TABLE 4*b*—*Loading rates prior to load-holding in the servohydraulic machine.*

	$\dot{P}$, kN/min	
Steel	"Conventional"	"Slow"
C-Mn	(4) [1.3][a]	(7×10^{-2}) $[7 \times 10^{-3}]$
150-mm A533B-1	(36) [3.2]	(0.4) $[12.5 \times 10^{-3}]$

( ) = Loading rate to the knee of the force/displacement curve, that is, ~ general yield.
[ ] = Loading rate from knee of curve to load-hold condition.

[a]Only one C-Mn steel specimen tested at "conventional" rates was loaded in load control. The remaining C-Mn steel specimens (and all Co-op A533B-1 specimens) tested on the servohydraulic machine were loaded in displacement control at 0.3 mm/min and transferred to load control at the selected load-hold condition.

APPENDIX II

Derivation of Initiation Toughness Values from Multiple Specimen Data and Load-Hold Conditions with Respect to the J_R-Curves

Values of J were calculated from the expression

$$J = Jo\left\{1 - \frac{|0.75f(ao/W) - 1|\,\Delta a}{b}\right\}$$

where

$Jo = A/Bb\,f\,(ao/W)$,
A = area under the force/displacement curve,
B = specimen thickness,
b = remaining ligament,
Δa = an average measurement of ductile crack extension derived from nine equidistant measurements across the full specimen thickness and taking the average of the two surface measurements as a single value, and
$f(ao/W)$ = a factor to correct for the tensile component of loading in a compact specimen.

The toughness at ductile crack initiation (Ji) was taken as the intersection of a $J = 4\sigma_f\Delta a$ blunting line with the linear regression line for $J - \Delta a$ data lying within the Δa range:

$$0.15 \text{ mm} \leq \Delta a \leq 0.06b$$

Size requirements for validity were taken to be

1. Initiation: $a, B, b \geq 50\,Ji/\sigma_f$
2. J-controlled growth: $b > 10\,J/dJ/da$, and
3. $a, B, b \geq 25\,J/\sigma_f$.

J_R-data and corresponding load-hold conditions are shown in Figs. 12*a* to 12*c*. The extent to which J_R-curve data met the foregoing requirements (1–3) is shown in Table 5.

TABLE 5—*Validity checks on R-curve data shown in Fig. 12.*

Steel	Compact Specimen Type	Initiation, $(50\,Ji/\sigma_f)$ m	J-Controlled Growth: $J_{allowable} = \frac{b}{10}\cdot\frac{dJ}{da}$ (MNm^{-1})	J-Controlled Growth: $J_{allowable} = \sigma_f \cdot b/25$ (MNm^{-1})
C-Mn	20-mm $a/W = 0.5$	0.022 (X)	0.70 (✓)	0.28[a] (X)
Co-op A533B-1	20-mm $a/W = 0.5$	0.012 (✓)	0.80 (✓)	0.42 (✓)
150-mm A533B-1	40-mm $a/W = 0.6$	0.025 (✓)	1.29 (✓)	0.64[a] (X)

✓ or X indicates whether or not all data from R-curve analyses met size requirement.
[a]Except for Condition G for the C-Mn steel ($J_S = 0.3$ MNm^{-1}), all other load-hold conditions were achieved under J-controlled crack growth.

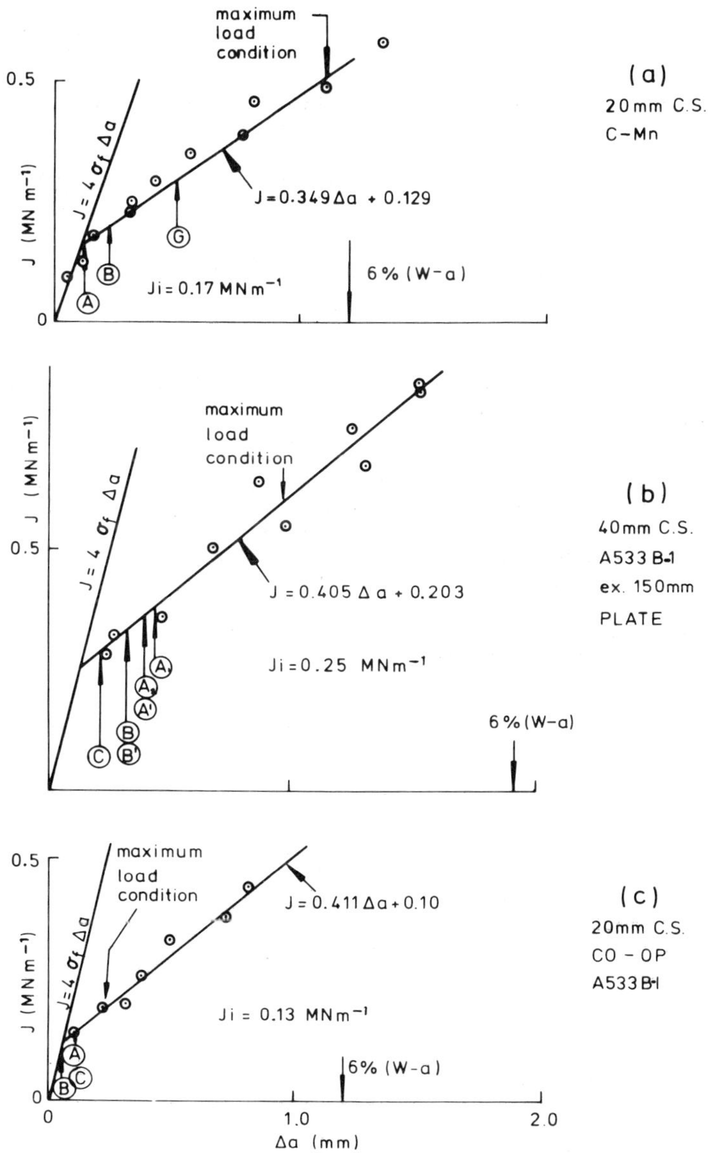

FIG. 12—*J_R-curve data and relative positions of load-hold for tests at 20°C.*

References

[*1*] Ingham, T. and Morland, E. in *Proceedings*, Committee on the Safety of Nuclear Installations, Specialists Meeting on Plastic Tearing Instability, Washington University, St. Louis, Mo., NUREG/CP-0010, Sept. 1979, pp. 330–358.

[*2*] Hutchinson, J. W. and Paris, P. C. in *Elastic-Plastic Fracture, ASTM STP 668*, J. D. Landes, J. A. Begley, and G. A. Clarke, Eds., American Society for Testing and Materials, 1979, pp. 37–64.

[*3*] Paris, P. C. and Zahalak, G. I. in *Proceedings*, Committee on the Safety of Nuclear Installations, Specialists Meeting on Plastic Tearing Instability, Washington University, St. Louis, Mo., NUREG/CP-0010, Sept. 1979, pp. 401–428.

[*4*] Shih, C. F. in *Proceedings*, Committee on the Safety of Nuclear Installations, Specialists Meeting on Plastic Tearing Instability, Washington University, St. Louis, Mo., NUREG/CP-0010, Sept. 1979, pp. 144–193.

[*5*] Chell, G. G. and Milne, I. in *Proceedings*, Committee on the Safety of Nuclear Installations, Specialists Meeting on Plastic Tearing Instability, Washington University, St. Louis, Mo., NUREG/CP-0010, Sept. 1979, pp. 376–401.

[*6*] Green, G., Smith, R. F., and Knott, J. F. in *Proceedings*, British Steel Corp. Symposium on the Mechanics and Mechanisms of Crack Growth, M. J. May, Ed., Cambridge, U.K., April 1973, Paper 5.

[*7*] Green, G. and Knott, J. F., *Journal of the Mechanics and Physics of Solids*, Vol. 23, No. 3, 1975, pp. 167–183.

[*8*] Tsuru, S. and Garwood, S. J. in *Proceedings*, Third International Conference on Mechanical Behavior of Materials, Cambridge, U.K., Vol. 3, Aug. 1979, pp. 519- 528.

[*9*] Schulze, D. in *Proceedings*, Fourth Materialprüfungsanstalt (MPA) Seminar on "Safety of the Pressure Boundary of Light Water Reactors, Staatliche Material prüfungsanstalt, Stuttgart University, Stuttgart, Germany, Oct. 1978, Paper 8 (in German).

[*10*] Fearnehough, G. D. in *Proceedings*, British Steel Corp. Symposium on the Mechanics and Mechanisms of Crack Growth, E. J. May, Ed., Cambridge, U.K., April 1973, Paper 16.

[*11*] Hahn, G. T., Hoagland, R. G., Rosenfield, A. R., and Barnes, C. R. in *Crack Arrest Methodology and Application, ASTM STP 711*, G. T. Hahn and M. F. Kanninen, Eds., American Society for Testing and Materials, 1980, pp. 248–269.

[*12*] Krempl, E., *Journal of the Mechanics and Physics of Solids*, Vol. 27, No. 5/6. Dec. 1979, pp. 363–375.

[*13*] Kujawski, D., Kalljanpur, V., and Krempl, E., *Journal of the Mechanics and Physics of Solids*, Vol. 28, No. 2, 1980, pp. 129–148.

Index

This Index combines entries for both volumes of STP 803. "I" denotes entries for Volume I, and "II" denotes entries for Volume II.

D

E

K

L

M

N

O

P

T

Schwager JD. Market wizards: interviews with top traders. New York: Harper-Collins; 1993.

Schwager JD. The new market wizards: conversations with America's Top traders. New York: Wiley; 1995.

Serban A. Combining mean reversion and momentum trading strategies in foreign exchange markets. J Bank Finance 2010;34:2720–2727.

Silber W. Technical trading: when it works and when it doesn't. J Derivatives 1994;1:39–44.

Simon H. A behavioral model of rational choice. Q J Econ 1955;69:99–118.

Sullivan R, Timmermann A, White H. Data–snooping, technical trading rule performance, and the bootstrap. J Finance 1999;54:1647–1691.

Sweeney RJ. Beating the foreign exchange market. J Finance 1986;41:163–182.

Sweeney RJ. Some new filter tests, methods and results. J Financ Quant Anal 1988;23:285–300.

Sweeney RJ, Lee EJQ. International dimensions of securities and currency markets. In: Aggarwal R, Lee C-F, editors. Volume 4, Advances in financial planning and forecasting series. Greenwich (CT): JAI Press; 1990. Part A.

Szakmary AC, Mathur I. Central bank intervention and trading rule profits in foreign exchange markets. J Int Money Finance 1997;16:513–535.

Taylor SJ. Rewards available to currency futures speculators: compensation for risk or evidence of inefficient pricing? Econ Rec 1992;68:105–116.

Taylor SJ. Trading futures using a channel rule: a study of the predictive power of technical analysis with currency examples. J Futures Mark 1994;14:215–235.

Taylor MP, Allen H. The use of technical analysis in the foreign exchange market. J Int Money Finance 1992;11:304–314.

White H. A reality check for data snooping. Econometrica 2000;68:1097–1126.